Genetic Diversity of Fruits and Nuts

Sustainable Utilization

Editor

Hosakatte Niranjana Murthy

Department of Botany, Karnatak University

Dharwad, India

CRC Press

Taylor & Francis Group

Boca Raton London New York

CRC Press is an imprint of the

Taylor & Francis Group, an **informa** business

A SCIENCE PUBLISHERS BOOK

First edition published 2025
by CRC Press
2385 NW Executive Center Drive, Suite 320, Boca Raton FL 33431

and by CRC Press
4 Park Square, Milton Park, Abingdon, Oxon, OXl 4 4RN

CRC Press is an imprint of Taylor & Francis Group, LLC

Library of Congress Cataloging-in-Publication Data (applied for)

ISBN: 978-1-032-77911-9 (hbk)
ISBN: 978-1-032-79468-6 (pbk)
ISBN: 978-1-003-49220-7 (ebk)

DOI: 10.1201/9781003492207

Typeset in Times New Roman
by Prime Publishing Services

Preface

The genetic diversity of crop plants is an important resource for crop improvement. In light of abiotic and biotic stresses, climate change, global warming, and increased human population these genetic resources have played greater importance in developing new cultivars with higher yields, nutritional values, stress, and disease-resistant characteristics. Fruits and nuts have demonstrated excellent genetic diversity and it has been estimated based on morphological, cytogenetic, biochemical, and molecular evaluation.

This volume contains scientific information generated by experts in the field who have analyzed and utilized the genetic diversity of fruits and nuts for their improvement. They have also discussed the degree of diversity found in each fruit and nut crop and how to use such diversity to release new varieties with exceptional yield and quality characteristics. Such data is also helpful in conserving a variety of germplasm for later use.

The book is divided into two sections, section I deals with genetic diversity and utilization of fruit plants involving breadfruit, coconut, tree tomato, wood apple, cocoa, pomegranate, guava, Malabar tamarind, bael, coffee, date palm, desert date, pequie, wild apple, peach, jamun, and citrus. Section II deals with genetic diversity and utilization of nut plants including walnut, cashew nut, and macadamia nut. The introduction, taxonomic background, genetic resource collection and characterization at the morphological, cytological/cytogenetic, biochemical, and molecular levels, use of genetic resources for breeding programs, cultivar release, and genetic resource conservation are all covered in each chapter. Tables and self-explanatory illustrations have been added to each chapter to supplement the main text. Each chapter contains material that will be helpful to academics, researchers, and students studying plant biology, horticulture, and industry. Horticulturists, pomologists, breeders, and conservationists can all benefit from this invaluable information.

I would like to thank and express my deepest gratitude to all the contributors who helped me in completing this book. I am thankful to Vijay Primlani, the Acquisition editor, and all the members editorial board of CRC Press for completing this assignment successfully.

July 2024 **Hosakatte Niranjana Murthy**

Contents

I. Genetic Diversity and Utilization of Fruit Plants

1

Genetic Diversity and Sustainable Utilization of Breadfruit [*Artocarpus altilis* (Parkinson) Fosberg]

Oral O. Daley,[1,*] *Angela T. Alleyne*,[2] *Seymour A. Webster*,[3] *Judy Rouse-Miller*[4] and *Laura B. Roberts-Nkrumah*[5]

Introduction

Breadfruit is a domesticated species that produces nutritious fruits which have made it an important food staple in the Pacific region for more than three millennia (Zerega et al., 2004). Recognizing its value as a food source due to its high productivity, almost year-round availability, and ease of cultivation, European explorers were the first to distribute it outside this region, most notably to the Caribbean (Powell, 1977; Leakey and Roberts-Nkrumah, 2016). Its more recent promotion as a crop with significant potential to contribute to food and nutrition security worldwide has been supported by renewed distribution to countries where it already existed and introductions to others (Ragone, 2016), so that it is currently grown in over 90 countries in all continents and, besides islands in the Pacific, those in the Indian Ocean and the Caribbean. As a tropical crop, its range extends between 17°N and 17°S and even to the Tropics of Cancer and Capricorn in tropical maritime climates as in Hawaii, USA, and Tropical Queensland, Australia, respectively (Ragone, 1997; Goebel, 2007; Jones et al., 2010).

1 Lecturer, Department of Food Production, Faculty of Food and Agriculture, The University of the West Indies, Trinidad and Tobago.

2 Senior Lecturer, Department of Biological and Chemical Sciences, Faculty of Science and Technology, The University of the West Indies, Barbados.

3 Head of Department, Plant, Soil Sciences and Engineering, Faculty of Agriculture, College of Agriculture, Science and Education, Jamaica.

4 Lecturer, Department of Life Sciences, Faculty of Science and Technology, The University of the West Indies, Trinidad and Tobago.

5 Retired Professor, Department of Food Production, Faculty of Food and Agriculture, The University of the West Indies, Trinidad and Tobago.

* Corresponding author

Breadfruit is one of few perennial tree species that is grown as a starchy food crop. The trees are typically tall, attaining heights up to 21 m at maturity with a growth rate typically between 0.5–1.5 m per year (Ragone, 2006; Deivanai and Bhore, 2010). This height was a competitive advantage in the agroforestry systems in which the trees were grown traditionally in the Pacific and, with their architecture and multiple branches, were able to achieve high yields (Elevitch and Ragone, 2018). Yield estimates vary widely. Based on studies on an agroforestry system, Fownes and Raynor (1993) reported an annual per-tree yield of 93–219 fruits with a maximum fruit number of up to 615, which produced an estimated yield of 6.7 t ha^{-1}. Liu et al. (2014) reported an expected average yield of 5–8 t ha^{-1} based on the yield of trees over 7 years at a plant density of 50 trees ha^{-1}. At higher densities of 83–100 trees ha^{-1}, annual tree yields up to 299 fruits and 500 kg in 4–5-year-old trees, and up to 521 fruits and 584 kg in 16-year-old trees were recorded (Roberts-Nkrumah, 1998), and it was estimated that in pure stands or as a major component in crop mixtures, breadfruit could consistently achieve yields as high as 50 tha^{-1}. Bowers (1981) indicated that breadfruit dry weight yield of 6 tha^{-1} compared favorably with the yield of meal from corn; this dry weight yield surpasses that of the global average dry weight yields of 3.7, 4.7, and 5.7 t ha^{-1} for the predominant staples, wheat, rice, and corn, respectively, in 2022 (FAO, 2024). Despite its demonstrated potential as a food source, breadfruit is a neglected and underutilized species (NUS) that remains a primary subsistence crop wherever it is grown today.

Given its potential, since 2009, breadfruit has been one of 35 crop species recognized by the International Treaty on Plant Genetic Resources for Food and Agriculture (IPGRFA) as an underutilized crop to be conserved for food and nutrition security (FAO, 2009). This agreement was a major step forward in underscoring that NUS and their genetic diversity were critical natural resources for sustainable agriculture and alleviation of hunger and, ultimately, poverty. Most breadfruit production is found in tropical developing countries, some of which are small island developing states (SIDS) and others, low-income, and least-developed countries. Efforts to achieve the Sustainable Development Goals (SDGs) set out by the United Nations (UN, 2015) are of particular significance for these countries, designed as they are to address historical inequities that have hindered development. Vulnerability to external market shocks is a key characteristic given their weak, open economies and high trade imbalances represented by high dependence on imports, including food. Economic vulnerability is linked to generally limited resources for production such as land and to inappropriate development models and agricultural production systems, for example, large-scale monocultures for export crops and under-developed supply chains of food crops for domestic markets (Beckford, 1985). The SIDs and several other countries are constrained further by fragile ecosystems and high levels of vulnerability to natural disasters such as hurricane or cyclone damage, floods, drought, and coastal degradation that are attributable to climate change. These adverse environmental conditions negatively impact agricultural production and productivity, thereby exacerbating food and nutrition insecurity and trade deficits.

Plant genetic resources (PGR) and, ultimately, the agro-biodiversity presented by NUS found in the wild or traditional agricultural production systems offer some of the best opportunities for sustainable agriculture and development (Galluzi et al., 2015). Through its genetic diversity, breadfruit has adapted to a wide range of tropical environments. It has demonstrated its ability to contribute significantly to the achievement of several of the SDGs in the Pacific and Caribbean regions and is expected to do so in others to which the crop was recently introduced. With high yields of very nutritious fruit displayed by several varieties (Broomes et al., 2009; Jones et al., 2011; Liu et al., 2015; Daley et al., 2019), breadfruit can continue, to play an important role in the achievement of zero hunger, improved domestic food and nutrition security (SDG 2), and human health and well-being (SDG 3). The crop also contributes to better livelihoods, foreign exchange earnings, and economic growth (SDG 8) based on higher farm income through fresh and processed fruit sales on domestic and export markets and involving more than one cultivar (McGregor, 2007; Roberts-Nkrumah et al., 2013). Use of the crop's genetic diversity on farms increases agro-biodiversity, particularly in mixed cropping systems (Fownes and Raynor, 1993) and agroforestry systems (Elevitch, 2015), which contributes to sustainable utilization of genetic resources and sustainable agriculture and directly addresses SDGs 2, 8, and 15.

Through their longevity, architecture, canopy size, mycorrhizal relationships, and leaf fall, breadfruit trees provide shade, soil cover, and mulch to reduce soil erosion, recycle nutrients, and sequester carbon which are all important ecosystem services contributing to their compatibility in production systems with many other tree crops and annual species (Coates-Beckford and Perreira, 1992; Ragone et al., 2016; Elevitch and Ragone, 2018; Roberts-Nkrumah, 2018; Livingston and Lincoln, 2023). These attributes address SDG 15 which aims at the protection, restoration, and sustainable use of terrestrial ecosystems and includes reversing land degradation and loss of biodiversity (UN, 2015). Climate change is expected to exert negative effects on agriculture in the tropics, therefore SDG 13 calls for urgent action to combat climate change and its impact. The effect of climate change on breadfruit has been assessed and consideration of its possible usefulness in resilient cropping systems has been recommended (Mausio et al., 2020; Yang et al., 2022; Livingston and Lincoln, 2023). The genetic diversity of breadfruit facilitates this array of functions in agriculture and justifies its identification as a crop to be conserved for food and agriculture described in greater detail below.

Breadfruit has traditionally been a major source of nutrition for the people of the Pacific region for millennia (Zerega et al., 2006; Jones et al., 2011) and to a lesser extent for other parts of the tropics where it has been introduced at different times (Medagoda, 2007; Moustache and Moustache, 2007; Omobuwajo, 2007; Roberts-Nkrumah, 2007). The fruit contains high levels of carbohydrates, all the essential amino acids, and all the essential fatty acids (Golden and Williams, 2007; Liu et al., 2015). It is also a good source of potassium and calcium and contains a fair amount of iron, ascorbic acid, and niacin (Jones et al., 2011; Bawa and Webb, 2016). It has a low to intermediate glycaemic index (Ramdath et al., 2004; Bahado-Singh et al., 2006) and has been promoted as one of the top 25 superfoods that are useful for the management and prevention of diet-related lifestyle diseases such as Type 2 diabetes (Magnus, 1997; Turi et al., 2015; Bawa and Webb, 2016). Availability is an important criterion for food and nutrition security and a study in Jamaica showed that breadfruit was available at lower prices than most other highly nutritious staples (Henry, 2016). Therefore, increased breadfruit consumption can be directly beneficial, at the individual or household levels in terms of its dietary value and could also contribute significantly to public health savings from having a healthier population (Ferguson et al., 2010).

The fruit is most often consumed fresh at the mature, firm stage as a starchy staple. However, it may also be consumed at the immature stage or when ripe and no longer firm. Common methods of preparation of breadfruit in producing regions include roasting, boiling, frying, and baking. The fruit is now also used in novel and creative ways to make many non-traditional products including snacks, baked goods, beverages, and desserts. The fruits are highly perishable and will become unacceptable to most consumers within just a few days after harvesting under ambient tropical conditions (Maharaj and Sankat, 1990; Sankat and Maharaj, 2007). This high perishability also limits the marketing of breadfruit to international markets. Controlled atmosphere storage at 16°C with 5% oxygen, and 5% carbon dioxide has been used to extend shelf life for 3–4 weeks (Sankat and Maharaj, 1993). However, these measures have not been widely adopted, possibly because these facilities are not available in many producing countries. Pit fermentation was used in many Oceanic countries to extend the supply of breadfruit in seasons of low supply (Atchley and Cox, 1985). This method of preservation has not been adopted in new breadfruit-producing regions and even in Oceania it has been declining. In contrast, the production of flour from breadfruit pulp has been growing in many producing areas because it has a long shelf life, is easier to store and can easily be incorporated into other products (Broomes et al., 2009; Liu et al., 2020).

Besides its food and environmental benefits, this multi-purpose crop is used as livestock feed, while non-food uses include fiber, medicine, wood for construction material, and adhesives. The potential exists to improve livelihoods and, indirectly, food and nutrition security through increased market development for these products, which can promote germplasm conservation for the sustainable use of breadfruit germplasm. As an NUS crop, loss of genetic diversity is a serious threat. In some countries genetic erosion has already occurred and many valuable breadfruit cultivars are disappearing due to loss of indigenous knowledge, changes in agricultural systems, urbanization, and natural disasters (Jones et al., 2011). These losses and insufficient efforts to conserve breadfruit germplasm are also indicators that its value has been underestimated and underexploited. Consequently, the negative impact of the loss of cultivars on food security and the health of future generations has been inadequately addressed to secure it as a dietary mainstay and a component of a resilient food system. Little is known about the agronomy of the species and how to improve cultivars, yield, and quality. Therefore, there is a need to develop the most effective commercialization, marketing and policy framework on food and nutrition security to promote its uses and maximize its socio-economic value. Understanding and conserving the diversity of breadfruit germplasm is important for better maintenance, development, and use for the benefit of present and future generations. Therefore, this chapter provides a comprehensive review of the collection, characterization, and evaluation of breadfruit genetic diversity to support its sustainable utilization.

Taxonomic Background

The cultivated breadfruit belongs to the monophyletic family Moraceae (Mulberry) (Kanzaki et al., 1997). The monophyly of Moraceae was supported by morphological data and more recently by molecular data (Zerega et al., 2005; Clement and Weiblen, 2009). The taxonomy of the tribe and lower taxa has been problematic, however, a monophyletic Artocarpeae tribe which includes the largest genus Artocarpus the third largest in the Moraceae family with more than 60 species, and two other small genera was circumscribed by Zerega et al. (2010). Species in this genus are monoecious, lactiferous, and mostly trees and a few shrubs (Sharma, 1963; Zerega et al., 2006).

Breadfruit and its relatives have usually been considered by some taxonomists as either a single species exhibiting wide variation (Jarrett, 1959) or as separate species (Blanco, 1837; Trécul, 1847; Quisumbing, 1940). The latter view has been supported by morphological and phylogenetic analyses (Zerega et al., 2004, 2005). This breadfruit complex is recognized as *A. altilis* (Parkinson) Fosberg and its closest relatives, *A. camansi* Blanco, and *A. mariannensis* Trecul (Zerega et al., 2004, 2005). *A. camansi*, breadnut, is the seeded wild progenitor of *A. altilis* and native to New Guinea where domestication of *A. altilis* began (Lebot, 1999). Previously, *A. camansi* was widely regarded as a seeded breadfruit cultivar and was classified as *A. communis* and *A. altilis* (Aurore et al., 2014). The species *Treculia africana*, also called breadfruit or African breadfruit, is a member of the Moraceae family and can be mistaken for *Artocarpus altilis* in the literature (Oduro et al., 2007; Zerega et al., 2010). Breadfruit and jackfruit (*Artocarpus heterophyllus* Lamarck) are the two most widely cultivated *Artocarpus* species throughout the tropics (Jarrett, 1976; Zerega et al., 2005).

The taxonomic classification for breadfruit is given in Table 1. The botanical name for breadfruit is *Artocarpus altilis* (Parkinson) Fosberg, formerly *Artocarpus communis* J.R.Forst. & G.Forst. *Sitodium altile* Parkinson, *Artocarpus incisus* (Thunb.) L.f., and *Radermachia incisa* Thunb (Leakey, 1977; Ragone, 1997; Ferrer-Gallego and Boisset, 2018). The common name breadfruit is an English name or word that has also been used for other species including *Artocarpus camansi* (Blanco) also known as breadnut or chataigne and *Artocarpus mariannensis* also known as dug-dug. In the Caribbean, other common names include cow, panbwa, pain bois, frutapan, and fruta de pan (Roberts-Nkrumah, 2007). Other common names from other countries and regions include beta (Vanuatu) bia, bulo, nimbalu (Solomon Islands), kapiak (Papua New Guinea), kuru (Cook Islands), meduu (Palau) mei, mai (Federated States of Micronesia [FSM], Kiribati, Marshalls, Marquesas, Tonga, Tuvalu), mos (Kosrae), ‘ulu (Hawai‘i, Samoa, Rotuma, Tuvalu) ‘uru (Society Islands) uto, buco (Fiji), árbol a pan (Spanish), brotfruchtbaum (German), fruta pão (Portuguese), khanun-sampalor, sa-ke (Thai), khnaôr

Table 1. Taxonomic classification of breadfruit.

Level	**Classification**
Kingdom	Plantae (Plants)
Subkingdom	Tracheobionta – Vascular plants
Superdivision	Spermatophyta (Seed plants)
Division	Magnoliophyta (Flowering plants)
Class	Magnoliopsida (Dicotyledon)
Subclass	Asteridae
Order	Urticales
Family	Moraceae (Mulberry family)
Genus	*Artocarpus*
Species	*Artocarpus altilis and A. altilis* x *A. mariannensis*

sâmloo, sakéé (Khmer) kulur, kuror, sukun (Indonesia), l'arbre à pan (French), laba pen, vèritab (Haiti, Creole), mshelisheli (Swahili), rata del (Sri Lanka), rimas (Philippines), sake (Vietnamese) (Ragone, 2011). This chapter focusses on the species *Artocarpus altilis* referred to as breadfruit unless otherwise stated.

Breadfruit's two closest relatives, *Artocarpus camansi* (Blanco) and *Artocarpus mariannensis* (Trécul) are both fertile diploids (2n = 2x = 56) and occur naturally in different parts of Oceania. *A. camansi* is native to New Guinea where it occurred in the wild and dominated the secondary forest in lowland areas. This native range may extend north-easterly to the Moluccas and the Philippine islands. *A. mariannensis* is native to the Mariana archipelago which extends from Guam in the southernmost part, to the Northern Mariana Islands. Breadfruit occurs in the wild in Guam and is dominant in some forests on limestone terraces, plateaus, and volcanic soil (Fosberg, 1960). Its native range may also include the neighboring island of Rota, in the Northern Mariana Islands. However, although *A. mariannensis* is very popular in Palau and the Caroline Islands, it seems doubtful that it occurred naturally in those islands (Fosberg, 1960).

A. altilis cultivars in Melanesia, a sub-region of Oceania that includes New Guinea, are mostly fertile diploids (2n = 2x = 56). Amplified Fragment Length Polymorphism (AFLP) analysis showed that a majority (91%) of these varieties, which include some sterile diploids (2n = 2x = 56) and triploids (2n = 3x = 84), contain only *A. camansi* specific markers and no *A. mariannensis* specific markers, which indicate that they evolved only from *A. camansi* (Zerega et al., 2004). They are also believed to be the oldest *A. altilis* varieties and are morphologically and physiologically akin to *A. camansi* compared to varieties from other regions (Ragone, 1997; Jones et al., 2013). The evolution of *A. altilis* from *A. camansi* most likely happened over several generations and started in Melanesia as humans spread across that sub-region and started selecting breadfruit types with specific traits which caused a change in morphology.

Breadfruit types in Polynesia are like those of Melanesia in terms of the level of *A. camansi* specific markers present suggest that most of these cultivars evolved only from *A. camansi*. However, the seeded diploid cultivars that are dominant in Melanesia are less prevalent in Western Polynesia and are almost absent in Eastern Polynesia where triploid cultivars are the dominant type. It has been hypothesized that these changes in breadfruit morphology were linked to the west-to-east human migration of the Lapita people through Melanesia onward to Polynesia (Zerega et al., 2004). The use of seed to propagate breadfruit in Melanesia and close areas was quite feasible. However, for longer distances, vegetative propagation would have taken precedence because breadfruit seeds are recalcitrant and do not store very well. Few-seeded and seedless cultivars may also have developed due to somatic mutation and meiotic defects caused by using vegetative propagation over several generations (Zerega et al., 2004).

Although both Melanesia and Polynesia were dominated by breadfruit types having only *A. camansi*-specific markers, other hybrid types also existed in those regions. Hybridization occurred between *A. camansi*-derived breadfruit and *A. mariannensis* which indicated the spread of *A. mariannensis* genes outside of Micronesia into Melanesia and Polynesia and contributes to a fuller explanation of the variation observed in breadfruit. The region of Micronesia which includes Guam and Rota, has mostly (78%) hybrid varieties containing at least one *A. camansi* specific marker and one *A. mariannensis* specific marker based on AFLP analysis. These included both sterile and fertile diploids and triploids. AFLP analysis further showed that 20% of the samples from Micronesia contained only *A. camansi* specific markers and 2% had only *A. mariannensis* specific markers. Therefore, the distribution of *A. camansi* into Micronesia can also be confirmed (Zerega et al., 2004). Different hypotheses have been proffered on the role that *A. camansi* and *A. mariannensis* played in the development of hybrid breadfruit cultivars. Zerega et al. (2004), postulated that *A. camansi* derived breadfruit was likely introduced to Micronesia through a direct northerly Lapita migration from Melanesia into the high mountains of Micronesia. This also helped to explain the absence of *A. camansi* in Yap and Palau which are islands in Micronesia but closer to Melanesia compared to Guam and Rota. It was also plausible that the Lapita voyagers moved between the two regions and must have introduced hybrids back to Melanesia. Hybrids found in Melanesia would also have been distributed to Polynesia in a similar way to *A. camansi* derived breadfruit. Both fertile and sterile hybrids are found in Micronesia. Introgression was therefore possible between fertile diploid hybrids and *A. mariannensis.* Knowledge of the origin of breadfruit has identified the need to conserve its wild relatives and to this end, accessions of *A. camansi* and *A. mariannensis* were established at the breadfruit germplasm collection held at the National Tropical Botanical Gardens Hawaii (Zerega et al., 2004).

Collection and Assessment of Germplasm/Genetic Resources

Most *in situ* and *ex situ* breadfruit gene banks exist in the Pacific where the greatest genetic diversity of the species emerged. The early people of the Pacific region were the first to collect breadfruit types that were suitable for food and for propagation by methods that facilitated long-distance travel as they journeyed across the region (Zerega et al., 2004). In traditional arboricultural systems, several breadfruit cultivars were grown on the same holding which extended the availability of the crop throughout the year, as in Pohnpei and Tahiti (Fownes and Raynor, 1993; Lincoln et al., 2018). Therefore, a diverse range of cultivars selected, planted, and managed on farms or in the wild on abandoned farmland constituted the first *in situ* breadfruit collections. Following the initial lists of cultivars that recorded breadfruit diversity in the Pacific, including 24 from Tahiti in 1860, 53 from the Caroline Islands in 1899, and 34 from the Marquesas Islands in 1910 (Wester, 1924), the need for documentation and conservation was recognized because of cultivar loss due to declining cultivation and use. *Ex-situ* breadfruit gene banks were first established in Hawaii during the 1920s (Ragone, 2016). The South Pacific Commission (SPC) established a regional collection in Samoa with 260 accessions from 14 countries and also started collections with duplicates in Fiji and Tahiti (Global Crop Diversity Trust, 2007). Based on a survey undertaken by the SPC in 1960 to characterize the breadfruit germplasm in seven Pacific territories, the Samoan and Fijian collections were expanded (Parham, 1966). Subsequently, smaller collections were established in the Federated States of Micronesia, the Solomon Islands, Papua New Guinea, and Kiribati to evaluate and select cultivars with desirable characteristics, including seedlessness, good fruit quality, pest and disease resistance, long bearing season, and suitability for use as dwarfing rootstock (Lorens and Englberger, 2007; Quartermain, 2007; Ragone et al., 2007; Redfern, 2007). No data were collected from these early SPC collections and except for those in the Solomon Islands, the others did not survive which reflected the continuing loss of genetic material. This loss was attributable not only to increasing westernization which led to cultural changes noted previously, but also to natural disasters, pests and diseases, and limited replanting of damaged trees (Global Crop Diversity Trust, 2007). Furthermore, a lack of resources and interest in managing, studying, and curating the collections were also reported as contributors to their decline (Ragone, 2016). In 2001, the Pacific Agricultural Plant Genetic Resource Network (PAPGREN), an initiative of the Secretariat of the Pacific Community, formerly the South Pacific Commission (SPC) identified breadfruit as a priority tree crop, second only to coconut, for access to and conservation of its genetic resources and by 2002, developed an action plan for implementation. Subsequently, the SPC established a regional breadfruit collection at the Centre for Pacific Crops and Trees and supported collections in French Polynesia and Vanuatu for characterization, evaluation, and conservation by increasing utilization (Navarro et al., 2007, Taylor and Tuia, 2007, Taylor et al., 2013).

In the 1970s, the National Tropical Botanical Garden (NTBG), a not-for-profit institution established in Hawaii, USA, for tropic plant research, education, and conservation, decided to establish a definitive collection of breadfruit and breadnut (Ragone, 2016) and supported collection activities in the Pacific and also received donated plants from the region, and the Seychelles. The NTBG collection was established on Maui at the Kahanu Gardens, the major site, while a small number of accessions were held at the Mc Bryde Garden and Limahuli Garden on Kauai. This has become the largest collection in the world with 340 accessions and 120 cultivars from 34 Pacific countries and the Philippines, Indonesia, the Seychelles, and Honduras (Ragone, 1997; Ragone et al., 2016).

European explorers collected breadfruit from the Pacific for their colonies, most notably in the Caribbean, where botanic gardens such as those in St. Vincent and Jamaica were established during the 18th century to receive the plants and propagate them for distribution throughout that region to boost local food supply (Bligh, 1976; Leakey and Roberts-Nkrumah, 2016). Among the collections outside the Pacific region, one of the largest and most diverse is held at the University of the West Indies (UWI) in Trinidad and Tobago with 40 accessions consisting of existing types in the Caribbean and recent introductions from the Pacific which comprise the majority. Like collections elsewhere, the main objectives are to assess the local germplasm, describe the accessions for ease of identification, and assess their suitability for commercial production and utilization. Thirty-three (33) accessions, including *A. camansi*, collected in Trinidad, and three *A. altilis* x *A. mariannesis* hybrids from the NTBG collection have been described (Roberts-Nkrumah, 2018).

Assessment of breadfruit genetic diversity is a key element of both the development and utilization of these collections or gene banks. This is important to avoid unnecessary duplication of accessions primarily because of the maintenance expense that may contribute to eventual neglect and loss of breadfruit collections (Ragone 1991, 1997). Synonymity is a major challenge during collection exercises as many vernacular names exist, some of which may refer to the same cultivar, while the same name may be used for different cultivars in the same or different locations (Ragone 1995, 2004; Roberts-Nkrumah 1997, 2018; Daley et al., 2022). Description of the leaf, fruit, and tree morphology has been the traditional approach to cultivar identification, with varying levels of attention to the evaluation of characteristics pertinent to increased production and consumption, for example, seasonality, pest and disease resistance, fruit yield, eating quality, nutritional content, flour yield and starch characteristics (Wester, 1924; Wilder, 1928; Koroiveibau, 1966; Parham, 1966; Ragone, 2007; Jones et al., 2011; Jones et al., 2013; Liu et al., 2014; Roberts-Nkrumah and Daley, 2014; Daley et al., 2016; Daley et al., 2019). Bioversity International and the NTBG provided an initial list of access and utilization descriptors to facilitate information sharing and utilization of the accessions held by gene banks (IPGRI, 2009). Besides these morphological and biochemical approaches, cytological characterization and, more recently, molecular characterization have been employed to assess breadfruit genetic diversity. These various approaches are reviewed in this section.

Characterization and Evaluation

Morphological Characterization

Breadfruit cultivars and types have been characterized based on morphological features of the tree, leaf, inflorescence, fruit, and seeds that were used initially as the main method for identification. During the collection of breadfruit in Tahiti, Bligh indicated that cultivars could be distinguished by differences in leaf and fruit form, and both in the Pacific and the Caribbean, some cultivar names reflect fruit and leaf morphological features (Ragone, 1991; Roberts-Nkrumah, 1998; Daley et al., 2020). Several early studies, based mainly on field-grown plants of differing numbers of cultivars in one or more countries or sub-regions of the Pacific, have pointed to high morphological variability within the species. Wilder (1928), described 32 breadfruit cultivars from Tahiti, with photographs of leaves and fruit, based on herbarium specimens. Sasuke (1953), grouped 50 cultivars from Pohnpei, Federated States of Micronesia into 12 leaf shapes ranging from entire to deeply dissected, five fruit skin textures, and the presence or absence of seeds. From a survey conducted in the South Pacific Region, Parham (1966) provided morphological data and drawings from 166 trees grouped in 12 leaf shapes and five skin texture classes, with documentation of the presence or absence of seeds, four fruit shapes, and variation in the hirsuteness of the midrib of leaves. Based on a survey in Fiji, Koroiveibau (1966), recorded and grouped 70 varieties by leaf shape, leaf apex and base shape, presence and color of hairs on midribs, petiole length, length and width of male inflorescence, fruit shape, with drawings. In surveys of the much smaller breadfruit germplasm in St Vincent, leaf, and fruit morphological descriptors for 22 specimens were applied to assist with identification (Roberts-Nkrumah, 1997). Five leaf lobe number groups and four fruit shape groups were used to classify the cultivars with other descriptors in each group, but like the surveys in the Pacific, there was a high level of variability even within groups, demonstrating the inadequacy of this approach (Roberts-Nkrumah, 1997). These studies on trees growing *in situ* provided descriptors that were useful for identifying some cultivars but not for the majority among which differences were unclear or inconsistent. Apart from the challenge of synonymity, the reliability of the data for identification purposes may have been compromised by the effects of environmental conditions at different locations on phenotype, variation within and among trees of the same clones at the same locations, tree age, and sampling procedures for the organs described. These inadequacies in the data collection strategies highlighted the need for systematic characterization and description of breadfruit germplasm (Ragone et al., 2007; Roberts-Nkrumah, 2018).

Several later studies in the Pacific and the Caribbean were conducted using standardized procedures to describe cultivars *in situ* during surveys and accessions at *ex situ* germplasm collections. Navarro et al. (2007) used standardized passport and morphological descriptors for the tree, mature leaves (n = 5), male flowers, and one mature fruit where this was available, as well as photographs, to describe 69 breadfruit cultivars collected during a survey of six islands in Vanuatu. The morphological descriptors were provided by the NTBG, but the number of descriptors used was not reported. Subsequently, the accessions were established in a gene bank (Navarro et al., 2007). Seven leaf and 11 fruit morphological descriptors, both quantitative and qualitative, were used to describe the leaves of 114 trees and the fruit of 106 trees, respectively, in surveys of breadfruit cultivars in five islands of four countries, in the western, eastern, and southern Caribbean, Jamaica, St. Kitts and Nevis, St. Vincent and the Grenadines, and Trinidad and Tobago (Figure 1) (Daley, 2018). Ethnographic information on cultivars collected during surveys was also used to help distinguish among cultivars (Ragone, 1995; Navarro et al., 2007; Daley, 2018; Daley et al., 2022).

A larger number of descriptors have been used to describe breadfruit in *ex situ* collections. At the breadfruit repository held at the Tropical Plant Genetic Resources and Disease Research Unit in Hawaii, the 32 accessions have been characterized using 42 morphological descriptors using a sample size of five (Ragone et al., 2007). Based on descriptors for other crops available from the Genetic Resources Information Network (GRIN) of the US Agricultural Research Service, Ragone

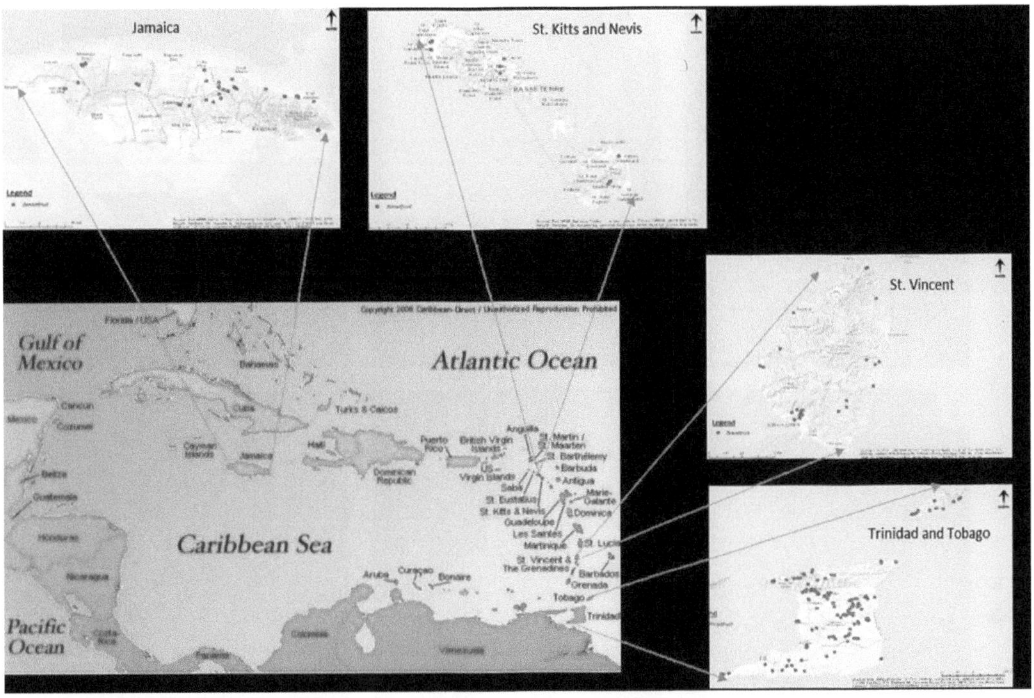

Figure 1. Map of the Caribbean highlighting countries surveyed for breadfruit characterization. (Photo credit: Oral Daley)

et al. (2007) described the use of this approach at the NTBG to collect data for 60 quantitative and categorical descriptors from 172 accessions comprising 220 trees planted before 2003. The same sample size (n = 5) was used for 60 quantitative and categorical descriptors to facilitate statistical analysis. Photographic records were also taken. Subsequently, Biodiversity International and NTBG published an initial list of quantitative and categorical descriptors for breadfruit like those used for other fruit crops by the US Department of Agriculture Genetic Resources Information Network and the International Board for Plant Genetic Resources (IPGRI, 2009). Thirty-two accessions in the Vanuatu collection have been characterized by 42 morphological descriptors. Subsequently, Jones et al. (2013) applied 57 previously standardized morphological descriptors to compare 222 accessions of breadfruit and closely related species and reported clear morphological differences among accessions according to the region of origin and species (Jones et al., 2013). Working with the UWI collection, Daley et al. (2020) used 44 morphological descriptors to compare 27 existing Caribbean accessions and newly introduced accessions. The study revealed that while descriptors related to leaf lobing were important for distinguishing a few accessions (Figure 2), fruit characters such as, skin texture, fruit shape, fruit skin color, and pulp colour (Figures 3, 4) were the most useful for distinguishing among accessions, and the recently introduced accessions expanded the range of morphological features for selection (Daley et al., 2020).

Cytological and Cytogenetic Characterization

Breadfruit genetics is complex due to the presence of seeded and seedless diploid (2n = 2x = 56) and triploid (2n = 3x = 84) types, with hybrids of both also seen in breadfruit growing regions around the globe (Ragone, 2001; Audi et al., 2023). A full draft genome of the species was only recently reported and presents an opportunity for further investigations into the detailed genetic structure of the species. Currently, there is a draft breadfruit genome assembly alongside its nearest similarly

Figure 2. Leaf morphological variations observed among accessions at the University of the West Indies, St. Augustine Campus breadfruit germplasm collection. Accessions: (A) Timor; (B) Kashee Bread; (C) Cassava; (D) Meitehid; (E) Puou and (F) Yellow. (Photo credit: Oral Daley)

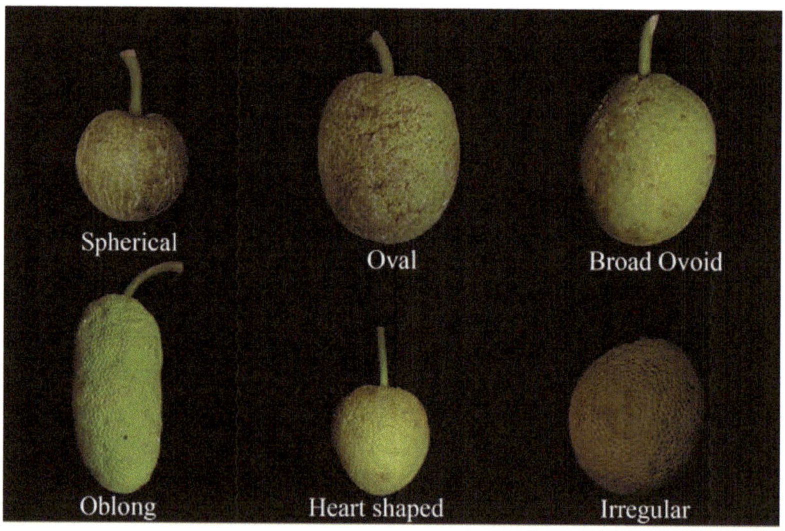

Figure 3. Fruit shape variations observed among accessions at the University of the West Indies, St. Augustine Campus breadfruit germplasm collection.

***Source*:** L.B. Roberts-Nkrumah 2018. The Breadfruit Germplasm Collection at the University of the West Indies, St Augustine Campus. The University of the West Indies Press. Copyright 2018 by Laura B. Roberts-Nkrumah. Reprinted with permission.

Figure 4. Fruit skin texture variations observed among accessions at the University of the West Indies, St. Augustine Campus breadfruit germplasm collection.

Source: L.B. Roberts-Nkrumah 2018. The Breadfruit Germplasm Collection at the University of the West Indies, St Augustine Campus. The University of the West Indies Press. Copyright 2018 by Laura B. Roberts-Nkrumah. Reprinted with permission.

domesticated relative, the Jackfruit (Sahu et al., 2020), although original wild relatives, such as *A. camansi* and *A. marianensis*, still exist (Witherup et al., 2013; Audi et al., 2023).

The available genome assembly for *A. altilis* suggests it is approximately 833 Mb, with 34,010 genes and 2034 gene families (Sahu et al., 2020). A high GC ratio of 68% and approximately 90% coverage in the draft genome indicates that the current assembly is helpful for future sequencing experiments where gene mapping to a reference is required. Over 50% of the *A. altilis* genome comprises repeat sequences of many long terminal repeats or LTRs (46%) (Sahu et al., 2020). This substantial number of repeats is typical of eukaryotic genomes and is responsible for plant genome expansion (Havecker et al., 2004; Du et al., 2010). For example, the potato genome consists of approximately 62% repeat sequences (Xu et al., 2011) and about 85% in rice (Lee and Kim, 2014). Repeat sequences such as LTRs are transposable elements, adding significant mutability to eukaryotic genomes (Havecker et al., 2004). Given that there have been repeated introductions across the globe from Oceania driven by migration expeditions (Daley et al., 2020; Audi et al., 2023), it is not surprising that these isolated gene pools allowed for the propagation of domestic types with their genetic variation expressed as a wide range of phenotypes with varying names, and morphologies based on their fruits (size and shape), leaf (size and shape), and flesh textures and colors (Zerega et al., 2015; Daley et al., 2020; Audi et al., 2023). Further, LTRs allow for gene regulation, gene translocation, and chromosome rearrangements in plant genomes, with most class 1 LTRs functioning by "copy and paste" mechanisms that can lead to genome expansion (Yang et al., 2023).

On an average, the length of genes in the *A. altilis* genome is 1,253 base pairs (bp) long with approximately 5–6 exons per gene (Sahu et al., 2020). Comparative genome analysis showed that *A. altilis* and *A. heterophyllus* share a core of approximately 9,462 gene families with *Arabidopsis thaliana*, and their duplication events occurred at similar times (Sahu et al., 2020). Most of the protein-coding genes are available and annotated as homologs in the Nr database (98%) and in other searchable protein databases such as TrEMBL (98%), InterPro (78%), and SWISS-PROT (79%) (Sahu et al., 2020). Approximately 1% (136) of unique gene families belonging to *A. altilis* were among the more than 15,000 gene families. While this is small, it represents a resource for further genetic exploration of the observed variation among breadfruit genotypes with more precise genome information. In functional databases such as KEGG and GO, functional pathways related to carbon fixation and photosynthesis were prevalent (Sahu et al., 2020).

The much smaller chloroplast (cp) genome of breadfruit and other *Artocarpus* species, including *A. heterophyllus*, has also been produced (Daniell et al., 2016; Li et al., 2020; de Souza et al., 2021; Lin et al., 2022). These all range around 160,000 bp in size. These chloroplast genome assemblies have assisted in confirming the origin of *A. altilis* as a direct descendant from *A. camansi* using phylogenetics (Zerega et al., 2004). Chloroplasts are highly conserved and represent a much smaller set of genes so genome reconstruction is more readily available and precisely assists in establishing phylogenetic histories (Daniell et al., 2016).

The cp genome of *A. altilis* encompasses 160,822 bp and consists of four parts: a large single-copy region (LSC-88,692 bp), a much smaller single-copy region (SSC-19,290 bp) and two inverted repeat regions (IRs) (IRA and IRB-26,420 bp each) (Wei et al., 2023). This arrangement is typical of most chloroplasts. More than 120 genes, including 83 protein-coding genes, were reported (Daniell et al., 2016; Wei et al., 2023). These recent findings in both the whole genome and the chloroplast genome of *A. altilis* and similar *Artocarpus* species (Li et al., 2020; Lin et al., 2022) are welcome additions to explore further regional variation, as well as the structure and function of single or community genes in future genetic research on *A. altilis*.

Biochemical Characterization

The biochemical composition of breadfruit has been analyzed for its major and minor nutrients, including the carbohydrate, protein, starch, and lipid content of its fruit, along with isoflavonoids and carotenoids from its leaves and peel (Golden and Williams, 2001; Ragone and Cavaletto, 2006; Beccaro et al., 2020; da Silva and Narain, 2021; Mehta et al., 2023). The vast array of compounds that both breadfruit and its nearest genetic relatives contain, make it a rich energy food source and a nutraceutical, because of its biochemical and biophysical properties. Several studies have examined breadfruit's nutrients because it is both a food source and functional food in its center of origin in the Pacific and other tropical countries in the Caribbean and Africa, where it was introduced (Umezuruike and Nwabueze, 2017; Clark and Aramouni, 2018; Beccaro et al,. 2020; Gonçalves et al., 2020; Sari et al., 2020; Amorim et al., 2021; da Silva and Narain, 2021).

Breadfruit contains approximately 27% carbohydrates, 1–2% proteins, 1.5% fat, and all the essential amino acids (Liu et al., 2015; Tarai et al., 2017; Mehta et al., 2023). Breadfruit contains exceptionally high levels of leucine, lysine, isoleucine, phenylalanine, and valine (Golden and Williams, 2001; Mehta et al., 2023) although this may vary by variety. When compared to more well-known starch-rich tropical crops, breadfruit contains more protein per dry weight than cassava, banana, and sweet potato (Liu et al., 2015; Huang et al., 2019) and high amounts of fiber (Carrington and Sankat, 2011). Breadfruit also has a high concentration of carbohydrates and starch, but recent studies indicate that it also contains high levels of resistant starch (Daley et al., 2019), which supports earlier findings of its low to intermediate glycemic index classification (Ramdath et al., 2004). In the Caribbean, studies have shown some differences in starch composition based on growth patterns and habitats (Latchoumia et al., 2014), with fruits grown in the dry season having higher starch content. Interestingly, one of the most widely cultivated varieties, Ma'afala, has high protein and carbohydrate levels (Liu et al., 2015; Daley et al., 2019) and is a good source of flavonoids (Soifoini et al., 2018). Breadfruit flour also compares well with other flour from tropical crops and is a good source of fiber and vitamins (Soifoini et al., 2018; Liu et al., 2020). Besides morphological and genetic diversity, there is also biochemical diversity among breadfruit, with cultivars showing variation in biochemical composition, which is also impacted by preparation and cooking methods (Ramdath et al., 2004; Bahado-Singh et al., 2006). Cooking practices are diverse and versatile, ranging from baking, roasting, and boiling to dehydrating, pickling, and fermenting (Roberts-Nkrumah et al., 2005; Ragone and Cavaletto, 2006; Liu et al., 2015; Mehta et al., 2023).

As a functional food, *A. altilis* has extensive nutraceutical activity to treat several conditions, including cancer, hypertension, and anxiety (Badrie and Broomes, 2010; Soifoini et al., 2018; Gonçalves et al., 2020). This is due to the presence of critical secondary compounds, including phenols, flavonoids, chalcones, terpenoids, stilbenes, and xanthones (Soifoini et al,. 2018), which are present in many *Artocarpus* species (Lan et al., 2013; Vianney et al., 2020). Breadfruit leaves contain antioxidants and valuable nutraceutical product additives (Vianney et al., 2020). Artocarpin $C_{2}6H_{2}8O_{6}$ is a prenylated flavone compound called [6-(3-methyl-1-butenyl)-5,2',4'-trihydroxy-3-isoprenyl-7-methoxy flavone] found in breadfruit (Soifoini et al., 2018; Beccaro et al., 2020). Artocarpin has many pharmacological properties, including dermatological (skin-whitening, photo-protection), antioxidant, cytotoxic, anti-inflammatory, antibacterial, antiandrogen, antitubercular, antiplasmodial, neuraminidase inhibitory, termiticidal and wound healing (Lan et al., 2013; Chan et al., 2018). Given these biochemical activities, its widespread ethnobotanical usage is not surprising among people who also consume the fruit. In some countries, therapeutic remedies exist for treating skin disorders, cancer, diabetes, and hypertension (Patil et al., 2002; Soifoini et al., 2018).

Besides its nutraceutical and nutritional value, some breadfruit tissues such as the skin and peels have been a good source for industrial applications (Anwar et al., 2020), for example, the absorption of dyes (Priyantha et al., 2018). Breadfruit starch has been shown to be a good coating for fruits and vegetables at higher temperatures to prevent spoilage (Worrell et al., 2002; Bezerra et al., 2019). Moreover, breadfruit starch is useful as a stabilizer in the food industry and therapeutics using breadfruit have been boosted by technological advancements in nanotechnology to produce starch nanoparticles (Harsanto et al., 2022). Starch nanoparticles are biological compounds with the ability to stabilize CO-bonded molecules and improve the emulsification properties of other compounds. They also have good biodegradable properties and absorption properties. Starch nanoparticles from breadfruit show potential for many applications in the food, biomedical, aesthetic, and pharmaceutical industries with better absorption and biodegradable properties (Andrade et al., 2020).

The biochemical variation in breadfruit is also observed in the isozyme composition of cultivars where six enzyme systems produced well-resolved bands among 150 breadfruit accessions, with diploid cultivars showing highest variations and triploid cultivars showing identical banding patterns for all enzymes used (Ragone, 1990). However, isozyme variations were not sufficient to distinguish among closely related cultivars.

Molecular Characterization

In the Caribbean, the various cultivars of breadfruit were the results of one major introduction in the early 18th century to St. Vincent and Jamaica, with subsequent distributions from these islands to other islands in the region (Daley et al., 2020; Daley et al., 2022;, Audi et al., 2023). Consequently, there are many morphological variations based on size, shape, texture, fruit color, and a variety of agronomic features (Jones et al., 2013; Daley et al., 2020). Not surprisingly, the range of cultivars that characterize the various morphological types in breadfruit is accompanied by reports examining various molecular markers. These include Random amplified polymorphic DNA (RAPDs), microsatellites (SSRs), amplified fragment polymorphisms (AFLPs), and next-generation sequencing and transcriptome analysis (Table 2). RAPDs are non-discriminant molecular markers in relation to lineage studies because they are unable to identify ancestral origins precisely. However, when used in small datasets, they can identify genetic relatives and groups within a small population. Before the first report of a representative genome (Zerega et al., 2004; De Bellis et al., 2016), several microsatellite marker sets were developed. A set of 25 was developed for *A. altilis* (Witherup et al., 2013) and later, another set of 50 SSR markers was developed using Next Generation sequencing (De Bellis et al., 2016). AFLP markers have helped establish the origin of breadfruit (Zerega et al., 2004; Audi et al., 2023).

While the reference genome is important for analyzing specific genes in breadfruit germplasm, molecular markers have been useful in assessing the identity and genetic diversity of breadfruit

Table 2. Published studies using various molecular tools for the characterization of breadfruit.

Type of Molecular Characterization	Year of Publication	Region/Location	Number of Samples Used	References
AFLPs	2004	Pacific	200	(Zerega et al., 2004)
AFLPs	2005	Oceania	261	(Zerega et al., 2005)
AFLPs	2007	India/Western Ghats	60	(Sreekumar et al., 2007)
Amplification and SSR markers	2013	Global/ Pacific, Caribbean,	600	(Witherup et al., 2013)
SSRs	2014	Oceania	349	(Zerega et al., 2015)
NGS of SSR markers	2016	Polynesia	41	(De Bellis et al., 2016)
SSRs	2016	Polynesia, Melanesia, Micronesia, and the Caribbean	423	(Zerega and Ragone, 2016)
RAPD PCR	2021	Indonesia	9	(Ifah et al., 2018)
Transcriptome	2018	Polynesia	24	(Laricchia et al., 2018)
NGS and read assembly	2021	Global	179	(Gardner et al., 2021)
AFLPs, SSRs, and NGS	2023	Global/ Pacific, Caribbean,	349	(Audi et al., 2023)
Sequencing of SSR markers	2024	Caribbean	95	(Daley et al., 2024)

cultivars and managing gene banks. These markers were concentrated on the Pacific germplasm and are generally useful in maintaining and analyzing the genetic diversity in that region. However, it is difficult to distinguish among triploid accessions using this approach (Daley, 2018; Audi et al., 2023; Daley et al., 2024), although studies have shown the morphological and genetic plasticity (Zerega and Ragone, 2016; Daley et al., 2020; Audi et al., 2023). A recent study confirms that two genotypes are related to Eastern Polynesian relatives that have been established (Audi et al., 2023). Notwithstanding this recent study, the specific genetic identity of several Caribbean cultivars or landraces is still unknown because of the intense morphological/cultural selection in the past 200 years. Nonetheless, the various molecular markers have established that breadfruit cultivars share a singular origin in Oceania but have since devolved into local genotypes through intensive self-selection and introgression over time.

Molecular transcriptome analysis of *A. altilis* diploids shows an overrepresentation of catalytic activity and binding as the most representative Gene ontology (GO) terms in this species (Laricchia et al., 2018; Zhou and Underhill, 2021). When grown on rootstocks or dwarfed, several key genes involved in sugar transport such as SWEET and SUT genes were downregulated in source tissues, while antioxidant genes were upregulated. These patterns of gene disruption reflect the importance of sugar and starch metabolism as well as antioxidant biosynthesis in breadfruit as reflected by its biochemical composition (Zhou and Underhill, 2021).

Utilization of Germplasm

The ITPGRFA is supported by two global action plans for the conservation and sustainable utilization of PGRFA which have advocated for the complementarity between conservation and utilization. The Second Global Plan of Action, which also responds to concerns for the sovereign rights of nations over their genetic resources and international interdependence on PGRFA, identified the following five activities to promote sustainable utilization of PGRFA (FAO, 2011):

- expanded characterization, evaluation, and development of specific subsets of collections to facilitate germplasm use
- support for plant breeding, genetic improvement, and base-broadening efforts
- promotion of crop production diversification and greater crop diversity for sustainable agriculture

- promotion of the development and commercialization of all varieties, mainly farmers' varieties or landraces and underutilized species
- support for planting material production and distribution

As indicated above, the natural genetic variation evident in *A. altilis* diploids and triploids, both seeded and unseeded, has been influenced by *A. camansi* and *A. mariannensis*. Types with desirable agronomic and consumption traits have been selected over time and indicate the key characteristics that growers and consumers prefer and should be considered in any breadfruit crop improvement program. Issues to be addressed include the alleviation of hunger and poverty, nutritional deficiencies, non-communicable diseases due to inappropriate diets, greater resilience of the crop to biotic and abiotic stresses that can reduce supplies, best production and post-production practices, information, and human capacity building. The objectives of germplasm collections, whether publicly or privately funded, and the genetic variability of the accessions will influence the extent of their use for the identification and selection of superior types for distribution. The collections held in the Pacific typically have a wider genetic base than those outside the region and would be more suitable for crop improvement research geared towards the development of new types. A sustainable approach to utilization should also avoid narrowing the genetic base which could increase susceptibility to biotic and abiotic stresses and reduce the resilience of the crop and production systems (Salgotra and Chauhan, 2023). Important lessons can be learned in this regard from the recent introductions of breadfruit cultivars in the Caribbean. In the case of breadfruit, attention to the broader agro-biodiversity, for example, micro-organisms, that support crop production, availability for consumption, and novel products. Additionally, stakeholder consultation and participatory approaches to crop improvement and crop production system research in the producing regions are key to ensuring optimum integration of local knowledge for increased multiplication, production, and consumption of PGRGA (Koohafkan and Altieri, 2011; Koohafkan and Cruz, 2011; Palanivel and Shah, 2021). This section will assess the sustainable utilization of breadfruit genetic resources held *in situ* and especially in *ex-situ* collections.

Expanded Characterization, Evaluation, and Development of Core Collections

While the characterization of accessions for agronomic and consumer-related traits has been in progress at several collections, only one has reported the identification of a core selection. Of the 120 cultivars held at the NTBG, 20 core or 'utility' (elite) cultivars were selected for more detailed study and distribution based on their widespread cultivation, consumer preference, year-round production, yield, fruit quality, and nutritional composition (Ragone, 2007, 2016). The characterization and evaluation being done at the collections in Vanuatu (Navarro et al., 2007) and St. Augustine (Roberts-Nkrumah, 2018) also aim to select economically promising cultivars to facilitate local use. At the St. Augustine Campus, the objective is to select from the recently introduced cultivars to add to the already available local germplasm.

Utilization in Breeding and Improvement Programs

The key constraints to increased breadfruit production and consumption to address the relevant SDGs are related to species-specific or inherent characteristics, environmental factors, and socio-economic challenges. The major inherent characteristics that adversely affect supply include planting material availability for commercial orchards, height of mature trees, seasonal bearing susceptibility to disease, and short shelf life (Leakey, 1977; Roberts-Nkrumah, 2007). Susceptibility to wind damage, drought, waterlogging, and storm surges that increase soil salinity are the key environmental threats. Socio-economic deterrents include competition from imported staples with changing lifestyles and food tastes, as well as traditional consumer preferences (Roberts-Nkrumah, 2007; Taylor and Tuia, 2007). The solution to many of these problems lies within the genetic diversity present in breadfruit.

Breeding programs intended to address these constraints to breadfruit production and consumption have not been reported. However, of the 120 cultivars held at the NTBG, 20 core or 'utility' (elite) cultivars were selected for more detailed study and distribution based on their widespread cultivation, consumer preference, year-round production, yield, fruit quality, and nutritional composition (Ragone, 2007; McGregor et al., 2016; Ragone, 2016). Most of the germplasm collection activities within the Pacific and in the Caribbean have been geared towards expanding the germplasm base by the introduction of new cultivars and many of them were obtained from the NTBG. With respect to the production constraints, both inherent and environmental, research has continued to evaluate introduced and previously existing cultivars within the local context prior to release to producers. For example, in Kiribati, cultivars were screened for growth habit, seasonality, fruit yield, and quality (Redfern, 2007). In Trinidad and Tobago, cultivars were evaluated for early tree growth and yield (Roberts-Nkrumah, 1998), phenology (McKenzie, 2000), pattern of fruit set on canopies (Roberts-Nkrumah, 2002), grafting on *A. camansi* rootstock for tree-decline control (Solomon Jr et al., 2012; Solomon Jr, 2018), fruit rots (Roberts-Nkrumah and Daley, 2014) and response to pruning and growth regulators to control tree height (McKenzie, 2000).

Breadfruit germplasm has been used for crop improvement using non-plant breeding techniques such as traditional grafting. Tree dieback which can result in tree loss and reduced yield (Coates-Beckford and Perreira, 1992) was addressed by grafting stem cuttings of selected, susceptible Caribbean cultivars on 6-week-old seedlings of breadnut which are unaffected by this disease. Grafted plants established well in the field, however, grafting success varied among cultivars (Solomon Jr et al., 2012). 'Ma'afala', an introduced cultivar, is also very susceptible to the disease (Roberts-Nkrumah, 2014). Solomon Jr et al. (2016) demonstrated epicotyl grafting of *in vitro* generated 'Ma'afala' scion (micro cuttings) on 1–4-week-old breadnut seedling rootstock produced high grafting success rates. This merging of biotechnological and grafting techniques to produce planting material can be applied to other economically valuable but susceptible accessions. A dwarf breadfruit phenotype is desirable to reduce tree loss from strong tropical winds and to facilitate high-density planting (Zhou and Underhill, 2022). These researchers reported the induction of a dwarfing phenotype when the breadfruit cultivar 'Noli' scion grown from stem cuttings was grafted onto *Artocarpus lakoocha* seedling rootstock; the grafted breadfruit scion showed reduced stem height and thickness and reduced internode length (Zhou and Underhill, 2022). This indicates that *A. lakoocha* rootstock has the potential to reduce tree height by reducing tree vigor (Zhou and Underhill, 2022). Additionally, *in vitro*-generated plantlets of breadfruit were used by Shandil and Tuia (2016) to assess the salt tolerance of the breadfruit cultivar 'Uto Dina', again this demonstrates that biotechnological approaches can be used to accelerate the identification of cultivars with desirable traits which can then be used for crop improvement. Breadfruit cultivars have been used to conduct field trials in several locations in Hawaii, which vary based on climate and soil type in an effort to determine how cultivars perform under varied conditions, this information will identify the best growing areas for each of the cultivars tested in the trial (Lincoln et al., 2019).

Conventional breeding may not be applicable for triploid or sterile breadfruit types which produce little or no viable pollen and seeds. Furthermore, conventional breeding can be a very slow process for fertile breadfruit types due to their long juvenile phase and heterogeneous genetic background (Zhou et al., 2014). Alternatively, molecular breeding and mutation breeding can be considered and have been successfully used in other tropical fruit tree crops. Regardless of the method of crop improvement and breeding used, there should be a clear goal to support utilization while conserving or improving the genetic diversity of the species. Based on existing constraints to breadfruit commercialization and activities for sustainable utilization, there should be a focus on identifying a range of cultivars that can be grown together to reduce the seasonality of supply. Additionally, cultivars should be appropriate for, agro-processing, have good post-harvest potential, high-yielding and have good eating quality and taste (Ragone, 1997).

Cultivars Released

In recent times, efforts have been made to promote breadfruit production and utilization in both traditional and new markets through the distribution of planting materials to several countries in Africa, Asia, Latin America, and the Caribbean. The NTBG has played a crucial role in these distributions. Starting in 1990, 30 accessions representing 22 cultivars of breadfruit were brought to the Caribbean to expand the range of cultivars available for commercialization (Roberts-Nkrumah, 2018). Some cultivars from the NTBG collection were also sent to Kiribati in 1993 (Redfern, 2007) and New Caledonia in 1999 (Lebegin and Desprez, 2007) to establish field trials to assess their productivity and fruiting patterns compared to local cultivars. These distributions relied on root shoots/cuttings, stem cuttings, or seeds and experienced low survival rates. Furthermore, it was difficult to source adequate quantities of roots at the optimal stage of development and condition for successful propagation and often resulted in damage to trees' root systems. More recently, there have been several global distributions of micropropagated breadfruit plants. Between 2009 and 2023 the Trees That Feed Foundation distributed more than 30,000 tissue-cultured breadfruit plantlets to more than 30 countries across the globe. Approximately 83% of the plantlets distributed were cultivar 'Ma'afala' (Mary Mclaughlin, personal communication, 08 Feb. 2024). The remainder comprises cultivars 'Paua'a', 'Otea', 'Pii Pii', and 'Ulu Fiti'. These plantlets were produced by Cultivaris, Global Breadfruit, and the University of British Columbia.

In the Caribbean, sustainable use of breadfruit germplasm can address key SDGs especially, SDGs 2 and 3, namely zero hunger and food and nutrition security, consequently, commercial stands and nurseries of recently introduced and existing cultivars for increased cultivation and consumption were encouraged. However, recent efforts relying on recently imported cultivars showed the need for evaluation of agronomic and consumption-related traits before release to farmers to ensure that the goals are met. Among producers and processors in Barbados, Jamaica, and Trinidad and Tobago, recently introduced cultivars that are highly susceptible to local diseases or adverse environmental conditions or do not meet consumer requirements in terms of size, cooking methods, and eating quality have not met with expected levels of acceptance. Furthermore, although micropropagated plantlets are more amenable to long-distance and cross-border transport, the lack of proper facilities and technical expertise to harden and acclimatize plants in the receiving country resulted in low survival rates. Consequently, orchard establishment with local cultivars continue and is being investigated in Barbados (A.T. Alleyne, personal communication, Roberts-Nkrumah, 2014; S.A. Webster, personal communication). However, the new introductions represent a broadening of the germplasm base in the region, and in suitable environments, with desirable traits such as a shorter time to bearing, bearing in off-season, and more compact habit, they can prove useful in-home gardens and in mixed cropping or agroforestry systems. It is reported that as part of the Trees that Feed project (Barbados - Trees That Feed Foundation), breadfruit trees imported into Barbados in 2012 have been successfully used in the Barbados trailways project for both horticulture and nutritional purposes.

In accordance with the System-wide Genetic Resources Program (SGRP) of the Convention on Biological Diversity (CBD) and the ITPGRFA protocols for access to and benefit sharing system, breadfruit genetic resources are available for research and breeding. conservation and training for food and agriculture (SGRP, 2010). The Standard Transfer Agreement (STA) with fixed conditions governs access and all benefits arising from their use are to be shared, including benefits from commercialization. The NTBG earns a royalty on every plant sold by Cultivaris and through a Memorandum of Understanding with the Government of Samoa, 50% of this income is shared with the government (Ragone, 2016). Information obtained from research on the germplasm held at the UWI St Augustine Campus is freely available and planting materials from the collection have been shared, without charge, with the Mona Campus on behalf of Jamaica or within Trinidad and Tobago, therefore, there was no need for the STA or revenue sharing. There have been no requests from other countries for planting material, but any future requests will be treated in compliance with the existing protocols, other legal provisions of countries from which the requested materials were obtained, and plant quarantine regulations.

Conservation of Useful Germplasm/Genetic Resources

The inclusion of breadfruit on the ITPGRFA Annex 1 list of crops for conservation highlights the essential link between the ability to continue to engage in the utilization of its genetic resources towards fully exploiting the crop for food and nutrition security, directly or indirectly and the protection of these resources as a biological good and service for present and future generations. Given the origin and existence of the greatest breadfruit diversity in the Pacific, Diversity International (formerly, the International Plant Genetic Resources Institute), assigned the SPC the responsibility for the development of a strategy for the conservation and use of breadfruit genetic resources (Global Crop Diversity Trust, 2007). Subsequently, a Breadfruit Conservation Strategy was developed in 2007 based on collaboration with the Breadfruit Institute and curators of existing breadfruit collections globally, mainly in the Pacific, the Caribbean, and Africa. Curators participated both in a survey on the status of existing collections and the First International Breadfruit Symposium, which was held in Fiji and co-convened by the SPC, the EU-ACP Technical Centre for Agricultural and Rural Cooperation (CTA), a major funder, and the e Breadfruit Institute. At the symposium, breadfruit diversity and conservation were also major themes of the presentations, and a focus session was held to develop a global strategy for the conservation, use, and sharing of breadfruit germplasm (Global Crop Diversity Trust, 2007; Ragone and Wiseman, 2007). Currently, the previously described *ex-situ* collections curated in the Pacific and the Caribbean, the two regions with the greatest interest in breadfruit utilization, are the main areas in which its genetic resources are conserved.

Breadfruit conservation involves both *in situ* and *ex situ* conservation strategies. Figure 5 provides a proposed model for the general conservation of breadfruit. Given the importance of the genetic resources of the breadfruit, as reflected under the multilateral agreement of the IPGRFA, the full range of genetic resources of the breadfruit complex needs to be preserved and safeguarded against all forms of biotic and abiotic hazards. *In situ* conservation, although not foolproof, permits the continuation of the evolutionary processes that undergird the genetic diversity and adaptability of populations of plant species to be maintained in their natural or cultivated habitat (Braverman, 2014). SPC has been entrusted with the responsibility to develop and implement a strategy to guide the development of a global conservation network (Global Crop Diversity Trust, 2007).

Ex-Situ Conservation Strategies

Seedless breadfruit is propagated vegetatively. Since the economically important cultivars do not produce viable seeds, the conservation of genetic resources is traditionally done by establishing accessions in open-field conservation systems. Open-field conservation systems are labor-intensive and compete with other crops for land with the best land-use capabilities. Furthermore, open-field conservation systems are exposed to the vagaries of abiotic and biotic hazards. Considering these disadvantages, even with duplication or triplication, present and future conservation systems should not rely on open-field systems as the sole means of conserving breadfruit genetic resources (Global Crop Diversity Trust, 2007; Murch et al., 2007). Recent advancements in plant biotechnology have opened new opportunities for propagating and conserving breadfruit *in vitro*. Plant tissue culture techniques have been applied to both breadfruit and breadnut. The 'Yellow- heart' breadfruit cultivar was propagated using shoot tip culture using explants collected from field-grown mature trees (Rouse-Miller and Duncan, 2000) and breadnuts (Rouse-Miller and Duncan, 2007). Micropropagation of breadfruit has also been reported by Murch et al. (2007) using axillary shoot tips from mature and juvenile trees. The Breadfruit Institute in collaboration with Cultivars, Global Breadfruit has developed *in vitro* systems for propagating and conserving breadfruit (Ragone, 2013). The *in vitro* culture of two Samoan varieties has been reported by Shandil and Tuia (2015). Through micropropagation 45 breadfruit cultivars and two related species have been successfully cultured for mass production and distribution as well as conservation of genetic material (Ragone, 2016). CePaCT has also established a network of laboratories for the conservation of several species of plant genetic resources including breadfruit, which are of food and nutrition security interest (Pacific Community, 2019). The *ex situ* field collection held at The University of the West Indies, St. Augustine, Trinidad and Tobago is not duplicated as an *in vitro* collection. However, globally an *in vitro* held germplasm collection for breadfruit and the seeded breadnut will become increasingly important as climate change impacts such as increased drought and waterlogging as well as more frequent hurricanes which can impact field collections occur more frequently.

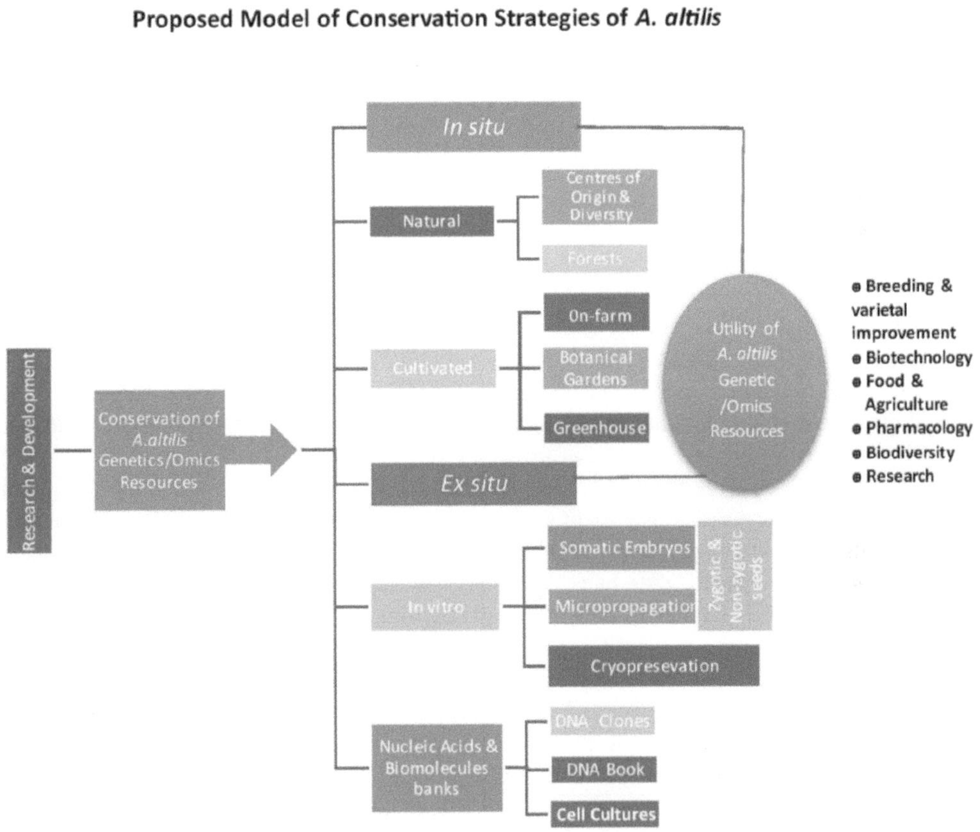

Figure 5. A proposed model for conservation of breadfruit (*A. altilis*) (Photo credit: Seymour Webster)

All *in vitro* protocols thus far have involved organogenic pathways using explant with pre-formed meristems. Despite these biotechnological advancements, the manipulation of the *A. altilis in vitro* lags when compared to the progress made in other tree crop species of nutritional and economic significance (Fang et al., 2004; Kamle et al., 2011; Guan et al., 2016; Mignon and Werbrouck, 2018; Ghadirzadeh-Khorzoghi et al., 2019; Wang et al., 2023). For example, reports on the exploration of protocols for the elicitation of somatic embryogenic competence from explants of the breadfruit, have been absent, except for the unsuccessful attempt reported by Shandil and Tuia (2015).

The development of a successful somatic embryogenesis system would provide a more efficient and economical approach to propagating and conserving the breadfruit genetic resources (Rouse-Miller and Duncan, 2000; Ikeda et al., 2006; Murch et al., 2007; Shandil and Tuia, 2015). With regards to the cryopreservation of varieties of the breadfruit, no published account was identified. Cryogenics is a powerful system for preserving or conserving vulnerable plant germplasm that are seedless or produce recalcitrant, or infertile seeds. Somatic embryos or non-zygotic seeds are amenable to cryopreservation and considering all the other advantages, more efforts should be invested in developing an efficient repetitive somatic embryogenesis system for breadfruit. The advent of 'omics' technologies also presents emerging opportunities to exploit the genetic resources of the breadfruit complex. For example, the setting of a breadfruit multiomic/bioinformatics database for research geared towards unraveling the structural and functional biological systems, conserving, and sharing genetic resources amongst plant breeders and biotechnologists, would represent a watershed achievement for breadfruit as a sustainable food and nutrition security crop. In general, the development of efficient, multimodal propagation and conservation systems for breadfruit and other *Artocarpus* species, could also provide a gateway for the application of novel, unconventional crop improvement technologies for exploiting the wide range of genetic diversity of breadfruit.

Conservation of Breadfruit in the Pacific

The largest and most established ex-situ conservation of breadfruit is located at Kahana Gardens, NTBG, Hawaii, USA, which is managed by the Breadfruit Institute (Jones et al., 2013). The NTBG collection comprises more than 150 different species and cultivars in the breadfruit complex, which are represented by more than 300 accessions (NTBG, 2019). These accessions were collected from more than 30 islands of the Pacific, Seychelles, Indonesia, the Philippines, and Honduras (NTBG, 2019). The Breadfruit Institute, an affiliate of the NTBG, core responsibility is to promote the conservation, and use the diversity in breadfruit through science-based research, curation and preservation of accessions, sustainable agroforestry, and education (NTBG, 2019). Besides being the largest and most diverse, the accessions have been characterized at the morphological and molecular levels, each accession mapped to its original habitat. Trees are identified with standard labeling procedures and have a full-time staff of Conservators and Curators.

Conservation efforts by individual countries within the Pacific Regions, due to their small sizes and number of candidate accessions present in gene banks, are pooled together and are represented as a spatially contiguous genescape. More specifically, the Centre for Pacific Crops and Trees (CePaCT) is a regionalized multi-species gene bank of the Pacific Region (Pacific Community, 2019). The foremost role of CePaCT is to assist countries and territories of the Pacific Region to sustainably conserve and utilize indigenous Plant Genetic Resources as well as to source extraterritorial crops with improved genetic diversity to address gaps in food and nutrition security that might exist in the regional food systems and to improve resilience to abiotic and biotic stressors (Pacific Community, 2019). CePaCT's objective-driven activities within Pacific Islands and Territories are implemented and executed through the Pacific Agricultural Resources Network, which includes approximately 22 countries of the Pacific Region and four other participating countries—USA, Australia, France, and the UK—outside of the Pacific Region (Pacific Community, 2019). In terms of breadfruit, approximately 150 accessions were reported from eight countries and territories of the Pacific Region (Global Crop Diversity Trust, 2007). In another report, approximately 110 accessions were identified from eight countries, in the region (Regional Strategy for the *ex-situ* Conservation and Utilization of Crop Diversity in the Pacific Islands (Ragone, 2006).

Conservation of Breadfruit in the Caribbean

The UWI, St Augustine Campus in Trinidad and Tobago is the major site at which breadfruit germplasm is curated in the Caribbean. This collection is an example of the advantage of duplication of field collections at alternative sites as a conservation strategy. With funding provided by the JADF for breadfruit improvement in Jamaica, the intention of the UWI, Faculty of Agriculture, was to establish the major gene bank in Jamaica consisting of cultivars existing in Jamaica and those introduced from the NTBG, through collaboration between the faculty office at the Mona Campus, the Department of Life Sciences of the Faculty of Science and Technology which had propagation facilities and the Ministry of Agriculture, which had committed to provide at least 2 ha of land at its research station at Bodles, and other stations such as at Orange River, if necessary. Therefore, 75% of the planting material from the NTBG was sent to the Mona Campus for propagation and subsequent field establishment. The remainder of the material was sent to St. Augustine to establish a duplicate collection of the introduced germplasm, with cultivars collected locally and in the Eastern Caribbean, mainly from St. Vincent and the Grenadines. When the root cuttings at the Mona Campus did not survive, rooted adventitious shoots, without soil, were sent to the Mona Campus where they were planted in a small field managed by the Faculty of Agriculture office. Duplicates were sent to the College of Agriculture, Science, and Education in Portland, where some accessions were established while others were lost. The trees at these two sites grew to maturity, but even with the provision of a data collection instrument, no data were collected and the trees at the Mona Campus were eventually lost in a fire. Similarly, early attempts to engage public institutions in Trinidad and Tobago such as the Tobago House of Assembly in collaborative, multi-locational evaluation of germplasm and development of alternate locations for conserving the collection or the more promising cultivars for local needs were unsuccessful, as no data were collected.

The main advantages of the St. Augustine Campus as a site for conservation of breadfruit germplasm were the lower risk of loss due to hurricane damage because Trinidad is south of the hurricane belt in the region, the availability of trained and experienced staff and suitable facilities for plant propagation at the Faculty of Food and Agriculture, access to land at the nearby University Field Station of the faculty and direct supervision of all activities related to propagation, field maintenance and data collection by the curator who was also the principal researcher. Among the success factors, access to physical and human resources and sustained research interest were critical to the successful conservation of the collection at this site and its utilization for germplasm evaluation, training, and information sharing. A database with characterization information based on morphological and evaluation descriptors based on traits of agronomic importance has been developed and this information, as well as utilization descriptors for 33 of the accessions, has been published (Roberts-Nkrumah, 2018). Other publications on morphological and molecular characterization, phenology, pruning, susceptibility to fruit rots, nutrient composition, resistant starch, starch characteristics, and sensory evaluation based on this collection have been identified under Section 4.0.

However, this field collection faces important constraints to the conservation of the germplasm. Tree decline and death associated with pathogens, including *Phytophthora* spp., termites, and soil water management, especially drought, lead to tree loss. Propagation facilities allow maintenance and regeneration of the collection as adventitious shoots that are air-layered to facilitate replacements in the field, however, an *in vitro* collection is not available. The income from the sale of fruits contributes towards disease management, but inadequate funding is the key limitation for field maintenance, including irrigation equipment, timely weed control and pruning, and dedicated staff for field and laboratory operations. Funding is also needed to address suitable space for the preservation of herbarium specimens as samples are subject to deterioration and must be replaced.

While more funding is required to advance the conservation and use of breadfruit genetic resources both in the Pacific and the Caribbean, one critical difference that has contributed to the lack of funding from external sources for the improvement of the crop at national or regional levels in the Caribbean is that breadfruit has not been specifically identified as a major crop for food and nutrition security, with two exceptions. Consistent with its existing agricultural policy which identified breadfruit and breadnut as crops to contribute towards agricultural diversification for food and nutrition security, St.Kitts and Nevis, obtained funding from the FAO between 2011 and 2012 to support the development of a breadfruit and breadnut industry (FAO, 2016). Existing cultivars were identified in a field survey and training in propagation and orchard establishment was provided but there was no request to expand the germplasm base. In 2018, the National Agricultural Research and Extension Institute of Guyana requested FAO assistance in identifying existing cultivars and training in propagation to support the government's expressed interest in expanding breadfruit production for food security and value addition. A subsequent report indicated that the country had received donations of germplasm for several crops, including breadfruit, to broaden the genetic base but progress was slow. Conservation of *ex-situ* collections was usually unsuccessful, and efforts were being made to establish an *in vitro* breadfruit collection (FAO, 2021). It is unclear to what extent these efforts are being pursued currently as both countries have experienced changes in government which often result in changes in policy or crops of focus or both. Therefore, while the absence of research interest, and inadequate physical and human resources at public institutions to participate in germplasm conservation and utilization may be due to lack of government funding, the core problem appears to be the absence of consistent policies that explicitly identify breadfruit as an important crop for food and nutrition security or any other SDG, which limits access by the UWI to funding for conservation activities, such assistance may be available from international agencies. Possibly, partnerships with private sector companies with an interest in breadfruit commercialization for funding the maintenance and germplasm collections under terms that uphold the ITPGRFA for access and benefit-sharing and meet other legal requirements of the entities involved.

The USDA/ARS has a collection of breadfruit accessions, which are maintained by the National Plant Germplasm System at the Pacific Basin Tropical Plant Genetic Resources Management Unit, Hilo, Hawaii, and the National Germplasm Repository, Mayaguez, Puerto Rico. The collection comprises 33 accessions of breadfruit, including 24 duplicate accessions from the NTBG collection (Global Crop Diversity Trust, 2007).

Conclusion

Conservation of breadfruit genetic resources is paramount to the sustainable utilization of this underutilized crop towards the achievement of SDGs. However, several constraints limit both conservation and sustainable utilization. These constraints include planting material availability for commercial orchards, height of mature trees, seasonal bearing, susceptibility to disease, short shelf life, susceptibility to wind damage, drought, waterlogging, and storm surges that increase soil salinity, competition from imported staples with changing lifestyles and food tastes, as well as traditional consumer preferences. Breadfruit's wide genetic diversity is very encouraging in finding solutions to many of these problems. However, breadfruit is generally understudied and efforts to promote its commercialization and utilization lags relative to other crops in most producing countries. Therefore, clear and consistent regional and national goals, policies, and actions, particularly to fund physical and human resource requirements, are needed to facilitate the conservation and sustainable utilization of this very important tropical crop. Several collections have been reported, but there is still a need to expand characterization, evaluation, and identify core or 'utility' cultivars in producing regions to facilitate germplasm use. Furthermore, many breadfruit cultivars are seedless and cannot be improved through conventional breeding methods. Therefore, crop improvement efforts by using grafting and biotechnological approaches will rely heavily on having in-depth characterization and evaluation of existing germplasm. Some breadfruit cultivars have shown resilience to the effects of climate change and other environmental issues. Breadfruit has also been used in many different cropping systems with the best results reported for cropping systems with high biodiversity. This must be promoted to show breadfruit as an important crop for sustainable agricultural systems such as agroforestry and food forest.

References

Amorim, T.S., I.H.P. Andrade, C.G. Otoni, G.P. Camilloto, and R.S. Cruz. 2021. Tailoring breadfruit (*Artocarpus altilis*) starch: Cross-linking starch from this non-conventional source towards improved technologically relevant properties and enabled food applications. *Starch - Stärke, 73*: 1–6.

Andrade, I.H.P., C.G. Otoni, T.S. Amorim, G.P. Camilloto, and R.S. Cruz. 2020. Ultrasound-assisted extraction of starch nanoparticles from breadfruit (*Artocarpus altilis* (Parkinson) Fosberg). *Colloids Surf. A: Physicochem. Eng. Asp., 586*: 124277.

Anwar, S.H., D. Hasni, S. Rohaya, M. Antasari, and C. Winarti. 2020. The role of breadfruit osa starch and surfactant in stabilizing high-oil-load emulsions using high-pressure homogenization and low-frequency ultrasonication. *Heliyon, 6*: 1–11.

Atchley, J., and P.A. Cox. 1985. Breadfruit fermentation in Micronesia. *Econ. Bot., 39*: 326–335.

Audi, L., G. Shallow, E. Robertson, D. Bobo, D. Ragone, E.M. Gardner, B. Jhurree-Dussoruth, J. Wajer, and N.J.C. Zerega. 2023. Linking breadfruit cultivar names across the globe connects histories after 230 years of separation. *Curr. Biol., 33*: 287–297.

Aurore, G., J. Nacitas, B. Parfait, and L. Fahrasmane. 2014. Seeded breadfruit naturalized in the Caribbean is not a seeded variety of *Artocarpus altilis*. *Gen. Resour. and Crop Evol., 61*: 901–907.

Badrie, N., and J. Broomes. 2010.Beneficial uses of breadfruit (*Artocarpus altilis*): Nutritional, medicinal and other uses. *In*: Watson, R.R., and Preedy, V.R. [Eds.], *Bioactive Foods in Promoting Health*, 491–505. Elsevier, Academic Press.

Bahado-Singh, P.S., A.O. Wheatley, M.H. Ahmad, E.Y.S.A. Morrison, and H.N. Asemota. 2006. Food processing methods influence the glycaemic indices of some commonly eaten West Indian carbohydrate-rich food. *B.J. of Nutri., 96*: 476–481.

Bawa, S.H., and M. Webb. 2016. Nutritional and health effects of the consumption of breadfruit. *Trop. Agric., 93*: 52–69.

Beccaro, G.L., T. Soifoini, D. Donno, V. Jeannoda, N.E. Rakotoniaina, S. Hamidou, S.M. Achmet, K. Afraitane, and C. Giacoma. 2020. *Artocarpus altilis* (Parkinson) Fosberg (breadfruit) nutritional characterization and local uses in Comoros Islands. *Acta Hortic.*, Article 1267: 27–32.

Beckford, G.L. 1985. Caribbean peasantry in the confines of the plantation mode of production. *Int. Soc. Sci. J., 37*: 401–414.

Bezerra, E.d.A., E.d.N. Santos, N.S.d. Farias, and M.T. Cavalcanti. 2019. Coating based on breadfruit starch (Artocarpus *altilis*) for fruit conservation: Influence of glycerol, sorbitol, and mannitol as plasticizers. *Food Sci. and Technol., 39*: 389–405.

Blanco, F.M. 1837. *Flora de filipinas: Según el sistema sexual de linneo Imprenta de Sto.* Thomas, Manila.

Bligh, W. 1976. *The Log of H.M.S. Providence 1791–1793.* Genesis Publications Limited, England.

Bowers, R.D. 1981. Breadfruit: A low energy requirement source of carbohydrate for the wet tropics. *Entwicklung und Ländlicher Raum, 15*: 11–13.

Braverman, I. 2014. Conservation without nature: The trouble with *in situ* versus *ex situ* conservation. *Geoforum, 51*: 47–57.

Broomes, J., N. Badrie, and L.B. Roberts-Nkrumah. 2009. Effect of cultivar and maturity on yield and nutrient composition of breadfruit (*Artocarpus altilis*) flour. *In*: Lugo, W.I., and Colón, W. [Eds.], *The Proceedings of the 45th Annual Meeting of the Caribbean Food Crops Society*, Frigate Bay, Federation of St. Kitts and Nevis, 138–140.

Carrington, C.M.S., and C.K. Sankat. 2011. Breadfruit (*Artocarpus altilis* (Parkinson) Fosberg). *In*: Yahia, E.M. [Ed.], *Postharvest Biology and Technology ofTtropical and Subtropical Fruits*: Vol 2: *Acai to Citrus*, 251–271. Woodhead Publishing Limited, Cambridge.

Chan, E.W.C., S.K. Wong, J. Tangah, and H.T. Chan. 2018. Chemistry and pharmacology of artocarpin: An isoprenyl flavone from *Artocarpus* species. *Syst. Rev. in Phar., 9*: 58–63.

Clark, E.A., and F.M. Aramouni. 2018. Evaluation of quality parameters in gluten-free bread formulated with breadfruit (*Artocarpus altilis*) flour. *J. of Food Qual., 2018*: 1–18.

Clement, W.L., and G.D. Weiblen. 2009. Morphological evolution in the mulberry family (Moraceae). *Syst. Bot., 34*: 530–552.

Coates-Beckford, P.L., and M.J. Perreira. 1992. Survey of root-inhabiting microorganisms on declining and non-declining breadfruit (*Artocarpus altilis*) in Jamaica. *Nematropica, 22*: 55–65.

Daley, O.O., L. B. Robert-Nkrumah, and A.T. Alleyne. 2016. Sensory and instrument assessment of colour and texture among breadfruit [*Artocarpus altilis* (Parkinson) Fosberg] cultivars. *Trop. Agric., 93*: 92–108.

Daley, O.O. 2018. *Ethnobotanical, Morphological, Genetic, and Physicochemical Studies on Breadfruit [Artocarpus altilis (Parkinson) Fosberg] in the Caribbean.* Ph.D. Thesis, University of the West Indies, St. Augustine, Trinidad and Tobago.

Daley, O.O., A.T. Alleyne, L.B. Roberts-Nkrumah, and L.A. Motilal. 2024. Microsatellite sequence polymorphisms reveals substantial diversity in Caribbean breadfruit [*Artocarpus altilis* (Parkinson) Fosberg] germplasm. *Horticulturae, 10*: 1–18.

Daley, O.O., L.B. Roberts-Nkrumah, and A.T. Alleyne. 2020. Morphological diversity of breadfruit [*Artocarpus altilis* (Parkinson) Fosberg] in the Caribbean. *Sci. Hort., 266*: 1–12.

Daley, O.O., L.B. Roberts-Nkrumah, A.T. Alleyne, I. Francis-Granderson, J. Broomes, and N. Badrie. 2019. Assessment of breadfruit (*Artocarpus altilis* (Parkinson) Fosberg) cultivars for resistant starch, dietary fibre, and energy density. *Afr. J. of Food, Agri. and Nutr. Dev., 19*: 15060–15076.

Daley, O.O., L.B. Roberts-Nkrumah, A.T. Alleyne, and M.C. Gloster. 2022. Folk nomenclature and traditional knowledge of breadfruit [*Artocarpus altilis* (Parkinson) Fosberg] diversity in four Anglophone Caribbean countries. *J. Ethnobiol. Ethnomed., 18*: 1–14.

Daniell, H., C. Lin, M. Yu, and W. Chang. 2016. Chloroplast genomes: Diversity, evolution, and applications in genetic engineering. *Genome Biol., 17*: 1–29.

da Silva, C.E., and N. Narain. 2021. Physicochemical characterization and bioactive compounds in breadfruit (*Artocarpus altilis*) and its dried products. *Res. Soc. Dev., 10*: 1–16.

De Bellis, F., R. Malapa, V. Kagy, S. Lebegin, C. Billot, and J. Labouisse. 2016. New development and validation of 50 SSR markers in breadfruit (*Artocarpus altilis*, Moraceae) by next-generation sequencing. *Appl. Plant Sci., 4*: 1–7.

de Souza, U.J.B., R.N. dos Santos, R.N. de Araújo Filho, G.R. dos Santos, R. Almeida Sarmento, F. De Bellis, and F.S. Campos. 2021. The complete chloroplast genome of *Artocarpus altilis* (Moraceae) and phylogenetic relationships. *Mitochondrial DNA Part B, 6*: 2291–2293.

Deivanai, S., and S.J. Bhore. 2010. Breadfruit (*Artocarpus altilis*): An underutilized and neglected fruit plant species. *Middle East J. of Sci. Res., 6*: 418–428.

Du, J., D. Grant, Z. Tian, R.T. Nelson, L. Zhu, R.C. Shoemaker, and J. Ma. 2010. Soytedb: A comprehensive database of transposable elements in the soybean genome. *BMC Genomics, 11*: 113.

Elevitch, C. 2015. *Agroforestry Landscapes for Pacific Islands: Creating Abundant and Resilient Food Systems*. Permanent Agriculture Resources, Hawaii, USA.

Elevitch, C.R., and D. Ragone. 2018. *Breadfruit Agroforestry Guide: Planning and Implementation for Regenerative Organic Methods*. Breadfruit Institute, Hawaii, USA.

Fang, J., A. Wetten, and P. Hadley. 2004. Cryopreservation of cocoa (*Theobroma cacao* l.) somatic embryos for long-term germplasm storage. *Plant Science, 166*: 669–675.

[FAO], Food and Agriculture Organization of the United Nations. 2009. *International Treaty on Plant Genetic Resources for Food and Agriculture*. FAO, Rome, Italy.

[FAO], Food and Agriculture Organization of the United Nations. 2011. *Second Global Plan of Action for Plant Genetic Resources for Food and Agriculture*. (January 20, 2024), https://www.fao.org/agriculture/crops/thematic-sitemap/theme/seeds-pgr/gpa/en/.

[FAO], Food and Agriculture Organization of the United Nations. 2016. *Evaluation of FAO's Contribution to Members of the Organization of Eastern Caribbean States and Barbados, 2010–2015*. Office of Evaluation, FAO, Rome, Italy.

[FAO], Food and Agriculture Organization of the United Nations. 2021. *Second Reporting Cycle Report on the Implementation of the International Treaty on Plant Genetic Resources for Food and Agriculture (ITPGRFA) – Guyana*. Rome: FAO.

[FAO], Food and Agriculture Organization of the United Nations. 2024. *Faostat Statistical database*. Food and Agriculture Organization of the United Nations 2024 [cited January 18, 2024 2024]. Available from https://www.fao.org/faostat/en/#data.

Ferguson, T.S., M.K. Tulloch-Reid, and R.J. Wilks. 2010. The epidemiology of diabetes mellitus in Jamaica and the Caribbean: A historical review. *West Indian Med. J., 59*: 259–264.

Ferrer-Gallego, P.P., and F. Boisset. 2018. The naming and typification of the breadfruit, *Artocarpus altilis*, and breadnut, *A. camansi* (Moraceae). *Willdenowia, 48*: 125–135.

Fosberg, F.R. 1960. Introgression in *Artocarpus* (Moraceae) in Micronesia. *Brittonia, 12*: 101–113.

Fownes, J.H., and W.C. Raynor. 1993. Seasonality and yield of breadfruit cultivars in the indigenous agroforestry system of Pohnpei, Federated States of Micronesia. *Trop. Agric., 70*: 103–109.

Galluzi, G., M. Halewood, I. López Noriega, and R. Vernooy. 2015. *Twenty-five Years of International Exchanges of Plant Genetic Resources Facilitated by the CGIAR Genebanks: A Case Study in International Interdependence*. International Treaty on Plant Genetic Resources for Food and Agriculture, Rome, Italy.

Gardner, E.M., M.G. Johnson, J.T. Pereira, A.S.A. Puad, D. Arifiani, Sahromi, N.J. Wickett, and N.J.C. Zerega. 2021. Paralogs and off-target sequences improve phylogenetic resolution in a densely sampled study of the breadfruit genus (*Artocarpus*, Moraceae). *Syst. Biol., 70*: 558–575.

Ghadirzadeh-Khorzoghi, E., Z. Jahanbakhshian-Davaran, and S.M. Seyedi. 2019. Direct somatic embryogenesis of drought resistance pistachio (*Pistacia vera* l.) and expression analysis of somatic embryogenesis-related genes. *S. Afr. J. Bot., 121*: 558–567.

Global Crop Diversity Trust. 2007. *Breadfruit Conservation Strategy*. Global Crop Diversity Trust, Bonn, Germany. [cited January 28, 2024]. Available from https://cdn.croptrust.org/wp/wp-content/uploads/2019/05/Breadfruit_Strategy_FINAL_14Sept07.pdf.

Goebel, R. 2007. Breadfruit: The Australian scene. *Acta Hortic., 757*: 141–148.

Golden, K.D., and O.J. Williams. 2001. Amino acid, fatty acid, and carbohydrate content of *Artocarpus altilis* (breadfruit). *J. Chromatogr. Sci., 39*: 243–250.

Golden, K.D., and O.J. Williams. 2007. The amino acid, fatty acid, and carbohydrate content of *Artocarpus altilis* (breadfruit): The white heart cultivar from the West Indies. *Acta Hortic., 757*: 201–208.

Gonçalves, N.G.G., J.I.. de Araújo, F.E.A. Magalhães, F.R.S. Mendes, M.D. P. Lobo, A.C. d.O.M. Moreira, and R.d.A. Moreira. 2020. Protein fraction from *Artocarpus altilis* pulp exhibits antioxidant properties and reverses anxiety behavior in adult zebrafish via the serotoninergic system. *J. of Funct. Foods, 66*: 103772.

Guan, Y., S.-G. Li, X.-F. Fan, and Z.-H. Su. 2016. Application of somatic embryogenesis in woody plants. *Front. in Plant Sci., 7*: 1–12.

Harsanto, B.W., Supriyanto, I. Kartini, and Y. Pranoto. 2022. The ability of breadfruit starch nanoparticle-stabilized pickering emulsion for encapsulating cinnamon essential oil. *Malays. Appl. Biol., 51*: 83–90.

Havecker, E.R., X. Gao, and D.F. Voytas. 2004. The diversity of ltr retrotransposons. *Genome Biol., 5*: 1–6.

Henry, F.J. 2016. Agriculture, health, and development in the Caribbean: The case for breadfruit. *Trop. Agric., 93*: 10–17.

Huang, S., M.M. Martinez, and B.M. Bohrer. 2019. The compositional and functional attributes of commercial flours from tropical fruits (breadfruit and banana). *Foods*, 8: 1–15.

Ifah, A.A., E. Yuniastuti, and Parjanto. 2018. Analysis of breadfruit plant diversity (*Artocarpus altilis* p.) by random amplified polymorphic DNA (RAPD) in DIY. *AIP Conf. Proc., 2021*: 1–6.

Ikeda, M., M. Umehara, and H. Kamada. 2006. Embryogenesis-related genes: Its expression and roles during somatic and zygotic embryogenesis in carrot and arabidopsis. *Plant Biotechnol J., 23*: 153–161.

[IPGRI], International Plant Genetic Research Institute. 2009. *Key Access and Utilization Descriptors for Breadfruit Genetic Resources*. IPGRI, Rome, Italy.

Jarrett, F.M. 1959. Studies in *Artocarpus* and allied genera, III. *Artocarpus* subgenus *Artocarpus. J. Arnold Arbor., 40*: 113–155.

Jarrett, F.M. 1976. The syncarp of *Artocarpus*: A unique biological phenomenon. *Gard. Bull. Singapore, 29*: 35–39.

Jones, A.M.P., R. Baker, D. Ragone, and S.J. Murch. 2013. Identification of pro-vitamin a carotenoid-rich cultivars of breadfruit (*Artocarpus*, Moraceae). *J. Food Compos. Anal., 31*: 51–61.

Jones, A.M. P., S.J. Murch, and D.C. Ragone. 2010. Diversity of breadfruit (*Artocarpus altilis*, Moraceae) seasonality: A resource for year-round nutrition. *Econ. Bot., 64*: 340–351.

Jones, A.M.P., S.J. Murch, J. Wiseman, and D.C. Ragone. 2013. Morphological diversity in breadfruit (*Artocarpus*, Moraceae): Insights into domestication, conservation, and cultivar identification. *Genet. Resour. Crop Evol., 60*: 175–192.

Jones, A.M.P., D.C. Ragone, K. Aiona, W.A. Lane, and S.J. Murch. 2011. Nutritional and morphological diversity of breadfruit (*Artocarpus,* Moraceae): Identification of elite cultivars for food security. *J. Food Compos. Anal., 24*: 1091–1102.

Jones, A.M.P., D.C. Ragone, N.G. Tavana, D.W. Bernotas, and S.J. Murch. 2011. Beyond the bounty: Breadfruit (*Artocarpus altilis*) for food security and novel foods in the 21st century. *Ethnobot. Res. Appl., 9*: 129–149.

Kamle, M., A. Bajpai, R. Chandra, S. Kalim, and R. Kumar. 2011. Somatic embryogenesis for crop improvement. *GERF Bull Biosci., 2*: 54–59.

Kanzaki, S., K. Yonemori, A. Sugiura, and S. Subhadrabandhu. 1997. Phylogenetic relationships between the jackfruit, the breadfruit and nine other *Artocarpus spp*. From rflp analysis of an amplified region of cpDNA. *Sci. Horticulturae, 70*: 57–66.

Koohafkan, P., and M.A. Altieri. 2011. *Globally Important Agricultural Heritage Systems: A Legacy for the Future*. Food and Agriculture Organization of the United Nations, Rome, Italy.

Koohafkan, P., and M.J.D. Cruz. 2011. Conservation and adaptive management of globally important agricultural heritage systems (giahs). *J. Resour. Ecol., 2*: 22–28.

Koroiveibau, D. 1966. *Some Fiji Breadfruit Varieties*. Department of Agriculture, Suva, Fiji.

Lan, W., C. Tzeng, C. Lin, F. Yen, and H. Ko. 2013. Prenylated flavonoids from *Artocarpus altilis*: Antioxidant activities and inhibitory effects on melanin production. *Phytochem., 89*: 78–88.

Laricchia, K.M., M.G. Johnson, D. Ragone, E.W. Williams, N.J.C. Zerega, and N.J. Wickett. 2018. A transcriptome screen for positive selection in domesticated breadfruit and its wild relatives (*Artocarpus* spp.). *Am. J. Bot., 105*: 915–926.

Latchoumia, J.N., S. Adenet, G. Aurore, K. Rochefort, A. Buléon, and L. Fahrasmane. 2014. Composition and growth of seedless breadfruit *Artocarpus altilis* naturalized in the Caribbean. *Sci. Hortic., 175*: 187–192.

Leakey, C., and L.B. Roberts-Nkrumah. 2016. The introduction of the breadfruit [*Artocarpus altilis* (Parkinson) Fosberg] to the West Indies: The role of Sir Joseph Banks. *Trop. Agric., 93*: 32–40.

Leakey, C.L.A. 1977. *Breadfruit Reconnaissance Study in the Caribbean Region.* Cali, Colombia: CIAT/Interamerican Development Bank.

Lebegin, S., and Z.L. Desprez. 2007. Horticultural evaluation of five introduced and one local breadfruit cultivar in New Caledonia. *Acta Hortic., 757*: 89–92.

Lebot, V. 1999. Biomolecular evidence for plant domestication in sahul. *Genet. Resour. Crop Evol., 46*: 619–628.

Lee, S.I., and N.S. Kim. 2014. Transposable elements and genome size variations in plants. *Genomics Inform., 12*: 87–97.

Li, D., J. Gul, and J. He. 2020. The complete chloroplast genome sequence of *Artocarpus* hypargyreus. *Mitochondrial DNA Part B 5*: 910–911.

Lin, X., C. Feng, T. Lin, A.J. Harris, Y. Li, and M. Kang. 2022. Jackfruit genome and population genomics provide insights into fruit evolution and domestication history in China. *Hortic. Res., 9*: 1–13.

Lincoln, N.K., A. Cho, G. Dow, and T. Radovich. 2019. Early growth of breadfruit in a variety × environment trial. *Agron. J., 111*: 3020–3027.

Lincoln, N.K., D. Ragone, N. Zerega, L.B. Roberts-Nkrumah, M. Merlin, and A.M. Jones. 2018. Grow us our daily bread: A review of breadfruit cultivation in traditional and contemporary systems. *Hortic. Rev.*, *46*: 299–384.

Liu, Y., P.N. Brown, D. Ragone, D.L. Gibson, and S.J. Murch. 2020. Breadfruit flour is a healthy option for modern foods and food security. *PloS One*, *15*: 1–19.

Liu, Y., A.M.P. Jones, S.J. Murch, and D.C. Ragone. 2014. Crop productivity, yield, and seasonality of breadfruit (*Artocarpus spp*. Moraceae). *Fruits*, *69*: 345–361.

Liu, Y., D.C. Ragone, and S.J. Murch. 2015. Breadfruit (*Artocarpus altilis*): A source of high-quality protein for food security and novel food products. *Amino Acids*, *47*: 847–856.

Livingston, C., and N. K. Lincoln. 2023. Determining allometry and carbon sequestration potential of breadfruit (*Artocarpus altilis*) as a climate-smart staple in Hawai'i. *Sustainability*, *15*: 1–12.

Lorens, A.S., and L. Englberger. 2007. The importance and use of breadfruit cultivars in pohnpei, federated states of Micronesia. *Acta Hortic.*, *757*: 101–108.

Magnus, M.H. 1997. The top 25 superfoods. *Cajanus*, *30*: 3–13.

Maharaj, R., and C.K. Sankat. 1990. The shelf-life of breadfruit stored under ambient and refrigerated conditions. *Acta Hortic.*, *269*: 411–424.

Mausio, K., T. Miura, and N.K. Lincoln. 2020. Cultivation potential projections of breadfruit (*Artocarpus altilis*) under climate change scenarios using an empirically validated suitability model calibrated in Hawai'i. *PLoS One*, *15*: 1–17.

McGregor, A. 2007. The export of horticultural and high value agricultural products from the Pacific Islands. *Pacific Economic Bulletin*, *22*: 81–99.

McGregor, A.M., L.D. Tora, and V. Lebot. 2016. Planting breadfruit orchards as a climate change adaptation strategy for the Pacific Islands. *Acta Hortic.*, *1128*: 55–66.

McKenzie, C. 2000. *Breadfruit (Artocarpus altilis Park. Fosberg) Phenology and Growth Responses to Selected Plant Growth Regulators*. M.Phil. Thesis, University of the West Indies, St. Augustine, Trinidad and Tobago.

Medagoda, I. 2007. Breadfruit in the Asian region: Focus on Sri Lanka. *Acta Hortic.*, Article 757: 65–70.

Mehta, K.A., Y.C.R. Quek, and C.J. Henry. 2023. Breadfruit (*Artocarpus altilis*): Processing, nutritional quality, and food applications. *Front. Nutr.*, *10*: 1–13.

Mignon, E., and S. Werbrouck. 2018. Somatic embryogenesis as key technology for shaping the rubber tree of the future. *Front. Plant Sci.*, *9*: 1–6.

Moustache, A.M., and M. Moustache. 2007. History and use of breadfruit in the Seychelles. *Acta Hortic.*, *757*: 135–140.

Murch, S.J., D. Ragone, W.L. Shi, A.R. Alan, and P.K. Saxena. 2007. *In vitro* conservation and micropropagation of breadfruit (*Artocarpus altilis*, Moraceae). *In*: Jain, S. M., and Häggman, H. [Eds.], *Protocols for Micropropagation of Woody Trees and Fruits*, 279–288. Springer, Netherlands.

Navarro, M., S. Malres, J.P. Labouisse, and O. Roupsard. 2007. Vanuatu breadfruit project: Survey on botanical diversity and traditional uses of *Artocarpus altilis*. *Acta Hortic.*, *757*: 81–88.

[NTBG], National Tropical Botanical Gardens 2024. The breadfruit institute of the National Tropical Botanical Garden promotes conservation, study, and use of breadfruit for food and reforestation. National Tropical Botanical Gardens 2019 [cited January 10, 2024]. Available from https://ntbg.org/breadfruit/.

Oduro, I., W.O. Ellis, and S.T. Narh. 2007. Expanding breadfruit utilization and its potentials for pasta production. *Discovery and Innovation*, *19*: 243–247.

Omobuwajo, T.O. 2007. Overview of the status of breadfruit in Africa. *Acta Hortic.*, *757*: 61–64.

Pacific Community. 2024. Centre for Pacific crops and trees (CePaCT). [cited January 28, 2024]. Available from https://www.spc.int/resource-centre/centre-for-pacific-crops-and-trees-cepact.

Palanivel, H., and S. Shah. 2021. Unlocking the inherent potential of plant genetic resources: Food security and climate adaptation strategy in Fiji and the Pacific. *Environ. Dev. Sustain.*, *23*: 14264–14323.

Parham, J.W. 1966. Coconut and breadfruit surveys in the South Pacific region: Report. In *Technical Information Paper No. 1*. South Pacific Commission: Noumea, New Caledonia.

Patil, A.D., A.J. Freyer, L. Killmer, P. Offen, P.B. Taylor, B.J. Votta, and R.K. Johnson. 2002. A new dimeric dihydrochalcone and a new prenylated flavone from the bud covers of *Artocarpus altilis*: Potent inhibitors of cathepsin K. *J. Nat. Prod.*, *65*: 624–627.

Powell, D. 1977. The voyage of the plant nursery, H.M.S.Providence, 1791–1793. *Econ. Bot.*, *31*: 387–431.

Priyantha, N., L.B.L. Lim, D.T.B. Tennakoon, E.T.Z. Liaw, C.H. Ing, and A.B. Liyandeniya. 2018. Biosorption of cationic dyes on breadfruit (*Artocarpus altilis*) peel and core. *Appl. Water Sci.*, *8*: 1–11.

Quartermain, A. 2007. Breadfruit in Papua New Guinea. *Acta Hortic.*, *757*: 109–114.

Quisumbing, E. 1940. The validity of *Artocarpus camansi* Blanco. *Philipp J. Sci.*, *72*: 331–337.

Ragone, D. 1990. Isozyme variation in Pacific Island cultivars of breadfruit. *HortScience*, *25*: 1153 –1154.

Ragone, D. 2001. Chromosome numbers and pollen stainability of three species of Pacific Island breadfruit (*Artocarpus*, Moraceae). *Am. J. Bot.*, *88*: 693–696.

Ragone, D. 2006. Conserving traditional varieties of economic plants: Breadfruit as a case study. Agroforestry.net [Online] Hawaii. [cited January 28, 2024]. Available from http://www.google.com.au/url?sa=t&rct=j&q=&esrc=s&source=web&cd=1&ved=0C 22
C4QFjAA&url=http%3A%2F%2Fwww.agroforestry.net%2Fevents%2Fafwksp2006

%2Fpres%2FRagone_Conserving_Trad_Varieties_script.pdf&ei=E4UmU4LaCabmi Aent4CADA&usg=AFQjCNEYH4QBmDo5cV8as3wkbzMf5BQQw&sig2=kojKIouF4pBgFALQPZy1Q&bvm=bv.62922401,d.aGc.

Ragone, D. 2007. Breadfruit: Diversity, conservation, and potential. *Acta Hortic.*, *757*: 19–30.

Ragone, D. 2024. *Progress report.* Breadfruit Institute 2013 [cited January 28, 2024 2024]. Available from https://ntbg.org/wp-content/uploads/2020/02/2013-BFI-Progress-Reports.pdf.

Ragone, D. 2016. Breadfruit for food and nutrition security in the 21st century. *Trop. Agric.*, *93*: 18–29.

Ragone, D., and C.G. Cavaletto. 2006. Sensory evaluation of fruit quality and nutritional composition of 20 breadfruit cultivars. *Econ. Bot.*, *60*: 335–346.

Ragone, D., C.R. Elevitch, and A. Dean. 2016. Revitalizing breadfruit in Hawaii: A model for encouraging the cultivation and use of breadfruit in the tropics. *Trop. Agric.*, *93*: 213–224.

Ragone, D.C. 1991. *Collection, Establishment, and Evaluation of a Germplasm Collection of Pacific Island Breadfruit.* PhD dissertation, University of Hawaii, Hawaii.

Ragone, D.C. 1995. Description of Pacific Island breadfruit cultivars. *Acta Hortic.*, *413*: 93–98.

Ragone, D.C. 1997. *Breadfruit: Artocarpus altilis (Parkinson) Fosberg-promoting the Conservation and Use of Underutilized and Neglected Crops.* Bioversity International, Rome, Italy.

Ragone, D.C. 2006. *Artocarpus altilis* (breadfruit). *In*: Elevitch, C.R. [Ed.]. *Species Profiles for Pacific Island Agroforestry*, 85–100. Permanent Agriculture Resources (PAR) Hōlualoa, Hawai'i.

Ragone, D.C. 2011. Farm and forestry production and marketing profile for breadfruit (*Artocarpus altilis*). *In*: *Specialty Crops for Pacific Island Agroforestry*. UH Press.

Ragone, D.C., and J. Wiseman. 2007. Developing and applying descriptors for breadfruit germplasm. *Acta Hortic.*, *757*: 71–80.

Ramdath, D.D., R.L.C. Isaacs, S. Teelucksingh, and T.M.S. Wolever. 2004. Glycaemic index of selected staples commonly eaten in the Caribbean and the effects of boiling v. Crushing. *Br. J. Nutr.*, *91*: 971–977.

Redfern, T. 2007. Breadfruit improvement activities in Kiribati. *Acta Hortic.*, *757*: 93–100.

Roberts-Nkrumah, L., and O. Daley. 2014. Cultivar differences in the incidence of fruit rot in breadfruit (*Artocarpus altilis*). *Acta Hortic.*, *1047*: 75–85.

Roberts-Nkrumah, L.B. 1997. Towards a description of the breadfruit germplasm in St. Vincent. *Fruits*, *52*: 27–35.

Roberts-Nkrumah, L.B. 1998. A preliminary evaluation of the imported breadfruit germplasm collection at the University of the West Indies, Trinidad. *In*: Harrison, R., and Young, F.A. [Eds.]. *The Proceedings of the Enhancing Regional Food Security and Exports by Integrating National Strategies: Proceedings of the Caribbean Food Crops Society 34th Annual Meeting*, 12–18 July 1998, Jamaica.

Roberts-Nkrumah, L.B. 2002. Management implications of the distribution of fruit set on breadfruit (*Artocarpus altilis*) trees. *Acta Hortic.*, *632*: 201–207.

Roberts-Nkrumah, L.B. 2007. An overview of breadfruit (*Artocarpus altilis*) in the Caribbean. *Acta Hortic.*, *757*: 51–60.

Roberts-Nkrumah, L.B. 2014. A review of the potential of breadfruit cultivar 'Ma'afala' for commercial production in Trinidad and Tobago. *Trop. Agric.*, *91*: 284–299.

Roberts-Nkrumah, L.B. 2018. *The Breadfruit Germplasm Collection at the University of the West Indies, St. Augustine Campus*. The University of the West Indies Press, Kingston, Jamaica.

Roberts-Nkrumah, L.B., and N. Badrie. 2005. Breadfruit consumption, cooking methods and cultivar preference among consumers in Trinidad, West Indies. *Food Qual. Prefer.*, *16*: 267–274.

Roberts-Nkrumah, L.B., and G. Legall. 2013. Breadfruit (*Artocarpus altilis*,
Moraceae) and chataigne (*A. camansi*) for food security and income generation: The case of Trinidad and Tobago. *Econ. Bot.*, *67*: 324–334.

Rouse-Miller, J., and J.E. Duncan. 2000. *In vitro* propagation of *Artocarpus altilis* (Park.) Fosberg (breadfruit) from mature plant material. *In Vitro Cell Dev. Biol. Plant*, *36*: 115–117.

Rouse-Miller, J., and J.E. Duncan. 2007. *In vitro* regeneration of *Artocarpus camansi* and *Artocarpus altilis*. *Acta Hortic.*, *757*: 153–160.

Sahu, S. K., M. Liu, A. Yssel, R. Kariba, S. Muthemba, S. Jiang, B. Song, P.S. Hendre, A. Muchugi, R. Jamnadass, S. Kao, J. Featherston, N.J.C. Zerega, X. Xu, H. Yang, A. Van Deynze, Y.V. de Peer, X. Liu, and H. Liu. 2020. Draft genomes of two *Artocarpus* plants, jackfruit (*A. heterophyllus*) and breadfruit (*A. altilis*). *Genes*, *11*: 1–17.

Salgotra, R.K., and B.S. Chauhan. 2023. Genetic diversity, conservation, and utilization of plant genetic resources. *Genes*, *14*: 1–20.

Sankat, C.K., and R. Maharaj. 1993. Refrigerated storage of breadfruit and the effect of waxing, packaging and storage in water. *Indian Packers*, *47*: 135–145.

Sankat, C.K., and R. Maharaj. 2007. A review of postharvest storage technology of breadfruit. *Acta Hortic.*, *757*: 183–192.

Sari, D.R., F.F. Ahmad, Y.Y. Djabir, and R. Yulianty. 2020. Breadfruit leaves extract (*Artocarpus altilis*) effect on pancreatic damage in Diabetic Type 11 animal model induced by alloxan–nicotinamide. *Medicina Clínica Práctica*, *3*: 1–4.

Sasuke, N. 1953. Breadfruit, yams, and taros of Ponape Island. [Ed.]. *In*: *The Proceedings of the 7th Pacific Science Congress*, Christchurch, New Zealand.

[SGRP], System-wide Genetic Resources Program 2010. The importance of recognizing the international treaty in the CBD's protocol on access and benefit-sharing. Edited by SGRP. System-wide Genetic Resources Program. Rome, Italy.

Shandil, A.S., and V.S. Tuia. 2015. Micropropagation of breadfruit (*A. altilis*) enhanced using a bioreactor system. *Acta Hortic., 1101*: 159–164.

Shandil, A.S., and V.S. Tuia. 2016. Using an improved micropropagation system combined with access to field evaluation information can support sustainable development of breadfruit. *Trop. Agric., 93*: 116–126.

Sharma, M.R. 1963. Morphological and anatomical investigations on *Artocarpus* Forst. The inflorescence. *Proc. Natl. Acad. Sci. India. Sect. B. Biol. Sci. 29B*: 77–92.

Soifoini, T., D. Donno, V. Jeannoda, E. Rakotoniaina, S. Hamidou, S.M. Achmet, N.R. Solo, K. Afraitane, C. Giacoma, and G.L. Beccaro. 2018. Bioactive compounds, nutritional traits, and antioxidant properties of *Artocarpus altilis* (Parkinson) fruits: Exploiting a potential functional food for food security on the Comoros Islands. *J. Food Qual., 2018*: 1–12.

Solomon Jr, J.F., L. Roberts-Nkrumah, and J. Rouse-Miller. 2016. Breadfruit propagation: Merging plant tissue culture technology and traditional grafting. *Trop. Agric., 93*: 225–229.

Solomon Jr, F.K. 2018. *An Evaluation of Factors Influencing Grafting Breadfruit, Artocarpus altilis (Parkinson) Fosberg, on Chataigne, Artocarpus camansi* (*Blanco*). M.Phil. Thesis, University of the West Indies, Trinidad and Tobago.

Solomon Jr, F.K., L.B. Roberts-Nkrumah, and J.A. Rouse-Miller. 2012. Development of a grafting protocol for the commercial propagation of three West Indian breadfruit cultivars. *Trop. Agric., 89*: 85–98.

Sreekumar, V.B., A.M. Binoy, and S.T. George. 2007. Genetic and morphological variation in breadfruit (*Artocarpus altilis* (Park.) Fosberg) in the western ghats of India using AFLP markers. *Genet. Resour. Crop. Evol., 54*: 1659–1665.

Tarai, R., D. Sahoo, and A.K. Panda. 2017. Breadfruit. *In*: Rana, M.K. [Ed.]. *Vegetable Crop Science*. 1023–1034. CRC Press.

Taylor, M., R. Paroda, H. Jaenicke, and P. Mathur. 2013. Implementing the strategy for the conservation through use of underutilised crops in the Pacific. *Acta Hortic.*, Article 979: 533–539.

Taylor, M.B., and V.S. Tuia. 2007. Breadfruit in the Pacific region. *Acta Hortic., 757*: 43–50.

Trécul, A. 1847. Mémoire sur la famille des Artocarpées. *Ann. Sci. Nat. Bot., 3*: 38–157.

Turi, C.E., Y. Liu, D. Ragone, and S.J. Murch. 2015. Breadfruit (*Artocarpus altilis* and hybrids): A traditional crop with the potential to prevent hunger and mitigate diabetes in Oceania. *Trends Food Sci. Technol., 45*: 264–272.

Umezuruike, A.C., and T.U. Nwabueze. 2017. Characterization of some food formulation functional properties of flour processed from roasted African breadfruit (*Treculia africana*) seeds. *J. Sci. Agric., 1*: 110–116.

[UN], United Nations. 2015. Transforming our world; the 2030 agenda for sustainable development. https://sdgs.un.org/2030agenda.

Vianney, Y.M., S.E.D. Putra, and M.G.M. Purwanto. 2020. Antioxidant and toxicity activity of aqueous extracts from various parts of breadfruit and breadnut. *Int. J. Fruit Sci., 20*: S1639–S1651.

Wang, J., X. Xia, G. Qin, J. Tang, J. Wang, W. Zhu, M. Qian, J. Li, G. Cui, Y. Yang, and J. Qian. 2023. Somatic embryogenesis and plant regeneration from stem explants of pomegranate. *Horticulturae, 9*: 1–14.

Wei, L., H. Wang, D. Chen, X. Ke, Z. Zhu, and H. Wang. 2023. The complete plastome sequence of *Artocarpus altilis* (Parkinson ex F.A. Zorn) Fosberg, (Moraceae). *Mitochondrial DNA Part B, 8*: 154–156.

Wester, P.J. 1924. The seedless breadfruits of the Pacific archipelagoes. *Philippine Ag. Rev., 17*: 24–39.

Wilder, G.P. 1928. *The Breadfruit of Tahiti*. Bulletin 50. Bernice P. Bishop Museum, Honolulu, Hawaii, USA.

Witherup, C., D. Ragone, T. Wiesner-Hanks, B. Irish, B. Scheffler, S. Simpson, F. Zee, M.I. Zuberi, and N.J.C. Zerega. 2013. Development of microsatellite loci in *Artocarpus altilis* (Moraceae) and cross-amplification in congeneric species. *Appl. Plant Sci., 1*: 1–6.

Worrell, D.B., C.M.S. Carrington, and D.J. Huber. 2002. The use of low temperature and coatings to maintain storage quality of breadfruit, *Artocarpus altilis* (Parks.) Fosb. *Postharvest Biol. Technol., 25*: 33–40.

Xu, X., S. Pan, S. Cheng, B. Zhang, D. Mu, P. Ni, G. Zhang, S. Yang, R. Li, J. Wang, G. Orjeda, F. Guzman, M. Torres, R. Lozano, O. Ponce, D. Martinez, G. De la Cruz, S.K. Chakrabarti, V.U. Patil, K.G. Skryabin, B.B. Kuznetsov, N.V. Ravin, T.V. Kolganova, A.V. Beletsky, A.V. Mardanov, A. Di Genova, D.M. Bolser, D.M.A. Martin, G. Li, Y. Yang, H. Kuang, Q. Hu, X. Xiong, G.J. Bishop, B. Sagredo, N. Mejía, W. Zagorski, R. Gromadka, J. Gawor, P. Szczesny, S. Huang, Z. Zhang, C. Liang, J. He, Y. Li, Y. He, J. Xu, Y. Zhang, B. Xie, Y. Du, D. Qu, M. Bonierbale, M. Ghislain, M. del Rosario Herrera, G. Giuliano, M. Pietrella, G. Perrotta, P. Facella, K. O'Brien, S.E. Feingold, L.E. Barreiro, G.A. Massa, L. Diambra, B.R. Whitty, B. Vaillancourt, H. Lin, A.N. Massa, M. Geoffroy, S. Lundback, D. DellaPenna, C. Robin Buell, S.K. Sharma, D.F. Marshall, R. Waugh, G.J. Bryan, M. Destefanis, I. Nagy, D. Milbourne, S.J. Thomson, M. Fiers, J. M.E. Jacobs, K. L. Nielsen, M. Sønderkær, M. Iovene, G.A. Torres, J. Jiang, R. E. Veilleux, C.W.B. Bachem, J. de Boer, T. Borm, B. Kloosterman, H. van Eck, E. Datema, B. te Lintel Hekkert, A. Goverse, R.C.H.J. van Ham, R.G.F. Visser, C. The Potato Genome Sequencing, C. The Potato Genome, B. G.I. Shenzhen, U. Cayetano Heredia, I. Central Potato Research, R.A.S. Centre Bioengineering, U.d.C. Cgr-Cmm, U.o.D. College of Life Sciences, S.A.o.A.S. High Technology Research Center, U. Huazhong Agriculture, U. Hunan Agricultural, L. Imperial College, A. Instituto de Investigaciones, B. Institute of, Biophysics, V. Institute of, C.A.o.A.S. Flowers, C. International Potato, E. Italian National Agency for New Technologies, D. Sustainable, J. C. V. Institute, I. N. d. T. A. Laboratorio de Agrobiotecnología, U. N. d. L. Laboratorio de Biología de Sistemas, U. Michigan State, I. Scottish Crop Research,

C. Teagasc Crops Research, P. The New Zealand Institute for, L. Food Research, A. University of, W. University of, I. Virginia Polytechnic, U. State, U. Wageningen, and C. Research. 2011. Genome sequence and analysis of the tuber crop potato. *Nature*, *475*: 189–195.

Yang, L., X. Zhang, L. Wang, Y. Li, X. Li, Y. Yang, Q. Su, N. Chen, Y. Zhang, N. Li, C. Deng, S. Li, and W. Gao. 2023. Lineage-specific amplification and epigenetic regulation of LTR-retrotransposons contribute to the structure, evolution, and function of Fabaceae species. *BMC Genom.*, *24*: 1–15.

Yang, L., N. Zerega, A. Montgomery, and D. Horton. 2022. Potential of breadfruit cultivation to contribute to climate-resilient low latitude food systems. *PLoS Clim.*, *1*: 1–19.

Zerega, N.J.C., W.L. Clement, S.L. Datwyler, and G.D. Weiblen. 2005. Biogeography and divergence times in the mulberry family (Moraceae). *Mol. Phylogenet. Evol.*, *37*: 402–416.

Zerega, N.J.C., and D. Ragone. 2016. Toward a global view of breadfruit genetic diversity. *Trop. Agric.*, *93*: 77–91.

Zerega, N.J.C., D.C. Ragone, and T.J. Motley. 2004. Complex origins of breadfruit (*Artocarpus altilis*, Moraceae): Implications for human migrations in Oceania. *Am. J. Bot.*, *91*: 760–766.

Zerega, N.J.C., D.C. Ragone, and T.J. Motley. 2005. Systematics and species limits of breadfruit (*Artocarpus*, Moraceae). *Syst. Bot.*, *30*: 603–615.

Zerega, N.J.C., D.C. Ragone, and T.J. Motley. 2006. Breadfruit origins, diversity, and human-facilitated distribution. *In*: Motley, T.J., Zerega, N.J.C., Ragone, D.C., and Cross, H. [Eds.], *Darwin's Harvest: New Approaches to Origins, Evolution, and Conservation of Crop Plants*, 213–238. Columbia University Press, New York, USA.

Zerega, N.J.C., M.N.N. Supardi, and T.J. Motley. 2010. Phylogeny and recircumscription of Artocarpeae (Moraceae) with a focus on *Artocarpus*. *Syst. Bot.*, *35*: 766–782.

Zerega, N.J.C., T. Wiesner-Hanks, D. Ragone, B. Irish, B. Scheffler, S. Simpson, and F. Zee. 2015. Diversity in the breadfruit complex (*Artocarpus*, Moraceae): Genetic characterization of critical germplasm. *Tree Genet. and Genomes*, *11*: 1–26.

Zhou, Y., M.B. Taylor, and S.J.R. Underhill. 2014. Dwarfing of breadfruit (*Artocarpus altilis*) trees: Opportunities and challenges. *Am. J. Exp. Agric.*, *4*: 1743–1763.

Zhou, Y., and S.J.R. Underhill. 2021. Differential transcription pathways associated with rootstock-induced dwarfing in breadfruit (*Artocarpus altilis*) scions. *BMC Plant Biol.*, *21*: 1–21.

Zhou, Y., and S.J.R. Underhill. 2022. Characterisation of breadfruit (*Artocarpus altilis*) plants growing on lakoocha (*A. lakoocha*) rootstocks. *Horticulturae*, *8*: 1–12.

2

Genetic Diversity of Coconut (*Cocos nucifera* L.) and Sustainable Utilization

Kobra Yousefi,[1] *Siti Nor Akmar Abdullah,*[1,2,*] *Umaa Malani Nadarajah,*[1] *Muhammad Asyraf Md Hatta*[1] and *Kong Lih Ling*[2]

Introduction

Coconut (*Cocos nucifera* L.), a perennial plant, is important economically in tropical and subtropical regions of the world. The coconut tree produces various commercially valuable products that serve both the residential and industrial sectors in edible and non-edible forms. Using all different parts of the coconut palm gives lucrative opportunities for farmers of tropical and subtropical areas, earning its title "tree of life". Coconut evolutionary history dates back to ancient periods when natural selection and human migrations altered the coconut's population dynamics and historical destiny. Recent genetic studies using DNA-based techniques revealed that coconut farming occurred independently in two geographically distinct locations, notably the land areas surrounded by the Pacific and Indian Oceans. Surprisingly, the coconut has a distinct and broad gene pool with no known wild equivalents, highlighting its uniqueness and genetic autonomy (Gunn et al., 2011). Nonetheless, the lack of high-yielding cultivars, aged palms, and the imposition of biotic and abiotic environmental stresses have hampered coconut's sustainable development (Yousefi et al., 2023).

The genus *Cocos* comprises a singular species, *nucifera*, categorized into two distinct groups: tall and dwarf coconuts. The long productive period of the coconut plant and the reliance of the tall variety on cross-pollination widens diversity within natural coconut populations across diverse geographical regions. This has been utilized to enhance and ensure the long-term sustainability of this valuable palm (Arunachalam and Rajesh, 2017; Preethi et al., 2020). The nutritional content of the coconut fruit components, such as the coconut endosperm, tender coconut water (TCW), and coconut oil, varies.

[1] Department of Agriculture Technology, Faculty of Agriculture, Universiti Putra Malaysia, 43400 UPM Serdang, Selangor, Malaysia.
Email: kobrayousefi65@gmail.com; umaamalani98@gmail.com; m.asyraf@upm.edu.my

[2] Laboratory of Sustainable Agronomy and Crop Protection, Institute of Plantation Studies, Universiti Putra Malaysia, 43400 UPM Serdang, Selangor, Malaysia.
Email: lihling@upm.edu.my

* Corresponding author: snaa@upm.edu.my

More than 90% of the fatty acids in coconut oil are saturated, and 63.5% are medium and short-chain fatty acids (SMCFAs). SMCFAs are more soluble than long-chain fatty acids (LCFAs). This property facilitates the action of various lipases and aids in the quick and complete digestion of glycerides (Arunima and Rajamohan, 2014). Coconut oil's fatty acids have antibacterial, anti-inflammatory, anti-cancer, and antioxidant effects. TCW is the clear liquid within the coconut fruit that is picked and consumed when the fruit is 6–8 months old. TCW is highly nutritious as it contains minerals, vitamins, carbohydrates, amino acids, electrolytes, and phytohormones (Rajamohan and Archana, 2018). TCW also has anti-thrombotic and antioxidant properties. It can improve blood pressure, cardiovascular function, and cholesterol levels. Vitamins, minerals, tocopherol, phytohormones, fiber, and protein are found in coconut endosperm (Padmakumaran Nair et al., 1998).

The primary focus of genetic diversity research, which plays an essential role in coconut breeding programs, is the systematic grouping of accessions into genetic clusters to maximize their potential in breeding strategies and the identification of redundant or replicated ecotypes within germplasm repositories. This aspect is critical for developing efforts to improve crops and maintain the coconut farming industry's sustainable existence (Arunachalam and Rajesh, 2017; Preethi et al., 2020). Genetic diversity, defined as a range of genetic traits within a crop or species, is one of the most important aspects of genetic research explored with a scientific and transparent approach. Genetic diversity analysis in coconut has previously been conducted utilizing various morphological and biochemical features of the germplasm. Molecular markers have emerged as superior tools for characterizing plant germplasm, surpassing many limitations associated with the morphological and biochemical approaches (Preethi et al., 2020). Genetic variety refers to the differences in the sequence of one or more regions of DNA linked with a specific feature. Comparisons of DNA sequences among individuals in a population can reveal information about that group's genetic variation (Choudhury et al., 2014; Swarup et al., 2021). Plant genomes' dynamic and evolutionary instability causes genetic and epigenetic changes in plants, which are potential sources of intraspecies genetic and phenotypic diversity. Phenotypic variation in a plant population directly results from genetic variation. Plants can respond to rapid environmental changes because of the genetic variety they generate (Raza et al., 2019). It is deemed highly critical to have access to information on the genetic diversity of a crop to carry out breeding programs in such a way as to achieve an effective and methodical level of exploitation of the crop. One of the most significant phenotypic qualities that demonstrates the coconut's genetic diversity is the coconut palm's stature and fruit composition (Arunachalam and Rajesh, 2017). Additionally, the economic performance of coconut hybrids created due to the crossing of dwarf and tall palms is between 80–100% greater than the economic performance of palms formed by crossing between the tall coconut palms only (Jain and Priyadarshan, 2009).

The coconut palm has major challenges in implementing development and sustainability programs owing to the cross-pollination nature of the widely grown tall variety, a limited genetic basis, and a long juvenile period. In addition, price volatility, lack of export competitiveness in the local market, low levels of added value, and a wide range of coconut products have posed severe challenges to the coconut business. Furthermore, since coconut is monotypic, the development of acceptable variations under varied ecological circumstances is restricted. Therefore, several of these constraints must be overcome to create a genetic improvement worthy of consideration in coconut (Perera and Manimekalai, 2021). On the other hand, coconut's genetic diversity is in major danger owing to genetic erosion caused by natural and human influences, putting the conservation of these unique genetic resources of foremost importance. It is critical to identify coconut germplasm to utilize its genetic diversity. Molecular markers based on DNA and protein have been employed to tackle coconut's ecological, phylogenetic, and categorization challenges to preserve and develop this valuable plant (Perera and Manimekalai, 2021). This chapter aims to provide an overview of coconut genetic diversity and its sustainable usage. The relevance of genetic variation in coconut and its function in guaranteeing the long-term survival and adaptation of coconut farms is discussed. It also looks at the many ways to assess and conserve coconut genetic diversity and the advantages and problems connected with its long-term usage.

Taxonomic Background

The coconut, a diploid with 2n = 32 chromosomes, is the solitary species in the *Cocos* genus of the Arecaceae family (Dransfield et al., 2008). The coconut palm is taxonomically classified as follows.

Family – Arecaceae/Palmae
Subfamily – Arecoideae
Tribe – Cocoseae
Subtribe – Attaleinae
Genus – *Cocos*
Species – *nucifera*

The first attempt to categorize coconuts meticulously may be traced back to Narayana and John (1949), who categorized coconuts into two distinct groups: tall and dwarf palms (Figure 1). The start of flowering in tall palms typically occurs between 5–7 years after planting, while their lifespan ranges between 80 and 100 years. Tall palms can reach between 20–30 meters in height (Figure 1). Furthermore, tall coconut is distinguished by cross-pollination, in which the individuals involved have heterozygous genetic compositions. The copra oil content has favorable features, and the diameters of the coconut fruits range from medium to large. Various cultivars can be identified within the tall category, including West Coast Tall, West African Tall, Malayan Tall, Fiji Tall, and Panama Tall.

Fruit production normally begins 3–4 years after cultivation in dwarf coconut trees which have an average age of 40–50 years. Palms classified as dwarfs have average heights ranging from 8 to 10 meters (Figure 1). Furthermore, these palms are self-pollinated and homozygous, which improves genetic purity. The size of the fruit ranges from small to medium. Dwarf palms exhibit high productivity, with internal fruits of excellent quality and softness. These palm plants are resistant to some diseases. Dwarf palm cultivars include the Malay Yellow Dwarf, Malay Red Dwarf, Malay Green Dwarf, Chowghat Green Dwarf, and Chowghat Orange Dwarf (Table 1) (Niral and Jerard, 2019). The germination rate is an essential factor determining genetic variability within coconut species (Harries, 2012). The dwarf palm species have a remarkable tendency for leaf fissuring and a high rate of germination. To obtain hybrids possessing the desirable traits of early maturity from dwarf varieties and high yield from tall parents, a combination of tall and dwarf varieties is employed. Observing petiole coloration is an additional trait used to identify hybrid seedlings (Fernando et al., 2010). The production of high-vigor hybrids inherently ensures the sustainability of cultivating the coconut plant.

Given the lack of wild varieties in coconuts and the uniformity of characteristics identified in coconuts (Baudouin et al., 2014), the evolutionary history of this palm has been hypothesized using historical sources and fossil data (Prasad et al., 2013) and its present-day distribution (Niral and Jerard, 2019). The scientific investigation of agricultural plant origins demands a thorough examination of four dimensions: the geographical site of origin, the chronological aspect of origin, the ancestral species or taxon of agricultural plants, and the domestication process (Chowdappa et al., 2017). Previous studies on the origin of coconuts primarily focused on their geographical origin, neglecting other aspects such as the time of origin in differentiating them from their wild ancestors.

Place of Origin for Cocos nucifera

The origin of coconuts is a subject of debate, with two distinct perspectives. Most scientific researchers widely acknowledge South Asia as the primary region of origin for the coconut (Beccari, 1917; Child, 1974; de Candolle, 1886; Mayuranathan, 1938), while some scholars attribute its presence to the west coast of Central America and the islands of the Pacific Ocean (Cook, 1901; Martius, 1850; Pursglove, 1985). Harries (1990) suggested Malaysia, specifically the coastal regions and islands between Southeast Asia and the Western Pacific Ocean, is the most probable location for coconut domestication (Harries, 1990). It is hypothesized that the ancestor of the modern coconut and its wild

Figure 1. Comparing the height of a dwarf coconut tree (Malayan Yellow Dwarf) and a tall coconut tree (Malayan Tall).

Table 1. Geographical distribution of some coconut fruit cultivars around the world.

Variety	Trait	Geographical Distribution	References
Makapuno	gelatinous endosperm	Laguna Province, Philippine	Zuniga (1953)
Kopyor	brittle endosperm	Java & Lampung Island, Indonesia	Asmah (1999); Mahmud (2000)
Nam Hom	aromatic endosperm	Damnoen Saduak BanPhaeo, Amphawa Sampran, Thailand	Nartvaranant (2019)
Pandan	aromatic endosperm	Selangor, Kedah, and Perak states in Malaysia	
The King Coconut	bright colored coconut fruit + tender nut	Kurunegala, Gampaha and Galle districts, Seri Lanka	Jayasinghe and Hewajulige (2021)
Wulung Coconut	pink mesocarp + medicinal benefits	Java Island, Indonesia	Rethinam and Krishnakumar (2022)
Chowghat Dwarf (green & orange)	tender nut + early bearing	Thrissur districts of Kerala State, India	Niral et al. (2014)
Kalparaksha	tender nut + resistance to root (wilt) disease	State of Kerala, India	Niral et al. (2014)
Malayan Dwarf	colorful fruits (yellow, red, and green) + high yield+ early bearing	Malaysia, Brazil, Thailand, Fiji and Florida	Rethinam and Krishnakumar (2022)
Green Xiem	super sweet solid and liquid endosperm	Mekong Delta Province, Vietnam	Rethinam and Krishnakumar (2022)
Mohachao Narel	sweet and soft endosperm with less fiber	Guhaghar taluk of Ratnagiri district of Maharashtra State, India	Kumar and Samsudeen (2018)
Brazilian Green	tender nut + high yield	Brazil and tropical areas throughout the world	Lédo et al. (2019)

counterparts may have gone extinct as a result of unfavorable geological and climatic occurrences in previous times. Based on phylogenetic research, the likely source of the coconut palm can be traced to a wide geographical area that includes the Indian archipelago, the Northern Indian Ocean, Southeast Asia, the Philippines, New Guinea, and the Western Pacific that covers Northern Indian Ocean, Malesia, Melanesia, and Micronesia (Niral et al., 2021).

Time of Origin for Cocos nucifera

An examination of anthropological and archaeological evidence establishes a plant's precise time of origin. More recently, the archaeological remnants of the most ancient coconuts in the Northeastern region of New Guinea, Micronesia, Solomon, Society Islands, and Cook Islands were discovered suggesting their existence before the arrival of modern human habitation (Chowdappa et al., 2017). Prebble and Dowe (2008) documented a significant decline in palm populations during the late Quaternary epoch in the Pacific Oceanic regions. Several additional extinctions and dispersals took place in diverse regions of the Pacific and Indian Oceans (Chowdappa et al., 2017). The initial documentation of coconuts in India and Sri Lanka can be traced back to 545 AD when an Egyptian, Cosmos, reported their presence. Marco Polo referenced the coconut palm during his 1280 travels in Indonesia (McCormack, 2005). The precise range of dispersal of coconut remains uncertain due to its inherent capacity for long-distance dispersion. Coconut nuts exhibit a survival strategy where they remain buoyant in water, facilitating their dispersal and germination upon reaching coastal territories. Fossil analysis conducted in Queensland, Australia in 1995 indicates that coconut palms have existed for approximately 2 million years, allowing ample time for their global dispersal (McCormack, 2005). Given the limited knowledge available, it is challenging to pinpoint a particular species as the ancestral predecessor of *Cocos nucifera*. The exact differentiation of domestic and cultivated coconuts is uncertain due to insufficient knowledge about their ancestral species. Early humans most likely cultivated and domesticated coconuts in coastal regions as a response to limited food resources and population growth (Cowan and Watson, 1992).

Evolution of Cocos nucifera

The evolution of today's coconut, which is derived from Mesozoic Guanda palms, is unknown (Lebrun et al., 1999). Coconut has a two-stage evolutionary history, the first involving geological eras and natural selection, and the second involving human selection (Teulat et al., 2000; Gunn et al., 2011). Coconut is an ancient plant that was able to adapt 80 million years ago with the assistance of natural selection and the dispersal of seeds in ocean water despite the number of challenges, frequent fluctuations, and intense storms. Coastal tropical regions with plenty of rainfall and no severe seasonal or periodic restrictions are important for coconut adaptability and proliferation. Coconut seeds cannot migrate to other coastal locations because of their enormous size which prevents their transfer by birds and other animals. Only through the ocean, it can spread to neighboring beaches with marginal sand pits (Nair, 2021). The coconut seed's specific components, such as the big endosperm and high husk ratio, allow this fruit to float in the oceans for an extended period, then land and germinate on the lands along the coastlines. The coconut seed can provide the energy and nutrients required for the seedling's growth for more than a year, allowing the roots to expand to the lower soil layers to search for water where they can play their role in water uptake for the growing plant. Endosperm, or coconut kernel, contributed to the colonization and adaptation of the coconut palm and humans to their new habitat and has played a significant role in the success of human civilizations (Batugal et al., 2005).

More than 100,000 years ago, when human colonization began on the tropical coastlines, can be regarded as the beginning of the evolution stage of the coconut population in response to human selection (Ingicco et al., 2018). Coconut kernels provided early colonists with a supply of high-energy food. Furthermore, coconut kernels were utilized for a long time to offer water and healthful nourishment to sea mariners. As a result, larger fruits were chosen to give water and nourishment. Sea passengers then cross-pollinated coconut seeds left over from the ships provided with wild coconut populations on the New Coast. As a result, the parents of the following coconut populations on the new coast were established (Harries, 1978). As the human population increased in the tropics, coconut became a staple in the diet and culture of the tropical region (Child, 1964). Furthermore, in recent years, there has been demand for coconuts outside of the tropical region for the use of coconut oil and water. Its beneficial use in the health, cosmetic and the pharmaceutical industry, has increased, which has had a significant impact on the distribution and production of this valuable product.

Collection and Assessment of Germplasm/Genetic Resources

The coconut, a species with a lengthy evolutionary history and significant genetic variety, grows in a broad range of tropical and subtropical climates across the world. A crop plant's germplasm comprises indigenous races, wild races, various species, and breeding lines, among other components (Perera, 2020). To monitor and maintain coconut genetic resources, the International Plant Genetic Resources Institute (IPGRI) developed the International Coconut Genetic Resources Network (COGENT) in 1991. COGENT addresses the global decline in coconut production to boost sustainable productivity. COGENT focuses on the economic security aim of 96% of coconut farmers, classified as smallholders (Karun et al., 2022). COGENT has created five International Coconut Gene Banks (ICGs) in India, Papua New Guinea, Ivory Coast, Brazil, and Indonesia (Perera, 2020) as well as 28 national gene banks in 24 countries to fulfill these objectives. These banks develop *ex-situ* plant species collections to preserve coconut genetic diversity and guarantee the safe dissemination of germplasm. In addition to preserving unique coconut subpopulations, these gene banks also preserve some mutant varieties that could benefit breeding. COGENT also promotes the establishment of an international network of research and training facilities that focus on preserving coconut variety. These efforts are made at national and regional levels to create worldwide initiatives to preserve the recognized variety, identify and assess existing diversity, conduct germplasm assessment, and deliver timely information to member nations.

Characterization of Evaluation

The absence of any closely related wild species of coconut has persisted, and no such relative has been definitively identified thus far. Nevertheless, the intraspecific diversity of this particular palm species is high as the different coconut varieties have exhibited considerable variation in their morphological traits (Niral and Jerard, 2019). Variations were observed in the leaf shape, palm height (dwarf and tall), vertical development of the stem, number of bunches, number of fruits in each bunch, inflorescence characteristics as well as fruit color, weight, quality, and composition of coconut products including water, oil and meat. In addition, variation in key economic traits including flowering time (early ripening) and resistance to pests and diseases has been observed. These features are required for germplasm conservation centers to identify and evaluate germplasm quickly (Batugal et al., 2005; Perera et al., 2010a).

Coconut germplasm evaluation is carried out as a thorough procedure to enhance utilization and develop coconut palm based on potentially beneficial characteristics. This approach is important as there is growth in economic demand for coconut and several environmental issues need to be addressed to ensure a sustainable supply of coconut products. It involves discovering and conserving genetic variety, enhancing product quality, and ensuring the best approach to reap the benefit from this abundant coconut resource. Germplasm evaluation deals with the comprehensive analysis of this palm's many components and gives important information in several sectors. It covers wide aspects from the morphological description of the plant and its appearance to the internal structure of the cells in the cytological description and from the study of plant biochemical substances to its genetic information.

Morphological Characterization

The coconut tree is a tropical perennial monoecious and protandrous. The roots of this palm are fibrous. The primary roots are around 6 meters in length and 8 mm in diameter, and their color ranges from white to brown. The primary thick main roots grow from the stem's base, and the rootlets grow from the primary roots. The total number of main roots rises as the palm becomes older. A mature palm tree has between 2,000–3,000 active main roots (Menon and Pandalai, 1958). The coconut stem is unbranched, light grey to brown, and straight or slightly curled. The base of the stem takes up to 4 years to mature in tall and 2–3 years in dwarf palms. The diameter of the cylindrical stem varies between 80–85 cm in tall form and 75 cm in dwarf, with a maximum height of 30 meters (Figure 1). The stem's width does not expand or vary after formation (Child, 1974). The crown's leaves are arranged in five rings, one above the other, and tall palms often have 30–36 leaves. Each leaf axil, with the exception of the aged leaves, bears an inflorescence or a fruit bunch (Ratnambal et al., 1995). Coconut leaves are pinnate and last 2–3 years (Patel, 1938). The leaf's length ranges from 4.5–6 meters (Davis, 1954). Tall palm produces 14–16 leaves annually, whereas dwarf produces up to 21 leaves. The number of leaflets per leaf ranges between 200 and 300. The maximum length of the leaflets is 130 cm, while the minimum length of the leaflets at the peak of the leaf is 25 cm. A leaflet has 220–170 stomata per square meter on average, and dwarfs have a higher stomatal density than tall palms (Satyabalan, 1993).

Male and female flowers appear on the same inflorescence, which is wrapped in a spathe. Male flowers are more numerous and smaller than female flowers (Menon and Pandalai, 1958; Dransfield et al., 2008). The inflorescence is pale yellow to cream in color, and the male and female flowers develop to an orange-brown color. The inflorescence ranges in length from 0.8–1.0 meters. Under normal conditions, the coconut palm produces 10–14 flowers per year (Perera et al., 2010b)

The coconut fruit consists of several layers. The outermost layer is a hard and thin epicarp, followed by a fibrous mesocarp called the husk. And then the hard endocarp (shell) preserves the seed and the solid and white endosperm (flesh), which is the edible and economic part of the fruit. The liquid endosperm or coconut water is a clear liquid that is rich in nutrients (Figure 2). Coconut fruits mature between 10–13 months following fertilization. The amount of solid endosperm increases as the fruit matures, and the amount of liquid endosperm or coconut water reduces drastically owing to absorption by the solid endosperm tissue or evaporation (Menon and Pandalai, 1958). There are three eyes at the end of the coconut shell. The embryo is located under the functional eye or soft eye, which matures and can sprout and form a new coconut tree. Two non-functional eyes remain undeveloped (Sudha et al., 2021).

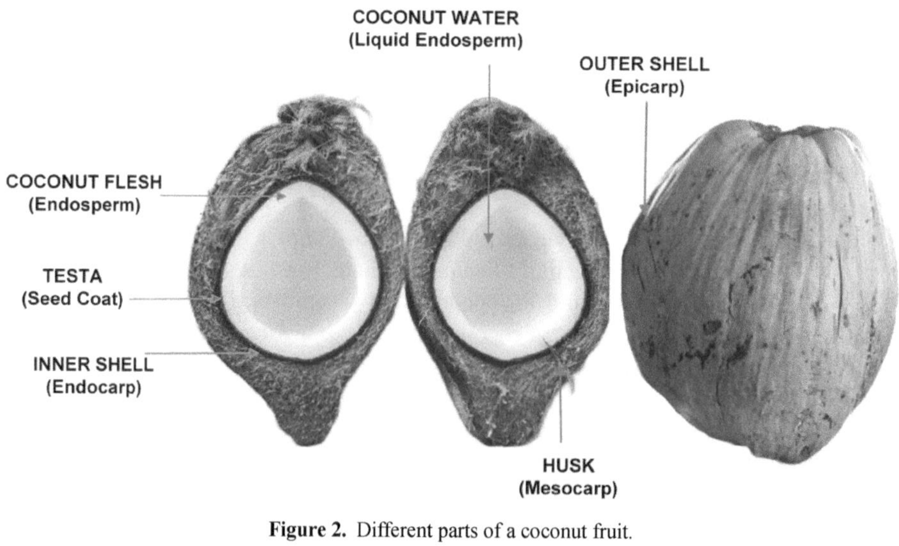

Figure 2. Different parts of a coconut fruit.

The variations of coconut can be observed in its quantitative and qualitative attributes. Qualitative characteristics in the fruits including fruit shape, color range, and dimensions, alongside resistance to biotic and abiotic stresses. Quantitative traits encompass the number of fruit bunches, flowering time, fruit ripening period, and characteristics of the different fruit components, such as nut weight, flesh thickness, and amount of water (Karun et al., 2022). The hue of coconut fruit exhibits a range of chromaticity, spanning from green and golden tones to tinges of orange and shades of brown. The fruit morphology typically includes spherical, elongated, or elliptical forms. It can be classified according to the presence of protuberances and the degree of curvature. The curvature can manifest as either round, flat, or angular. The internal nuts of the coconut fruit exhibit varying shapes, including oval, rectangular, and round forms (Niral and Jerard, 2019). Unripe fruits of the tall coconut variety typically exhibit a green color with hints of tan and occasional; yellow tones (Foale, 2003). The coconut displays significant genetic variations in the husk, shell, and kernel biomass ratio (Whitehead, 1964). Wild coconut populations have evolved to have thicker shells and husks with longer and more angular fruit shapes. The wild coconut population successfully expanded in aquatic environments and began growing in coastal and inland areas. The dense shell protects the developing embryo in aquatic environments (Harries, 1978).

Cytological Characterization

The somatic chromosome number of coconut is 2n = 32 based on the first reported meiosis study in the Philippines by Santos (1929) as the haploid chromosomes were n = 16 which was later proven by a few other studies in India. Darlington and Janaki (1945), Nambiar and Swaminathan (1960), Abraham et al. (1961) studied the karyotype, microsporogenesis, pollen sterility, and genetic behavior of different coconut varieties, such as Laccadive Ordinary (tall), Apricot (semi-tall), and Dwarf Red (dwarf), and dwarf palms were proposed as excellent candidates for developing high-yielding hybrid trees. Nambiar and Swaminathan (1960) revealed various meiotic irregularities among the tall variety, semi-tall variety, apricot variety, and dwarf varieties, including the Laccadive ordinary, red dwarf, and green dwarf and suggested that the dwarf coconut origin could be attributed to inbreeding among the tall varieties, rather than a sudden mutation.

Tan (1969) discovered how the meiosis behavior has directly affected the pollen sterility in three different coconut varieties. 'Coconino' (*nana*) and 'Tambolilid' (*japonica)* displayed irregular meiosis with higher pollen sterility rates of 30% and 41%, respectively; the 'Laguna' (typical) variety demonstrated regular meiosis with 11% pollen sterility. The chromosomal counts (2n = 2x = 32) and spindle orientations of all three cultivars were irregular, but 'Coconino' and 'Tambolilid' exhibited different numbers of chromosomes and were multivalent, leading to genetic instability. Pereira et al. (2017) conducted a detailed chromosomal characterization specifically on the Brazilian green dwarf. It was observed that the Brazilian dwarf varieties have some irregularities in meiosis, but the pollen viability was good, which indicated that these irregularities do not affect the fertility of the genotype. The Makapuno is a rare type of coconut jelly. Its endosperm changes over time from mostly triploid cells in the early stages to high levels of polyploidy (6n, 9n, 12n, 24n, 48n, and above) as it matures. The latter stages are characterized by indefinite growth, high multiplication, increased abnormalities, cessation of normal mitosis, onset of amitosis, and nuclear degeneration. This suggests that disruptions in growth regulation may be to blame for the abnormality in the buttery kernel of Makapuno coconuts (Abraham et al., 1965). The coconut palm also has some cytological changes due to external stressors. A study carried out by Nambiar and Prasannakumari (1964) has found that cytological aberrations abnormalities in the process of cell division were rare. They discovered that palms affected by root wilt disease exhibited successful seed development and high levels of pollen fertility, indicating that the disease had no discernible effect on the reproductive characteristics of the palms.

In the most recent study carried out by Wang et al. (2021) emphasis was placed on the evolution of its karyotype through intensive genomic analysis, such as the gene collinearity and comparative genomics, to the chromosomal evolution by analyzing the genomes of *Cocos nucifera* and other related species. Before the study of divergence of the monocot's studies, there were 10 proto chromosomes (n = 10) at node A, and these underwent Pre-τ Whole Genome Duplication (WGD) at (Node B) end-to-end joins (EEJ) and other rearrangements, reduction to (n = 8) and after the post-τ WGD (Node C), the chromosome number double leading to the ancestral genome having (n = 16). Then, the chromosome numbers were further deducted to (n = 9) due to complex reorganization involving the five EEJs, arm-exchanging crossovers, and nested chromosome fusions. With the most recent WGD, which occurred at Node E, the chromosome number was doubled to n = 18. And the final 5 crossovers and nested chromosome fusions, eventually resulted in the extant chromosome number of *C. nucifera* being n = 16.

Biochemical Characterization

The evaluation of coconut germplasm through biochemical characterization is important due to the notable changes in the biochemical components within the palm. Such changes happen during various critical stages and events, such as during seed germination, and plant development stages, and when exposed to biotic and abiotic stressors. The coconut endosperm mainly contains lipids (68–70%), sugars (6–7%), and protein (6–9%) (Nathanael, 1967). According to Nampoothiri et al. (2018) the germination period, which is longer in tall cultivars, is one of the distinguishing features between dwarf and tall varieties. This is because in the dwarf coconut varieties, the lipids that serve as the main source of reserve food, solubilize quickly (within 28–42 days of sowing) due to the significant threefold increase in lipase activity. This leads to a higher reduction of oil content (4.1–4.8%) and as a result faster germination in the dwarfs than the talls and hybrids (2.3–3.2%).

Félix et al. (2023) found that biochemical processes involved in the maturation of coconuts vary between two distinct coconut cultivars: the Yucatan Green Dwarf (YGD) and the Mexican Pacific Tall (MPT), through protein analysis that identified 244 proteins in YGD and 347 in MPT, with 155 proteins shared between both. This indicates significant biochemical diversity between the two cultivars. In terms of the carbohydrate metabolism in MPT, proteins involved in carbohydrate metabolism (e.g., glycolysis) were more accumulated in the mature stage. This suggests a shift towards synthesizing precursors for fatty acids in the later stages of development, but YGD showed higher carbohydrate metabolism in the immature stage but reduced activity in the intermediate and mature stages. This aligns with YGD being valued for its tasty coconut water in the immature stage. For fatty acid metabolism, MPT exhibited a greater accumulation of proteins involved in fatty acid metabolism during intermediate and mature stages compared to YGD, indicating a higher efficiency in producing lipids and fatty acids. This varietal difference in the biochemical composition is important as the distinct proteomic signatures of these cultivars can guide selective breeding programs aimed at enhancing desired traits, such as higher lipid content in MPT or improved taste qualities in YGD.

Apart from that, the Makapuno variety, which exhibits the unique features of the jelly endosperm in Indonesia is due to the deficiency of the alpha-D galactosidase enzyme. Initially, Mujer et al. (1983) reported that this coconut endosperm's unique viscosity is due to the presence of a higher level of polysaccharides due to the overgrowth of microcells and a higher concentration of galactomannans compared to the normal coconut endosperm. Mujer et al. (1983) and Mendoza et al. (1987) further confirmed that, in comparison with the normal coconut endosperm, the activity of α-D-galactosidase increases with age, reaching its peak at the final stage of maturity, which plays a role in the natural degradation of the galactomannans. However, in Makapuno's endosperm, a notable deficiency in α-D-galactosidase and β-mannosidase activity is observed. This deficiency is linked to the high levels of galactomannans, which correlate with the variety's distinctive characteristics, such as its gelatinous endosperm. Hence, it is important to preserve the diverse genetic profile to focus on improving the variety or maintaining unique traits, such as the gelatinous endosperm.

Another biochemical characterization of coconut germplasm involved the Guam Tall variety with pink husks that showed significantly higher total phenol content in its tender nut water compared to normal husked coconuts (Thomas et al., 2021). These phenolic compounds which are known for their antioxidant properties, can neutralize harmful free radicals in the cell. Guam Tall coconut water also showed a higher protein content, with 86 mg per 100 mL, compared to 58.7 mg per 100 mL in normal husked types. Protein is essential for many bodily functions, and this increased content could enhance the nutritional value of coconut water. Anthocyanins are pigments that belong to the flavonoid class of polyphenols and are known for their antioxidant properties. The presence of these pigments in the exocarp of the pink-husked coconuts is likely responsible for the distinctive pink color of the husks and contributes to the overall antioxidant activity, which can serve the interests of the pharmaceutical industry in the development of health supplements and natural remedies.

Konan et al. (2011) examined the biochemical characteristics of the coconut water from three important cultivars: West African Tall (WAT), Malaysian Yellow Dwarf (MYD), and hybrid PB121+. The researchers obtained in-depth information about the germplasm collection by measuring weight, dry matter, ash, protein, sugar, pH, and dry extract refractometrically. Specifically, they found that the coconut water from the "WAT" and hybrid PB121+ was richer in sugar and ashes compared to the MYD. In the most recent study, Amritha et al. (2023) highlighted key hormones like BA, TDZ, SA, IAA, GA4, JA, cisJ, and 24-epiBL and their roles in growth and germination across various coconut cultivars and hybrids. They also carried out the biochemical characterization of the coconut liquid endosperm of two coconut cultivars, which were the tall (West Coast Tall) and dwarf (Gangabodam), as well as their reciprocal hybrids. This has revealed that different maturity stages have distinct hormonal profiles, with significant implications for optimal harvest timing and breeding strategies.

Response to Abiotic and Biotic Stresses

The biochemical changes also contribute to the resistance or susceptibility of the different coconut varieties to biotic and abiotic stressors. Izaitu Aida et al. (2020) carried out a study to determine the genetic basis of color attraction in coconut palms which influences rhinoceros beetles' infestation. MYD varieties were more affected than MRD and Tagnanan which suggested that the appearance of plants, specifically their color, may impact the population of rhinoceros beetles. MYD palm which has yellow stems and fruits is highly preferred compared to MRD and Tagnanan by the female adults. These different colors (yellow for MYD, orange for MRD, and greenish for Tagnanan) of the palms may be linked to chemical compounds or secondary metabolites that attract or repel the beetles. However, more research is needed to analyze the chemical components of related compounds from both palm varieties to understand their roles in attracting beetles.

A study was conducted on two popular coconut varieties, MATAG and Pandan, to examine their biochemical response to red palm weevil (RPW) infestation. RPW causes rapid death to the coconut tree through the destruction of tissues in the trunk of the coconut tree (Figure 3). The study found that the Pandan cultivar is more tolerant to RPW infestation than the MATAG. The levels of hydrogen peroxide (H_2O_2), malondialdehyde (MDA), and total ion leakage in the leaves of the coconut trees at various stages of infestation were measured. MATAG cultivar had higher levels of H_2O_2, MDA, and total ion leakage compared to the Pandan cultivar, especially after 21 days of infestation. These markers indicate oxidative stress, which is linked to plant defense against RPW infestation.

Coconuts also showed genetic diversity in several physiological and morphological adaptations that help tolerate water deficits, including a deep root system, high protoplasmic tolerance, and enhanced antioxidant activity. The genotypic differences in these traits can largely explain the agronomic performance of coconut trees under drought conditions (Gomes and Prado, 2007). In the early 1990s, 23 coconut varieties (10 tall, 6 dwarfs, and 7 hybrids) were used to determine the most tolerant variety to drought using parameters, such as stomatal resistance, leaf water potential, and epicuticular wax content. Their key finding revealed the desirable drought-tolerant genotypes, such as the hybrid from tall genotypes, especially the West Coast Tall x West Coast Tall and LO X

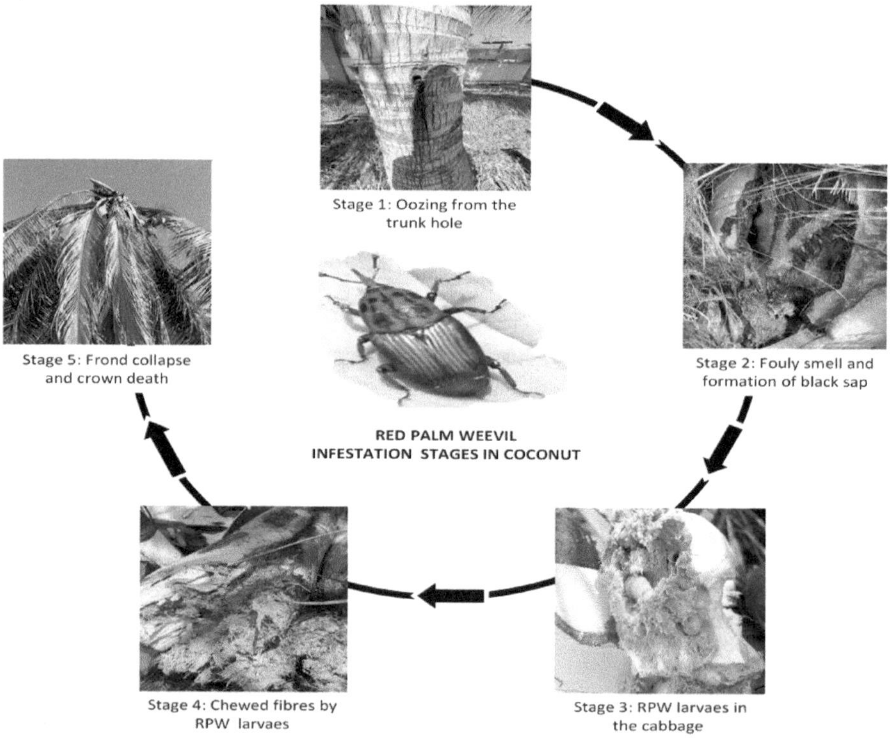

Figure 3. Field observation of red palm weevil (RPW) infestation stages in the Pandan variety.

Chowghat Orange Dwarf which had higher leaf water potential and epicuticular wax content than other dwarf varieties, and some hybrids that resembled their female parents to be less tolerant to drought (Rajagopal et al., 1990).

The genetic basis of the drought-responsive trait was also studied in 2 x 4-line to understand drought tolerance in coconut cultivars using two dwarfs (CGD and MYD) and four tall (ECT, PHOT, LCT, and FMST) cultivars which include parameters, such as leaf water potential, transpiration rate, net photosynthetic rate (Pn), and lipid peroxidation. Transpiration rates and leaf water potentials showed greater specific combining ability (SCA) effects, indicating non-additive gene action and indicating heterosis for these traits. Pn under stress had good combining ability with additive gene action during times of no stress and recovery. Heterosis could be used to improve drought tolerance through breeding. In non-stress conditions, lipid peroxidation gene action was unpredictable, with low dominance and insignificant additive gene action. Non-additive gene action was observed during stress and recovery. This study suggests using gene action in breeding initiatives to develop drought-tolerant coconut varieties (Rajagopal et al., 2007). The most recent transcriptomic study involved two populations called the Ambakelle special (AS) and unselected variety CRIC60 and six-year yield data. It was found that during low rainfall both varieties faced reduced nut yield, but the AS variety maintained its superiority indicating better adaption to low rainfall conditions than the CRIC60 (Samarasinghe et al., 2021).

Molecular Characterization

Coconut has a wide range of quantitative and qualitative phenotypic features, but the molecular foundation for most of them needs to be clarified. There have been a few reports of studies on the genetic basis of somatic embryogenesis, fatty acid biosynthesis, and solid endosperm development.

Regarding somatic embryogenesis, Pérez-Núñez et al. (2009) discovered an ortholog of *SOMATIC EMBRYOGENESIS RECEPTOR-LIKE KINASE* (*SERK*) gene named *CnSERK* in the coconut genome. Gene expression research revealed that this gene can be observed in embryogenic tissues before embryo development. However, it was not found in non-embryogenic tissues or at considerably lower levels, indicating a link between *CnSERK* gene expression and the initiation of somatic embryogenesis. Rajesh et al. (2016) identified fourteen genes and described the gene expression pattern during somatic embryogenesis in coconuts using the transcriptomic approach. Expression of *SERK, AP2, SAUR, ECP, MAPK, AGP, LEA,* and *ANT* genes increased in the embryonic callus stage. *GST, PKL, GLP, WUS*, and *WRKY* genes were expressed during the somatic embryogenic stage. Another study employed gene transcription data from coconut embryogenic calli to find conserved miRNA. A total of 27 miRNA sequences from 15 different miRNA families were discovered (Sabana et al., 2018).

MADS-box transcription factors are involved in the development of coconut solid endosperm, particularly in the modification and maturity of solid endosperm (Liang et al., 2014; Sun et al., 2020). Gao et al. (2022) found that the transcription factor *CnGATA20* and the downstream regulatory gene *CnOLE18* influenced endosperm development (Gao et al., 2022). The transcription factor CnMADS1 is involved in cell proliferation and lipid metabolism in coconut endosperm (Sun et al., 2020). Several genes and enzymes are associated with the biosynthesis of coconut endosperm fatty acids, the most important of which are acyl-acyl carrier protein thioesterases (FatA/B) and fatty acyl-ACP thioesterases and lysophosphatidic acid acyltransferase (LAAPT), which are involved in the accumulation of medium-chain fatty acids in the endosperm (Fan et al., 2013; Xia, 2021).

Molecular Marker for Assessing Coconut Genetic Diversity

A plant's morphological, cytological, biochemical, and molecular characteristics can be used to investigate and analyze diversity. Based on these, morphological, cytological, and biochemical markers were used to evaluate the genetic diversity of plants, and then with the advent of genomic tools, molecular markers became the preferred way to evaluate genetic diversity. The field of molecular markers focuses on analyzing genetic variation at the DNA/RNA level. The classification of molecular markers primarily encompasses hybridization-based and PCR-based techniques. Markers for genetic variation can be derived from different sources, such as genomic DNA/RNA, ribosomal RNA, or organelle genome sequences. Molecular markers are widely used to assess genetic diversity due to their high variability, broad genomic representation, reproducibility, automation potential, and independence from environmental factors (Bhandari et al., 2017). Restriction fragment length polymorphism (RFLP), random amplified polymorphic DNA (RAPD), amplified fragment length polymorphism (AFLP), microsatellites or simple sequence repeats (SSR), and single nucleotide polymorphisms (SNPs) are widely recognized as crucial molecular markers in various fields of research. These markers have proven to be highly effective and efficient tools for the evaluation of genetic diversity in plant research. In the mid-1990s, coconut research embarked on utilizing molecular marker technologies. The first efforts were made to analyze the diversity of coconut germplasm using RFLP, RAPD, and AFLP markers, with the current trend of growing usage of SSR (Perera et al., 2020).

In a recent study, Riangwong et al. (2020) employed genotyping by sequencing (GBS) to obtain SSR markers from 38 coconut accessions. Genetic diversity and genetic relationships among 40 coconut species in Thailand were assessed using principal coordinate analysis (PCoA) as well as population structure and phylogenetic tree analyses with 74 polymorphic SSR markers. The results identified three distinct coconut groups.

Muñoz-Pérez et al. (2022) utilized GBS to develop SNPs across the coconut genome. The objective was to assess the genetic diversity, population structure, and linkage disequilibrium (LD) within a diverse panel of 112 coconut accessions from Colombia's Atlantic and Pacific coasts. A total of 40,614 SNPs were identified, with 19,414 mapped to coconut chromosomes. Out of the total, 10,338 SNPs were found exclusively in the Atlantic gene pool, while 4,606 SNPs were exclusive to the Pacific gene pool. Additionally, 3,432 SNPs were identified as capable of distinguishing between both gene pools. The Pacific groups exhibited a gradual decline in linkage disequilibrium (LD), a low Fixation Index (Fst), and low nucleotide diversity (π). In contrast, the Atlantic group displayed slightly higher genetic diversity and a faster LD decay.

Sudha et al. (2023) examined the genetic diversity and population structure of 16 coconut varieties (96 palms) from the International Gene Bank of South Asia and the Middle East (ICG-SAME), Kidu, India. They utilized 35 SSR markers for their analysis. Significant genetic diversity in the coconut germplasm was revealed using these markers. The genotypes were classified into two main clusters with two sub-groups based on their geographic origins using hierarchical clustering analysis. The initial cluster consisted of the dwarf accessions, while the second cluster included all the tall accessions. This data enhances the efficiency of safeguarding the genetic materials for coconut breeding initiatives.

Non-coding RNA (ncRNA) SSR markers are significant and growing scientific tools for measuring plant genetic variation. These potential markers do not encode proteins, yet they serve a crucial function in gene expression regulation and reacting to biotic and abiotic stressors (Summanwar et al., 2020; Bhatia et al., 2021). These markers are highly polymorphic and cross-transferable (Singh et al., 2023). Sabana et al. (2022) discovered 597 ncRNA-derived SSR markers in coconut, consisting of 509 lncRNASSRs and 88 miRNASSRs. A total of 20 primers, with 10 derived from lncRNA-SSR and 10 from miRNA-SSR, were chosen and tested on six coconut accessions. The results showed that 50% of the primers generated polymorphic fragments. The average heterozygosity and Shannon index were 0.5 and 1.1, respectively, indicating high levels of polymorphism in ncRNA-SSRs. Through distance-based cluster analyses, it was observed that the tall and dwarf accessions exhibited distinct differentiation and were classified into separate clusters. These markers may identify hidden genetic variations in the non-coding areas of the genome. Researchers can detect more significant genetic variation in the crop by analyzing these genomic locations. As these markers are based on RNA, they may give information on the functional characteristics of non-coding areas as well as putative regulatory factors that may influence the functional properties of coconut and its adaptability to diverse environments. This greatly adds to the long-term management and use of coconut genetic resources.

Continuous advancements in next-generation sequencing (NGS) technology and bioinformatics have enabled the mining of SSR markers from the entire genome, efficiently generating numerous SSR markers. In a recent study, Caro et al. (2022) successfully identified and characterized 7,139 novel SSR markers from the genome assembly of the coconut variety 'Catigan Green Dwarf' (CATD). This study effectively synthesized, optimized, and amplified 131 new SSR markers in coconut. These markers were developed using four different genotypes: 'Catigan Green Dwarf' (CATD), 'Laguna Tall' (LAGT), 'West African Tall' (WAT), and SYNVAR (a hybrid of LAGT and WAT). Among the analyzed coconut genotypes, 113 of the 131 SSR markers exhibited polymorphism. These newly developed SSR markers have significant potential as a beneficial tool for assessing coconut's genetic diversity and population structure, as well as for application in breeding programs that rely on molecular markers for their analyses.

A total of 39,002 SSR loci were found by Lantican et al. (2019) with di-repeats like AT/TA being the most common motif. The frequency of SSR loci was determined to be 25.11 per million base pairs in the Catigan Green Dwarf' (CATD) genome. There were 22,031 distinct SSR markers designed from these loci, with a final set of 7,139 markers excluding AT/TA motifs. In addition, 13 gene-linked SSRs for genes involved in drought response, insect and pest resistance, and oil production were generated. The physical attachment of markers to target genes is essential for downstream applications, such as precise mapping and genetic enhancement.

Lantican et al. (2019) analyzed sequence variations, such as SNPs and InDels between the reference Dwarf coconut genome (CATD) and the reference tall coconut genome (Hainan Tall) (Xiao et al., 2017) to build SNP genotyping arrays. Following the mapping and recalibration of base quality scores, 58,503 variations were discovered, with 98.9% of them being SNPs in non-repeat areas. Most of the SNPs (A/G, C/T) were single-base transitions, with A/G being the most frequent. SNPs were found at a rate of one every 7,881 base pairs in the non-repeat area. Furthermore, 631 InDels, mostly single-base, were discovered, and the nucleotide diversity between the 'Dwarf' and 'Tall' coconut genomes was calculated to be 128 base variations per million base pairs.

Utilization of Germplasm

In general, genetic materials, particularly those found in the form of seeds and tissues that are accessible in populations as genetic resources, are referred to as germplasm. Therefore, it is critical to investigate and diversify the germplasm that is now available. Scientists and farmers can identify and reproduce coconut varieties with superior qualities through the study and use of coconut germplasm. This provides pathways to enhance productivity, secure food security, and support the livelihood of farmers worldwide who depend on this vital crop. The process of preserving and choosing the available germplasm materials in the coconut population specifically to develop improved coconut varieties with the desired features like high-yielding and improved tolerance to biotic stresses requires a well-coordinated breeding program.

Utilization in Breeding and Improvement Programs

High-yielding coconut variants have long been the focus of breeding and improvement projects using coconut germplasm. Through genetic selection and hybridization, targets like disease and insect resistance, faster maturation, enhanced nut quality and size, and environmental adaptability seek to increase the coconut productivity per tree. By identifying which traits are heritable, how they correlate with yield, and their direct and indirect effects, Natarajan et al. (2010) have provided strategic use of coconut germplasm. These findings which were further supported by Suchithra and Paramaguru (2018) guide breeders in choosing which genotypes from the germplasm to crossbreed or select for targeted breeding that can successfully develop high-yielding coconut varieties.

The breeding initiative known as Dwarf × Dwarf (D × D), as highlighted by Batugal et al. (2009), has shown promising results. The T × T hybrids developed in Thailand have great potential for long-term genetic improvement. These hybrids are particularly well-suited for intercropping due to their tall nature, as noted by (Anupap et al., 1992). The D × D hybrids studied in Côte d'Ivoire and Malaysia have shown exceptional performance in terms of precocity, nut production, and copra content. These hybrids have outperformed the local varieties in both regions. In Côte d'Ivoire, the PB series stands out for its high copra content compared to the local West African Tall (WAT) varieties. Similarly, in Sri Lanka, the CRIC 60 and CRIC 65 varieties are renowned for their excellent copra and nut yields. According to recent research by the PCA (2016), the germplasm of Dwarf x Tall F1 has become increasingly favored for its early fruit-bearing, high yield, and stress resilience. Due to the high cost of their seeds, regular farmers face limited access to them. The Philippine Coconut Authority (PCA) has taken steps to address the challenge by developing a genetically multi-stored (GMA) coconut variety. This variety offers several advantages, such as high yield, vigor, early maturation (precocity), and genetic stability across generations. This variety is the result of combining the agronomic qualities of two foreign tall types (West African and Rennel) with four indigenous tall kinds (Laguna, Bago Oshiro, Baybay, and Tagnanan).

Coconut germplasms serve multiple purposes, including the development of high-yielding cultivars and enhancing the coconut palm's ability to withstand various biotic and abiotic stresses. Coconut germplasm plays a crucial role in germplasm breeding for the development of resistant cultivars. By harnessing its potential, we can effectively create hybrids that possess enhanced tolerance or resistance to various stresses. This approach holds great promise in the quest for more resilient coconut varieties. The coconut germplasm also offers promising opportunities for advancing coconut production and addressing specific needs in the industry. Preserving germplasm enables genetic diversity to be utilized which is crucial for ensuring the long-term viability and productivity of the coconut industry.

Breeding for Resistance

Breeding programs to produce cultivars that are resistant to pathogen infection or pest infestation, often rely on the utilization of germplasms. The germplasm materials are used in selective hybridization to enhance their potential. A study was conducted by Nair et al. (1996) aimed to combat coconut root wilt disease (RWD), a severe disease caused by phytoplasma. The study focused on breeding initiatives and hybridizing different coconut varieties, including the disease-free elite WCT and CGD coconuts. The hybrids were used as planting material to assess their reaction to the root disease. After eight years of planting, it was found that only 16% of the 31 hybrids displayed symptoms of the disease. This indicates that there is a potential source for genetic resistance or tolerance to root wilt in the germplasm materials. The reduced incidence of root wilt disease is a positive outcome of this study. In a study conducted by Nair et al. (2006), it was found that a specific hybrid had a disease incidence of 65%. However, despite this, the hybrid still demonstrated a good average yield of 78 nuts per palm per year. This suggests that the hybrid exhibits tolerance to the disease. The lethal yellowing disease (LYD) has been a significant concern. Fortunately, researchers discovered that the Maypan hybrids, which are a combination of MYD and Panama Tall, showed resistance to LYD. This finding has proven to be highly effective in combating the disease in affected areas. The effectiveness of their methods gradually declined, particularly in Jamaica (Harries and Romney, 1974; Lebrun et al., 2010).

After 20 years of screening in Ghana, it has been observed that the Sri Lanka Green Dwarf (SGD) and Vanuatu Tall (VTT) cultivars exhibited resistance or tolerance to LYD. The SGD x VTT hybrid variety has shown exceptional performance, especially in areas affected by the disease. This makes it a highly promising option for wider adoption in regions prone to LYD. Hybridization has shown promise in developing mite-resistant varieties. The inclusion of hybrids like the Ordinary Tall from the Philippines, San Blas Tall, and Cochin China Tall, which exhibit different levels of resistance, highlights their potential in combating mite infestations. Researchers have discovered that by combining genetic traits from different varieties, they can create new hybrids that have enhanced resistance profiles (Jerard et al., 2021). This exciting finding opens up possibilities for developing crops that are better equipped to withstand various challenges. By harnessing the power of genetic diversity, scientists can create hybrids that possess improved resistance to diseases, pests, and environmental stressors. This breakthrough has the potential to revolutionize agriculture and contribute to more sustainable and resilient food production systems.

In addition, the purpose of the available coconut germplasm is to identify and screen for resistant cultivars that can be used in selective breeding. To address the issue of root wilt, a total of four seed gardens have been set up across five different locations. These seed gardens have been specifically designed to yield a substantial quantity of planting material that possesses resistance or tolerance to root wilt. In the gardens, there are a total of 6472 artificially pollinated seedlings. These seedlings belong to three different species: WCT, CGD, and COD. Three cultivars, namely Kalpasree, Kalpa Raksha, and Kalpa Sankara, have been specifically developed to exhibit resistance against RWD. Breeding populations were created to study the coconut disease caused by phytoplasma, as demonstrated by Nair et al. (2004), and Chowdappa et al. (2016) These populations proved to be highly effective in combating the pathogen responsible for RWD, thus enhancing the overall effectiveness of the breeding program. A 19-year study conducted in Florida by Broschat et al. (2002), and Odewale et al. (2013) focused on the LYD disease. The study found that certain dwarf germplasms, like the Fiji Dwarf and dwarf varieties in Nigeria (MGD, MYD, and MOD), exhibited resistance to LYD and showed higher nut yield compared to the West African Tall (WAT) variety. These findings suggest that these disease-tolerant dwarf x tall hybrids have the potential to be developed. Southeast Asian and Malaysian genotypes showed a certain level of tolerance to LYD, as they had an infection rate of less than 35%.

The potential for exploring these cultivars for resistance genes (R genes) is significant. This could play a crucial role in the development of new LYD-resistant hybrids (Jain and Priyadarshan, 2021).

Finally, the coconut germplasm can be utilized for breeding that focuses on enhancing specific traits to improve resistance against pests, such as mites and environmental stresses, such as drought and cold. Ramanujam et al. (1998) examined 26 different coconut germplasm for mite resistance where the study revealed none of the cultivars showed resistance, however different coconut germplasm materials exhibited different damages sizes where the Banawali Green Round displayed the smallest lesions, and Malayan Green Dwarf showcasing larger ones. The study was further supported by Muthiah and Bhaskaran (2000), and Arunachalam et al. (2013) which revealed that the damage was correlated to the morphological traits that contribute to its different levels of resistance towards mite infestations. Such morphological traits are the color of the fruit/tepal, tepal weight, and the presence of resin. Chowghat Orange Dwarf (COD), Laccadive Micro, and Spicata have been identified as having higher resistance to mites (Chowdappa et al., 2016).

Rajagopal et al. (2007) found that the dwarf genotypes absorbed more water than other types and hybrids. Tall forms which are West African Tall (WAT) were found to be drought tolerant. These characteristics are highly correlated with the morphological trait which is the deep root system. Drought-tolerant genotypes had more VAM-colonized roots and higher infection intensities than sensitive ones. It also has lower lipid peroxidation and better catalase, superoxide dismutase, and peroxidase activity exhibiting more scalariform thickening on tracheid and large sub-stomatal cavities (Gomes and Prado, 2007).

The coconut germplasm can be effectively utilized in the breeding and crop improvement program by selection, hybridization, and breeding for selective traits which lays the genetic foundations for creating a resistant/tolerant cultivar (Figure 4).

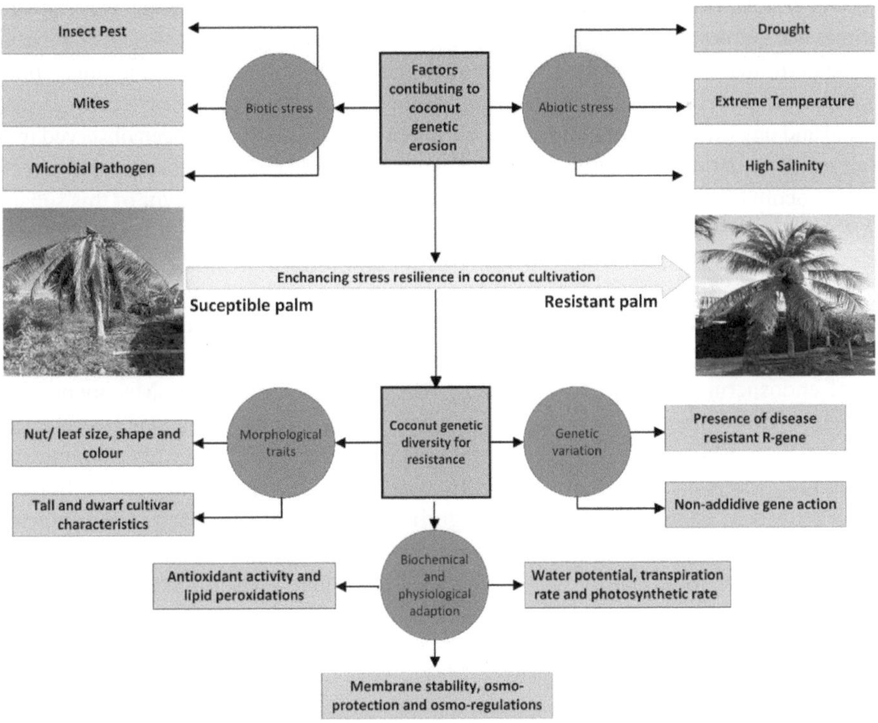

Figure 4. Highlights the importance of genetic diversity for enhancing the resilience of coconut palms to biotic and abiotic factors that contribute to genetic erosion.

Different Coconut Fruit Cultivars and Their Genetic Characterization

Tender coconut refers to coconut varieties with special characteristics, such as soft, sweet, and moist husks as well as juicy, jelly-like, and soft endosperm at the early stages of development before maturity. Because of these special features, it has gained popularity as a source of healthy, energizing cold refreshments throughout the hot season. The fruits of these varieties are harvested before they reach maturity. Tender coconut can be used to provide flavors to culinary and confectionaries. It is also an excellent source of vitamins and minerals (Rajamohan and Archana, 2018), which are in high demand in worldwide markets. This enables tender coconut to fetch higher prices than other varieties and at the same time encourages the creation of employment. This improves the lives of farmers' families, increases exports, strengthens the national economy, and supports the sustainable development of the coconut industry. The following section discusses a few aspects of tender coconut varieties in terms of genetic diversity and characterization.

Jelly-like Endosperm

Coconut fruits with gelatinous, sticky, and tasty endosperm are highly popular in tropical countries. This coconut fruit, known by various local names, is found in different countries. *Makapuno* in the Philippine (Table 1), *Dua Sap* in Vietnam, *Dikiri Pol* in Sri Lanka, *Kopyor* in Indonesia, *Dong Kathy* in Cambodia, *Maphrao Kathi* in Thailand, *Dahi Nariyel* in Myanmar, *Thairu Thengai* in India and *Niu Garuk* in Papua New Guinea.

Makapuno

Wicaksono et al. (2021) compared Makapuno and Kopyor, stating that the endosperm of Makapuno is gelatinous while that of Kopyor is often solid and brittle. Makapuno possesses a gelatinous and sticky endosperm that is soft buttery, and fragrant, and the fruit may vary in the juice content. The palm appears normal in terms of its physical appearance and embryo structure. However, it lacks the biochemical and physical abilities required for germination. Makapuno has been observed in tall palm trees (Nguyen et al., 2016). The endosperm of Makapuno exhibits higher moisture and protein levels than regular coconut fruits (Adriano and Manahan,1931). The delightful flavor of this substance can be attributed to the abundant presence of easily digestible sugars, along with citric acid, malic acid, and potassium (Santoso et al., 1996). The fatty acid composition of Makapuno endosperm differs from that of regular endosperm. Due to its inherent resistance to spoilage, this particular substance holds great promise in the food processing sector (Luengwilai et al., 2014).

A recent study by Lado et al. (2019) examined genes expressed differentially in Makapuno and normal endosperm using RNA sequencing. They discovered that the Makapuno endosperm exhibits high levels of glucomannan due to the early synthesis of glucomannan that remains stable in undegraded form. Makapuno endosperm thickness is greater than the normal coconut endosperm owing to its high levels of cytokinins, which aid in cell growth (Islam et al., 2009). In the absence of the α- galactosidase enzyme in Makapuno, significant quantities of water-soluble galactomannan accumulate compared to water-insoluble mannan, which is present in normal coconut endosperm, causing stickiness and modification in the cell wall structure. These effects cause Makapuno to produce endosperm with jelly and sticky characteristics (Mujer et al., 1984).

Kopyor

Kopyor coconut is found in Indonesia (Table 1) with jelly-like endosperm and comes in tall, dwarf, and hybrid varieties. The fruits of the dwarf coconuts have a variety of colors including green, yellow, orange, red, brownish green, reddish brown green, and greenish-brown. Tall Kopyor coconut can be green, brown, or red (Sukendah and Priyadarshini, 2023). Kopyor coconuts are grown in a bunch of common tall and dwarf coconut trees. The ratio of Kopyor in the tall palm is minimal, with one or two fruits per bunch on average; however, in the dwarf palm, this ratio may approach 50% (Sukendah and Priyadarshini, 2023). The texture of the endosperm, the proportion of endosperm volume filling the fruit cavity, and the amount of coconut water all contribute to the differences and are categorized into 3 groups. Type A that has been coagulated without water, crumbs, and endosperm is loose from the shell. Type B has little fruit liquid, and some endosperm is linked to the shell, but type C contains much more water, and the endosperm is highly soft and glued to the shell (Sukendah and Priyadarshini, 2023).

The phenotype of Kopyor coconut is determined by the xenia effect, wherein the pollen affects the seeds and fruits of the fertilized plant. The effect is influenced by a recessive mutant gene called 'k' at the K locus (Sukendah et al., 2009; Maskromo et al., 2016). Kopyor coconut is classified into three classes based on the phenotype, endosperm genotype, and zygotic embryo. The phenotype of KK nuts is normal, and they have a homozygous KKK endosperm genotype. The nuts with a normal phenotype are heterozygous for the Kk genotype, with either KKk or Kkk endosperm genotype. The Kopyor phenotype mutant has a homozygous kk genotype for nuts and a homozygous kkk genotype for endosperm (Sudarsono et al., 2013). Puan Kalianda Kopyor coconut (*Cocos nucifera* L.) is a newly released tall Kopyor coconut from Kalianda, South Lampung, Indonesia. Rahayu et al. (2022) aimed to analyze the genetic diversity of Puan Kalianda Kopyor coconut using SSR markers. The Puan Kalianda coconut population showed a 70% expected heterozygosity (He) and 60% observed heterozygosity. The phylogenetic analysis formed two main clusters, each consisting of three sub-clusters. The Genetic structure analysis showed that the population probably derived from two ancestral origins (K = 2) and can further be clustered into six sub-clusters (K = 6). Therefore, since genetic diversity within the population is relatively high, the Puan Kalianda tall Kopyor coconut population can be considered an essential genetic resource for future Kopyor coconut development.

Aromatic Endosperm

The fragrant coconut is a unique variety of dwarf coconut that emits a pleasant aroma reminiscent of *pandan* (Chongpraditnun et al., 2001). This specific variety of coconut is incapable of natural germination due to a genetic mutation affecting endosperm development. The primary component of the aromatic coconut scent is 2 acetyl-1-pyrroline (2-AP), also found in other plants like rice, jasmine, and soybean (Saensuk et al., 2016; Nartvaranant, 2019). Saensuk et al. (2016) employed a *de novo* transcriptome assembly of Thai aromatic coconut endosperm and identified a single nucleotide mutation G-to-C in the coding sequence of *CnAMADH2* in exon 14, which resulted in an amino acid change from alanine to proline at position 442 (P442A). This mutation was found exclusively in aromatic coconut among the sequences of AMADH/BADH proteins from diverse plant species.

The fruits are spherical and weigh between one to one and a half kilograms (Somsri and Vichitrananda, 2007). The fragrant coconut, known for its aromatic and refreshing water, has emerged as a lucrative and economically valuable commodity in countries that cultivate coconuts. The optimal time for harvesting the aromatic fruit is eight months, when it contains the highest sugar content, lowest acidity, and exhibits a potent aroma (Puchakawimol and Jangchud, 2003). Various local names refer to the aromatic coconut cultivars from the different Southeast Asian countries including, *Pandan* in Malaysia and Indonesia, *Maphrao Num Hom* in China, Philippines and Thailand, and *Dua Dua* in Vietnam.

Nam Hom

Thailand cultivates two dwarf coconut varieties, Nam Hom and Nam Wan, which yield fresh water and soft nuts. Nam Wan is recognized for its high yield and small-size fruits, and it is commonly used to make fresh juice. Nam Hom (Table 1) produces coconut water that is highly aromatic, and sweet with a vibrant pink color due to beneficial polyphenols (Rethinam and Krishnakumar, 2022). It exhibits a more pronounced aromatic fragrance due to the presence of 2-AP and is recognized as a variety with a lower yield but larger fruit size (Pablito, 2014). It is Thailand's most common aromatic coconut cultivar, with minimal morphological variations. There are two distinct types of nuts: one with three loops at the end of the fruit, known as Kon Chip, and the other with a rounded shape at the end, called Luk Klom (Chomchalow, 1999). This aromatic variety is called Young Aromatic Coconut in Thailand due to its dwarf stature (Nartvaranant, 2019).

A study examined the genetic diversity of fragrant coconut cultivars, explicitly focusing on various dwarf accessions, such as the "green dwarf fragrant" in the Philippines (Noël et al., 2011). Hierarchical cluster analysis revealed a genetic distinction between fragrant green dwarf and other tested accessions in this study. The study revealed that aromatic cultivars had a genetic diversity index of 0.067, indicating relatively low genetic diversity. In contrast, other accessions had at least three times higher genetic diversity. In another study conducted by Nartvaranant (2019), 10 AFLP markers were used to examine the genetic diversity and relationship of Nam Hom coconuts cultivated in Western Thailand. The findings revealed that the group consisting of all dwarf coconuts including "Nam Hom" ("Kon Chip" strains and "Kon Klom") and "Nam Wan" from various production areas had similarity coefficients ranging from 1.00 to 0.895, indicating a limited genetic variation. Furthermore, the findings indicated that the coconut samples of the "Nam Hom" variety, obtained from both the 'Kon Chip" and "Kon Klom" strains, exhibited a high degree of genetic similarity across various locations, as evidenced by a cophenetic correlation coefficient (r) of 0.9347.

Pandan

Aromatic green dwarf (Pandan) coconuts are planted in the states of Selangor, Kedah, and Perak (Table 1) in Malaysia for the production of edible nuts. The pandan tree may grow to be 6 meters in height. Pandan fruits are green, small, and spherical, with 10–15 fruits produced in each bunch (Figure 5). This variety contains water that is pleasant and aromatic, similar to the fragrance of pandanus leaves. The endosperm is soft and has a jelly-like texture. At 7–8 months, the fruits are harvested when the endosperm is still tender. The weight of the fruit can range from less than 1.5 kg to more than 2 kg, with an average of 300–400 ml of water per nut.

Pandan coconut plantations cover 1.7% of coconut farms in Malaysia. This variety has an average production of 27,000–30,000 coconuts per hectare (Man and Shah, 2020). In Malaysian marketplaces, each pandan fruit sells at double the price of other coconut varieties. Most smallholders obtain economic value and profit from selling mild-drinking nuts with an appealing flavor and scent. As a result of their faster fruiting and shorter height for optimum harvest, these dwarf coconuts provide more economic benefits to farmers.

Sweet Endosperm

Mohachao Narel is a coconut variant reported in India (Table 1) with a highly delicious and delicate endosperm. This variant is spread at random in the population. The percentage of nuts with delicious endosperm in various palms ranges from 10–77% per bunch. This fruit is more expensive than the normal type and is consumed fresh. Sweet endosperm has less copra and oil, lower fruit and husk weight but more total sugar content. Consequentially, it has less fiber and endosperm weight, however, the shell and endosperm thickness are not affected based on fruit components analysis (Kumar and Samsudeen, 2018). Samsudeen et al. (2013) examined the characteristics of Mohachao Narel fruit by SSR markers. They found significant variability in traits, such as the weight of the husk, embryo and the copra, and oil content. In this study, 14 SSR loci were identified, revealing the presence of 35 alleles. The average number of alleles per primer was found to be 2.5. The analysis of microsatellites

Figure 5. A Pandan palm tree, its fruit bunch and fruit (Teluk Bharu Agricultural Center, Perak, Malaysia)

revealed a 45% similarity among the sweet endosperm palm type, suggesting a substantial genetic diversity within this group of palms.

The province of Ben Tre in Vietnam is renowned for its production of Green Xiem Coconut, known for its exceptional sweetness and delicious taste (Table 1). As a result, the region has developed a specialized coconut confectionery business. The palm begins flowering three years after planting, yielding an average annual productivity of 140–150 fruits. The fruit exhibits a green coloration, is round in shape, and features a three-pointed knob at its lower portion. It weighs approximately 1.36 kg, while the coconut water content ranges from 250–350 ml. The fruit is typically harvested after 8–9 months when it reaches its peak flavor and sweetness (Rethinam and Krishnakumar, 2022).

Different Characteristics of Elite Coconut Husk

The utilization of coconut husk (mesocarp) as a natural fiber has been steadily increasing in economic significance, particularly in organic farming. The husk thickness and weight of coconuts exhibit significant diversity, which is influenced by factors, such as the coconut variety and its level of maturity (Pandiselvam et al., 2018). The coconut husk also displays diversity in terms of consistency, thickness, color, taste, and texture. The thickness of coconut husks can vary, ranging from thin to thick (ICAR-CPCRI, 2017). Regular coconut husks have a higher fiber content and a more rigid texture. In certain exceptional instances, coconut husks that are softer where the fibers are easily separated have been observed, whereas the husks of regular coconuts are characterized by their firmness.

The color of tender coconut fruits varies, with white husks being the most common and pink husks being rare. The presence of soft husks with lighter-colored fiber has been reported in the Lakshadweep and Nicobar Islands (Niral and Jerard, 2017). The pink husk (Wulung and Cungap Red in Indonesia) (Table 1) contains higher levels of phenol, protein, flavonoid, and anthocyanin, resulting in an enhanced free radical scavenging activity. This characteristic in pink husk holds potential for medical and pharmaceutical applications (Thomas et al., 2021). The coconut husk astringent taste is attributed to the tannin compound's presence. Husks with elevated tannin and sugar levels, leading to increased sweetness, have been documented in India and Indonesia (Menon and Pandalai, 1958; Tampake et al., 1982; Niral and Jerard, 2017).

Conservation of Useful Coconut Germplasm

Genetic diversity is a fundamental prerequisite in plant breeding for acquiring desirable traits and germplasm conservation is critical to provide genetic resources for crop improvement through various breeding strategies. The coconut has significant economic value and is considered a culturally and socially important commodity. Improvement of coconut genetic varieties contributes to their heightened resilience to climate change and genetic erosion, drought, pests, and diseases, mitigating species extinction risk. Such efforts are crucial for the cultivation of high-yielding and economically viable cultivars. Hence, the evaluation, restoration, and safeguarding of genetic resources are essential for global food security and the long-term viability of natural ecosystems. The main objective of a plant genetic resources (PGR) conservation program is to preserve optimal genetic diversity within the gene pool of a specific crop, encompassing its natural habitat and germplasm collection. The preservation of PGR is commonly achieved through *in situ* and *ex situ* techniques (Dulloo et al., 2005).

In Situ Conservation

Millions of hectares of agricultural area dedicated to coconut farming and household gardens are the primary locations for conserving coconut genetic diversity. The *in situ* conservation of coconut germplasm involves preserving the genetic diversity of coconuts in their natural environment. This includes cultivating traditional stands and species in agricultural ecosystems that have developed unique traits through human selection. Implementing conservation *in situ* relies on rigorous research, scientific expertise, and fruitful farmer collaboration. It is imperative to diligently observe the rights of farmers, along with their agricultural expertise and techniques, while preserving the natural ecosystem for the propagation and dissemination of coconut seeds and seedlings. Furthermore, modern techniques, such as satellite data and machine learning (Bourdeix et al., 2020) may aid in the implementation of germplasm conservation by informing farmers about the performance and condition of their fields, as well as providing the required feedback to enhance the process of conserving and maintaining seeds and seedlings. This strategy may not only help coconut cultivars survive for extended periods, but it can also assist farmers in strengthening their dedication and engagement in sustaining coconut biodiversity.

Several efforts have been undertaken to enhance the *in situ* conservation approach, which allows for the dynamic conservation of coconut populations (Sthapit and Jarvis, 1999). The Polymotu idea is an experimental technique and alternative conservation strategy for coconuts in which diverse coconut accessions are farmed individually on small islands and isolated valleys (the Polynesians' ancient method). These isolation conditions in geographically different places enable the reproduction essential for true breeding of each accession via natural pollination. The benefits of this technique include lower field genebank maintenance costs, easier access to germplasm, and optimal protection of coconut germplasm. This strategy is utilized to improve field genebanks' effectiveness (Bourdeix et al., 2011).

Ex situ Conservation

Ex situ conservation involves preserving species outside their natural habitats; facilities, such as seed banks, farm gene banks, botanical gardens, and laboratory collections (*in vitro* culture and cryopreservation) play an important part in conserving germplasm in the form of plants, seeds, pollen, tissues, and DNA. *Ex situ* conservation is significant in various aspects. This protection method is crucial for maintaining the genetic diversity of coconuts and preventing the loss of different coconut varieties in their natural environment. This collection offers the opportunity to investigate and comprehend the genetic traits of various coconut cultivars. Researchers can utilize conserved varieties and their diversity to develop improved cultivars resistant to biotic and abiotic stresses. International Coconut GeneBanks (ICGs) and national genebanks were created in major coconut-producing countries for this reason (Perera, 2020). *Ex situ* germplasm management and protection takes a lot of labor and evaluation and includes steps like collecting suitable samples, determining the identity of different coconut varieties, evaluating more preferable traits, and organizing related information. In addition, it involves efforts to ensure safe germplasm exchange with other international seeds and gene banks, removal of older accessions, as well as establishment and maintenance of field collections (Yang et al., 2018).

Seed Bank Conservation

The recalcitrant nature of coconuts, the vast size of the nuts, the various shapes of the seed, and the allogamous character of coconuts make conserving these plants' seeds challenging. As a result, other techniques of coconut conservation prevail over seed protection (Dulloo et al., 2005).

Field GeneBank Conservation

Large areas of land are required to establish field genebanks based on the biological properties of coconut. Even though the development of this gene bank is time-consuming and expensive, it is a reference approach for coconut conservation and has been developed in all coconut-producing countries

Field gene banks have a straightforward microstructure and are simple to characterize. The following key points should be explored in the establishment and creation of this gene bank:

- the minimum number of palms required to maintain genetic diversity
- updating and rationalizing the collections and removing old and duplicate data
- estimating the costs of maintaining the coconut collection
- developing the core collection
- Standardization of field plotting techniques to create sufficient diversity germplasm evaluation (Dulloo et al., 2005).

Natural disasters, pests, and diseases can all threaten field collections.

In Vitro Conservation

Tissue culture is an *ex situ* conservation method for coconut. Coconut germplasm is primarily transmitted in the form of embryos developed in laboratory conditions and occasionally in the form of seed nuts. Farmers primarily receive coconut planting materials in the form of seeds or seedlings (Yang et al., 2018). Coconut seeds are abundant in remote coral islands. The bulky size of coconut seeds makes collection challenging in these areas (Chin and Roberts, 1980). To address this issue, various embryo culture protocols are used in the laboratory to allow for the production of complete seedlings from coconut zygotic embryos. Embryo rescue culture is an appropriate method for protecting the expensive mutant coconuts, such as Kopyor in Indonesia (Sukendah and Priyadarshini, 2023) and Makapuno in the Philippines (Wicaksono et al., 2021) (the jelly-like endosperm prevents zygote embryo germination). Due to the long seed germination time, embryo culture is an appropriate method for germinating and conserving horned coconut (thick pericarp, India) (Jerard et al., 2014).

The *in vitro* method has limitations, such as slow propagation of disease-free planting material, low regeneration rate, somaclonal changes in planting material, and regular subculture requirement of plant material. Due to the seed's recalcitrant nature, such coconut tissue culture is not entirely standardized yet (Yang et al., 2018).

Cryopreservation

Bajaj (1984), was the first to assess the possibility of long-term preservation of embryos that can restart normal development after being frozen at extremely low temperatures. Cryopreservation is the storage of living plant materials at extremely low temperatures (–196) and is used to preserve the germplasm of resistant and vegetatively propagated plant species for long periods. Given seed bank and field bank conservation limitations, cryopreservation can be an important strategy for the long-term preservation of coconut germplasm (Welewanni and Bandupriya, 2017).

Other long-term conservation efforts in coconut include embryo and pollen preservation, which generates appropriate and reliable backups for field gene banks. If valuable materials preserved in farm gene banks are lost, they can be recovered from preserved germplasm. An efficient cryopreservation strategy for coconut zygotic embryos has been developed. It includes inoculating embryos in Petri dishes with glucose medium and placing them in a sealed container with silica gel to reduce humidity. The embryos are then transferred to cryotubes for freezing by immediately immersing them in liquid nitrogen. They can then be transferred to a culture medium containing Eeuwens mineral elements for fast embryo growth. This approach was reported for 10 diverse accessions with germination percentages ranging from 13.7–74.7% (N'Nan et al., 2012). Coconut pollen only survives for a few days in natural environments. Karun et al. (2014) proposed methods for long-term cryopreservation of coconut pollen, which maintained its fertility and viability in liquid nitrogen for four years. Germination was 100% in embryos obtained from hybrid nuts produced with frozen pollen. The resulting seedling grew normally as well. Thus, the possibility of establishing a pollen cryobank in coconut was confirmed (Karun et al., 2014).

Reserving coconut genetic resources is crucial for ensuring the long-term sustainability of the crop. Immediate conservation of the available coconut variability is necessary to safeguard them from the effects of expansion in high-yielding selections and production of hybrids, genetic erosion, climate change, sea level rise, and pests and diseases

Conclusion and Future Perspective

Coconut palm is an important economic crop in tropical regions. Coconut contains large genetic resources that can be utilized in agricultural improvement programs. The exponential rate of population growth and the adverse effects of climate change have made it critical for genetic resources to be exploited as efficiently as possible to meet new challenges in increasing agricultural output and providing food for the growing population. Genetic diversity not only contributes to the development of diverse and high-quality products that satisfy consumers but also serves as an important source of genetic resistance to diseases, pests, and abiotic stresses. Genetic diversity and characterization of genetic resources are important and primary requirements in crop improvement research to select parents for efficient breeding. With the rise in demand for coconuts, it is critical to find prospective superior genotypes that fulfill the needs of today's coconut market. These genotypes can then be employed to create profitable coconut hybrids for replanting programs. International collaboration and sharing efforts in the management of genetic resources and the optimal use of coconut genetic varieties are essential in sustainable agriculture development. Support obtained in the conservation of coconut diversity by major coconut-producing countries led to the establishment of farm banks in 24 countries. Breeding programs have focused mainly on quantitative traits like high yield. However, in addition to yield, other traits, such as resistance to pests and diseases, adaptation to varied climatic conditions, and tolerance to difficult growth conditions should be considered. With the availability of coconut genomic resources and bioinformatics tools, the opportunity to study and develop molecular markers grows. As a result, it is possible to analyze the molecular mechanisms of genes related to key traits, such as yield, early fruit ripening, copra quality, and quantity and quality of tender coconut water, which will aid in ensuring sustainable coconut production, resource management, and coconut product improvement. On the other hand, to conserve the genetic variety of coconuts, germplasm conservation techniques in dataset evaluation, complete access to data, standardization, data management, and documentation must be addressed and strengthened. Furthermore, *in situ* and *ex-situ* conservation methods should be linked with environmental conditions and specialized needs to preserve genetic diversity as effectively as feasible. Enhancing farmers' knowledge of the value of genetic variety and boosting their agricultural skills can impact the preservation of many coconut varieties, as well as the trade and long-term economic development of this precious agricultural product. Attention should be given to tender coconut varieties, which have distinct qualities and are approved by the coconut market sector. This can help provide financial support for smallholders and has the potential for employment creation. Although the coconut industry is growing at a slow pace, it has a bright future because all parts of the coconut tree have economic value. The availability of quality planting materials, proper crop management practices, breeding strategies, germplasm protection, and preservation of coconut biodiversity are important and beneficial for the coconut industry. In addition, there should be access to new technologies in the development of research projects to increase coconut productivity and support policies should be in place to ensure implementation of the different programs for sustainability of the coconut industry.

Acknowledgement/ Funding

This work was funded by the Ministry of Higher Education Malaysia (MoHE) through Long-Term Research Grant Scheme (LRGS/1/2020/UPM/01/2/2).

References

Abraham, A., C.A. Ninan, and P. Gopinath. 1965. Cytology of development of abnormal endosperm in Philippine Makapuno coconuts. *Caryologia, 18*: 395–408.

Abraham, A., P.M. Mathew, and C.A Ninan. 1961. Cytology of *Cocos nucifera* L. and *Areca catechu* L. *Cytologia*. (Tokyo), *26*: 327–332.

Adriano, F.T. and M. Manahan. 1931. The nutritive value of green, ripe, and sport coconut (buko, niyog, and Makapuno). *Philipp Agric., 20*: 195–198.

Amritha, V.V., K.G. Ajith Kumar, S. Sreedevi, S. John, and K. Prathapan. 2023. Changes in Relative Proportions of Plant Growth Regulators in the Liquid Endosperm during Seed Development in Coconut Cultivars/Hybrids. *Russian Journal of Plant Physiology, 70*: 135. https://doi.org/https://doi.org/10.1134/S1021443723601428.

Anupap, T., C. Petchipiroon, and M. Rattanapruk. 1992. *Existing Status of Coconut Genetic Resources of Thailand.* Report of a working session for implementation of the International Coconut Database held at CIRAD, Montpellier, 19–22. France.

Arunachalam, V., and M.K. Rajesh. 2017. *Coconut Genetic Diversity, Conservation and Utilization, Biodiversity and Conservation of Woody Plants*, 3–36. Springer, Cham.

Arunachalam, V., B.A. Jerard, S.E. Apshara, C. Jayabose, K. Subaharan, N. Ravikumar, and C. Palaniswami. 2013. Digital phenotyping of coconut and morphological traits associated with eriophyid mite infestation. *Journal of Plantation Crops, 41*(3): 417–424.

Arunima, S. and T. Rajamohan. 2014. Influence of virgin coconut oil-enriched diet on the transcriptional regulation of fatty acid synthesis and oxidation in rats: A comparative study. *British Journal of Nutrition, 111*: 1782–1790. https://doi.org/doi:10.1017/S000711451400004X.

Asmah, N. 1999. *Analisis protein spesifik sebagai penanda sifat kopyor pada kelapa. Skripsi Jurusan Kimia FPMIPA*. IPB. Bogor.

Bajaj, Y.P.S. 1984. Induction of growth in frozen embryos of coconut and ovules of citrus. *Current Science (Bangalore) 53*: 1215–1216.

Batugal, P., R. Bourdeix, and L. Baudouin. 2009. Coconut breeding. *In*: *Breeding Plantation Tree Crops: Tropical Species*, 327–375. Springer.

Batugal, P., V. Ramanatha Rao, and J. Oliver. 2005. *Coconut Genetic Resources*. International Plant Genetic Resources Institute – Regional Office for Asia, the Pacific and Oceania (IPGRI-APO), Serdang, Selangor DE, Malaysia.

Baudouin, L., B.F. Cunn, and K.M. Olsen. 2014. The presence of coconut in southern Panama in pre-Columbian times: clearing up the confusion. *Annals of Botany, 113*(1): 1–5.

Beccari, U. 1917. The origin and dispersal of *Cocos nucifera. Philipp. J. Sci Series C, 12*: 27–43.

Bhandari, H., A. Bhanu, K. Srivastava, M.N. Singh, Shreya, and A. Hemantaranjan. 2017. Assessment of Genetic Diversity in Crop Plants: An Overview. *Advances in Plants and Agriculture Research, 7*: 00255. https://doi.org/10.15406/apar.2017.07.00255.

Bhatia, G., S.K. Upadhyay, A. Upadhyay, and K. Singh. 2021. Investigation of long non-coding RNAs as regulatory players of grapevine response to powdery and downy mildew infection. *BMC Plant Biol., 21*: 1–16. https://doi.org/https://doi.org/10.1186/s12870-021-03059-6.

Bourdeix, R., S. Adkins, V. Johnson, L. Perera, Sisunadar. 2020. *In situ* and *ex-situ* conservation of coconut genetic resources. *In*: Adkin S., M. Foale, R. Bourdeix, Q. Nguyen, and J. Biddel [Eds.], *Coconut Biotechnology: Towards the Sustainability of the 'Tree of Life'*. Springer, Cham. https://doi.org/10.1007/978-3-030-44988-9_4

Bourdeix, R., V. Johnson, L. Baudouin, V.S. Tuia, T. Kete, S. Planes, C. Lusty, and S. Weise. 2011. Polymotu: A new concept of island-based germplasm bank based on an old Polynesian practice. *Ogasawara Research, 37*: 33–51. https://doi.org/https://hdl.handle.net/10568/35778.

Broschat, T.K., N.A. Harrison, and H. Donselman. 2002. Losses to lethal yellowing cast doubt on coconut cultivar resistance. *Palms, 46*: 185–189.

Caro, R.E.S., J. Cagayan, R.R. Gardoce, A.N.C. Manohar, A.O. Canama-Salinas, R.L. Rivera, D.V. Lantican, H.F. Galvez, and C.E. Reano. 2022. Mining and validation of novel simple sequence repeat (SSR) markers derived from coconut (*Cocos nucifera* L.) genome assembly. *Journal of Genetic Engineering and Biotechnology, 20*: 71. https://doi.org/10.1186/s43141-022-00354-z.

Child, R. 1964. *Coconuts*. Longmans, Green and Co. Ltd, London, UK.

Child, R. 1974. *Coconuts*. 2nd Edn. Longman: London, UK.

Chin, H.F., and E.H. Roberts. 1980. *Recalcitrant Crop Seeds*, 152pp. Tropical Press SDN BHD, Kuala Lumpur, Malaysia.

Chomchalow, N. 1999. Coconut varieties in Thailand. *Assumption University Journal of Technology, 3*(1): 19–30.

Chongpraditnun, P., P. Limsmuthchaiporn, H. Hattori, S. Nakamura, A. Watanabe, and M. Chino. 2001. Nutrient concentrations of phloem sap of aromatic coconut grown under various nutrient conditions. *In*: Horst, W.J. et al. [Eds.], *Plant Nutrition. Developments in Plant and Soil Sciences*. Springer, Dordrecht. *92*: 232–233. https://doi.org/10.1007/0-306-47624-X_112.

Choudhury, B.I., M.L. Khan, and S. Dayanandan. 2014. Patterns of nucleotide diversity and phenotypes of two domestication-related genes (OsC1 and Wx) in indigenous rice varieties in Northeast India. *BMC Genet., 15*: 71. https://doi.org/10.1186/1471-2156-15-71.

Chowdappa, P., J. George, H. Muralikrishna, and M.K. Rajesh. 2016. *CPCRI: 100 Years of Scientific Excellence*. ICAR - Central Plantation Crops Research Institute Kasaragod - 671 124, Kerala, India.

Chowdappa, P., V. Niral, B.A. Jerard, and K. Samsudeen. 2017. *COCONUT*. Daya Publishing House. A Division of Astral International Pvt. Ltd. NewDelhi-110002.

Cook, O.F. 1901. Origin and distribution of cocoa palm. *Contr. U.s Nat. Herb., 7*: 57–97.

Cowan, C.W., and P.J. Watson. 1992. *The Origins of Agriculture*. Smithsonian Institution Press, Washington, DC, USA.

Darlington, C.D., and E.K. Janaki Ammal. 1945. *Chromosome Atlas of Cultivated Plants.* G. Allen & Unwin Ltd, London.

Davis, T.A. 1954. Hermaphroditism in *Coconus nucifera* L. *Indian Coconut Journal, 7*: 133–142. De Candolle, A. 1886. *Origin of Cultivated Plants* (2nd Edn.), 468pp. Hafner Publ. Co., New York,

Dransfield, J., N.W. Uhl, C.B. Asmussen, W.J. Backer, and M.M. Harley. 2008. *Genera palmarum: The Evolution and Classification of the Palms,* 732pp. Royal Botanic Gardens, Kew, UK, London: https://doi.org/10.34885/92.

Dulloo, M.E., V.R. Rao, F. Engelmann, and J. Engels. 2005. Complementary conservation of coconuts. *In:* Batugal, P., V.R. Rao, and J. Oliver [Eds.], *Coconut Genetic Resources,* 75–90. International Plant Genetic Resources Institute–Regional Office for Asia, the Pacific and Oceania (IPGRI-APO), Serdang, Selangor DE, Malaysia,

Fan, H., Y. Xia, Y. Yang, Y. Yang, Xia W, A.S. Mason, Z. Xia, F. Qiao, S. Zhao, and H. Tang. 2013. RNA-Seq analysis of *Cocos nucifera*: Transcriptome sequencing and *de novo* assembly for subsequent functional genomics approaches. *PLoS One, 8*(3): e59997. https://doi.org/10.1371/journal.pone.0059997.

Félix, J.W., M.I. Granados-Alegría, R.R. Gómez-Tah, M. Tzec-Sima, E. Ruiz-May, B. Canto-Canche, J.A. Zamora-Bariseno, E. Bojorquez-Velazquez, C. Oropeza-Salin, and I. Islas-Flores. 2023. Proteome Landscape during Ripening of Solid Endosperm from Two Different Coconut Cultivars Reveals Contrasting Carbohydrate and Fatty Acid Metabolic Pathway Modulation. *Int. J. Mol. Sci., 24*: 10431. https://doi.org/https://doi.org/10.3390/ijms241310431.

Fernando, W.M.U., T.S.G. Peiris, and M.R.T. Wickramaratne. 2010. Variation in some seedling characters of three different coconut cultivars and their use in identification in the nursery. *Cocos, 9*: 23–29. https://doi.org/DOI:10.4038/*cocos*.v9i0.2126.

Foale, M. 2003. *Coconut Odyssey: The Bounteous Possibilities of the Tree of Life,* 132pp. Australian Centre for International Agricultural Research Monograph No. 101.

Gao, L.W., Y. Wang, Q. Guo, and D. Li. 2022. Identification and functional analysis of transcription factors related to coconut (*Cocos nucifera* L.) endosperm development based on ATAC-seq. *Trop. Plants, 1*: 8. https://doi.org/doi: 10.48130/TP-2022-0008.

Gomes, F.P., and C.H.B.A. Prado. 2007. Ecophysiology of coconut palm under water stress. *Braz. J. Plant Physiol., 19*: 377–391. https://doi.org/https://doi.org/10.1590/S1677-04202007000400008.

Gunn, B.F., L. Baudouin, and K.M. Olsen. 2011. Independent origins of cultivated coconut (*Cocos nucifera* L.) in the Old-World tropics. *PLoS One, 6*(6): e21143 https://doi.org/10.1371/journal.pone.0021143.

Harries, H.C. 1978. The evolution, dissemination and classification of *Cocos nucifera* L. *Bot. Rev., 44*: 265–319.

Harries, H.C. 1990. Malesian origin for a domestic *Cocos nucifera. In*: Baas P., K. Kalkman, R. Geesink [Eds.] *The Plant Diversity of Malesia*, 351–357. Springer, Dordrecht. https://doi.org/10.1007/978-94-009-2107-8_29.

Harries, H.C. 2012. Germination rate is the significant characteristic determining coconut palm diversity. *AoB Plants, 2012*: pls045. https://doi.org/https://doi.org/10.1093/aobpla/pls045.

Harries, H.C., and D.H. Romney. 1974. Maypan: An F1 hybrid coconut variety for commercial production in Jamaica. *Farmer, Jamaica, 79*: 57–60.

ICAR-CPCRI. 2017. *Annual Report 2016–2017*, 236pp. ICAR Central Plantation Crops Research Institute, Kasaragod, Kerala, India.

Ingicco, T., G.D. van den Bergh, C. Jago-on, J.J. Bahain, M.G. Chacon, H. Forestier, C. King, K. Manalo, S. Nomade, A. Pereira, M.C. Reyes, A.M. Semah, Q. Shao, P. Voinchet, C. Falgueres, P.C.H. Albers, M. Lising, G. Lyras, D. Yurnaldi, P. Rochette, A. Bautista, and J. de Vos. 2018. Earliest known hominin activity in the Philippines by 709 thousand years ago. *Nature, 557*: 233–237. https://doi.org/https://doi.org/10.1038/s41586-018-0072-8.

Islam, M.N., M.L.O. Cedo, L.O. Namuco, T.H. Borromeo, E.A. Aguilar. 2009. Effect of fruit age on endosperm type and embryo germination of Makapuno coconut. *Gene Conserve, 8*: 708–722.

Izaitul Aida, I., Z. Rasdi, R. Ismail, M.B. Ismeazilla, K. Mohd Faizol, Z. Muhammad Shakir, A. Nur Fakriyah, Noor and S.M. Shuhina. 2020. Susceptibility and resistance of different host varieties of oil palm and coconut palm towards pest, rhinoceros beetle (*Oryctes rhinoceros*). *Asian Journal of Agriculture and Rural Development, 10*: 56. https://doi.org/DOI:10.18488/journal.1005/2020.10.1/1005.1.56.67.

Jain, S.M., and P.M. Priyadarshan. 2009. *Breeding Plantation Tree Crops. Tropical Species*. Springer, New York. 978-0-387-71201-7.

Jain, S.M. and P.M. Priyadarshan. 2021. *Cash Crops: Genetic Diversity, Erosion, Conservation and Utilization*. Springer, Cham. https://doi.org/10.1007/978-3-030-74926-2.

Jayasinghe, M.D. and I.G.N. Hewajulige. 2021. King coconut: The golden nut of Sri Lanka. *Sri Lanka Journal of Food and Agriculture, 7*(9). doi: 10.4038/sljfav7i1.90.

Jerard, B.A., V. Niral, and M.K. Rajesh. 2021. Breeding Strategies. *In*: Rajesh, M.K., S.V. Ramesh, L. Perera, and C. Kole [Eds.], *The Coconut Genome. Compendium of Plant Genomes*, 47–76. Springer, Cham. https://doi.org/10.1007/978-3-030-76649-8_4.

Jerard, B.A., V. Niral, R. Dhanapal, V. Damodaran, V. Arunachala, M.K. Rajesh, K. Devakumar, K. Samsudeen, R.V. Nai, P.M. Kumarn, V. Thomas Geroge. 2014. IND 221–Andaman horned Cocos (IC0598221; INGR13063), a coconut (*Cocos nucifera*) germplasm with distinct character of horny nuts. *Indian J. Plant Genet. Resour., 27*: 76–77.

Karun, A., K.K. Sajini, V. Niral, C.H. Amarnath, P. Remya, M.K. Rajesh, K. Samsudeen, B.A. Jerad, and F. Engelmann. 2014. Coconut (*Cocos nucifera* L.) pollen cryopreservation. *Cryo Letters, 35*: 407–417.

Karun, A., S.V. Ramesh, M.K. Rajesh, V. Niral, R. Sudha, and K.S. Murlikrishna. 2022. Conservation and utilization of genetic diversity in coconut (*Cocos nucifera* L.). *In*: *Cash Crops: Genetic Diversity, Erosion, Conservation and Utilization*, 197–250.. Springer, Cham. https://doi.org/10.1007/978-3-030-74926-2.

Konan, B.R., J.L. Konan, F. Tetchi, R.R. Assa, and G. Amani. 2011. The biochemical characteristics of coconut (*Cocos nucifera* L.) water during germination. *Int. J. Biol. Chem. Sci., 5*: 2214–2223. https://doi.org/10.4314/ijbcs.v5i6.4.

Kumar, A., and K. Samsudeen. 2018. Seasonal variation and developmental changes in the biochemical composition of coconut kernel in Mohachao Narel, a sweet endosperm coconut (*Cocos nucifera* L.) population from Maharashtra. Conference paper, *Trends in Biosciences. 11*(7): 1599–1603, ref.16. Kerala, India.

Lado, J.P., B.Z. Arellano, C.A.L. Osio, B.J.T. Tendero, C.B.D. Vina, M.A.O. Torio, R.P. Laude, and M.G.Q. Diaz. 2019. Early differential expression of galactomannan biosynthesis genes in 'Makapuno' coconut (*Cocos nucifera* L.) revealed by the *de novo* assembly and analysis of endosperm transcriptome. *Philipp. Agricul. Sci., 102*: 6–24.

Lantican, D.V., S.R. Strickler, A.O. Canama, R.R. Gardoce, L.A. Mueller, and H.F. Galvez. 2019. *De Novo* Genome Sequence Assembly of Dwarf Coconut (*Cocos nucifera* L. 'Catigan Green Dwarf') provides insights into genomic variation between coconut types and related palm species. *G3 (Bethesda), 9*: 2377–2393. https://doi.org/10.1534/g3.119.400215.

Lebrun, P., L. Baudouin, W. Myrie, A. Berger, and M. Dollet. 2010. Recent lethal yellowing outbreak: Why is the Malayan yellow dwarf coconut no longer resistant in Jamaica? *Tree Genet. Genomes, 4*: 125–131. https://doi.org/10.1007/s11295-007-0093-1.

Lebrun, P., L. Grivet, and L. Baudouin. 1999. Use of RFLP markers to study the diversity of the coconut palm. *In:* Oropeza, C., J.L. Verdeil, G.R. Ashburner, R. Cardena, and J.M. Santamaria [Eds.], *Current Advances in Coconut Biotechnology*, 73–87. Springer. https://doi.org/10.1007/978-94-015-9283-3_6.

Lédo, A. da S., E.E.M. Passos, H.R. Fontes, J.M.S. Ferreira, V. Talamini and W.A. Vendrame. 2019. Advances in Coconut palm propagation. *Revista Brasileira de Fruticultura, 41*(2). https://doi.org/10.1590/0100-29452019159.

Liang, Y., Y. Yuan, T. Liu, W. Mao, Y. Zheng, D. Li, J. Sun, X. Peng, Y. Chen, Y. Yang, H. Zhang, Y. Wang, Z. Li, Y. Liu, Z. Lin, Y. Wei, Y. Yang, L. Liu, S. Wang, X. Li, Y. Fang, W. Liu, J. Li, Y. Zhang, Y. Zhang, J. Zhang, Y. Yang, Y. Liu,

Y. Wei, Y. Zhang, Y. Liu, Y. Wang, Y. Li, and Y. Fang. 2014. Identification and computational annotation of genes differentially expressed in pulp development of *Cocos nucifera* L. by suppression subtractive hybridization. *BMC Plant Biology*, *14*: 205. https://doi.org/10.1186/s12870-014-0205-

Luengwilai, K., D.M. Beckles, O. Pluemjit and J. Siriphanich. 2014. Postharvest quality and storage life of 'Makapuno' coconut (*Cocos nucifera* L.). *Sci. Hortic.*, *175*: 105–110. https://doi.org/10.1016/j.scienta.2014.06.005.

Mahmud, Z. 2000. *Petunjuk teknis budidaya kelapa kopyor*. Departemen Kehutanan dan Perkebunan Dirjen Perkebunan Jakarta, *29*: 41.

Man, N., and J. Shah. 2020. Acceptance of New Coconut Seed Matag Among Coconut Growers in Bagan Datoh, Perak and Bachok, Kelantan. *International Journal of Academic Research in Business and Social Sciences*, *10*: 16, 1–26. https://doi.org/10.6007/IJARBSS/v10-i16/8286.

Martius, C.F.P. 1850. *Historia Naturalis Palmarum*, 1823–1850. T O Weigel, Leipzig. https://doi.org/10.5962/bhl.title.506.

Maskromo, I., S.H. Larekeng, H. Novarianto, and S. Sudarsono. 2016. Xenia negatively affecting kopyor nut yield in Kalianda Tall kopyor and Pati Dwarf kopyor coconuts. *Emir J. Food Agric.*, 644–652. https://doi.org/10.9755/ejfa.2015-07-552.

Mayuranathan, P.V. 1938. Original home of the coconut. *J. Bombay Nat. His. Soc.*, *40*: 174–182. https://doi.org/10.5962/bhl.title.506.

McCormack, G. 2005. *The Origin of the Coconut Palm*. Cook Islands Natural Heritage Trust, Rarotonga. http://cookislands.bishopmuseum.org/showarticle.asp?id=15. (Accessed 31 Dec 2023).

Mendoza, E.M.T., J.L. Samonte, D.A. Ramirez, N.B. de la Cruz and C.V. Mujer. 1987. Galactomannans in developing normal and makapuno coconut endosperms. *Bulletin of the Philippine Biochemical Society*, *7*: 15–19. https://doi.org/10.5962/bhl.title.506.

Menon, K.P.V., and K.M. Pandalai. 1958. *The Coconut Palm: A Monograph*. Indian Central Coconut Committee. S. India. 384pp.

Mujer, C.V., A.S. Arambulo, and E.M.T. Mendoza. 1983. The viscous component of the mutant (makapuno) coconut endosperm. I. isolation and characterization. *Philippine Journal of Biology*, *12*: 42–50. https://doi.org/10.5962/bhl.title.506.

Mujer, C.V., D.A. Ramirez, and E.M.T. Mendoza. 1984. Coconut α-d-galactosidase isoenzymes: Isolation, purification and characterization. *Phytochemistry*, *23*: 1251–1254. https://doi.org/10.1016/S0031-9422(00)80435-5.

Muñoz-Pérez, J.M., G.P. Cañas, L. López, and T. Arias. 2022. Genome-wide diversity analysis to infer population structure and linkage disequilibrium among Colombian coconut germplasm. *Sci. Rep.*, *12*: 2958. https://doi.org/10.1038/s41598-022-07013-w.

Muthiah, C., and R. Bhaskaran. 2000. Major outbreak of eriophyid mite of coconut in India. *Planter, Kuala Lumpur*, *76*: 243–246.

N'Nan, O., M. Borges, J.L.K. Konan, V. Hocher, J.-L. Verdeil, J. Tregear, A.S.P. N'guetta, F. Engelmann, and B. Malaurie. 2012. A simple protocol for cryopreservation of zygotic embryos of ten accessions of coconut (*Cocos nucifera* L.). *In Vitro Cellular & Developmental Biology – Plant, 48*: 160–166. https://doi.org/10.1007/s11627-012-9425-4.

Nair, K.P. 2021. *Tree Crops: Harvesting Cash from the World's Important Cash Crops*. Springer International Publishing. https://doi.org/10.1007/978-3-030-62140-7.

Nair, M.K., P.K. Koshy, P.M. Jacob, R.V. Nair, E.V.V. Bhaskara Rao, K.U.K. Nampoothiri, and R.D. Iyer. 1996. A root (wilt) disease resistant coconut hybrid and strategy for resistance breeding. *Indian Coconut Journal*, *1*(27), 2–5.

Nair, R.V., P.M. Jacob, R.J. Thomas, and M. Sasikala. 2004. Development of varieties of coconut (*Cocos nucifera* D resistant. *Journal of Plantation GGGG*, *32*(33).

Nair, R.V., P.M. Jacob, R.J. Thomas, S.P. Pillai, and C. Mathews. 2006. Performance of CGD x WCT hybrid in the coconut root (wilt) disease prevalent tract. *Journal of Plantation Crops*, *34*(1):15.

Nambiar, M.C., and M.S. Swaminathan. 1960. Chromosome morphology, micro-sporogenesis and pollen fertility in some varieties of coconut. *Indian Journal of Genetics*, *20*: 200–211.

Nambiar, M.C., and T.O. Prasannakumari. 1964. Effect of root (wilt) disease on microsporogenesis in coconut (*Cocos nucifera* L.). *Indian Cocon. J.*, *17*: 93–100.

Nampoothiri, V., V. Krishnakumar, P.K. Thampan, and M.A. Nair. 2018. *The Coconut Palm (Cocos nucifera L.):esearch and Development Perspectives*. Springer, Singapore. ISBN 978-981-13-2753-7.

Narayana, GV., and CM. John. 1949. Varieties and forms of coconut. *Madras Agric. J. India*, 36: 349–366.

Nartvaranant, P. 2019. Analysis of markers of the genetic variants of (Nam Hom coconut *Cocos nucifera* L.) grown in western Thailand. *Interdisciplinary Research Review*, *14*(1): 12–20. https://doi.org/https://ph02.tci-thaijo.org/index.php/jtir/article/view/176694.

Natarajan, C., K. Ganesamurthy, and M. Kavitha. 2010. Genetic variability in coconut (*Cocos nucifera*). *Electronic Journal of Plant Breeding*, *1*(5): 1367–1370. https://doi.org/https://www.ejplantbreeding.org/index.php/EJPB/article/view/1653.

Nathanael, W.R.N. 1967. The Application of fertilizers to adult coconut palms in relation to theoretical concepts. *Ceylon Cocon. Q.*, *18*(5): 30.

Nguyen, Q.T., H.D.D. Bandupriya, M. Foale, and S.W. Adkins. 2016. Biology, propagation, and utilization of elite coconut varieties (makapuno and aromatics). *Plant Physiology and Biochemistry*, *109*: 579–589. https://doi.org/10.1016/j.plaphy.2016.11.003.

Niral, V., and B.A. Jerard. 2019. Botany, origin, and genetic resources of coconut. *In*: Krishnakumar, V., P. Thampan, and M. Nair [Eds.], *The Coconut Palm (Cocos nucifera L.): Research and Development Perspectives*, 57–111. Springer, https://doi.org/10.1007/978-981-13-2754-4_3

Niral, V., B. Augustine, J.K. Samsudeen, R.J. Thomas, and G.V. Thomas. 2014. *Coconut Varieties and Hybrids*. CPCRI Technical Bulletin No. 87, Indian Council of Agricultural Research: Kasaragod, India.

Niral, V., and B.A. Jerard, 2017. Botany, growth, and development. *In*: P. Chowdappa, V. Niral, B.A. Jerard, K. Samsudeen. *Coconut*, 73–129. Daya Publishing House, A Division of Astral International Pvt. Ltd, New Delhi, India.

Niral, V., B.A. Jerard, and M.K. Rajesh. 2021. Germplasm Resources: Diversity and Conservation. *In*: Rajesh, S.V. Ramesh, L. Perera, C. Kole. *The Coconut Genome: Compendium of Plant Genomes*, 27–47. Springer, Cham. https://doi.org/10.1007/978-3-030-76649-8_3.

Noël, K.K.J., K.K. Edmond, K.J.L. Konan, and K.K. Eugene. 2011. Microsatellite gene diversity within Philippines dwarf coconut palm (*Cocos nucifera* L.) resources at Port-Bouët, Côte d'Ivoire. *Scientific Research and Essays*, *6*: 5986–5992. https://doi.org/10.5897/SRE11.877.

Odewale, J.O., M.N. Okoye, G. Odiowaya, J.M. Ahanon, and C.I. Agho. 2013. Bunch and nut production of surviving coconut palms in lethal yellowing disease endemic area of Nigeria. *International Journal of Plant Breeding and Genetics*, *7*(1): 57–64. https://doi.org/10.3923/ijpbg.2013.57.64.

Pablito, P.P. 2014. Opportunities in growing coconut for sweet aromatic water and young, tender nuts. *Agriculture Monthly Magazine. Philippine*. https://agriculture.com.ph/2018/03/07/opportunities-in-growing-coconut-for-sweet-aromatic-water-and-young-tender-nuts/. Accessed on 11/-1/2024.

Padmakumaran Nair, K.G., T. Rajamohan, and P.A. Kurup. 1998. Coconut kernel protein modifies the effect of coconut oil on serum lipids. *Plant Foods for Human Nutrition*, *53*: 133–144. https://doi.org/10.1023/A:1008078103299.

Pandiselvam, R., M.R. Manikantan, A. Kothakota, G.K. Rajesh, S. Shameena Beegum, S.V. Ramesh, V. Niral, and K.B. Hebbar. 2018. Engineering properties of five varieties of coconuts (*Cocos nucifera* L.) for efficient husk separation. *Journal of Natural Fibers*, *17*(4), 589–597. https://doi.org/10.1080/15440478.2018.1507863.

Patel, J.S. 1938. The coconut. A monograph. Government Press, Madras, India.

PCA. 2016. *Genetically Multi-ancestored Farmers' Composite Variety: A Sustainable Option for the Farmers*. Philippine Coconut Authority. https://pca.gov.ph/pdf/techno/gmasynvar.pdf. Accessed on 11/01/2024.

Pereira, TNS., M.F. Neto, M.M. de Souza, I.G. da Costa Geronimo, C.A.F. de Melo, and M.G. Pereira. 2017. Cytological characterization of Brazilian green dwarf coconut (*Cocos nucifera* L.) via meiosis and conventional and differential karyotyping. *Cytologia* (*Tokyo*), *82*: 167–174. https://doi.org/https://doi.org/10.1508/cytologia.82.167.

Perera, C., H.D.D. Bandupriya, R.J. Thomas, and R. Bourdeix. 2020. Diversity Studies Using Molecular Markers. *In*: Adkins, S.M. Foale, R. Bourdeix, Q. Nguyen, and J. Biddle [Eds.], *Coconut Biotechnology: Towards the Sustainability of the 'Tree of Life'*, 101–122. Springer, Cham. https://doi.org/10.1007/978-3-030-44988-9_6.

Perera, L., and R. Manimekalai. 2021. Characterization of Genetic Diversity Using Molecular Markers. *In*: M.K. Rajesh, S.V. Ramesh, L. Perera, and C. Kole [Eds.], *The Coconut Genome: Compendium of Plant Genomes*, 77–90.. Springer, Cham. https://doi.org/10.1007/978-3-030-76649-8_5.

Perera, L., S.A.C.N. Perera, C.K. Bandaranayake, and H.C. Harries. 2010a. Coconut. *In*: *Oil Crops*, 369–396. Springer. https://doi.org/10.1007/978-0-387-77594-4.

Perera, P.I.P., V. Hocher, L.K. Weerakoon, D.M.D. Yakandawala, S.C. Fernando, and J.-L. Verdeil. 2010b. Early inflorescence and floral development in *Cocos nucifera* L. (Arecaceae: Arecoideae). South African Journal of Botany, *76*: 482–492. https://doi.org/10.1016/j.sajb.2010.03.006.

Perera, S.A.C.N. 2020. Genetic Improvement for Sustainability of Coconut Production: The Sri Lankan Experience. *In*: Marambe, B., J. Weerahewa, and W. Dandeniya [Eds.], 149–169. *Agricultural Research for Sustainable Food Systems in Sri Lanka*. Springer, Singapore. https://doi.org/10.1007/978-981-15-2152-2_7.

Pérez-Núñez, M.T., R. Souza, L. Sáenz, J.L. Chan, J.J. Zúñiga-Aguilar, and C. Oropeza. 2009. Detection of a SERK-like gene in coconut and analysis of its expression during the formation of embryogenic callus and somatic embryos. *Plant Cell Rep.*, *28*: 11–19. https://doi.org/10.1007/s00299-008-0616-8.

Prasad, M., E.G. Khare, and S.K. Singh. 2013. Plant fossils from the Deccan Intertrappean sediments of Chhindwara district, Madhya Pradesh, India: Their palaeoclimatic significance. *Jour. Palaeontol. Soc. India*, *58*: 229–240.

Prebble, M., and J.L. Dowe. 2008. The late Quaternary decline and extinction of palms on oceanic Pacific islands. *Quat. Sci. Rev.*, *27*: 2546–2567.

Preethi, P., S. Rahman, S. Naganeeswaran, A.A. Sabana, K.P. Gangaraj, B.A. Jerard, V. Niral, and M.K. Rajesh. 2020. Development of EST-SSR markers for genetic diversity analysis in coconut (*Cocos nucifera* L.). *Mol. Biol. Rep.*, *47*: 9385–9397. https://doi.org/10.1007/s11033-020-05981-8.

Puchakawimol, P., and K. Jangchud. 2003.Study on the quality changes of aromatic coconut at different maturity. *In*: *Proceedings of 41st Kasetsart University Annual Conference*. *Subject: Agro-Industry,* 166–174. Kasetsart University.

Pursglove, J.W. 1985. Tropical Crops: Monocotyledons. *In*: 5th Edn. *Germplasms Resources: Diversity and Conservation*. Longman, London.

Rahayu, M.S., A. Setiawan, I. Maskromo, A.A. Sabana, K.P. Gangaraj, B.A. Jerard, V. Niral, and M.K. Rajesh. 2022. Genetic diversity analysis of Puan Kalianda kopyor coconuts (*Cocos nucifera*) from South Lampung, Indonesia based on SSR markers. *Biodiversitas Journal of Biological Diversity*, *23* (1). https://doi.org/https://doi.org/10.13057/biodiv/d230126.

Rajagopal, V., K.V. Bai, S.N. Kumar, and V. Niral. 2007. Genetic analysis of drought responsive physiological characters in coconut. *Indian Journal of Horticulture*, *64*: 181–189.

Rajagopal, V., K.V. Kasturibai, and S.R. Voleti. 1990. Screening of coconut genotypes for drought tolerance. *Oleagineux*, *45*: 215–223.

Rajamohan, T., and U. Archana. 2018. Nutrition and health aspects of coconut. *In*: *The Coconut Palm (Cocos nucifera L): Research and Development Perspectives*, 757–777.

Rajesh, M.K., T.P. Fayas, S. Naganeeswaran, K.E. Rachana, U. Bhavyashree, K.K. Sajini, and A. Karun. 2016. De novo assembly and characterization of global transcriptome of coconut palm (*Cocos nucifera* L.) embryogenic calli using Illumina paired-end sequencing. *Protoplasma*, *253*: 913–928. https://doi.org/10.1007/s00709-015-0856-8.

Ramanujam, B., K.K.N. Nambiar, and M.J. Ratnambal. 1998. Screening of coconut cultivars/hybrids against *Thielaviopsis paradoxa* (de Seynes) Hohnel, using petiole inoculation technique. *Development in Plantation Research Crop*, 284–286.

Ratnambal, M.J., M.K. Nair, K. Muraleedharan, P.M. Kumaran, E.V.V.B. Rao, and R.V. Pillai. 1995. *Coconut Descriptors*, Part I. 198pp. CPCRI. Kasaragod, Kerala, India.

Raza, A., A. Razzaq, S.S. Mehmood, X. Zou, X. Zhang, Y. Lv, and J. Xu. 2019. Impact of climate change on crops adaptation and strategies to tackle its outcome: A review. *Plants*, *8*(2): 34. https://doi.org/10.3390/plants8020034.

Rethinam, P., and V. Krishnakumar. 2022. Tender Coconut Varieties. *In*: Rethinam, P., and V. Krishnakumar [Eds.], *Coconut Water: A Promising Natural Health Drink-Distribution, Processing and Nutritional Benefits*, 37–76. Springer International Publishing, Cham.

Riangwong, K., S. Wanchana, W. Aesomnuk, C. Saensuk, P. Nabankoh, V. Ruanjaichon, T. Kraithong, T. Toojinda, A. Vanavichit, and S. Arikit. 2020. Mining and validation of novel genotyping-by-sequencing (GBS)-based simple sequence repeats (SSRs) and their application for the estimation of the genetic diversity and population structure of coconuts (*Cocos nucifera* L.) in Thailand. *Hortic. Res.*, *7*: 156. https://doi.org/https://doi.org/10.1038/s41438-020-00374-1.

Sabana, A.A., G. Antony, C.U. Rahul, and M.K. Rajesh. 2018. *In silico* identification of microRNAs and their targets associated with coconut embryogenic calli. *Agri. Gene*, *7*: 59–65. https://doi.org/10.1016/j.aggene.2018.01.002.

Sabana, A.A., G. Antony, M.K. Rajesh, K.P. Gangaraj, V. Niral, R. Sudha, and B.A. Jerard. 2022. Development and characterization of non-coding RNA-derived simple sequence repeat markers in coconut (*Cocos nucifera* L.). *Funct. Integr. Genomics*, *22*: 1243–1251. https://doi.org/10.1007/s10142-022-00911-2.

Saensuk, C., S. Wanchana, K. Choowongkomon, S. Wongpornchai, T. Karaithong, W. Imsabai, E. Chaichoompu, V. Ruanjaichon, T. Toojinda, A. Vanavichit, and S. Arikit. 2016. *De novo* transcriptome assembly and identification of the gene conferring a "pandan-like" aroma in coconut (*Cocos nucifera* L.). *Plant Sci.*, *252*: 324–334. https://doi.org/10.1016/j.plantsci.2016.08.014.

Samarasinghe, C.R.K., M.K. Meegahakumbura, D.P. Kumarathunge, H.D.M.A.C. Dissanayaka, P.R. Weerasinghe, and L. Perera. 2021. Genotypic selection approach made successful advancement in developing drought tolerance in perennial tree crop coconut. *Sci. Hortic.*, *287*: 110220. https://doi.org/https://doi.org/10.1016/j.scienta.2021.110220.

Samsudeen, K., M.K. Rajesh, D.D. Nagwaker, R. Reshmi, P. Ajith Kumar, K. Devadas, and K. Anita. 2013. Diversity in *Mohachao Narel*, a Sweet Endosperm Coconut (*Cocos nucifera* L.) Population from Maharashtra, India. *Natl. Acad. Sci. Lett.*, *36*: 319–330. https://doi.org/10.1007/s40009-013-0128-0.

Santos, J. 1929. Cytological study of *Cocos nucifera* L. *Philippine J. Sci.*, *17*: 417–437.

Santoso, U., K. Kubo, T. Ota, T. Tadokoro, and A. Maekawa. 1996.Nutrient composition of kopyor coconuts (*Cocos nucifera* L.). *Food Chem.*, *57*: 299–304. https://doi.org/https://doi.org/10.1016/0308-8146(95)00237-5.

Satyabalan, K. 1993. *The Coconut Palm: Botany and Breeding*. Jakarta: Asian and Pacific Coconut.

Singh, R.S., P. Singh, S. Sinha, U. Kumar, R. Kumari, and S. Kumar. 2023. Non-coding RNA Based Marker: A New Weapon in Armory of Molecular Markers. *In*: *Molecular Marker Techniques: A Potential Approach of Crop Improvement*, 117–125. https://doi.org/10.1007/978-981-99-1612-2_7.

Somsri, S., and S. Vichitrananda. 2007. Tropical Fruit Production in Thailand. *In*: *International Workshop on Tropical and Subtropical Fruits*, *787*: 33–46.

Sthapit, B., and D. Jarvis. 1999. Participatory plant breeding for on-farm conservation. *LEISA*, Leusden, *15*: 40–41.

Suchithra, M., and P. Paramaguru. 2018. Variability and correlation studies for vegetative, floral, nut and yield characters in indigenous and exotic coconut genotypes. *Int.J. Curr. Microbiol. App. Sci.*, *7*(7): 3040–3054 https://doi.org/10.20546/ijcmas.2018.707.355

Sudarsono, H.N., M.L.A. Hossang, D. Dinarti, I. Maskromo, M.R. Rahayu, and D. Sopandie. 2013. Produksi bibit kelapa kopyor true to type dengan persilangan terkontrol dan peningkatan produksi buah kelapa kopyor dengan polinator lebah madu. *Laporan Tahunan Prosiding Seminar Hasil-Hasil Penelitian*, IPB University, Bogor, Indonesia.

Sudha, R., M.K. Rajesh, V. Niral, K. Samsudeen, S. Shil, S. Ramchander, and Y. Diwakar. 2023. Analysis of genetic diversity and population structure in worldwide coconut germplasm (*Cocos nucifera* L.) using microsatellite markers. *Sci. Hortic.*, 309: 111681. https://doi.org/10.1016/j.scienta.2022.111681.

Sudha, R., V. Niral, and K. Samsudeen. 2021. Botanical Study and Cytology. *In*: *The Coconut Genome*, 13–25. Springer.

Sukendah, and R. Priyadarshini. 2023. Morphological Diversity of Kopyor Coconut in Indonesia. *Plant Diversity in Biocultural Landscapes*, 205–215. https://doi.org/10.1007/978-981-19-8649-9_10.

Sukendah, S., H. Volkaert, and S. Sudarsono. 2009. Isolation and analysis of DNA fragment of genes related to kopyor trait in coconut plant. *Indones. J. Biotechnol.*, *14*(2). https://doi.org/https://doi.org/10.22146/ijbiotech.7814.

Summanwar, A., U. Basu, N.N.V. Kav, and H. Rahman. 2020. Identification of lncRNAs in response to infection by *Plasmodiophora brassicae* in *Brassica napus* and development of lncRNA-based SSR markers. *Genome*, *64*: 547–566. https://doi.org/10.1139/gen-2020-0062.

Sun, R., L. Gao, Z. Mi, Y. Zheng, and D. Li. 2020. CnMADS1, a MADS transcription factor, positively modulates cell proliferation and lipid metabolism in the endosperm of coconut (*Cocos nucifera* L.). *Planta*, *252*: *83*. https://doi.org/10.1007/s00425-020-03490-3.

Swarup, S., E.J. Cargill, K. Crosby, L. Flagel, J. Kniskern, and K.C. Glenn. 2021. Genetic diversity is indispensable for plant breeding to improve crops. *Crop Sci.*, *61*: 839–852. https://doi.org/10.1002/csc2.20377.

Tampake, H., T. Kuswara, and T.A. Davis. 1982. Coconut germplasm survey of Nusa Tenggara, Timur Province, the initial step towards inducing drought resistance in coconut strains. *Indonesian Agri. Res. Devt. J.*, *4*: 52–61.

Tan, AQ. 1969. *Cytological Study of Three Cultivars of Coconuts, Tambolilid, Laguna, and Coconiño*. Undergraduate Theses. 8679. https://www.ukdr.uplb.edu.ph/etd-undergrad/8679.

Teulat, B., C. Aldam, R. Trehin, P. Lebrun, J.H.A. Barker, G.M. Arnold, A. Karp, L. Baudouin, and F. Rognon. 2000. An analysis of genetic diversity in coconut (*Cocos nucifera*) populations from across the geographic range using sequence-tagged microsatellites (SSRs) and AFLPs. *Theoretical and Applied Genetics*, G*100*: 764–771. https://doi.org/https://doi.org/10.1007/s001220051350.

Thomas, R.J., M. Shareefa, H. Harsha, and A. Karun. 2021.Biochemical characterization of pink husked coconut types. *Indian Society of Plantation Crops*, *49*: 72–76. https://doi.org/DOI:10.25081/jpc.2021.v49.i1.7062.

Wang, S., Y. Xiao, Z.W. Zhou, J. Yuan, H. Guo, Z. Yang, J. Yang, P. Sun, L. Sun, Y. Deng, W.Z. Xie, J.M. Song, M.T. ul Qamar, W. Xia, R. Liu, S. Gong, Y. Wang, F. Wang, X. Liu, A.R. Fernie, X. Wang, H. Fan, L.L. Chen, and J. Luo. 2021. High-quality reference genome sequences of two coconut cultivars provide insights into evolution of monocot chromosomes and differentiation of fiber content and plant height. *Genome Biol.*, *22*: 304 https://doi.org/10.1186/s13059-021-02522-9.

Welewanni, I., and D. Bandupriya. 2017.Coconut cryopreservation: Present status and future prospects. *Cord.*, *33*: 21. https://doi.org/10.37833/cord.v33i1.54.

Whitehead, R.A. 1964. *Sample Survey and Collection of Coconut Germplasm in the Pacific Islands*. Ministry of Overseas Development, London.

Wicaksono, A., R. Raihandhany, and J.A. Teixeira da Silva. 2021. Kopyor versus macapuno coconuts: Are these two edible mutants of Southeast Asia the same? *Planta*, *254*(5): 86. https://doi.org/10.1007/s00425-021-03740-y.

Xia, W. 2021. Genes Involved in Lipid Metabolism in Coconut. *In*: *Innovation in the Food Sector Through the Valorization of Food and Agro-Food By-Products*, Ch. 12. Intech Open, Rijeka, https://doi.org/10.5772/intechopen.90998.

Xiao, Y., P. Xu, H. Fan, L. Baudouin, W. Xia, S. Bocs, J. Xu, Q. Li, A. Guo, L. Zhou, J. Li, Y. Wu, Z. Ma, A. Armero, A.E. Issali, N. Liu, M. Peng, and Y. Yang. 2017. The genome draft of coconut (*Cocos nucifera*). *Gigascience*, *6*: 1–11. https://doi.org/10.1093/gigascience/gix095.

Yang, Y., A. Iqbal, and R. Qadri. 2018. Breeding of Coconut (*Cocos nucifera* L.): The Tree of Life. *In*: Al-Khayri, j. S. Jain, and D. Johnson [Eds.], *Advances in Plant Breeding Strategies: Fruits*, 673–725. Springer, Cham. https://doi.org/10.1007/978-3-319-91944-7_17.

Yousefi, K., S.N.A. Abdullah, M.A.M. Hatta, and K.L. Ling. 2023. Genomics and Transcriptomics Reveal Genetic Contribution to Population Diversity and Specific Traits in Coconut. *Plants*, *12*: 9–13. https://doi.org/10.3390/plants12091913.

Zuniga, L.C. 1953. The probable inheritance of the Makapuno character of the coconut. *Philipp. Agric.*, *36*: 402–413.

3

Genetic Diversity of Tree Tomato Tamarillo (*Solanum betaceum* Cav.) and Sustainable Utilization

Fernando Ramírez

Introduction

The Solanaceae family is found throughout the world, with close to 90 genera and between 3,000–4,000 species (Gebhardt, 2016). *Solanum* includes close to 1,500 species worldwide (Lamas et al., 2022). Several species within this genus have economic relevance such as potato (*Solanum tuberosum* L.), lulo (*Solanum quitoense* Lam.) (Ramírez and Davenport, 2020), cocona (Ramírez, 2020), tomato (*Solanum lycopersicum* L.) (Meier, 2001), eggplant (*Solanum melongena* L.) (Mangino et al., 2020), etc. Among this genus, *Solanum betaceum* Cav. commonly known as tree tomato, cholito, or tamarillo is an important crop in Latin American countries. In the late XIX century, it was introduced to Southeast Asia (Vietnam, Indonesia Malaysia, India, and Thailand,), Oceania (New Zealand and Australia), Africa (South Africa, Uganda, and Rwanda), and Europe (Italy, Germany, Spain, Portugal, France, and the Netherlands) (Diep et al., 2020; Diep et al., 2022). *S. betaceum* has been cultivated in countries, such as Colombia, Brazil, Venezuela, Ecuador, Argentina, the United States, Australia, New Zealand, and Kenya where it is produced commercially for export to other countries (Lim, 2013). It is also cultivated in the highlands of Costa Rica, Puerto Rico, Jamaica, Haiti, and Guatemala (Morton, 1987).

Tree tomato grows from 1,000–3,000 m in tropical and subtropical regions (Lim 2013) (Figures 1–3). The best growth temperature ranges from 18–22°C and between 600–800 mm of annual precipitation. It is a tree that can reach from 2–5 meters and rarely 7 m (Carrillo-Perdomo et al., 2015). It has a grey monopodial trunk that is branched with simple leaves that are groped on the distal portion of branches. The flowers are pink or cream-colored to white, 1.3–1.5 cm across, bisexual, pentamerous, and occur in groups of 3–10 or more in axillary cymes near the branch apex (Figure 1c–e) (Lim, 2013). The fruits are ovoid, obtuse, or acute at their ends, ranging in length from 4–10 cm and 3–5cm in width, glabrous, and can vary in color from yellow to orange, purple, or red. It is an appealing fruit with a delicious taste.

Independent Researcher, Bogotá, Colombia.
E-mail: fernando.ramirez.lopez.p@gmail.com

Tree tomato fruit is eaten raw, but it is most often used to make juice (Schotsmans et al., 2011). It is also cooked as a vegetable and used to make hot sauce, desserts, jellies, and jams (Duarte and Paull 2015). Fruits can be grilled and seasoned or baked (Lim, 2013). Fruit sliced can be mixed with ice cream. Slices can be used on chutneys, curries, compote, soups, and stews (Schotsmans et al., 2011; Lim, 2013). The peeled fruits can be mixed with lemon juice, gelatin, sugar, and milk to make a dessert that can be decorated with tree tomato fruit slices (Lim, 2013).

Tree tomato fruits are low in fat, a source of dietary fibers, vitamin C, B1-3, B6, E, provitamin A, proteins, soluble sugars (sucrose, glucose, and fructose), minerals, such as iron, copper, zinc, magnesium, phosphorus, calcium, potassium, organic acids (e.g., malic and citric acids), 70 volatile compounds, anthocyanins, carotenoids, and phenolics (Pratt et al., 2008; Hurtado et al., 2009; Schotsmans et al., 2011; Wang and Zhu, 2020; Diep et al., 2022; Coyago-Cruz et al., 2023). It is well known for its tasty fruits, antioxidant, antiobesity, and anticancer properties (Diep et al., 2020; Diep et al., 2022).

Tree tomatoes have been used to treat buccal infections. Indigenous groups use tree tomato fruit to treat anemia and respiratory diseases (Lim, 2013). Tree tomato fruit pulp when cooked is used as a bandage for sore tonsils in Colombia. It has been used to treat liver maladies in Jamaica (Lim, 2013). Warm leaves are used as a medicine against sore throat when wrapped around the neck in Ecuador (Bohs, 1989). Tene et al. (2007) reported that tree tomato juice is used to treat stomach ache, high cholesterol, and tonsillitis in Ecuador. Sliced fruits have been used as a medicine to treat mumps and tonsillitis (Gutiérrez et al., 2000). The leaves of tree tomatoes have been used as cough medicine in Uganda, Africa (Lacroix et al., 2011). Fruits are used in inflammatory diseases, tonsillitis, and liver sicknesses in Mizoram, India (Ralte et al., 2021). Its juice has been used to treat high blood pressure in the Peren district of Nagaland, India (Jamir et al., 2022).

Figure 1. Tree tomato fruits and flowers. (a) Orchard with reddish cultivar near Rondón, Vereda, Pirnaguata, Boyacá State, Colombia, (b) detail of leaves, (c) Mango cultivar flower from Bogotá, Cundinamarca State, Colombia, (d) detail of inflorescence, and (e) close-up view of buds. Photographs by Fernando Ramírez. Reproduced with permission.

Figure 2. Tree tomato fruits. (a) reddish cultivar near Rondón, Vereda, Pirnaguata, Boyacá State, Colombia, (b) detail of unripe reddish fruit, (c) mature reddish fruit, and (d) Mango cultivar fruits from Bogotá, Cundinamarca State, Colombia. Photographs by Fernando Ramírez. Reproduced with permission.

Figure 3. "Mango" cultivar of tree tomato. (a) tree bearing leaves, mature and immature fruit, (b) inflorescence, (c) detail of flower-bearing anthers and carpel, (d) fruit, (e) fruit cut in half showing pulp and seeds, and (f) seeds extracted from the pulp. Photographs by Fernando Ramírez. Photographs taken in La Mana, near Bogotá, Cundinamarca State, Colombia.

The genetic diversity of tree tomatoes is an important aspect that has not been thoughtfully investigated. Relatively few investigations have covered this topic, particularly in Colombia (Enciso-Rodríguez et al., 2010), Ecuador (Ordoñez A. et al., 2009; Peñafiel et al., 2009; Acosta-Quezada et al., 2012) and in the wild-type *S. betaceum* in Argentina, the claimed site of origin of tree tomato (Broglia and Caruso, 2022). More research is needed to elucidate the center(s) of origin and dispersion of both the wild and cultivated forms.

The current chapter aims to discuss the latest information on tree tomato genetic diversity, morphological, biochemical, and molecular characterization, and its sustainable utilization through germplasm repositories, grafting with wild *Solanum* species to provide disease resistance and increase yield, and fruit size.

Taxonomic Background

The tree tomato was originally named *Solanum betaceum* Cav. by the Spanish botanist Antonio José Cavanilles, who made a description of a tree cultivated in the Royal Garden in Spain (Cavanilles, 1799). *Solanum betaceum* Cav. was then moved to the genus *Cyphomandra* by Otto Sendtner in 1845 and named *Cyphomandra betacea* (Cav.) Sendtn., (Sendtner, 1845) and finally transferred to the original *Solanum betaceum* (Cav.), which is the currently accepted name.

The tree tomato (*Solanum betaceum* Cav.) is an important fruit from South America. Although its origin remains unknown (Morton, 1982; Guimarães et al., 1996; Schotsmans et al., 2011), it has been claimed that wild populations occur in northwestern Argentina and Bolivia (Bohs, 1989; Lamas et al., 2022), as well as Colombia, Ecuador, Peru, Brazil, and Bolivia (Duarte and Paull, 2015; Ramírez and Kallarackal, 2019). Fruits are small, yellow, or pale red in the wild (Schotsmans et al., 2011). Guimarães et al. (1996) reported that it could be native to the Andean region from Peru to Argentina. Morton (1987) reported that tree tomato is native to Peru, and possibly Chile, Ecuador, and Bolivia. Ecuador and Colombia have been claimed to be the centers of diversity of tree tomatoes (Enciso-Rodríguez et al., 2010; Acosta-Quezada et al., 2012). Bolivia and northeastern Argentina have been determined as the center of origin of tree tomatoes (Bohs, 1991; Bohs and Nelson, 1997). Northern Peru, Ecuador, and Colombia are also considered domestication centers for this tree species (Enciso-Rodríguez et al., 2010; Acosta-Quezada et al., 2012; Viera et al., 2019). A close relative of the tree tomato, called *Solanum maternum* Bohs was originally described from Bolivia as its closest wild relative (Bohs and Nelson, 1997). This tree is self-incompatible and infertile with tree tomatoes. Tree tomato is closely associated with *S. unilobum* and *S. roseum* from Bolivia (Bohs, 1994, 1995).

Collection and Assessment of Germplasm/Genetic Resources

Characterization of Evaluation

Morphological Characterization

Tree tomato has shown a high morphological variability among cultivars. Acosta-Quezada et al. (2011) examined 39 morphological traits in 24 tree tomato accessions from six countries and five cultivar groups. The countries were Ecuador, Colombia, Peru, Portugal, Bolivia, and New Zealand. They used the following cultivars: red, red conical, purple, orange, and orange pointed. They found that the best morphological descriptors with the highest heritability values were associated with seed, fruit, and infructescence features except in the red conical group. The morphological features studied revealed a high variation among cultivars. Floral and inflorescence-related traits showed very little variation. Plant architecture showed a considerable variation among tree tomato cultivars except in the red conical group. Seed-linked traits showed very little morphological variation among tree tomato accessions. They conducted a Principal Component Analysis (PCA) and a cluster analysis among the morphological traits. They found that cultivars were grouped based on infructescence and fruit number and also fruit shape, and size. There was a grouping absence among the tree cultivars derived from different countries.

Lamas et al. (2022) investigated the morphological variation among wild tree tomato populations from northwestern Argentina in Salta and Jujuy provinces. They analyzed nine populations with 657 fruits from 125 plants; However, the seed characterization was conducted in seven populations for a total of 543 fruits derived from 91 plants. The morphological variables included fruit pericarp thickness, equatorial diameter, length, weight, and seed weight. They also measured the pH, and total soluble solids. Overall, they obtained a high morphological variability among tree tomato populations (4.97–24.26 CV%) mostly associated with fruit size. Features associated with fruit acidity and flavor showed less variability. They conducted a principal component analysis (PCA) among the morphological variables and found that fruit length, equatorial diameter, and weight, showed a high variability and separated populations. They found no association between geographical location and morphological traits. The morphological variation among populations was greater than within populations. The environment caused a high phenotypic variance among the tree tomato population. This is due to humidity precipitation, and temperature differences.

Cytological Characterization

Tree tomato has two sets of chromosomes 2n = 2x = 24 (Pringle and Murray, 1993; Currais et al., 2013). Several variants have been obtained, such as tetraploids, spontaneous triploids, and aneuploids (Standring et al., 1990; Pringle and Murray, 1992). Furthermore, Pringle and Murray (1992) produced tetraploids by applying colchicine during seed germination while aneuploids and triploids were generated by crosses. Standring et al. (1990) reported tree tomato accession with different chromosome numbers, for example, 2n = 24 (B24, B29, and Peru), aneuploids bearing 25, 26, and 47 chromosomes, autotriploids 2n = 3x =36 and autotetraploids 2n = 4x = 48. Tree tomato triploids, tetraploids, and aneuploids could be used for breeding purposes (Currais et al., 2013).

Correia et al. (2023) suggested that tree tomato ploidy levels could have an important role in drought responses. They subjected both diploid (2n = 2x = 24) and tetraploid (4n = 4x = 48) tree tomatoes to water stress conditions. They found that water stress induced stomatal closure and reduced CO_2 assimilation in both ploidy-level plants. Tetraploid trees behaved better under water shortage conditions when taking into account water potential. Diploid trees showed an increased proline accumulation when subjected to water stress conditions.

Biochemical Characterization

Somatic embryogenesis is a relevant tool that has been used in the biochemical characterization of tree tomatoes. Alves et al. (2017) used somatic embryogenesis to establish cell suspensions and characterize the extracellular protein profiles in embryogenic and non-embryogenic cells. Embryogenic suspension cultures were effectively established beginning with 40 mg cells over 20 mL of liquid. Mass fluorometric and spectrometry tools were used to identify extracellular proteins, main classes of proteases, and hydrolytic activity between embryogenic and non-embryogenic cell mediums. They found that extracellular protein profiles showed qualitative and quantitative variation between embryogenic and non-embryogenic cell mediums. These differences were due to hydrolytic enzymes, i.e., xylanases and glucanases. Proteolytic activity analysis reported the occurrence of metalloproteases aspartic proteases, and serine proteases, in embryogenic cultures. Non-embryogenic cell mediums were rich in serine proteases.

Several compounds have been identified on the pulp and mesocarp of tree tomatoes by biochemical methods. Gannasin et al. (2015) characterized the hydrocolloid fractions of tree tomato fruit. They discovered two new hydrocolloids in the tree tomato mesocarp. The hydrocolloid was mainly made of arabinogalactan protein-associated pectin and the pulp hydrocolloid was composed mainly of hemicellulosic polysaccharides coupled with neutral polysaccharides and proteins. do Nascimento et al. (2013) extracted a polysaccharide called galactoarabinoglucuronoxylan from the pulp of a tree tomato. This compound showed antinociceptive activity in inflammatory pain models. do Nascimento et al. (2016) studied the polysaccharides found in tree tomato pulp. They found polysaccharides in the mucilage of tree tomatoes. They reported differences in the ratio and concentration of polysaccharides between pulp and mucilage. Diep et al. (2020) found a high dietary fiber content (3%) in tree tomato cultivars, Mulligan Laird's Large, and Amber cultivars from New Zealand. They reported also the occurrence of 22 amino acids, two that were essential and five nonessential. They found that γ-aminobutyric acid, l-aspartic acid l-glutamic acid were the dominant amino acids found in tree tomato peel except in the Amber cultivar. They reported differences among the amino acid profiles and soluble sugar concentrations in the tree cultivars. Orqueda et al. (2017) studied lyophilized powder from pulp seed and skin from tree tomatoes from Yungas, Argentina. They reported that the powders were a source of vitamin C, fiber phenolics, carotenoid, and potassium, and had a low sodium and carbohydrate content. The phenolic-enriched extracts had an inhibitory effect on key enzymes, such as lipase, α-glucosidase, and α-amylase. The enriched polyphenol extract of skin and seed could have nutraceutical applications.

Carotenoids have been analyzed through different biochemical methods in tree tomatoes. Giuffrida et al. (2018) compared different biochemical techniques for the extraction characterization and detection of carotenoids and apocarotenoids from yellow tree tomato fruits. They used a combination of methods, such as an online-based method linking the supercritical fluid chromatography with triple quadrupole mass spectrometry detection and supercritical fluid extraction. They also used an offline liquid extraction and LC-PDA-MS analysis, which used the conventional C30 (3 μm I.D. particles) and an innovative C30 column including the sub-2 micron particles. They found that the SFE-SFC-MS methodology was quicker in contrast to offline conventional chromatography and liquid extraction. SFE-SFC-MS methodology was optimized for the consistent detection of apocarotenoids.

Molecular Characterization

Genetic diversity is a measure of the genetic variability occurring among individuals of a cultivar or a population in a species (Salgotra and Chauhan, 2023). Brown (1983) defines genetic diversity as the result of the recombination of genetic material (DNA) throughout the inheritance gene flow, process, genetic drift, and mutations. Plant genetic resources comprise total hereditary material, that involves all the multiple alleles of genes that occur within a particular crop species (Salgotra and Chauhan, 2023), i.e., it involves medicinal, horticultural plants, and wild-type species. Furthermore, plant genetic resources comprise cultivated varieties, new varieties, cultivars, landraces, old cultivars, wild-type cultivars, weedy forms, breeding stocks, genetic stocks, and wild species of cultivated crops.

Relatively few investigations have focused on the genetic diversity of tree tomatoes. Acosta-Quezada et al. (2012) used Amplified Fragment Length Polymorphism (AFLP) markers to study the genetic diversity of 25 tree tomato accessions from four cultivar groups from Colombia, Ecuador, Peru, Bolivia, Portugal, Spain, and New Zealand. They used a total of 11 combinations of markers. They found that 43% of the AFLP fragments were polymorphic. The combined genetic diversity among cultivated accessions was 0.29. The total genetic diversity in the orange group ranged between 0.18–0.24. There was a large genetic variation within cultivar groups but not among groups. Based on their results, Ecuador is considered a center of diversity for tree tomatoes. They found a low genetic differentiation among geographical provenances meaning that there is a flow of seeds and fruits among regions. Peñafiel et al. (2009) studied the genetic variability by SSR markers of 50 tree tomato accessions from Imbabura, Pichincha, and Tungurahua provinces in Ecuador. They classified the accessions by fruit color, size, and pulp color into common yellow, yellow gigant, yellow-ball, and purple gigant. They tested 53 primers but only 33 (62%) were amplified and the number of polymorphic primers was 7.5%. They found a low genetic variability among tree tomato accessions. The genetic similarity among accessions was 80–100% and within each cultivar was above 80%. Ordoñez et al. (2009) characterized the genetic variability of tree tomatoes by AFLP markers from Imbabura, Pichincha, and Tungurahua provinces in Ecuador. They studied 26 tree tomato accessions. They obtained 183 markers and from these, 110 (60%) were polymorphic. They found a limited genetic variability among tree tomato accessions. Based on these results, the authors conclude that there is a high degree of inbreeding among varieties. Furthermore, wild-type tree tomatoes should be included in future studies. Tapia (2005) used Random Amplified Polymorphic DNA (RAPD) markers to characterize the genetic diversity of 44 tree tomato accessions from Ecuador. They found a narrow genetic diversity among accessions. Bolagay et al. (2015) used RAPD markers to study de genetic diversity among 15 tree tomato accessions from Ecuador. They found a low genetic diversity.

Enciso-Rodríguez et al. (2010) characterized the genetic structure of 30 tree tomato accessions mostly from Colombia and other countries, such as Ecuador, and Kenya. They used the PCR-based markers developed from conserved orthologous genes (COSII) from other Solanaceae species. They successfully amplified 26 out of 41 tree tomatoes. Five polymorphic markers were found in tree tomatoes. Both observed heterozygosity (0.003) and expected heterozygosity (0.003) values were low among accessions. They found that the genetic differentiation among populations was high. This could be due to geographical distribution, which does not benefit the genetic flow among populations. They propose that Colombia is a center of diversification for tree tomatoes based on the high genetic differentiation obtained. Broglia and Caruso (2022) evaluated tomato (*Solanum lycopersicum* L.) molecular markers that could be transferred and used to study the genetic variability of wild-type tree tomatoes from Yungas, northwestern Argentina. They studied two wild-type tomato populations, one from San Lorenzo, Salta (24° S–65° W), and the other one from San Francisco, Jujuy (23° S–64° W). They evaluated 30 simple sequence repeat (SSR) markers and selected 13 that were consistent. Only 10 out of the 13 markers were polymorphic (77%) and 53% of the SSRs of tomato could be transferred to tree tomato. The SSR105, SSR223, SSR192, SSR295, and SSR23 were the most effective markers due to their ease of use and consistency during the amplification process. The greatest genetic variability was found in San Lorenzo followed by San Francisco.

Lagos-Burbano et al. (2022) induced mutations in tree tomatoes by applying diethyl sulfate. They used Random Amplified Polymorphisms (RAMs) to evaluate the genetic changes caused by diethyl sulfate in 'Manzano' tree tomato seedlings. They used 0,5, 1,0, 1,5 y 2% diethyl sulfate concentrations and seedlings were incubated for 8 and 16 hours. They found that polymorphic loci showed 87.5–100% variability. The genetic variability of seedlings treated with diethyl sulfate was greater than that of non-treated plants. However, the genetic changes caused by diethyl sulfate were low.

Transcriptome analysis and real-time sequencing have proved essential for future tree tomato genetic diversity studies. Deng et al. (2022) used the PcaBio SMRT sequencing to build the first full-length transcriptome atlas of tree tomatoes. This could be an important contribution to the study of post-transcriptional events. They have made available the first simple sequence repeat markers (SSRs), which are key for future genetic diversity studies. Pacheco et al. (2021) conducted a *de novo* transcriptome assembly of two tree tomato cultivars and found a high number of polymorphisms that could be used in future genetic diversity analyses. Wild relatives of *Solanum* species play a key role as germplasm donors for the improvement of economically important Solanaceae species (Pratt et al., 2008). Transfer of intended genes among domestic donors could cause fewer problems due to unwanted horticultural features that occur in wild relatives.

Utilization of Germplasm

Utilization in Breeding and Improvement Programs

The sustainable utilization and conservation of plant genetic resources are key to enabling food security for future generations (Salgotra and Chauhan, 2023). Genetic resources must be ready to use by plant breeders to implement the continuous development of new cultivars and varieties. Salgotra and Chauhan (2023) reported that plant genetic resources are a key pool for genes that deal with insect and disease-pest-resistant genes. Furthermore, genetic resources can be utilized to improve fruit quality features and yield (Ramírez and Davenport, 2021). Given the current climate change scenario, plant genetic resources play an important role in producing climate-resistant crops.

In the case of tree tomato, the sustainable utilization and conservation of this crop have been done by (1) genetic diversity studies, (2) hybridization, (3) grafting (asexual hybridization) with wild-*Solanum* species, (4) germplasm repositories, (5) studies dealing this analyze the wild relatives and site or origin and diversification.

Germplasm repositories are important due to their role as sites for preserving plant genetic resources. Particularly, they maintain a "living collection" of different accessions, and related wild species and contribute to both interspecific and intraspecific hybridization studies. They also play an important role in genetic and morphological studies as well as disease resistance, yield, and fruit quality improvement. Colombia´s National Germplasm Repository housed 54 tree tomato accessions by 2002 (Lobo et al., 2002). Furthermore, Valencia et al. (2010) reported that Colombia´s National Germplasm repository had 61 accessions of which 53 were tree tomatoes from Colombia and 8 were from other countries. Pratt et al. (2008) reported that the Colombian Agricultural Research Corporation Agrosavia (Formerly Corpoica) housed one of the largest tree tomato germplasm collections. This cluster has been analyzed through genotypic (AFLP marker) phenotypic tools. COS are a group of genes that are preserved during plant evolution in copy number and sequence (Lyons et al., 1997). Tree tomato COSII primers have produced 70% PCR amplification products (Pratt et al., 2008). Moreover, wild relatives of tree tomatoes have been characterized by COSII markers.

The Fruit Program of the National Institute of Agricultural Research (INIAP) in Ecuador, has researched the breeding and genetics of tree tomatoes (Viera et al., 2019). This program aims to increase the resistance to diseases and pests, yield, and enhance fruit quality features for consumption and industrial use. Ecuador´s National Germplasm Repository housed 43 tree tomato accessions, of which 32 were cultivated and 12 wild by 2005 (Tapia, 2005).

Ramírez and Kallarackal (2019) reviewd the hybridization of thee tomato. Tree tomato crosses among different cultivars are possible as there are no technical difficulties or genetic barriers (Prohens and Nuez, 2000; Shah et al., 2022). However, due to the self-fertilization of tree tomatoes, individuals bear a high degree of homozygosis, which leads to crosses with individuals that are genetically different and produce genetically similar offspring (Prohens and Nuez, 2000).

Crosses between tree tomatoes and related species have been successful, particularly, with *Solanum roseum* Bohs and *Solanum unilobum* (Rusby) Bohs (Bohs, 1991). However, Pringle and Murray (1991) conducted nine crosses between tree tomatoes and other eight relative species, such as *Solanum corymbiflorum* (Sendtn.) Bohs, *Solanum diploconos* (Mart.) Bohs, *Solanum diversifolium* Humb. & Bonpl. ex Dunal, S. *unilobum*, *Solanum obliquum* Ruiz & Pav., *Solanum cajanumense* Kunth, *Solanum splendens* (Dunal) Bohs, and *Solanum sibundoyense* (Bohs) Bohs and noted that fruit set was low and no viable seed was obtained. They also used polyploid *S. betaceum* trees but failed to obtain consistent interspecific hybridization. Bohs (1991) showed that the F_1 as a result of crosses between *S. betaceum* x *S. roseum*, *C. unilobum* x *betaceum* had reduced fertility in contrast to their parents. Interspecific hybridization of the tomato with wild relatives for disease resistance has been conducted in the experimental station "La Selva", which is part of Colombia´s National Institute of Agronomic Research (AGROSAVIA) (Lobo et al., 2000). For example, the cross between tree tomato and *S. unilobum* produced offspring with resistance against fruit anthracnose (*Colletotrichum gloeosporioides* (Penz.) Penz. & Sacc.). Lobo et al. (2002) obtained interspecific hybrids between *Solanum maternum* Bohs x S. *betaceae* and *S. unilobum* x *S. betacea*. However, the former cross was under evaluation in 2002.

S. corymbiflorum (Sendtn.) Bohs, *Solanum diversifolium* Humb. & Bonpl. ex Dunal, and *Solanum amotapense* Svenson, are self-pollinated tree species, which could be used for hybridization with tree tomatoes in Colombia (Lobo, 2006). *Nicotiana glauca*, which is compatible with *S. betaceum* could be utilized as rootstock for commercial production in Ecuador (Navarrete et al., 2018).

Since 2006, the National Institute of Agricultural Research (INIAP) in Ecuador has investigated the crosses between *S. betaceum* × *S. maternum* and *S. betaceum* × *S. unilobum* (Viera et al., 2019). These species have some degree of resistance to anthracnose but show poor fruit size and flavor. To solve this issue, backcrosses were held with tree tomatoes to improve the fruit quality features. Arroyo et al. (2016) continued the selection process to obtain anthracnose-resistant plants. Several viable hybrids have been obtained through grafting a form of asexual hybridization.

Grafting is an effective way to improve fruit size and yield in tree tomatoes. Wild-type *Solanum* species have been grafted onto tree tomatoes to produce better features, such as size, taste, longevity, yield, and nutrient content (Bohs, 1989; Viteri et al., 2010). Ramírez (2021) successfully grafted mora, mango, and común tree tomato cultivars onto cucubo (*Solanum stellatiglandulosum* Bitter) a wild Solanaceae species that grows from Costar Rica to Ecuador. A cleft graft was used by introducing a 15 cm long tree tomato scion onto cucubo. Grafted trees showed a high survival rate above 94%. Grafted-obtained fruits were heavier longer, and wider than their ungrafted counterparts (Table 1). The longest fruits occurred in the grafted mango cultivar (7.9 cm). The greatest fruit width occurred in the mora cultivar (6.7 cm). Grafted trees produced more fruits (30–40 fruits/tree) in contrast to ungrafted ones (10–15 fruits/tree) (Table 1).

Numerous *Solanum* wild-type species offer resistance against diseases (Bradeen, 2021). Tree tomato grafting has also been used to provide resistance against diseases. Vargas et al. (2018) reported that some wild-type Solanace species are graft-compatible with tree tomatoes, such as SN5, which provided resistance against root-knot nematode (*Meloidogyne incognita* Kofoid and White). Viteri-Díaz et al. (2020) reported that *Nicotiana glauca* Graham, a wild-type Solanaceae species from the Andes showed the best potential rootstock for *S. betaceum* due to its 100% resistance against the pathogen *Fusarium solani* (Mart.) and low tolerance to the nematode *M. incognita*. Tree tomatoes are grafted on rootstocks such as *Nicotiana glauca* or *S. auriculatum*, which offer resistance against *F. oxysporum* and *M. incognita* in Ecuador (Viera et al., 2019). Grafring experiments have shown that wild tobacco (*Solanum mauritianum* Scop.) has been effective to provide tolerance to humid soil conditions by decreasing the susceptibility to root rot in New Zealand and Australia (Slack, 1976; Guimarães et al., 1996).

Species/Cultivars Released

Tree tomato has numerous cultivars in South America and different countries. However, Morton (1987) reported that there are no established cultivars but preferences based on fruit color. It is difficult to know the crosses and pedigree information as well as the general background of established cultivars. Many of the currently established cultivars were developed by growers and or scientists who failed to document their provenience. Furthermore, cultivar features are poorly understood (Lamas et al., 2022). Named cultivars occur in New Zealand (Table 2) (Sale and Pringle, 1999; Sale, 2006; Schotsmans et al., 2011). Currently, tree tomato cultivars are distinguished by color, either yellow, purple, and red (Diep et al., 2022) or orange, cultivar groups (Bohs, 1994; National Research Council, 1989; Prohens and Nuez, 2000). Acosta-Quezada et al. (2011) proposed five cultivar groupings from Ecuador based on genebank description, the grouping of cultivars by farmers, and fruit features, such as fruit color, apex shape, and the color of the seeds and the pulp. These included the red, red conical, orange pointed, orange, and purple. The tomato cultivars from Ecuador can be classified by chemical composition and pulp color (Viera et al., 2019). Ordoñez et al. (2009) recognized tree tomato cultivars from Ecuador as giant purple, giant yellow, purple-pointed, yellow-pointed, yellow ball-shaped. In New Zealand, tree tomatoes can be distinguished by their skin color into yellow (separated into amber and gold) and purple-red (separated into purple and red) (Prohens and Nuez, 2000; Schotsmans et al., 2011). Farmers from Argentina recognize tree tomato varieties based on shape and skin color (Lamas et

Table 1. Fruit features of tree tomato cultivars grafted onto cucubo. N = 100 fruits per each cultivar. Reproduced with permission after Ramírez, F. (2021) Tree tomato (*Solanum betaceum* Cav.) grafted with a wild Solanum species. *Genet. Resour. Crop Evol.*, 68: 2265–2271. doi: 10.1007/s10722-021-01194-5.

Cultivar/Graft	Mean Number of Fruits/Tree	Mean Fruit Weight (g)	SD	Mean Fruit Length (cm)	SD	Mean Fruit Width (cm)	SD	Number of Trees/Cultivars	Graft Success (%)	Graft Failure (%)
Mango (grafted)	40	160.1 ± 0.6	1.5	7.9 ± 0.3	0.8	6.4 ± 0.3	0.7	30	97	3
Mango (ungrafted)	15	92.3 ± 0.6	1.4	6.7 ± 0.2	0.6	5.9 ± 0.1	0.3	30		
Mora (grafted)	37	181.4 ± 0.8	2	7.2 ± 0.4	1	6.7 ± 0.3	0.8	30	94	6
Mora (ungrafted)	14	130.2 ± 0.8	1.9	6.5 ± 0.4	0.9	5.6 ± 0.3	0.7	30		
Común (grafted)	33	193.7 ± 1.25	2.8	7.7 ± 0.2	0.5	6.5 ± 0.4	0.9	30	99	1
Común (ungrafted)	12	103.9 ± 1.32	3.7	7.2 ± 0.5	1.3	5 ± 0.4	1	30		

SD = standard deviation
± = standard error

Table 2. New Zealand tree tomato cultivars.

Fruit features of tree tomato cultivars grafted	Skin Color	Notes	Source
'Red Beau'	red	Common cultivar but declined by virus and replaced by other cultivars	Schotsmans et al. (2011)
'Ted's Red'	red	Replaced 'Red Beau'	Schotsmans et al. (2011)
'Laird's Large'	red	Replaced 'Red Beau'	Schotsmans et al. (2011)
'Oratia'	red		Sale and Pringle (1999)
'Red Delight'	red		Sale and Pringle (1999)
'Kerikeri Red'	red		Sale and Pringle (1999)
'Andy's Sweet Red',	red		Sale and Pringle (1999)
'Red Beauty'	red		Sale and Pringle (1999)
'Red Chief'	red		Sale and Pringle (1999)
'Seccombe Red'	red		Sale and Pringle (1999)
'Bold Gold'	yellow	Common cultivar	Sale (2006)
'Goldmine'	yellow		Sale (2006)
'Amberlea Gold'	yellow		Sale (2006)
'Kaitaia Yellow'	yellow		Sale (2006)

al., 2022). They classify them as round and pointed orange-purple, round elongated red "sangre de toro", small rounded and orange fruits 'cherry'. and orange-reddish "bola de chivo". However, these cultivars are different from those described for other South American countries (Acosta-Quezada et al., 2011). In Colombia, tree tomato cultivars are recognized by fruit color and shape. However, more research is warranted to characterize the cultivars in different regions of Colombia. Cultivars include the orange skin fruit, such as 'mango' (Figure. 2 and 3), red "common red", 'red-purple' and yellow "common yellow", and "rounded yellow", and orange with red 'parthenocarpic' (Bernal, 1994).

Hay & Sons introduced and developed the purple and yellow cultivars of tree tomatoes in New Zealand in 1891(Morton, 1987; Diep et al., 2022). The purple-red cultivar was developed in New Zealand in the 1920s (Prohens and Nuez, 2000).

Conservation of Useful Germplasm/Genetic Resources

Conservation initiatives to protect native tree tomatoes have been conducted in the provinces of Tucuman Catamarca, Jujuy Salta, in Northwest Argentina between 400–700 m. a. s. l. (Orqueda et al., 2020). This has been done by a sustainable utilization of the red and orange tree tomatoes. This is also an alternative wild crop that is important for forest sustainability and conservation (Lamas et al., 2022).

Cryopreservation techniques have been applied to tree tomatoes. Graça et al. (2018) developed a consistent protocol for the cryopreservation of embryonic cultures. First, they obtain the embryonic masses from leaf sections and zygotic embryos in suitable culture media. Then the embryonic masses are prepared for cryopreservation at 4°C for 5 days in dark conditions. Subsequently, a dehydration process is conducted by introducing the cells in MS salts with a high sucrose concentration. Next is to verify the Plant Vitrification Solution 2 (PVS2) as suggested by Graça et al. (2018) and the embryonic masses receive 500 µl loading solution for 30 minutes. Then a 500 µl PVS2 solution is added at 0°C for 60 minutes and subsequently, the embryonic mass is transferred to liquid nitrogen at –196°C for 60 minutes. Once the embryonic masses are frozen, they need to be removed from the liquid nitrogen and placed immediately in a 40 °C bath for 1 minute and then they should be placed in an appropriate growth media at 25 °C in dark conditions. The time for embryonic mass recovery is between 2–4 months. The embryonic masses will subsequently turn into somatic embryos.

Conclusions

Relatively few investigations about the genetic and morphology of tree tomatoes have been conducted worldwide. The current understanding of tree tomato genetic and morphological diversity is quite fragmented and centered in a few countries in South America, i.e., Ecuador Colombia and Argentina. Within these countries, studies must focus on more regions and cultivars. It is also important to include other countries where tree tomato is grown, such as Venezuela, Peru, Brazil, and Bolivia to obtain a complete picture of tree tomato genetic diversity in South America. Moreover, it is essential to analyze the genetic diversity of tree tomatoes outside South America such as India, etc.

Tree tomato genetic and morphological diversity studies are key to determining genetic variability, which can be applied to breeding programs to improve yield and disease resistance. Furthermore, the center(s) of the diversity of tree tomatoes could provide an array of variable genotypes that could play an important role in reducing adverse genetic effects such as genetic drift and the founder effect. Genetic diversity studies are relevant for conservation purposes, particularly for wild populations in Bolivia and northwestern Argentina. These wild populations are relevant to obtain desirable traits that can be incorporated into cultivated forms and thus, improve yield and quality features.

Germplasm repositories are important for keeping wild and cultivated tree tomato genotypes. However, not many of them have been created and there is a need to increase their number as well as to collect more cultivars from different regions and ecological conditions. Governments and scientific groups should focus more on determining the genetic diversity of tree tomatoes and their possible benefits for conservation, growers, and sustainable use.

Hybridization of tree tomatoes with other related Solanaceae species has been used to obtain better fruit quality features, yield, and disease resistance. Relevant hybrids have been obtained that play a commercial advantage. Regarding disease resistance tree tomato hybrids could show benefits by reducing pesticide applications and being more beneficial to the environment and human consumption. However, the hybridization of tree tomatoes with other species is often difficult, and much effort is recured to obtain a suitable hybrid with desirable characteristics.

Tree tomato grafting has been an effective mechanism to increase fruit quality feature and yield. However, more research is warranted to obtain successful results with grafts with other Solanaceous species. Graft studies are currently in their infancy and the mechanisms that drive grafting need to be elucidated. Growers have used grafting often by trial and error and their personal experience to obtain desirable fruit quality and yield features in countries like Colombia. However, these experiences are often forgotten or remain anecdotal and can be lost as they are not passed on to other growers or researchers. It is thus important to conduct investigations on grafting by growers and indigenous groups to document possible benefits for tree tomato breeding programs.

References

Acosta-Quezada, P.G., J.B. Martínez-Laborde, and J. Prohens. 2011. Variation among tree tomato (*Solanum betaceum* Cav.) accessions from different cultivar groups: implications for conservation of genetic resources and breeding. *Genet. Resour. Crop Evol., 58*: 943–960. https://doi.org/10.1007/s10722-010-9634-9.

Acosta-Quezada, P.G., S. Vilanova, J.B. Martínez-Laborde, and J. Prohens. 2012. Genetic diversity and relationships in accessions from different cultivar groups and origins in the tree tomato (*Solanum betaceum* Cav.). *Euphytica, 187*: 87–97. https://doi.org/10.1007/s10681-012-0736-7.

Alves, A., A. Caeiro, S.I. Correia, P. Veríssimo, and J. Canhoto. 2017. Establishment and biochemical characterization of tamarillo (*Solanum betaceum* Cav.) Embryogenic cell suspension cultures. *In Vitro* Cellular & Developmental Biology. *Plant, 53*: 606–618. https://doi.org/10.1007/s11627-017-9864-z.

Arroyo, William Fernando Viera, Andrea Verónica Sotomayor Correa, María Viviana Tamba Sandoval, Wilson Arturo Vásquez Castillo, Aníbal Martínez, Pablo Francisco Viteri Díaz, and Lenin Ron. 2016. "Estimación de Parámetros de Calidad Del Fruto Para Segregantes Interespecíficos de Tomate de Árbol (Solanum Betaceum Cav.) En Respuesta de Resistencia a La Antracnosis (Colletotrichum Acutatum J.H. Simmonds)." *Acta Agronomica 65*(3). doi:10.15446/acag.v65n3.49771.

Bernal, J.E. 1994. El cultivo del tomate de árbol. *In*: *Memorias Del Curso Regional de Actualización En Frutales Tropicales, ICA, Corpoica, Espinal, Tolima*, 223–230.

Bohs, L. 1994. *Cyphomandra* (Solanaceae). *Flora Neotrop.*, *63*: 1–175. https://doi.org/10.2307/2805360.

Bohs, L. 1995. Transfer of *Cyphomandra* (Solanaceae) and its species to *Solanum*. *Taxon*, *44*: 583–587.

Bohs, L. 1989. Ethnobotany of the genus *Cyphomandra* (Solanaceae). *Economic Bot.*, *43*: 143–163. https://doi.org/10.1007/BF02859855.

Bohs, L. 1991. Crossing studies in *Cyphomandra* (Solanaceae) and their systematic and evolutionary significance. *Am. J. Bot.*, *78*: 1683–1693. https://doi.org/10.2307/2444847.

Bohs, L, and A. Nelson. 1997. *Solanum maternum* (Solanaceae), a new Bolivian relative of the tree tomato. *Novon*, *7*: 341–345. https://doi.org/10.2307/3391759.

Bolagay, I., D. Rueda, N. Soria, M. Tulcan, B. Manjunatha, R.R.R. Kundapur, S. Mulla, et al. 2015. Agronomic and molecular characterization of four genotypes of *Solanum betaceum*. Cav tolerant to *Colletotrichum acutatum* from the region of Patate in Ecuador. *J. Chem. Pharm. Res.*, *7*: 933–939.

Bradeen, J.M. 2021. On the value of wild *Solanum* species for improved crop disease resistance: resistances to nematodes and viruses. *In:* D. Carputo, R. Aversano, and M.R. Ercolano. [Eds.], *The Wild Solanums Genomes*, 95–118. Springer International Publishing. Cham, Switzerland. https://doi.org/10.1007/978-3-030-30343-3_5.

Broglia, V.G., and G.B. Caruso. 2022. Transferibilidad de marcadores SSR de tomate (*Solanum lycopersicum*) al chilto (*Solanum betaceum*) para la evaluación de la diversidad genética de dos poblaciones del Noroeste Argentino. *Lhawet Nuestro Ent.*, *8*: 15–23.

Brown, W.L. 1983. Genetic diversity and genetic vulnerability -an appraisal. *Economic Bot.*, *37*: 4–12. https://doi.org/10.1007/BF02859301/METRICS.

Carrillo-Perdomo, E., A. Aller, S.M. Cruz-Quintana, F. Giampieri, and J. M. Alvarez-Suarez. 2015. Andean berries from Ecuador: a review on botany, agronomy, chemistry, and health potential. *J. Berry Res.*, *5*: 49–69. https://doi.org/10.3233/JBR-140093.

Cavanilles, AJ. 1799. *Solanum betaceum*. *Anal. Hist. Natural*, *1*: 44–45.

Correia, S., A. Braga, J. Martins, B. Correia, G. Pinto, and J. Canhoto. 2023. Effects of polyploidy on physiological performance of acclimatized *Solanum betaceum* Cav. plants under water deficit. *Forests*, *14*: 208. https://doi.org/10.3390/f14020208.

Coyago-Cruz, E., A. Guachamin, G. Méndez, M. Moya, A. Martínez, W. Viera, J. Heredia-Moya, E. Beltrán, E. Vera, and M. Villacís. 2023. functional and antioxidant evaluation of two ecotypes of control and grafted tree tomato (*Solanum betaceum*) at different altitudes. *Foods*, *12*: 3494. https://doi.org/10.3390/foods12183494.

Currais, L., J. Loureiro, C. Santos, and J.M. Canhoto. 2013. Ploidy stability in embryogenic cultures and regenerated plantlets of tamarillo. *Plant Cell, Tissue Organ Cult.*, *114*: 149–159. https://doi.org/10.1007/s11240-013-0311-5.

Deng, H., L. Zhang, M. Liao, J. Wang, D. Liang, H. Xia, X. Lv, et al. 2022. A PacBio single molecule real-time sequencing-based full-length transcriptome atlas of tree tomato (*Solanum betaceum* Cav.) and mining of simple sequence repeat markers. *Front. Plant Sci.*, *13*: 1–6. https://doi.org/10.3389/fpls.2022.1052817.

Diep, T.T., C. Pook, and M.J.Y. Yoo. 2020. Physicochemical properties and proximate composition of tamarillo (*Solanum betaceum* Cav.) fruits from New Zealand. *J. Food Compos. Anal.*, *92*: 103563. https://doi.org/https://doi.org/10.1016/j.jfca.2020.103563.

Diep, T.T., E.C Rush, and M.J.Y. Yoo. 2022. Tamarillo (*Solanum betaceum* Cav.): A review of physicochemical and bioactive properties and potential applications. *Food Rev. Int.*, *38*: 1343–67. https://doi.org/10.1080/87559129.2020.1804931.

Duarte, O., and R.E. Paull. 2015. *Exotic Fruits and Nuts of the New World*. CABI, Wallingford. UK.

Enciso-Rodríguez, F., R. Martínez, M. Lobo, and L.S, Barrero. 2010. Genetic variation in the solanaceae fruit bearing species lulo and tree tomato revealed by conserved ortholog (COSII) markers. *Gen. Mol. Biol.*, *33*: 271–278. https://doi.org/10.1590/s1415-47572010005000016.

Gannasin, S.P., N.M. Adzahan, M.Y. Hamzah, S. Mustafa, and K. Muhammad. 2015. Physicochemical properties of tamarillo (*Solanum betaceum* Cav.) hydrocolloid fractions. *Food Chem.* *182*: 292–301. https://doi.org/https://doi.org/10.1016/j.foodchem.2015.03.010.

Gebhardt, C. 2016. The historical role of species from the solanaceae plant family in genetic research. *Theor. Appl. Genet.*, *129*: 2281–2294. https://doi.org/10.1007/s00122-016-2804-1.

Giuffrida, D., M. Zoccali, A. Arigò, F. Cacciola, C. Osorio Roa, P. Dugo, and L. Mondello. 2018. Comparison of different analytical techniques for the analysis of carotenoids in tamarillo (*Solanum betaceum* Cav.).*Arch. Biochem. Biophys.*, *646*: 161–167. https://doi.org/https://doi.org/10.1016/j.abb.2018.03.011.

Graça, D., S. Correia, E.A. Ozudogru, M. Lambardi, and J.M. Canhoto. 2018. Cryopreservation of tamarillo (*Solanum betaceum* Cav.) embryogenic cultures. *In*: S.M. Jain and P. Gupta. *Step Wise Protocols for Somatic Embryogenesis of Important Woody Plants* (Vol. II), 95–101. [Eds.], Springer International Publishing, Cham, Switzerland https://doi.org/10.1007/978-3-319-79087-9_7.

Guimarães, M.L., M.E. Tomé, and G.S. Cruz. 1996. *Cyphomandra betacea* (Cav.) Sendtn. (tamarillo). *In*: YPS Bajaj [Ed.] *Biotechnology in Agriculture and Forestry*, *35*: 120–137. *Trees IV*, Springer-Verlag, Berlin, Germany.

Gutiérrez, L., Vidal, N. Barrera, and R. Cadena. 2000. Sinopsis de las especies colombianas del genero *Cyphomandra* (Solanaceae) transferidas a *Solanum*. *Acta Agron.*, *50*: 7–19.

Hurtado, N.H., A.L. Morales, M.L. González-Miret, M.L. Escudero-Gilete, and F.J. Heredia. 2009. Colour, pH stability and antioxidant activity of anthocyanin rutinosides isolated from tamarillo fruit (*Solanum betaceum* Cav.). *Food Chem.*, *117*: 88–93. https://doi.org/https://doi.org/10.1016/j.foodchem.2009.03.081.

Jamir, K., K. Seshagirirao, and M.D. Meitei. 2022. Indigenous oral knowledge of wild medicinal plants from the Peren District of Nagaland, India in the Indo Burma Hot-Spot. *Acta Ecol. Sinica.*, *42*: 206–223. https://doi.org/10.1016/j.chnaes.2021.04.001.

Lacroix, D., S. Prado, D. Kamoga, J. Kasenene, J. Namukobe, S. Krief, V. Dumontet, E. Mouray, B. Bodo, and F. Brunois. 2011. Antiplasmodial and cytotoxic activities of medicinal plants traditionally used in the village of Kiohima, Uganda. *J. Ethnopharmacol.*, *133*: 850–855. https://doi.org/10.1016/j.jep.2010.11.013.

Lagos-Burbano, T.C., L.K. Lagos-Santander, D.E. Duarte-Alvarado, and J. Garcia-Alzate. 2022. Induced mutagenesis in *Solanum betaceum* Cav. plants by the use of diethyl sulphate (DES). *Revista U.D.C.A.*, *25*: 1–10. https://doi.org/10.31910/rudca.v25.n1.2022.1956.

Lamas, C.Y., M.M. Urtasun, E.M. Giamminola, G.R. Pratta, G.B. Caruso, M.N. Morandini, and M.L. de Viana. 2022. Fruit and seed characterization of wild populations of a traditional Andean crop: *Solanum betaceum* Cav. (Solanaceae) in the Argentinian Yungas. *Genet. Res. Crop Evol.*, *69*: 231–244. https://doi.org/10.1007/s10722-021-01223-3.

Lim, TK. 2013. *Edible Medicinal and Non-Medicinal Plants* (Vol. 6), *Fruits*. Springer, New York. U.S.A. https://doi.org/10.1007/978-94-007-5628-1.

Lobo, AM, C.I. Medina, and M. Cardona. 2000. Resistencia de Campo a la antracnosis de los frutos (*Colletotrichum gloeosporioides*) en tomate de árbol (*Cyphomandra* (*Solanum*) *betacea* (*betaceum*) Cav. Sendt.). *Rev. Fac. Nal. Agron. Medellín*, *53*: 1129–1142.

Lobo, M. 2006. Recursos genéticos y mejoramiento de frutales Andinos: una visión conceptual. *Rev. Corpoica*, *7*: 40–54.

Lobo, M., C.I. Medina, O.A. Delgado, M.L. Zuluaga, M. Cardona, and A. Osorio. 2002. Recursos genéticos en frutales Andinos en el sistema de bancos de germoplasma del Estado Colombiano, 43–48. *In*: *IV Seminario Nacional de Frutales de Clima Frio Moderado*, Medellín, Colombia. Corpoica.

Lyons, L.A., T.F. Laughlin, N.G. Copeland, N.A. Jenkins, J.E. Womack, and S.J. O'Brien. 1997. Comparative anchor tagged sequences (CATS) for integrative mapping of mammalian genomes. *Nat. Genetics*, *15*: 47–56. https://doi.org/10.1038/ng0197-47.

Mangino, G., M. Plazas, S. Vilanova, J. Prohens, and P. Gramazio. 2020. Performance of a set of eggplant (*Solanum melongena*) lines with introgressions from its wild relative *S. incanum* under open field and screenhouse conditions and detection of QTLs." *Agronomy*, *10*: 467. https://doi.org/10.3390/agronomy10040467.

Meier, U. 2001. *Growth Stages of Mono-and Dicotyledonous Plants: BBCH Monograph*. Federal Biological Research Centre for Agriculture and Forestry, Berlin. Germany. http://www.reterurale.it/downloads/BBCH_engl_2001.pdf.

Morton, J. 1982. The tree tomato, or 'tamarillo', a fast-growing, early-fruiting small tree for subtropical climate. *Proc. Florida State Hort. Soc.*, *95*: 81–85.

Morton, J. 1987. Tree Tomato. *In: Fruits of Warm Climates*. 437–440. Echo Point Books & Media, Miami. U.S.A.

Nascimento, G.E., L.A. Hamm, C.H. Baggio, M.F. de P. Werner, M. Iacomini, and L.M.C Cordeiro. 2013. Structure of a galactoarabinoglucuronoxylan from tamarillo (*Solanum betaceum*), a tropical exotic fruit, and its biological activity. *Food Chem.*, *141*: 510–516. https://doi.org/https://doi.org/10.1016/j.foodchem.2013.03.023.

Nascimento, G.E., M. Iacomini, and L.M.C Cordeiro. 2016. A comparative study of mucilage and pulp polysaccharides from tamarillo fruit (*Solanum betaceum* Cav.). *Plant Physiol. Biochem.*, *104*: 278–283. https://doi.org/https://doi.org/10.1016/j.plaphy.2016.04.055.

National Research Council. 1989. *Lost Crops of the Incas: Little-known Plants of the Andes with Promise for Worldwide Cultivation*. National Academy Press.

Navarrete, X., L. Ron, P. Viteri, and W. Viera. 2018. Parasitism of the root knot nematode *Meloidogyne incognita* (Kofoid and White) chitwood in five wild solanaceae species. *Rev. Fac. Nal. Agron. Medellin*, *71*: 8367–8373. https://doi.org/10.15446/rfna.v71n1.67122.

Ordoñez, S.R., M. Lourdes Torres and V.S. Arahana. 2009. Diferenciación molecular en variedades cultivadas de tomate de árbol (*Solanum betaceum*), mediante la técnica molecular de AFLP. *ACI Av. Cien, Ingenierías*, *1*: 63–68. https://doi.org/10.18272/aci.v1i1.12.

Orqueda, M.E., M. Rivas, I.C. Zampini, R. Alberto, S. Torres, S. Cuello, J. Sayago, et al. 2017. Chemical and functional characterization of seed, pulp and skin powder from chilto (*Solanum betaceum*), an Argentine native fruit. Phenolic fractions affect key enzymes involved in metabolic syndrome and oxidative stress. *Food Chem.*, *216*: 70–79. https://doi.org/10.1016/j.foodchem.2016.08.015.

Orqueda, M.E., S. Torres, I.C. Zampini, F. Cattaneo, A. Fernandez Di Pardo, E.M. Valle, F. Jiménez-Aspee, et al. 2020. Integral use of Argentinean *Solanum betaceum* red fruits as functional food ingredient to prevent metabolic syndrome: Effect of *in vitro* simulated gastroduodenal digestion. *Heliyon*, *6*: e03387. https://doi.org/10.1016/j.heliyon.2020.e03387.

Pacheco, J., S. Vilanova, R. Grillo-Risco, F. García-García, J. Prohens, and P. Gramazio. 2021. *De novo* transcriptome assembly and comprehensive annotation of two tree tomato cultivars (*Solanum betaceum* Cav.) with different fruit color. *Horticulturae*, *7*: 431. https://doi.org/10.3390/horticulturae7110431.

Peñafiel, N., V.S. Arahana, and M. Lourdes Torres. 2009. Evaluación de la variabilidad genética del tomate de árbol (*Solanum betaceum* Cav.) en los cultivos de tres Provincias del Ecuador por medio de marcadores microsatélites. *Av. Cien. Ingenierias, 1*: 69–74.

Pratt, R.C., D.M. Francis, and L.S. Barrero Meneses. 2008. Genomics of tropical solanaceous species: established and emerging crops. *In:* P.H. Moore and R. Ming [Eds.], *Genomics of Tropical Crop Plants*, 453–67. Springer, New York. U.S.A. https://doi.org/10.1007/978-0-387-71219-2_19.

Pringle, G.J., and B.G. Murray. 1991. Interspecific hybridisation involving the tamarillo, *Cyphomandra betacea* (Cav.) Sendt. (Solanaceae). *N. Z. J. Crop Hortic. Sci., 19*: 103–111. https://doi.org/10.1080/01140671.1991.10421787.

Pringle, G.J., and B.G. Murray. 1992. Polyploidy and aneuploidy in the tamarillo, *Cyphomandra betacea* (Cav.) Sendt. (Solanaceae). I. Spontaneous polyploidy and features of the euploids. *Plant Breed., 108*: 132–138. https://doi.org/10.1111/j.1439-0523.1992.tb00112.x.

Pringle, G.J., and B.G. Murray. 1993. Karyotypes and c-banding patterns in species of *Cyphomandra* Mart. Ex Sendtner (Solanaceae). *Bot. J. Linn. Soc., 111*: 331–342. https://doi.org/10.1111/j.1095-8339.1993.tb01907.x.

Prohens, J., and F. Nuez. 2000. The tamarillo (*Cyphomandra betacea*): A review of a promising small fruit crop. *Small Fruits Rev., 1*: 43–69. https://doi.org/10.1300/J301v01n02_06.

Ralte, L., U. Bhardwaj, and Y.T. Singh. 2021. Traditionally used edible solanaceae plants of Mizoram, India have high antioxidant and antimicrobial potential for effective phytopharmaceutical and nutraceutical formulations. *Heliyon, 7*: e07907. https://doi.org/10.1016/j.heliyon.2021.e07907.

Ramírez, F., and J. Kallarackal. 2019. Tree tomato (*Solanum betaceum* Cav.) reproductive physiology: A review. *Sci. Hortic., 248*: 206–215. https://doi.org/10.1016/j.scienta.2019.01.019.

Ramírez, F. 2020. Cocona (*Solanum sessiliflorum* Dunal) reproductive physiology: A review. *Genet. Resour. Crop Evol., 67*: 293–311. https://doi.org/10.1007/s10722-019-00870-x.

Ramírez, F. 2021. Tree Tomato (*Solanum betaceum* Cav.) grafted with a wild *Solanum* species. *Genet. Resour. Crop Evol., 68*: 2265–2271. https://doi.org/10.1007/s10722-021-01194-5.

Ramírez, F., and T.L. Davenport. 2020. The development of lulo plants (*Solanum quitoense* Lam. Var. *septentrionale*) characterized by BBCH and landmark phenological scales. *Int. J. Fruit Sci., 20*: 562–585. https://doi.org/10.1080/15538362.2019.1613470.

Ramírez, F. 2021. Uchuva *(Physalis peruviana L.)* reproductive biology. Springer, Cham. Switzerland.

Sale, P. 2006. Some interesting developments and the odd hiccup in the 2006 tamarillo season. *Orchardist, 79*: 52 – 55.

Sale, P., and G. Pringle. 1999. *The Tamarillo Handbook: A Guide for New Zealand Growers and Handlers.* New Zealand Tamarillo Growers Association Inc, Kerikeri. New Zealand.

Salgotra, R., and B.S. Chauhan. 2023. Genetic diversity, conservation, and utilization of plant genetic resources. *Genes, 14*: 174. 10.3390/genes14010174. https://doi.org/10.3390/genes14010174.

Schotsmans, W.C., A. East, and A. Woolf. 2011. Tamarillo (*Solanum betaceum* (Cav.)). *In:* E.M. Yahia. [Eds.], *Postharvest Biology and Technology of Tropical and Subtropical Fruits*, 427–441. Woodhead Publishing, Oxford. UK. https://doi.org/10.1533/9780857092618.427.

Sendtner, O. 1845. *Cyphomandra betacea. Flora, 28*: 172–173.

Shah, R., P. Bakshi, H. Itoo, and G. Kour. 2022. Tamarillo (*Cyphomandra betacea* (Cav.)) origin, cultivation, breeding, and management. *In:* M.S. Khan. *Tropical Plant Species and Technological Interventions for Improvement.* IntechOpen, Rijeka. Croatia. https://doi.org/10.5772/intechopen.106601.

Slack, JM. 1976. Growing Tamarillos. *Agric. Gaz. N.S.W., 86*: 2–4.

Standring, L.S., G.J. Pringle, and B.G. Murray. 1990. The control of chloroplast number in *Solanum muricatum* Ait. and *Cyphomandra betacea* (Cav.) Sendt. and its value as an indicator of polyploidy. *Euphytica, 47*: 71–77. https://doi.org/10.1007/BF00040366.

Tapia, C. 2005. Conservación complementaria y uso sostenible de cultivos subutilizados en Ecuador: Rescate, promoción y uso de recursos fitogenéticos interandinos del Ecuador. *Agrociencia, 9*: 147–160. http://www.fagro.edu.uy/~agrociencia/index.php/directorio/article/view/288.

Tene, V., O. Malagón, P. Vita Finzi, G. Vidari, C. Armijos, and T. Zaragoza. 2007. An ethnobotanical survey of medicinal plants used in Loja and Zamora-Chinchipe, *Ecuador. J. Ethnopharmacol., 111*: 63–81. https://doi.org/10.1016/j.jep.2006.10.032.

Valencia, R.A., M. Lobo, and G.A. Ligarreto. 2010. Estado del arte de los recursos genéticos vegetales en Colombia: sistema de bancos de germoplasma. *Corpoica Cien. Tecnol. Agrop., 11*: 85–94. https://scholar.google.es/scholar?start=10&q=lulo+genetics+&hl=es&as_sdt=0,5.

Vargas, Y., J. Nicolalde, W. Alcívar, L. Moncayo, C. Caicedo, J. Pico, L. Ron, and W. Viera. 2018. Response of wild solanaceae to *Meloidogyne incognita* inoculation and its graft compatibility with tree tomato (*Solanum betaceum*). *Nematropica, 48*: 126–135.

Viera W.F., A.V. Sotomayor Correa, M.V. Tamba Sandoval, W.A, Vásquez Castillo, A. Martínez, P.F. Viteri Díaz, and L. Ron. 2016. Estimación de parámetros de calidad del fruto para segregantes interespecíficos de tomate de árbol (*Solanum betaceum* Cav.) en respuesta de resistencia a la antracnosis (*Colletotrichum acutatum* J.H. Simmonds). *Acta Agron., 65*: 304–311. https://doi.org/10.15446/acag.v65n3.49771.

Viera, W., A. Sotomayor, and P. Viteri. 2019. Breeding of three Andean fruit crops in Ecuador. *Chronica Hort., 59*: 20–29.

Viteri-Díaz, P., W. Vásquez-Castillo, M. Racines-Oliva, and W. Viera-Arroyo. 2020. Andean solanaceae species with resistance to biotic factors, such as tree tomato (*Solanum betaceum* Cav.) rootstocks. *Manglar, 17*: 347–352.

Viteri, P., J. León, W. Vásquez, C. Encalada, A. Martínez, J. Revelo, M. Posso, and M. Hinojosa. 2010. Solanáceas silvestres utilizadas como portainjertos de tomate de árbol (*Solanum betaceum* Cav.) con alto rendimiento, resistencia a enfermedades y mayor longevidad. *Boletín Div., 371*, INIAP, Quito. Ecuador

Wang, S., and F. Zhu. 2020. Tamarillo (*Solanum betaceum*): chemical composition, biological properties, and product innovation. *Trends Food Sci. Technol., 95*: 45–58. https://doi.org/https://doi.org/10.1016/j.tifs.2019.11.004.

4

Genetic Diversity of Wood Apple (*Limonia acidissima* L.) and Sustainable Utilization

Shrinivas Lamani[1,2] and *Hosakatte Niranjana Murthy*[1,3,*]

Introduction

Family Rutaceae includes 2,070 species of woody herbs, shrubs, and trees that are found all over the world, primarily in warm temperate and tropical climates and have glandular punctuate, frequently pungent foliage. Rutaceae is the largest family of the order Sapindales and is well-known for its economic importance (Kubitzki et al., 2011). The family is found all over the world, but it is most common in the tropics and subtropics. Southern Africa, Australasia, and America all have high levels of generic and species richness, as well as endemicity (Engler, 1896; Kubitzki et al., 2011). The presence of schizogenous secretory cavities filled with essential oils is the family's most unique phenotypic character. The pores can be observed as pellucid dots in the leaves, but they can also be observed in other parts of the plant, such as the pericarp, flowers, and young axes (Turner et al., 1998; Kubitzki et al., 2011; Groppo et al., 2012). Fruits are considered a vital source of proteins, carbohydrates, essential fats, amino acids, vitamins, mineral elements, dietary fiber, and reducing and non-reducing sugars. Fruits are not only the primary source of our food but they also have immense therapeutic potential due to the presence of bioactive compounds, including antioxidants, such as fruit acids, sugars, organic acids, etc. (Joseph et al., 2016; Lamani et al., 2021; Lamani et al., 2022).

Wood apple (*Limonia acidissima* L.) of the Rutaceae family is one of the nutritional and medicinally important underutilized fruit-yielding tree species that belongs to the monotypic genus *Limonia* and is native to India and Sri Lanka. It is widely cultivated in Southeast Asian countries. Wood apple is a deciduous, slow-growing, erect tree (Figure 1). It can be planted under almost all conditions of soils and climates and it bears about 800–1000 fruits per annum (Ghosh et al., 2012). The tree is subtropical and it prefers dry climatic conditions for flowering and fruit set. It is found widely in dry deciduous and semi-arid regions of South India, majorly in the Northwestern part of Tamil Nadu and the Eastern Ghats of Andhra Pradesh, central and Southern parts of Karnataka, Northern region of Telangana, and central region of Kerala (Lamani and Murthy, 2022). Fruits of *Limonia acidissima* are well known by various vernacular names, such as wood apple, elephant apple, and monkey fruit.

[1] Department of Botany, Karnatak University, Dharwad - 580003, India.
[2] Department of Botany, DVS College of Arts and Science, Shimoga - 577201, India.
[3] Department of Horticulture Science, Chungbuk National University, Cheongju 28644, Republic of South Korea.
* Corresponding author

Figure 1. *Limonia acidissima* (Wood apple): (A) Habit, (B) Leaf, (C) Flower, (D) Fruits, (E) Ripened fruit with rind, (F) Pulp and (G) Seeds.

Wood apple is used by indigenous people for food and medicine. It is an evergreen tree that grows widely in states of South India and is considered to have great potential as a medicinal plant. Both the pulp and leaves extract for various medicinal uses. Wood apple as a whole and its parts, such as unripe fruit, ripened fruit, leaves, roots, and bark have been used in the indigenous system of medicine for the treatment of various ailments. Indonesians used to eat wood apple ripe fruits by mixing them with sugar for breakfast. Indians eat the pulp of ripe fruit with sugar or jaggery. It is used for making chutney and pickles or is blended with coconut milk and palm sugar syrup and consumed as a beverage. Sri Lankan people prepare a popular drink known in Simhalese as Dimbulkiri by mixing ripened wood apple pulp with coconut milk and palm sugar (Morton, 2002). Leaves of wood apple consist of an essential oil, used as an astringent, and the gum of wood apple is used as an ointment to calm an irritated skin (Chowdhury et al., 2008). The fruits are used in the treatment of diarrhea, dysentery, tumors, asthma, injuries, and hepatitis. In the recent past, various chemicals have been isolated from various parts of the plant and are composed of phenols, flavonoids, alkaloids, triterpenoids, coumarin, lignans, quinones, sterols, fatty acids, and volatile compounds. A fruit with nutritional values and immense health potential indicates its potentiality for processing commercially valuable products (Chopra and Singh, 2001).

The shape of the fruit is round to oval, 5–12.5 cm wide with a hard, woody, greyish white, scurfy rind about 6 mm thick; the pulp is brown, odorous, resinous, astringent taste (sour or sweetish) with numerous small, white seeds scattered throughout and are highly nutritive (Orwa et al., 2009; Lamani et al., 2021). Seeds of wood apple contain a bland, non-bitter oil that is high in unsaturated fatty acids (Morton, 1987). These seeds showed an abundance of protein, carbohydrate, and amino acid content, and a high amount of mineral elements (Ratnayake et al., 2009; Sonawane et al., 2016).

Wood apple fruits have significant importance in traditional medicine; their parts—leaves, fruit pulp, seed, and seed oil—are utilized as famine foods since they are rich in nutritional and phytochemicals, both in cake and seed oil. In this regard, systematic cultivation of this plant is critical to achieving its full potential as a source of the above-mentioned high-value products from India. It is important to discuss the recent updates regarding genetic diversity, morphological, biochemical, and molecular characterization, and most importantly the sustainable utilization through germplasm conservation and breeding of promising genotypes for cultivation programs to provide disease resistance and increase yield, and fruit size.

Distribution and Morphology of Wood Apple

Wood apple is indigenous to India and is widely cultivated in Southeast Asia and Northern part of Malaya, and mainly in the dry deciduous regions of Burma, Sri Lanka, Pakistan, and Bangladesh (Morton, 1987). In India and Sri Lanka, it is cultivated along roads and edges of fields and occasionally in orchards. In Malaysia and Indonesia, it is being cultivated in villages and parks (Verheij and Coronel, 1992). Wood apple is a dry plains moderate drought-resistant tree species, which prefers mainly a monsoon climate with distinct dry or deciduous forests and it prefers dry climatic conditions for flowering and fruit set which is commonly found throughout the Southern part of India. The wood apple tree may grow on a variety of soil types; the decreasing order of soil concerning growth richness is red soil, black cotton soil, sandy loamy soils, and red-black mixed soil. In India, it is found throughout the plains of forests, particularly in drylands up to an elevation of 450 m. It is found abundantly in the wild forests, neglected areas, and majorly on plains; whereas the density decreases as the altitude increases (Lamani and Murthy, 2022).

Wood apple is a deciduous, slow-growing, erect tree (Figure 1A) that reaches up to a height of about 12–15 meters with a few upward-reaching branches bending outwards near the summit where they are subdivided into slender branchlets drooping at the tips and bears fruits (Ghosh et al., 2012). The spines are axillary, short, straight, and 2–3 cm long on some of the zig-zag twigs. The leaves are deciduous dark green, alternate, pinnate with 5–7 leaflets and each leaflet size of 25–35 mm length and 10–20 mm thickness, with oil glands and a slight lemon scent when crushed (Figure 1B). The flowers are bisexual, small, numerous, dull-red, or greenish in colour (Figure 1C). The shape of the fruit is round to oval, 5–12.5 cm wide with a hard, woody, greyish white (Figure 1D), scurfy rind about 6 mm thick; the pulp is brown (Figure 1E), odorous, resinous, astringent taste (sour or sweetish) with numerous small, white seeds scattered throughout and are highly nutritive (Figure 1G) (Orwa et al., 2009). The leaves shed from November to January; a flush of new leaves appears from February to March. Subsequently, flowering is initiated in March and continues up to May. Fruit development starts from May to June and the ripening of fruits begins from October to March; they remain attached even after the commencement of flowering in the next season (Murrinie et al., 2017). Wood apple is a seasonal fruit; in India, the fruit matures in October and is available until March. The maximum yield per plant varies from 40–70 kg (Singh et al., 2014).

Traditional Medicinal Values of Wood Apple

Wood apple plant sources are well known for their medicinal properties. As a whole, unripened and ripened, wood apples and their parts, such as leaves, bark, roots, and fruits have been used as medicine for the treatment of various diseases. Ethnopharmacological properties have been documented in various studies. A mixture of young leaf juices, milk, and candies is given as a remedy for children's biliousness and intestinal disorders. Powdered gum mixed with honey, is given to control dysentery and diarrhea in children. The leaves of the plant contain an essential oil, which is used as an astringent and the wood apple gum is used as an unguent to calm an irritated skin (Chowdhury et al., 2008). The leaves are astringent and carminative when mixed with a pinch of pepper, making them an excellent medicine for hiccups, vomiting, indigestion, and dysentery (Ilango and Chitra, 2009). Traditionally, the constituents from the stem bark of wood apples are mixed with water and applied mainly to the face; it is believed that regular application on the skin helps to keep the complexion of the skin cool, smooth, fair, and well-textured (Bandara et al., 1988).

The fruit pulp has a wide range of medicinal benefits, including free radical scavenging, antibacterial, antioxidant, anticancer, anti-ulcerative, gastro-protective, antiviral, anti-diarrheal, hepatoprotective, and actions that prevent lipid peroxidation (Ilango and Chitra, 2009). Extracts of wood apple are used traditionally for curing various diseases. It is antimicrobial and antifungal, a liver and cardiac tonic, helps in dysentery, hiccough, and sore throat problems. It is also used as a good antidote for snake bites and as a face cream to remove small spots and lesions on the skin (Pandey

Table 1. Nutritional composition in pulp and seed cake of *Limonia acidissima.*

Parameters	Composition		References
	Pulp	Seed cake	
Proximate composition			
Moisture (%)	58.89 ± 1.21	4.83 ± 0.01	Lamani et al. (2021; 2022)
Oil (%)	0.99 ± 0.00	32.02 ± 0.08	
TSS (°Brix)	19.52 ± 0.17	1.52 ± 0.29	
pH	3.61 ± 0.09	1.76 ± 0.19	
Acidity (%)	4.61 ± 0.13	1.83 ± 0.02	
Ash (%)	2.73 ± 0.16	3.76 ± 0.18	
Crude fiber (%)	3.32 ± 0.02	-	
Total Protein (%)	9.30 ± 0.02	25.24 ± 0.07	
Total Carbohydrate (%)	24.74 ± 0.17	31.66 ± 0.04	
Total phenolics (mg GAE/ 100 g)	500.00 ± 0.01	135.42 ± 1.47	
Mineral elements (mg/ 100 g)			
Sodium	8.50	9.55 ± 0.20	Joshi et al. (2008); Sonawane et al. (2016)
Potassium	-	494.3 ± 9.84	
Magnesium	-	703.78 ± 14.08	
Manganese	0.70	1.66 ± 0.03	
Copper	0.80	7.18 ± 0.14	
Iron	3.50	8.49 ± 0.17	
Zinc	386.30	5.76 ± 0.12	
Phosphorus	46.50	1227.6 ± 24.55	
Calcium	15.90	-	
Amino acids (mg/ 100 g)			
Aspertic acids	104.16	1.4 ± 0.05	Priya Darsini et al. (2013); Sonawane et al. (2016)
Glutamic acid	Negligible	14.77 ± 0.68	
Histidine	414.95	1.24 ± 0.04	
Glycine	152.23	6.90 ± 0.28	
Arginine	206.49	4.59 ± 0.18	
Alanine	169.82	4.99 ± 0.20	
Tyrosine	605.19	2.23 ± 0.08	
Valine	471.53	1.09 ± 0.04	
Phenylalanine	984.15	2.87 ± 0.08	
Isoleusine	2870.41	2.44 ± 0.06	
Leucine	176.06	5.19 ± 0.21	
Proline	939.04	-	
Tryptophan	1154.51	-	
Free-sugar composition (% DW)			
Fructose	16.40 ± 0.23	-	Lamani et al. (2022)
Glucose	14.23 ± 0.10	-	
Organic acid composition (% DW)			
D-Tartaric acid	4.01 ± 0.03	-	Lamani et al. (2022)
Ascorbic acid	4.51 ± 0.05	-	
Citric acid	4.27 ± 0.04	-	
Succinic acid	1.83 ± 0.03	-	

GAE-Gallic acid equivalent; DW-Dry weight

et al., 2014). Although the unripe fruit is used as an astringent, the seeds can help diagnose cardiac disorders (Senthilkumar et al., 2010). Fruits are also used as a refrigerant, stomachic, stimulant, astringent, aphrodisiac, diuretic, cardiotonic, tonic for liver and lungs, to cure cough, hiccup, asthma, leucorrhoea, diarrhea, and in wound healing (Ilango and Chitra, 2010). Wood apple is used by the tribal and rural populations of the developing world which contributes to the traditional health system (Murthy and Dalawai, 2020).

Nutritional Values of Wood Apple

Wood apple fruits are considered an important source of nutrients that include proteins, carbohydrates, fatty acids, sugars, dietary fiber, vitamins, minerals, and amino acids. Fruits are abundant in nutrients and contain various bioactive components, thus serving as a good source of antioxidants (Sonawane and Arya, 2013). Table 1 depicts the overall nutritional composition of fruit pulp and seed cake. Nutritional compositions of wood apple have 58.89% of total moisture content in the matured fruits along with 9.30% of protein, 24.74% of carbohydrate, 0.99% of crude fat, 3.32% of crude fiber, and 2.73% of ash. The principal factors used to check the quality of fruits included total soluble solids, titratable acidity, and pH, shown in the fruit pulp were 19.52 ± 0.17 °Brix, 4.61 ± 0.13%, and 3.61 ± 0.09%, respectively (Lamani et al., 2022). Wood apple fruits can be preserved through processing and used during the off-season as this belongs to underutilized tree species (Vidhya and Narain, 2011). According to Patel (2013), wood apple pulp can be transformed into a beverage with an outstanding flavor and taste, and can be kept in bottles using flow-over pasteurization and it can also be canned after being sweetened with sugar to produce a satisfying product with high-keeping qualities.

Wood apple seeds are a good source of protein, carbohydrate, amino acid content, and a high amount of mineral elements (Ratnayake et al., 2009; Sonawane et al., 2016). Moisture content in seeds was 4.83%. The total ash content in seed cake is 3.76 ± 0.18%, which is the major indication of the presence of mineral elements. Seeds were recognized for their higher content of total protein (25.24%) and the carbohydrate content was 31.66%. Seed cake showed 135.42 mg gallic acid equivalent/100g of total phenolic content (Table 1) (Lamani et al., 2021). Nutritional and nutraceutical analysis of fruits revealed that the rich content of protein, carbohydrates, and total phenols helped maintain good health. Utilization of such underutilized fruits provides a significant source of livelihood support for many rural communities.

Taxonomic Background

Limonia acidissima L., (synonyms *Feronia limonia* (Linn.) Swingle, *Feronia elephantum* Correa, *Limonia acidissima* Groff and *Schinus limonia* L.) member of the order Sapindales belongs to the monotypic genus Limonia, in the family Rutaceae. It is referred to as a wood apple since the fruit shape resembles an apple and the fruits possess a hard shell. It is a neglected seasonal fruit-yielding species; fruits mature in the month of October and are available till March. The wood apple is native to India and Sri Lanka and spreads over India's dry deciduous regions.

Collection and Assessment of Germplasm

Morphological Characterization

Characterization and assessment of diversity are critical for identifying distinct genotypes, recognizing genetic relationships, including parentages and efficient germplasm management and utilization. Quantitative or qualitative characteristics can be used to distinguish genetic variability from phenotypic variability. Genetic variability will be examined using multiple approaches, according to Govindaraj et al. (2015), allowing breeders to select genotypes with contrasting traits. A comprehensive investigation of morphological diversity, on the other hand, provides an excellent basis for evaluating promising genotypes for cultivation programs (Schlautman et al., 2020). Hence an assessment of genetic diversity is needed to cultivate sustainable exploitation and the production of value products to meet global demands.

Wood apple tree and fruit growth attributes are associated with the yield and quality of the fruits and have revealed a significant variability among the populations and results obtained from a few studies are depicted in Table 3. Ghosh et al. (2012) conducted a thorough study to characterize the desirable genotypes from the different areas of West Bengal. As a result of the survey, a total of 20 plants were identified from 20 locations. They found a great deal of variability among the genotypes, with respect to the yield (5,000–6,000 fruits/tree) and fruit size ranging from 7.3–8.9 cm, and 7.2–8.4 cm in length and breadth, respectively. The weight of the fruit ranged from 130 g–225 g. The percentage of pulp and seed was observed maximum of 60.6% and 40.4%, respectively.

An extensive survey was carried out at several places in Gujarat by Singh et al. (2016). In total, 25 genotypes were collected based on the morphological traits of wood apple. Wide diversity was recorded for the agronomical characteristics like fruit shape, i.e., out of 25 genotypes eight were round, 10 were flattened round, two were triangular, and five were oblong in shape. Fruit size was measured as length, width, and girth and these ranged between 5.25–9.46, 5.10–9.29, and 20.14–29.77 cm, respectively. The weight of fruit pulp and seeds ranged from 80.04–319.32 g among all the studied genotypes, while the weight of seeds ranged from 2.36–7.93 g.

Dowarah et al. (2021) studied fruit physicochemical studies of 12 local wood apple genotypes collected from various regions of West Bengal. They explained the physical characteristics including fruit size, shape, shell, seed, pulp percentage, etc., and it showed a lot of variability within the studied genotypes. Same as above, they analyzed qualitative characters and got significant results in all the studied genotypes. Fruit shapes observed were flattened round, round, and oblong. The size of the

Table 2. Physicochemical and Biochemical composition of pulp and seed oil of *Limonia acidissima.*

Parameters	Composition		References
	Seed oil	Pulp oil	
Physicochemical properties (at 30 °C)			
Color	Golden yellow	-	Lamani et al. (2021)
Physical state at 4 °C	Liquid	-	
Viscosity (cP)	42.33 ± 0.57	-	
Wax content (mg/g)	0.13 ± 0.00	-	
Refractive index	1.44 ± 0.01	-	
Specific gravity	0.92 ± 0.00	-	
Free fatty acid (%)	1.38 ± 0.02	-	
Acid value (mg KOH/g)	2.12 ± 0.01	-	
Iodine value (g I_2/100g)	116.16 ± 0.28	-	
Saponification value (mg KOH/g)	186.50 ± 0.86	-	
Unsaponifiable matter (%)	1.15 ± 0.01	-	
Fatty acid composition (%)			
Palmitic acid (16:0)	17.68 ± 0.65	18.52 ± 0.12	Lamani et al. (2021; 2022)
Stearic acid (18:0)	14.15 ± 0.27	9.02 ± 0.08	
Oleic acid (18:1n9c)	21.56 ± 0.57	23.89 ± 0.06	
Vaccenic acid (18:1n7)	1.70 ± 0.04	1.78 ± 0.23	
Linoleic acid (18:2)	10.02 ± 0.43	9.23 ± 0.35	
α-Linolenic acid (18:3)	16.28 ± 0.29	16.55 ± 0.26	
Σ Saturated fatty acids	33.38 ± 0.60	32.17 ± 0.35	
Σ Monounsaturated fatty acids	25.51 ± 0.67	26.20 ± 0.33	
Σ Polyunsaturated fatty acids	26.86 ± 0.13	25.78 ± 0.61	
Tocopherol composition (mg/100g)			
α-Tocopherol	3.77 ± 0.00	-	Lamani et al. (2021)
γ-Tocopherol	39.27 ± 0.07	-	
δ-Tocopherol	12.64 ± 0.01	-	

Σ-Sum

fruits ranged from 7.9–10.7 cm in length and 7.1–9.8 cm in diameter. The weight of the fruit varied from 97.4–225.7g, whereas the weight of the pulp varied from 50.7–117.5g. The average weight of seeds ranged from 6.5g–37.9g. They found less variation in rind thickness and the rind weight ranged from 53.45–78.40 g.

Lamani and Murthy (2022) examined 16 morphological traits in 32 wood apple populations from Southern states of India. They selected the wild naturally grown populations with contrasting characteristics, such as tree height, canopy spread, fruit yield and size, pulp and seed percentage, etc. The morphological features studied revealed a high variation among populations. The plant height, stem girth, and bole length of studied populations ranged from 7.68–16.69 m, 0.98–2.62 m, and 1.47–3.57 m, respectively. Wood apple fruits exhibited a wide range of variability in concern with the fruit's shape and size. The round-shaped fruits were dominant and represented 53.12% of the total population (Figures 2B, F, G), followed by 21.87% of flattened round (Figures 2C, H), and 18.75% of oblong (Figures 2A, D, E). The size of the fruit including length, width, and girth of populations ranged from 6.39–8.99 cm, 6.09–9.74 cm, and 18.25–29.01 cm respectively. The characters of fruits including fruit, pulp, seed, and rind weight showed wide variability that ranged from 118.30–233.42 g, 70.19–151.82 g, 5.90 g–14.89 g, and 34.96–73.19 g, respectively. Their study revealed that the populations (KA-01, KA-04, KA-11, AP-02, TS-07, TN-03) under evaluation ranged greatly in terms of fruit physical traits, such as size, shape, rind, seed percentage, and pulp content.

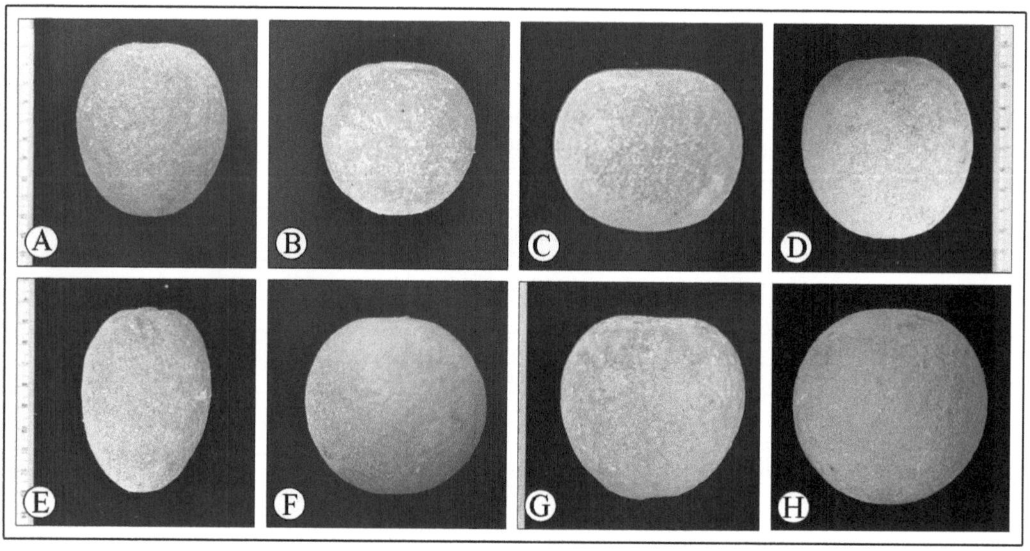

Figure 2. Morphological variability of wood apple fruits: (A) KA-01, (B) KA-06, (C) AP-03, (D) TS-01, (E) TS-06, (F) TS-07, (G) TN-01, and (H) TN-03.

Cytological Characterization

Cytological characterization has been considered to be necessary for enhancing the chromosomal database and preserving significant information for future usage. *Limonia acidissima*, is a moderate-sized medicinally important deciduous fruit-yielding plant of the family Rutaceae, which is reported that the species contains n = 9 and 2n = 18 chromosomes (Singhal et al., 1980; Gunaseeli and Sampathkumar, 1986). Jha and Nath (2015) reported the chromosomes in a wood apple and it was confirmed to contain 2n = 18 chromosomes with median and sub-median chromosomes. Chromosomes are relatively smaller in size (1.53–2.83 µm) and the length of individual chromosomes varies from 1.53–2.83 µm and total chromatin length varies from 32–43.5 µm among observed metaphase plates (Jha and Yamamoto, 2012; Zhao-Yang et al., 2002).

Biochemical Characterization

Due to the increasing awareness among people towards natural products, natural medicine is attracting more attention than the allopathic system. However, the collection of medicinal plants from natural habitats poses a serious threat to their existence and they would disappear. Therefore, medicinal plants are grown, to obtain the active compounds. Breeding has an advantage over a collection of natural populations, i.e., a controlled and regular supply of materials (Ramesha et al., 2008). The preliminary phytochemical analysis of wood apple plant parts extracted using ethanol exhibited antioxidant and antibacterial activity due to the presence of biochemicals and also helpful in preventing the diseases associated with oxidative stress and infectious diseases (Nithya and Saraswathi, 2010; Vijayvargia et al., 2014). Reports of Ilaiyaraja et al. (2015) revealed wood apple fruits are good at antioxidant activity. They extracted polyphenols by response surface methodology and reported 7.37 g gallic acid equivalents/g. The polyphenolic extract also demonstrated very good antioxidant activities. The unripe fruits contain stigmasterol. Root bark yielded osthenol, geranyl umbelliferone, marmin, marmesin, aurapten, bergapten, isopimpinellin, etc. (Gupta et al., 1979). Three volatile flavor components were obtained from fresh wood apple fruits, namely, methyl hexanoate, ethyl-3-hydroxyhexanoate, and butanoic acid (Patel, 2013). The fruit contains flavonoids, saponins, tannins, and glycosides. Coumarins and tyramine derivatives were isolated from the fruits. The leaves and fruit shells have antifungal compounds, such as psoralene, xanthotoxin, 2, 6-dimethoxybenzoquinone, and osthenol (Pradhan et al., 2012). The fruit pulp represents 36% of the whole fruit and the pectin content of the pulp is 3–5%, so it can be an excellent source for the preparation of jelly (Narsing Rao et al., 2011). Phapale and Thakur (2010) revealed that repeated extractions of ripe fruit pulp produced extracts of both the free and bound phenolic components. The highest total phenolic content (229.0 mg/g, gallic acid equivalent) was found in the phenolic glycoside extract, which was accompanied by the phenolic ester (37.5 mg/g) and free phenolics (11.0 mg/g).

Gas chromatography-mass spectroscopy (GC-MS) reports of wood apple essential oil revealed a total of 24 compounds, in which some major compounds, such as trans-anethole (57.72%), methyl chavicol (37.48%), and p-anisaldehyde (1.66%) are reported in aerial parts (Pande et al., 2010). According to Ahmad et al. (1989), methyl chavicol (27.20%) and thymol (24.40%) are the major constituents obtained from aerial parts of wood apples. Lakshmi et al. (2015) studied the biochemical changes in fruits at three different ripening stages. The aqueous ethanol extracts were used to quantify the free sugar content and the results observed revealed an increase in total sugar content from 3.2g–4.9 g/ 100g and a decrease in acidity from 3.5–3.2 g/100g. The total soluble sugar content was reduced upon ripening from 20.4–14.0 °Brix. Total phenolic content was increased from 8.7–9.6 g/ 100g upon fruit ripening and they contain maximum content of phenolic acids in bound-form than free-form. According to Jayakumar and Geetha (2012), oxalic, malic, citric, and tannic acids are present abundantly in wood apple fruits. The percentage of organic acids decreases from 3.5 g (unripe) to 3.2g/100g (ripe), which means that the percentage of organic acid content will decrease as the fruit reaches the maturity stage (Lakshmi et al., 2015).

Ghosh et al. (2012) conducted a thorough study to characterize the desirable genotypes from the different areas of West Bengal. Biochemical studies including total sugar, reducing sugar, acidity, total soluble solids, and Vitamin-C ranged from 5.6–14.3%, 4.1–8.8%, 1.2–4.6%, 15.0–18.4 °Brix, and 2.7–13.0 mg/100g pulp, respectively, among the studied accessions and this showed significant variability (Table 3). An extensive survey was carried out at several places in Gujarat by Singh et al. (2016) concerning the biochemical characteristics like total soluble solids and Vitamin-C characterized in all the genotypes. The TSS, observed ranged from 11.07–19.36 °Brix in the genotypes and the vitamin-C content ranged from 7.08–19.60 mg/100g in pulp (Table 3).

Lamani et al. (2021, 2022) studied fruit biochemical studies of local wood apple populations. According to their investigation fruit pulp showed interesting results, the quantity of fructose and glucose were 16.40% and 14.23%, respectively, while the total content of sugar recorded was 31.59%. The organic acid profile of wood apple revealed greater variability, the content of tartaric, citric, ascorbic and succinic acid was 4.01%, 4.27%, 4.51%, and 1.83%, respectively. The total polyphenol content of the pulp showed a significant value and it was 500 ± 0.01 mg GAE/100g (Table 1). Wood apple pulp is a good source of essential fatty acid compositions, the average content of unsaturated fatty acids, such as oleic acid, linoleic, and α-linolenic acid was 23.89%, 9.23%, and 16.55%, respectively, whereas the saturated fatty acids, such as palmitic and stearic acid were 18.52% and 9.02%, respectively.

In wood apple seed oil, unsaturated fatty acids including oleic acid, linoleic, and α-linolenic acid account for 21.56%, 10.02%, and 16.28% of the total fatty acids, respectively, whereas the saturated fatty acids, such as palmitic and stearic acid was 17.68% and 14.15%, respectively. Seeds are a rich source of tocopherols, in these three tocopherols (α, γ, and δ) were identified in wood apple genotype seed oils. For α, γ, and δ-tocopherols, respectively, the average tocopherol amount was 3.77, 39.27, and 12.64 mg/ 100g oil (Table 2).

Molecular Characterization

The genetic diversity of plants provides plant breeders an opportunity to develop new and improved cultivars with desired traits. Plant DNA markers have been widely employed for several applications, such as genotype identification, taxonomic and phylogenetic connection inference, assessment of genetic diversity, and phylogeographic study (Weising et al., 2005). The degree of genetic variability found in individuals of a variation or a population within a species is known as genetic diversity. It is the result of mutations, gene flow, genetic drift, and the recombination of genetic material (DNA) throughout the inheritance process (Brown, 1983). Plant genetic resources are made up of all of the various gene alleles that are present in a certain crop species, which is known as total hereditary material, which includes wild-type species as well as horticultural and therapeutic plants. Genetic diversity is the main driving force for the selection and evolution of populations (Nevo et al., 1984).

Studies on the genetic diversity of wood apples have been extremely limited. The genetic diversity of wood apple has been reported by Vyas et al. (2008). Using Inter-short sequence repeat (ISSR) markers, they investigated the genetic variation observed in the samples from three different localities of South-east Rajasthan. To develop ISSR markers, eight individual and 17 pair combinations of primers were selected from all the individuals from the three populations. They reported that out of eight individual and 17 primer pair combinations, 307 loci were scored in all populations. It was found that 50.81% of these ISSR segments were polymorphic. The overall amount of variance between populations was estimated to be 0.20. The total genetic diversity of the wood apple ranged from 0.16–0.19. Their findings represent that gene flow is mostly influenced by the topography of the area or logging, which divides populations, lowering the efficient interchange of germplasm and diversity among regions.

Wood apples have considerable genetic diversity in fruit shape, size and many other qualitative characteristics and they are currently underexploited in their natural habitats. Because of the considerable genetic diversity, Bhargava et al. (2019) randomly collected 11 genotypes of wood apple from Indore and Vadodara and evaluated the genetic diversity using the Random Amplified Polymorphic DNA (RAPD) approach. To detect the RAPD polymorphisms, 25 independently random primers were utilized, out of which 22 primers gave 192 amplified bands, 139 (based on 25 RAPD markers they detected 72.39% of polymorphism) of which were polymorphic. The polymorphic information contents (PIC) value of the primers ranged from 0.08–0.50, while the low genetic distance was (0.31) among the genotypes studied. The results of this study proved that this technique could potentially used to develop the specific markers needed in the wood apple crop improvement program.

Genetic diversity analysis in wood apple by Vyas et al. (2008) and Bhargava et al. (2019) was only based on dominant DNA markers, which were not suitable for large-scale genotyping and marker-assisted breeding because of their low information content and reliability. The reports of Lata et al. (2023) showed the use of simple sequence repeats (SSRs). SSR markers are suitable markers for the genetic analysis for linkage map constructions, assessment of germplasm diversity, and recognition of molecular markers for marker-assisted selection (Wang et al., 2010). By applying SSR markers, Lata et al. (2023) assessed the genetic diversity of 96 wood apple samples collected from 16 states of India and found a significant amount of genetic diversity among them. To evaluate the genetic diversity, they used 24 polymorphic simple sequence repeats (SSRs), developed *de novo* through RNA sequencing and got 11,401 SSRs and 2,15,859 unigenes. The most common SSRs (56.89%) were dinucleotide repeat motifs, which were followed by trinucleotide (37.61%) and tetranucleotide (3.63%) repeats. The most prevalent SSRs (56.89%) were dinucleotide repeat motifs followed by trinucleotide and tetranucleotide repeats. With an average of 5.37 alleles per locus and an average PIC value of 0.62, SSR profiling revealed 129 alleles. In the 96 samples Jaccard's genetic dissimilarity varied from 0.39–1.00. The overall analysis indicated moderate to high genetic diversity among the wood apple samples. They suggested nine highly informative SSRs and 14 wood apple samples to conserve genetic resources and develop breeding programs to improve them.

Utilization of Germplasm

Utilization in Breeding and Improvement Programs

The basic and essential first step towards enhancing any crop is collecting germplasm (Mani et al., 2017). Genetic diversity is the desired factor for the characterization of the selection of efficient parents for utilization in breeding programs. It is a great challenge to the horticulturists to conserve the genetic purity by the seed propagation method as it is a cross-pollinated tree species. So far, no thorough investigation has been conducted on breeding and improvement in wood apple. The plants that have grown so far are of seed origin and exhibit lots of variation. Not much data is available on wood apple germplasm to select for superior traits, but some have attempted to characterize wood apple germplasm, viz., Hiwale et al. (2008) due to the wide range of diversity present in the wood apple species. They attempted to identify the superior types that could be replicated vegetatively true to type on a large scale and the suitable technique as well as the time for the large-scale multiplication. Through the use of softwood grafting techniques, patch budding and seeding, the gathered sources were multiplied. The rate of seed germination ranges from 22.9%–62.40% (CHES-2). When the propagation approach was used in February, they had 53.35% success in softwood grafting as opposed to 29.83% success in other collections using budding. Hence, they indicated that softwood grafting was the best technique for propagating one-year-old desi rootstock. The *in situ* soft wood grafting technique is better than nursery grafting in terms of growth parameters when transplanted in the field.

Mani et al. (2017) reported the elite wood apple germplasm from the collected natural populations from different locations in the Nadia district of West Bengal. From 20 screened populations, 10 elite germplasm were selected for further studies on flowering, fruiting, and specific physio-chemical characteristics. Their research revealed that the two accessions, Acc. 5 (Karimpur) and Acc. 9 (Chakdah), with abundant flowers, large-sized fruits (8.89 cm and 7.96 cm), maximum percentage of pulp (63.08% and 65.33%), and lesser shell thickness (0.33 cm and 0.36 cm) (Table 3). Both accessions can overcome a condition of market oversupply by exhibiting late fruit maturity, ranging from early May to late November and they are excellent even in the biochemical contents. Consequently, they suggested that this germplasm could potentially be preserved through asexual reproduction for use in future breeding initiatives.

The results of Debbarma et al. (2019) showed the implementation of biotechnological methods for germplasm conservation and long-term utilization of neglected fruits of India. By keeping into consideration, the wood apple's nutritional and biochemical importance as well as its adaptability in various kinds of climates. Because almost all fruit crops tend to be grown vegetatively or clonally, it is difficult to maintain them for extended periods and they also rapidly lose genetic stability. Large planting materials need to be stored in a cold, vast area; which is a lengthy and expensive process. According to their outcomes, the *in vitro* conservation method provides many advantages for annual

Table 3. Major morphological traits of morphologically and biochemically superior genotypes of *Limonia acidissima*.

Genotype	State	Morphological traits								Biochemical composition				References
		Fruit shape	Tree height (m)	Bole length (m)	Canopy spread (m^3)	Fruit weight (g)	Pulp weight (g)	Seed weight (g)	Rind thickness (cm)	TSS (°Brix)	**Acidity (%)**	**Reducing sugar (%)**	**Vitamin-C (mg/100g)**	
KA- 01	Karnataka	Oblong	12.28	1.88	26.45	185.69	108.65	8.40	0.42	-	-	-	-	Lamani and Murthy (2022)
KA- 03		Round	12.38	2.64	33.30	146.64	79.42	5.90	0.41	-	-	-	-	
KA- 04		Round	10.12	2.34	31.23	171.49	94.24	9.71	0.42	-	-	-	-	
KA- 06		Round	12.56	2.26	32.62	160.35	90.66	14.55	0.42	-	-	-	-	
KA- 11		Triangular	12.73	3.40	30.21	159.65	89.90	9.18	0.44	-	-	-	-	
AP- 02	Andhra Pradesh	Round	14.88	3.57	23.40	138.73	87.04	7.28	0.41	-	-	-	-	
AP- 03		Flat round	11.81	1.87	33.55	175.09	102.87	11.10	0.42	-	-	-	-	
AP- 04		Round	13.28	2.52	32.81	118.30	70.19	8.20	0.41	-	-	-	-	
TS- 01	Telangana	Oblong	14.25	3.29	23.93	203.92	119.44	14.06	0.44	-	-	-	-	
TS- 06		Oblong	13.01	2.50	24.48	160.52	100.47	10.05	0.39	-	-	-	-	
TS- 07		Round	10.28	2.28	32.96	186.30	112.40	14.89	0.41	-	-	-	-	
TN- 01	Tamil Nadu	Round	13.35	2.70	34.19	153.16	111.63	9.06	0.38	-	-	-	-	
TN- 03		Flat round	11.13	1.47	35.10	233.42	151.82	9.98	0.41	-	-	-	-	
TN- 04		Oblong	13.19	3.43	31.12	138.06	82.41	7.77	0.39	-	-	-	-	
WA-4	West Bengal	Oblong	-	-	-	206.33	112.26	15.66	0.27	22.08	1.51	0.85	7.45	Dowarah et al. (2021)
WA-5		Oblong	-	-	-	195.13	109.26	19.63	0.26	18.50	1.31	1.38	10.53	
WA-10		Round	-	-	-	161.80	81.50	9.07	0.33	11.16	1.60	0.83	16.02	
WA-11		Round	-	-	-	225.73	117.54	37.94	0.25	18.50	1.80	1.02	13.17	
WA-12		Oblong	-	-	-	184.26	110.30	16.47	0.33	18.50	1.79	0.93	14.24	
Acc.5	West Bengal	Flat round	-	-	-	281.30	177.33	-	0.33	17.87	5.15	6.22	17.76	Mani et al. (2017)
Acc.9		Round	-	-	-	283.08	184.67	-	0.36	18.43	4.33	6.49	18.43	

GW-6	Gujarat	Flat round	-	-	-	320.53	214.45	5.71	0.46	15.65	4.27	-	7.08	Singh et al. (2016)
GW-10		Flat round	-	-	-	314.83	213.63	5.20	0.47	19.05	4.18	-	13.20	
GW-13		Flat round	-	-	-	298.78	221.88	3.01	0.38	17.53	4.51	-	7.96	
GW-15		Oblong	-	-	-	418.35	319.32	2.98	0.36	17.17	3.42	-	17.68	
GW-16		Round	-	-	-	400.03	295.87	6.29	0.42	13.83	4.11	-	12.57	
GW-17		Flat round	-	-	-	384.50	270.55	6.87	0.43	11.07	3.53	-	13.89	
GW-18		Round	-	-	-	432.93	317.74	6.27	0.39	16.83	4.27	-	14.78	
BS-1	West Bengal	-	-	-	-	202	122.41	17.57	-	16.4	1.2	8.8	4.9	Ghosh et al. (2012)
JS-4		-	-	-	-	225	103.95	15.52	-	18.4	3.9	5.0	13.0	
JS-7		-	-	-	-	130	52.52	14.56	-	16.8	3.6	4.1	8.5	
LS-1		-	-	-	-	217	123.90	27.55	-	16.0	1.7	6.1	2.7	
LS-2		-	-	-	-	140	69.44	10.78	-	15.0	4.6	2.6	11.2	

fruit plants, such as decreased storage space demands, the preservation of genetic stability without affecting viability, extended storage periods, and germplasm that is free from pests, diseases and insects. As a result, they propose that biotechnological methods might be an alternate approach to conservation, which promotes the utilization of genetic diversity and enables the future use of genetic diversity.

The phenotypic variability in wood apple, including canopy size of the tree, fruit weight, rind weight, number of seeds, and percentage composition of pulp and seeds, was recorded by Lamani et al. (2022) through the analysis of 32 populations from South India based on selected qualitative and quantitative traits. The height of the tree ranged from 7.68–16.69 m, the canopy area was 18.78–36.36 m^3 whereas, the fruit weight ranged from 118.30–233.42 g, the pulp percentage was 53.91–72.93% and the seed percentage was 4.02–9.07% (TN-03). The variability noted among all the populations was used for multivariate statistical analysis and formed into 4 clusters. They recorded that the results of populations considered under clusters 1 and 2 are more diverse as compared to clusters 3 and 4. TN-03 is the most promising population based on the overall assessment, while the populations grouped under clusters 1 (AP-03, TS-01, TS-07, and TN-05) and cluster 2 (TN-03) revealed elite fruit-related traits by having the highest fruit, pulp, and seed weight despite having average tree morphology (Table 3). Hence, the population that showed greater variability could be considered as a germplasm and conserved for utilization for cultivation and future breeding programs. Hence to conserve genetic purity it is suitable to multiply by the vegetative propagation techniques, especially the grafting in the months of June to September. As suggested in the above-cited research approaches it is better to go for vegetative propagation techniques to retain the superior traits without any variations and such populations can be used as germplasm for the utilization of cultivation and future breeding programs.

Cultivars Released

Wood apple is abundant in arid plains and monsoon climates with distinct dry seasons. These are extremely drought-tolerant and grow on light soils. Morton (2002) observed different varieties of wood apple based on morphological and sensory evaluation of trees and these were classified into two varieties, one with larger fruits sweeter in taste and another one is smaller fruits that are acidic/sour in taste. Some of the wood apple varieties are differentiated based on the density of thorns, such as branches with high and low density of thorns. Apart from some basic reports no known report was available on cultivars/varieties available in wood apple. Wood apple are neglected since they are not available on the market for edible or commercial purposes, and there is a lack of improved cultivars/varieties. In terms of their economic potential, many of these are beneficial to human health in various places, such as offering food/fodder for animals, nutritionally balanced diets, and enhancing local revenue (Murthy and Dalawai, 2020). Wild fruit species contribute significantly to the sustainability of ecosystems.

Conservation of Useful Germplasm

To maintain the diversity observed and suggested by the prior research investigations, it is critical need of protect and conservation wood apple species both *in situ* and *ex situ* to maintain the germplasm to promote sustainable development based on the use of available genetic diversity and protect the habitats where they usually exist. Conservation of germplasm promotes crop improvement aiding more ways for varietal development by plant breeding techniques; hence it is necessary to conserve the wood apple plant species for future utilization (Pujari, 2015). Through the observation of species variability, it is possible to identify elite types that can be propagated true to type on a wide scale through vegetative propagation. After evaluating one-year-old desi rootstock for propagation, they concluded that the best approach was softwood grafting. The comparative study among *in situ* soft wood grafting and nursery grafting, followed by field transplantation revealed that *in situ* grafting proved more potent in terms of growth characteristics than nursery grafting (Hiwale et al., 2008). According to Pujari (2015) biotechnological approaches of germplasm conservation have great potential over the conventional method for underutilized fruit species. These approaches include cryopreservation, slow-growth cultures, artificial seeds/synthetic seeds and DNA clones. Hence the method of *in vitro* conservation of explants of wood apple by cryogenic storage of explants under controlled conditions can lead to sustained storage of germplasms without losing viability and genetic stability. The *Limonia acidissima* populations, which include the superior population reported by Ghosh et al. (2012), Singh et al. (2016), Dowarah et al. (2021), and Lamani and Murthy (2022), are all found in natural forests and protected areas that are under the control of the respective state governments. The commercial cultivation and preparation of these plants for value addition could provide financial benefits for small and marginal farmers as well as tribe members. This could inspire these people to cultivate or maintain forests and other areas as a way of supporting on-farm conservation. These genetic resources must be conserved, improved, diversified, modified, and distributed to farmers for them to remain sustainable.

Conclusion

Wood apple genotypes were selected based on horticultural traits including early bearing, reduced spine intensity, earliness, regular bearing, dwarf tree structure, fruit with a strong aroma and fruit characters. Wood apple fruits contain 58.73% pulp, 1% oil, and 6.44% seeds, containing 32.02% seed oil. Fruits are rich in proteins, lipids, carbohydrates, amino acids, minerals and other bioactive components, making them an essential source of nutrients and natural antioxidants. The studies on biochemical analysis revealed that fruits are rich in essential fatty acids, sugars, organic acids and polyphenols, all of which contribute to their taste and quality. Furthermore, seeds are an excellent source of important lipids and vitamin E, making them useful for cooking, frying, and preparation of cosmetics. Wild fruits are especially significant for individuals who reside in rural, forest, or tribal locations. Wood apple, an indigenous tropical and subtropical fruit plant are still underutilized due to a lack of knowledge about their potential, accessibility and geographical specificity. Climate change, increasing urbanization, and deforestation caused by land use changes are all affecting the genetic diversity of these fruits. No named cultivars are available to date. However sour and sweet types, high yielders with big fruit sizes exist in the variable seedling progenies in nature. Fruits of selected potential genotypes were selected for propagation through vegetative means. At present though seed propagation is done mainly, budding has been reported to be successful if done during late summer and early monsoon. This chapter not only addresses growth traits and biochemicals, but also genetic variability and promising genotypes for breeding programs.

There have been only a few investigations performed in India on both the genetics and morphology of wood apple. The genetic and morphological diversity of wood apples is currently understood to exist in a few states, namely West Bengal, Gujarat, Karnataka, Andhra Pradesh, Tamil Nadu, Kerala, and Telangana. The investigation needs to be focused on further regions and genotypes within the nation. In addition to this, further research into the life cycle, floral biology, and fruit setting of these varieties may be included in the future research program. The advanced selection criteria can be used to select desirable high-yielding vigorous genotypes, early flowering and more biotic and abiotic stress-resistant wood apple genotypes.

References

Ahmad, A., L.N. Mishra, and R.S. Thakur. 1989. Composition of the volatile oil from *F. limonia* leaves. *Planta Med., 5*: 55–99.

Bandara, B.M., A.A.L. Gunatilaka, E.M.K. Wijeratne, and N.K.B. Adikaram. 1988. Antifungal constituents of *Limonia acidissima*. *Planta Med., 54*: 374–375.

Bhargava, R., K. Gurjar, A.K. Singh, and S. Singh. 2019. Genetic diversity of wood-apple (*Feronia limonia* L.) revealed by random amplified polymorphic. *Int. J. Chem. Stud.*, *7*(1): 208–212.

Brown, W.L. 1983. Genetic diversity and genetic vulnerability: An appraisal. *Economic Botany*, *37*: 4–12.

Chowdhury, M.G.F., M.N. Islam, M.S. Islam, A.F.M.T. Islam, and M.S. Hossain. 2008. Study on preparation and shelf-life of mixed juice based on wood apple and papaya. *J. Soil Nat.*, *2*(3): 50–60.

Dalawai, D., and H.N. Murthy. 2021. Chemical Profile and Antioxidant Properties of *Andrographis producta* (C.B. Clarke) Gamble. *Pharmacogn. J.*, *13*: 475–485.

Darsini, D.T.P., V. Maheshu, M. Vishnupriya, S. Nishaa, and J.M. Sasikumar. 2013. Antioxidant potential and amino acid analysis of underutilized tropical fruit *Limonia acidissima* L., Free Radic. *Antioxid.*, *3*: 62–69.

Debbarma, R., S. Diengngan, and R. Kambale. 2019. Utilization of biotechnological approaches for germplasm conservation and sustainable use of underutilized fruit species in India. *Acta Hortic.*, *1241*: 159–164.

Dowarah, B., P. Deb, G. Rana, and K. Sushmitha. (2021). Fruit physico-chemical studies of some local wood apple (*Limonia acidissima* L.) genotypes. *International Journal of Minor Fruits, Medicinal and Aromatic Plants*, *7*: 53–56.

Engler, A. 1896. Rutaceae. *In*: Engler, A. and Prantl, K. [Eds.], *Die natürlichen Pflanzenfamilien, III* (*4*): 95–201. Leipzig: Engelmann.

Ghosh, S.N., A.K. Banik, B.C. Banik, B. Bera, S. Roy, and A. Kundu. 2012. Conservation, multiplication and utilization of wood apple (*Feronia limonia*): A semi-wild fruit crop in West Bengal (India). *Acta Hortic*., Article 948: 279–283.

Govindaraj, M., M. Vetriventhan, and M. Srinivasan. 2015. Importance of genetic diversity assessment in crop plants and its recent advances: An overview of its analytical perspectives. *Genet. Res. Int.* https:// doi.org/10.1155/2015/431487.

Groppo, M., J.A. Kallunki, J.R. Pirani, and A. Antonelli. 2012. Chilean pitavia more closely related to Oceania and old World Rutaceae than to neotropical groups: Evidence from two cpDNA non-coding regions, with a new subfamilial classification of the family. *PhytoKeys*, *19*: 9–29.

Gunaseeli, B.P. and R. Sampathkumar. 1986. On the karyomorphological features between *Limonia acidissima* Linn. and Poncirus trifoliate (L) Raf. *Proc. Indian Sci. Congr. Assoc.*, *73*: 88.

Gupta, R., S. Johri, and A.M. Saxena. 2009. Effect of ethanolic extract of *Feronia elephantum* Correa fruits on blood glucose levels in normal and streptozotocin induced diabetic rats. *Nat. Prod. Radiance*, *8*: 32–36.

Hiwale, S.S., D.G. Dhandar, and B.G. Bagle. 2008. Vegetative propagation of Wood Apple (*Feronia limonia* Correa). *Indian J. Agrofor*., *10*(2): 58–61.

Ilaiyaraja, N., K.R. Likhith, G.R. Sharath Babu, and F. Khanum. 2015. Optimisation of extraction of bioactive compounds from *Feronia limonia* (wood apple) fruit using response surface methodology. *Food Chem.*, *173*: 348–354.

Ilango, K., and V. Chitra. 2009. Hepatoprotective and antioxidant activities of fruit pulp of *Limonia acidissima* Linn. *Int. J. Health Res.*, *2*: 361–367.

Ilango, K., and V. Chitra. 2010. Wound healing and anti-oxidant activities of the fruit pulp of *Limonia acidissima* Linn (Rutaceae) in rats. *Trop. J. Pharma. Res.*, *9*: 223–230.

Jayakumar, A., and K. Geetha. 2012. Wood apple uses and benefits. *Facts for You*, *12*: 23–24.

Jha, T.B., and S. Nath. 2016. EMA based chromosome analysis in *Bixa orellana* L. and *Limonia acidissima* L. *Nucleus*, *59*: 93–98.

Jha, T.B., and M. Yamamoto. 2012. Application of EMA fluorescent staining and FISH of rDNA in analysis of *Aloe vera* (L.) Burm. f. chromosomes. *Bull. Fac. Agric. Kagoshima Univ.*, *62*: 83–89.

Joseph, K.S., S. Bolla, K. Joshi, M. Bhat, K. Naik, S. Patil, S. Bendre, G. Brunda, V. Haibatti, S. Payamalle, S. Shinde, Y.H. Dewir, and H.N. Murthy. 2016. Determination of chemical composition and nutritive value with fatty acid compositions of African Mangosteen (*Garcinia Livingstonei*). *Erwerbs-Obstbau.*, *59*: 195–202.

Joshi, P., and S. Jain. 2008. Products from wood apple (*Limonia acidissima*) fruit and their quality evaluation. *J. Food Sci. Tech*., *45*: 270–271.

Kubitzki, K., J.A. Kallunki, M. Duretto, and P.G. Wilson. 2011. Rutaceae. *In*: Kubitzki K. [Ed.], *The Families and Genera of Vascular Plants* (Vol. 10), 276–356. Berlin: Springer.

Lakshmi, Y., A. Ushadevi, and R. Baskaran. 2015. Post-harvest ripening changes in wood apple (*Feronia elephantum* Corr), an underutilized fruit. *Int. J. Fruit Sci*., *15*(4): 425–441.

Lamani, S., K.A. Anu-Appaiah, H.N. Murthy, Y.H. Dewir, and H.Z. Rihan. 2021. Fatty acid profile, tocopherol content of seed oil, and nutritional analysis of seed cake of wood apple (*Limonia acidissima* L.), an underutilized fruit-yielding tree species. *Hortic.*, *7*: 275.

Lamani, S., K.A. Anu-Appaiah, H.N. Murthy, Y.H. Dewir, and J.J. Rikisahedew. 2022. Analysis of free sugars, organic acids, and fatty acids of wood apple (*Limonia acidissima* L.) fruit pulp. *Hortic.*, *8*: 67.

Lamani, S., and H.N. Murthy. 2022. Diversity of wood-apple (*Limonia acidissima* L., Rutaceae) genetic resources in South India. *Genet. Resour. Crop Evol.*, *69*: 2929–2946.

Lata, A., H.K. Yadav, and N.K. Nair. 2023. Genetic diversity and population structure analysis in wood-apple (*Limonia acidissima* L.) using simple sequence repeat (SSR) markers. *Genet. Resour. Crop Evol.* https://doi.org/10.1007/s10722-023-01699-1.

Mani, A., A.K. Singh, N. Jain, and S. Misra. 2017. Flowering, Fruiting and Physio-chemical Characteristics of Bael. *Curr. Appl. Sci. Technol.*, *23*(3): 1–8.

Morton, J. 1987. Wood-Apple. *In*: *Fruits of Warm Climates*, 190–191. Julia F. Morton, Miami, FL. Morton, J. 2002. Wood apple. *Fruits of Warm Climates*. 190–191.

Murrinie, E.D., P. Yudono, A. Purwantoro, and E. Sulistyaningsih. 2017. Morphological and physiological changes during growth and development of wood apple (*Feronia limonia* (L.) Swingle) fruit. *Int. J. Botany*, *–13*: 75–81.

Murthy, H.N., and D. Dalawai. 2020. Bioactive compounds of wood apple (*Limonia acidissima* L.) *In*: Murthy, H.N., Bapat, V.A. [Eds.], *Bioactive Compounds in Underutilized Fruits and Nuts,* 544–569. Springer Nature, Cham, Switzerland.

Narsing Rao, G., P.G. Prabhakara Rao, and D. Govardhana Rao. 2011. Preparation of wood apple (*Feronia limonia* L.) seed protein concentrate and evaluation of its nutritional and functional characteristics. *Int. Food Res. J.*, *18* (3): 949–955.

Nevo, E., A. Beiles, and R. en-Shlomo. 1984. The evolutionary significance of genetic diversity: Ecological, demographic and life history correlates. *Lect. Notes Biomath.*, *53*: 13–21.

Nithya, N., and U. Saraswathi. 2010. *In vitro* antioxidant and antibacterial efficacy of *Feronia elephantum* Correa fruit. *Indian J. Nat. Prod. Resour.*, 301–305.

Orwa, C., A. Mutua, R. Kindt, R. Jamnadass, and S. Anthony. 2009. *Agroforestry Database: A Tree Reference and Selection Guide Version 4.0*. World Agroforestry Centre, Nairobi, Kenya.

Pande, C., G. Tewari, C. Singh, S. Singh, and R.C. Padalia. 2010. Chemical composition of the essential oil of Feronia elephantum Correa. *Nat. Prod. Res.*, *24*(19): 1807–1810.

Pandey, S., S. Gouri, and R.K. Gupta. 2014. Evaluation of nutritional, phytochemical, antioxidant and antibacterial activity of exotic fruit *Limonia acidissima*. *J. Pharmacognosy Phytother.*, *3*(2): 81–88.

Patel, D., R. Kumar, D. Laloo, and S. Hemalatha. 2012. Natural medicines from plant source used for therapy of diabetes mellitus: An overview of its pharmacological aspects. *Asian Pac. J. Trop. Dis.*, *2*(3): 239–250.

Patel, H.B. 2013. Master of technology. Production technology of wood apple (*Feronia limonia*) juice, 95pp. College of Food Processing Technology and Bioenergy, Anand Agricultural University, Anand 388 110.

Phapale, R., and S.M. Thakur. 2010. Antioxidant activity and antimutagenic effect of phenolic compound in *Feroia limonia* (L.) fruit. *Int. J. Pharm. Sci.*, *2*(4): 68–73.

Pradhan, D., G. Tripathy, and S. Patanaik. 2012. Anticancer activity of *Limonia acidissima* Linn (Rutaceae) fruit extracts on human breast cancer cell lines. *Trop. J. Pharm. Res.*, *11*: 413–419.

Pujari, S. 2015. Germplasm conservation of different plant species. www.yourarticlelibrary.com/. /33255.

Ramesha, B.T., T. Amna, G. Ravikanth, R.P. Gunaga, R. Vasudeva, K.N. Ganeshaiah, R. Uma Shaanker, R.K. Khajuria, S.C. Puri, and G.N. Qazi. 2008. Prospecting for camptothecines from *Nothapodytes nimmoniana* in the Western Ghats, South India: Identification of high-yielding sources of camptothecin and new families of camptothecines. *J. Chromatogr. Sci.*, *46*: 362–368.

Ratnayake, R.M.R.N.K., H.J. Sumithra, M.D. Fernando, and K.B. Palipane. 2009. Effect of GRAS compounds on Aspergillus rot of wood-apple (*Feronia limonia*). *Phytoparasitica*, *37*: 431–436.

Schlautman, B., L. Diaz-Garcia, and S. Barriball. 2020. Morphometric approaches to promote the use of exotic germplasm for improved food security and resilience to climate change: A kura clover example. *Plant Sci.*, *2019*: 110319. https:// doi. org/ 10.1016/j.plant sci. Senthilkumar, A., V. Venkatesalu, K. Kannathasan, and M. Chandrasekaran. 2010. Chemical constituents and antibacterial activity of the leaf essential oil of *Feronia limonia*. *Indian J. Microbiol.*, *50*(1): 70–73.

Singh, A.K., S. Sanjay, and T.A. More. 2014. Preliminary evaluation of bael varieties under rainfed conditions of hot semi-arid ecosystem of western India. *Indian J. Hortic.*, *71*(2): 264–268.

Singh, A.K., S. Singh, V. Yadav, and B.D. Sharma. 2016. Genetic variability in wood apple (*Feronia limonia*) from Gujarat. *Indian J. Agric. Sci.*, *86*(11): 1504–1508.

Singhal, V.K., B.S. Gill, and S.S. Bir. 1980. Chromosome number reports LXVII. *Taxon.*, *29*: 355–337.

Sonawane, S.K., M.B. Bagul, J.G. LeBlanc, and S.S. Arya. 2016. Nutritional, functional, thermal and structural characteristics of *Citrullus lanatus* and *Limonia acidissima* seed flours. *J. Food Meas. Charact.*, *10*: 72–79.

Sonawane, S., and S.S. Arya. 2013. Antioxidant activity of Jambhul, wood apple, Ambadi and Ambat Chukka: An indigenous lesser known fruits and vegetables of India. *Adv. J. Food Sci. Technol.*, *5*(3): 270–275.

Turner, G.W., A.M. Berry, and E.M. Gifford. 1998. Schizogenous secretory cavities of *Citrus limon* (L.) Burm.f. and a revaluation of the lysigenous gland concept. *Int. J. Pl. Sci.*, *151*: 75–88.

Verheij, E.W.M., and R.E. Coronel. 1992. Plant Resources of South-East Asia No. 2 Edible Fruits and Nuts, 190–191. Prosea, Bogor, Indonesia.

Vidhya, R. and A. Narain. 2011. Formulation and evaluation of preserved products utilizing under exploited fruit, wood apple (*Limonia acidissima*). *American-Eurasian J. Agric. Environ. Sci.*, *10*(1): 112–118.

Vijayvargia, P., and R. Vijayvergia. 2014. A review on *Limonia acidissima* L.: multipotential medicinal plant. *Int. J. Pharm. Sci. Rev. Res.*, *28*(1): 191–195.

Vyas, S., A. Gaur, A.K. Tyagi, S.D. Purohit. 2008. Assessment of genetic diversity in *Feronia Limonia* (L.) Swingle using inter simple sequence repeats. *J. Sustain. For.*, *27*: 328–342.

Wang, Z., B. Fang, J. Chen, X. Zhang, Z. Luo, L. Huang, X. Chen, and Y. Li. 2010. *De novo* assembly and characterization of root transcriptome using Illumina paired-end sequencing and development of cSSR markers in sweet potato (*Ipomoea batatas*). *BMC Genomics*, *11*: 726.

Wang, J.M., J.M. Yang, J.H. Zhu, Q.J. Jia, and Y.Z. Tao. 2010. Assessment of genetic diversity by simple sequence repeat markers among forty elite varieties in the germplasm for malting barley breeding. *J. Zhejiang Univ. Sci. B.*, *11*(10): 792–800.

Weising, K., H. Nybom, K. Wolff, and G. Kahl. 2005. *DNA Fingerprinting in Plants: Principles, Methods, and Applications* (2nd Edn.), 472 pp. CRC Press, Boca Raton.

Zhao-Yang, Z., S. Hang, and G. Zhi-Jian. 2002. Karyomorphological study of the *Spiraea japonica* complex (Rosaceae). *Brittonia*, *54*: 168–174.

5

Genetic Diversity of Cocoa (*Theobroma cacao* L.) and Sustainable Utilization

Elain Apshara, S.[1,*] and *Anuradha Sane*[2,*]

Introduction

Cocoa (*Theobroma cacao* L.) is known as the 'Food of the Gods'. It yields beans, which are the economic part and, when fermented and dried, constitute the primary ingredient in chocolate. The indigenous people of the Amazon basin of South America, the Mayas, Aztecs, and Pipil-Nicaraos, roasted and crushed cocoa beans to make an energy drink called 'xocoatl', which was then mixed with sugar and milk by the Spanish in Europe and changed into 'chocolate', the world's favourite item. According to the World Cocoa Foundation (WCF), cocoa is widely distributed and provides livelihood security for around 40–50 million people, with a total production of 5.8 million tonnes, of which Africa contributes 70%, Latin America 11%, and Asia and Oceania 19%. Ivory Coast, Ghana, Cameroon, Nigeria, and Brazil are the major cocoa producers in the world, followed by Indonesia and Malaysia in the Asian region (www.wcf.org.). Cocoa is a renowned beverage crop and happy food, with its nutritional and health benefits from dark chocolates (Wood, 1964).

Cocoa was introduced into India in 1798 at Courtallam in Tirunelveli district of the old Madras state (Ratnam, 1961). It spread mainly in the Western Ghats Hills and Plains region, where the climate was subtropical humid with many rainy days and brief dry periods (Wood, 1964). Due to the limited land in the lower hills, commercial cultivation of cocoa expanded in the 1970s under palms across Kerala, Karnataka, Tamil Nadu, and Andhra Pradesh. The area under cocoa cultivation is now 1.5 lakhs ha, but the production is only 24 mt. (www.dccd.nic.in) far below the 60,000 mt. demand of the Indian chocolate industry. Cocoa is a perennial tree crop having distinct growth habits, adapted to different cropping systems in different regions, has diverse fruit characteristics, is highly responsive to changing climates and growing environments and so dynamic, systematic, and thus it needs long-term breeding strategies (Bekele, 2001).

[1] ICAR – Central Plantation Crops Research Institute-CPCRI, Regional Station, Vittal- 574 243, Karnataka. Email: elainapshara@gmail.com

[2] ICAR – Indian Institute of Horticultural Research-IIHR, Bangalore.

* Corresponding author

Taxonomic Background

Cocoa belongs to the kingdom Plantae, sub-kingdom Tracheobionta, division Magnoliophyta, subclass Delleniidae, and order Malvales. The only cultivated species of cocoa is *Theobroma cacao* L., a diploid (2n = 20), originally classified under Sterculiaceae and reclassified under the Malvaceae family, with a small genome of 380 Mbp (Lanaud et al., 1992; Figueira et al., 1992). There are 22 identified species in this genus, including *Theobroma angustifolium, T. bicolor, T. grandiflorum, T. mammosum, T. microcarpum, T. simiarum, T. speciosum,* and *T. subincanum. Theobroma grandiflorum* is widely consumed as 'cupuacu' in Brazil for its flavored mucilage. Only a few species are used in breeding programs.

Bartley (2005) documented the diversity among cocoa populations based on the degree of human involvement in the establishment of cocoa groups. Cocoa populations are categorized into (i) natural or spontaneous, (ii) sub-spontaneous, where the population is established by human action but has no evidence of exploitation, and (iii) cultivated or domesticated populations formed by human action and under continuous utilization. There are three basic types of cocoa available: Criollo, Forastero, and Trinitario, which have specific pod and bean characteristics and form the old plantations as well as the new varieties developed by researchers (Wood and Lass, 1955). *Theobroma cacao* consists of Criollo populations of Central and South America and *Theobroma cacao* subsp. *sphaerocarpum* includes other populations such as Forastero and Trinitario (Cuatrecasas, 1964).

Criollos

Criollos have red or purple pods, elongated fruits with sharp apex, rough surfaces, and hard husks. The fresh beans are white or pinkish, dry beans are cinnamon colored after three days of fermentation, large and thick, 20–30 in number. Trees are slow-growing and susceptible to pests and diseases. Pure Criollos are found only in isolated groups, characterized by specific flavors, and are called 'fine' cocoa. Colombian, Mexican, Nicaraguan Criollo or Cacao Real, and Pentagona or Lagarto were described based on their origin and distinct characteristics (Soria, 1970).

Forasteros

Forasteros have green to yellow pods, rounded smooth fruits with hard husks, and fresh beans that are dark purple, turning to dark chocolate brown after six days of fermentation, flat, and more than 30 in number. Trees are fast-growing and tolerant to pests and diseases. Forasteros are generally referred to as basic cocoa or 'bulk' cocoa, suitable for the preparation of mild chocolates. Amelonado, Comum, West African Amelonado, Cacao Nacional, Matina or Ceylan, Guiana Wild Amelonado, and Amazon populations are grouped under Forasteros (Toxopeus, 1985).

Trinitarios

Trinitarios are natural hybrids of Criollos and Forasteros from Trinidad. They are mostly cultivated types and have characteristics of both types (Cheesman, 1944). They are vigorous and distributed across many countries. Trinitarios in Southeast Asia and Oceania have largely contributed to the cocoa economy in the region.

The diversity available with cocoa throughout the world has been widely investigated. Cocoa breeding is one of the oldest crop improvement programs, beginning in 1930 with the Cocoa Research Unit (CRU) in Trinidad and continuing until 1944 with the West African Cocoa Research Institute (WACRI), Cocoa Research Institutes of Ghana (CRIG), and Nigeria (CRIN). Cameroon (IRAD) and France (CIRAD) began cocoa improvement in the 1950s, and in 1963, Brazil and Ecuador (INIAP) strengthened resistance breeding programs with the establishment of the Executive Commission for the Cocoa Development Plan (CEPLAC). The Malaysian Cocoa Board (MCB), Indonesian Coffee and Cocoa Research Institute (ICCRI), and Papua New Guinea's Coconut and Cocoa Research Institute (CCRI) all began systematic breeding programs in the 1980s.

Collection and Assessment of Germplasm

The basic step in any breeding program is the collection of germplasm from both primary and secondary centers of origin and distribution. Cocoa in its native zones of South and Central America and other major producing countries of Africa suffers from many debilitating diseases caused by both viral and fungal pathogens and serious pests. To safeguard germplasm exchange programs and ensure continuous breeding activities, cocoa quarantine centers were established, one of which is the Intermediate Cocoa Quarantine Centre (ICQ, C) in Europe. The University of Reading took over the responsibility for cocoa quarantine and created the facility in 1987 (Hadley and Lee, 1992). The centre collects important clones and wild collections from international and national gene banks, performs virus indexing, and conducts strict quarantine for other major diseases and pests as per the guidelines for the safe movement of germplasm described by FAO/Bioversity International (Frison and Feliu, 1989; End et al., 2014). Clones that have cleared the quarantine are being supplied as bud sticks to cocoa researchers from over 20 countries.

In India, under the Indian Council of Agricultural Research (ICAR), the Central Plantation Crops Research Institute (CPCRI) is the national institute dedicated to cocoa research. It initiated systematic cocoa research in 1969 by introducing germplasm from Malaysia and Nigeria and later through ICQ, C. ICAR-NBPGR (National Bureau of Plant Genetic Resources) is the nodal agency within the country for all imports, whereas the export of germplasm is governed by the National Biodiversity Authority (NBA, Chennai). More than 500 collections are being maintained in the National Active Germplasm Site (NAGS) for cocoa at CPCRI, Regional Station, Vittal, Karnataka. The germplasm collections include clones from the Amazon, Brazil, Ecuador, England, Ghana, Jamaica, Mexico, Nigeria, and Peru and local collections from the hills of Kallar in Tamil Nadu, Wayanad, Idukki in Kerala, and Shiradi Ghats of Karnataka. All these are conserved in field gene banks (FGBs) and assessed and evaluated for their precocity, adaptability, compatibility reaction, yield stability, productivity, and quality in the introduced environment.

Morphological Characterization

The morphological expression of diversity is estimated from different indicators of variability, especially with respect to observable traits. Cocoa has a specific plant habit of branching in tiers. Epigeal germination continues with vertical or orthotropic growth of the main shoot. At a certain height, five lateral plagiotropic branches or fans are formed, and this process is called 'jorquetting'. Further vertical growth continues on the main stem by the activation of orthotropic buds in the leaf axils below the whorl of fan branches, which are called 'chupons'. This pattern continues until the tree reaches an average height of 20 m in the wild in multiple tiers. However, in inter/mixed cropping systems, a single-tier canopy is maintained and managed with proper training through formation and structural pruning measures. Genotypes with medium vigor, small stature, and canopy spread with optimum pod yield are being selected for cultivation.

Zhang et al. (2009) classified cocoa into 12 groups based on geographical origin and distinct morphological traits. Bioversity International has standardized the descriptor status for cocoa, comprising 60 characteristics. Turnbull and Eskes (2010) developed a visual aid to identify widely distributed cocoa accessions with a minimal descriptor of 20 characters. UPOV developed DUS (Distinct Uniform Stable) testing guidelines for cocoa in 2013 with 30 characteristics. Cataloging, characterization, and passport data documentation have been undertaken with collections at CPCRI (Elain Apshara and Rajan, 2009), and national identity numbers (IC-Indigenous Collection No./

EC-Exotic Collection No.) were obtained from the National Bureau of Plant Genetic Resources (NBPGR), New Delhi for further utilization in the breeding program. ICAR-CPCRI also developed DUS guidelines for cocoa with a total of 25 characteristics comprised of three each of leaf and flower characters, 12 fruit and six bean traits and fat content as qualitative parameters and with 69 reference varieties with Protection of Plant Varieties and Farmers Rights Authority (PPV&FRA) in 2022 and notified in the Gazette of India which help in registration of cocoa varieties by breeders and farmers (Table 1).

Table 1. Selective indicators of morphological variability in cocoa.

Plant habit	Erect, Intermediate, and Pendulous growth habits
Leaves	Young flush color, leaf apex, and leaf base
Flowers	Color, petal diameter, pedicel length, pollination behavior
Fruits	Cherelle color, ripe pod color Pod shapes: Orbicular/Ovate/Obovate/Oblong/Elliptic
Basal constriction	Absent/Intermediate/Present
Apex form	Indented/Rounded/Obtuse/Acuminate/Acute/Mammilate/Attenuate
Surface rugosity	Absent/Medium/Rugose/Highly rugose
Prominence of ridges	Absent/Shallow/Deep
Pairing of ridges	Single/Fused/Paired
Pod size	Small/Medium/Large
Husk thickness	Thin/Medium/Thick
Beans	Shape, Color, Size
Wood and Lass (1985); Bartley (2005); PPVFRA (2023).	

Cytological Characterization

The cytology of cocoa is typical of the Sterculiaceae family. Cytogenetic investigations indicated that all of the examined *T. cacao* accessions are diploid (2n = 2x = 20). The genome of *T. cacao* is diploid. Davie (1935) and Carletto (1946) were the first to record the chromosomal number of several species in the *Theobroma* genus, with Cacau Catongo, Cacau Criollo, and Cacau Tigre having 2n = 20. The "Cupuaçu tree" (*Theobroma grandiflorum* Schum.) has an asymmetrical karyotype that includes metacentric, submetacentric, telocentric, and acrocentric chromosomes, as well as a satellite chromosome pair (Santos, 2002). *T. cacao* exhibits chromosomal variability, with chromosomes classified by length. The chromosomal banding held by *T. cacao* exhibits chromosomal variability, with chromosomes classified by length (Muñoz, 1948). A comparative investigation of *T. cacao* and *T. grandiflorum* found that both species have symmetrical karyotypes with tiny metacentric chromosomes. Until now, the presence of a satellite chromosome pair was regarded as a cytological marker for the species *Theobroma*, although it has also been observed in meiotic chromosomes and conjunction with nucleoli (Glicenstein and Fritz, 1989; Dantas and Guerra, 2010).

Within the genus *Theobroma*, cytogenetic analyses have shown karyotypes with modest chromosome lengths: 1.19–2.0 µm in *T. cacao*, 1.15–2.21 µm in *T. grandiflorum* (Dantas and Guerra, 2010), and 9.1–1.69 µm in *T. speciosum*. Between 1.14–3.23 µm, there was more variation in chromosomal lengths among the accessions that were studied. In the diploid, bivalent formation was seen, and the disjunction was regular despite the low apparent chiasma frequency. Bivalents were more common than trivalents and quadrivalents in the two established auto tetraploids under study (Martinson, 1975). Reassociation kinetics analysis of *T. cacao* revealed a genome size of 2.01 x 108, which is small in comparison to previously studied temperate and tropical plants. Additionally, information was provided on the *T. cacao* DNA's melting point, base composition, and relative degree of methylation (Couch et al., 1993).

Biochemical Characterization

Trait-specific improvements are being considered in current breeding programs. A bean index of 100 beans/100 g, i.e., dry bean weight of 1 g, is considered standard for grade I beans (GOI, 1997); therefore, selections are aimed at larger bean sizes of 1 g and above, which will have high butter content suitable for the chocolate industry. Bean size is influenced by genotype, environment, and processing methods. Variation in bean indices was observed across collections, and the single dry bean weight ranged from 0.7–2.5 g. To improve the qualitative parameters, the Criollo and Trinitario types are used in hybridization programs. Cocoa is considered a functional food; therefore, dark chocolates are consumed for their health benefits. Catechin, epicatechin, and procyanidins are the main polyphenols present in cocoa that contribute to bitterness, astringency, and the organoleptic quality of cocoa (Counet et al., 2004). Cocoa beans of different clones evaluated at CPCRI for polyphenols and antioxidant activity exhibited distinct differences. Polyphenols ranged from 82–136 mg/g, procyanidin 49–64 mg/g, fat content from 24%–55%, and antioxidant activity from 77%–98% among the cocoa clones studied. In general, cocoa beans with high polyphenol and procyanidin contents exhibited high antioxidant activities that can be used for qualitative improvement. Cocoa butter with free fatty acid content (FFA) of 1% or less together with an acceptable flavor is the best indication of good quality beans. Prolonged storage may cause fat degradation and an increase in FFA concentration. The type of fatty acids in 18 hybrids and 10 elite clones was assessed, and 11 fatty acids were found to be involved in the quality of cocoa beans. The fatty acids palmitic, stearic, oleic, linoleic, and arachidic are in all the genotypes invariably. The percentage of stearic acid was the highest, ranging from 30.5% in VTLCP-7 to 44.2% in VTLCP-1. The proportion of expression for the other fatty acids varied between genotypes (Senthil Amudhan and Elain Apshara, 2015). The pre-processing procedures (harvest, breaking, fermentation, and drying) are critical to producing high-quality beans. Chocolate producers favored cocoa bean quality based on flavour, cocoa butter hardness, purity, and yield. The quality evaluation of beans obtained from farmers' gardens in various locations continues to identify regional variability. Fat content varied from 32–50% in Karnataka and Kerala, 35–52% in Tamil Nadu, and 30–60% in Andhra Pradesh. The micronutrient profile of cocoa beans cultivated in Karnataka and Kerala states was also evaluated. Cocoa beans in Vittal, Karnataka contained Fe, Zn, and Cu at an average level of 22.25, 30.66, and 29.36 ppm, while in Kasaragod, Kerala, they had 40.43, 33.44, and 20.10 ppm, respectively. The soil micronutrient status of Vittal was 73.6, 38.1, 5.3, and 53.8 ppm and at Kasaragod, it was 22.7, 16.7, 0.29, and 0.44 ppm of Fe, Mn, Zn, and Cu, respectively. Though soil Zn content was more in Vittal, the translocation of Zn to bean is low, which might be due to the very high soil available Cu. The nutrient balance in the soil also influences the nutritional quality of cocoa beans.

Molecular Characterization

Molecular markers are more efficient in detecting genetic variation than morphological markers. Good quality DNA is essential for cocoa genome-related studies. Several DNA extraction methods have been proposed for DNA extraction from cocoa conventional (Sane et al., 2003), rapid, high-throughput, low-cost extraction (Bhattacharjee et al., 2004), and cocoa-derived products by several workers (Ha et al., 2015). Over the years, several types of molecular markers have been developed and widely used in *T. cacao*, including Isozymes, restriction fragment length polymorphism (Sounigo et al., 2005), random amplified polymorphic DNA (Russell et al.,1993; Sane and Sekar, 2006), amplified fragment length polymorphism (Perry et al., 1998) and microsatellites or simple sequence repeats (Zhang et al., 2009; Smulders et al., 2008). The structure of the genetic diversity in cacao has been extensively studied using morphological traits (Bartley, 2005), and molecular markers (Marita et al., 2001). Among, three major populations, the Forastero from the Amazon area, are the rich and diverse source of resistance to diseases, yield, and quality; the Criollos which are native to Central and North America (Mexico) are known for their rich flavor and the Trinitario which is a result of the natural cross of the two original populations. In addition to the major populations, there are several subpopulations of cacao and molecular markers suggest that there are 10 populations instead of the three previously identified (Motamayor et al., 2008). SSR markers have become increasingly common in DNA fingerprinting of cacao germplasm. These markers have been utilized for individual clone identification, parentage analysis, diversity assessment, the genetic identity of cocoa germplasm (Saunder et al., 2004), and determination of the origin and domestication of cocoa (Motamayor and Lanaud, 2002), linkage map (Crouzillat et al., 1964). Single nucleotide polymorphism (SNP) markers have become increasingly popular in cacao germplasm management due to their compatibility with high-throughput systems, universal data comparability, and lower genotyping costs. The cacao genome was first sequenced in 2011, and since then, the number of cacao genomes sequenced, as well as the availability of sequence information, has increased significantly, contributing to the identification of SNPs and use. Researchers have attempted to develop a core set of the most informative SNPs to identify off-types, parental and population identification, and determine admixture levels of different genetic groups in various cacao collections. These small SNP panels, which range from 48–192 SNPs, have been used to generate multi-locus profiles for individual cacao trees using the "multi-locus matching" method, which assesses the genetic integrity of genotyped germplasm (Gutiérrez et al., 2021).

Utilization of Germplasm

Attempts are underway in all cocoa-producing countries to exploit the available variability, which is the stock for breeders to produce varieties with good economic characteristics. Systematic and long-term cocoa improvement programs are being implemented with the following strategies: (1) Introduction of germplasm, conservation-cataloguing characterization, and evaluation of collections; (2) Breeding through selection and hybridization; (3) Establishment of seed gardens/clonal orchards through vegetative multiplication; (4) Comparative yield of CYTs; (5) Multilocation MLT; (6) Screening for biotic and abiotic stress and resistance breeding; (7) Biotechnology and bioinformatics approaches and 8.Demonstration gardens (Elain Apshara et al., 2005).

Breeding through Selection

The presence of genetic variability offers a wide scope for the selection of potential types. In cocoa, in a seedling population, approximately 75% of yields are obtained from 25% of the trees. Yield is the main selection criterion. An easy approach is to select superior plants and develop them into clones. The major selection criteria followed in cocoa are trees yielding 100 pods/tree/year, pods weighing 350–400 g or more, and 35–40 beans having a fermented single dry bean weight of 1 g and with favorable compatibility reaction (Table 2). Throughout the world, initial selections have been made from both native types and landraces. Traditionally, trunk girth has been used to measure yield efficiency in cocoa, which ultimately represents the Vigor of genotypes (Thong and Ng, 1980) to produce more pods. Dry bean production is generally considered to be a combination of bean weight per pod, number of pods per tree, and number of trees per hectare. It is expected to be 1 kg per tree per year, and productivity is usually assessed over 5–8 years in varietal trials. Although individual tree selections are made from seedling progenies, they must be further evaluated in clonal trials for confirmation. Heritability for yield is low or average when based on single-tree harvests (Soria, 1975) but higher when based on yields from plots containing several trees (Lockwood and Pang, 1995). Thus, clonal varieties are gaining importance. At the beginning of this century, clonal selection programs were initiated with pod index, bean size, and disease resistance as selection criteria.

To assess the phenotypic value of a genotype even in hybrid selection programs, clonal trials are advised (Eskes and Lanaud, 2001). In palm-based intercropping systems, the pod yield is generally expressed with respect to the canopy area, which is mainly maintained as a single-tier architecture. In the evaluation trials of the Trinitario type of cocoa and Wayanad collections, five clones each were

Table 2. Breeding criteria and traits involved.

Yield	Quality
Vigor of young trees Correlation between yield and vigor of adult trees Interaction with planting density Ability to adapt in the field Productivity of young trees (earliness) Productivity of adult trees No. of pods per tree Bean wt. per pod (pod index: no. of pods needed for 1 kg of dry beans)	**Physical Criteria:** dry bean size (1 gram and above), shell percentage (10–15%), nib recovery (85–90%), butter content (> 50%) hardness, the color of cotyledons **Organoleptic Characters:** flavor, viscosity, taste (bitter, astringent)
Resistance to Diseases	Resistance to Insects
Intrinsic Resistance: resistance to infection, colonization or pathogen multiplication, grading of damage levels **Escape:** period of fruiting or vegetative growth, duration of pod ripening, low productivity	Attractiveness **Screening:** grading with the level of infection, husk characters, Intrinsic resistance (antibiosis), Antixenosis, Tolerance
Eskes et al. (1998); Eskes and Lanaud (2001); Bekele et al. (2005); Eskes and Efron (2006).	

Table 3. Best cocoa clones selected with desirable characteristics.

Clone	Canopy Area (m^2)	Pod No./ Tree/yr.	Bean No./Pod	SBW (g)	DBY Kg/Tree/yr.	Shell %	Fat %
VTLC-25	11.81	61.9	42.7	1.03	2.45	10.9	51.0
VTLC-15	20.40	53.3	42.3	1.12	2.53	11.5	51.3
VTLC-18	12.26	49.4	43.0	1.17	2.48	13.1	53.2
VTLC-36	10.74	48.4	43.3	1.00	1.45	12.7	50.6
VTLC-13	11.34	45.1	43.0	1.00	1.93	11.2	54.5
VTLC-37	10.11	44.2	43.3	1.00	1.83	13.9	50.1
VTLC-17	13.39	43.1	40.3	1.24	2.16	13.1	53.2
VTLCS-1	15.25	55.0	41.0	1.13	2.52	11.0	52.0
VTLCS-2	14.60	54.5	41.0	1.21	2.70	15.0	53.0

SBW- Single dry bean weight DBY- Dry bean yield

selected for high pod and dry bean yields ranging from 2.2–3.3 kg/tree/year (Elain Apshara, 2015; 2016) with an optimal canopy of 15–20 m^2. From the clonal trials, ICAR-CPCRI has developed three varieties, VTLCC-1, VTLCS-, and VTLCS-2, that are available for commercial cultivation in different agroclimatic zones in India. Genetic analysis of 17 plant, pod, and bean characteristics in 44 Nigerian cocoa clones resulted in the selection of superior genotypes for higher performance with traits of high heritability and genetic advance. Based on pod yield, VTLC-25, VTLC-15, VTLC-18, VTLC-36, VTLC-13, VTLC-37, and VTLC-17 were identified as heavy bearers with optimal canopy (Table 3). These clones recorded a single dry bean weight of more than 1 g, 10–15% shell, high nib recovery, and more than 50% fat, making them suitable for industries as well (Elain Apshara et al., 2009; Elain Apshara and Nair, 2011).

Breeding through Hybridization

Hybrid vigor between parents showing good combining ability is readily exploited in cocoa improvement programs. Initial crosses involving Pound's seedling collections showed exceptional vigor precocity and high yield in Ghana. These observations and similar ones in Trinidad are attributed to hybrid vigor (Bell and Rogers, 1956). The most commonly adopted method is to develop hybrids between two distant genotypes and select hybrid varieties based on heterosis and hybrid vigor

(Toxopeus, 1972). At CPCRI, seven high-yielding clones, VTLC-1, VTLC-5, VTLC-7, VTLC-8, VTLC-9, VTLC-11, and VTLC-30, were selected and used as parents in breeding programs (Bhat, 1999; Elain Apshara et al., 2005).

Pollination Behavior in Cocoa

For hybridization programs to be effective, knowledge of cocoa pollination behavior is important. Cocoa is a cross-pollinated crop. Cocoa flowers are small and hermaphrodite and appear in large numbers on floral branches called 'cushions'. The flowering is referred to as 'cauliflorous' or truncate bearing, and the flowers are approximately 15 mm in diameter, borne on long pedicels, have five free sepals, five free petals, 10 stamens, and an ovary of five united carpels. The petals have an interesting shape; they are very narrow at the base but expand into a cup-shaped pouch and end in a broad tip or ligule. The 10 stamens that form the androecium are in two whorls: the outer whorl, consisting of five long non-fertilized staminodes, and the inner whorl with five fertile stamens. The stamens bear two anthers that lie in the pouch of the corresponding petal. The ovary has five parts containing many ovules arranged around a central axis. The style has the appearance of a single style. The ovary contains 30–60 ovules and is a highly heritable characteristic of cocoa. Fruit setting and ovular fertility depend on pollination and nutrition conditions (Lachenaud, 1995). The flowers are generally pink with darker tissue in the staminodes and petals, but there is variation between cultivars in the size and color of the flowers. The inflorescence primordia arise from old leaf axils, and it takes approximately 30 days from initiation for the bud to emerge through the bark and mature. Although cocoa trees produce many flowers, only 1–5% of the flowers are successfully pollinated to produce a pod.

When a bud matures, the sepals split in the afternoon and continue to open during the night. Early in the following morning, the flowers are fully open and the anthers release their pollen, i.e., the anthesis will be between 14.00–16.00 h, which will be completed 2.00–4.00 h the next day. Anther dehiscence will be at 4.00–6.00 h and will be completed between 8.00–10.00 h. The style matures a little later. This is the best day for pollination, and failure of fertilization on this day will cause the flower to abscise the following day (Zamora et al., 1960). Pollination in cocoa is affected by many small insects, particularly midges of the genus *Forcipomyia*, which pollinate approximately 5–10% of all flowers produced. A self-incompatibility system was recorded among cocoa populations, with allogamy ranging from 50 –100%. Up to 97% pollen viability is recorded in cocoa, and 66% *in vitro* germinability is observed (Rajamony and Mohan Kumaran, 1995). Pollen germination studies have shown that 20% sucrose + 40% PEG + 20 mM $CaCl_2$ will result in a good percentage of germination (Anonymous, 2023).

Incompatibility Mechanism in Cocoa

Self-incompatibility in cocoa was first reported by Pound (1932), who showed that some trees in Trinidad could not set fruit with their pollen nor with one another. Since then, the existence of self-compatible and self-incompatible trees has been established in most cocoa-growing countries. In cocoa, the pollen tubes develop normally, but when mating is incompatible, the male gamete does not fuse with the female gamete and is mostly considered as 'gametophytic' incompatibility (Bartley and Cope, 1973). Cope (1962) in Trinidad proposed five different S-factors or S-alleles to explain the results of extensive selfing and crossing studies between many cultivars. The dominance relations of these alleles are expressed in the following formula: Sa=Sb=Sc>Sd>Sf. Previously, Knight and Rogers (1955) studied the compatibility relations within a few families of Amazon material in Ghana and explained the dominance relations as S1>S2=S3>S4>S5. In an incompatible pollination, the proportion of non-fusion between the gametes will be 25, 50, or 100%. On failure of setting, the flower falls off three or four days after pollination. The degree of incompatibility varies among different populations. Self-compatible genotypes are found in the Lower Amazon Forastero,

Criollo, and Trinitario, whereas the Upper Amazon Forasteros are generally self-incompatible (Eskes and Lanaud, 2001). Amazon cultivars are all self-incompatible but are generally cross-compatible. Trinitario cultivars have a high proportion of self-incompatible trees that will not cross with other self-incompatible trees, requiring pollen from self-compatible trees for successful pollination. The Amelonado population is entirely self-compatible. Enzymic markers have enabled verification of the origin of seeds collected depending on gene combinations, and approximately 0–100% of seeds resulted from selfing (Lanaud, 1987). Self-incompatible but cross-compatible genotypes have become advantageous in hybrid development programs as well as in the design and operation of seed gardens to ensure the desired parentage and production of true hybrid seeds (Wood and Lass, 1985).

Method of Hand Pollination and Fruit Set

For the production of true hybrids with specific objectives and to confirm the compatibility reaction, hand pollination is being practiced. In artificial pollination, a flower bud that will open the following day, recognized by its whitish color and swollen appearance, is selected. The bud is covered with a hood of a plastic tube/hose pipe piece 5 cm 1.5–2 cm, which is sealed to the bark using materials such as plasticine/glaze putty. The tube is covered with a muslin cloth at the top and held in place with a rubber band. This ensures the circulation of air and the exclusion of insects. Opened flowers are collected from the desired male parent, and stamens are carefully removed by pushing the corresponding petal. One entire anther with a part of the filament is deposited on the stigma. The style is surrounded by a ring of staminodes, and if these are long, the removal of two or three staminodes should be performed for easy access to the style. Emasculation is not necessary because of the presence of self-incompatibility. For selfing, hand pollination is performed using stamens from the same flower. Pollinated flowers are labeled using tin foil pieces fixed in the cushion using ball pins. The hoods were removed 24 hours after pollination, and in 3–5 days, fertilization was confirmed by visual swelling of the ovary. To prevent undue shedding and wilting of fruits from hand pollination, it is usual to remove all developing fruits on the tree produced by open pollination (Vikraman Nair et al., 2002).

During natural pollination, too many fruits are normally set, but they do not reach maturity. The young developing pod is called 'Cherelle' and cocoa has its fruit thinning mechanism of "Cherelle wilt" which occurs by yellowing, blackening, and shriveling. The incidence of wilt increases from pollination to a peak at approximately 50 days, then falls off and rises to a second peak at 70 days. After 95–100 days, no further wilting occurs; the cherelles then become immature pods, reach full size, mature, and finally ripe. The ovule is filled with jelly-like endosperm. It takes 5–6 months from pollination to ripening. The mature fruit or pod is botanically an indehiscent drupe that remains on the tree unless harvested. Seeds are recalcitrant, without dormancy, and should therefore be sown on the day of extraction to raise seedlings. Pods developed through hand pollination should be protected from mammalian pests and timely harvest should be ensured because cocoa has vivipary.

Progeny Trials

Various cross-combinations have been attempted with specific objectives to develop hybrids. In the Asia Pacific region, the Malaysian Cocoa Board (MCB) and the Indonesian Coffee and Cocoa Research Institute (ICCRI) have developed many hybrids. At CPCRI, five progeny trials were evaluated with 76 cross combinations during 1983–1994 at Vittal, Kidu, and Kasaragod centers with the objectives of increasing the number of pods, high dry bean yield, large bean size, and drought tolerance. Of these, 20 hybrids were identified as the best and were further evaluated as clones. Among them, 17 progeny exhibited high vigor and cropping efficiency even in the early years of development (Bhat and Ananda, 1997; Elain Apshara et al., 2007, 2008). From the progeny trials, four hybrids, VTLCH-1, VTLCH-2, VTLCH-3, and VTLCH-4, were released as improved varieties for cultivation in the country in 2006, the Golden Jubilee year of CPCRI. RS, Vittal, and VTLCH-5 were released as Netra Centura for the centenary year of CPCRI, Kasaragod.

Comparative Yield Trial (CYT)

The clones and progenies developed through selection and hybridization programs were multiplied as clones and evaluated under comparative yield trials in different situations. Under high-density planting in arecanut gardens, five hybrids, VTLCP-6, VTLCP-20, VTLCP-15, VTLCP-1, and VTLCP-19 have been identified as the best performers even in their initial years of growth (Elain Apshara et al., 2008). A comparative study of parents and progenies as clones resulted in the identification of VTLCP-6, VTLCP-2, and VTLCP-20 and parents VTLC-1 and VTLC-56 as potential high-yielders (Elain Apshara, 2013a). In another trial, clones suitable for both arecanut and coconut canopies were identified (Elain Apshara, 2013b) and released as varieties. Under coconut in the double hedge system of planting, hybrids VTLCP-22, VTLCP-18, and VTLCP-1 showed the best performance with optimal canopy and high yield. Evaluation of clonal and seedling progenies of selected genotypes has resulted in the identification of four hybrids and two clones for multiplication both as clones and seedlings for utilization in the area expansion program (Elain Apshara, 2014). Selective clones are being evaluated under arecanut, coconut, and oil palm gardens as well.

Multi-location Trial (MLT) and Demonstration Plots

To assess the adaptability and stability of hybrids and clones in different agro-climatic conditions, multi-location trials are important. Elite clones of cocoa are under evaluation in both traditional and non-traditional states, namely Karnataka, Kerala, Tamil Nadu, Andhra Pradesh, Gujarat, Maharashtra, West Bengal, and Assam, for studying genotype x location x environment interactions. Further, 94 front-line demonstration plots are established under the participatory research cum demonstration plots scheme funded by the Directorate of Cashewnut and Cocoa Development (DCCD), Kochi in eight taluks in Karnataka and one taluk of northern Kerala and five AICRPP (All India Co-ordinated Research Project on Palms) centers for identification of location-specific varieties and to tackle the climate change effects.

Resistance Breeding

The perennial crop stand of cocoa in the native and productive zones is facing severe setbacks with debilitating diseases, which causes an estimated loss of 30% of world production. The most important diseases of cocoa are black pod rot caused by several *Phytophthora* species, frosty pod rot caused by *Moniliophthora roreri,* and witches' broom caused by *Moniliophthora perniciosa* (Ploetz, 2007). *P. palmivora* is present in the entire cocoa growing area and *P. megakarya* is the most aggressive species with losses up to 50%, but is limited to West Africa. In Brazil, *M. perniciosa* is responsible for the drastic yield loss. Vascular Streak Die-back (VSD) caused by *Oncobasidium theobromae* is found in Southeast Asia causing 20–30% loss in Papua New Guinea. In Africa, the cocoa economy is extensively affected by several viral diseases caused by the Cacao swollen shoot virus (CSSV), Cacao yellow mosaic virus (CYMV), and Cacao necrosis virus (CNV) which necessitated strict quarantine measures in germplasm exchange programs (Michelle End et al., 2014). Integrated disease and pest management by use of resistant materials, and cultural and biological methods is probably the best way to contain pathogens and pests in the longer run for sustainable resistance. And so, disease resistance became the primary trait targeted by cocoa breeders worldwide. Sources of resistance have been identified for black pod rot (Iwaro et al., 2006), witches' broom (Umaharan et al., 2005), and frosty pod rot (Philips-Mora and Wilkinson, 2007). Since the cocoa genome is sequenced, it is expected to provide models for plant-pathogen interactions and also facilitate the identification of resistance genes. The insects causing severe damage in cocoa are mirids (*Sahlbergella* sp., *Distantiella* sp., *and Helopeltis* sp.) and thrips (*Selanothrips rubrocintus*). In Africa, mirids pose a severe threat and in South East Asia, the cocoa pod borer (*Conopomorpha cramerella*) causes heavy loss (Azhar, 1986). Many cocoa pests like mealy bugs, aphids, caterpillars, and borers are also reported in India (Daniel, 2002). Screening of available germplasm for prevailing diseases, and existing, emerging pests is very important in light of the seasonal weather variations.

Black Pod Rot (BPR)

India is free of most of the debilitating viral and fungal diseases of cocoa. Since the current cocoa growing area comes under a high rainfall zone and the main harvest season coincides with the monsoon, the incidence of black pod rot caused by *Phytophthora palmivora* is comparatively higher. Though the main harvest is safeguarded with systematic annual pruning, in the post-monsoon period, the second-season crop is still affected by pod rot. In Costa Rica, through screening nine cultivars, viz., EET 59, EET 376, UF 713, UF 715, SCA 6, SCA 12, Pound 7, Catongo, and Diamantes 800 have been found to have promising degrees of resistance (Lawrence, 1978). In Java, SCA 6, SCA 12, and ICS 6 were found to exhibit resistance. At CPCRI based on field screening, clones have been categorized into <10%, 10–25%, and >25% damage levels. *In vitro* screening using isolates of *P.palmivora, P.capsici,* and *P. citrophthora* indicated that collections of Nigerian origin exhibit a certain degree of tolerance (Chandra Mohanan, 1982). Further, 21 exotic clones collected exclusively for *Phytophthora* pod rot resistance are identified for utilization in the cocoa hybridization program.

Vascular Streak Die Back (VSD)

In India, Vascular streak die-back (VSD) caused by *Oncobasidium theobromae*, was first reported in Kottayam, Kerala by Abraham (1981). Since this disease is contained in the region, Kerala Agricultural University (KAU) took upbreeding programs specific to VSD resistance (Mallika et al., 2002). The hybridization program at KAU was initiated in 1983 and continued up to 1993. During this period, a total of 159 parents were included to produce 239 crosses, 187 pods, and 21,819 hybrid seedlings. A total of 3,126 superior seedlings were field planted in series I, II, III, IV, and progeny trials I, II, III, IV, 163 superior hybrids were selected, utilized in various breeding program for the release of hybrids and used as better combiners in clonal gardens. Varieties and hybrids CCRP 1–10 with high yield and good levels of tolerance to VSD have been released as varieties for cultivation in the country (Minimol et al., 2011). Malaysia, Indonesia, Philippines, Vietnam, and Papua New Guinea in the Asia-Pacific region have taken up VSD resistance programs and most of the Trinitario hybrids are resistant to this disease.

Tea Mosquito Bug (TMB)

Tea mosquito bug (TMB) (*Helopeltis* sp.) incidence became severe during the last three years in summer and post-monsoon seasons. *Helopeltis antonii, H. theivora,* and *H. bradyi* are reported in cocoa in south India. The insect population is influenced by many factors like temperature, humidity, water stress, the condition of cocoa trees, etc. The development and use of mirid-resistant cocoa varieties mostly concentrated on the assessment of field damage (N' Guessan et al., 2004). Tolerance to *Sahlbergella singularis* and *Distantiella theobromae* has been measured by observing cumulative and recent damage in field trials in Cameroon and Ivory Coast. This method permits the ranking of genotypes according to their global reaction to mirid attacks (Anikwe et al., 2009). Damage on flushes, cherelles, and pods of individual trees and different grade levels of infection are assessed to work out the TMB tolerance among genotypes. Penetrometer readings for determining the hardness of the sclerotic layer and thickness at primary and secondary furrows of pod husk have been recorded at CPCRI in 100 cocoa genotypes and interpreted with reference to insect resistance (Elain Apshara, 2013a). Husk characteristics have been given special importance in recent years in the Asia-Pacific region, especially for pod borer resistance. The mechanism of plant resistance to insects is a complex phenomenon. Red-colored pods with smooth surfaces have been identified as tolerant to TMB damage.

Low Moisture Stress

Cocoa plants are susceptible to environmental conditions, especially temperature, and drought, and considerably influence pod yield (Daymond et al., 2000). Cocoa is very sensitive to water scarcity and undergoes a period of low moisture stress for 5–6 months in its current growing condition in India. Breeding for drought tolerance is unique to our country and is taken up with systematic screening of available germplasm as well as hybridization programs. Screening of accessions is conducted for physiological parameters like stomatal resistance, chlorophyll fluorescence, proline accumulation under stress, and studies on seed germination under low osmotic potential, etc. A total of 216 cocoa genotypes have so far been screened for physiological and biochemical parameters under different trials (Balasimha, 1999). In all these studies, field measurements were taken during unstressed (October) and stressed (March) conditions. Few Nigerian collections have been identified as drought-tolerant and used for hybridization with high-yielding Malaysian collections under two progeny trials. Two hybrids, VTLCH-3 and VTLCH-4, have been released as varieties suitable for cultivation under water-limited conditions in the country. Studies on leaf morphology, stomatal behavior, water relation components, and biochemical factors indicated that thick leaves, higher wax content, efficient stomatal closure, and high tissue elasticity are responsible for the better adaptation of cocoa plants to drought conditions. The application of chlorophyll fluorescence as a tool to screen cocoa for drought tolerance has been confirmed with a series of genotypes. Recently, photosynthesis, chlorophyll fluorescence, and water potential under stress and non-stress conditions were estimated in 11 genotypes from different geographical origins: Columbia, Brazil, Peru, Mexico, and Ecuador (Elain Apshara et al., 2013a). Seasonal and varietal differences were found, and transpirational water loss was found to be reduced with increased stomatal closure, which is considered a favorable drought trait in any crop. Among the 52 new introductions, 5 Amazon and Pound collections are adaptable to water-limited conditions (Balasimha et al., 2013) with high yields, which will be further utilized in the breeding program. Genotypic differences for morpho-physiological criteria, potential antioxidant enzymes, and biochemical components depicting drought tolerance in young seedlings were determined with cocoa clones and hybrids under controlled low-moisture stress conditions (M'bo Kacou et al., 2015, 2015a, 2016). In these trials, standards, and thresholds for several physiological parameters related to cocoa were standardized.

To make the breeder's task easier Bioversity International, the Common Fund for Commodities (CFC), and the International Cocoa Organisation (ICCO) have also designed the working procedures for field screening and physiological evaluation of cocoa germplasm (Eskes et al., 1998). Hadley (2000) detailed the visual estimates of physiological traits in cocoa, which are to be taken in clonal plants on a plot basis with 6–8 trees and in hybrid plants on an individual tree basis. Morpho-physiological parameters include measurements of flowering, fruiting, cherelle wilting, leaf flushing, branching and pruning intensity, canopy shape, density, and light transmission on different point scales. To understand and elucidate the optimum canopy shape and structure of cocoa, different spacing and canopy sizes have been studied (Balasimha, 2001), which showed significant differences in crop yield. In an experiment with grafts, the photosynthetically active radiation (PAR) and light interception varied significantly over the years with two spacings (2.7 m x 2.7 m and 2.7 m x 5.4 m) and three canopy sizes (small, medium, and large), and similar results were noticed with transpiration rate and stomatal conductance as well (Balasimha, 2006). It is important to note that the maximum leaf area should be maintained, self-shading of leaves should be avoided, and pruning should be done to the extent of retaining 20–30 leaves or developing pods to ensure the yielding potential of the genotype.

Species/Cultivars Released

From the progeny trials, five hybrids, VTLCH-1, VTLCH-2, VTLCH-3, VTLCH-4, and VTLCH-5, and the clonal trials, three varieties, VTLCC-1, VTLCS-1, and VTLCS-2, have been released by ICAR-CPCRI for commercial cultivation in different agro-climatic zones in the country (Tables 4, 5, and Figure 1). These are mainly suitable for the cocoa-growing states of southern India, Karnataka, Kerala, Tamil Nadu, and Andhra Pradesh, in addition to the states of Gujarat, Maharashtra, and the northaastern states. Seedling and clone standards were also developed for the production of quality planting materials (Anonymous, 2023a).

Table 4. CPCRI cocoa varieties developed through the selection.

Characters	VTLCC 1	VTLCS 1	VTLCS 2
Canopy area m^2	12	12	15
No. of pods/tree/year	55	55	55
No. of beans/pod	35	42	41
Single dry bean weight (g)	0.9–1.05	1.3	1.21
Dry bean yield kg/tree/year	1.3	2.5	2.7
Dry bean yield kg/ha	890	1700	1840
Shelling %	12	11	15
Nib recovery %	88	88	85
Fat content %	50	52	53
Features	Self-compatible line suitable for all regions including North Eastern Zones	High yielder withstands both biotic and abiotic stress	High bean index, stable yielder withstands both biotic and abiotic stress

Table 5. CPCRI varieties developed through hybridization.

Characters	VTLCH 1	VTLCH 2	VTLCH 3	VTLCH 4	VTLCH 5
Canopy area m^2	16	15	18	18	16
No. of pods/tree/year	50	50	41	40	66
No. of beans/pod	40	40	41	40	43
Single dry bean weight (g)	1–1.11	1–1.5	1–1.05	1–1.07	1–1.11
Dry bean yield kg/tree/year	1.4	1.5	1.7	1.6	2.5–3.0
Dry bean yield kg/ha	959	1030	1150	1090	1500–1800
Shelling %	13	11	15	15	11
Nib recovery %	87	89	87	87	88
Fat content %	54	54	51	51	52
Features	Early stable heavy bearer	Heavy bearer, tolerant to black pod rot	Suitable for water-limited conditions	Suitable for water-limited conditions	Suitable for high-density planting both under arecanut and coconut

Conservation of Useful Germplasm/Genetic Resources

Cocoa germplasm is mainly conserved as field gene banks (FGBs) in research institutes. Alternate gene bands are also being maintained as safety duplication in similar and different agro-climatic regions. In India, germplasm collections are being maintained under palm-based cropping systems. Under an arecanut garden of 2.7 m x 2.7 m, cocoa is planted in the center of four areca palms at 2.7 m x 5.4 m spacing. Under the coconut garden, collections are being maintained in two systems. In the single hedge system, in a coconut garden spaced at 7.5 m x 7.5 m, cocoa is planted at 3 m x 7.5 m, and in

Figure 1. Cocoa varieties released by CPCRI, India.

the double hedge system, cocoa is spaced in two rows of 2.5–2.7 m. Under an oil palm of 9 m x 9 m square planting, cocoa is conserved at 2.5–3 m in the alternate genebank at Andhra Pradesh. All these collections are provided with 100:40:140 g of NPK per tree/year and 20 liters of water through a drip. Annual systematic pruning is practiced to maintain the canopy architecture at 15–20 m^2.

The Outcome of Cocoa Breeding Programs

1. High-yielding varieties to increase productivity and double farmers' income.
2. Region-specific varieties and common varieties for multiple locations.
3. Qualitative improvement exploring health benefits.
4. Varieties suitable for climate change effects, tolerant to biotic and abiotic stresses, and with nutrient and water use efficiency for marginal lands.
5. Vegetative multiplication to maintain true to type, reduce perennial cropping cycle, and overcome self-incompatibility problems.
6. Seedling/clonal standards, regional nurseries, and model nurseries for quality planting material production.

Conclusion

Cocoa improvement has attained a positive phase with the sequencing of its genome. Identifying genes responsible for incompatibility and disease resistance is the main concern. Expression of genes for resistance and quality parameters and their validation with trait-specific germplasm is very important for future cocoa improvement programs. The past few decades of research have been restricted to single-cross hybrids and future development should focus on double and triple crosses, backcrosses, and the possible use of inbred lines. Development of early selection, detection, and diagnostic methods for resistance will enable rapid screening of plant material and permit preselection activities. Because of the health benefits of dark chocolates, biochemical constituents and antioxidant properties of cocoa are to be given greater attention in the breeding programs. Farmers' participatory plant breeding, *in situ* conservation of landraces, exploitation of flavor components from genotypes belonging to specific geographic regions, varieties for changing climatic conditions, and environment-friendly management strategies are to be considered. At the National level, the expansion of cocoa cultivation with quality planting material of elite clones, and a collaborative approach between research institutes, universities, state horticulture departments, and developmental agencies are required. At the International level, participation in cocoa genetic resources networking and regional breeding groups of both developed and developing countries is important. Collaboration of India with Asia Pacific regional countries, Malaysia, Indonesia, Philippines, Vietnam, and Papua New Guinea is essential with their common coconut-based cropping systems with known pest and disease problems. It will enhance the region-specific sustainability of cocoa cultivation.

References

Abraham, C.S. 1981. Vascular streak dieback of cocoa in India. *Indian Arecanut, Spices, and Cocoa Journal, 4*: 119–120.

Anikwe, J.C., A.A. Omoloye, P.O. Aikpokpodion, F.A. Okelana, and A.B. Eskes. 2009. Evaluation of resistance in selected cocoa genotypes to the brown cocoa mirid, *Sahlbergella singularis* Haglund in Nigeria. *Crop Protection, 28*(4): 350–355.

Anonymous. 2023. *Annual Report ICAR- CPCRI.*

Anonymous. 2023a. Cocoa (Seed/Clone Standard). *Indian Minimum Seed Certification Standards Part – II,* Chapter 3: Seed Certification Standards for Trees, 57–62. *In*: Malhotra, S.K., and Dilip Kumar Srivastava. [Eds.]. The Central Seed Certification Board, Dept. of Agriculture & Farmers Welfare, Govt. of India.

Azhar, I., 1986. A threat of cocoa pod borer (*Conopomorpha cramerella*) infestation to the Malaysian cocoa industry. 1. On the biology and damage. *Teknologi Koko-Kelapa MARDI, 2*: 53–60.

Balasimha, D. 1999. Stress physiology of cocoa. *Journal of Plantation Crops, 27*(1): 1–8.

Balasimha, D. 2001. Growth and yield of cocoa grown under areca in relation to plant density and canopy architecture. *In*: *Proceedings of 13th International Cocoa Research Conference*, 365–372. Kota Kinabalu, Malaysia.

Balasimha, D. 2006. Performance of cocoa in relation to spacing and pruning regimes. *In*: *Proceedings of 15th International Cocoa Research Conference*, 265–272. Costa Rica.

Balasimha, D., S. Elain Apshara, and C.T. Jose. 2013. Genotypic variations in chlorophyll fluorescence and stomatal conductance of cocoa in relation to drought tolerance. *Journal of Plantation Crops, 41*(1): 40–45.

Bartley, B.G. 2005. *The Genetic Diversity of Cacao and Its Utilization.* 341pp. CABI publishing, UK.

Bartley, B.G.D., and F.W. Cope. 1973. Practical aspects of self-incompatibility in *Theobroma cacao* L. *Agric. Genet. Sel. Top. R. Moav.*,109–134.

Bekele, F.L. 2001. *Proceedings of the International Workshop on New Technologies and Cocoa Breeding–INGENIC, 16–17th October 2000*, 190. Kota Kinabalu, Sabah, Malaysia,

Bekele, F.L., J. Michelle End, and A.B. Eskes. 2005. *Proceedings of the International Workshop on Cocoa Breeding for Improved Production Systems- INGENIC*, 198. Accra, Ghana.

Bell, G.D.H., and H.H. Rogers. 1956. Cacao breeding at WACRI. *In*: *Proceedings of the Cacao Breeding Conference, West African Cocoa Research Institute*, 31–49. Tafo, Ghana.

Bhat, V.R. 1999. Cocoa germplasm characterization, conservation, and utilization. *In*: Ratnambal, M.J., P.M. Kumaran, K. Muralidharan, V. Niral, and V. Arunachalam, V. [Eds.], *Improvement of Plantation Crops*, 77–85. ICAR- CPCRI, Kasaragod, India.

Bhat, V.R., and K.S. Ananda. 1997. Evolving high yielding varieties by selection and hybridization in cocoa. *Ann. Rep. of CPCRI*: 71–72.

Bhattacharjee, R., M. Kolesnikova-Allen, P. Aikpokpodion, S. Taiwo, and I. Ingelbrecht. 2004. An improved semiautomated rapid method of extracting genomic DNA for molecular marker analysis in cocoa. *Plant Molecular Biology Reporter, 22*: 435–436.

Chandra Mohanan, R. 1982. Studies on the reaction of pods of different cocoa accessions to *Phytophthora palmivora. Planter, 58*: 99–103.

Cheesman, E.E. 1944. Notes on the nomenclature, classification, and possible relationships of cacao populations. *Tropical Agriculture, 21*: 144–159.

Cope, F.W. 1962. The mechanism of pollen incompatibility in *Theobroma cacao* L. *Heredity, 17*(2):157–182.

Couch, J.A., H.A. Zintel, and P.J. Fritz. 1993. The genome of the tropical tree *Theobroma cacao* L. *Molecular and General Genetics, 237*:123–128.

Counet, C., C. Ouwerx, D. Rosoux, and S. Collin. 2004. Relationship between procyanidin and flavour contents of cocoa liquors from different origins. *Journal of Agricultural and Food Chemistry, 52*(20): 6243–6249.

Crouzillat, D., E. Lerceteau, V. Petriard, J. Morena, H. Rodgriguez, D. Walker, W. Phillips, C. Ronning, R. Schnell, J. Osei, and P. Fritz. 1996. *Theobroma cacao* L.: A genetic linkage map and quantitative trait loci analysis. *Theoretical and Applied Genetics, 93*: 205–214.

Cuatrecasas, J. 1964. Cacao and its allies: A taxonomic revision of the genus Theobroma. *Systematic Plant Studies, Contributions to the US National Herbarium*, *35*(6): 379–614.

Daniel, M. 2002. Pest management of cocoa. *In*: Ravi Bhat, D. Balasimha and S. Jayasekhar [Eds.], *Proceedings of National Seminar on Technologies for Enhancing Productivity in Cocoa*, 87–94. ICAR-CPCRI, Regional Station, Vittal, India.

Dantas, L.G., and M. Guerra. 2010. Chromatin differentiation between *Theobroma cacao* L. and *T. grandiflorum* Schum. *Genet. Mol. Biol.*, *33*: 94–98

Daymond, A.J., P. Hadley, R.C.R. Machado, and E. Ng. 2000. An investigation of physiological parameters underlying yield variation in cocoa. *In*: *Proceedings of 13th International Cocoa Research Conference*, 331–340. Kota Kinabalu, Sabah, Malaysia.

Elain Apshara, S. 2013. Comparative performance of selected cocoa parents and progenies as clones in South India. 25. *In*: Sabariah Samsudin, Samuel Yap Kian Chee, and Badrul Hisyam Zainudin [Eds.], Abstracts of *Malaysian International Cocoa Conference on Innovation and Technology Driving Cocoa Productivity/Sustainability,* 7–8 October *2013*, Sunway Pyramid Convention Centre, Kuala Lumpur.

Elain Apshara, S. 2013a. Performance of selected cocoa (*Theobroma cacao* L.) clones under arecanut and coconut. *Journal of Plantation Crops*, *41* (2): 242–246.

Elain Apshara, S. 2013b. *In*: Sabariah Samsudin, Samuel Yap Kian Chee, and Badrul Hisyam Zainudin [Eds.], Abstracts of *Malaysian International Cocoa Conference on Innovation and Technology Driving Cocoa Productivity/Sustainability*, 7–8 October 2013, Sunway Pyramid Convention Centre, Kuala Lumpur.

Elain Apshara, S. 2014. Performance of elite cocoa clones under coconut in double hedge system of planting. *The Cashew and Cocoa Journal*, *3*(2): 13–16.

Elain Apshara, S. 2015. Growth and yield performance of Trinidad cocoa collections in Karnataka. *Journal of Plantation Crops*, *43*(3): 237–240.

Elain Apshara, S. 2016. Selection of potential clones from Wayanad cocoa collections. *Journal of Plantation Crops*, *44*(2): 124–128.

Elain Apshara, S., and P. Rajan. 2009. *Profile of Cocoa Collections at CPCRI, Research Centre, Kannara.* 61pp. ICAR- CPCRI, Kasaragod, India.

Elain Apshara, S., and R.V. Nair. 2011. Genetic analysis in cocoa collections obtained from Nigeria. *Journal of Plantation Crops*, *39*(1): 200–206.

Elain Apshara, S., K.S. Ananda, and D. Balasimha. 2005. Cocoa breeding at CPCRI, India. *INGENIC Newsletter*, *10:* 34–37.

Elain Apshara, S., V.R. Bhat, and R.V. Nair. 2008. Comparative studies on elite cocoa progenies in their initial years of growth. *Journal of Plantation Crops*, *36*(1): 38–44.

Elain Apshara, S., M.K. Rajesh, and D. Balasimha. 2013. Assessment of morphological, physiological, and molecular characteristics of cocoa accessions from Central and South America in relation to drought tolerance. *Journal of Plantation Crops*, *41*(3): 389–397.

Elain Apshara, S., V.R. Bhat, K.S. Ananda, and R.V. Nair. 2007. High Yielding Cocoa Varieties of the Central Plantation Crops Research Institute, India. *INGENIC Newsletter*, *11*: 12–15.

Elain Apshara, S., V.R. Bhat, K.S. Ananda, R.V. Nair, and D. Suma. 2009. Evaluation and identification of high yielding trees in Nigerian cocoa germplasm. *Journal of Plantation Crops*, *37*(2):111–116.

End, M.J., A.J. Daymond, and P. Hadley. 2014. *Technical Guidelines for the Safe Movement of Cacao Germplasm.* Revised from the FAO/IPGRI Technical Guidelines No. 20. Global Cacao Genetic Resources Network (CacaoNet), Bioversity International, Montpellier, France.

Eskes, A.B., J.M.M. Engels, and R.A. Lass. 1998. Working procedures for cocoa germplasm evaluation and selection. *In*: *Proceedings of the CFC/ICCO/IPGRI Project Workshop*, *1*: 6.

Eskes, A.B., and C. Lanaud. 2001. Cocoa. In: Andre Charrier, Michel Jacquot, Serge Hamon and Dominique Nicolas [Eds.], *Tropical Plant Breeding*, 78–105. Science Publishers Inc., USA and CIRAD, France.

Eskes, A.B., and Y. Efron. 2006. *Global Approaches to Cocoa Germplasm Utilization and Conservation.* CFC Technical Paper No. 50, 224. Bioversity International.

Figueira, A., J. Janick and P. Goldsbrough.1992. Genome size and DNA polymorphism in *Theobroma cacao* L. *Journal of the American Society for Horticultural Science*, *117*(4): 673–677.

Frison, E.A., and E. Feliu. 1989. *FAO/IBPGR Technical Guidelines for the Safe Movement of Cacao Germplasm.* Food and Agricultural Organisation of the United Nations, International Board for Plant Genetic Resources, Rome.

Glicenstein, L.J., and P.J. Fritz. 1989. Ploidy level in *Theobroma cacao. Heredity*, *80*: 464–467.

GOI.1997. Cocoa beans grading and marking rules. AGMARK standards. *Gazette of India Part II*, *3*: 1.

Gutiérrez, O.A., K. Martinez, D. Zhang, D.S. Livingstone, C.J. Turnbull, and J.C. Motamayor. 2021. Selecting SNP markers reflecting population origin for cacao germplasm identification. *Beverage Plant Research*, *1*(1): 1–9.

Ha, L.T.V., L. Vanlerberghe, L, H.T. Toan, K. Dewettinck, and K. Messens. 2015. Comparative evaluation of six extraction methods for DNA quantification and PCR detection in cocoa and cocoa-derived products. *Food Biotechnology*, *29*(1): 1–19.

Hadley, P. and T. Lee. 1992. Current cocoa quarantine facilities of the University of Reading.. *In*: *International Workshop on Conservation, Characterization and Utilization of Cocoa Genetic Resources in the 21st Century*, *1317*: 65–68, Trinidad.

Iwaro, A.D., D.R. Butler, and A.B. Eskes. 2006. Sources of resistance to *Phytophthora* pod rot at the International Cocoa Genebank, Trinidad. *Genetic Resources and Crop Evolution, 53*(1): 99–109.

Knight, R., and H. Rogers. 1955. Incompatibility in *Theobroma cacao. Heredity, 9*(1): 69–77.

Lachenaud, P. 1995. Variations in the number of beans per pod in *Theobroma cacao* L. in the Ivory Coast. III. Nutritional factors, cropping effects, and the role of boron. *Journal of Horticultural Science, 70*(1): 7–13.

Lanaud, C. 1987. *Novelles Data on Cocoa Biology: Diversity of Populations of System Incompatibility, Spontaneous Haploid, and Their Consequences on Amelioration of Genetics of This Species.* Ph.D. Thesis, University of Paris XI. Orsay, France.

Lanaud, C., P. Hamon, and C. Duperray. 1992. Estimation of nuclear DNA content of *Theobroma cacao* L. by flow cytometry. *Coffee Cacao Tea, 36*(1).

Lawrence, J.S. 1978. Evaluation of methods for assessing resistance of cacao cultivars and hybrids to *Phytophthora palmivora* (Butler). *Cocoa Research Centre Brazil Technical Bulletin 62.*

Lockwood, G., and J.T.Y. 1994. Cocoa breeding at BAL Plantations. Genetic analysis and its implications for breeding strategy. *In*: Proceedings of the International Workshop on Cocoa Breeding Strategies, *6680*, Kuala Lumpur.

Mallika, V.K., S. Prasanna Kumari Amma, and R. Vikraman Nair. 2002. Crop improvement in cocoa. *In*: Proceedings of National Seminar on Technologies for Enhancing Productivity in Cocoa, 19–27. CPCRI, Regional Station, Vittal.

Martinson, V.A. 1975. Cytological studies of diploid and tetraploid *Theobroma cacao. Genetica, 45*(3): 341–348.

M'bo Kacou Antoine Alban, S. Elain Apshara, KB. Hebbar, Tahi G.Mathias, and AkéSévérin. 2015. Potential of antioxidant enzymes in depicting drought tolerance in cocoa genotypes at young age. *African Journal of Science and Research, 4*(5): 18–23.

M'bo Kacou Antoine Alban, S. Elain Apshara, K.B. Hebbar, Tahi G. Mathias, and AkéSévérin. 2015a. Morpho-physiological criteria for assessment of two-month-old cocoa genotypes for drought tolerance. *Indian Journal of Plant Physiology, 21*(1): 23–30.

M'bo Kacou Antoine Alban, S. Elain Apshara, KB. Hebbar, K.S. Ananda, Tahi G.Mathias, and AkéSévérin. 2016. Change in epicuticular wax and biochemical secondary metabolites in cocoa under hydric stress. *International Journal of Current Research, 8*(4): 28988–28999.

Minimol, J.S., S. Prasanna Kumari Amma, S., and E.K. Lalitha Bai. 2011. Three decades of research on breeding of cocoa in Kerala Agricultural University. *In*: *Proceedings of Seminar on Strategies for Enhancing Productivity of Cocoa*, 10–14. CPCRI, Regional Station, Vittal.

Munoz, O.J.M. 1948. *Chromosomic Studies in the Genus Theobroma.* M.Sc.Thesis. Faculty of the Inter American Institute of Agricultural Sciences, Turrialba.

Perry, M.D., M.R. Davey, J.B. Power, K.C. Lowe, H.F.J. Bligh, P.S. Roach, and C. Jones. 1998. DNA Isolation and AFLP™ Genetic Fingerprinting of shape *Theobroma cacao* (L.). *Plant Molecular Biology Reporter, 16:* 49–59.

Phillips-Mora, W., and M.J. Wilkinson. 2007. Frosty pod of cacao: A disease with a limited geographic range but unlimited potential for damage. *Phytopathology, 97*(12): 1644–1647.

Ploetz, R.C. 2007. Cacao diseases: Important threats to chocolate production worldwide. *Phytopathology, 97*: 1634–1639.

PPVFRA. 2023. *Cocoa DUS Guidelines.* The Gazette of India Notification Extraordinary (S.No.163) - Part II(3ii) (S.O.683(E), Ministry of Agriculture and Farmers Welfare.

Pound, E.J. 1932. The genetic constitution of cocoa crop. *First Ann. Rep. Cocoa. Res. Trinidad*: 9–25.

Rajamony, L., and N. Mohanakumaran. 1995. Studies on the floral biology of cocoa. *Planter, 71*: 119–127.

Ratnam, R. 1961. Introduction of Criollo cacao into Madras state. *South Indian Horticulture, 9*(4): 24–29.

Russell, J.R., F. Hosein, E. Johnson, R. Waugh, and W. Powell. 1993. Genetic differentiation of cocoa (*Theobroma cacao* L.) populations revealed by RAPD analysis. *Molecular Ecology, 2*(2): 89–97.

Sane, A., Upadhyay, A., Manimekalai, R., and V.A. Parthasarathy. 2003. An efficient method of DNA extraction from *Theobroma cacao* and *Cola nitida* leaves. *Agrotrópica, 127.*

Sane, A., and A.C. Sekar. 2006. Molecular characterization of cocoa (*Theobroma cocoa* L.) Germplasm Using RAPD Markers. *Plant Cell Biotechnology and Molecular Biology*,151–154.

Santos, O.C. 2002. *Chromosomic Characterization of Cupuacu, Theobroma grandiflorum (Willd. ex. Spreng.) Schum. (Sterculiaceae) Cultivadona Amazônia.* M.Sc. Thesis, Center for Biological and Health Sciences, Federal University of Sao Carlos, Sao Carlos.

Saunders, J.A., S. Mischke, E.A. Leamy, and A.A. Hemeida. 2004. Selection of international
molecular standards for DNA fingerprinting of *Theobroma cacao. Theoretical and Applied Genetics, 110*: 41–47.

Senthil Amudhan, M., and S. Elain Apshara. 2015. A comparative study on antioxidant activity and biochemical profile of exotic cocoa clones. *Journal of Plantation Crops, 43*(3): 231–235.

Smulders, M.J.M., D. Esselink, F. Amores, G. Ramos, D.A. Sukha, D.R. Butler, B. Vosman, and E.N. Van Loo. 2008. Identification of cocoa (*Theobroma cacao* L.) varieties with different quality attributes and parentage analysis of their beans. *IGENIC Newsletter, 12*: 1–13.

Soria, V.J. 1970. Principal varieties of cocoa cultivated in tropical America. *Cocoa Growers' Bulletin, 15*: 12–21.

Soria, V.J. 1975. The genetics and breeding of cacao. 18–24. *In*: 5th International Cocoa Research Conference.

Sounigo, O., R. Umaharan, Y. Christopher, A. Sankar, and S. Ramdahin. 2005. Assessing the genetic diversity in the International Cocoa Genebank, Trinidad (ICG, T) using isozyme electrophoresis and RAPD. *Genetic Resources and Crop Evolution, 52:* 1111–1120.

Thong, K.C., and W.L.Ng. 1980. *Growth and Nutrients Composition of Monocrop Cocoa Plants on Inland Malaysian Soils*, 262–286.

Toxopeus, H. 2001. Botany, types, and populations. *In*: *Cocoa*,11–37. Longman Ltd., London.

Turnbull, C.S., and A.B. Eskes. 2010. A Visual Aid to Identify Widely Distributed Cocoa Accessions. 43pp. CFC, Amsterdam, The Netherlands/ICCO, London, UK/Bioversity International, Italy.

Umaharan, R., J.M. Thévenin, S. Surujdeo-Maharaj, and D.R. Butler. 2005. Identification of resistance to Witches' Broom Disease. *In*: *The International Cocoa Genebank*, 161–169, Trinidad. Cocoa Producers' Alliance.

Vikraman Nair, R., V.K. Mallika, and S. Prasanna Kumari Amma. 2002. Breeding and Genetics. *In*: Balasimha D. [Eds.], *Cocoa*, 18–47. ICAR-CPCRI, Kasaragod, India.

Wood, G.A.R.1964. *Cocoa Growing in India*. Bournville, UK.

Wood, G.A.R, and R.A. Lass. 1955. *Cocoa*. Longman Group Limited, Longman House, England.

Wood, G.A.R, and R.A. Lass. 1985. *Cocoa*. (4th Edn.) Longman Group Limited, Longman House, Burnt Mill, Harlow, Essex CM20 2JE, England.

www.dccd.nic.in. Directorate of Cashewnut and Cocoa Development, Ministry of Agriculture and Farmers Welfare, Govt. of India.www.wcf.org.World Cocoa Foundation.

Young, A.M. 1994. *The Chocolate Tree: A Natural History of Cacao*. Smithsonian Institution Press.

Zamora, P.M., N.M. Orlido, and J.M. Capinpin. 1960. Ontogenetic and embryological studies in Theobroma cacao Linn. *Philippine Agriculturist*, *43*: 613–636.

Zhang, D., S. Mischke, E.S. Johnson, W. Phillips-Mora, and L. Meinhardt. 2009. Molecular characterization of an international cacao collection using microsatellite markers. *Tree Genetics & Genomes*, *5*: 1–10.

6

Genetic Diversity of Pomegranate (*Punica granatum* L.) and Sustainable Utilization

Linta Vincent[1,*] and *Sonam Meena*[2]

Introduction

The pomegranate (*Punica granatum*), historically consumed as a medicinal fruit in the Middle East for thousands of years, is now gaining global popularity. Whether consciously or not, consumers have embraced it as a superfood, and marketing has captured this opportunity, promoting it with labels, such as antioxidant, anti-inflammatory, antiviral, antibacterial, and antifungal properties (Johanningsmeier and Harris, 2011). The industrialization of pomegranate cultivation for various products such as juices, dehydrated seeds, dried rind, and value-added items has further amplified the demand for this versatile crop. In response to the escalating consumer interest, the cultivation area and production of pomegranate have witnessed a significant surge, reaching an expansive 300,000 hectares with an annual production of 3,000,000 tonnes (Melgarejo-Sánchez et al., 2015). From a producer's perspective, pomegranate cultivation is considered a highly profitable and rewarding agricultural venture. The crop's resilience, capacity for high dividends, adaptability to hi-tech systems, and potential for export contribute to its attractiveness.

Pomegranate is renowned for its rich array of nutritional compounds. These include hydrolyzabletannins, condensed tannins, flavonols, anthocyanins, and phenolic and organic acids, all of which have been associated with numerous health benefits (Nuncio-Jáuregui et al., 2015). Additionally, the fruit boasts a low pH value (< 4.0), high acidity, and a sugar content primarily composed of fructose and glucose. Apart from the juicy arils, the non-edible peel of the fruit contains valuable phenolics, minerals, and complex polysaccharides. Furthermore, the seeds are a plentiful source of proteins, crude fibers, vitamins, minerals, pectins, sugars, polyphenols, and isoflavones (Kandylis and Kokkinomagoulos, 2020). Notably, the oil extracted from the seeds is gaining popularity due to its high content of polyunsaturated fatty acids, such as linolenic and linoleic acids, along with punicic acid, oleic acid, stearic acid, and palmitic acid (Viuda-Martos et al., 2010).

[1] Scientist, Division of Fruit Crops, ICAR-IIHR, Bengaluru-560089.
[2] Ph.D. Scholar, Division of Fruit Crops, ICAR-IIHR, Bengaluru-560089.
* Corresponding author: linta.vincent@icar.gov.in

However, the impact of climate change poses challenges for pomegranate cultivation. Issues, such as fruit cracking, especially prevalent in arid and semi-arid regions, along with sun scald and aril browning, have emerged as serious threats. In non-traditional cultivation areas, new pests and diseases like bacterial blight and wilt have become significant concerns. The monoculture of specific varieties in each region exacerbates the spread of these stresses, prompting the scientific community to explore the diversity within pomegranates for gene introgression.

Given its status as a superfood crop and the unforeseen challenges arising from climate change, understanding the spread and domestication of the crop, its diversity, and the global genetic resources available is crucial. This knowledge is essential for devising conservation strategies and improving the crop's resilience to address current challenges.

Taxonomic Background

Pomegranate belongs to the order Myrtales and likely originated from Saxifragales. The Punicaceae and Sonneratiaceae families originated from the Lythraceae family. Morphological (Graham and Graham, 2014), molecular (Berger et al., 2016), and genomic (Yuan et al., 2018) evidence widely supports classifying pomegranate within the Lythraceae family rather than Punicaceae. Its well-resolved phylogenetic position is crucial for plant phylogenomics (Zhang et al. 2020). Graham and Thorne (1998) positioned the genus *Punica* in the Lythraceae, previously placed in Punicaceae. However, confusion persists among taxonomists regarding the pomegranate family. However, literature commonly places it in the Punicaceae family, consisting of a single genus *Punica* with two species (*Punica granatum* L. and *P. protopunica* Balf.f.), *P. protopunica* is endemic to Socotra island (Yemen). *P. protopunica* is suggested as the genus ancestor based on xylem anatomy.

Chromosomal studies are essential in crop breeding programs. The haploid number of chromosomes is n = 7 in *P. protopunica*, considered a primitive characteristic in an evolutionary context compared to *P. granatum*, which has n = 8 (Levin, 2006a). There were reports of 2n = 16, 18 in *P. granatum* (Smith, 1976), and it varied among cultivars. 'Dholka', 'Ganesh', 'Kandhari', "Muscat White", and 'Patiala' exhibited a chromosome count of 2n = 16, while "Double Flower",' 'Vellodu', and 'Kasmiri' showed a count of 2n = 18 (Nath and Randhawa, 1959; Raman et al., 1963). Singh et al. (2019) stated that its chromosome number can range from 2n = 14 to 2n = 18.

The species *P. granatum* has two subspecies, namely chlorocarpa and porphyrocarpa, distinguished by their geographical distribution. The former is found in the Transcaucasus region, while the latter is native to Central Asia (Sharma et al., 2009). The variations in karyotypes among different varieties and forms, evident in the shape and size of chromosomes, not only contribute to evolutionary phylogenetic studies (Levin, 2006a) but also offer valuable insights for modifying hybridization programs.

Currently, NCBI provides access to four reference genomes, which include the two traditional Chinese cultivars: 'Taishanhong' (2n = 18), characterized as semi-soft seeded (Chater et al., 2020; Gasic et al., 2020); 'Dabenzi' (2n = 18), known for its hard-seeded nature; and the soft-seeded Tunisian cultivar 'Tunisia' (2n = 16), widely adopted in China (Luo et al., 2020; Chen et al., 2022). Additionally, the collection comprises the soft-seeded Indian cultivar 'Bhagwa' (2n = 16) (Roopa Sowjanya et al., 2022).

History

The pomegranate has a rich and ancient history, spanning various cultures and religions. Long before its medicinal properties were recognized, the fruit held sacred significance in many of the world's major religions. References to the pomegranate can be found in Greek mythology, Persian mythology, Judaism, Buddhism, Christianity, and Islam. Across these religions, the pomegranate symbolizes fertility, sanctity, and abundance (Langley, 2000). In Greek mythology, the pomegranate symbolized the indissolubility of marriage; in Persian mythology, Isfandiyar gained invincibility by consuming a pomegranate; in Judaism, it is believed that a single pomegranate contains 613 seeds, representing each of the Bible's 613 commandments; in Buddhism, the pomegranate is considered one of three blessed fruits; in Chinese ceramic art associates the pomegranate with fertility and abundance; in Christianity and among Bedouin tribes, it is linked to fertility; in Islam, Korean descriptions depict a heavenly paradise featuring pomegranates (Langley, 2000).

The pomegranate might have been cultivated in Egypt before the time of Moses and was prominently featured in Egyptian mythology and the Old Testament of the Bible. An ancient Egyptian proverb even attests to the rejuvenating qualities of the fruit: "Eat a pomegranate and visit a bath, your youth will haste back." Even Egyptians were laid to rest with pomegranates. The Babylonians held the belief that chewing the seeds before battle granted invincibility. In Jewish tradition, the crown of the emperor was crafted in the shape of the calyx of a pomegranate, and it served as the personal emblem of the Roman emperor. (Chandra et al., 2010). The pomegranate, symbolizing life, fertility, and regeneration, derived from the myth of Persephone, has been adopted as the distinctive heraldic bearing (coat of arms) by institutions, such as the British Medical Association, the Royal College of Physicians of London, and three other colleges (Langley, 2000).

The term 'pomegranate' finds its roots in medieval Latin, specifically "pomum granatum" translating to "apple of many grains or seeds". Carl Linnaeus, formally named the pomegranate as *Punica granatum*. Notably, 'granatum' emphasizes the numerous seeds within the fruit, while the term 'Punica' may be misleading. In classical Latin, it referred to the Carthaginians, inhabitants of Carthage in modern Tunisia. Historical records indicate a rivalry between the Phoenicians and Carthaginians from 264–146 BC, followed by an amalgamation. The classical Latin term 'Punici' evolved from 'Phoenici' during this period, and concurrently, 'Punicus' emerged, suggesting an Eastern origin for the pomegranate. In ancient Egypt, the pomegranate was known as 'Arhumani'. The old Semitic term for it was 'Rimmon', from which the Hebrew name 'Ramon' and the Arabic name 'Rumman' originated. In the Near East (including Persia, Turkey, etc.), Central Asia (such as Afghanistan), and India, the fruit is commonly referred to as 'Anar' (Evreinoff. 1949; Diry, 1975).

Origin

Even though Linnaeus designated the pomegranate fruit, French botanist de Candolle identified its origin in Persia (current Iran) in 1882. Later, Vavilov suggested the pomegranate's origin westward to Asia Minor (Turkey) in 1992. However, Russian botanist Armen. L. Takhtajan expanded the origin from the Balkan Peninsula to the Western Himalayas. Recently, based on wild pomegranates in Transcaucasia and Central Asia from Iran and Turkmenistan to Northern India, these areas were considered the center of origin (Levin, 2006a; Holland et al., 2009).

More precisely, the origin and genetic diversity of pomegranate were well described with three mega-centers (primary, secondary, and tertiary) and five macro-centers (Middle Eastern, Mediterranean, Eastern Asian, American, and South African), according to Levin (2006b). The maximum diversity was observed in the primary mega-center (Middle Eastern mega-center) within the natural habitat boundaries of pomegranate (Iran, Afghanistan), included by N.I. Vavilov into the Front Asian center of origin and evolution of cultivated plants.

The secondary mega-centers (Mediterranean and Eastern Asian) formed during the introduction of pomegranates to the east and west. There was a better classification of local varieties by Arabs during the 13th century in Arabia and Mesopotamia, considered the secondary mega-center. Valuable varieties might have been collected from secondary centers and slowly evolved into tertiary centers. However, the formation of macro-centers inside the primary and secondary mega-centers took place nonsynchronously during the domestication process.

The mega-centers/macro-centers evolved during domestication, as evidenced by modern biotechnological tools. Fluorescent-AFLP (Amplified Fragment Length Polymorphism) markers analysis was performed on 85 pomegranates sourced from six geographical populations in Shandong, Anhui, Shaanxi, Henan, Yunnan, and Xinjiang Provinces in China (Yuan et al., 2007). The findings suggest that population-level diversity is lower than species-level diversity, with significant genetic differences between populations. Molecular variation analysis indicates that the majority of genetic variation in 49 accessions of wild Indian pomegranate resides within populations (54%) rather than between populations. Furthermore, gene flow in wild pomegranate accessions is lower compared to Chinese pomegranate (Yuan et al., 2007; Narzary et al., 2010). In Tunisia, the limited diversity of pomegranate, as revealed by SRR (Simple Sequence Repeats) markers analysis, highlights its narrow genetic background stemming from restricted origins (Sarkhosh et al., 2009).

In addition to mega/macro centers, Levin also mentioned endemic micro-genetic centers of pomegranate, including the Kandahar oasis, Afghanistan, Dashnabad, Uzbekistan, Azerbaijani, and the Central Asian region. The Mediterranean region varieties are local types selected and maintained by unknown individuals and propagated vegetatively, considered the primary gene pool, still growing wild in the Near East, Transcaucasia, Dagestan, and Asia Minor. Hybridization between cultivated and wild forms in these areas leads to secondary gene pool development. However, the tertiary gene pool consists of forms of *Punica protopunica*. The ancestor, *Punica protopunica*, was believed to be located in the Socotro Archipelago (Yemen) by Takhtajan (1966). A confusion arises with the double process of domestication, one involving wild species in Albania and Montenegro, and the other in Eastern Turkey and the South of the Caspian Sea.

Domestication

It is among the earliest fruit species to undergo domestication, with the process estimated to have commenced in the Neolithic era, around 8000–5000 BC (Levin, 2006a; Still, 2006). This timeframe aligns with the advent of modern humans and the initiation of agricultural activities between 2000–6000 BC, placing it in the cradle of Humankind as per a historical interpretation of the biblical narrative. The ancient Sumerian civilization, flourishing in the Tigris and Euphrates region (present-day southern Iraq) from 2500–2600 BC, played a pivotal role in introducing the pomegranate crop to the Zagros mountains (Iran, Northeastern Iraq, and Southeastern Turkey) during the third dynasty of Ura (4100–4200 BC). Historical literature also references the discovery of pomegranates in Jericho (modern Israel) around 5000 BC.

The expansion of pomegranates is substantiated by archaeological findings, including carbonized seeds and fragments discovered in early Bronze Age sites, such as Jericho and Arad in the third millennium BC. Remains of *Punica* species have been identified in Nimrud (Zohary and Spiegel-Roy, 1975). The Linnean Latin term *Punica* indirectly suggests the tree's dissemination throughout the Mediterranean, possibly linked to Phoenician trade routes that traversed the Mediterranean region (Touwaide and Appetiti, 2023). However, the crop was already cultivated in the Carthago (900 BC), the Pyrenean peninsula, Greece, and Italy (400 BC).

By the Middle Ages, pomegranates were domesticated throughout the Levant and in Egypt. At the beginning of the century, wild groves were still found thriving at the bottom of gorges throughout the Mediterranean and Central Asian regions (IPGRI, 2001). It is estimated that its cultivation started in Central Asia during 3000 BP. There is evidence of a double domestication of the pomegranate. Interestingly, it underwent a second, independent process of domestication, possibly between the 3rd and 2nd millennia, in the region corresponding to present-day Albania. Zohary and Spiegel-Roy (1975) proposed Albania as the first center of domestication of pomegranate (Hammer et al., 1996). Albania can be considered a 'Non-center' for domestication that took place autonomously since 3500 BC. Etymology also suggests the same; the pomegranate's name in Albania is 'Shege', which is an ancient name evolved during the 3rd–2nd millennium BC by the predecessors of Albanians (Xhuveli, 2012).

In the Mediterranean, pomegranate was introduced to India and, further east, to China. It was assimilated into the local medical systems, appearing in Ayurvedic and ancient traditional Chinese medicine. According to traditional narratives, the pomegranate has a linear history from Greek medical literature, translated into Arabic between the 9th–11th centuries. The Arabians then spread the knowledge eastward, passing through India to China. Genomic evidence indicates that Tunisia, a significant cultivar in China, is genetically closest to Bhagwa, an Indian cultivar. There is a possibility that it was geographically transferred from the Bhagwa population (Akparov et al., 2023). In contrast, Indian literature like Sushruta Samhita mentions pomegranate in Ayurvedic medicine, written during the 3rd–4th century. This could be another route of introduction of pomegranate to India by Alexander the Great, who arrived in Kashmir (327–325 BC), and Greeks settled in this region, possibly introducing the crop and its medicinal properties.

During the age of great geographic discoveries (AD 1600–1700) and later, pomegranate spread to many subtropical and tropical countries worldwide (Levin, 2006). By the early 1400s, the pomegranate had reached Indonesia (Morton, 1987). In the 1500s and 1600s, the Spanish introduced this species to Central America, Mexico, and South America (LaRue, 1980). The presence of pomegranate in California predates its official inclusion into the United States. The first clear evidence that the pomegranate was in the area that would become the United States dates back to the early 1700s. The cultivation of pomegranate in the southern and northern coastal regions of California was established in AD 1769 by Franciscan missionaries from Spain, led by Father Junipero Serra, who founded Roman Catholic Christian centers (Morton, 1987).

During the domestication process, morphological changes might have occurred while one can distinguish them from non-domesticated progenitors. Majorly, pomegranate selections are made based on flower, rind, and aril color, fruit size, sugar, and acid contents, keeping quality, yield, seed hardiness, and also resistance to biotic and abiotic stresses (Zhukovskij, 1950; Holland et al., 2009).

Collection and Assessment of Germplasm/Genetic Resources

Alphonse de Candolle, Zhukovsky, Harlan, and De Wet are notable figures who pioneered the exploration and recognition of diversity in crop plants, contributing to the development of concepts, such as the center of origin, the center of diversity, and the concept of gene pools (Vavilov, 1992; Zhukovsky, 1965; Harlan, 1971). Subsequently, the International Union for Conservation of Nature (IUCN), the Crop Ecology and Genetic Resources Unit of the FAO, and the International Board of Plant Genetic Resources (IBPGR) played pivotal roles in systematically conserving genetic resources.

In the realm of pomegranate conservation, concerted efforts have been made to avoid genetic erosion and vulnerability. The Mediterranean region considered the native homeland of pomegranate, harbors a multitude of local varieties selected and maintained through vegetative propagation by unknown individuals, as documented by various explorers (Levin, 1995; Ozyguven, 1996; Mars, 1996).

Significantly, the Garrygala Research Station in Turkmenistan boasts the world's largest *ex situ* pomegranate germplasm collection, established in 1934 and currently housing 1117 accessions from 26 countries across four continents, including cultivars obtained from commercial firms, scientific institutions, wild populations, and landraces (Levin, 1994, 1995). The Vavilov Research Institute of Plant Industry in Russia also maintains a substantial collection of 800 accessions.

A more modest germplasm collection of approximately 225 accessions is curated at the Agricultural Experiment Station of Elche (EEA-Elx) in Alicante, Spain, overseen by the Regional Government of Valencia. Notably, a systematic pomegranate breeding program commenced in 2008 at the EEA-Elx and the Valencian Institute of Agrarian Research, Spain (Zuriaga et al., 2022).

The US National Germplasm System's pomegranate collection comprises 192 accessions predominantly sourced from Turkmenistan. Additionally, other Mediterranean countries, such as Morocco, Tunisia, Greece, Turkey, and Egypt maintain significant collections of pomegranate germplasm (Mars, 1996).

India also plays a role in pomegranate conservation efforts, with Levin documenting wild material in Turkmenistan and wild pomegranates referred to as 'Daru' in Northern India (Levin, 1976, 1977,

1981). By the 1980s, wild collections were reported in India, prompting the initiation of breeding programs targeting yield, quality parameters, and pest and disease resistance (Sharma and Sharma, 1990). The Indian Himalayas hold a good diversity of wild pomegranates. The country has also introduced various pomegranate varieties from different countries through the efforts of the National Bureau of Plant Genetic Resources (NBPGR) in New Delhi. Additionally, the National Research Centre on Pomegranate in Solapur and the Indian Institute of Horticultural Research in Bengaluru are actively engaged in pomegranate collection and improvement programs across India. Notably, the NRC-Pomegranate Solapur holds approximately 350 collections, the ICAR-Indian Institute of Horticultural Research, Bengaluru maintains around 269 collections, and the ICAR-Central Institute for Arid Horticulture in Bikaner curates approximately 92 collections.

China has a national pomegranate germplasm repository in the Yicheng district of Shandong Province with a collection of 298 pomegranate accessions (Yuan and Zhao, 2019). Afghanistan has an *ex situ* national collection in Nangarhar and Kandahar provinces holds 59 Afghan varieties and 20 imported cultivars (Saeedi et al., 2012).

Characterization and Evaluation

A thorough characterization of any germplasm collection is essential for maximizing the potential of these resources. It enables the identification of promising accessions and facilitates the optimization of future breeding efforts through informed crossbreeding designs.

Morphological Characterization

The International Union for the Protection of New Varieties of Plants, headquarters in Geneva, has developed descriptors for pomegranate for distinctness, uniformity, and stability (DUS) testing. Similarly, the Protection of Plant Varieties and Farmers' Rights Authority (PPVFRA) in India has established guidelines for conducting DUS tests on pomegranates. These assessments encompass various plant growth characteristics, including plant vigor, growth habit, intensity of grey coloration of main branches, number of one-year-old shoots ending in thorns, and predominant number of leaves per node on young shoots. Leaf characteristics, such as length, width, ratio of length to width, apex shape, intensity of green color, petiole length, and anthocyanin coloration are also evaluated. Floral attributes, including calyx and corolla dimensions, color, surface smoothness, and number of flowers per node on one-year-old shoots, are examined. Fruit-related traits are scrutinized, focusing on length, width, length-to-width ratio, crown length, fruit over color, extent of over color, cross-sectional shape, skin thickness, sweetness, acidity, and juiciness. Aril characteristics, such as length, width, and color, as well as seed traits including length, width, and hardness, are also assessed.

Researchers have additionally utilized morphological features, such as anthocyanin presence in branches, petiole color, fruitful flower size, fruit albedo color, fruit heel and bottom shape, aril juice content, seed color, and size, number of fruits per tree, mean fruit weight, and mean tree yield to study Iranian germplasm (Farsi et al., 2023). Similar characterizations have been conducted on Indian germplasm consisting of 350 accessions (Parashuram et al., 2018) and Iranian germplasm (Khadivi et al., 2018). Furthermore, fruit volume, aril percentage, number of arils per fruit, 100-aril weight, 100-seed weight, and aril firmness have been employed by researchers to explore genetic diversity in 40 Indian *ex situ* collections (Parashuram et al., 2022). Among the 18 morphological descriptors used to characterize Himalayan indigenous collections, traits, such as the number of arils and aril weight were found to have significant relationships within all populations (Dar et al., 2022). In Tunisian germplasm, genetic diversity assessments focused on fruit size, color, and juice characteristics, which were identified as the most determinant criteria for cultivar clustering (Mars and Marrakchi, 1999). Pomegranate tegmen size and bundles present on its external surface have been utilized to characterize 13 Italian genotypes, serving as indirect indicators of seed hardiness (Beghe et al., 2022).

Cytological Characterization

Studies have investigated genetic diversity at the chromosomal level in various pomegranate cultivars. Sheidai and Noormohammadi (2005) reported 2n = 2x = 16 chromosomes in several cultivars of pomegranate collected from Iran. Furthermore, research conducted on 21 pomegranate cultivars in Iran found significant differences in chiasmata and chromosome pairing, shedding light on chromosomal diversity (Sheidai et al., 2012; Sheidai and Noormohammadi, 2005).

Furthermore, the identification of all the SUT gene family members present in the pomegranate genome, totaling ten SUT genes, has been accomplished in the Tunisian germplasm. This endeavor aids in understanding the xylem structure, which has implications for seed hardiness in pomegranates (Poudel et al., 2020).

Biochemical Characterization

As pomegranate is renowned for its nutraceutical and therapeutic properties, the review on biochemical compounds in pomegranate is widely available. However, the characterization of germplasm through biochemical approaches remains relatively limited, despite its immense potential to characterize germplasm in terms of nutraceutical and therapeutic aspects.

In Iranian pomegranate germplasm, significant attention has been given to studying major biochemical properties, such as the pH of juice, Total Soluble Solids (TSS), acidity, total anthocyanin content, polyphenol content, and antioxidant activity (Farsi et al., 2023). Additionally, the evaluation of genetic diversity in Indian *ex situ* pomegranate collections incorporates the assessment of ascorbic acid content along with the aforementioned biochemical parameters (Parashuram et al., 2022).

Moreover, oil content and fatty acid profiling have been utilized to characterize 13 genotypes of Italian pomegranate accessions, contributing to their valorization (Beghe et al., 2022). These biochemical approaches offer valuable insights into the diverse nutritional and therapeutic properties of pomegranate germplasm.

Molecular Characterization

Amplified Fragment Length Polymorphism (AFLP) markers were employed to characterize pomegranate germplasm in Northeastern Turkey (Ercisli et al., 2011). The genetic diversity of 94 accessions from the Spanish pomegranate collection at the Agricultural Experiment Station of Elche was analyzed using 19 microsatellite markers (Zuriaga et al., 2022). Microsatellite markers were also utilized to characterize Indian germplasm, revealing high genetic diversity and a broad genetic background (Perveen et al., 2023). Additionally, 29 Simple Sequence Repeats (SSR) markers were applied to understand the diversity in 40 pomegranate accessions in the Indian ex-situ collection (Parashuram et al., 2022). SSR markers were also employed to identify genetic diversity in 63 genotypes, including 55 accessions (landraces) from Italy, Turkmenistan, Japan, and the USA, and eight cultivars from Israel, established at the pomegranate repository of the Fruit Tree Unit of the Department of Soil, Plant, and Food Science at the University of Bari "Aldo Moro", Italy. The study identified polymorphic markers suitable for varietal identification and traceability (Giancaspro et al., 2023). ISSR markers were used to characterize the Indian Himalayan indigenous collection at PirPanjal Himalaya, revealing 58% variation among populations and 42% variation within populations (Dar et al., 2022).

Comparative analysis between two *Punica* species, *P. protopunica* Balf. fil., native to Socotra, and *P. granatum* L., was conducted using sequence-related amplified polymorphism (SRAP), target region amplification polymorphism (TRAP), and intron-targeted amplified polymorphism (ITAP). The study concluded that *P. protopunica* could be an ancestral species of *P. granatum* (Youssef et al., 2018).

Utilization of Germplasm

Utilization in Breeding and Improvement Programs

Floral Biology

The flowering pattern of pomegranate is influenced by prevailing climatic conditions, with tropical climates fostering year-round flowering, while subtropical regions typically experience annual flowering. In regions where continuous flowering occurs, three waves of flowering are observed, such as in South India coinciding with June, October, and March (Babu, 2010).

Pomegranates have a moderate juvenile period of 3–4 years, and a high number of seeds in a single fruit is a desirable trait for hybridization (Jalikop, 2010; Holland et al., 2009). The floral biology of pomegranate is distinctive, featuring two flower types—functional male and hermaphrodite (Wetzstein et al., 2013), or sometimes three types—functional male, intermediate, and hermaphrodite (Babu, 2010; Teixeira et al., 2013). This varies from genotype to genotype and environmental conditions (Chaudhary and Desai, 1993).

The unique aspect of pomegranate flower morphology is that all types contain both male and female organs. Male flowers have a negligible female part and do not set fruit, while intermediate types have somewhat developed female organs, resulting in small or dropped fruit. Hermaphrodite flowers, with fully developed female organs, produce the highest quality fruit. Intermediate flowers often possess degenerative ovaries despite a well-developed androecium, leading to infrequent fruit setting (Wetzstein et al., 2011).

Hermaphrodite flowers typically exhibit heterostyly (style raised above anthers), making pomegranate often cross-pollinated, and the presence of heterostyly varies among genotypes (Babu, 2010; Teixeira et al., 2013). Visual selection parameters like flower size, ovary width, and heterostyly prove successful in certain genotypes, influencing fruit set percentage in cross combinations (Ikram et al., 2023). Morphological studies conducted in the growing season are crucial for distinguishing between intermediate and hermaphrodite flower types before hybridization, with standards established to improve fruit set percentage.

The first sign of unfolding in the floral bud is the appearance of cracks at the apex, where the calyx lobes unite. Subsequently, sepals separate, and scarlet petals emerge. Anthesis timings vary among varieties, and the optimum temperature for anthesis also varies accordingly. Anther dehiscence in most varieties begins after full bloom and is influenced by maximum temperature and atmospheric humidity. Pomegranate pollen grains are abundant, with hermaphrodite flowers having more viable pollen than male flowers (Dhiman et al., 2021). Stigma maturity precedes anthesis by one day, reaching peak receptivity on the same and the second day after anthesis. Detailed floral biology descriptions can be found in the literature of Babu (2010).

Hybridization

Hybridization procedures involve collecting mature, unopened flower buds at the balloon stage of male flowers (identified by their smaller size). After removing sepals, petals, stigma, and style, the flower buds are placed under a 100-watt lamp overnight. The shredded pollen grains are then collected and utilized for pollination the following morning. For about-to-open flower buds of hermaphrodite flowers (recognized by their large size and urn-shaped ovary) at the partially open stage, emasculation is carried out by removing sepals, petals, and stamens. Following emasculation, pollen grains from the desired male plants, collected the previous day, are dusted onto the stigma of emasculated flowers using a camel hairbrush. Stigma receptivity is confirmed by observing a sticky stigmatic solution on the stigma. If the stigma is not receptive, it is covered with a butter paper bag and crossed the next day. After pollen dusting, the flower buds are covered with butter paper bags and tagged. To enhance crossing success, the remaining flowers on the branch are removed (Ikram et al., 2023).

Breeding Objectives

In any pomegranate breeding program, the primary focus is on achieving high yields. Additionally, breeding objectives for pomegranate include enhancing organoleptic qualities, juice content, seed mellowness, and the overall appearance of fruits, alongside maximizing fruit yield (Jalikop, 2010).

Consumer preferences for bold, juicy arils with soft seeds can vary and are influenced by indigenous markets and the standards of each country. Breeding objectives are tailored accordingly, considering whether the purpose is fresh consumption or processing. Given the export-oriented nature of pomegranate, breeders should align with international market standards, emphasizing factors, such as fruit size, shape, skin color, aril color, easily separable bold arils, seed size, sugar-acid blend, and beneficial nutrients in the juice. Other crucial considerations include ripening time, shelf life, and post-harvest quality. Over the years, there has also been a focus on breeding for physiological disorders like fruit cracking and aril browning/internal breakdown of arils (Jalikop, 2010).

Breeding objectives might differ from a cultivation perspective, as the pomegranate crop is adapted to different climatic and soil conditions. Ideal traits include an evenly spreading canopy with thorn-free, strong branches (Jalikop, 2010). A dwarf plant habit is also preferred for easy mechanical operations and harvesting (Glozer and Ferguson, 2008). In hot, dry areas, breeding objectives may shift towards developing drought and high-temperature-tolerant varieties.

Addressing region-specific diseases, such as bacterial blight and wilt in India and neighboring countries, is a crucial aspect of breeding programs (Jalikop et al., 2005). In frost-prone regions like Turkmenistan and Azerbaijan, breeding for frost resistance becomes a major objective (Levin, 1979). Resistance to fruit borers, including the carob moth (*Ectomyelois ceratoniae*) and anar butterfly (*Viracholais ocrates*), is a significant focus.

Rootstock breeding has recently been initiated in pomegranate with multiple objectives, including the development of precocious, size-controlling rootstocks, reduced suckering, compatibility with a wide spectrum of commercial varieties, and resistance to abiotic stresses, such as waterlogging, root, and salinity. Resistance to biotic stresses, including root-knot nematode, wilt (*Ceratocystis fimbriata*), and shot hole borer (*Xyleborus perforans*), is also a key consideration in these breeding efforts (Sarkhosh and Chaparro, 2022).

Breeding Methods

The majority of pomegranate cultivars cultivated worldwide have arisen from the selection of seedlings in wild populations, demonstrating intra-species variation (Mars, 2000). This is attributed to the pomegranate's frequent cross-pollination and high heterozygosity. The development of new varieties can be achieved through conventional breeding methods, such as seedling selection, hybridization with subsequent selection, and mutation breeding.

Seedling Selection

Creating a seedling population is relatively straightforward due to the abundance of seeds, good germination, and high heterozygosity in pomegranate. Numerous promising varieties have been developed through simple seedling selection, often named based on the cultivation location or distinctive fruit characteristics. For instance, in Israel, 'Emek' was chosen from open-pollinated seeds, and 'Kamel', similar to 'Wonderful' but with dark red skin and earliness, was another selection (Holland et al., 2007). In Tajikistan, 'Apsheronsk' and 'Melesi' were selected for their flavor and yield (Rozanov, 1972). Wang et al. (2006) selected 'Zaoxuan 018' from 'Dahongpao' and 'Zaoxuan 027' from 'Daqinpi' in China as chance seedlings. The Indian cultivar 'Ganesh', known for its soft seeds, originated from seedling selection in 'Alandi' (Jalikop, 2003), and subsequently, 'G-137' was identified from the open-pollinated progeny of Ganesh.

Hybridization and Selection

In pursuit of specific breeding objectives, suitable donor parents can be identified from the germplasm collection. Following meticulous screening, numerous pomegranate varieties, as highlighted in Table 1, emerge as valuable candidates for hybridization and subsequent selection.

Once superior selections are made, they have to be multiplied clonally, so that the genetic integrity of the selected progeny is not lost. The varieties developed have to be tested against the ruling variety in multi-locations.

Table 1. Source of germplasm for various traits in pomegranate.

Sl. No.	Trait	Source/Donor Parent
1.	Large fruit, juicy, sweet, pink arils	Ganesh, Teipitian
2.	Large fruit size	Bopi, Teipitian
3.	Red skin, pink arils	Duanzhihong
4.	Earliness	87-Qing 7, Alack
5.	Sweet-sour, green skin	Dabaitian, Heyinruanzi
6.	Resistance to fruit splitting	ApsheronskiiKrasnyi, Frantsis, KyrmyzKabukh
7.	High juice content	Slunar, Pirosmani, Vedzisuri
8.	Attractive thick skin, late maturing	Mridula, Bhagwa
9.	Thin skin, early maturing	Ruby, Muskat, Dholka
10.	Thick red rind, red arils, late ripening	Malas-e-saveh, Rabab-e-Neyriz, Malas-e-yazdi, Naderi-e-bud-rood
11.	Rich Vitamin C	Siyah
12.	Yellow skin	Lefan, Kabul Yellow
13.	Red skin, red arils, hard seeds, sweet-sour	Janarnar
14.	Red glossy skin, sweet-sour	Wonderful
15.	Sweet pink arils	Rosh Hapered, Malisi
16.	Sweet fruits with soft seeds and red arils	Mollar de Elche, ME1, ME5, ME6, ME14
17.	Red color, high yield	Kamel, Akko
18.	High yield	ME15, ME16, ME17, Agria de Blanca, Albar de Bianca, Borde de Albatera, BA1, Borde de Blanka, Casta del Reino de Ojos, CRO1, Mollar de Albatera, Mollar de Orichula, Lnxuan-, Hicaznar
19.	Resistance to moisture stress	TaishanDahongshiliou, Nana
20.	Adaptation to alkaline soils	Yushiliu 1, Yushiliu 2
21.	Bacterial blight resistance	99A

Source: Jalikop (2010).

Mutation Breeding

While mutation breeding proves to be a valuable tool for generating variability, it necessitates the cultivation of a large population to identify desirable traits, making it a somewhat unpredictable method. Furthermore, a significant drawback is that the majority of mutants tend to be recessive, rendering them limited or no practical use, thereby reducing the accessibility of the technique. This might explain why limited experiments have been conducted in pomegranate breeding, given the inherent challenges.

Experiments involving seeds and cuttings exposed to Gamma (γ) radiations within the range of 1–40 kR have been conducted, leading to the identification of types that surpass the initial material in terms of fruit yield, size, and quality. Notably, varieties bearing sweet fruit exhibited higher sensitivity to radiation compared to those with acid-sweet or acid fruit. Akhund-zade et al. (1977) delved into the effects of irradiation on cuttings, seeds, and pollen, revealing significant variability at doses ranging from 5–10 kR. In China, natural mutants (bud sports) like Hongma-naozi, Taihanghong, and Hongyushizi were isolated from different varieties. Levin (1990) explored the treatment of pomegranate seeds with chemical mutagens and x-rays, resulting in forms with desirable fruit quality and good juice, exhibiting superior keeping quality. The most promising results were achieved with x-rays at doses ranging from 10–20 kR. Mutants were also successfully bred through N, N-dimethyl N-nitrosourea treatment, with one notable form being the soft-seeded seedling Sverkhrannil, signifying 'super-early'. The application of chemical mutagenesis in pomegranate breeding has also been considered.

Conservation of Useful Germplasm/Genetic Resources

The conservation of local pomegranate germplasm is a vital undertaking at the primary gene pool level, with established collections in various Mediterranean countries, such as Spain, Morocco, Tunisia, Greece, Turkey, and Egypt (Mars, 1999). Azerbaijan, Ukraine, Uzbekistan, and Tajikistan boast relatively extensive collections, each comprising 200–300 accessions, including wild, semi-wild, and cultivated forms. Remarkably, the Turkmenistan Experimental Station of Plant Genetic Resources maintains the world's largest collection, with 1,117 accessions gathered from 26 countries on four continents, showcasing the morphological diversity of cultivars and materials obtained from commercial firms, scientific institutions, and wild populations (Levin, 1994, 1995).

Notably, a core collection, representing 10% of the main collection, has been established. The unique wild relative, *P. protopunica*, was successfully located on Socotra Island (Yemen) during a 1989 expedition, resulting in the collection of seeds with high germination success. The germplasm exhibits substantial morphological diversity, encompassing variations in shape, color (petal, skin, and aril color), resistance to splitting, ripening date, taste, flavor, juice content, and seed size. A comprehensive list of reported germplasm is provided in Table 2.

International groups dedicated to pomegranate biodiversity collaborate to collect, conserve, and evaluate germplasm globally (Frison and Servinsky, 1995; Mars, 2000; Fadavi et al., 2006; Levin, 2006b; Still, 2006; Zamani et al., 2007). Verma et al. (2010) extensively documented genetic diversity in pomegranates for several key countries, offering detailed listings of important cultivars and germplasm on a country-by-country basis.

Although the genetic base in pomegranate is limited, India exhibits significant variability due to seedling trees, landraces, primitive cultivars, and diverse agroclimatic conditions (Jalikop and Kumar, 1990). The National Bureau of Plant Genetic Resources (NBPGR) has collected wild pomegranate

Table 2. Major varieties are cultivated in major pomegranate producing countries.

Major Countries Growing Pomegranate	Major Varieties Cultivated
India	Alandi or Vadki, Dholka, Kandhari, Kabul, Muskati Red, Paper Shelled, Spanish Ruby, Ganesh (GB I), G 137, P 23, P 26, Mridula, Aarakta, Jyoti, Ruby, IIHR Selection, Yercaud 1
Afghanistan	Kandahari and Bedana
Iran	Shisheh-e cape Ferdous, Malas-e Sava, Robab-e Neyriz, Ghojagh-e Qom, Malas-e Yazd, Khazar-e Bardaskan, Nadari-e Badrood, Ardestani-e Mahalat, and Bejestani-e Gonabad
Turkey	Hicaznar, Çekirdesksiz VI, Silifkeaşısı, Katırbaşı, and Lefan
China	'Daqingpitian', 'Dahongpao', 'Dahongpitian', 'Damayatian', 'Taishanhong', 'Taishansanbai', 'Taishanjinhong, Dahongtian', 'Dabaitian', 'Dagangmazi', 'Heyintiepi', 'Heyintongpi', 'Yudazi' and 'Dongyan', Mantianhong', Taihanghong', 'Tianlvzi', 'Tianguangyan', 'Houpitianshazi', 'Suanlvzi', 'Huahongpi', 'Huopao', Qingpiruanzi
Kazakhstan	Ashyk-anar, Shah-Nar
USA	Wonderful
Iraq	Salakhani', 'Zivzik'
Pakistan	Desi, Kandhari, and Bedana
Syria	al-Lafan, al-Faransy, al-Asfoury
Italy	Shany, Emek, Ako, and Wonderful
Spain	Mollar de Elche, Mollar de Valencia, and Wonderful

germplasm from Western Himalaya, while the Central Institute of Arid Horticulture (CIAH) and, later, the National Research Centre on Pomegranate (NRCP) have played key roles in maintaining collections. NRCP established a National Field Gene Bank in 2007, currently holding 187 accessions, with 31.56% being exotic and 68.44% indigenous. Several Indian agricultural universities and ICAR institutes also contribute to maintaining pomegranate germplasm in their repositories.

Verma et al. (2010) extensively documented the genetic diversity in pomegranates across several important countries, along with detailed listings of significant cultivars and germplasm on a country-by-country basis. Vincent et al. (2019) compiled a comprehensive list of germplasm collected worldwide from various regions.

Conclusion

The challenges facing pomegranate cultivation are intensifying at a rapid pace due to climate change. Biotic stresses, such as bacterial blight and wilt, have become major concerns in Southeast Asian countries, while various physiological issues and pests are emerging as significant threats in the northern hemisphere. Therefore, understanding the availability of germplasm and the sources of desirable traits will be invaluable for future breeding programs. Despite the abundance of available germplasm, the characterization and utilization of pomegranate germplasm remain limited. However, there is significant potential for diversifying the use of pomegranate in various aspects.

References

Akhund-zade, I.M., E.E. Fedorova, G.M. Mamedov, and Z.D. Iskenderova. 1977. Study of the cytogenetic characteristics of pomegranate. *Ispolz-biofiz-meto-dov-v-genet-selektsioneksperimente*, *17*: 8–9.

Akparov, Z., S. Hajiyeva, M.Abbasov, S.Kaur, A.Hamwieh, A.M. Alsamman, and A. Jighly. 2023. Two major chromosome evolution events with unrivaled conserved gene content in pomegranate. *Front. Plant Sci.*, *14*: 103–9211.

Babu K.D. 2010. Floral biology of pomegranate (*Punica granatum* L.). *Fruit Veg. Cereal Sci. Biotechnol.*, *4*(2): 45–50.

Beghè, D., C. Giordano, B. Pizzo, V. Lolli, A. Caligiani, F. Bonzanini, and R. Petruccelli. 2022. Characterization of Pomegranate (*Punica granatum* L.) through the Morphoanatomical and Lipid Profile of Seeds for the Valorisation of Italian Accessions. *Sustainability*, *14*(22): 14780.

Berger, B.A., R. Kriebel, D. Spalink, and K.J. Sytsma. 2016. Divergence times, historical biogeography, and shifts in speciation rates of Myrtales. *Mol. Phylogen. Evol.*, *95*: 116–136.

Chandra, R.K., DhineshBabu, and T.V. Jadhav. 2010. Origin, history, and domestication of pomegranate. *Fruit Veg, Cereal Sci. Biotech.*, *4*(2): 1–6.

Chater, J.M., A. Yavari, A, Sarkhosh, Z, Jia, D.J. Merhaut, and J.E. Preece. 2020. World pomegranate cultivars. *Pomegranate*, 157–195.

Chaudhary, S.M., and U.T. Desai. 1993. Effects of plant growth regulators on flower sex in pomegranate (*Punica granatum* L.). *Indian Journal of Agriculture Science*, *63*: 34–35.

Chen, Y. H., H.F. Gao, W.A.N.G. Sa, X.Y. Liu, Q.X. Hu, Z.H. Jian et al. 2022. Comprehensive evaluation of 20 pomegranate (*Punica granatum* L.) cultivars in China. *J. Integr. Agric.*, *21*(2): 434–445. doi:10.1016/S2095-3119(20)63389-5.

Dar, T.U.H., W.W. Tantray, S.U. Islam, Z.A. Mangral, A.A. Khuroo, R. Ahmad, and B.A. Bhat. 2022. Characterizing wild germplasm of neglected and underutilized crops: A case study of pomegranate (*Punica granatum* L.) from remote PirPanjal Himalaya. *Biochemical Systematics and Ecology*, *105*: 104524.

Dhiman, B., Y. Kumar, and Y.C. Hu. 2021. A general purpose multi-fruit system for assessing the quality of fruits with the application of recurrent neural network. *Soft Computing*, *25*(14): 9255–9272.

Diry, N. 1975. *The Pomegranate. Fruit Orchards*, 302–307. Publ. Aleppo University, Fac. of Agriculture, Syria.

Ercisli, S., E. Kafkas, E. Orhan, S. Kafkas, Y. Dogan, and A. Esitken. 2011. Genetic characterization of pomegranate (*Punica granatum* L.) genotypes by AFLP markers. *Biol. Res.*, *44*(4): 345–350.

Evreinoff, V.A. 1949. Le grendier. *Fruits Journal of Agriculture Science*, *19*: 320–336.

Fadavi, A., M. Barzegar, and M. H. Azizi. 2006. Determination of fatty acids and total lipid content in oilseed of 25 pomegranate varieties grown in Iran. *Journal of Food Composition and Analysis*, *19*(6–7): 676–680.

Farsi, M., M. Kalantar, M. Zeinalabedini, and M.R. Vazifeshenas. 2023. First assessment of Iranian pomegranate germplasm using targeted metabolites and morphological traits to develop the core collection and modeling of the current and future spatial distribution under climate change conditions. *Plos One*, *18*(2) : e0265977.

Frison, E.A., and Servinsky. 1995. *Directory of European Institutions Holding Crop Genetic Resources Collections* (*4th*), (Vol. 1), 499. International Plant Genetic Resources Institute, Rome.

Gasic, K., J.E. Preece, and D. Karp. 2020. Register of new fruit and nut cultivars list 50. *Hort. Science*, *55*(7): 1164–1201. doi: 10.21273/HORTSCI50register-20.

Giancaspro, A., S. L.Giove, I.Marcotuli, G. Ferrara, and A. Gadaleta. 2023. Datasets for genetic diversity assessment in a collection of wild and cultivated pomegranates (*Punica granatum* L.) by microsatellite markers. *Data in Brief*, *49*: 109346.

Glozer, K., and L. Ferguson. 2008. *Pomegranate Production in Afghanistan*. UCDAVIS College of Agricultural & Environmental Sciences, (s32).

Graham, S.A., and A. Graham. 2014. Ovary, fruit, and seed morphology of the Lythraceae. *Int. J. Plant Sci.*, *175*: 202–240.

Graham, S.A., and R. Thorne R. 1998. Validation of subfamily names in Lythraceae. *Taxon.*, *47*: 435–436.

Hammer, K., G. Laghetti, G. Olita, P. Perrino, and L. Xhuveli. 1996. Collecting in the Albanian mountains in 1995. *Plant Genet. Resour. Newslett.*, *107*: 36–40.

Harlan, J.R. 1971. Agricultural Origins: Centers and Noncenters: Agriculture may originate in discrete centers or evolve over vast areas without definable centers. *Science*, *174*(4008): 468–474.

Holland, D., K. Hatib, I. Bar-Yaakov, E. Yonay and F. Abd El Hadi. 2007. Shani-Yonav Pomegranate. Hort. Science, *42*(3): 710–711.

Holland, D.K., Hatib, and I. Bar-Ya'akov. 2009. Pomegranate: botany, horticulture, breeding. pp. 127–191 *In*: Janick, J. [Ed.], *Horticultural Reviews* (Vol. 35). John Wiley & Sons, New Jersey.

Ikram, S., W. Shafqat, S.U. Rehman, M.A. Qureshi, S.U. Din, S. Ikram, and M.J. Jaskani. 2022. Floral Characterization of Pomegranate Genotypes to Improve Hybridization Efficiency. *Plants*, *12*(1): 165.

IPGRI (International Plant Genetic Resources Institute). 2001. *Regional Report*, CWANA, Rome, 1999–2000, 145–178.

Jalikop, S.H., and P.S. Kumar. 1990. Use of a gene marker to study the mode of pollination in pomegranate (*Punica granatum* L.). *Journal of Horticultural Science*, *65*(2): 221–223.

Jalikop, S.H. 2003. Rosetted siblings in F2 of a cross in pomegranate (*Punica granatum* L.) can be useful model for rosetting investigations. *Euphytica*, *131*: 333–342.

Jalikop, S.H. 2010. Pomegranate breeding. *Fruit Veg. Cereal Sci. Biotechnol.*, *4*(2): 26–34.

Jalikop, S.H., R.D. Rawal, and R. Kumar. 2005. Exploitation of sub-temperate pomegranate Daru in breeding tropical varieties. *Acta Horticulturae*, *696*: 107–112.

Johanningsmeier, S.D., and G.K. Harris. 2011. Pomegranate as a functional food and nutraceutical source. *Annu. Rev. Food Sci. Technol.*, *2*: 181–201.

Kandylis, P., and E. Kokkinomagoulos. 2020. Food applications and potential health benefits of pomegranate and its derivatives. *Foods*, *9*(2): 122.

Khadivi, A., D. Ayenehkar, M. Kazemi, and A. Khaleghi. 2018. Phenotypic and pomological characterization of a pomegranate (*Punica granatum* L.) germplasm collection and identification of the promising selections. *Sci. Hortic.*, *238*: 234–245.

Langley, P. 2000. Why a pomegranate? *BMJ.*, *321*(7269): 1153–1154.

LaRue, J.H. 1980. *Growing Pomegranates in California*. Pamphlet of the University of California Division of Agricultural Sciences, 2459, 1977.

Levin, G.M. 2006a. *Pomegranate* (1st Edn.), 1–129. Third Millennium Publishing, East Libra Drive Tempe, AZ.

Levin, G.M. 2006b. *Pomegranate Roads: A Soviet Botanist's Exile from Eden*, 15–183. *In*: Baer, B.L. [Ed.], Floreat Press. Forestville, CA.

Levin, G.M. 1990. Induced mutagenesis in pomegranate. *Dostizheniya-naukiv-praktiku:-Kratkie-tezisy-dokladov-k-predstoyashchei-nauchnoi-konferentsii:-Puti-uskoreniya selektsionnogo-protsessa-rastenii*, 126–128.

Levin, G.M. 1976. The pomegranate of the western Kopetdag as a genetic resource for breeding and a target for nature conservation. *Tezisy-dokl-1-oiNauch-konf-po-okhrane-prirody-Turkmenistana*, 71–72.

Levin, G.M. 1977. Variation in wild pomegranate in the western Kopetdag. *Turkmenistan-SSR-Ylymlar-Akademijasynyn-Habarlary,-Biologik-Ylymlaryn*, *5*: 40–45.

Levin, G.M. 1979. Polymorphism of pomegranate seeds and the prospects of using it in breeding, *50*: 921. *In*: *Referrativnyl Zhurnal*, 11.65.349 vide Plant Breeding Abstracts 1075.

Levin, G.M. 1981. Wild pomegranate (*Punica granatum* L.) in Turkmenistan. *Turkmenistan SSR ylymlar Akademijasynyn Habarlary Biologik Ylymlaryn*, *2*: 60–64.

Levin, G.M. 1994. Pomegranate (*Punica granatum*) plant genetic resources in Turkmenistan. *Plant Genet. Resour. News*, *97*: 31–36.

Levin, G.M. 1995. Aspects of pomegranate culture in Turkmenistan. *Plant Genetic Resources Newsletter*, *102*: 29–31.

Luo, X., H. Li, Z. Wu, W. Yao, P, Zhao, D. Cao, et al. 2020. The pomegranate (Punica granatum l.) draft genome dissects genetic divergence between soft and hard-seeded cultivars. *Plant Biotechnol. J.*, *18*(4): 955–968. doi: 10.1111/pbi.13260.

Mars, M. 1996. Pomegranate genetic resources in the Mediterranean region, *In*: *Proceedings of Ist MESFIN Plant Genetic Resources Meeting*, 345–354. Tenerife, Spain.

Mars, M. 2000. Pomegranate plant material: Genetic resources and breeding: A review. *Options Mediterraneennes Serie A*, *42*: 55–62.

Mars, M., and M. Marrakchi. 1999. Diversity of pomegranate (*Punica granatum* L.) germplasm in Tunisia. *Genetic Resources and Crop Evolution*, *46*: 461–467.

Melgarejo-Sánchez, P., J.J. Martínez, P. Legua, R. Martínez, F. Hernández, and P. Melgarejo. 2015. Quality, antioxidant activity, and total phenols of six Spanish pomegranates clones. *Sci. Hortic.*, *182*: 65–72.

Morton, J.F. 1987. *Fruits of Warm Climates*. JF Morton, Miami.

Narzary, D., T.S. Rana, and S.A. Ranade. 2010. Genetic diversity in inter-simple sequence repeat profiles across natural populations of Indian pomegranate (*Punica granatum* L.). *Plant Biology*, *12*(5): 806–813.

Nath, N. and G.S. Randhawa. 1959. Studies on Floral Biology in the Pomegranate (*Punica granatum* L.) I. Flowering Habit, Flowering Season, Bub Development, and Sex-Ratio in Flowers. *Indian Journal of Horticulture*, *16*(2): 61–68.

Nuncio-Jauregui, N., P. Nowicka, S. Munera-Picazo, F. Hernández, Á. Carbonell-Barrachina, and A. Wojdyło. 2015. Identification and quantification of major derivatives of ellagic acid and antioxidant properties of thinning and ripe Spanish pomegranates. *J. Funct. Foods*, *12*: 354–364.

Ozguven, A.I. 1996. The genetic resources of pomegranate (*Punica granatum*) in Turkey. *In*: Proceedings of Ist MESFIN Plant Genetic Resources Meeting, 269–284. Tenerife, Spain.

Parashuram, S., N.V. Singh, N.N. Gaikwad, G. Corrado, P. Roopa Sowjanya, B. Basile, and P.G. Patil. 2022. Morphological, Biochemical, and Molecular Diversity of an Indian *Ex Situ* Collection of Pomegranate (*Punica granatum* L.). *Plants*, *11*: 3518.

Parashuram, S., N.V. Singh, P. Roopasowjanya, K.D. Babu, V. Sangnure, S, Singh, and R.K. Pal. 2018. Germplasm conservation and phenotypic characterization of pomegranate (*Punica granatum* L.) germplasm accessions for various morphological and physico-chemical characters. *Journal of Pharmacognosy and Phytochemistry*, *7*(2S): 114–118.

Perveen, N., S.S. Cholin, K. Hipparagi, G. Prabhuling, B.N.S. Murthy, and D. Peerjade. 2023. Molecular diversity assessment among the pomegranate genotypes belonging to diverse genetic background using microsatellite markers. *Acta Physiol. Plant.*, *45*(7): 92.

Poudel, K., X. Luo, L. Chen, L. Jing, D. Xia, X. Tang, and S. Cao. 2020. Identification of the SUT gene family in pomegranate (*Punica granatum* L.) and functional analysis of PgL0145810. 1. *Int. J. Mol. Sci.*, *21*(18): 6608.

Raman, V.S., P.C. Kesavan, G. Manimekalai, W.M. Alikhan, and S.R. Rangaswami. 1963. Cytological studies in some tropical fruit plants: Banana, annona, guava, and pomegranate. *South Indian Horticulture*, *11*: 27–33.

Roopa Sowjanya, P., P. Shilpa, P.G. Patil, K.D. Babu, J. Sharma, V.R. Sangnure, et al. 2022. Reference quality genome sequence of Indian pomegranate cv. 'Bhagawa' (*Punica granatum* L.). *Front. Plant Sci.*, *13*: 947164. doi: 10.3389/fpls.2022.947164.

Rozanov, B.S. 1972. Results of varietal trials and breeding of pomegranate in the south of Tajikistan. *Tr-Vakhmsk-zonal'-n-opyt-st-subtrop-kul'-tur.*: 57–68.

Saeedi, A.M., G. Mohammad, G.R. Samadi, S. Abdiani, and E. Giordani. 2012. The pomegranate national collection of Afghanistan. *In*: Melgarejo P., and Valero D. [Eds.], *II International Symposium on the Pomegranate*, 57–60. CIHEAM, Universidad Miguel Hernández.

Sarkhosh, A., and J.X. Chaparro. 2022. Rootstock breeding in pomegranate (*Punica granatum* L.), *In*: *Vth International Symposium on Pomegranate and Minor Mediterranean Fruits*, *1349*: 39–46.

Sarkhosh, A., Z. Zamani, R. Fatahi, and H. Ranjbar. 2009. Evaluation of genetic diversity among Iranian soft-seed pomegranate accessions by fruit characteristics and RAPD markers. *Sci. Hortic.*, *121*(3): 313–319.

Sharma, S.D., and V.K. Sharma. 1990. Variation for chemical characters in some promising strains of wild pomegranate (*Punica granatum* L.). *Euphytica*, *49*(2):131–133.

Sheidai, M., Noormohammadi, Z. 2005. Chromosome pairing and unreduced gamete formation in 19 pomegranate (*Punica grantum* L.) cultivars. *Cytologia*, *70*(3), 257–265.

Sheidai, M., S., Kolahizadeh, Z. Noormohammadi, N. Azani, and M. Nikoo. 2012. Correlation between geography and cytogenetic diversity in Pomegranate (*Punica granatum* L.) cultivars in Iran. *Acta Botanica Brasilica*, *26*: 953–965.

Singh, T.J., Gupta, T., and Sharma, S. 2019. Development and purity identification of hybrids by using molecular marker in wild pomegrante (Punica granaum L.). *Scientia Horticulturae*, *247*: 436–448.

Smith, P.M. 1976. Minor crops. *In*: N.W. Simmonds [Ed.], *Evolution of Crop Plants*, 301–324. Longman, London.

Still, D.W. 2006. Pomegranates: A botanical perspective. *In*: N.P Seeram, R.N. Schullman, and D. Heber [Eds.], *Pomegranates: Ancient Roots to Modern Medicine*, 199–209. CRC Press Taylor and Francis Group, Boca Raton.

Takhtajan A.L. 1966. *Systema et phylogenia Magnoliophytorum*. Mosqua; Leningrad: Nauka. 611 p. [In Russian] (Тахтаджян А. Л. 1966. *Система и филогения цветковых растений*. М.; Л.: Наука. 611 с.).

Teixeira da Silva J.A., and T.S. Rana. 2013. Pomegranate biology and biotechnology: A review. *Sci. Hortic.*, *160*: 85–107.

Touwaide, A., and E. Appetiti. 2023. *Herbs in History*. American Herbal Products Association, Silver Spring, MD.

Vavilov, N.I. 1992. *Origin and Geography of Cultivated Plants*. Cambridge University Press, Cambridge, UK, Transl. by D. Love.

Verma, N., A. Mohanty, and A. Lal. 2010. Pomegranate genetic resources and germplasm conservation: A review. *In*: Chandra, R. [Ed.], *Pomegranate*, 120–125. *Fruit Veg. Cereal Sci. Biotechnol.*, *4*(Special Issue 2).

Vincent, L., P.L. Anushma, C. Vasugi, A. Rekha, and Shiva, B. 2019. Genetic resources of tropical fruits. *In*: *Conservation and Utilization of Horticultural Genetic Resources*, 79–116. Springer, Singapore.

Viuda-Martos, M., J. Fernández-López, and J.A. Pérez-Álvarez. 2010. Pomegranate and its many functional components as related to human health: A review. *Compr. Rev. Food Sci. F.*, *9*(6): 635–654.

Wang Y, X.L. Yin, and L..F. Yang. 2006. Breeding of Zaoxuan 018 and 027 pomegranate selections. *China Fruits*, *4*: 6–8.

Wetzstein, H. Y., W. Yi, J.A. Porter, and N. Ravid. 2013. Flower position and size impact ovule number per flower, fruit set, and fruit size in pomegranate. *Journal of the American Society for Horticultural Science*, *138*(3): 159–166.

Xhuveli, L. 2012. Albania, the domestication country for pomegranate (*Punica granatum* L.). *Genet. Resour. Crop Evol.*, *59*: 1605–1610.

Youssef M., A.S. Alhammadi, and J.H. Ramírez-Prado. 2018. Remarks on genetic diversity and relationship of *Punica protopunica* and *P. granatum* assessed by molecular analyses. *Genet. Resour. Crop Evol.*, *65*: 577–590.

Yuan, Z., and X. Zhao. 2019. Pomegranate genetic resources and their utilization in China. *Acta Hortic.*, *1254*: 49–56. https://doi.org/10.17660/ActaHortic.2019.1254.8.

Yuan, Z., Y. Fang, T. Zhang, Z. Fei, F. Han, C. Liu, M. Liu, W. Xiao, W. Zhang, S. Wu, M. Zhang, Y. Ju, H. Xu, H. Dai, Y. Liu, Y. Chen, L. Wang, J. Zhou, D. Guan, M. Yan, Y. Xia, X. Huang, D. Liu, H. Wei, and H. Zheng. 2018. The pomegranate (*Punica granatum* L.) genome provides insights into fruit quality and ovule developmental biology. *Plant Biotechnol J.*, *16*(7): 1363–1374. doi: 10.1111/pbi.12875.

Yuan, Z., Y. Yin, J. Qu, L. Zhu, and Y. Li. 2007. Population genetic diversity in Chinese pomegranate (*Punica granatum* L.) cultivars revealed by fluorescent-AFLP markers. *Journal of Genetics and Genomics*, *34*(12): 1061–1071.

Zamani, Z., A. Sarkhosh, and R. Fatahi. 2007. Genetic relationships among pomegranate genotypes studied by fruit characteristics and RAPD markers. *J. Hortic. Sci. Biotechnol.*, *82*(1): 11–18.

Zhang, X., Y. Zhao, Y. Ren, Y. Wang, and Z. Yuan. 2020. Fruit Breeding in Regard to Color and Seed Hardness: A Genomic View from Pomegranate. *Agronomy*, *10*(7): 991. https://doi.org/10.3390/agronomy10070991.

Zhukovskij, P.M. 1950. *Punica. Cultivated Plants and Their Wild Relatives*, 60–61. State Publishing House Soviet Science, Moscow.

Zhukovsky, P.M. 1965. Main gene centres of cultivated plants and their wild relatives within the territory of the USSR. *Euphytica*, *14*: 177–188.

Zohary, D., and P. Spiegel-Roy. 1975. Beginnings of Fruit Growing in the Old World: Olive, grape, date, and fig emerge as important Bronze Age additions to grain agriculture in the Near East. *Science*, *187*(4174): 319–327.

Zuriaga, E., J. Pintová, J. Bartual, and M.L. Badenes. 2022. Characterization of the Spanish pomegranate germplasm collection maintained at the Agricultural Experiment Station of Elche to identify promising breeding materials. Plants. *11*(9): 1257.

7

Genetic Diversity of Guava (*Psidium guajava* L.) and Sustainable Utilization

Nagaraja, A., Nayan Deepak, G., Madhubala Thakre, Vasugi, C., Raman Selvakumar* and *Singh, S.K.*

Introduction

As a perennial crop grown in tropical and subtropical climates across the world, the guava tree (*Psidium guajava* L.), sometimes known as the "apple of the tropics" holds significant importance. The guava fruit is well-known for its significant mineral makeup as well as its high concentration of vitamins C, A, B_1, and B_2. According to Hayes (1970), the main nutrients in red-fleshed guava include antioxidant nutraceuticals such as β-carotene, lycopene, and phenolic compounds. Red-fleshed guava is also a good source of vitamin C. Guava is now recognized in many tropical and subtropical countries as a crop of substantial economic importance due to the combination of these factual aspects and the comparatively low production costs involved (Rodriguez et al., 2010b). Each part of the tree serves a distinct function/use. For instance, the decoction of the bark is used to treat stomach aches, while the broth of the leaves is used as a treatment for diarrhea. Additionally, the leaf extract exhibits antibacterial qualities effective against *Staphylococcus*. Those with diabetes are advised to drink leaf tea, while ripe fruits are used to alleviate diarrhea and other gastrointestinal conditions (Kothagoda and Rao, 2009). Due to the astringent qualities demonstrated by the roots and leaves, gastroenteritis, diarrhea, and dysentery are treated using these parts along with immature fruits. Crushed leaves are used topically for treating wounds, ulcers, and rheumatic disorders. According to Morton (1987), providing a leaf decoction to cholera patients has been proven to be effective in stopping vomiting and diarrhea symptoms. The plant is native to tropical America and ranges in size from Mexico to Peru. It later gained economic significance in several nations due to its hardy traits, generous output, and profitable returns, even with modest cultivation efforts. Due to its hardy and prolific characteristics, guava has expanded significantly in more than 60 countries (Negi and Rajan, 2007). According to Knight (1980), the capacity of this specific fruit crop to survive saline sets it apart from other tropical

Division of Fruits and Horticultural Technology, Indian Council of Agricultural Research- Indian Agricultural Research Institute, New Delhi, India.

* Corresponding author: anrciah@gmail.com

All the figures in this chapter are at the end of the chapter.

and subtropical fruit crops. It can also be grown on unproductive soils with little effort. India, Mexico, Brazil, Cuba, Venezuela, the United States (particularly Hawaii, Florida, and California), Australia, New Zealand, South Africa, China, Thailand, the Philippines, Indonesia, Malaysia, Sri Lanka, the Dominican Republic, Myanmar, and Haiti are the nations with the highest guava production.

This fruit was introduced to India by the Portuguese and has now become established in certain regions with great variety (Purseglove, 1968). Several guava clones have been imported to India, including Beaumont and Indonesian seedless from Australia, Acerapera from Brazil, and Verchie from the United States. Among the several wild species brought from the United States are: *P. sartorianum*, *P. littorale* var. *longipes*, and *P. longipes* (Singh and Rana, 1993; Ray, 2002). A number of potentially promising guava cultivars have been imported from Thailand (Solanki et al., 2011). Despite its beneficial qualities and large nutritional makeup, guava is susceptible to a variety of diseases, most notably *Fusarium oxysporum* f. sp. *psidii* induced guava wilt. Furthermore, guava exhibits restricted fruit development, a short viability period, a high seed content, and a high susceptibility to stress (Rai et al., 2010). The primary objective of germplasm development projects, whether through introductions or breeding procedures, is to produce a medium-sized tree with colorful fruit, Good Shelf life, and a low seed count. Additionally, the development of a guava wilt disease-resistant tree is critical (Chandra and Mishra, 2007). The Indian Council of Agricultural Research (ICAR)-Central Institute for Subtropical Horticulture in Lucknow, India, has developed a wilt-resistant interspecific hybrid of *P. molle* and *P. guajava*. In addition to graft compatibility with commercial varieties of *P. guajava*, this hybrid may serve as a source of guava wilt resistance.

Origin and Evolution

Oviedo, a Spanish historian, first recorded the guava tree during his expedition to Haiti between 1514 and 1557. Oviedo referred to the vegetative activity of plants present in the West Indies region as guayabo (Ruehle, 1964). The oldest known reference to guava may be found in Gonzalo Fernandez de Oviedo's Natural History of the Indies, published in 1526 (Popenoe, 1920). Linnaeus formally named it *P. guajava* in 1753 (Ellshoff et al., 1995). The guava fruit is native to tropical America and was brought to India in the early 17th century (Hayes, 1957). Don Francisco de Paula Marin, a Spanish botanist who arrived in the early 1800s and chose to stay in Hawaii during the reign of King Kamehameha I, is credited with introducing guava to Hawaii (Shigeura and Bullock, 1976). It was later hypothesized that Spanish invaders transported guava trees native to the Pacific area to India and the Philippines. This phenomenon expanded from India and the Philippines to the Malay Archipelago, Hawaii, and South Africa (Soubihe Sobrinho, 1951). The Aztec culture was known as *xal xocotl* in the aforementioned country, which translates to "sand plum", alluding to the abrasive texture of its flesh (Gates, 1952). The production of this crop in the United States is predominantly centered in the states of Florida and Southern California. According to historical records, the arrival of this specific entity in the former state may be traced back to 1847, when it originated in the country of Cuba. The fruit in question is known by several names in different parts of the world. It is known as guayaba in Spanish-speaking portions of tropical America. The French call it goyave, whereas the Germans and Portuguese call it guajava and goiaba, respectively. The tree is known in Spanish as guayabo or guayavo, in Dutch as *guyaba* or *goeajaaba*, in Surinamese as *guave* or *goejab*, and in Portuguese as *goiaba* or *goaibeira*. The indigenous Hawaiians refer to the fruit widely known as guava or kuawa as such. The current situation in the Guam region may be classified as abas. This fruit is usually known as guava or jambu batu in the Malayan region. It is crucial to note, however, that it also has other dialectal names, comparable to its equivalents in India, tropical Africa, and the Philippines, where it is commonly referred to as bayabas. Indigenous communities in Mexico, Central America, and South America use a variety of tribe designations such as pichi, posh, enandi, and others. In India, the fruit known as 'Amrud' is noted for its similarity to a pear, but the name "safari aam" refers to a mango connected with a voyage (Toro, 1993).

Taxonomy

The guava fruit is classified within the Myrtaceae family, which encompasses a diverse range of approximately 130 genera and 3,000 species of trees and shrubs. These species are predominantly found in tropical and subtropical regions across the globe (Luiz et al., 2011). Eucalyptus, cloves, nutmeg, allspice, cinnamon, java plum, and jamun are among the significant species belonging to the Myrtaceae family. It encompasses four genera that are known for their production of edible fruits, namely *Psidium, Eugenia, Syzygium*, and *Feijoa.* The genus *Eugenia* encompasses several significant species, including *E. braziliensis, E. uniflora, E. victoriana, E. axillaris, E. aggregate, E. dysenterica, E. lutescens,* and *E. luschnathiana*, among others. In contrast, the genus *Syzygium* comprises approximately 500 species, with notable species such as *S. jambos, S. malacense, S. suborbiculare, S. paniculatum, S. aqueum, S. cordatum, S. cumini, S. forte, S. samarangense*, and others. The genus *Feijoa* encompasses the species *Avea sellowiana*, which is predominantly cultivated in Brazil and Uruguay (Mitra et al., 2012). The Myrtaceae family is characterized by its shining bark and greying brown color, which peels off in scales (Figure 1).

Additionally, the family is distinguished by its angular young shoots that bear flowers containing numerous stamens with an inferior ovary. The fruit of this family consists of many-seeded berries (Hayes, 1957). *Psidium* belongs to the taxonomic classification of tribe Myrteae, specifically within the subtribe Myrtinae. This genus consists of a total of 70 species, many of which are extensively cultivated, such as *P. guajava* and *P. guineense*. The *Psidium* genus can be distinguished from other genera based on discreet characteristics pertaining to the gynoecium, including a trilocular ovary, multiseriate ovules arranged in a peltate placenta, and a durable bony seed coat. Linnaeus classified *P. guajava* into two subspecies, namely *pomiferum*, which encompasses individuals bearing round-shaped fruits, and *pyriferum*, which encompasses individuals bearing pear-shaped fruits (De Candolle, 1908). In addition to *Psidium guajava*, several other species of relatively limited significance can be found in specific regions. These include *P. littorale*, commonly known as strawberry guava, found in Brazil, *P. guineense* in the West Indies, *P. montanum* in Jamaica, *P. microphyllum* in Puerto Rico, and *P. friedrichsthalianum* in Malaya (Pontikis, 1996). According to Atchinson (1947), cytological investigations have revealed that all cultivated guava varieties exhibit either a diploid state with a chromosome count of $2n = 2x = 22$ or a triploid state with a chromosome count of $2n = 3x = 33$. The guava plant is a type of crop that relies on cross-pollination for reproduction. Among the various pollinators, the honeybee plays a significant role in facilitating the pollination process for guava plants. The degree of often cross pollination instead of cross pollination varies between 25.7%–41.3%. According to Morton (1987), the existing diversity in seedling populations, which has led to the identification of promising genotypes, can be attributed to a combination of high heterozygosity and frequent cross-pollination (Dinesh and Vasugi, 2010b).

Botanical Origin

According to Hayes (1953), the guava plant is said to have originated in tropical America. Purseglove (1968) believes that its origins may be traced back to Mexico, whilst De Candolle (1908) believes they are in Brazil. The plant has a wide distribution over the equatorial regions, flourishing in tropical and subtropical temperatures. It was first introduced into India in the 17th century. Through the process of elimination of the Old World, De Candolle inferred that the plant's origin was in America. De Candolle also underlined the significance of pinpointing the particular place inside America where guava originated. Furthermore, Gonzaga Neto (1990) stated that the origin of guava might be traced back to Mexico, Colombia, Peru, or Brazil. According to Risterucci et al. (2005), the guava plant is indigenous to northern South America, despite its vast distribution across numerous tropical locales in the Americas. The plant is widely farmed for its fruit and has naturalized in tropical and subtropical locations across the world. The guava fruit has been cultivated and dispersed for a long time by people, bird species, and other creatures, making it difficult to pinpoint its original origin. However, Morton

(1987) suggests that the guava's roots may be traced to a geographical range stretching from southern Mexico through Central America. The species is found in warm areas of tropical America as well as the West Indies. It has been observed in the Bahamas, Bermuda, and southern Florida, where it was first imported in 1847. By 1886, it had spread over more than half of the state of Florida (Yadav, 2006).

Botany of Guava

The guava plant is distinguished by its arboreal shrub-like form, which may reach heights of 25–30 feet. It has a thin trunk that is ornamented with greenish-brown scaly bark that peels off in exquisite flakes. According to Orwa et al. (2009), the tree has a taproot system with significant branching, giving rise to low-hanging and suckers from the roots. The leaves have an oblong form that can range from elliptical to oval in shape. Their length ranges from 3–6 inches, with a sharp to rounded tip. The bottom epidermis of the leaves has coarse pubescence, but the top epidermis has pronounced depressed venation. The blooms are arranged axillary and solitary. Crane and Balerdi (2016) state that the young branches yield white bisexual or perfect blooms around one inch across, growing as solitary blooms or in bunches on a peduncle. The calyx has gamosepalous features, which means that 4–5 sepals are linked and shorter in length. The flower's corolla is made up of 4–5 petals and has gamopetalous features. Polyandry is exhibited by the androecium, which has an infinite number of stamens. The Gynoecium is made up of 4–5 carpels, each with its tiny style and stigma. The ovary is located beneath, with quadra-penta-locules that are syncarpous. The fruit is described as a berry with several little brown seeds by Crane and Balerdi (2016). The color of the fruit's outer layer/peel ranges from yellow to orange. According to Soetopo (1991), aromatic pulp comes in a variety of hues, including white, yellow, pink, and red, with flavors ranging from acidic to sweet. According to Weinert and Wyk (1998), the fruit seeds are 3–5 mm long and have a distinct mesocarp made up of minute fibrous stone cells. The presence of blooms on freshly growing vegetative terminals is a prominent botanical feature of guava. However, the onset of flower buds and subsequent fruit development is highly variable, influenced by factors like as rainfall patterns and fertilizer availability. The aforementioned trait allows cultural adaptation of blooming and fruit development processes (Shigeura and Bullock, 1976).

Spread and Distribution

Guava is a highly esteemed cultivated species within the Myrtaceae family. It originated in tropical America and is now found in many tropical and subtropical places across the world (Samson, 1986; Morton, 1987): In India, South Africa, Florida, Hawaii, Egypt, Brazil, Colombia, West Indies, Cuba, Venezuela, New Zealand, the Philippines (Wilson, 1980; Yadava, 1996), Vietnam (Le et al. 1998), and Thailand (Tate 2000). South Africa has a vast range of weather patterns and climatic conditions, allowing for the production of a large variety of fruits. This has given the country a worldwide reputation as a major grower and exporter of subtropical fruits, most notably guava. Suitable conditions for guava cultivation have led to the Western Cape province, with Fan Retief being the dominant cultivar.

Guava cultivation is also reported in the Northern Provinces of Limpopo and Mpumalanga, according to Middle East Regional Cooperation (MERC, 2013). Commercial guava farming in the United States is mostly located in Hawaii, Puerto Rico, and Florida.

The fruit crop has been recorded growing without purposeful cultivation on numerous residential properties spanning a substantial area of South Florida (Crane and Balerdi, 2016). Guava, along with the mango from India, was first brought into Egypt in 1825. In 1927, a seedless variant was created. Guava farming in Egypt is predominantly concentrated in the country's lower area, notably in the governorates of El-Beheira, Damietta, Kafr El Sheikh, Alexandria, and Qaliubiya (ISHS, 2016). Guava seedling populations in different locations of India's Indo-Gangetic plains show great heterogeneity. Bose et al. (1999) discovered considerable heterogeneity in the Uttar Pradesh districts of Adilabad, Mirzapur, Kanpur, Unnao, and Fatehpur, as well as the neighboring regions in India. Through GIS-based mapping of guava-growing areas in Uttar Pradesh, India, the districts of Allahabad and Kaushambi were identified to be important foci of rich variation. This mapping also assisted in identifying considerable variability in a variety of features (Rajan et al., 2007). Brazil is the world's third-largest producer of commercial guava due to ideal agroclimatic conditions, promoting the growth and cultivation of this fruit. According to Martins et al. (2014), guava is largely grown in Brazil's southeastern and northeastern areas. Guava was introduced into Queensland in 1850, and it has since been found to flourish abundantly as a wild plant in numerous regions. Guava varieties from Hawaii and Asia were introduced for study and selection. According to Menzel (1985), the guava fruit is geographically dispersed in the southern area of Queensland, namely in Nambour, as well as in the northern region of New South Wales, specifically in Alstonville. Thailand's guava-producing regions are predominantly located in the country's central and western regions, with considerable concentrations in the provinces of Samat Sakhon, Ratchaburi, and Nakhon Pathom (Thaipong et al., 2017).

Domestication and Dispersal

Despite the fact that the guava plant underwent domestication over 2,000 years ago, the initial instance of its commercial cultivation, originating from the Caribbean Islands, was documented in 1526. Subsequently, explorers disseminated it to the Philippines and India (Yadav, 2006). The Spanish and Portuguese colonizers promptly facilitated the transfer of the Spanish and Portuguese languages from the New World to the East Indies and Guam. Subsequently, the cultivation of this crop gained popularity in Asia and the tropical regions of Africa. The cultivation of this crop has been observed in Egypt for an extended period, suggesting the possibility of its dissemination from Egypt to Palestine. The species has been observed sporadically in Algeria and along the Mediterranean coastline of France. De Candolle postulated that the distribution of the species occurred from Mexico to Peru during the pre-Columbian era. The guava fruit was introduced to India during its early history and subsequently underwent domestication and cultivation in various regions. In the United States, the cultivation and domestication of [plant species] primarily occurs in the states of Florida and Southern California. This practice was initiated by P.W. Reasoner, who introduced the plant from Cuba to the former state in 1847. According to Popenoe (1920), successful cultivation of this plant has been observed as far north as the Pinellas Peninsula on the west coast and Cape Canaveral on the east coast. The species in question was among the initial introductions to the Galapagos Islands and is currently found across the primary Hawaiian Islands, thriving in various soil conditions (Mauchamp, 1997). The Aborigines introduced it to the West Indies prior to the arrival of Christopher Columbus. According to Gonzalo Fernandez de Oviedo's account in 1535, this particular species was commonly found in the West Indies and was cultivated by indigenous populations for its fruit. According to Solanki et al. (2011), guava cultivation has been observed in a total of 58 countries, such as Brazil, Burma, China, Columbia, Cuba, Egypt, India, Israel, Mexico, Pakistan, Peru, South Africa, Sri Lanka, United Arab Republic, the USA, Venezuela, and the West Indies.

Collection and Assessment of Germplasm/Genetic Resources

Status of Guava Genetic Resources

Guava improvement began a century ago, but the lack of vegetative propagation techniques at the time hampered the retention of better germplasm (Iyer and Subramanyan, 1988). It is believed that there are about 400 guava varieties globally; however, just a few dozen are principally responsible for the considerable growth of orchards on a global scale. Despite its origins in tropical America, this species has spread throughout the world's tropical and subtropical climates. According to Orwa et al. (2009), the species has successfully adapted to the tropical parts of the Old World as well as the West Indies. The vast majority of commercially cultivated varieties are classified as *P. guajava* L. Although Linnaeus classified *pyriform* and *pomiferum* as separate species, they were later reclassified as trees with pear-shaped and round-shaped fruits, respectively. These are only two of the countless variants reported within this species (Popenoe, 1920). Aside from the aforementioned species, there are numerous more landraces and wild species that are important in breeding efforts. *P. cattleianum* is a botanical species distinguished by its small stature as a tree or shrub and its smooth-textured bark. This species' leaves are obovate and elliptical, with a glabrous surface. The berries are obovate to roundish in form and range in hue from purple-red to pale yellow. They have a delicate outer peel and soft flesh that is rich in seeds. Because of its pleasant taste and aromatic qualities reminiscent of strawberries, this fruit is commonly known as "strawberry guava". *P. guajava,* according to Barbosa et al. (2016), has less tolerance to low temperatures than the species under consideration. *P. molle* Bertd is a plant species that consists of shrubs or small trees. Its leaves are obovate and have a reddish velvety texture on the underside. The fruits of this species undergo a color change, turning light yellow as they mature. The fruit's pulp is distinguished by its white appearance, numerous seeds, and a characteristic acidic flavor. The plant in question is well-known for its resistance to wilting and has been successfully used in the development of wilt-resistant interspecific hybrids in collaboration with *P. guajava* (Negi and Rajan, 2007; Rajan et al., 2005; Vishwakarma et al., 2023). *P. guineense,* sometimes known as Brazilian guava, exhibits a high degree of variety in its traits. The plant grows in a shrub-like manner most of the time, yet it can occasionally approach the size of a small tree. The plant grows at a rather moderate rate and yields fruits of significant size and quality, with a pleasant flavor. The tree is grown at moderate elevations in tropical regions to produce its fruits, which are consumed as food. This species of plant has been observed to have a high level of tolerance to a wide range of soil conditions. It has been established that the production of dwarf hybrids exhibits cross-compatibility with the commonly found guava species, *Psidium guajava. P. friedrichsthalianum* Niedenzu (1893) is a Central American native. It is distinguished by its tall trees, which can reach heights of 7–10 meters, and its slender branches with a smooth texture. This species' leaves are oval and have a glossy appearance on the upper surface, while the lower surface is covered in fine hairs. Fruits have a globose morphology, are small in size, and have a sour taste. Because of their high levels of acidity, the fruits are ideal for the production of jelly. Furthermore, this is critical for conferring wilting resistance in commercially cultivated varieties. This fruit is also known as Chinese guava and Costa Rican guava (Popenoe, 1920). *P. cujavilis* consists of shrubs or small, contorted trees that grow to heights of 1–4 meters. These plants produce spherical berries with a sour flavor that measures about 2–3 centimeters in diameter. The species is distributed at altitudes ranging from low to medium and is believed to have originated in the Antilles. It was subsequently introduced and cultivated in Indonesia and the Philippines (Backer et al., 1963).

The plant species *P. montanum* Swartz, commonly referred to as mountain guava, is typically observed in the elevated regions of Jamaica. The plant produces spherical fruits that contain white plup and a higher quantity of seeds. In addition to the guava varieties mentioned above, there is another type of guava known as para guava. Because it does not belong to the guava genus, the fruit known as para guava should not be referred to as guava. However, because it shares many characteristics with guavas, it is appropriate to classify it alongside guavas. This fruit is known as araca or para guava in Brazil, and it was recently introduced to the state of Florida (Morton, 1987). There is a distinct group of species known as aracazeiros in Brazil, which refers to native *Psidium* species, *viz., Psidium cattleianum* Sat., *Psidium incandescens* Martius, *Psidium grandiflorum* Martius, and *Psidium arboretum* Vell. These species are recognized for their early maturity, resistance to some pests and illnesses, and foreign fruit flavor (Luiz et al., 2011).

Collection

The key responsibilities for the effective usage and improvement of genetic resources are the exploration, acquisition, preservation, assessment, and strategic management of such resources. Numerous scientific institutions are involved in the collection and protection of guava genetic resources through the development of field genebanks. The Active Germplasm Bank of Embrapa Tropical Semi-Arid, located in Petrolina, Pernambuco state, and the Active Germplasm Bank of Universidade Estadual Paulista School of Agricultural and Veterinary Studies (FCAV) UNESP/FCAV, located in Jaboticabal, Brazil, are two significant institutes responsible for maintaining guava genetic variety. Open-pollinated seedlings were gathered and shipped from both America and India to the Universidade Estadual Paulista (UNESP) in Jaboticabal, Brazil. As recorded by Boora in 2012, the cultivars Rica and Paluma were produced through a selection procedure from imported germplasm. The Foundation of Cuban Fruit Tropical and Subtropical Germplasm Bank, commonly known as the Tropical Fruit Research Institute, developed the biggest guava collection in Cuba in 1965 (Fernandez and Pelea, 2015). The University of Hawaii College of Tropical Agriculture and Human Resources in the United States has actively contributed to the improvement of guava agriculture. It has considerably aided in the introduction and evaluation of numerous guava cultivars coming from seven different nations (Morton, 1987). Due to the non-indigenous character of the crop, numerous guava cultivars from Thailand (Kazi Piyara), Hawaii (BAU Piyara-5), and India (Allahabad L-49) are planted in different parts of Bangladesh (Rahim et al., 2012). Multiple national institutions in India are conserving a wide collection of roughly 160 guava types in their field gene banks (Dinesh and Vasugi, 2010a; Figure 2).

Characterization and Evaluation

Crop diversity must be examined and assessed to determine the efficacy of germplasm in crop promotion. The morphological characterization is a simple activity because of its ease of use, low cost, and value in examining the relationships and numerous factors among distinct varieties/accessions. Fruit is an important component of production, varying in form, size, and weight. As a result, it is regarded as an important morphological trait used in the assessment of relatedness and genetic cluster analysis. The Mahalanobis D2 statistics were used in the study to assess the amount and importance of genetic differences in fruit weight and seed properties. When Ward's approach was used for cluster analysis, it revealed a discernable pattern of groups that proved effective in discovering optimal cross combinations of genetically diverse clones for obtaining low seed content (Rajan et al., 2007, 2012). According to Jana et al. (2015), using Ward's minimal variance technique based on growth yield aided in the selection of superior segregants in the breeding program. Cluster and discrimination analysis was shown to be useful in identifying the key factor contributing to variety within the group with the highest level of variability (Pelea et al., 2016). The authors performed varietal analyses to identify the relative contributions of genetic and environmental variables to the variation in fruit quality characteristics seen across diverse agroclimatic conditions (Dinesh and Reddy, 2001; Thaipong and Boonprakob, 2005). Singh et al. (2015) evaluated numerous commercial cultivars for their compatibility with certain agroecological zones, concentrating on growth, yield, and quality characteristics. This assessment resulted in the selection of potential cultivars. Kidaha et al. (2015) investigated the morphological properties of guava landraces from Kenya's western and coastal areas. The study found substantial physical variety across Kenyan landraces, both within and between areas, using both quantitative and qualitative descriptors. Similarly, Santos et al. (2011) studied the genetic divergence of *Psidium* accessions from several Brazilian states. The data revealed that there was no distinct grouping depending on state of origin, implying that there are no obstacles to guava availability. Guava collections from throughout the world have been thoroughly examined in order to ascertain their agronomic and morphological properties (Periera et al., 2003; Luiz et al., 2011; Dinesh and Vasugi, 2010b). The study of biochemical features of guava fruits focuses particularly on functional chemicals such as lycopene, carotenoids, and other compounds (de Lima et al., 2002; Correa et al., 2012). The existence of genetic heterogeneity within germplasm aids in the discovery of superior parental individuals that may be introduced into the breeding process to create innovative cultivars. The evaluation of genetic diversity in relation to fruit and seed-related properties is critical. As a result, a study is required to investigate the potential for utilizing favorable variations in guava seed-related traits, such as the number of seeds per fruit, seed weight proportion, seed weight per fruit, seeds per 100 grams of pulp, 100 seed weight, and seed content in open-pollinated selections. The goal of this research is to find genotypes with acceptable seed properties in the progeny of half-sibling crossings (Rajan et al., 2007). Rajan et al. (2008) investigated the role of seed-related features in the evolution of small-seeded cultivars. The researchers examined the direct and indirect impacts of these factors using correlation and path coefficient analyses. The findings emphasized the need to use additive gene effects throughout the selection process, as mentioned before by Rajan et al. (2005). The researchers assessed the appropriateness of guava accessions with high acidity and pectin content for cider production (Garg et al., 2005). Because of their high acidity levels, these guava accessions were shown to have a poor table value. A wide range of morphological characteristics are used to differentiate genotypes, but their efficiency decreases when discriminating closely related genotypes. As a result, it is critical to create efficient molecular markers that can distinguish between closely related *Psidium* spp. and their wild counterparts (Saxena et al., 2007). Guava biotechnology has gotten less attention in the past than other crops, and it is now in its early phases of research. The use of molecular markers in guava began with the identification of cultivars and the assessment of biodiversity within germplasm. The use of morphological and molecular characterization techniques enabled the demonstration of varied degrees of diversity (Figures 3 and 4).

Nagueira et al. (2012) and Nasution and Hadiati (2014) investigated the diversity of wild guava in Brazil. This study used morphological descriptors in concert with microsatellites to identify groups of guava having common ancestors. Valera-Montero et al. (2016) assessed genetic diversity using a mix of RAPD and morphological markers. The results showed that the samples were 80% comparable. Furthermore, the use of clads for the combined data helped to resolve any ambiguity and aided in the identification of various groups based on fruit form and banding pattern.

RAPD markers were used to identify *P. guajava* germplasm from diverse geographical regions (Ahmed et al., 2011; Manna et al., 2015). Nasution and Hadiati also undertook the characterization of guava germplasm accessions based on leaf and fruit characteristics in 2014. The discovery of a substantial level of similarity across farmed guava varieties from diverse places was aided by the study of AFLP markers in guava accessions from 15 nations (Thaipong et al., 2017). The use of microsatellite loci to assess genetic variability in guava accessions and wild *Psidium* spp. germplasm to guide genetic resources and breeding programs revealed the presence of genetic variability in guava across nine Brazilian states (Costa and Santos, 2013). The use of microsatellites for molecular characterization aided in the finding of cultivars with unusual alleles, as well as wild genotypes that might serve as a useful gene reservoir for conservation goals (Valdes-Infante et al., 2007). The use of SSR markers aided in the unambiguous differentiation of *P. guava* accessions and their related species, as well as the clarification of genetic linkages among the majority of accessions depending on their geographical origins (Sitter et al., 2014). Valdes-Infante et al. (2003) published the first genetic linkage map that demonstrated the relationship between multiple quantitative trait loci (QTLs) in guava. The first linkage map was made using the mapping population "Enana Roja Cubana" 'N6' reported by Valdes-Infante et al. (2003). The researchers created and used SSR markers specific to guava in order to create an AFLP linkage map. Under the European Commission-funded 'GUAVAMAP' project, the mapping populations were increased by the study of AFLP and SSR markers. AFLP and SSR markers were used to create parental linkage maps in three guava mapping populations (Ritter et al., 2010; Ritter, 2012). These linkage maps are useful for investigating potential links between linkage groups from other maps. Following that, linkage maps in diverse genetic backgrounds were created, followed by quantitative trait locus (QTL) studies for numerous beneficial characteristics (Luiz et al., 2011).

Conservation of Guava Genetic Resources

Guava germplasm conservation is accomplished through the development and management of field gene banks in various places across the world. The Bangladesh Agricultural University Germplasm center (BAU) Germplasm Centre (GPC) is the most comprehensive germplasm center specializing in fruit crops. The center was founded in 1991 with the assistance of the Swiss Agency for Development and Cooperation (SDC). Its principal goal was to establish a seed bank and to make agricultural technology more accessible to farmers. The institution presently handles around 85,000 mother plants from 161 different species, including 57 different types of guava. Significant progress has been made in the previous two decades in the field of science-based conservation resources relevant to the propagation and management of small fruit trees, such as guava (Rahim et al., 2012).

Brazil is largely regarded as a major source of guava variety, with research organizations housing gene banks specialized to *Psidium* species. These genebanks have successfully maintained 310 *Psidium* spp. accessions, as well as an additional 174 accessions from Brazilian guava. Several noteworthy cultivars have been generated through the careful selection of open-pollinated seed orchards, according to Luiz et al. (2011). Paluma, Rich, Sassaoka, and Pedro Sato are among these varieties. In India, around 160 genotypes of several *Psidium* spp. are now maintained in collections and administered in field genebanks at multiple sites across the nation (Dinesh and Vasugi, 2010a). Several notable genebanks in India are in charge of preserving a large number of guava germplasm accessions. These include ICAR-CISH (105 accessions), ICAR-IIHR (43 accessions), Fruit Research Station in Sangareddy (27 accessions), Fruit Research Station in Rewa (60 accessions), Regional Centre for Agriculture in Udaipur (11 accessions), HC and RI Periyakulam (18 accessions), and Govind Ballabh Pant University of Agriculture and Technology (GBPUAT) in Pantanagar (22 accessions). Tissue culture is an important tool in both basic and applied research, contributing significantly to species preservation through *in vitro* procedures. In the establishment, multiplication, and rooting stages, a technology for *in vitro* multiplication of wild guava was standardized. This approach plays an important role in species propagation, conservation, and breeding. For the short- to medium-term conservation of guava, a standardized technique was developed. The experiment involved growing lateral buds on a basal medium known as Murashige and Skoog (MS) at a temperature of 21°C. The medium's mineral content was lowered by half and supplemented with 15 g/L sucrose, 0.25 mg/L 6-benzylaminopurine, 0.025 mg/L 1-naphthylacetic acid, and 1% activated charcoal. Guava was found to be sensitive to conservation temperatures below 9–10°C (Rodriguez and Velasaquez, 1999; Rodriguez et al., 2010b).

Documentation

Documentation plays a crucial role in facilitating germplasm exchange and the equitable sharing of benefits, serving as a fundamental component of fruit genetic resources management. The information can be compiled in diverse formats, such as published articles, directories, and gene bank documentation systems. The information can be disseminated through various means, such as traditional print media including books and journals, as well as through easily accessible electronic formats. The International Plant Genetic Resources Institute (IPGRI) is a worldwide organization dedicated to the management and updating of databases containing concise data pertaining to *ex situ* germplasm collections. The utilization of descriptors can aid in addressing the challenges associated with the identification of a larger quantity of plants within a germplasm collection. The guidelines for conducting tests to determine distinctness, homogeneity, and stability in guava were published by the International Union for the Protection of New Varieties of Plants (UPOV) in 1987. These guidelines serve as a valuable tool for the global characterization of germplasm, providing a standardized and efficient approach to the rapid assessment of guava varieties. The 1999 publication by the International Board for Plant Genetic Resources (IBPGR) provided a comprehensive list of tropical and subtropical fruits and tree nuts. This directory documented a total of 269 accessions, which encompassed 191 genera and 879 species. These accessions were further categorized into 14 crop sections. Currently, these collections are being preserved and managed by 242 research institutions worldwide, spanning across 69 countries.

The directory provided in this collection includes comprehensive information pertaining to each crop, including the responsible curator, details regarding holdings, duplicate sites, quarantine regulations, evaluation sites, and the maintenance of the collection. It should be noted that this directory has replaced the previous version published by IBPGR in 1984 (Bettencourt et al., 1992). Rodriguez developed an illustrated bilingual (English and Spanish) descriptor for the characterization of guava in Cuban collections. This document provides an analysis of 395 accessions, including a catalog of 18 of them, accompanied by newly developed illustrated descriptors consisting of over 70 characteristics.

This method provides a convenient and efficient means of distinguishing between phenotypes, while also serving as a valuable tool for establishing a global standard for guava characterization (Rodriguez et al., 2010a, b). The initial utilization of guava germplasm characterization in India involved the use of illustrated guava descriptors developed by the All India Coordinated Research Project on Subtropical Fruits, ICAR-CISH, Lucknow. This approach was employed until the publication of the standard Distinctness, Uniformity, and Stability (DUS) test guidelines in 2016 by a task force established by the Protection of Plant Varieties and Farmers' Rights (PPV&FRA) authority (Rajan et al., 2011; PPV and FRA, 2016). Table 1 shows the achievements of guava breeding in India (Figures 5, 6, 7, 8, 9, and 10).

Conclusion

The effective collection, maintenance, and strategic management of genetic resources from both wild and farmed species are critical activities to ensure the right usage of these resources. Prospecting, the introduction of foreign materials, and the selection of superior individuals as a consequence of open and controlled pollination have greatly expanded the spectrum of germplasm. Nonetheless, the original variety has been degraded due to many biotic and abiotic causes. Therefore, preserving indigenous germplasm is critical to increasing the species' genetic diversity. Fruit morphology variability presents a significant problem for farmers because of the relevance of qualities such as size, skin and flesh color, fruit shape, flesh thickness, and other factors in determining guava quality. There is now a scarcity of compact, highly productive plant types with features like soft-seeded and colored fruits in the germplasm. When maintaining germplasm, it is critical to focus on key properties such as tree height, fruit shape, uniformity, seed count, vitamin C content, acidity level, pulp thickness, and other essential characteristics.

Guava growing in agriculture production systems with poor soil fertility zones offers a significant benefit in terms of ease of cultivation, encouraging its spread. Therefore, cultivating guava genotypes with higher yield potential in such environmental circumstances is an important step toward improving the economic sustainability of guava orchard producers in these locations. It is critical to develop guava model ideotypes that contain a thorough understanding of the interaction between plant development processes and yield. Wild species have economic value because they include uncommon genes that confer tolerance and resilience to both abiotic and biotic stressors. The availability of uncontrolled genetic resources in small clusters inside remote areas limits the viability of *in situ* conservation initiatives. In such circumstances, *ex situ* conservation and the creation of seed orchards might be pursued as viable conservation approaches. The genetic pool of regenerated plants can be expanded by a mix of spontaneous regeneration and the introduction of seedlings from seed orchards. Furthermore, this occurrence contributes to the species' adaptability. Another key concern is the establishment of a genetic marker that may be used to identify local genotypes from others. This would allow duplicate samples in different field gene banks to be removed, maximizing resource efficiency.

Table 1. Breeding Achievements in Guava.

Varieties	Breeding Methods	Specific Features
Pant Prabhat	Selection	White pulp
Harijha	-	A popular variety in Bihar
Allahabad Safeda	Selection	The most popular cultivar in UP
Allahabad Surkha	-	Pink flesh and large-fruited variety
Chittidar	-	Highest TSS guava variety (small red dots present on the peel)
Ganesh Khind Fruit Research Station (GKFRS), Pune by GS Cheema and Deshmukh		
Lucknow-46	-	Pear or pyriform fruit shape
Lucknow-49	Sardar Guava: A selection from Allahabad Safeda	Highest Vitamin C variety; national variety
Nagpur Seedless	-	-
Safed Jam	Allahabad Safeda × Kohir	-
Kohir Safeda	Kohir × Allahabad Safeda	-
ICAR-Indian Institute of Horticultural Research (IIHR), Bangalore		
Arka Amulya	Allahabad Safeda × Triploid-Seedless	-
Arka Mridula	Seedling selection from Allahabad Safeda	White pulp
Arka Kiran	Kamsari × Purple local	Pink pulp soft-seeded, high lycopene variety
Arka Rashmi	Kamsari × Purple local	Pink flesh dual purpose, soft- seeded, suitable for HDP
ICAR-Central Institute of Subtropical Horticulture (CISH), Lucknow, Uttar Pradesh		
CISH-G-1	Seedling selection from Allahabad Safeda	Red color and longer shelf life
Lalit	Seedling selection from Allahabad Safeda	Red color pulp
Shweta	Selection from a half-sib population of Apple color	Color variety (pink pulp)
Hissar Agricultural University, Hisar		
Hisar Safeda	Allahabad Safeda × Seedless	White pulp
Hisar Surkha	Apple color × Banarasi Surkha	Pink pulp hybrid
TRY (G)-1	Selection	Off-season, drought, sodicity tolerant
Apple Colour		Red-peel guava variety
Hafshi	-	Red pulp
ICAR-Indian Agricultural Research Institute, New Delhi		
Pusa Srijan (2n=24) (Aneuploid-82)	Potential dwarfing rootstock for HDP Tetrasomic type of aneuploidy (2n+2)	Tolerance to wilt Recommended for dwarf rootstock from IARI
Others		
Giant Thailand Guava	Introduced from Bangladesh	
Behat Coconut		
Seedless Varieties: Saharanpur Seedless, Nagpur Seedless		

References

Ahmed, B., M.A. Mannan, and S.A. Hossain 2011. Molecular characterization of guava (*Psidium guajava* L.) germplasm by RAPD analysis. *Intnl. J. Nat. Sci.*, *1*(3): 62–67.

APEDA. 2016. http://apeda.gov.in/apedawebsite/

Atkinson, E. 1947. *Chromosomes Atlas of Flowering Plants*. London: Allen and Unwin Ltd.

Backer, C.A., V.D.B. Bakhuizen, and C. Reinier. 1963. *In*: C.A. Backer, and V.D.B. Bakhuizen (Eds.), *Flora of Java (Spermotophytes only)*. Groningen: P. Noordhoff.

Backer, C.A., and R.C. Bakhuizen van den Brink. 1968. *Flora of Java*. (3 Vols.). Groningen: Noordhoff.

Bailey, L.H. 1976. *Hortus Third: A Concise Dictionary of Plants Cultivated in the United States and Canada*. New York: Macmillan Publishers.

Barbosa, J.M., G.P. Asner, R.E. Martin, C.A. Baldeck, F. Hughes, and T. Johnson. 2016. Determining subcanopy *Psidium* cattleianum invasion in Hawaiian forests using imaging spectroscopy. *Remote Sensing, 8*: 33.

Bettencourt, E., T. Hazekamp, and M.C. Perry. 1992. *Directory of Germplasm Collections. Tropical and Subtropical Fruits and Tree Nuts*, 337pp. Rome: IBPGR.

Boora, R.S. 2012. Improvement in guava (*Psidium guajava* L.): A review. *Agricultural Rev*., *33* (4): 341–349.

Bose, T.K., S.K. Mitra, A.A. Farooqui, and Sandhu, M.K. 1999. *Tropical Horticulture* (1st Edn.), 297pp. Kolkata: Nava Prokash Publication.

Catalogue of Life. 2010. WCSP in species 2000, and ITIS Catalogue of Life.

Chandra, R., and M. Mishra, 2007. Biotechnological interventions for improvement of guava (*Psidium guajava* L.). *Acta Hort*., *735*: 117–125.

Correa, L.C., C.A.F. Santos, and G.P.P. Lima, 2012. Chemical and biochemical characterization of guava and areca fruits from different regions of Brazil. *In*: *3rd International Symposium on guava and other Myrtaceae*, 20. Petrolina, PE, Brazil.

Costa da, S.R., and C.A.F. Santos. 2013. Allelic database and divergence among *Psidium* accessions by using microsatellite markers. *Genetics and Molecular Research*, *12*(4): 6802–6812.

Crane, H.C., and C.F. Balerdi. 2016. *Guava Growing in the Florida Home Landscape,* 1–7. HS4, one of a series of the Horticultural Sciences Department, UF/IFAS Extension,. Website at http://edis.ifas.ufl.edu.

De Candole, A. 1908. *Origin of Cultivated Plants*. New York: Appleton.

de Lima, M.A.C., de Assis, J.S., and Gonzaga, N.L. 2002. Caracterização dos frutos de goiabeira e seleção de cultivares na Região do Submédio São Francisco. *Rev. Bras. Frutic*., *24*: 273–276.

Dinesh, M.R., and B.M.C. Reddy. 2001. Evaluation of *Psidium guajava* L. accessions and some other *Psidium* species for fruit characters. *J. Appl. Hortic*., *3*(1): 41–43.

Dinesh, M.R., and C. Vasugi. 2010a. Guava improvement in India and future needs. *J. Hortic. Sci*., *5*: 94–108.

Dinesh, M.R., and C. Vasugi. 2010b. Phenotypic and genotypic variations in fruit characteristics of guava (*Psidium guajava* L). *Indian J. Agric. Sci*., *80*(11): 62–63.

Ellshoff, Z.E., D.E. Gardner, C. Wikler, and C.W. Smith. 1995. *Annotated Bibliography of the Genus Psidium, with Emphasis on P. Cattleianum (strawberry guava) and P. Guajava (common guava), Forest Weeds in Hawaii* (Technical Report 95). Cooperative National Park Resources Study Unit, University of Hawaii at Manoa.

FAOSTAT. 2016. *FAOSTAT Database Collections*. Rome: Food and Agriculture Organisation of United Nations. http://faostat.fao.org.

Fernández de Oviedo, G. 1535. *Historia General y Natural de las Indias: Islas y Tierra-firme del Mar Océano*, pp. 770pp. ([English title not available]). Madrid: Real Academia de la Historia.

Fernández, B.E., and P.L. Pelea, 2015. Genetic breeding of guava (*Psidium guajava* L.). *Cult*., *36*(1): 96–110.

Gangappa N.D., C. Singh, M.K. Verma, M. Thakre, A.M. Sevanthi, R. Singh, R, M. Srivastav, K. Raghunandan, C. Anusha, V. Yadav, and A. Nagaraja. 2022. Assessing the genetic diversity of guava germplasm characterized by morpho-biochemical traits. *Front. Nutr*., *9*: 1017680.

Garg, N., P. Yadav, N. Goel, A.K. Bhattacharjee, S. Rajan, and R. Kumar. 2005. Screening of acidic guava accessions for cider preparation. *International Guava Symposium*, *735*: 647–650.

Gates, W. 1952. *An Aztec Herbal, the Classic Codex of 1552* (Series publication no. 23). Baltimore: Maya Society 1939.

Gill, K.S. 2016. Guavas. *In*: B. Caballero, P.M. Finglas, and F. Toldra (Eds.), *Encyclopedia of Food and Health*. Amsterdam: Academic.

Gonzaga Neto, L. 1990. Cultura da goiabeira. *Petrolina, pe: Embrapa-Cpatsa*, *3*(26).

Hamilton, R.A., and H.S. Smith. 1959. *Growing Guava for Processing*. Extension Bulletin 63. University of Hawaii.

Hayes, W.B. 1953. *Fruit Growing in India*. Allahabad: Kitabistan.

Hayes, W.B. 1957. The guava and its relative. *In*: *Fruit Growing in India*, 286–303. Allahabad: Kitabistan.

Hayes, W.B. 1970. *Fruit Growing in India [M]*, 72. Allahabad: Kitabistan.ISHS. 2016. International Society for Horticultural Science Section Tropical and Subtropical Fruits. *Newsletter*, *14*: 1–9.

ITIS. 2017. ITIS, catalogue of life. *In*: *WCSP species 2000*. Digital resource at www.catalogueoflife.org/col.

Iyer, C.P.A., and M.D. Subramanyan. 1988. IIHR. Selection-8 and improved guava cultivar. *South Indian Horticulture*, *36*(5): 258–259.

Jagtiani, J., H.T. Chan, and Sakai, W.S. 1988. *Tropical Fruit Processing*. San Diego: Academic.

Jana, B., P.S. Munsi, and D.C. Manna. 2015. Correlation study of yield, flowering duration, and fruit physico-chemical characters of guava (*Psidium guajava* L.). *World J. Agric. Res*., *3*(2): 91–93.

Kidaha, L.M., A.E. Alakonya, and A.B. Nyende. 2015. Morphological characters of guava landraces in western and coastal Kenya. *Am. J. Exp. Agric*., *9*(6): 1–11.

Knight, R.J. 1980. Origin and world importance of tropical and subtropical fruit crops. *In*: Nagy, S. and Shaw, P.E (Eds.),*Tropical and Subtropical Fruits*, 1–120. AVI Publishing Westport.

Kothagoda, N., and A.N. Rao. 2009. Medicinal use and fruit development of ten tropical fruit species. *J. Trop. Med. Plants*, *10*(2): 231–235.

Le, T.H., J.F. Hancokc, and T.T. Trinh. 1998. The fruit crops of Vietnam: Introduced species and their native relatives. *Fruit Var. J*., *52*(3): 158–168.

Luiz, C.C., A.F. Carlos, F.V. Santos, and P.L. Gep. 2011. Antioxidant content in guava (*Psidium guajava*) and araca (*Psidium* spp.) germplasm from different Brazilian regions. *Plant Genetic Resources: Characterization and Utilization, 9*: 384–391.

Manna, D.C., H. Sarkar, and B. Das. 2015. Genetic divergence studies in guava (*Psidium guajava* L.). *Electron. J. Plant Breed.*, *6*(1): 161–166.

Martins, M.V.V., S.F. Silveira, and L.A. Maffia. 2014. Guava fruit loss caused by rust. *Summa Phytopathol.*, *40*(2): 107–113.

Mauchamp, A. 1997. Threats from alien plant species in the Galapagos Islands. *Conservation Biology*, *11*(1): 260–263.

MERC. 2013. Market and economic research centre. *South African Fruit Trade Flow*, *10*: 1–37.

Merrill, E.D. 1925. *An Enumeration of Philippine Flowering Plants* (4 Vols.). Government of the Philippine Islands, Department of Agriculture and Natural Resources, Bureau of Printing, Manila.

Missouri Botanical Garden. 2007. *Tropicos Database*. St. Louis: Missouri Botanical Garden. http://www.tropicos.org/

Mitra, S. 1997. *Postharvest Physiology and Storage of Tropical and Subtropical Fruits*. New York: CAB International.

Mitra, S.K., T.K.S. Irenaeus, M.R. Gurung, and P.K. Pathak. 2012. Taxonomy and importance of Myrtaceae. *In*: *3rd International Symposium on Guava and Other Myrtaceae,* 33. Petrolina, PE, Brazil. Morton, J. 1987. Guava. *In*: F. Julia, and J. Morton [Eds.], *Fruits of Warm Climates*, 356–363. Miami: Creative Resource Systems.

Nasution, F.Y., and S. Hadiati, S. 2014. Characterization and clustering of some guava germplasm collections based on leaf and fruit characters. *Agrivita*, *36*(1): 91.

Negi, S.S., and S. Rajan. 2007. Improvement of guava through breeding. *Acta Hort.*, *735*: 31–37.

Niedenzu, F.J. 1893. Nova Genera et Species Pantarum seu Prodromus. *In*: H.G.A Engler, and K.A.E. Prantll (Eds.), *Naturlichen Pflanzenfamilien*, *3*(7): 69.

Nisha, K., G. Shetty, and A. Chaturvedi. 2013. *Psidium* guajava a fruit or medicine: An overview. *J. Pharm. Innov.*, *2*(8): 63.

Nogueira, A.M., J.H.S. Guilhen, E. Mangaravite, A. Ferreira, and M.F.S. Ferreira. 2012. Discriminant analysis of wild guava tree by morphological descriptors and microsatellites. *In*: *3rd International Symposium on Guava and Other Myrtaceae*, 15.Petrolina, PE, Brazil. Orwa, C., A. Mutua, R. Kindt, R. Jamnadass, and S. Anthony. 2009. *Agroforestry Database: A Tree Reference and Selection Guide Version 4.0*. World Agroforestry Centre, Kenya.

Pelea, L.P., A.S. Gonzalez, E.B. Fernandez, N.N.M. Rodriguez, M.T. Cornide, and J.E.S. Garcia. 2016. Characterization of genetic diversity in three guava (*Psidium guajava* L.) populations. *Cult.*, *37*(2): 115–126.

Pereira, F.M., C.A. Caryalho, and J.C. Nachtigal. 2003. S'ecullo XXI: Nova cultivar de goiaberiade dupla finalidade. *Rev. Bras. Frutic.*, *25*: 498–500.

Plant information Database. 2017. http://www.tradewindsfruit.com/content/fruitsscientific.htm.

Pontikis, C.A. 1996. *Psidium guajava* L. (Guava). *In*: Y.P.S. Bajaj (Ed.), *Trees IV. Biotechnology in Agriculture and Forestry*, 35pp Berlin/Heidelberg: Springer.

Popenoe, W. 1920. *Manual of Tropical and Subtropical Fruits*. New York: Macmillan.

PPV and FRA. 2016. *Guidelines for the Conduct of Tests for Distinctiveness, Uniformity, and Stability. Guava* (*Psidium guajava L.*), 25.PPV & FRA. Department of Agriculture & Cooperation Ministry of Agriculture and Farmers Welfare. GOI, New Delhi, India, Purseglove, J.W. 1968. *Tropical Crops: Dicotyledons*. New York: Wiley.

Rahim, M.A., M.S.A. Fakir, M.M. Hossain, M.S. Alam, M.M. Anwar, A.K.M.A. Alam, F. Islam, N. Naher, K.S. Islam, K. Fatema, and Ghosh, S.N. 2012. BAU germplasm center: The largest fruit repository in Bangladesh one stop service for quality planting materials of fruits, conservation, development, production, diversity, research, and extension. *In*: *Proceedings of the International Symposium on Minor Fruits and Medicinal Plants for Health and Ecological Security* (*ISMF & MP*), 3–6. West Bengal, India.

Rai, M.K., P. Asthana, V.S. Jaiswal, and U. Jaiswal. 2010. Biotechnological advances in guava (*Psidium guajava* L.): Recent developments and prospects for further research. *Trees*, *24*: 1–12.

Rajan S., and C.M.U. Hudedamani Menzel. 1985. Guava: An exotic fruit with potential in Queensland. *Queensland Agric. J.*, *111*(2): 93–98.

Rajan, S., L.P. Yadava, R. Kumar, and S.K. Saxena. 2005. Selection possibilities for seed content- A determinant of fresh fruit quality in guava (*Psidium guajava* L.). *J. Appl. Hortic.*, *7*: 52–54.

Rajan, S., L.P. Yadava, R. Kumar, and S.K. Saxena. 2007. Genetic divergence in seed characteristics of guava: A basis for parent selection in hybridization. *Indian J. Hort.*, *64*(3): 290–293.

Rajan, S., L.P. Yadava, R. Kumar, and S.K. Saxena. 2008. Direct and indirect effects of seed related characters on number of seeds in guava (*Psidium guajava* L.) fruits. *Sci. Hortic.*, *116*: 47–51.

Rajan, S., A.K. Misra, and H. Ravishankar. 2011. *Guava Descriptor*, 19 pp. Lucknow: AICRP (STF), CISH.

Rajan, S., L.P. Yadava, R. Kumar, and S.K. Saxena. 2012. Variation among guava (*Psidium guajava* L.) accessions in seed hardness and its association with fruit characteristics. *Int. J. Innov. Hortic.*, *1*(1): 18–23.

Ray, P.K. 2002. Guava. *In*: *Breeding Tropical and Subtropical Fruits*, 143–154. New Delhi: Springer.

Risterucci, A.M., M.F. Duval, W. Rohde, and N. Billottte. 2005. Isolation and characterization of microsatellite loci from *Psidium guajava* L. *Mol. Ecol. Notes*, *5*: 745–748.

Ritter, E. 2012. Guava biotechnologies, genomic achievements and future needs. *In*: *3rd International Symposium on Guava and Other Myrtaceae*, 15. Petrolina, PE, Brazil.

Ritter, E., A. Herran, J. Valdes-Infante, N.N. Rodriguez-Medina, A. Briceno, G. Fermin, F. Sanchez-Teyer, A. Connor-Sanchez, J. Muth, J. Boike, D. Prufer, C.A. Santos, I.C. Nunes dos Santos, M.A. Rodriguez, A.M. Risterucci, N.

Billotte, D. Becker, and W. Rohde. 2010. Comparative linkage mapping in three Guava mapping populations and construction of an integrated reference map in Guava. *In*: W. Rohde, and G. Fermin (Eds.), *Proceedings of the IInd IS on Guava and other Myrtaceae*. *Acta Hort*., 849.

Rodriguez, N.N., and B. Velasaquez. 1999. *Propagación del guayabo. Informe final de proyecto. Propagación, mejoramiento y conservación de frutales cultivados bajo condiciones tropicales*, 40 pp. CITMA. Ciudad de La Habana.

Rodriguez, M.N.N., G.A. Fermin, J. Valdes-Infante, B. Velasquezi, D. Rivero, D.F. Martinez, J. Rodriguez, and W. Rohde. 2010a. Illustrated descriptors for guava (*Psidium guajava*). *Acta Hort*., *849*: 103–110.

Rodriguez, M.N.N., J.J. Valdes, J.A. Rodriguez, J.B. Velasquez, D.F. Rivero, F. Martinez, G. Gonzalez, D.G. Sourd, L. Gonzalez, and J. Canizares. 2010b. Genetic resources and breeding of guava (*Psidium guajava* L.) in Cuba. *Appl. Biotechnol*., *27*: 238–241.

Ruehle, G.D. 1964. El cultivo de la guayaba en la Florida. *Trop. Agric*., *10*, 555–564.

Samson, J.A. 1986. The minor tropical fruits. *In*: *Tropical Fruits* (Vol. 2) 270. New York: Longman.

Santos, C.A.F., L.C. Correa, and S.R. Costa 2011. Genetic divergence among *Psidium* accessions based on biochemical and agronomic variables. *Crop. Breed. Appl. Biotechnol*., *11*: 149–156.

Saxena, S.K., S. Rajan, R. Chandra, N. Srivastava, and A. Bajpai. 2007. Molecular characterization of closely related open pollinated seedling selections in guava. *Acta Hort*., *735*: 49–55.

Shigeura, G.T., and R.M. Bullock. 1976. Flower induction and fruit production of guava (*Psidium guajava* L.). *Acta Hort*., *57*: 247.

Singh, B.P., and R.S. Rana. 1993. Promising fruit introductions. *In*: K.L. Chadha, and O.P. Pareek (Eds.), *Advance in Horticulture* (Vol. 1), 43–66. New Delhi: Malhotra Publishing House.

Singh, T.K., A. Kumar, G. Mahajan, and P. Kumar. 2015. Evaluation of commercial guava (*Psidium guajava* L.) varieties for growth, yield, and quality attributes under rainfed conditions of Vindhya regions. *Environ. Ecol*., *33*(4B): 1867–1870.

Sitther, V., Zhang, D., Harris, D.L., Yadav, A.K., Zee, F.T., Meinhardt, L.W., and S.A. Dhekney. 2014. Genetic characterization of guava (*Psidium guajava* L.) germplasm in the United States using microsatellite markers. *Genet. Resour. Crop Evol*., *61*: 829–839.

Soetopo, L. 1991. *Psidium guajava* L. *In*: E.W.M. Verheij, and R.E. Coronel (Eds.), *Record from Proseabase*. Bogor: PROSEA (Plant Resources of South-East Asia) Foundation.

Solanki, I.S., V.P. Ahlawat, A.S. Charia, S.K. Sehrawat, and D.S. Dahiya 2011. Crop improvement scenario in guava. *Haryana J. Hort. Sci*., *40*(1, 2): 50–58.

Soubihe Sobrinho, J. 1951. *Estudos b'asicos para o melhoramento da goiabeira (Psidium guajava* L.), 166pp. Sao Paulos.

Species 2000. Naturalis, Leiden, The Netherlands. ISSN 2405-8858.

Swartz, O.P. 1788. *Procr*. 77.

Tate, D. 2000. *Tropical Gruit of Thailand*. Bangkok: Asia Books Co. Ltd.

Thaipong, K., and U. Boonprakob. 2005. Genetic and environmental variance components in guava fruit qualities. *Sci. Hortic*., *104*: 37–47.

Thaipong, K., S. Promchot, A. Auvuchanon, and Boonprakob, U. 2017. Genetic analysis of guava Germplasm using AFLP markers. *Int. J. Agric*., *13*(5): 741–752.

Toro Toro, E. 1993. *El Cultivo de la Guayaba*.

UN Comtrade. 2017. *United Nations Commerce Trade and Development*. New York and Geneva. https://comtrade.un.org/

UPOV. 1987. *Guidelines for the Conduct of Tests for Distinctness, Homogeneity, and Stability*. TG/110/3. Geneve.

(Duplication)USDA. 2010. *USDA ARS, Germplasm Resources Information Network. Psidium friedrichsthalianum* in the Germplasm Resources Information Network (GRIN), U.S. Department of Agriculture Agricultural Research Service.

USDA, NRCS. (2009, August 28). *The Plants Database*. National Data Center. Baton Rouge, LA, USA. http://plants.usda.gov.

Usman, M., W.A. Samad, B. Fatima, and M.H. Shah. 2013. Pollen parent enhances fruit size and quality in inter varietal crosses in guava (*Psidium guajava*). *Int. J. Agric. Biol*., *15*: 125–129.

Valdes-Infante, J., D. Becker, N. Rodriguez, B. Velazquez, G. Gonzalez, D. Sourd, L. Rodriguez, E. Ritter, and W. Rohde. 2003. Molecular characterization of Cuban accession of guava (*Psidium guajava* L.), establishment of a first molecular linkage map and mapping of QTLs for vegetative characters. *J. Genet. Plant Breed*., *57*: 349–358.

Valdes-Infante, J., N. Rodriguez, D. Becker, G. Velazquez, D. Sourd, J. Espinosa and W. Rohde. 2007. Microsatellite characterization of guava (*Psidium guajava* L.), germplasm collections in Cuba. *Cult*., *28*(3): 61–67.

Valera-Montero, L.L., P.J. Munoz-Rodriguez, H. Silos-Espino, and S. Flores-Benitez. 2016. Genetic diversity of guava (*Psidium guajava* L.) from Central Mexico revealed by morphological and RAPD markers. *J. Exp. Bot*., *85*: 176–183.

Weinert, I.A.G., and V.P.J. Wyk. 1998. Guava puree with reduced stone cells content: Preparation and characteristics of concentrated and nectars. *Int. J. Food Sci. Technol*., *23*(5): 501–510.

Wilson, C.W. 1980. Guava. *In*: S. Nagy, and P.E. Shaw (Eds.), *Tropical and Subtropical Fruits*, 279–299. Westport: Corm.

Yadav, A.K. 2006. Guava for valley state agricultural research station. http://www.ag.fvsu.edu/publicat/commoditysheets/fvsu003.htm.

Yadava, U.L. 1996. Guava production in Georgia under cold-protection structure. *In*: J. Janick (Ed.), *Progress in New Crops*, 451–457. Arlington: ASHS Press.

Figure 1. Genetic resources of guava-wild species.

Figure 2. Variability among guava germplasm for fruit shape, size and colour.

Figure 3. Different guava germplasm

a(a) Allahabad Safeda, (b) Allahabad Safeda variant, (c) Arka Kiran, (d) Arka Mridula, (e) Black guava (f) Hisar Safeda, (g) Hisar Surkha, (h) Hisar Surkha variant, (i) Kasipur collection, (j) Lalit, (k) Lucknow-49, (l) *P. friedrichstalianum*, (m) *P. pumilum*, (n) Pant Prabhat, (o) Punjab Pink, (p) Red type I, (q) Red type II, (r) Sasni, (s) Sasri, (t) Shweta, (u) Snow White, (v) Soft Seeded variant, (w) Sour type, (x) Thai, (y) Thai variant, (z) Trichy, (aa) VNR & (bb) Yellow type.

(Gangappa et al., 2022)

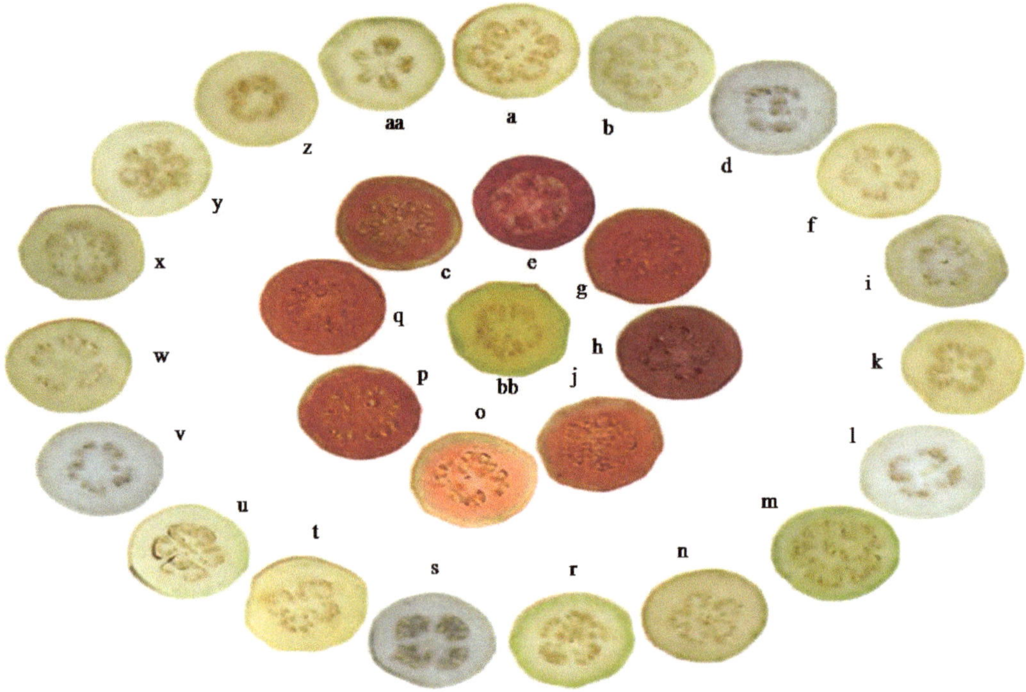

Figure 4. Transverse section of guava germplasm

a(a) Allahabad Safeda, (b) Allahabad Safeda variant, (c) Arka Kiran, (d) Arka Mridula, (e) Black guava (f) Hisar Safeda, (g) Hisar Surkha, (h) Hisar Surkha variant, (i) Kasipur collection, (j) Lalit, (k) Lucknow-49, (l) *P. friedrichstalianum*, (m) *P. pumilum*, (n) Pant Prabhat, (o) Punjab Pink, (p) Red type I, (q) Red type II, (r) Sasni, (s) Sasri, (t) Shweta, (u) Snow White, (v) Soft Seeded variant, (w) Sour type, (x) Thai, (y) Thai variant, (z) Trichy, (aa) VNR & (bb) Yellow type.

(Gangappa et al., 2022)

Figure 5. Different guava germplasm.

Figure 6. Improved guava varieties developed from ICAR-IIHR, Bengaluru.

Figure 7. Improved guava varieties developed from ICAR-CISH, Lucknow.

Figure 8. Improved guava varieties developed from ICAR-IARI, New Delhi.

Figure 9. Improved guava varieties developed from ICAR-Research complex for NEH Region, Umian Meghalaya.

Figure 10. Varieties released from PAU, Ludhiana.

8

Genetic Diversity of *Garcinia gummi-gutta* and Sustainable Utilization

Shilpa Susan Scaria,[1] *Kadanthottu Sebastian Joseph,*[1,*]
Vijayalaxmi S. Dandin[2] and *P.S. Shameer*[3]

Introduction

Garcinia gummi-gutta (L.) N. Robson is a fruit-bearing tree in the Clusiaceae family, better known by the names Gamboge and Malabar tamarind. This species is native to Southeast Asia. It has become established in the Western Ghats of India. The Clusiaceae family has about 200 species, but *G. gummi-gutta* stands out for its special qualities and highlights its range of uses. The *Garcinia* genus contains a wide range of plant species, each with distinct botanical characteristics and frequently significant effects on human health. The complexities of *G. gummi-gutta*'s genomic landscape and its implications for sustainable consumption become more apparent as we set out on a quest to investigate its genetic diversity. The Clusiaceae family member *G. gummi-gutta* is evidence of the botanical treasures found in Southeast Asia's tropical and subtropical climates (Sweeney, 2008). Admired for its unique tart fruit, the species has branched out to appeal to a variety of markets, including traditional medicine, food preparation, and the rapidly growing nutraceutical industry. Since ancient times, communities have recognized the plant's potential for curing illnesses and boosting well-being, and have included various portions of the plant in their traditional pharmacopeias (G. Joseph, Nair, and Kuttappan, 2023). Specifically, the fruit has been a mainstay in cooking, adding its distinct flavor to local cuisine. Homesteads grow and care for this tree, and its dried fruit rind is used for a variety of purposes, especially as a condiment and souring agent in Kerala, South India. *G. gummi-gutta* has a rich phytochemical profile in addition to its culinary and medical uses. Of these, hydroxycitric acid (HCA) is one component of particular interest. The species' significance has increased worldwide due to the inclusion of HCA in dietary supplements due to its alleged impact on metabolic processes (Raina et al., 2021). As a result, the tree is now acknowledged as a plantation crop in South India in response to the growing demand for both the industrial extraction of natural HCA and the use of its fruit rind in regional cuisine.

[1] Department of Life Science, School of Sciences, Christ University, Bangalore, India, 560029.
[2] Department of Biology, JSS RSH PU College, Dharwad 580004, India.
[3] ICAR- Central Tuber Crops Research Institute, Sreekariyam, Thiruvananthapuram, Kerala.
* Corresponding author: joseph.ks@christuniversity.in

In addition to its culinary uses, the Malabar tamarind has nutritional and therapeutic benefits. Its fruit rind is frequently used as a food flavoring, food bulking, and food preservative. *G. gummi-gutta* is used in traditional medicine to treat a variety of conditions, including rheumatism, irregular menstruation, oedema, constipation, and intestinal parasites, which are common in many Asian nations. Several organic acids, benzophenones, and xanthones have been identified as the plant's main ingredients by prior phytochemical investigations, which may have contributed to the plant's medicinal qualities. Its biological activity, including hypolipidemic, anti-obesity, and anticancer qualities, has been further supported by scientific investigations, making it a valuable and adaptable resource for the scientific community, herbal medicine, and culinary traditions alike (Tomar et al., 2019).

There are various variants within the species, and each has special qualities of its own. The chapter explains the botanical subtleties and geographic limitations that characterize these variations, from the common and economically significant *var. gummi-gutta* to the uncommon *var. papilla* and *G. conicarpa*, providing a nuanced perspective on their ecologies and relevance. The central focus of this investigation is the genetic complexity of *G. gummi-gutta*. Concerns regarding the long-term viability of *G. gummi-gutta* use have been raised considering the increased exploitation that has resulted from the surge in global demand for the plant in recent decades (Pramanik et al., 2021). The growing attention emphasizes the species' economic potential but also emphasizes how critical it is to understand and protect its genetic variety. The need for conservation methods grows as human activity continues to alter the environments in which *G. gummi-gutta* is found. This chapter explores the various conservation approaches and how field gene banks, seed banks, and *in vitro* banks contribute to the preservation of this remarkable botanical specimen's genetic resources. It also considers the wider ramifications of conservation initiatives, highlighting the significance of local community engagement and global cooperation.

In order to better understand the genomic complexities that underlie *G. gummi-gutta*'s adaptability and usefulness, this chapter will explore the species' genetic diversity. Our goal is to investigate genetic diversity assessment approaches and genetic diversity's implications for sustainable consumption. This chapter aims at the conservation and responsible management of this plant treasure by summarizing current knowledge and highlighting research gaps. Our voyage through the genetic realms of *G. gummi-gutta* is not just an academic investigation; rather, it is a call to action for the conservation of biodiversity and the wise use of natural resources. This book chapter aims to provide readers with a full overview of *G. gummi-gutta* by exploring its range, variations, traditional uses, biological activities, phytochemicals, toxicity studies, genetic diversity, and conservation efforts. It does this by revealing the species' remarkable biological splendor as well as its profound cultural significance.

Botanical Description and Distribution

G. gummi-gutta, colloquially recognized as *G. cambogia* (commonly known as Malabar tamarind), is a tropical evergreen tree belonging to the family Clusiaceae. Native to Southeast Asia, it is renowned for both its cultural significance and its association with potential health benefits. The unique appearance and utility of this species are attributed to its botanical features. At a height of 5–20 meters, *G. gummi-gutta* usually has a symmetrical canopy that is well-branched. The bark is smooth and greyish, and the trunk is straight. The leaves have a leathery texture and are glossy, dark green, and elliptic. They can range in length from 5–12 cm and are placed along the branches in opposing pairs. On slender stalks, tiny, unisexual flowers appear in groups. The main flower parts are sepals and stamens, and the flowers are usually greenish-yellow in color and lack petals. *G. gummi-gutta* is primarily known for its pumpkin-shaped fruit, which resembles a tiny yellow or greenish-yellow pumpkin. With a diameter of roughly 5–8 cm, the fruit can range in size from an orange to a grapefruit. The pericarp, or outer rind, of the fruit, is thick and rich in HCA, a substance that may help explain some of the possible health advantages of *G. gummi-gutta*. The flesh of the fruit is sweet, sour, and divided, containing the seeds. Usually oval, seeds are encased in a white, gelatinous aril. The aril of the seeds is occasionally employed in customary culinary techniques, but the seeds themselves are not eaten. The tree grows slowly, and it takes about 7–9 years to distinguish between male and female trees until they begin to flower (Kalia, Malik, and Chaudhury, 2012).

G. gummi-gutta exhibits a restricted native global range, primarily confined to India, Nepal, and Sri Lanka. It thrives in the humid, tropical climates of this region. These trees thrive predominantly in the semi-evergreen to evergreen forests of southwest India, with a notable presence in the Western Ghats, encompassing the states of Kerala, Maharashtra, Karnataka, and Tamil Nadu. Beyond its native range, *G. gummi-gutta* has been introduced to various other tropical and subtropical areas worldwide. This includes regions in Africa, Australia, and certain parts of South America. *G. gummi-gutta* demonstrates a preference for lowland evergreen forests, often found near riverbanks and in secondary forests. The species is adaptable and can thrive in a variety of soil types. The distribution of *G. gummi-gutta* is influenced by ecological factors such as temperature, precipitation, and soil characteristics. The species tends to favor areas with consistent rainfall and well-drained soils (Semwal et al., 2015).

Garcinia gummi-gutta (L.) Robs. holds significant economic importance as a fruit crop and is a crucial element of the forest ecosystem in the Western Ghats. Within this species, three varieties are recognized in India: *G. gummi-gutta* (L.) N. Robson *var. gummi-gutta, G. gummi-gutta var. papilla* (Wight) N.P. Singh, and *G. conicarpa Wight* (Shameer et al., 2016). Among these, var. *gummi-gutta* stands out as the most prevalent and economically vital, extensively cultivated across the Western Ghats, particularly in Kerala, spanning from coastal areas to highlands, and existing in the wild. In contrast, *G. conicarpa* and *var. papilla* are less common, primarily found in the highlands of evergreen forests with limited distribution. The widespread cultivation preference for *var. gummi-gutta* can be attributed to its favorable characteristics, including large fruit size, pulpy aril, and a higher seed count (4–8 seeds per fruit). *G. conicarpa* is morphologically distinct due to the absence of leaf ligules, the arrangement of stamens in a convex torus head, and the conical nature of its fruits (Shameer et al., 2016).

Varieties of *G. gummi-gutta*

Kerala, India stands out as a hub for the diversity of *G. gummi-gutta*, with notable variations observed in morphological characteristics such as leaves, flowers, fruits, and seeds. *G. gummi-gutta* (L.), an evergreen tree with a maximum height of 20 meters, is distinguished by its pale yellow, sticky exudation. The leaves of this tree are elliptic and elliptic-ovate, ranging from 6–13 x 2.5–6 cm. In terms of male flowers, they are tetramerous and appear in clusters of 3–8 on axillary fascicles, with individual flowers measuring 1–1.7 x 1–1.2 cm and pedicels spanning 7–12 mm. The sepals are orbicular, featuring a membranous margin with fringed-like projections, while the oblong petals are pale yellow or orange-yellow, Pink, and reddish brown, with a membranous margin. Stamens are arranged in a globose head, and a rudimentary pistil may be absent or, if present, bears a stigma with a discoid structure and a 4-lobed cleft. Female flowers, also tetramerous, occur solitarily or in fascicles of 1–3, measuring 1.5–2 x 1.5 cm, with yellow, yellow-orange, and red. Staminodes number 4–20, and the ovary is 4–12 locular, with one ovule in each locule. The sub-globose or ovoid ovules are grooved, and stigmatic rays are spreading, free nearly to the base, with a crenate, tuberculate margin. Fruits are globose, approximately 6–8 cm in diameter, with 6–10 grooves and a yellow or orange-yellow color upon ripening. The pericarp is notably thick and fleshy. Seeds, numbering 6–8, are ovoid, measuring 2–3.3 x 0.7–0.9 mm, compressed, and surrounded by a pulpy aril that can be either white or red. Field identification of *G. gummi-gutta* involves noting elliptic leaves measuring 6–13 cm in length, stigmatic lobes numbering 6–10, and a deeply grooved fruit with 6–10 grooves (Tharachand et al., 2015).

The morphological diversity within *var. gummi-gutta* is particularly pronounced among cultivated varieties in Kerala compared to their wild counterparts. In an examination of 18 accessions of var.

gummi-gutta cultivated across different regions in Kerala, ranging from coastal areas to midlands, the focus was on assessing variability in size and shape. Noteworthy traits of *var. gummi-gutta* include large fruit size, pulpy aril, and a higher seed count (4–8 seeds per fruit), making it a preferred choice for cultivation and contributing to its widespread distribution. The processed pericarp of *var. gummi-gutta* is highly esteemed for its delicate taste and flavor. Accessions were meticulously evaluated based on fruit size, rind thickness, acidity, and yield, with an average fruit weight of 173 g. Earlier investigations by (Abraham et al., 2006) on 13 fruit and five seed characters across 51 accessions of "Malabar tamarind" indicated maximum variability in nipple length (74.8%) and minimum for fruit girth (12.8%), with an average fruit weight of 161 g (Abraham et al., 2006). The branching pattern of *var. gummi-gutta* generally exhibits a horizontal orientation, occasionally with a pendulous drooping pattern. Leaf sizes range from 7–12 x 3.5–5 cm, displaying considerable variability in shapes from elliptic to broader forms. Leaf apices and bases are predominantly acute, occasionally obtuse. This morphological diversity extends to flowers, fruits, and seeds, with fruit shapes varying from globose to oblong, and rarely, discoid. The thickness of fruit rind, a crucial factor in the food sector, varies significantly among selected accessions, ranging from 6.25 mm–16.03 mm. Fresh fruit weights fall within the range of 45.7–173.3 g. The number of grooves on the fruit surface also demonstrates significant variation, ranging from 5–11 (Figure 1A) (Abraham et al., 2006).

Garcinia conicarpa (Wight), an evergreen tree reaching heights of up to 15 meters, is characterized by its distinct yellow exudation. The leaves are predominantly obovate-ovate, occasionally oblong, or broader beyond the middle, measuring 6–10 x 4–8 cm. Male flowers, arranged in a tetramerous fashion, appear either singly or in 2–5 flowered fascicles, featuring stout pedicels and ovate sepals with a membranous margin and fringed-like projection. Yellow, oblong-orbicular petals with a slightly membranous margin are complemented by stamens in a convex torus head, with the presence or absence of rudimentary pistils. Female flowers, also tetramerous, occur solitarily or in 2–3 flowered fascicles, with staminodes forming a ring. The ovary is 3–5 locular, housing one ovoid, grooved ovule per locule, and 3–5 stigmatic rays. Fruits typically exhibit a conical shape, occasionally ovoid, with a yellowish-green hue, around 5 cm in diameter, and 3–5 grooves, featuring a terminal mamilla. The pericarp is notably thick and fleshy, while seeds, numbering 2–4, are ovate-oblong, measuring 2–3 x 0.8–10 mm, and enclosed in a thin fibrous aril. Field identification rests on the absence of a leaf ligule on the petiole, the broader leaf shape beyond the middle, and the conical fruit shape with 3–5 grooves (Figure 1B) (Shameer et al., 2016).

Garcinia gummi-gutta var. papilla (Wight) is characterized as an evergreen tree, reaching heights of up to 15 meters, with distinctive yellow exudation. The leaves are elliptic, measuring 6–9 x 1.5–3 cm. In male flowers, organized in tetramerous clusters of 3–5, the pedicels are stout (5–7 mm), and sepals, ovate to oblong, possess a membranous margin. Brick-red oblong petals with a membranous margin complement stamens in a globose androphore, with a rarely present rudimentary pistil. Female flowers, also tetramerous, occur solitarily or in fascicles of 1–3, measuring 1–1.2 x 7–10 mm, with staminodes forming a ring. The ovary is 6–8 locular, containing 1 ovule in each locule, sub-globose,

Figure 1. Diversity in fruit varieties of *G. gummi-gutta* (A) var. *gummi-gutta*, (B) *G. conicarpa*, and (C) var. *papilla*.

grooved, and possessing 4–8 stigmatic rays. Fruits are sub-globose, yellowish-green, around 6 cm in diameter, featuring 4–8 grooves with a terminal mamilla, and a notably thick and fleshy pericarp. Seeds number 3–5, sub-triangular, measuring 2–3 x 0.8–10 mm, enclosed in a dense mass of fibrous aril. Field identification of *var. papilla* is facilitated by the reddish tinge on young shoots and leaf margins. The fruits are ovoid-oblong, presenting 4–8 grooves and a terminal mamilla (Figure 1C) (Shameer et al., 2016).

Ethnobotanical Uses

G. gummi-gutta is highly valued for its ethnobotanical uses, particularly as a spice tree. The sun-dried smoked rind, a key component, is widely embraced as a flavoring agent, notably enhancing the taste of fish curries. Renowned for its bacteriostatic properties, the dried rind, when combined with salt, becomes a preferred method for curing fish in India and Sri Lanka. The smoke-dried fruit rind of *G. gummi-gutta*, integral to the "Colombo curing" of fish, was utilized in pickling processes alongside brine. Additionally, it serves as a substitute for Kokum butter (derived from *G. indica*) and is a common additive for increasing the satiety of meals (Burdock, Bagchi, and Bagchi, 2005). In traditional medicine, the fruit rind is utilized for various purposes, including addressing bowel and rheumatism complaints and acting as a purgative, anthelmintic, and emetic. Its applications extend to veterinary medicine, where a rinse made from the plant is used to treat mouth diseases in cattle (Abraham et al., 2006). Although the fruits are edible, their high acidity generally deters raw consumption. Notably, a tonic rich in vitamin C, derived from *G. gummi-gutta*, is employed in India to manage diverse heart conditions. Beyond medicinal uses, the plant's versatility is evident. The fruit rind is employed in polishing silver and gold ornaments and serves as an alternative to acetic acid for coagulating rubber latex. The gum finds use as a varnish, while the resin serves as a pigment in miniature canvases and watercolors. This diverse range of applications underscores the plant's significance in traditional practices, as well as its contributions to various industrial and artistic endeavors (Pandey, Rastogi, and Rawat, 2013).

Phytochemical Composition and Biological Activities

Preliminary phytochemical investigations have revealed the existence of phenolic compounds, alkaloids, saponins, flavonoids, polysaccharides, tannins, and proteins in various plant components. Furthermore, a variety of substances, such as xanthones, organic acids, amino acids, and benzophenones have been extracted from various plant sections. These identified phytochemicals are important players with a variety of functions, including antidiabetics, immunomodulators, antimicrobials, antioxidants, and anti-inflammatories (Noreen et al., 2023). Table 1 represents the various phytocompounds present in *G. gummi-gutta.*

Table 1. Phytochemical composition of *G. gummi-gutta.*

Phytochemical Class	Plant Part	Compounds	References
Organic acids	Fruit	Hydroxycitric acid (HCA), Citric acid, Tartaric acid, Malic acid, HCA Lactone	(Jayaprakasha and Sakariah, 2000)
Xanthones	Roots, fruits	Oxy-guttiferone I, Garbogiol, Oxy-guttiferone K, Oxy-guttiferone K2, Oxy-guttiferone M, Rheediaxanthone A,	(Masullo et al., 2008)
Fatty acids	Seed	Stearic acid, Palmitic acid, Oleic acid, Elaidic acid, Arachidic acid, Linoleic acid	(Carratù et al., 2008)
Benzophenones	Fruit, bark, leaf	Garcinol, Guttiferone I, Guttiferone N, Guttiferone M, Isogarcinol, Guttiferone J, Guttiferone K	(Masullo et al., 2010)
Amino acids	Fruit	Arginine, Glutamine, Asparagine, Proline, Threonine, Glycine, γ-Aminobutyric Acid, Ornithine, Leucine, Lysine, Isoleucine	(Carratù et al., 2008)

A variety of biological functions shown by *G. gummi-gutta* have piqued scientific curiosity (Anilkumar et al., 2023). The fruit is well-known for its anti-obesity qualities, which are mostly linked to HCA and make it a popular dietary supplement. Studies show that HCA can help with weight management by lowering hunger and preventing the creation of fat. *G. gummi-gutta* has been shown to have hypolipidaemic effects, which affect serum and tissue lipid levels (Noreen et al. 2023). Numerous experiments have shown its antioxidant activity, which is related to phenolic components. The fruit has demonstrated potential in its anti-ulcer properties by guarding against the harm that some compounds cause to the gastrointestinal mucosa. Antimicrobial and anthelmintic actions, anticholinesterase qualities, antidiabetic qualities, and hepatoprotective effects are additional possible advantages (M.K. Joseph and Thangavel, 2023). While some research, particularly involving the molecule garcinol, suggests it has anti-cancer capabilities, a more thorough investigation is necessary to clarify its processes and therapeutic potential across other health dimensions.

Genetic Diversity Assessment

Morphological Diversity

The diversity of *G. gummi-gutta* varieties seems to be much, however, there have been significant studies on the var. *gummi-gutta* cultivars when compared to other varieties and wild relatives. 18 accessions of the var. *gummi-gutta,* are studied by Shameer et al. (2016), accessing the size and shape of fruits (Table 2) (Shameer et al., 2016). Accessions of *G. gummi-gutta* have been collected by various researchers and have been utilized for the improvement of the crop. However, very few of them have been utilized for crop improvement, especially IC 354028, IC 354019, IC 354047, IC 354063, and IC354070 collected from a germplasm pool of 56 accessions from Karnataka and

Table 2. Morphological diversity in accessions of *G. gummi-gutta* var. *gummi-gutta* (adapted from Shameer et al., 2016).

Accession	Number of Fruit Grooves	Fruit Rind Thickness (mm)	Fruit Shape	Fruit Weight (g)	No. of Seeds
Arammulla	8	13.8	Discoid	99.92	4
Calicut	8–9		Globose, grooves splitted	66.24	7–8
Karanikkunnu (ii)	7–10	7.2	Oblong	45.68	2–4
Karanikunnu (i)	7–8	11.71	Oblong, mamillae	89	5–7
Karannikunnu(iii)	7–10	6.25	Globose, oblong	54.86	4–6
Karimbam	8–11		Globose	58.94	6–9
Kurianippally	8	9.2	Oblong	58.42	5
Kotta	6–9	11.21	Globose, grooves splitted	52.16	2–4
Manipuzha	8		Globose, grooves splitted	124.82	5
Mezhuveli	7–9	10.31	Oblong	43.52	1–2
Podiyadi	6		Globose, oblong with depressed	85.48	3
Pulikezh	9	13.41	Globose, oblong with mamillae	46.98	7
TBG. G.g-1	6–8	16.03		198.8	4–5
TBG. G.g- 2	6–9			148.94	4–6
Ullanoor	8	11.51	Globose, mammilla te	85.28	7
Vaikom - 2	7–9		Globose, grooves splitted with mamillae		6–8
Vaikom 1	6–7		Globose, grooves splitted with mamillae	95.53	5–6
Wayanad	5–6	12.7	Oblong		3–4

Kerala, India (Abraham et al., 2006). Figure 2 represents the diversity of fruit size and shape in *G. gummi-gutta* var. *gummi-gutta.*

Physiological Characterization

Determining the photosynthetic pigments in *G. gummi-gutta* is an essential step in performing physiological research. The technique used to extract pigment is crucial, and in this study, a DMSO (dimethyl sulfoxide)-based approach outperformed the conventional acetone method in terms of efficiency. Total carotenoids and chlorophylls were recovered more fully using the DMSO technique. Although acetone (80%) is a common solvent, it is not as effective at removing certain less polar substances, such as beta-carotene and chlorophyll a. Reduced solvent loss was seen in pigment extraction using the DMSO approach, which entails direct immersion of leaf tissues followed by incubation. This result implies that the DMSO-based approach, which offers improved photosynthetic pigment recovery, is more appropriate for *G. gummi-gutta.*

In a study conducted between morphotypes GG-02 and GG-05, there are notable differences in parameters including petiole length, thickness, width, and length of leaves. Furthermore, there are significant variations in the amounts of carotenoid and chlorophyll, with GG-02 having larger quantities than GG-05. The detected differences provide important new information about the variety of the species. By comparing the ripe fruits from GG-02 and GG-05, it was possible to see notable variations in several measures, including weight, rind thickness, pedicle dimensions, polar and equatorial circumference, and grooves. GG-02 has a much thicker rind, which is indicative of the plant's economic component and contributes to increased rind recovery. The fruits of GG-02 are further distinguished from GG-05 by subtle differences in pulp content and seed characteristics (Bohra and Waman, 2019).

Biochemical Characterization

One major area of interest in the biochemical analysis of *G. gummi-gutta* was the measurement and identification of volatile oil components. Dried leaves from several *Garcinia* species were hydro-distilled to determine the volatile oil concentration. After that, GC-MS analysis was used to determine which kinds of secondary metabolites were in these oils. Interestingly, *G. gummi-gutta* had eight significant volatile compounds. Furthermore, secondary metabolites like trans-caryophyllene, gamma-murolene, gamma-gurjunene, beta-bourbonene, alpha-humulene, and gamma-cadinene, are present significant in the perfume industry. A further aspect of the biochemical characterization was determining the free fatty acid content of *G. gummi-gutta* seeds. Gas chromatography with flame ionization detection (GC-FID) was used to analyze the free fatty acids that were isolated from the seed kernels and transformed into fatty acid methyl esters (FAME). Stearic acid was the most common fatty acid found in both *G. gummi-gutta* samples, accounting for a significant percentage (28–40%). This result is consistent with other studies that showed stearic acid to be a significant component,

Figure 2. Diversity of fruit shapes and sizes in different accessions of *G. gummi-gutta* var. *gummi-gutta.*

with quantities rising with maturation. These results highlight the importance of Garcinia species as possible suppliers of beneficial methyl esters of fatty acids with a variety of uses (Parthasarathy et al., 2013).

In a study, the biochemical exploration of various *Garcinia* species centered on two crucial components: lycopene and Hydroxy Citric Acid (HCA). It is interesting to note that the amount of lycopene in each species under study varied very little. The study confirmed that *G. gummi-gutta* has the highest concentrations of HCA of all the Garcinia species tested. Samples from fresh and dry fruits, as well as leaves, demonstrated this. By contrast, HCA concentrations were lower in *G. pedunculata* and intermediate in *G. indica*. The biochemical diversity within the genus is highlighted by the observed range in HCA content among various Garcinia species. These results are important for future research on *G. gummi-gutta*, especially considering its high HCA concentration, which has drawn interest due to its purported anti-obesity qualities (Parthasarathy et al., 2013). In another study, the biochemical examination of *G. gummi-gutta* fruits shows the differences in pulp and rind ascorbic acid content, pH, and total soluble solids (TSS). The ascorbic acid level is the same in all morphotypes, despite notable variations in TSS and pH. The results offer insight into prospective uses and nutritional characteristics by providing useful information on the chemical makeup of the fruit's various parts (Bohra and Waman, 2019).

Even though *G. gummi-gutta* has a lot of promise for commercial production, no improved varieties have been introduced, which highlights the need for systematic study to find better genotypes. Because most of the current populations, which include both wild and homestead varieties are seedling progenies, there is a great deal of variation in morphological features. The investigation undertaken showed that the several morphotypes of *G. gummi-gutta* differed significantly in terms of morphological and biochemical characteristics related to leaves and fruits. For a species of medicinal significance, this variety is especially beneficial, as changes in pigment recovery are linked to variations in the amounts of the enzyme Chlorophyllase in the leaf tissues. This emphasizes how crucial it is to characterize morphological and biochemical criteria to comprehend the variety within *G. gummi-gutta* and establish the groundwork for future developments in the production and application of this priceless plant species.

Molecular Characterization

A range of molecular and genetic methods are used to evaluate the genetic diversity of populations of *G. gummi-gutta*. DNA fingerprinting, DNA sequencing, and molecular markers (such as SSRs and SNPs) are common techniques. *G. gummi-gutta* populations can have subgroups that can be found via population structure analysis. This approach sheds light on possible variables influencing population differentiation as well as the distribution of genetic diversity. Allelic variation within populations is evaluated using genetic markers like single nucleotide polymorphisms (SNPs) and microsatellites. Gaining knowledge about allelic variety advances our understanding of genetic richness as a whole. Evaluations of genetic diversity frequently investigate how regional and ecological factors affect the distribution of genetic variation. This covers elements that can influence the observed genetic diversity, such as altitude, climate, and geographic isolation. Nei's distance and other genetic distance measurements are used to measure the genetic relatedness of individuals or populations. These measures contribute to our understanding of *G. gummi-gutta*'s historical divergence and general genetic linkages (Tharachand et al., 2015).

In a comprehensive study, the researchers sought to assess the morphogenetic diversity within 35 accessions of *G. gummi-gutta* (L.) N. Robson by employing a dual approach, utilizing quantitative morphological traits and species-specific simple sequence repeat (SSR) markers. The recorded quantitative morphological data exhibited statistical significance ($p < 0.001$) based on One-way ANOVA, highlighting the diverse nature of the analyzed traits. Significant variation was observed among accessions, particularly in growth and yield-related characteristics such as the total number of fruits per tree, the mean number of fruits required to achieve a weight of 1 kg, and the mean fresh and dry rind yield per tree. Additionally, morphometric traits, including single fruit weight, showcased considerable diversity among the studied accessions. To complement the morphological assessment, molecular profiling was conducted using 24 SSR primers selected from a set of 32, based on their ability to generate precise, intense, and reproducible banding patterns. The SSR analysis yielded a total of 39 alleles across the 24 loci, with an average of 1.63 alleles per locus and an impressive average polymorphism rate of 86.11%. The "Polymorphic Information Content" (PIC) averaged 0.239, and the Marker Index averaged 0.005, indicating a moderate level of allelic heterozygosity and specificity of SSR primers for the selected accessions. Dendrograms constructed based on the "Unweighted Pair Group Method of Arithmetic Means" (UPGMA) and Jaccard's similarity coefficient highlighted substantial genetic variation among the accessions. These dendrograms illustrated the genetic links within the population visually, highlighting the significance of taking into account both morphological and molecular traits when assessing the diversity of *G. gummi-gutta* (Babu et al., 2021).

In a ground-breaking study entitled «Isolation and Characterization of Microsatellite Markers in *G. gummi-gutta* by Next-Generation Sequencing and Cross-Species Amplification», a comprehensive analysis of the plant's genetic makeup was conducted. They identified 27,313 microsatellites and designed primers for 9,964. PCR amplification of genomic DNA from 30 *G. gummi-gutta* genotypes showed high polymorphism (PIC ranging from 0.867–0.951). This marks the first generation of genomic SSR markers in *G. gummi-gutta*, with applications in mapping, identification, diversity assessment, and breeding studies. This work represents the first generation of genomic SSR markers in *G. gummi-gutta* through the application of state-of-the-art next-generation sequencing technology. These genomic SSR markers have broad applications since they have the potential to be very important in many different areas of genetic research, such as mapping, identification, diversity evaluations, and breeding studies. The multiplicity of alleles per locus, observed and expected heterozygosity, and strong polymorphic information content highlight the usefulness of these markers in identifying and describing the complex genetic diversity found in *G. gummi-gutta* populations (Ravishankar et al., 2017).

In a comprehensive study titled «Molecular Insights into the Genetic Diversity of *G. cambogia* Germplasm Accessions», the genetic relationships among 12 *G. cambogia* (Gaertn.) Desr. accessions were systematically evaluated using Random Amplified Polymorphic DNA (RAPD) markers. The germplasm under investigation originated from the collection maintained at "NBPGR Regional Station", Thrissur, India. Out of 30 RAPD primers, seven were chosen, revealing 128 polymorphic markers across the 12 accessions. The study evaluated genetic relatedness through metrics like Polymorphic Information Content (PIC), Marker Index (MI), and Jaccard's coefficient. The resulting dendrogram illustrated genetic relationships, validated by a positive correlation ($r = 0.49$) with morphological data. The significant variance in morphological features (flowers, fruits, leaves, and seeds) between the studied accessions provided additional evidence for this finding (Tharachand et al., 2015).

In another study, the primary objective was to conduct a comprehensive genetic evaluation of 50 accessions of *G. gummi-gutta* L. (Roxb.) collected from diverse geographical locations within the Kerala state, India. Employing ISSR (Inter Simple Sequence Repeat) markers, a set of 30 primers was initially tested, and 21 of them exhibited clear, reproducible, and scorable banding patterns. The analysis of these markers revealed a total of 157 ISSR fragments, with an average of 7.48 bands per primer. Notably, a considerable degree of genetic polymorphism was observed among the accessions, with an average of 59.52% polymorphic bands. UBC 809 stood out with the highest percentage of polymorphic bands at 87.5%. The Polymorphism Information Content (PIC) value, calculated at 0.355, indicated a high degree of allele divergence within the population. Moreover, the "Simpson and Shannon indices", with values of 0.99 and 1.02, respectively, underscored the significant level of polymorphism present within the studied population. The construction of a dendrogram using the neighbor-joining method with bootstrapping values (100) revealed the grouping of accessions into three distinct clusters. This clustering pattern suggested continuous gene flow among the accessions, irrespective of their geographical distances. The genetic relationships unveiled by this dendrogram provide valuable insights into the diversity and relatedness of *G. gummi-gutta* accessions within the studied population. Further analysis employing the STRUCTURE method based on Bayesian clustering identified the probability of eight subpopulations within the population. The idea that the *G. gummi-gutta* population under study contains discrete genetic subgroups is reinforced by this discovery. The discovery of these subpopulations has ramifications for conservation and breeding initiatives as it deepens our knowledge of the genetic variety and organization of this species (Babu et al., 2022).

Utilization of Germplasm

Germplasm of *G. gummi-gutta* has been collected by various researchers and has been utilized for the improvement of the crop. However, very few of them have been utilized for crop improvement, especially IC 354028, IC 354019, IC 354047, IC 354063, and IC354070 collected from a germplasm pool of 56 accessions from Karnataka and Kerala, India (Abraham et al., 2006). Similarly, the National Bureau of Plant Genetic Resources (Regional Station, Thrissur, India) has registered INGR No.04061 (IC244100-2) and INGR No. 04062 (IC244111-1) as early maturing germplasms. Accessions, GG-02 and GG-05 have been identified for growing in Andaman and Nicobar conditions (Bohra and Waman, 2019).

Conservation Strategies

Numerous research groups have conducted extensive studies on the collection, conservation, and utilization of Garcinia germplasm, particularly focusing on its center of diversity. Scientists affiliated with the National Bureau of Plant Genetic Resources (NBPGR) and various agricultural universities in India have actively participated in the comprehensive gathering of Malabar tamarind from diverse regions, including the Western Ghats, Northeastern India, and Andaman and Nicobar Islands. In the past, the primary objective of Garcinia germplasm collection centered around identifying trees with exceptional fruit characteristics, encompassing aspects such as attractiveness, fruit shape, size, rind thickness, number of arils per fruit, size, seed shape, color, and overall yield. However, in recent years, the research focus has shifted toward the selection of germplasm with traits conducive to early maturity, early and late blooming, as well as enhanced tolerance to drought and diseases. This shift reflects an evolving emphasis on not only the aesthetic and productive aspects of the fruit but also on factors crucial for sustainable cultivation and adaptability to challenging environmental conditions (Murthy et al., 2018).

In situ and *Ex situ* Conservation of Germplasm

In situ conservation approaches are instrumental in preserving plant populations within their natural distribution range, creating optimal conditions for the manipulation of genetic variability within species. This comprehensive method encompasses two primary methods. The first method involves establishing genetic reserves, characterized by the identification, management, and ongoing monitoring of genetically diverse wild populations within designated areas. This proactive and long-term conservation strategy aims to safeguard the dynamic genetic makeup of the species. The second method, on-farm conservation, has gained increasing significance in recent years, although it has been practiced by farmers for centuries. This method focuses on the preservation of traditional crop cultivars or landraces and the associated farming systems within the context of traditional agricultural practices. Notably, on-farm conservation allows for the integration of conservation efforts into the existing agricultural landscape, promoting the sustainable maintenance of valuable genetic resources. Exploration and collection missions conducted by the "NBPGR regional station" in Thrissur, India, have identified 110 sites across 13 districts in Kerala for the on-farm conservation of *G. gummi-gutta*. This strategic initiative is in line with the understanding of how critical it is to use conventional agricultural systems to preserve and propagate *G. gummi-gutta's* distinctive genetic diversity in its natural habitat.

Ex situ conservation is a crucial aspect of preserving genetic diversity in plant species, and the efforts undertaken by scientists at the NBPGR Regional Station in Thrissur, Kerala, for the conservation of Garcinia genetic resources and germplasm, are noteworthy. In this context, various conservation methods are employed, including field gene banks, seed banks, and *in vitro* banks. Keeping live plants in field gene banks entails either keeping them in their native environments or in specially created spaces that closely resemble those environments. 124 accessions of *G. gummi-gutta* have been gathered by scientists from the Western Ghats regions of Kerala, Karnataka, Maharashtra, and Goa. These accessions most likely comprise a wide variety of genetic material originating from several geographic regions. A specific plant species' genetic diversity is preserved by the collection and careful storage of seeds. For species that yield orthodox seeds, which are resistant to the freezing and drying procedures involved in seed banking, seed banks are invaluable. Plant tissues, such as meristems, shoots, embryos, or plantlets, are maintained in sterile, controlled environments at facilities known as "*in vitro* banks". It is particularly beneficial for species that produce resistant seeds, or seeds that are intolerant of conventional seed storage techniques, and for clonal crop germplasm. By guaranteeing a pathogen-free environment, *in vitro* conservation lowers the possibility of illnesses that could harm conventional fields or seed bank collections. It works especially well for clonal crops, where it's important to preserve the genetic identity of individual plants. Safe germplasm transfers under controlled phytosanitary control are supported by in vitro conservation. This is crucial for transferring plant material between nations or regions while lowering the possibility of bringing in diseases or pests (Murthy et al., 2018).

Toxicity Studies

The use of *G. gummi-gutta* and other herbs in weight-loss solutions has raised concerns about potential negative side effects, such as atrial tachycardia, palpitations, and hypertension. Overconsumption, of suggested dosages, caused testicular atrophy (shrinkage) in animals and reduced rat sperm production (Saito et al., 2005). The usage of *G. gummi-gutta* has been linked to major adverse events, including hepatotoxicity, according to the Brazilian Pharmacovigilance System (Choudhury et al., 2023). When taken by recommended dosage, the No Observed Adverse Effect Level (NOAEL) of up to 2800 mg/day indicates safety (Chuah et al., 2012). To precisely estimate the potential harm posed by *G. gummi-gutta* and related formulations, a focus on pre-marketing safety studies and strong post-marketing adverse impact tracking systems is required.

Future Perspective

Prospects for *G. gummi-gutta* include several important factors to take into account. We must step up our efforts to conserve biodiversity as we go forward, paying special attention to uncommon forms like *var. papilla* and *G. conicarpa*. This entails not just protecting their natural habitats but also putting *ex situ* conservation measures into practice, such as creating and maintaining gene banks. Further understanding of the genetic variety of *G. gummi-gutta* can be gained through developments in genetic research, particularly through the use of molecular techniques. By maintaining the species' resilience and flexibility in the face of environmental changes, a better understanding of the species at the molecular level can help establish conservation and breeding plans. It is imperative to involve local populations in the preservation and sustainable utilization of *G. gummi-gutta*. The effectiveness of conservation efforts can be increased by combining traditional knowledge with contemporary conservation techniques. This could entail awareness campaigns, community-based conservation initiatives, and instruction on sustainable harvesting methods. We ought to investigate sustainable methods of harvesting as well as non-traditional uses for *G. gummi-gutta*. This necessitates looking into its possible applications in the nutraceutical, pharmaceutical, and other industries while making sure that doing so won't jeopardize the long-term sustainability of natural populations.

Conclusion

In conclusion, this book chapter provides a comprehensive exploration of *G. gummi-gutta*, offering insights into its distribution, varieties, phytocompounds, biological activities, toxicity cases, genetic diversity, and conservation strategies. The examination of its distribution highlights its presence in the biodiverse Western Ghats, with particular emphasis on the rarity of *var. papilla* and *G. conicarpa*, confined to the highland regions of forests. The varieties, especially *var. gummi-gutta*, reveal a wealth of genetic diversity, notably manifested among the cultivars. The elucidation of genetic diversity not only enhances our understanding of the species but also underscores its ecological and economic significance. The rarity of certain varieties emphasizes the need for dedicated conservation efforts to safeguard this valuable genetic resource. Conservation strategies, including field gene banks, seed banks, and *in vitro* banks, play a pivotal role in ensuring the preservation and sustainable utilization of *G. gummi-gutta* genetic resources. The knowledge acquired from this chapter helps us negotiate the complex environment of *G. gummi-gutta*, which advances our understanding of the species and paves the way for responsible conservation measures. This chapter offers a comprehensive view by combining morphological and phytochemical characteristics. This viewpoint is crucial for the continued preservation and ethical management of *G. gummi-gutta*, guaranteeing the plant's position in both human culture and natural ecosystems.

References

Abraham, Z., S.K. Malik, G.E. Rao, S.L. Narayanan, and S. Biju. 2006. Collection and Characterisation of Malabar Tamarind [*Garcinia cambogia* (Gaertn.) Desr.]. *Genetic Resources and Crop Evolution*, *53*(2): 401–406. doi:10.1007/s10722-004-0584-y.

Anilkumar, A.T., S. Manoharan, S. Balasubramanian, and E. Perumal. 2023. *Garcinia gummi-gutta* : Phytochemicals and Pharmacological Applications. *BioFactors*, *49*(3): 584–599. doi:10.1002/biof.1943.

Babu, V., R. Ramachandran, L.V. Aswanilal, A. Gangaprasad, and E.A. Siril. 2022. Genetic Evaluation of *Garcinia gummi-gutta* L. (Roxb.) Accessions Based on Inter Simple Sequence Repeat Markers. *Ecological Genetics and Genomics*, *24*(September): 100130. doi:10.1016/j.egg.2022.100130.

Babu, V., R. Ramachandran, A. Gangaprasad, and E.A. Siril. 2021. Assessment of Morphogenetic Diversity in *Garcinia gummi-gutta* (L.) Roxb. Using Species-Specific Morphological and SSR Markers. *Ecological Genetics and Genomics*, *18*(February): 100081. doi:10.1016/j.egg.2021.100081.

Bohra, P., and A.A. Waman. 2019. Morphological and Biochemical Studies in *Garcinia gummi-gutta* (L.) Roxb. *Erwerbs-Obstbau*, *61*(3): 217–223. doi:10.1007/s10341-019-00419-3.

Burdock, G., M.I. Bagchi, and D. Bagchi. 2005. *Garcinia cambogia* Toxicity Is Misleading. *Food and Chemical Toxicology*, *43*(11): 1683–1684. doi:10.1016/j.fct.2005.05.011.

Carratù, B., C. Boniglia, S. Giammarioli, M. Mosca, and E. Sanzini. 2008. Free Amino Acids in Botanicals and Botanical Preparations. *Journal of Food Science*, *73* (5). doi:10.1111/j.1750-3841.2008.00767.x.

Choudhury, A., P. A. Singh, N. Bajwa, S. Dash, and P.Bisht. 2023. Pharmacovigilance of Herbal Medicines: Concerns and Future Prospects. *Journal of Ethnopharmacology*, *309*(June): 116383. doi:10.1016/j.jep.2023.116383.

Chuah, L.O., S.K. Yeap, W.Y. Ho, B.K. Beh, and Noorjahan Banu Alitheen. 2012. *In Vitro* and *In Vivo* Toxicity of *Garcinia* or Hydroxycitric Acid: A Review. *Evidence-Based Complementary and Alternative Medicine*, *2012*: 1–12. doi:10.1155/2012/197920.

Jayaprakasha, G.K., and K.K. Sakariah. 2000. Determination of (−) hydroxycitric acid in commercial samples of *Garcinia cambogia* extract by liquid chromatography with ultraviolet detection. *Journal of Liquid Chromatography & Related Technologies*, *23*(6): 915–923. doi:10.1081/JLC-100101498.

Joseph, G., A. Nair, and S. Kuttappan. 2023. *Garcinia cambogia*. *In*: *Herbs, Spices, and Their Roles in Nutraceuticals and Functional Foods*, 211–218. Elsevier. doi:10.1016/B978-0-323-90794-1.00013-2.

Joseph, M.K., and I. Thangavel. 2023. GC-MS/HPTLC Analysis, Antioxidant, Anti-Inflammatory and Antimicrobial Efficacy of *Garcinia cambogia*. *Free Radicals and Antioxidants*, *13*(1): 10–20. doi:10.5530/fra.2023.1.3.

Kalia, R.K., S.K. Malik, and R. Chauhury. 2012. *In Vitro* Morphogenetic Studies on Three Apomictic Species of Garcinia. *Journal of Plant Biochemistry and Biotechnology*, *21*(2): 279–285. doi:10.1007/s13562-011-0085-y.

Masullo, M., C. Bassarello, G. Bifulco, and S. Piacente. 2010. Polyisoprenylated Benzophenone Derivatives from the Fruits of *Garcinia cambogia* and Their Absolute Configuration by Quantum Chemical Circular Dichroism Calculations. *Tetrahedron 66*(1): 139–145. doi:10.1016/j.tet.2009.11.034.

Masullo, M., C. Bassarello, H. Suzuki, C. Pizza, and S. Piacente. 2008. Polyisoprenylated Benzophenones and an Unusual Polyisoprenylated Tetracyclic Xanthone from the Fruits of *Garcinia cambogia*. *Journal of Agricultural and Food Chemistry*, *56*(13): 5205–5210. doi:10.1021/jf800416j.

Murthy, H. N., V. S. Dandin, D. Dalawai, S.Y. Park, and K.Y. Paek. 2018. Breeding of Garcinia Spp. *In*: Jameel M. Al-Khayri, Shri Mohan Jain, and Dennis V. Johnson, [Eds.], *Advances in Plant Breeding Strategies: Fruits*, 773–809. Cham: Springer International Publishing. doi:10.1007/978-3-319-91944-7_19.

Noreen, S., M.K. Naizi, T. Tufail, F. Hassan, and C.G. Awuchi. 2023. Nutraceutical, Functional, and Therapeutic Properties of *Garcinia Cambogia* : A Review. *International Journal of Food Properties*, *26*(1): 729–738. doi:10.1080/10942912.2023.2178458.

Pandey, M.M., S. Rastogi, and A.K.S. Rawat. 2013. Indian Traditional Ayurvedic System of Medicine and Nutritional Supplementation. *Evidence-Based Complementary and Alternative Medicine*, *2013*: 1–12. doi:10.1155/2013/376327.

Parthasarathy, U., K.N. Babu, R.S. Kumar, G.R. Ashis, S. Mohan, and V.A. Parthasarathy. 2013. Diversity of Indian garcinia: A medicinally important spice crop in India. *Acta Horticulturae*, Article No. *979*(March): 467–476. doi:10.17660/ActaHortic.2013.979.50.

Pramanik, M., A.K. Diwakar, P. Dash, S. Szabo, and I. Pal. 2021. Conservation Planning of Cash Crops Species (*Garcinia gummi-gutta*) under Current and Future Climate in the Western Ghats, India. *Environment, Development and Sustainability*, *23*(4): 5345–5370. doi:10.1007/s10668-020-00819-6.

Raina, R., P.K. Verma, I.Taku, J.K. Malik, and R.C. Gupta. 2021. *Garcinia cambogia*. *In*: *Nutraceuticals*, 975–990. Elsevier. doi:10.1016/B978-0-12-821038-3.00058-6.

Ravishankar, K.V., R. Vasudeva, B. Hemanth, B.S. Sandhya, B.R. Sthapit, V.A. Parthasarathy, and V.R. Rao. 2017. Isolation and Characterization of Microsatellite Markers in *Garcinia gummi-gutta* by next-Generation Sequencing and Cross-Species Amplification. *Journal of Genetics*, *96*(2): 213–218. doi:10.1007/s12041-017-0756-0.

Saito, M., M. Ueno, S. Ogino, K. Kubo, J. Nagata, and M. Takeuchi. 2005. High Dose of *Garcinia cambogia* is Effective in Suppressing Fat Accumulation in Developing Male Zucker Obese Rats, but Highly Toxic to the Testis. *Food and Chemical Toxicology*, *43*(3): 411–419. doi:10.1016/j.fct.2004.11.008.

Semwal, R.B., D.K. Semwal, I. Vermaak, and A. Viljoen. 2015. A Comprehensive Scientific Overview of *Garcinia cambogia*. *Fitoterapia*, *102*(April): 134–148. doi:10.1016/j.fitote.2015.02.012.

Shameer, P., R. Koranappallil, T. Sabu, and N.N. Mohanan. 2016. Diversity of Malabar Tamarind (*Garcinia gummi-gutta* (L.) N. Robson) in the Western Ghats-Morphological and Phytochemical Evaluation. *In*: *Diversity of Garcinia Species in the Western Ghats: Phytochemical Perspective*, 1st Edn., 132–141. Jawaharlal Nehru Tropical Botanic Garden and Research Institute (JNTBGRI).

Sweeney, P.W. 2008. Phylogeny and Floral Diversity in the Genus *Garcinia* (Clusiaceae) and Relatives. *International Journal of Plant Sciences*, *169*(9): 1288–1303. doi:10.1086/591990.

Tharachand, C., C.I. Selvaraj, and Z. Abraham. 2015. Molecular Insights into the Genetic Diversity of *Garcinia cambogia* Germplasm Accessions. *Brazilian Archives of Biology and Technology*, *58*(5): 765–772. doi:10.1590/S1516-89132015050197.

Tomar, M., R.P. Rao, P. Dorairaj, A. Koshta, S. Suresh, M. Rafiq, R. Kumawat, R. Paramesh, U.V. Babu, and K.V. Venkatesh. 2019. A Clinical and Computational Study on Anti-Obesity Effects of Hydroxycitric Acid. *RSC Advances*, *9*(32): 18578–18588. doi:10.1039/C9RA01345H.

9

Genetic Diversity of *Aegle marmelos* and Sustainable Utilization

Madhumathi V.,[1,2] *M.K. Dhakar,*[3,*] *Bikash Das*[4] and *Ashok Yadav*[5]

Introduction

Bael (*Aegle marmelos* Correa), also known as Shri Phal, Baelpatra, and Bengal quince, holds significant cultural and historical importance as an indigenous fruit in India. Its presence dates back to prehistoric times, and it is widely cultivated across the country in subtropical, tropical, arid, and semi-arid regions (Yadav and Ram, 2021). Bael is cultivated across diverse regions of Southeast Asia, encompassing countries such as India, Sri Lanka, Pakistan, Burma, Bangladesh, Thailand, and various other Southeast Asian nations. In India, it is grown extensively, with Uttar Pradesh, Bihar, Jharkhand, Madhya Pradesh, West Bengal, Orissa (presently known as Odisha), Chattisgarh, and Rajasthan being the primary states where cultivation is prominent.

Notably, bael trees are commonly planted near temples dedicated to Lord Shiva. According to Hindu traditions, the leaves of the bael tree are considered sacred and are offered to Lord Shiva. The fruits play a crucial role in religious rituals, forming an integral part of offerings to deities through sacred pyres (havan). Historical references to the bael tree can be traced back to Vedic times (circa 2000 BC – circa 800 BC) in the Yajur Veda. Omprakash (1961) documented mentions of bael in early Buddhist and Jain literature (circa 800 BC – circa 325 BC), describing methods of ripening the bael fruit alongside other fruits. In the era of the Ramayana, bael fruit was well-known, with its trees reported to grow in the Chitrakuta hills and Panchvati. Sanskrit treatises such as the *Upavana Vinod*, focusing on silviculture, and the *Brihat Samhita* also make references to the bael fruit. Legend has it that in the forest, Lord Rama performed religious rites, offering various fruits, including the bael. The bael fruit is even depicted in the paintings of the Ajanta Caves, sharing space with other fruits in these artistic representations. Similar to numerous other members of the Rutaceae family, the bael

[1] ICAR-Indian Agricultural Research Institute, Barhi, Hazaribagh, Jharkhand – 825405, India.

[2] The Graduate School, Indian Agricultural Research Institute, New Delhi – 110012, India.

[3] ICAR Research Complex for Eastern Region, Farming System Research Centre for Hill and Plateau Region, Plandu, Ranchi, Jharkhand 834010, India.

[4] ICAR-National Research Centre on Litchi, Muzaffarpur - 842002 (Bihar), India.

[5] ICAR-Central Agroforestry Research Institute, 284003 Jhansi, Uttar Pradesh, India.

* Corresponding author: mahesh2iari@gmail.com

tree possesses aromatic flowers. There is a belief that this particular tree serves as an indicator plant, aiding in the identification of underground water, as suggested by Singh and Roy (1984).

Bael is a plant with both nutritional and medicinal value, particularly well-suited for cultivation in regions with limited water resources. The fruit of this plant is abundant in essential nutrients such as riboflavin, vitamin A, and carbohydrates. Various chemical components, including alkaloids, coumarins, and steroids, have been identified in different parts of the Bael plant (Maity et al., 2009). Renowned as a remedy for stomach ailments, bael is often referred to as a panacea for digestive issues. Marmelosin, found in bael fruit at varying concentrations (0.03–0.37%), is believed to be the bioactive compound responsible for therapeutic effects, with its levels dependent on the specific variety and geographical location. The mature fruit's pulp is consumed either in its fresh state or in the form of sharbat. Additionally, it is utilized in the creation of various value-added products, including jam, murabba, squash, powder, pickle, ice cream, slab, nectar, toffee, and more (Sharma et al., 2007).

The significance of the bael fruit lies in its therapeutic attributes, establishing the tree as one of the most valuable medicinal plants in India, as noted by Kirtikar and Basu (1935). The medicinal qualities of bael have been discussed in the *Charaka Samhita*, an ancient Sanskrit medical treatise, as documented by Aiyer (1956). Every part of the tree, including the stem, bark, root, leaf, flower, seed oil, and fruits at various stages of maturity, is utilized in diverse Ayurvedic medicines. In Indonesia, an infusion of bael flowers is enjoyed as a refreshing drink. The adhesive quality of the gummy substance surrounding the seeds makes it a valuable addition to waterpaints, enhancing their strength and brilliance. The hard fruit shells are crafted into pill and snuff boxes, often adorned with gold and silver, with greater abundance in young fruits. The gum derived from bael has been employed for stabilizing drilling fluids. Additionally, the stem contains a gum akin to gum Arabic (Hiwale, 2015). Due to its rutaceous nature, the leaves have a mild flavor, making them suitable for chewing as a natural mouth freshener.

Taxonomic Background

Bael is a native tree of India and is also prevalent in neighboring countries such as Nepal, Sri Lanka, Malaysia, Pakistan, Bangladesh, Myanmar, and Thailand, as well as across various Southeast Asian nations (Singh et al., 2019). Due to its revered status as a sacred tree, it is cultivated in regions like north Malaya, the arid zones of Java, and, to a limited extent, in northern Luzon of the Philippines, as well as in gardens in Egypt, Surinam, and Thailand. In India, bael is widely distributed throughout the country, with a concentration in the eastern parts of the Gangetic plains and nearby areas, particularly in Uttar Pradesh, Bihar, Madhya Pradesh, Chhattisgarh, and Jharkhand. Additionally, it can be observed growing in West Bengal, Punjab, and Odisha. In Gujarat, bael trees naturally thrive in forested areas, displaying significant diversity. While most genotypes in Gujarat's forest regions feature small-sized fruits, those with larger fruits are found in temple premises or courtyards of houses, likely brought by travelers, saints, and pilgrims from northern India, as documented by Singh et al. (2008, 2014).

Cytogenetics

Aegle marmelos Corr., a member of the Rutaceae family, shares its plant family with other species like Citrus, Catimiroa, Clausena, Eremocitrus, Limonia, Feroniella, Fortunella, Poncirus, Triphasia, and more. The botanical name *Aegle* has Greek origins, while the species name *marmelos* is derived from Portuguese. The chromosome configuration is $x = 9$, and $2n = 18$. While detailed cytogenetic studies on Bael are limited, reports from regions like the Punjab plains, Shiwalik hills, and Pachmarhi hills have indicated the presence of both diploid ($2n = 18$) and tetraploid cytotypes ($2n = 36$). According to Singhal et al. (2011), the wild and cultivated trees in these areas were predominantly diploid, exhibiting normal meiosis with 9 bivalents at metaphase-I and high pollen fertility ranging from 92%–98%. On the other hand, trees with larger leaves and fruits demonstrated a tetraploid chromosome constitution, featuring 18 bivalents at metaphase-I. The tetraploid trees also displayed regular microsporogenesis and a high pollen fertility rate of 95%, suggesting the potential for polyploidy in Bael.

Floral Biology

The success of a crop improvement program hinges largely on a thorough understanding of the floral biology of the specific crop. In the case of bael, the bisexual flowers are arranged in clusters, appearing as greenish-white axillary or terminal cymes. The calyx is shallow, featuring four short, broad teeth, with a pubescent exterior. The oblong-oval petals number four and exhibit a pale greenish-white hue. Stamens are numerous, hypogynous, and accompanied by short filaments. The ovary is oblong-ovoid, slightly tapering into a thick, short style that thickens upwards, with a capitate stigma having multiple cells (8–20) arranged in a circle, and numerous ovules in each cell (Pandey et al., 2008).

The open-type style possesses a variable number of stylar canals based on the number of carpels. The stigma is classified as a wet stigma, covered by a sticky secretion during pollination. Various floral characteristics such as flower bud emergence, flowering duration, time of anthesis, anther dehiscence, stigma receptivity, and pollen viability exhibit variability depending on the variety and location (Srivastava and Singh, 2000; Singh et al., 2008, 2011).

Studies conducted at Central Horticultural Experiment Station (CHES), Godhra, under rain-fed conditions in a semi-arid ecosystem, have evaluated the size and shape of floral organs, including bud size, flower size, and petal size for different varieties (Singh et al., 2008, 2014, 2016). Singh et al. (2014) have reported on the phenology, floral biology, and main pollinating agents for bael varieties. Insects are attracted to the nectar-secreting disc beneath the ovary, as indicated by Srivastava and Singh (2000). However, Singhal et al. (2011) found that diploid and tetraploid trees exhibited normal meiosis and high pollen fertility.

Phenological events such as leaf fall, leaf emergence, floral bud break, flowering, and fruiting were found to be consistent between wild and cultivated trees (Singh et al., 2013, 2014; Lal, 2002; Misra et al., 2000, 1999). Natural pollen transfer proved to be highly efficient, with high levels of fruit set following open pollination. However, fruit set significantly reduced following hand pollination, likely due to injury to the stigma during the experimental process. Despite the synchronous nature of anther dehiscence and stigmatic receptivity, selfing in a flower is avoided due to herkogamy. Despite the dull-colored appearance of the flowers, various insects, including honey bees, houseflies, butterflies, and ants, visit the flowers due to their sweet fragrance and the stamen/pollen serving as a food reward. Among insect pollinators, the honey bee (*Apisdorsata*) emerged as the primary and legitimate pollinator, while others played minor roles or were mere visitors (Singh et al., 2014).

The mature fruits of bael are sturdy, sizeable, and exhibit a spherical, oblong, or pear-shaped structure, attaining diameters of up to 23 cm. The surface is generally smooth, occasionally slightly tuberculate. This amphisarca fruit is multicarpellary and multichambered, developing from a superior ovary. Typically globose, the pericarp is nearly smooth, with a grayish-yellow color, a hardness of 1.6–2.7 mm, and an interior filled with soft, yellow, and orange, highly fragrant, and pleasantly flavored sweet pulp. From a botanical perspective, the fruit is classified as a berry with a hard pericarp. The number of cells in the fruit, arranged in a circular fashion, corresponds to the number of cells in the ovary. The inner portion of the pericarp and the placenta constitute the edible part of the fruit. The seed's testa is mucilaginous, and the numerous seeds are somewhat compressed, organized in closely packed tiers within the cells. They are surrounded by a tenacious, slimy, transparent mucus that solidifies upon drying. The testa is white and covered with woolly hairs immersed in the mucus. The embryo features large cotyledons and a short superior radicle, with no endosperm present (Singh and Roy, 1984).

The ripening process for these fruits spans about 11 months on the tree, resulting in sizes comparable to large grapefruits, and in some cases, even larger. Due to the hardness of the shell, it necessitates cracking with a hammer or machete for access to the delectable contents.

Species Diversity

Initially, the genus *Aegle* was considered monotypic, consisting solely of *Aegle marmelos* (L.) Correa. However, recent findings from the World Checklist of Species of Plants (WCSP) and global species resources, such as The Plant List (2010), acknowledge three species within the genus: *Aegle decandra* Fern-Vill., *A. glutinota* (Blanco) Merr., and *A. marmelos* (L) Correa. *Aegle marmelos* is characterized as a medium to tall deciduous tree, exhibiting slow growth and reaching heights of 5–10 meters. The aromatic trifoliate leaves are green, and the branches often feature spines, with a robust and stout trunk. The leaves are tripartite, composed of a pair and a terminal leaflet, with the terminal one typically being the largest. Some instances of leaf abnormalities have been observed in *Aegle marmelos*. The branches are distinctive, adorned with long, straight spines. The bark displays shallow furrows and a corky texture.

Germplasm Collection and Evaluation

Naturally occurring bael seedlings exhibit significant phenotypic diversity in regions such as Uttar Pradesh, Bihar, Jharkhand, Madhya Pradesh, Odisha, West Bengal, and Chattisgarh, warranting comprehensive documentation and utilization. In Uttar Pradesh, particularly in districts like Faizabad, Sultanpur, Jaunpur, Varanasi, Mirzapur, Allahabad, Pratapgarh, Etawah, Mathura, and Agra, numerous superior bael types are either naturally growing or planted in residential backyards.

The sensory attributes of bael fruit seem to be influenced by three key factors: mucilage, sugar content, and total phenolics. Optimal palatability is achieved when bael fruit exhibits high levels of sugars, especially non-reducing sugars, coupled with low phenolic content and mucilage (Pandey et al., 2008). Through the systematic collection and evaluation of bael trees with favorable fruit characteristics, several promising varieties have been developed via clonal propagation at Indian Council of Agriculture Research (ICAR) Institutes, Regional Stations, and State Agricultural Universities.

Traditionally, bael fruits were named based on their local availability. Existing reports on cultivars primarily focus on states such as Uttar Pradesh, Uttarakhand, Bihar, and West Bengal. Singh (1961) described six bael varieties from Uttar Pradesh, highlighting Mirzapuri as the superior type due to its thin, easily breakable rind, fine-textured pulp free of mucilage, excellent flavor, and low seed count. Other varieties included Darogaji, Ojha, Rampuri, Azamati, and Khamaria. Teaotia et al. (1963) identified five varieties from Uttar Pradesh, with KagziGonda standing out for its large fruit size (1412.82 g), thin rind (0.15 cm), and soft yellow pulp boasting excellent flavor.

In a survey conducted by Jauhari and Singh (1971) in significant fruit-growing areas of Uttar Pradesh and Bihar, KaziEtawah, Sewan Large, Mirzapuri, and Deoria Large were identified as excellent in taste and overall quality among the studied varieties. Mazumdar (1975) examined five bael fruit varieties in West Bengal and noted that the spherical flattened ones generally exhibited superior fruit and chemical composition.

Rai et al. (1991) identified six promising genotypes from the Tarai belt of Uttar Pradesh based on their yield potential and fruit quality. Within the same study, three wild genotypes with small, non-edible fruits but exceptionally high bearing capacity (up to 3,000 fruits per tree) were recognized, suggesting their potential use as donor parents in improvement programs.

Nath et al. (2003) conducted a study on 33 genotypes of bael collected from six districts of Jharkhand and 10 districts of Bihar in East Central India. Variability was observed in tree morphological characters, intensity of bearing, fruit characteristics, and physicochemical constituents. Five genotypes were identified as promising based on fruit characters and bearing behavior, with fruit weight ranging from 0.5–2.6 kg, fruit length from 16.8–30.5 cm, and rind thickness from 0.17–0.4 cm.

Pandey et al. (2005) collected and evaluated 17 germplasm accessions from Bihar and Uttar Pradesh, identifying four promising accessions based on tree and fruit characteristics. Selections made at Narendra Dev University of Agriculture And Technology (NDUA&T), Faizabad, and Govind

Ballabh Pant University of Agriculture and Technology (GBPUA&T), Pantnagar have also gained popularity for commercial cultivation.

Kumar and Nath (2010) characterized 21 genotypes of bael with natural diversity from Odisha for morphological and physicochemical parameters. CHB-l-13 and CHB-l-14 were found promising, showing favorable attributes such as high pulp weight, thin rind, superior fruit weight, minimum seed number, and optimal Total Soluble Solids (TSS) content.

More recently, an extensive exploration at the Central Institute for Subtropical Horticulture (CISH), Lucknow covered districts in Uttar Pradesh, Madhya Pradesh, Bihar, and Jharkhand, resulting in the collection of 108 accessions. Twenty-seven promising types were identified based on desirable attributes, including a dwarf and spreading tree canopy, profuse bearing, medium-sized fruits (0.8–1.25 kg), round to oblong shape, thin and papery shell, low seed count, mucilage and fiber content, attractive golden yellow pulp color, sweet taste, pleasant aroma, and tolerance to biotic and abiotic stresses. These promising types have been multiplied and planted in the field gene bank at CISH for further evaluation.

Germplasm Conservation and Utilization

The existing biodiversity of bael in India can be broadly categorized into two main types. The first type is characterized by small fruits with a bitter pulp, more seeds, and higher mucilage and fiber content. The second type consists of large fruits with a thin shell, fewer seeds, less fiber and mucilage, and a sweeter pulp. The former type is commonly used in ayurvedic preparations due to its high content of marmelosin and psoralen, while the latter type is typically utilized as a dessert fruit and for the preparation of processed products.

Extensive germplasm of bael has been systematically collected from various states across India, including Uttar Pradesh, Bihar, Gujarat, Haryana, Madhya Pradesh, and West Bengal. Institutions such as NDUAT, Faizabad; CHES, Godhra; CISH, Lucknow; Chaudhary Charan Singh Haryana Agricultural University (CCSHAU), Regional Research Station, Bawal; Central Arid Zone Research Institute (CAZRI), Jodhpur; and research studies in Jharkhand, Bihar, Rajasthan, and Gujarat have contributed significantly to this effort (Pandey et al., 2008; Rai et al., 1991; Pandey et al., 2005, 2008; Saroj et al., 2008; Singh et al., 2014, 2019).

Singh et al. (2008, 2010, 2016) have reported on the growth behavior of various bael varieties. Additionally, the characterization of bael genotypes has been conducted at CHES, Godhra, under rainfed hot semi-arid conditions in Gujarat (Singh et al., 2011)

Various bael genotypes display significant variations in vegetative characteristics, as noted by Rai et al. (1991), and Singh et al. (2012, 2015). Singh et al. (2015) have reported notable diversity among genotypes collected from Rajasthan. Fruit yield and associated traits, including fruit weight, length, width, and girth, exhibit considerable differences among genotypes. For instance, fruit yield ranges from 40.50–69.29 kg, fruit weight from 0.43–4.25 kg, length from 10.61–19.59 cm, width from 9.40–22.00 cm, and fruit girth from 29.10–70.00 cm. The physical composition of bael fruit also varies widely in shell weight, shell thickness, number of seeds per fruit, total fresh seed weight, number of seed sacs, fiber weight, and pulp weight per fruit.

The qualitative characteristics of bael fruit, such as Total soluble solids (TSS) mucilage, TSS pulp, total sugar, reducing sugar, non-reducing sugar, vitamin C, total phenols, acidity, and TSS to acid ratio, exhibit a broad range. These parameters vary between 37.00–49.50° Brix, 30.57–37.45° Brix, 16.15–19.98%, 3.30–4.95%, 12.85–15.13%, 17.13–21.03 mg/100g, 2.34–2.75%, 0.30–0.49%, and 68.88–124.83, respectively, as reported by Singh et al. (2011, 2014, 2016) and Sharma et al. (2013). Bael varieties also exhibit variations in leaf characteristics (Singh et al., 2015). Germplasm of bael shows variability in growth and maturity, and the diversity based on morphological characteristics of the fruit has been documented by Singh et al. (2015) and Rai and Misra (2005).

Physico-chemical characteristics of bael seed oil reveal a light yellow oil with a refractive index of 1.468. The oil's iodine value is 114.81 ± 0.07 mg iodine/g, saponification value is 183.69 ± 2.41 mg KOH/g, acid value is 19.05 ± 0.09 mg KOH/g, and peroxide value is absent in the analyzed oil. The seed oil composition is predominantly composed of unsaturated fatty acids, including linoleic acid (2452.06 ppm), oleic acid (961.52 ppm), and linolenic acid (37.55 ppm), as reported by Bajaniya et al. (2015).

Ghosh and Gayen (1990) have documented the application of Metroglyph grouping and association analysis to study the physical characteristics of bael fruit. The examination of genetic divergence in seventeen bael germplasm using Mahalanobis D2 statistics by Rai et al. (2002) revealed substantial genetic diversity. The genotypes were categorized into three clusters, with one distinct group (PB-3). Interestingly, the clustering pattern did not align with geographic distribution.

Critical contributors to genetic divergence included ascorbic acid content, fruit weight, fruit length, number and weight of seeds per fruit, fiber content, and petiole length. Rai and Mishra (2005) suggested that these traits could be valuable in selecting diverse parents for hybridization and intercrossing between clusters to develop superior bael clones with desirable traits.

The genetic resources of bael are safeguarded in various field gene banks of ICAR institutes and State Agricultural Universities. Collections are curated at multiple locations, including CCSHAU, Regional Research Station, Bawal (10), NDUAT, Faizabad (22), Central Institute for Arid Horticulture (CIAH), Bikaner (21), CISH, Lucknow (44), GBPUAT, Pantnagar (10), CAZRI, Jodhpur (5), and CHES, Godhra (190). The following genotypes developed through selection were found promising after long-term evaluation at various SAUs and ICAR institutes (Ghosh 2014).

Hindrance to Crop Improvement

1. The challenging aspect of predicting hybrid characteristics arises from the highly heterozygous nature and cross-pollination tendencies in bael, making it a complex task.
2. Bael trees exhibit a prolonged juvenile period, both from seed to maturity and between generations, leading to a considerable delay in the release of new varieties.
3. Despite the absence of reported hybridization for cultivar development in bael, crossability studies are currently lacking. Preliminary investigations suggest the presence of self-incompatibility, hindering the establishment of pure lines.
4. Bael plants are prolific pollen producers, yet persistent fruit drop results in low fruit sets. The primary cause of this issue is linked to delayed fertilization caused by slow pollen tube growth, leading to the shedding of flowers before the pollen tube can reach the ovary. Desai and Bhandwalkar (1995) proposed that reduced fruit retention may be attributed to lower pollen retention on selfed pistils compared to cross-pollinated ones.
5. Bael flowers are grouped in axillary or terminal cymes with delicate pedicels, making it challenging to bag individual flowers, necessitating the need to bag the entire cluster.
6. The recalcitrant nature of bael seeds, coupled with slow seedling growth, contributes to an extended breeding cycle.

Breeding Objectives

1. The objective is the development of bael varieties characterized by dwarf stature, compact canopies, early high yield, and consistent fruit bearing, with reduced susceptibility to fruit drop and cracking.
2. The goal is to create a bael variety featuring medium-sized fruits (1.0–1.5 kg per fruit) that are uniform in shape, have thin skin, exhibit bright yellow color upon maturity, and possess low seed and mucilage content.

3. The aim is to breed a bael variety with fruits suitable for processing, emphasizing larger size, higher pulp content, yellow coloration, excellent flavor, and taste, along with low levels of seed, mucilage, and fiber.
4. The objective is the development of high-yielding bael varieties that are resistant to fruit borers
5. The goal is to create a bael variety with fruits containing high sugar content and minimal levels of mucilage, tannin, and other phenolics, boasting a pleasant flavor and acceptable aroma.
6. An important objective is the development of high-yielding bael varieties that are either spineless or have low-intensity spines.

The existing population exhibits extensive genetic diversity, offering ample opportunities for directly selecting improved genotypes. This diversity can be leveraged in deliberate breeding programs with well-defined objectives.

Improved Bael Varieties

A description of some of the important varieties is given below:

Narendra Bael 5 (NB-5)

The growth habit is semi-spreading, characterized by early and abundant fruiting, although there is inconsistency in fruit size. The fruits are of medium size (12.50 cm x 11.50 cm), round in shape, with a smooth surface and an extremely thin rind measuring 0.16–0.17cm. At maturity, they adopt a straw-yellow hue and have low mucilage, a moderately fibrous texture, and a light yellow pulp with minimal seed content. The pulp, known for its softness, boasts an excellent taste (330 Brix), and the flavor is complemented by an ascorbic acid content of 18.63 mg/100g in the edible portion. On average, the fruit weighs between 700–1300 g, and the tree yields an average of 54 kg of fruits during the 8th year, particularly thriving under semi-arid conditions.

Narendra Bael9 (NB-9)

The plants exhibit a semi-vigorous and spreading nature, characterized by a compact canopy. They have a moderate height ranging from 4–7 meters, featuring a compact canopy. This variety is known for its precocity and prolific fruit-bearing capability, with an eight-year-old plant yielding an average of 69 kg of fruits. The fruits are medium to large (16.00 cm x 13.50 cm), with a roundish-oblong shape, a smooth surface, and a thick rind measuring 0.24 cm. Upon reaching maturity, they take on a light yellow color and have an average mucilage content, a moderately fibrous texture, and a slightly golden-yellow pulp with low seed content. In terms of taste, the fruits are deemed good, containing 380 Brix total soluble solids in the pulp, exhibiting a slight acidity, and having an ascorbic acid content of 19.20 mg/100 g in the edible portion. This variety is versatile, and suitable for consumption both fresh and processed into various post-harvest products. Notably, it boasts excellent keeping quality for storage.

Narendra Bael7 (NB-7)

The growth habit is spreading, featuring semi-compact foliage. The trunk has a grey color and exhibits rectangular splitting of bark. The medium-sized leaves are lanceolate in shape, and green in color, with a smooth surface and non-wavy margins. The plants are tall and semi-spreading, with sparse fruit-bearing but large-sized fruits. These fruits are round, with a smooth surface and a very thick rind, turning yellow upon maturity. They have low levels of mucilage and fiber, and their attractive yellow pulp contains minimal seeds. Known for their good taste (300 Brix) and flavor, the fruits also boast an ascorbic acid content of 19.78 mg/100 g. On average, each plant yields 54.34 kg of fruits during the 8th year. This variety is highly suitable for processing; however, the fruits do not ripen uniformly.

Narendra Bael16 (NB-16)

The plants display an upright growth habit and exhibit precocity in fruiting, with budded plants initiating fruiting as early as the 3rd year. An eight-year-old plant yields an average of 52.40 kg of fruits. The fruits are small in size (12.50 cm x 10.00 cm), round, featuring a smooth surface and an exceptionally thick rind measuring 0.35cm. Upon reaching maturity, they take on a straw-yellow hue, are low in mucilage, and possess a moderately fibrous, yellow pulp with a high seed content. The fruit pulp is distinguished by an excellent taste (36° Brix), appealing flavor, and an ascorbic acid content of 17.61 mg/100 g in the edible portion. The fruit weight falls within the range of 750–800 g, and due to its very thick rind, it is particularly well-suited for powder making.

Narendra Bael 17 (NB-17)

The plants exhibit an erect growth pattern with a semi-spreading habit. They are moderately and sparsely bearing, with budded plants commencing fruiting in the 4th year. The average fruit yield for an eight-year-old plant is 51.39 kg. The fruits are slightly larger (20.00 cm x 15.50 cm), round, with a smooth surface and an exceptionally thin rind measuring 0.24cm. Upon reaching maturity, they take on a straw-yellow color, are low in mucilage, moderately fibrous, and feature an attractive yellow pulp with low seed content. The taste of the pulp is excellent (34° Brix), and it contains an ascorbic acid content of 19.63 mg/100g in the edible portion. With a fruit weight ranging from 1.65–1.8 kg, it serves well for both table consumption and processing purposes.

Pant Aparna

Pant Aparna trees are characterized by their dwarf stature, featuring drooping foliage and nearly devoid of thorns. These trees are known for their precocity and abundant fruit bearing. The leaves are large, pear-shaped, and dark green. The fruit has a globose shape, with an average size of 13.00 cm x 12.00 cm and a weight ranging from 0.8–1.25 kg. The flesh of the fruit is yellow, sweet, and possesses a delightful flavor, with a thin rind. The fruit pulp is also yellow, with TSS (Total Soluble Solids) measuring at 35%, titratable acidity at 0.67%, and an ascorbic acid content of 17.15 mg/100 g of pulp. Notably, the levels of mucilage, seed, and fiber are low, and mucilage and seeds are enclosed in separate segments. The flavor and taste of Pant Aparna are highly regarded, and the yield during the 8th year reaches an impressive 61.06 kg per plant.

Pant Shivani

This variety is an early mid-season cultivar with tall, vigorous, and densely upright-growing trees that exhibit precocity and are prolific in bearing fruits. The fruit is characterized by an ovoid, oblong shape, measuring 18.50 cm x 15.00 cm, and a weight ranging from 1.5–2.0 kg. The color of the fruit is lemon-yellow, and it demonstrates good storage quality. The rind is of medium thickness, while the pulp is lemon-yellow, offering a pleasant flavor with low to medium levels of mucilage, seeds, and fiber. The taste is highly regarded, featuring 69% pulp, TSS of 36%, total titratable acidity at 0.47%, and an ascorbic acid content of 19.55 mg/100 g of flesh. The average yield is 49.12 kg per plant during the 7th year.

Pant Sujata

This variety falls within the mid-season category but faces an issue of fruit splitting. The trees are of medium-dwarf stature, featuring drooping and spreading foliage that is dense, precocious, and prolific in bearing. The thorns are stout and larger. The fruit has a globose shape, depressed at both ends, with an average size of 14.50 cm x 13.50 cm. Under rainfed conditions in the hot semi-arid ecosystem of western India, the fruit weight varies from 1.12–1.70 kg. The fruit and pulp are light yellow, boasting better storage life, a thin rind, and low levels of seeds, mucilage, and fiber. The flavor is pleasant, and the taste is highly regarded, with flesh constituting 72%, TSS measuring at 32° Brix, acidity at 0.44%, and an ascorbic acid content of 17.10 mg/100 g of flesh. The average yield is 65.57 kg per plant during the 8th year.

Pant Urvashi

This variety falls into the early mid-season category. The trees are tall and vigorous, and exhibit dense and upright growth, displaying precocity and being prolific in bearing fruits. The fruit is ovoid-oblong, with an average size of 14.50 cm x 17.20 cm, and the fruit weight ranges from 1.5–2.50 kg. The fruit itself is yellow, with a medium to thin rind and light yellow pulp. The fruit contains 62.35% pulp, characterized by a pleasant flavor. Seeds and mucilage are of medium quantity, and the fiber content is low. The TSS (Total Soluble Solids) measures at 33° Brix, with titratable acidity at 0.49%, and an ascorbic acid content of 17.15 mg/100g in the pulp. The average yield per plant is 60.15 kg during the 8th year.

CISH-B-1

This variety is characterized by its early maturation. The plants are semi-tall with a spreading growth habit. Budded plants commence fruiting in the 4th year. The fruits are small to medium in size, measuring 16.50 cm x 12.00 cm, and have an oval-oblong shape with a smooth surface. Upon maturation, they turn yellow, exhibiting low levels of mucilage and fibrous content. The attractive yellow pulp has a high seed content. Renowned for their excellent taste and flavor, the fruits boast 30–34° Brix total soluble solids in the pulp and 430 Brix in mucilage. The fruit weight varies from 0.8–1.40 kg, contributing to an average yield of 67.00 kg per plant during the 8th year.

CISH-B-2

The eighth-year-old plant yields an average of 56.78 kg of fruits. These fruits are of medium size, measuring 16.00 cm x 14.00 cm, with a rounded shape and a smooth surface. Upon reaching maturity, they turn yellow, and the rind is characterized by thickness. The fruits have low levels of mucilage and fibrous content, featuring light yellow pulp with minimal seed content. Renowned for their good taste and flavor, the fruits contain 31° Brix total soluble solids and titratable acidity at 0.41%. The fruit weight ranges from 1.7–2.6 kg per fruit. However, under natural conditions, the fruit does not ripen uniformly. This variety is particularly suitable for processing.

Goma Yashi

The Central Horticultural Experiment Station (CIAH-ICAR) in Vejalpur, Panchmahals (Godhra), Gujarat, developed this variety. A promising genotype was identified and collected from Badalpatti village in the Jaunpur district of Uttar Pradesh State. *In situ* budding was carried out in 2006. This genotype stands out due to its desirable characteristics, including earliness, compact growth, medium height, minimal spines, and a superior yield of quality fruits with a pleasant flavor and attractive pulp color. The plant exhibited flowering and fruiting from the 4th year after budding. During the 5th year of orchard life, the recorded measurements included a plant height of 4.2 m, stem girth of 29.42 cm, and plant spread of 4.1 m. In terms of yield, the average was 29 kg in the 5th year, with individual fruits weighing 1.71 kg and measuring 22.20 cm x 15.50 cm. The fruit girth was 46.21 cm, and the shell thickness measured 0.20 cm. A total of 152 seeds were observed, with a seed weight of 0.16 g and a total seed weight of 31 g. Fiber weight was recorded at 69 g, shell weight at 280 g, and 14–15 locules were observed in cross-section. The pulp content was 70%, with TSS in the pulp at 35° Brix, acidity at 0.30%, and vitamin C at 20 mg per 100 g of pulp. This genotype is less susceptible to sun scald due to its compact plant growth.

Thar Dhivya

The ripening process for this variety begins after 270 days of fruit setting, specifically under rainfed conditions in a hot semi-arid ecosystem. Fruits of this variety are relatively less affected, with only 40% susceptibility to sunscald, attributed to the dense canopy and luxuriant growth. During the 8th year, the average yield per tree is 7.50 kg, with individual fruits weighing 1.62 kg and measuring 18.60 cm x 14.80 cm. The fruit girth is 45.80 cm, and the shell thickness is 0.19 cm. In cross-section, there are 14.72 locules in the seed cavity. The Peel weight is 235.50 g, the pulp weight is 1.30 kg, the fiber weight is 61.10 g, the seed weight is 0.13 g, and the total seed weight is 32.00 g, with a total number of seeds at 120.75. The TSS (Total Soluble Solids) of the pulp is 38.50° Brix, the TSS of mucilage is 51° Brix, the acidity is 0.30%, and the TSS/acidity ratio is 128.33. This variety matures in February, categorizing it as belonging to the very early maturing group, and it thrives under rainfed semi-arid ecosystem conditions.

Thar Neelkanth

This superior genotype exhibits compact growth, medium height, and fewer spines, and delivers a superior yield of quality fruits with a pleasant flavor and an appealing pulp color. Flowering and fruiting commence from the 3rd year after budding. During the 8th year, the average yield per plant is 75.67 kg, with individual fruits weighing 1.45 kg and measuring 15.10 cm x 15.00 cm. The fruit girth is 47.30 cm, and the shell thickness is 0.18 cm. A total of 73 seeds are observed, with a seed weight of 0.21 g and a total seed weight of 15.46 g. The fiber weight is 110.17 g, and the shell weight is 265.00 g. In cross-section, the fruits exhibit 13–16 locules. The pulp content is 71.30%, with TSS (Total Soluble Solids) in the pulp measuring 40.1° Brix, TSS in mucilage at 51.50 Brix, acidity at 0.30%, and vitamin C at 19.90 mg per 100 g of pulp. The fruit of this genotype is characterized by its good flavor and aroma. It is highly suitable for drought-prone dry land conditions and is also well-suited for making sherbet, powder candy, and squash.

CHES-B-11

This genotype has been identified as promising based on horticultural traits at CHES, Godhra. In the 8th year, the average yield per plant is 84.10 kg, with individual fruits weighing 1.48 kg and measuring 14.10 cm x 15.20 cm. The fruit girth is 44.21 cm, and the shell thickness is 0.14 cm. A total of 75 seeds are observed, with a seed weight of 0.20 g and a total seed weight of 17.58 g. The fiber weight is 25.60 g, and the shell weight is 200.20 g. In cross-section, the fruits exhibit 14–17 locules. TSS (Total Soluble Solids) in the pulp is recorded at 38.13 Brix, TSS in mucilage at 49.8° Brix, acidity at 0.29%, and vitamin C at 22.83 mg per 100 g of pulp. This variety is considered a medium-maturing type, with maturity occurring in the 1st week of May. It is rich in antioxidant activity, and the fruits are distinguished by their good flavor and aroma. This genotype is highly suitable for making sherbet, murabba, and powder.

CHES-B-16

This genotype, identified as promising based on horticultural traits, was collected from Vidyanagar in 2011. In the 8th year, the average yield per plant is 74.20 kg, with individual fruits weighing between 0.9–1.20 kg and measuring 15.50 cm x 11.40 cm. The fruit girth is 42.20 cm, and the shell thickness is 0.20 cm. A total of 98.14 seeds are observed, with a seed weight of 0.19 g and a total seed weight of 19.51 g. The fiber weight is 31.42 g, and the shell weight is 185.20 g. In cross-section, the fruits exhibit 14–16 locules. TSS (Total Soluble Solids) in the pulp is recorded at 37.13° Brix, TSS in mucilage at 48.75 Brix, acidity at 0.34%, and vitamin C at 20.80 mg per 100 g of pulp. The growth habit is drooping, and it is considered a late-maturing variety, with maturity occurring in the 3rd week of May. It is rich in antioxidant activity, with CUPRA C (micro M TE/g) recorded at 127.87 in mucilage and 90.87 in fruit pulp. The fruits of this genotype are distinguished by their good flavor and aroma, making them highly suitable for RTS, candy, murabba, and powder making.

Swarna Vasudha

It is a medium vigorous, regular bearing, fruit weight ranging from 1100g–1800g, high yield (43.8 kg/plant) under rainfed conditions, pulp recovery (> 80%), TSS (> 40°B), acidity (0.54%). The variety has low seed content (< 2%) with a thin shell (< 2 mm). The plant is semi-vigorous having 34% less height than check variety Pant Sujata.

Molecular Characterization

Limited research has been conducted on the molecular characterization and enhancement of bael. Germplasm accessions from Thrissur, Kerala, were found to be distinguishable from those in Tenkasi, Attur, and Kolli, Tamil Nadu, based on protein markers and the isoperoxidase system. Meanwhile, accessions from Dharmapuri, Tenkasi, and Salem, Tamil Nadu, shared a common protein expression pattern (Johnson, 2010). Genetic uniformity between the mother plant and *in vitro* raised plants from axillary buds and callus culture in *Aegle marmelos* from Rasipuram, a Salem genotype, was confirmed using five isoenzyme systems, including peroxidase, esterase, acid phosphatase, alkaline phosphatase, and polyphenol oxidase (Smila et al., 2011). Pati et al. (2008) demonstrated the *in vitro* clonal propagation of bael cv. CISH-B-1 through enhanced axillary branching and ensured the genetic fidelity of the *in vitro* raised plants using 13 RAPD, 3 ISSR, and 2 DAMD primers. Another study on genetic diversity and relationships, based on sequence-related amplified polymorphism markers (SRAPs), revealed that *Aegle marmelos, Baltamocitrutdawei*, and *Aegloptitchevalieri* formed a distinct group with a similarity value of 0.28. This finding was consistent with previous studies (Morton et al., 2003). Within this clade, *A. marmelos* and *B. dawei* showed higher similarity, with a value of 0.45. These three species belong to the subtribe Balsamocitrinae, the tribe Citreae, and the subfamily Aurantioideae.

Future Thrust

Considering the abundant genetic diversity present in bael germplasm across various states such as Uttar Pradesh, Bihar, Jharkhand, Madya Pradesh, Odisha, Chhatisgarh, West Bengal, etc., it is imperative to carefully select superior genotypes with high-quality attributes and significant yield potential. Addressing prevalent issues like fruit drop, fruit cracking, and rotting requires robust research support and the identification of resistant or tolerant genotypes. Developing cultivars with reduced seed content and mucilage is crucial to expanding the cultivation of bael. Screening genotypes or cultivars for adaptability to problematic soils is also essential.

Further emphasis on post-harvest technology is necessary for the production of value-added and export-oriented processed products. The establishment and promotion of small-scale processing industries are essential for increasing bael processing. To boost fruit production in wastelands, screening genotypes and cultivars becomes imperative. Highlighting the medicinal properties of bael is essential for creating awareness about this fruit. Research efforts should be directed towards developing bael-based farming systems for efficient land utilization. Protecting traditional knowledge related to bael through patents is crucial. The encouragement and establishment of model nurseries and bael orchards on problematic soils can significantly contribute to the widespread commercial cultivation of this fruit crop.

References

Bajaniya, V. K., U.K. Kandoliya, N.H. Bodar, N.V. Bhadja, and B.A. Golakiya. 2015. Fatty acid profile and phytochemical characterization of bael seed (*Aegle marmelos* L.) oil. *International Journal of Current Microbiology and Applied Sciences, 4*(2): 97–102.

Desai, U.T., and S.M. Bhandwalkar. 1995. Compatibility and fruit set studies in some mango cultivars with Kesar as pollen parent. *South Indian Horticulture. 43*(3/4): 115–116.

Dixit, B.B.L., and S. Dutt. 1932. The constitution of marmelosin. *J. Indian Chem. Soc., 9*: 271–79.

Ghosh, D.K., and P. Gayen. 1990. Metroglyph grouping and association analysis on physical characters of bael fruit (*Aegle marmelos*). *Progressive Horticulture, 22*(1–4): 6–11.

Ghosh, S.N. 2014. *Tropical and Subtropical Fruit Crops: Crop Improvement and Varietal Wealth.* JAYA Publishing House.

Hiwale, S. 2015. Bael (*Aegle marmelos* Correa.). *In*: *Sustainable Horticulture in Semiarid Dry Lands*, 177–195. Springer, New Delhi. https://doi.org/10.1007/978-81-322-2244-6_12.

Jauhari, O.S., and R.D. Singh. 1971. Bael: A valuable fruit. *Indian Hort., 16*: 9–10.

Johnson, M. 2010. Biochemical variation studies in *Aegle marmelos* L. Corr: A medicinally important plant. *J. Chem. Pharm. Res., 2*(6): 454–462.

Kirtikar, K.R., and B.D. Basu. (1918). *Indian Medicinal Plants* (Vol. 2). Publisher not identified, Basu, BhuwaneśwariÂśrama.

Kumar, D., and Nath, V. (2010). Variability in bael (*Aegle marmelos* Correa) genotypes from Orissa. *Indian Journal of Plant Genetic Resources, 23*(3): 303–305.

Lal, G. 2002. Evaluation of bael (*Aegle marmelos* Correa) germplasm in semi-arid regions of Rajasthan. *Current Agriculture, 26*(1/2): 127–129.

Maity, P., D. Hansda, U. Bandyopadhyay, and D.K. Mishra. 2009. Biological activities of crude extracts and chemical constituents of Bael, *Aegle marmelos* (L.) Corr. *Indian Journal of Experimental Biology, 47*: 849–861.

Mazumdar, B.C. 1975. Physico-chemical analysis of some type of Bael (*Aegle marmelos* Correa) fruit growing in West Bengal. *Indian Agric., 19*(3): 295–98.

Misra, K.K., R. Singh, and H.R. Jaiswal. 1999. Studies on leaf characters, development pattern, and shoot growth in bael genotypes. *Progressive Horticulture, 31*(3/4): 144–150.

Morton, C.M., M. Grant, and S. Blackmore. 2003. Phylogenetic relationships of the Aurantioideae inferred from chloroplast DNA sequence data. *American Journal of Botany, 90*(10): 1463–1469.

Nath, V., D. Pandey, and B. Das. 2003. Diversity of bael (*Aeglemarmelos* Corr.) in east central India. *Indian Journal of Plant Genetic Resources, 16*(3): 222–225.

Om Prakash. 1961. Food and Drinks in Ancient India. Munshi Ram Manohar Lal Oriental Booksellers and Publishers, Delhi. pp 341.

Pandey, D., S.K. Shukla, and A. Kumar. 2008. Variability in bael accessions from Bihar and Jharkhand. *Indian Journal of Horticulture, 65*(2): 226–229.

Pandey, D., S.K. Shukla, and G. Pandey. 2008. Bael (*Aegle marmelos* Correa). *In*: [Ed.], Peter, K.V.], *Underutilized Horticultural Crops*, (Vol. 3), 201–224. New India Publishing Agency, New Delhi.

Pandey, D., S.K. Shukla, and V. Nath. 2005. *Diversity of Bael (Aegle marmelos Corr.) in Bihar and Uttar Pradesh.* ICAR, India.

Pati, R., R. Chandra, U.K. Chauhan, M. Mishra, and N. Srivastava. 2008. *In vitro* clonal propagation of bael (*Aegle marmelos* Corr.) CV. CISH-B1 through enhanced axillary branching. *Physiology and Molecular Biology of Plants, 14*: 337–346.

Rai, D., and Misra, K.K. 2005. Studies on genetic divergence in bael (Aegle marmelos Correa). Indian Journal of Horticulture, *62*(2): 152–154.

Rai, D., Misra, K.K., and Singh, V.P. 2002. Analysis of genetic divergence in Bael (Aegle marmelos Correa) germplasm. *Progressive Horticulture. 34*(1): 35–38.

Rai, M., R. Dwivedi, and P.N. Gupta. 1991. Variablity and PO1" entails identified germplasm in bael (Aegle marmelos Corr.). *Indian Journal of Plant Genetic Resources, 4*(02): 86–92.

Saroj, P.L., T.A. More, and U.V. Singh. 2008. Performance of bael (*Aegle marmelos*) cultivars under hot arid ecosystem of Rajasthan. *The Indian Journal of Agricultural Sciences, 78*(12).

Sharma, P.C., V. Bhatia, N. Bansal, and A. Sharma. 2007. A review on Bael tree. *Natural Products Radiance*, *6*(2): 171–178.

Sharma, S.K, R.S. Singh, and A.K. Singh. 2013. Bael. *In*: Biodiversity in Horticultural Crops Vol. 4, Peter K.V. (Ed.). Daya Publishing House, New Delhi, pp. 285–300.

Singh, A.K., H.K. Joshi, and S. Singh. 2013. Improving socio-economics through rainfed bael. *Indian Horticulture*, *58*(4).

Singh, A.K., S. Sanjay, and R.S. Singh. 2011. Popularization and marketing of bael in western India. SPJS and *In*: *National Conference on Hoti Business-linking Farmers with Market,* (103).

Singh, A.K., S. Sanjay, H.K. Joshi, B.G. Bagle, and T.A. More. 2008. Evaluation of bael genotypes for growth behaviour and floral traits under semi-arid ecosystem of western India. *The Horticultural Journal*, *21*(3): 140–142.

Singh, A.K., S. Sanjay, P. Makwana, and S.K. Sharma. 2016 Evaluation of bael germplasm under rainfed semi-arid environment of western India. *International Journal of Noni Research*, *11*(1, 2): 11–19.

Singh, A.K., S. Singh, R.S. Singh, B.G. Bagle, and B.D. Sharma. 2011. The Bael-fruit for dryland, Tech. *Bull*, (38): 46.

Singh, A.K., S. Singh, R.S. Singh, J.K. Joshi, and S.K. Sharma. 2014. Characterization of bael (*Aegle marmelos*) varieties under rainfed hot semi-arid environment of western India. *Indian J. Agric. Sci.*, *84*(10): 1236–1242.

Singh, A.K., S. Singh, P.L. Saroj, H. Krishna, R.S. Singh, and R.K. Singh. 2019. Research status of bael (Aegle marmelos) in India: A review. *The Indian Journal of Agricultural Sciences*, *89*(10): 1563–1571.

Singh, L.B. 1961. Some promising selections of Bael. Annual Report. Horticultural Research Institute, Saharanpur: 111–119.

Singh, R.N., and S.K. Roy. 1984. *The Bael*. ICAR, New Delhi, India.

Singhal, V.K., A. Salwan, P. Kumar, and J. Kaur. 2011. Phenology, pollination and breeding system of *Aegle marmelos* (Linn.) Correa (Rutaceae) from India. *New Forests*, *42*: 85–100.

Smila, K.H., R. Muthezhilan, S. Radhika, and M. Rajasekarapandian. 2011. Studies on genetic conformity of two micropropagated medicinal plants using isozymes. *Bioinformatics Bioeng.*, *1*: 325–330.

Srivastava, K.K., and H.K. Singh. 2000. Floral biology of beal (*Aegle marmelos*) cultivars. *Indian Journal of Agricultural Science*, *70*(11): 797–798.

Teaotia, S.S., V.N. Maurya, and B.N. Agnihotri. 1963. Some promising varieties of bael (*Aegle marmelos*) of eastern districts of Uttar Pradesh. *Indian Journal Horticulture*, *20*: 210–214.

The Plant List. 2010. www.theplantlist.org. Retrieved 7 July 2024.

Yadav, D. and D. Ram. 2021. Response of nutrient combinations on fruit cracking and its attributes in different cultivars of bael [*Aegle marmelos* (L.) Correa]. *Journal of Pharmacognosy and Phytochemistry*, *10*(2): 289–292.

10

Genetic Diversity of Coffee (*Coffea arabica* L.) and Sustainable Utilization

Milene Silvestrini

Introduction

Coffee, which includes Arabica coffee (*Coffea arabica* L.) and Robusta or Conillon coffee (*Coffea canephora* Pierre ex A. Froehner), is the second most valuable commodity of the world (after oil), with a total biennial production of 168.2 million bags in 2022/23 (Vega et al., 2003; Conway, 2024; ICO, 2024). Currently, *C. arabica* and *C. canephora* account for 60% and 40% of the traded worldwide coffee, respectively (ICO, 2024), being essential to the incomes of millions of families around the world (Vega et al., 2003; Poncet et al., 2024). To ensure this huge production and trade as well as their social benefits, it is necessary the maintenance, accessibility, and proper use of coffee genetic resources. Both wild (or subspontaneous) and cultivated accessions are key for the conservation of *C. arabica* genetic resources (Carvalho et al., 1962; Labouisse et al., 2008; Aerts et al., 2012; Tesfaye et al., 2014; Pruvot-Woehl et al., 2020; Montagnon et al., 2021; Mekbib et al., 2022), but according to the vast majority of studies, the most important reservoir of genetic diversity are plants from the forests of southwestern Ethiopia and South Sudan, the center of origin of the species (Anthony et al., 2001, 2002; Etienne et al., 2002; Aga et al., 2003, 2005; Chaparro et al., 2004; Silvestrini et al., 2007; Labouisse et al., 2008;, Aerts et al., 2012; Pruvot-Woehl et al., 2020; Scalabrin et al., 2020; Montagnon et al., 2021, 2024; Krishnan et al., 2021; Mekbib et al., 2022). In this chapter, we review the historical and up-to-date information about the speciation and domestication of Arabica coffee; its morphological, agronomic, and genetic diversity; the conservation, assessment, and use of global Arabica coffee genetic resources; and finally, we discuss the best strategies to keep producing coffee in face of possible environmental changes, mainly the current global warming and climatic challenges.

Taxonomic Background

C. arabica is one of 103 species of the genus *Coffea* (Davis et al., 2006) or, more precisely 124, if all species of genus *Psilanthus* Hook. f. are included as recommended initially by the phylogenetic

Universidade Estadual de Campinas (UNICAMP), Centro de Biologia Molecular e Engenharia Genética (CBMEG), Campinas, Brazil
Email: milene.silvestrini@gmail.com, milenes@unicamp.br

studies of Lashermes et al. (1997) and Cros et al. (1998), and later by Davis (2010, 2011) and Davis et al. (2019). As mentioned earlier, only *C. arabica* and *C. canephora* are commercially exploited. *C. arabica* produces the most appreciated beverage due to its superior quality, which is characterized by a rich flavor profile and complexity (Ky et al., 2001; Aguiar et al., 2005). *C. canephora* has higher soluble solids and caffeine content than *C. arabica*, and it is produced mainly for the soluble coffee industry (Ky et al., 2001; Aguiar et al., 2005). The remaining species are wild and are considered a crucial source of genetic diversity for traits such as plant architecture, resistance to diseases and pests, as well as various agronomic and industrial characteristics (Etienne et al., 2002; Silvestrini et al., 2007; Mekbib et al., 2022; Tapaça et al., 2023).

C. arabica is a predominantly self-pollinating species with cross-pollination rates varying from 7–15% depending on the type of recessive mutant evaluated (Carvalho and Krug, 1949; Carvalho and Monaco, 1962). However, the crop yield (amount and stability) seems to benefit from cross-pollination by bees as reported by Klein et al. (2003a, b) in agroforestry systems in Indonesia, in coffee plantations in Brazil (Saturni et al., 2016; Hipólito et al., 2018), and in Ethiopian coffee production forests (Geeraert et al., 2020). Also, contrasting with the common knowledge of predominant autogamy in the species, Berecha et al. (2014) recorded crossing rates of 76% in unmanaged Ethiopian highland forests using simple sequence repeat (SSR) markers. These rates might be valid only for natural populations of *C. arabica* or, more likely, indicate that mating system patterns are variable and possibly influenced by plant variety, the intensity of crop or forest management, and landscape conditions (Saturni et al., 2016; Hipólito et al., 2017; Geeraert et al., 2020). This variation was first observed by Krug and Carvalho (1951), who reported cross-pollination rates ranging from 7–9% and 40–93%, for cera and purpurascens mutants, respectively. Seed dispersal by birds, bats, rodents, and monkeys (Aerts et al., 2012) reaches long distances and, along with the relatively high aforementioned outcrossing pollen rates, appears to hinder the genetic structure of natural coffee populations (Berecha et al., 2014).

The species stands as the sole allotetraploid (with a chromosome count of 2n = 4x = 44) within the genus. In contrast, the other *Coffea* species are diploids, with a chromosome count of 2n = 2x = 22, and typically exhibit self-incompatibility (Krug and Carvalho, 1951). The allotetraploid was formed in the upland rain forests of Ethiopia (Sylvain, 1955; Friis et al., 1982), classified at the present as the Moist Evergreen Afromontane Forests and Transitional Rain Forests (Friis et al., 2010). Therefore, it is adapted to annual mean temperatures between 18–21°C and annual rainfall of 1,200–1,800 mm (Alègre, 1959) or 1,600 to more than 2,000 mm with only three or four dry months with less than 50 mm, according to Sylvain (1955). There are also observations of wild Arabica coffee on the Boma Plateau in Sudan (South Sudan) (Thomas, 1942), which were recently confirmed as a natural habitat of the species by Krishnan et al. (2021), and on Mount Marsabit in Kenya (Berthaud and Charrier, 1988; Davis et al., 2012; Montagnon et al., 2024). The species is considered an amphidiploid that originated from a unique site or event of hybridization also followed by a singular event of polyploidization (Lashermes et al., 1999; Tesfaye et al., 2007; Scalabrin et al., 2020). The hybridization occurred between an ancestor or close relative of *Coffea eugenioides* S. Moore as the female parent and *C. canephora*, or variants related to this species. The subsequent polyploidization resulted in fertile hybrid types (Lashermes et al., 1999; Tesfaye et al., 2007). Furthermore, the low divergence observed in chloroplast DNA and genomic DNA between these putative parents and *C. arabica* suggests a relatively recent origin for *C. arabica* (Cros et al., 1998; Lashermes et al., 1999;, Tesfaye et al., 2007; Scalabrin et al., 2020, 2024).

It is well-known that *C. arabica* has a narrow genetic basis due mainly to its mating system associated with its recent and particular formation and history of domestication (Lashermes et al., 1996; Anthony et al., 2001; Moncada and McCouch, 2004; Maluf et al., 2005; Silvestrini et al., 2007; Pereira et al., 2019; Scalabrin et al., 2020, 2024; Montagnon et al., 2021, 2022). Arabica coffee cultivation began in Yemen, Arabia, around five or six centuries ago (Montagnon et al., 2022). From Yemen, a significant number of seeds were introduced to India in 1670, and 25–30 years later, some progenies were transported to present-day Indonesia by the Dutch, giving rise to the Typica variety (Scalabrin et al., 2020). A smaller number of seeds from Yemen were taken to Bourbon Island (Réunion) in

1708, 1715, and 1721, leading to the origin of the Bourbon variety (Carvalho, 1945; Scalabrin et al., 2020). By the early 18th century, descendants of a singular Indonesian Typica plant, first cultivated in Amsterdam, were disseminated to South America. These plants subsequently became the genetic foundation for the primary coffee cultivars in Brazil and other countries (Chevalier and Dagron, 1928; Carvalho, 1945). The Bourbon variety only reached the Americas and East Africa later, in the mid-19th century (Carvalho, 1945; Scalabrin et al., 2020). Today *C. arabica* is cultivated under tropical and subtropical conditions in practically all of Latin America, its main producer, as well as some regions of Africa, Asia, and Oceania (Krug and Carvalho, 1951; ICO, 2024).

One aspect of the domestication of the species that has only recently been clarified (Montagnon et al., 2022) is the origin of cultivated plants in Yemen. Some authors claim that plants in Yemen originated from eastern Ethiopian forests and/or plantations (Montagnon and Bouharmont, 1996; Lashermes et al., 1996; Silvestrini et al., 2007; Tesfaye et al., 2014), while others suggest that they were directly derived from southwestern Ethiopia (Tadesse and Nigatu, 1996; Anthony et al., 2002; Scalabrin et al., 2020). A related question that arises is whether *C. arabica* seeds were naturally dispersed from the western to the eastern side of the Great Rift Valley by birds, bats, or other animal dispersers (see seed dispersal details above), being originally distributed across the Ethiopian highlands (as it seems to be in the eastern part of South Sudan) (Krishnan et al., 2021), or if the species was brought by humans, and thus, the first plants in southeastern Ethiopia were cultivated.

According to Tadesse and Nigatu (1996), *C. arabica* from Sidamo might be genuinely wild, while there is no spontaneous coffee in Harerge, and trees from the Harenna forest in Bale might have been introduced by the legendary 'Shaa-de-Liya' from Sidamo (Tadesse and Nigatu, 1996). Based on agro-morphological characteristics and records of wild coffee in Sidamo (Scott, 1953), Montagnon and Bouharmont (1996) first pointed out that plants in the southeast (Sidamo) might have been native but were destroyed or drastically reduced, and most likely, were the source of the domestication of *C. arabica* in Harerge and Yemen. These results were recently confirmed by Montagnon et al. (2022) using a much larger sample size and simple sequence repeats (SSR) DNA markers.

Montagnon et al. (2022) data indicates that during the historical movement of *C. arabica*, wild genotypes from the southwest went spontaneously to the southeast (Sidamo) and then were possibly carried to the east (Harerge) by humans before reaching Yemen. Alternatively, Arabica coffee could have been introduced in Yemen directly from Sidamo. Also, cultivation could have been initiated on different occasions and from different sources in Yemen and Harerge. In all possible scenarios, however, wild genes from southeastern Ethiopia were included in the domestication pathway (Montagnon et al., 2022). Note that the geographical location of the Sidamo region, adjacent to the narrowest part of the Great Rift Valley in Ethiopia (Figure 1), might have facilitated seed dispersal across the valley and consequent natural colonization, leading to the expansion of the species distribution into the southeastern highland forests, similarly to what happened in South Sudan (Krishnan et al., 2021). Hence, cultivated coffee in Yemen and Harerge originated from southeastern natural forests with likely an admixture of genotypes coming from Sidamo, Harerge, or Yemen, respectively, and even southwestern Ethiopia (Montagnon et al., 2022). These findings would help to explain the high genetic variability of some accessions from Sidamo (Silvestrini et al., 2007) and Bale (due to mainly the proximity to Sidamo) (Tesfaye et al., 2014) and also provide further tools to decision-makers concerned with conservation, use, and management of the genetic resources of species.

Collection and Assessment of Germplasm

Despite the initial uncertainty about the wild origin of southeastern coffee plants, there has always been a consensus that the foundation of the genetic diversity of *C. arabica* lies in southwest Ethiopia and South Sudan, while Yemen is considered the secondary center of dispersal or the center of domestication of the species, bearing its unique pool of genes (Anthony et al., 2001; Anthony et al., 2002; Aga et al., 2003; Chaparro et al., 2004; Silvestrini et al., 2007; Labouisse et al., 2008; Pruvot-

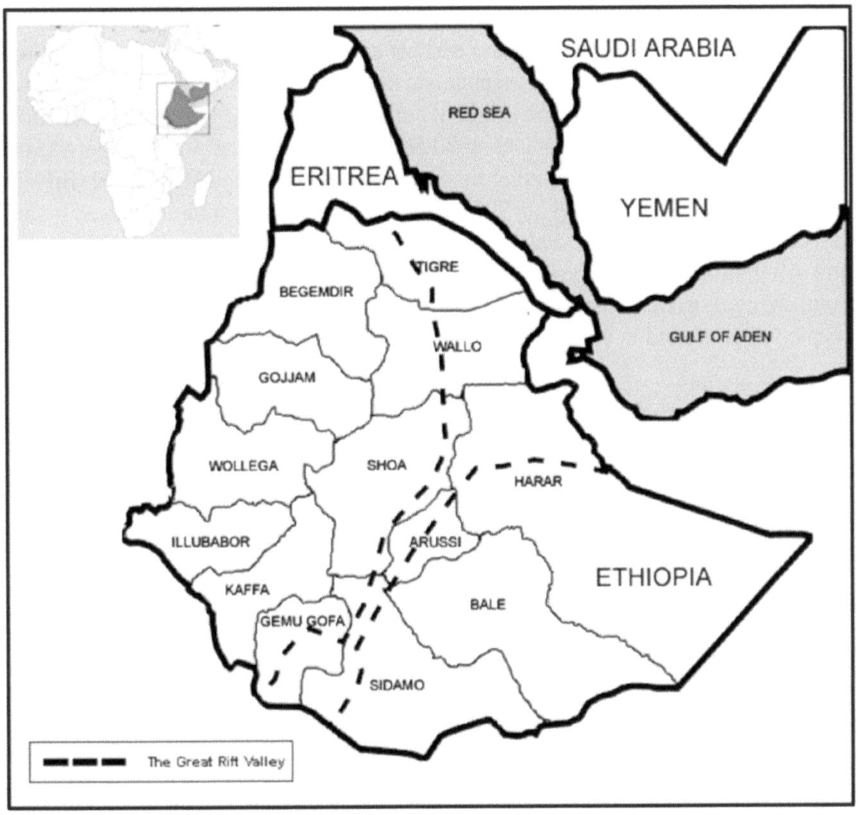

Figure 1. Map of Eritrea, Yemen, and Ethiopia, showing the main provinces according to FAO (1968). (*Source*: Silvestrini et al., 2007) and a schematic drawing of The Great Rift Valley escarpments.

Woehl et al., 2020; Scalabrin et al., 2020; Krishnan et al., 2021; Montagnon et al., 2021, 2024; Mekbib et al., 2022).

Conservation of all these genetic resources began on farms in Ethiopia and then after the domestication, on farms and field genebanks around the world (Elgelmann and Dulloo, 2007). In the early 20th century, the low genetic diversity within cultivars, mainly related to susceptibility to diseases and insects, raised the interest in collecting material from the center of origin of the species. In the second half of the 20th century, worldwide germplasm banks were significantly enriched with plant material from two main expeditions supported by FAO in 1965–1966 and Orstom in 1966 to Ethiopia, and in 1977 to Kenya (FAO, 1968; Anthony et al., 2007; Labouisse et al., 2008; Bramel et al., 2017).

Numerous studies have assessed the genetic diversity of *C. arabica* using botanical, morphological, and agronomic characteristics (Carvalho et al., 1962; Carvalho et al., 1991; Montagnon and Bouharmont, 1996; Yirga et al.; 2021); cytological (Pinto- Maglio et al., 2006), chemical and biochemical methods (Silvarolla et al., 2000; Ky et al., 2001; Aguiar et al., 2005), and, finally, molecular markers (Lashermes et al., 1996; Anthony et al., 2001; Aga et al., 2003; Maluf et al., 2005; Silvestrini et al., 2007; Aerts et al., 2012; Tesfaye et al., 2014; Sousa et al., 2017a,b; Pruvot-Woehl et al., 2020; Zhang et al., 2020, 2021; Krishnan et al., 2021; Bolívar-González et al., 2023; Ge et al., 2024). All approaches have their advantages and disadvantages, but geneticists and breeders agree that an association of methods is more powerful to characterize and successfully use the germplasm for coffee improvement. For example, morphological and cytomolecular studies were fundamental to elucidate the taxonomy of species (Carvalho and Krug, 1951), to establish genetic affinity and relationships between species of *Coffea* and interspecific hybrids (Pinto-Maglio et al., 2006) as well as to reveal the putative parents of *C. arabica* (Lashermes et al., 1999). Gene mutation studies based on morphological changes allowed the earliest knowledge of the genetic properties of coffee such as mode of inheritance, pollination mechanism and crossing rates, pleiotropic effects, and gene expression, in addition to insights into the origin of cultivated varieties: Typica and Bourbon (Krug and Carvalho, 1951; Carvalho et al., 1991). Chemical and biochemical characterization was particularly useful in describing cup quality and finding the first naturally decaffeinated coffee (Silvarolla et al., 2000; Silvarolla et al., 2004; Benatti et al., 2012). New insights from genomic studies, including those conducted by Combes et al. (2023), Salojärvi et al. (2024) and Scalabrin et al. (2024), suggest that sub-genome differential gene expression inherited from the paternal species and generated by recent chromosomal changes, such as aneuploidy, deletions, duplications, and exchanges; is responsible for the phenotypic variety found in cultivated *C. arabica*.

Molecular Characterization

Molecular markers have been playing an important role in measuring genetic diversity levels (see references below classified by the specific molecular marker used), revealing *C. arabica* speciation and domestication processes (Lashermes et al., 1999; Tesfaye et al., 2007; Scalabrin et al., 2020; Montagnon et al., 2022), and in assisted- selection in coffee breeding programs (Molino et al., 2010; Sousa et al., 2017b; Favoretto et al., 2017; Tran et al., 2018). Earlier studies employed dominant markers such as random amplified polymorphic DNA (RAPD) (Lashermes et al., 1996; Anthony et al., 2001; Aga et al., 2003; Chaparro et al., 2004; Maluf et al., 2005) and amplified fragment length polymorphism (AFLP) (Lashermes et al., 1996; Steiger et al., 2002; Anthony et al., 2002; Aga et al., 2003; Chaparro et al., 2004; Maluf et al., 2005), and afterward inter-simple sequence repeat (ISSR) (Aga et al., 2005; Masumbuko and Bryngelsson, 2006; Tesfaye et al., 2014). In the early 2000s, there was a boom in the use of co-dominant multiallelic simple sequence repeat (SSR) markers, which has continued to the present day (Anthony et al., 2002; Moncada and McCouch, 2004; Maluf et al., 2005; Silvestrini et al., 2007; Aerts et al., 2012; Sousa et al., 2017a; Pruvot-Woehl et al., 2020; Sánchez et al., 2020; Benti et al., 2021; Krishnan et al., 2021; Montagnon et al., 2021, 2022, Tapaça et al., 2023), followed by the current adoption of single nucleotide polymorphism (SNP) markers (Zhou et al., 2016; Sousa et al., 2017b; Scalabrin et al., 2020, 2024; Zhang et al., 2020, 2021;, Mekbib et al., 2022;, Bolívar-González et al., 2023; Tapaça et al., 2023; Ge et al., 2024).

Most studies revealed low genetic variability levels in *C. arabica*, especially in cultivated plants (Anthony et al., 2002; Maluf et al., 2005; Masumbuko and Bryngelsson et al., 2006; Silvestrini et al., 2007; Zhou et al., 2016; Zhang et al., 2020; Sánchez et al., 2020; Scalabrin et al., 2020, 2024; Tapaça et al., 2023, but see Aerts et al., 2012 and Benti et al., 2020, for Ethiopian cultivars) and relatively high variation in wild and semi-wild accessions from living collections as well as from forests in Ethiopia and south Sudan (Lashermes et al., 1996; Anthony et al., 2001; Aga et al., 2003, 2005; Chaparro et al., 2004; Silvestrini et al., 2007; Aerts et al., 2012; Tesfaye et al., 2014; Krishnan et al. 2021; Mekbib et al., 2022). Interestingly, Krishnan et al. (2021) have presently found a distinct pool of alleles in accessions from forests in south Sudan, clustering them apart from Ethiopia, Yemen, and global commercial cultivars. In fact, despite being extremely useful in quantifying genetic diversity levels and group wild, semi-wild, and cultivated material (Anthony et al., 2001; Aga et al., 2003, 2005; Silvestrini et al., 2007; Aerts et al. 2012; Tesfaye et al., 2014; Krishnan et al., 2021; Montagnon et al., 2021; Mekbib et al., 2022), it has been hard to identify molecular markers capable of discriminate commercial cultivars of Arabica coffee, see for example, Lashermes et al. (1996), Anthony et al. (2002), Steiger et al. (2002), Maluf et al. (2005), Zhou et al. (2016), Zhang et al. (2020), Sánchez et al. (2020). Zhou et al. (2016) using SNPs identified in expressed sequence tags (ESTs) did not separate cultivars Bourbon from Sumatra (Typica), Caturra from Catuaí, and Mundo Novo from Enrea. Using the same SNP set, Zhang et al. (2020) could not identify 20 cultivars of *C. arabica* even with a diverse developing African origin such as Ethiopia, South Sudan, Congo, Cameroon, Malawi, Kenya, and Tanzania.

Nevertheless, Orozco-Castillo et al. (1994) reported a clear separation between Bourbon and Typica cultivars using RAPD markers and Padilha et al. (2022) found an EST-SSR marker that separated the Bourbon group from the other *Instituto Agronômico de Campinas* (IAC) cultivars. In addition, Pruvot-Woehl et al. (2020), Benti et al. (2020) and Montagnon et al. (2021) demonstrated the effectiveness of a set of 8–14 SSR loci in varietal discrimination and authentication. It is important to highlight that, apart from Bourbon and typical varieties (Pruvot-Woehl et al., 2020), the majority of the cultivated accessions analyzed in these three studies were Ethiopian landraces, Yemeni varieties, and introgressed varieties derived from prior hybridization with *C. canephora*- Catimors and Sarchimors with a Bourbon parent (Caturra or Villa Sarchi) and a Timor Hybrid (Pruvot-Woehl et al., 2020; Benti et al., 2020; Montagnon et al., 2021). Therefore, there was no significant genetic bottleneck observed compared to the history of movement and improvement of cultivars in Europe and Latin America. Cultivars from these regions underwent the complete or traditional domestication pathway, which includes southwestern Ethiopia, southeastern Ethiopia, Yemen, India/Indonesia (Typica) or Réunion (Bourbon), then subsequently introduced to Europe and other countries (Montagnon et al., 2022). As a result, they exhibit very low genetic variability and have only been partially distinguished so far when traditional molecular markers or EST-SNPs are employed.

Nowadays, with the advance of DNA-sequencing technologies and the increasing use of SNP markers with a wide coverage of the genome, it is expected to find molecular descriptors that finally identify cultivars or inbred lines. Indeed, based on whole-genome resequencing (WGS), Bolívar-González et al. (2023) distinguished hybrids and geisha from other Arabica cultivars as well as Bourbon and Typica cultivars. By screening 672 SNP candidates from the 8.5 K SNP array developed by Merot-L'Anthoene et al. (2019), Zhang et al. (2021) could detect a collection of 96 SNPs that were well-suited for varietal identification. Preliminary results of Pereira et al. (2019), employing 1181 SNPs detected by genotyping-by-sequencing (GBS) technology, discriminated Bourbon cultivars and Icatus from all other commercial cultivars developed by IAC, and posterior analyses allowed the identification of almost all IAC cultivars, including the closely related Mundo Novo and Catuaí (L. Padilha, personal communication). These kinds of markers are imperative for assisted selection in breeding programs, varietal identification, accreditation of seeds and nurseries, traceability, and authentication of coffee beans (Zhang et al., 2020, 2021). In some cases, however, they are not available or published due to country-specific laws of varietal protection.

According to Scalabrin et al. (2020, 2024), *C. arabica* exhibits very low nucleotide diversity, attributed to a few recent mutations that arose only after the formation of the allopolyploid, as they are not shared with the progenitor species. The contrasting results obtained using EST-SNPs (Zhou et al., 2016; Zhang et al., 2020) and SNPs detected from wide SNP arrays, WGS and GBS (Pereira et al., 2019; Zhang et al., 2021; Bolívar-González et al., 2022) underline the need for much larger and deeper coverage of the genome to identify sufficient SNPs that provide varietal profiles within the limited variation of cultivated material. Furthermore, the phenotypic variation observed in Arabica commercial cultivars may result from differential gene expression patterns already present in the homeologous chromosomes inherited from the two parental species, as well as from expression differences caused by recent genomic rearrangements and subsequent alterations in gene dosage due to polyploidy (Combes et al., 2023; Scalabrin et al., 2024). Consequently, chromosomal changes and differential expression in sub-genomes can be considered today the main drivers of phenotypic diversity across the species. Genome and transcriptome sequencing, along with chromosome assemblies and broader SNP identification using high-throughput technologies, are helping bridge the gap between genotype and phenotype in *C. arabica*, with wide applications in coffee germplasm conservation and crop enhancement.

Utilization of Germplasm

Utilization in Breeding and Improvement Programs

The primary global breeding objective has been to transfer desired traits, particularly disease and pest resistance, from diploid relative species like *C. canephora* or *C. liberica* into *C. arabica* cultivars, while ensuring the preservation of coffee quality (Etienne et al., 2002). Crossings between related species have been used to increase the genetic basis for breeding purposes in *C. arabica* since the second half of the 20th century. Cultivars like IAPAR 59 and Icatus (vermelho, amarelo, precoce), for instance, were the first ones that resulted from natural or artificial hybridizations of *C. arabica* with *C. canephora* that are resistant to leaf rust disease (*Hemileia vastatryx*). The Icatus are derived from a crossing between a tetraploid *C. canephora* var. Robusta plant and Bourbon vermelho (*C. arabica*) followed by successive backcrossing with Mundo Novo (Carvalho and Fazuoli, 1993). IAPAR 59 originated from the crossing between the cultivar Villa Sarchi CIFC 971/10 and the Timor Hybrid CIFC 832/2, a natural hybrid found in a plantation of cultivar Typica in Timor Island in 1917 (Bettencourt, 1973). There are several cultivars derived from crossings with the Timor hybrid. Timor hybrid has been used as a source of resistance genes for economically important diseases and pests of coffee such as coffee leaf rust (CLR), coffee berry disease (CBD) caused by *Colletotrichum Kahawae*, root-knot nematode (*Meloidogyne exigua*), and bacteriosis caused by *Pseudomonas syringae* pv *garçae* (Bettencourt, 1973; Setotaw et al., 2010). Besides IAPAR 59, only in Brazil, there are cultivars Obatã, Tupi, and IAC 125 RN (Fazuoli et al., 2002; Fazuoli et al., 2018). Although Obatã IAC 1669-20 has become susceptible to leaf rust disease due to the emergence of new *H. vastatrix* races (Mendonça et al., 2016), the others are still resistant to this pathogen. IAC 125 RN was launched and registered in 2012 and includes also resistance to nematodes, more specifically to two races of *M. exigua* and one race of *M. incognita* (Fazuoli et al., 2018). The cultivar raises some concerns, though, regarding future environmental conditions due to its intolerance to drought, one of the significant challenges posed by global warming, making it suitable primarily for irrigated plantations.

Other *Coffea* species have been also crossed with *C. arabica* in breeding programs. Arabica coffee varieties with resistance to coffee leaf miner (*Perileucoptera coffeella*, Guérin-Méneville) have been developed in Brazil using genes from the *Coffea racemosa* Lour. species (Guerreiro Filho et al., 1999; Mendonça et al., 2016) and from *C. stenophylla* G. Don. *C. liberica* Bull. ex Hiern genes were used for leaf rust disease resistance, *C. salvatrix* Swynn. and Philipson for drought tolerance and *C. eugenioides* S. Moore to improve beverage quality (Carvalho, 1988; Van der Vossen et al., 2015).

Intraspecific genetic diversity has been utilized through traditional breeding methods and molecular marker-assisted selection to transfer wild genes conferring tolerance or resistance to diseases such as CLR, CBD, and coffee wilt disease (CWD), as well as genes involved in caffeine synthesis and degradation to produce decaffeinated coffee (Favoretto et al., 2017; Melese and Kolech, 2021). Moreover, it has been employed to enhance cup quality and increase coffee yield by incorporating genes from spontaneous and sub-spontaneous accessions of Ethiopia and South Sudan into cultivars (Van der Vossen et al., 2015; Favoretto et al., 2017; Melese and Kolech, 2021). However, only a small portion of the genetic variability available in these accessions has been explored so far, mainly due to limited accessibility and knowledge of the wild and semi-wild material (Labouisse et al., 2008; Van der Vossen et al., 2015).

Currently, with next-generation sequencing technology, genomic-assisted selection is the new promise to obtain cultivars resistant to diseases and pests, as well as to deal with the increasing climate change effects such as tolerance to drought and elevated temperatures. Molecular markers, especially SNP markers, have great potential for assisted-selection in coffee breeding programs. Indeed, some expressed SSR loci associated with leaf-miner resistance were found (Molino et al., 2010). Favoretto et al. (2017) described SNPs associated with caffeine content with potential use in assisted-selection and breeding of caffeine-free cultivars. Tran et al. (2018) found SNPs associated with cup quality. By combining whole-genome sequencing of wild and cultivated accessions of C. arabica (including cultivars derived from the Timor hybrid) with differential expression data, Salojärvi et al. (2024) made significant progress in identifying the genomic region conferring resistance to coffee leaf rust. In addition, intraspecific genetic diversity associated with drought and heat tolerance has been identified through RNAseq analyses in commercial cultivars (Oliveira et al., 2020; Fernandes et al., 2021) and wild accessions (Davis et al., 2012; Van der Vossen et al., 2015), offering new perspectives for near-future breeding strategies.

Conservation of Useful Germplasm/Genetic Resources

Coffee genetic resources can be conserved *in situ* through protected areas or on farm (Aga, 2003, 2005; Aerts et al., 2012; Labouisse et al., 2008). *In situ* conservation of *C. arabica* is primarily conducted in Ethiopia through (1.1) protected forests, and (1.2) forest coffee systems where coffee berries are conveniently picked; or (2) on farm, which include what Labouisse et al. (2008) called as 2.1) garden coffee systems, and 2.2) plantations of Ethiopia. Additionally, there is a traditional method of coffee cultivation in Ethiopia known as 3) semi-forest coffee, which falls between protected areas and on-farm cultivation. In this method, coffee is grown in forest fragments where farmers thin the canopy layer and annually slash the understory vegetation (Labouisse et al., 2008; Aerts et al., 2011). Coffee material from forests (1.1 and 1.2) and semi-forests (3) are considered wild or spontaneous and semi-wild or sub-spontaneous coffee, respectively (Labouisse et al., 2008). Although the name can be confused with botanical garden *ex situ* conservation, garden coffee systems (2.1) are on-farm conservation of genetic resources where seedlings are collected from forests and cultivated in agroforestry systems in different regions of Ethiopia. According to Labouisse et al. (2008) coffee plants from garden coffee or agroforestry systems are cultivated and commonly referred as landraces while coffee material from plantations can be consisted either of landraces or most frequently of inbred lines or cultivars developed by local national research institutions.

Conservation in forests (1.1 and 1.2), semi-forests (3) and agroforestry coffee systems (2.1) are highly valuable because they help preserve coffee genetic resources as well as some forest biodiversity (Labouisse et al., 2008; Arai et al., 2023; Cassamo et al., 2023; Poncet et al., 2024). In fact, all coffee production systems in Ethiopia are considered important for conservation because they are not isolated from each other and allow sufficient gene flow among populations (natural and cultivated) to avoid losses of genetic diversity (Labouisse et al., 2008). However, according to Aerts et al. (2012), commercial cultivars might pose a risk for wild material due to genetic introgression and loss of wild gene pools. This problem can be aggravated by the increased replacement of disease-susceptible landraces with resistant cultivars (Zewdie et al., 2022). Also, despite the well-recognized importance of the wild genetic resources of Arabica coffee in Ehiopia and South Sudan, little has been done in order to preserve the remaining natural forests in these countries (Aerts et al., 2011, 2012; Berecha et al., 2014; Van der Vossen et al., 2015;, Davis et al., 2019, Geeraert et al., 2019; Moat et al. 2019; Krishnan et al., 2021; Arai et al., 2023). Deforestation, fragmentation, and intense management of natural resources, mainly coffee berry extraction and planting in the forests appear to be the most serious threats (Barecha et al., 2014; Davis et al., 2019; Krishnan et al., 2021; Arai et al., 2023). Besides the genetic introgression commented earlier, the reduced pollen dispersal distances generated by deforestation and fragmentation and increased inbreeding may lead to additional loss of wild gene pools and genetic erosion (Barecha et al., 2014). Furthermore, when climate changes are included in the assessment of the International Union for Conservation of Nature's Red list of Threatened Species (IUCN Red List), *C. arabica* is considered as Threatened with Extinction (placed in the Endangered category) (Moat et al., 2019), moving up from Least Concern to the Endangered category, particularly considering full sun coffee systems (Cassamo et al., 2023).

The main changes in climate include elevated temperatures and altered precipitation (IPCC, 2023). Severe water deficit and long-term exposure to increased warming can have strong negative effects on quantity and quality of Arabica coffee production (DaMatta and Ramalho, 2006; Davis et al., 2012; Oliveira et al., 2020; Fernandes et al., 2021; Poncet et al., 2024). *C. arabica* is adapted to the mild temperatures and well-distributed rainfall regimes of the highlands of Ethiopia (Sylvain, 1955; Alègre, 1959) (with a short dry period important to stimulate blooming) (DaMatta and Ramalho, 2006). Thus, temperatures above 23°C can accelerate fruit ripening and decrease coffee bean quality (DaMatta and Ramalho, 2006; Tapaça, et al., 2023) as well as inhibit plant growth and development (DaMatta and Ramalho et al., 2006). Also, in regions with no irrigation, coffee yields may decrease as much as 80% in very dry years (DaMatta and Ramalho, 2006). Indeed, the historical climate fluctuations have already led to decreases in coffee production by over 20% in the southeastern region of Brazil (Koh et al., 2020).

Models by Davis et al. (2012) demonstrated a 38–90% reduction in the bioclimatically suitable area of the species distribution in Ethiopia and South Sudan by 2080, using the most favorable greenhouse gas emission scenario. Additionally, other studies have estimated that 39–59% of the current growing area under progressive global warming in Ethiopia (Moat et al., 2017), half and two-thirds (by 2041–2060), and up to 91% (by 2081–2100) of the country area in Mozambique (Cassamo et al., 2023), 60% (by 2050) in Brazil (Koh et al., 2020), and 90% (by 2050) in Nicaragua (Läderach et al., 2017), might become unsuitable for coffee farming in the absence of substantial interventions or alterations in the main determinants of climate change (Davis et al., 2012; Läderach et al., 2017; Moat et al., 2017; Koh et al., 2020; Cassamo et al., 2023).

For DaMatta et al. (2018, 2019), these effects might not be so harmful because elevated atmospheric carbon dioxide concentrations [CO2] may mitigate the negative outcomes of increasing temperatures by enhancing photosynthesis and crop yield. However, the authors recognized that they did not consider the interaction with other climatic factors such as extreme rainfall and drought. A highly recommended way to reduce impacts of elevated temperature and drought on Arabica coffee crop is the employment of agroforestry systems (Lasco et al., 2014; Moat et al., 2017; Gomes et al., 2020; Koh et al., 2020; Koutouleas et al., 2022; Cassamo et al., 2023; Tapaça et al., 2023; Poncet et al., 2024). Despite the current lower productivity (but see, Moreira et al., 2018) and short-term economic return, under the canopy of shade trees, coffee is protected from extreme heat and excessive loss of water, being buffered from extreme microclimatic fluctuations. Moreover, agroforestry systems enhance soil productivity and nutrient cycling, improve cup quality, and also help control pests and diseases (Lasco et al., 2014; Gomes et al., 2020; Koutouleas et al., 2022; Poncet et al., 2024).

Therefore, *in situ* conservation of coffee through forested systems and alternative, more diverse, and resilient cultivation methods such as agroforestry are crucial to ensure the sustainability of coffee production (Davis et al., 2012; Krishnan et al., 2021; Arai et al., 2023; Poncet et al., 2024). In addition to mitigating the local detrimental effects of climate change mentioned above, these methods might contribute to the conservation of global forests and biodiversity, ultimately helping sustain large-scale climatic conditions by reducing further greenhouse gas emissions and increasing carbon sequestration. The development and utilization of cultivars suited for agroforestry systems or those tolerant to drought and heat, and resistant to emerging diseases and pests resulting from these changing environmental conditions, should be the primary objectives of future coffee breeding programs (Davis et al., 2012; Lasco et al., 2014; Van der Vossen et al., 2015; Oliveira et al., 2020; Fernandes et al., 2021; Koutouleas et al., 2022; Tapaça et al., 2023; Poncet et al., 2024). Furthermore, as recommended by Poncet et al. (2024), socioeconomic demands, especially from local smallholder farmers, need to be concomitantly addressed in order to "fulfill the United Nations Sustainable Development Goals1 ("no poverty") and 8 ("decent work and economic growth"). However, these efforts alone may not suffice, and a substantial shift in the current agricultural and economic development models may be necessary to ensure genuinely sustainable coffee production and the maintenance of its natural genetic resources under climate change.

C. arabica genetic resources are also conserved *ex situ*, most commonly in field germplasm banks throughout the world, where the levels and quality of phenotypic and genetic diversity are more easily assessed (Labouisse et al., 2008; Bramel et al., 2017). There are, for example, the Choche germplasm bank of the Ethiopian Biodiversity Institute in Jimma zone, Goma district, Ethiopia evaluated by Mekbib et al. (2022); the Cameroon Collection, in Cameroon (Montagnon and Bouharmont, 1996); the International Coffee Collection at the *Centro Agronómico Tropical de Enseñanza e Investigación* (CATIE) and Germplasm collection at the *Centro de Investigación del Instituto del Café* (CICAFÉ), in Costa Rica (Lashermes et al., 1996; Anthony et al., 2001; Pruvot-Woehl et al., 2020; Sánchez et al., 2020; Zhang et al., 2021; Bolívar et al., 2023); the National Coffee Research Center (CENICAFE) in Colombia (Chaparro et al., 2004); the Coffee Germplasm Collection of *Instituto Agronômico de Campinas* (IAC) in Brazil (Silvestrini et al., 2007, 2008); and the Qima coffee breeding populations in Yemen (Montagnon et al., 2021).

These worldwide coffee field collections are in turn classified into two main types, according to Bramel et al. (2017): origin and users. The first holds plant materials from their region of origin, such as CATIE, and the second comprises wild and semi-wild accessions conserved for breeding purposes, as well as all the progenies, inbred lines, and crosses produced during the breeding programs (Bramel et al., 2017). In Bramel et al. (2017) survey, they showed that 57.2% of accessions in genebanks are of *C. arabica*. Most of them are conserved in field collections as trees. This type of collection presents some inconveniences such as high requirement of labor and land for site rotation and high cost, with the additional risk of genetic erosion due to diseases, pests, human error or vandalism, and poor adaptation to local environmental conditions (Eira et al., 2007; Engelmann and Dulloo, 2007). Thus, other *ex situ* methods have been evaluated for coffee such as *in vitro* collections by cryopreservation of seeds, embryos (zygotic and somatic), apices, and by tissue culture repositories (Etienne et al., 2002; Engelmann and Dulloo, 2007).

Even though *Coffea* intermediate recalcitrant seeds have little tolerance to low temperatures and desiccation for long periods (Eira et al., 1999a), cryopreservation of *C. arabica* seeds has been tested with gradual improvement and success (Dussert et al., 1997 1998, 2000; Eira et al., 1999a, 2005; Dussert and Engelmann, 2006). Initially, certain precooling conditions (Dussert et al., 1997, 1998), water content (Dussert et al., 1998; Eira et al., 1999a), and osmo-conditioning after cryopreservation (Dussert et al., 2000) were tested, and water sorption isotherms were determined (Eira et al., 1999b). Some survival of seeds after exposure to liquid nitrogen was found, but their intolerance to desiccation damaged them, decreasing seed viability (Eira et al., 1999a). After further tests, Eira et al. (2005) demonstrated that cryopreservation of coffee seeds is possible through a combination of factors, including seed water content of 20%, water sorption temperature of 15°C (Eira et al., 1999b), appropriate packing, and rapid thawing in a 40°C water bath after storage (as previously described by Dussert et al., 1997). Similar protocols were also reported by Dussert and Engelmann (2006) and Coelho et al. (2017), with an additional step consisting of slowly precooling seeds in a dry ice bath for 25 minutes before immersion in liquid nitrogen (Dussert and Engelmann, 2006). Seed cryopreservation or other *in vitro* collections are indeed costly and demand intensive labor, but they are highly recommended as complementary alternatives to field banks and *in situ* conservation of coffee, especially considering the new abiotic and biotic hazards that might threaten the coffee trees in the face of global warming.

Conclusion

In conclusion, the genetic resources of Arabica coffee are indispensable for sustaining its global production and trade. The intricate history of its speciation and domestication, particularly in southwestern Ethiopia, underscores the importance of conserving wild and cultivated accessions. Molecular markers have proven invaluable in assessing genetic diversity and guiding breeding efforts to develop resilient cultivars. Conservation efforts, encompassing both *in situ* and *ex situ* approaches, as well as more diverse and resilient coffee production systems are crucial for safeguarding the genetic diversity of *C. arabica* in the face of climatic challenges and anthropogenic pressures. Moving forward, a substantial shift in the current agricultural and economic development models along with continued collaboration and innovation in genetic research and conservation practices will be essential for ensuring the long-term sustainability of this crop.

References

Aerts, R., G. Berecha, P. Gijbels, K. Hundera, S. Van Glabeke, K. Vandepitte, B. Muys, I. Roldán-Ruiz, and O. Honnay. 2012. Genetic variation and risks of introgression in the wild *Coffea arabica* gene pool in southwestern Ethiopian montane rainforests. *Evolutionary Applications*, *6*: 243–252.

Aerts, R., K. Hundera, G. Berecha, P. Gijbels, M. Baeten, M. Van Mechelen, M. Hermy, B. Muys, and O. Honnay. 2011. Semi-forest coffee cultivation and the conservation of Ethiopian Afromontane rainforest fragments. *Forest Ecology and Management*, *261*: 1034–1041.

Aga, E., E. Bekele, and T. Bryngelsson. 2005. Inter-simple sequence repeat (ISSR) variation in forest coffee trees (*Coffea arabica* L.) populations from Ethiopia. *Genetica*, *124*: 213–221.

Aga, E., T. Bryngelsson, E. Bekele, and B. Salomon. 2003. Genetic diversity of forest arabica coffee (*Coffea arabica* L.) in Ethiopia as revealed by random amplified polymorphic DNA (RAPD) analysis. *Hereditas.*, *138*: 36–46.

Aguiar, A.T.E., L.C. Fazuoli, T.J.G. Salva, and J.L. Favarin. 2005. Chemical diversity in coffee species of genebank of Instituto Agronômico do Estado de São Paulo. *Crop. Breed. Appl. Biotechnol.*, *5*: 460–466.

Alègre, C. 1959. Climates et caféiers d'Arabie. *Agron. Trop.*, *14*: 23–58.

Anthony, F., B. Bertrand, O. Quiros, A. Wilches, P. Lashermes, J. Berthaud, and A. Charrier. 2001. Genetic diversity of wild coffee (*Coffea arabica* L.) using molecular markers. *Euphytica*, *118*: 53–65.

Anthony, F., M.C. Combes, C. Astorga, B. Bertrand, G. Graziosi, and P. Lashermes. 2002. The origin of cultivated *Coffea arabica* L. varieties was revealed by AFLP and SSR markers. *Theor. Appl. Genet.*, *104*: 894–900.

Arai, Y., K. Hundera, and T. Yoshikura. 2023. Challenges in conserving forest ecosystems through coffee certification: A case study from southwestern Ethiopia. *Front. Environ. Sci.*, 11: 1193242.

Bawin, Y., T. Ruttink, A. Staelens, A. Haegeman, P. Stoffelen, J.C.I. Mwanga Mwanga, I. Roldan-Ruiz, O. Honnay, and S.B. Janssens. 2021. Phylogenomic analysis clarifies the evolutionary origin of Coffea arabica. *J. Syst. Evol.* 59: 953–963.

Benatti, L.B., M.B. Silvarolla, and P. Mazzafera. 2012. Characterization of AC1: A naturally decaffeinated coffee. *Bragantia*, *71*: 143–154.

Benti, T., E.E. Gebre, K. Tesfaye, G. Berecha, P. Lashermes, M. Kyallo, and N. Kouadio Yao. 2021. Genetic diversity among commercial arabica coffee (*Coffea arabica* L.) varieties in Ethiopia using simple sequence repeat markers. *J. Crop Improv.*, *35*: 147– 168.

Berecha, G., R. Aerts, K. Vandepitte, S. Van Glabeke, B. Muys, I. Roldán-Ruiz, and O. Honnay. 2014. Effects of forest management on mating patterns, pollen flow, and intergenerational transfer of genetic diversity in wild Arabica coffee (*Coffea arabica* L.) from Afromontane rainforests. *Biol. J. Linn. Soc.*, *112*: 76–88.

Berthaud, J., and A. Charrier. 1988. Genetics resources of *Coffea*. *In*: Clarke R.J., and R. Macrae [Eds.], *Coffee* (Vol. 4): *Agronomy*, 11–42. Elsevier Applied Science, London.

Bettencourt, A. 1973. *Considerações gerais sobre o 'Híbrido de Timor'*. Secretaria da Agricultura do Estado de São Paulo, Campinas, 20pp. (Circular 23).

Bolívar-González, A., R. Molina-Bravo, W. Solano-Sánchez, E. ArayaValverde, S.T. IvamotoSuzuki, L.F.P. Pereira, and A. GaticaArias. 2023. SNP markers found in non-coding regions can distinguish among low-variant genotypes of Arabica and othercoffee species. *Genet. Resour. Crop. Evol.*, *70*: 1215–1228.

Bramel, P., S. Krishnan, D. Horna, B. Lainoff, and C. Montagnon. 2017. *Global Conservation Strategy for Coffee Genetic Resources*. The Crop Trust and World Coffee Research, Bonn. Germany.

Carvalho, A. 1945. Distribuição geográfica e classificação botânica do gênero *Coffea* com referência especial à espécie *arabica*. *Boletim da Superintendência dos Serviços do Café*, *21*:174–180.

Carvalho, A. 1988. Principles and practice of coffee plant breeding for productivity and quality factors: Coffea arabica. *In*: R.J. Clarke, and R. Macrae [Eds.] *Coffee*: (Vol. 4), *Agronomy*, 129–165. Elsevier Applied Science, London and New York.

Carvalho, A., and C.A. Krug. 1949. Agentes de polinização da flor do cafeeiro (*Coffea arabica L.*). *Bragantia*, *9*: 11–24.

Carvalho, A., and L.C. Fazuoli. 1993. Café.*In:* A.M.C. Furlani, and G.P. Viégas [Eds.], *O melhoramento de plantas no Instituto Agronômico. Instituo Agronômico de Campinas*, 29–76. Campinas, Brazil.

Carvalho, A., and L.C. Monaco. 1962. Natural cross pollination in *Coffea arabica*. L. *Int. Soc. Hort. Sci.*, *4*: 447–449.

Carvalho, A., L.C. Monaco, and H.J. Scaranari. 1962. Melhoramento do Cafeeiro: XXIV – Variação na produtividade de cafeeiros importados, com referência especial ao material da Etiópia e Sudão. *Bragantia*, *21*(13): 215–239.

Cassamo, C.T., D. Draper, M.M. Romeiras, I. Marques, R. Chiulele, M. Rodrigues, M. Stalmans, F.L. Partelli, A. Ribeiro-Barros, and J.C. Ramalho. 2023. Impact of climate changes in the suitable areas for *Coffea arabica* L. production in

Mozambique: Agroforestry as an alternative management system to strengthen crop sustainability. *Agric. Ecosyst. Environ., 346*: 108341.

Chaparro, A.P., M.A. Cristancho, H.A. Cortina, and A.L. Gaita´n. 2004. Genetic variability of *Coffea arabica* L. accessions from Ethiopia evaluated with RAPDs. *Genet. Resour. Crop. Evol., 51*: 291–297.

Chevalier, A. and M. Dagron. 1928. *Recherches historiques sur les débuts de la culture du caféier en Amérique.*, Communications et Actes de l'Académie des Sciences Coloniales, Paris.

Coelho, S.V.B., S.D.F. Rosa, and J.S. Fernandes. 2017. Cryopreservation of coffee seeds: A simplified method. *Seed Sci. Technol., 45*(3): 638–649.

Combes, M.C., T. Joët, A.K. Stavrinides, and P. Lashermes. 2023. New cup out of old coffee: Contribution of parental gene expression legacy to phenotypic novelty in coffee beans of the allopolyploid *Coffea arabica* L. *Ann. Bot., 131*: 157–170.

Conway, J. 2024. Export Volumes of Coffee-Producing Countries December. Available online: https://www.statista.com/statistics/268135/ranking-of-coffee-exporting- countries. (Accessed on 7 February 2024).

Cros, J., M.C. Combes, P. Trouslot, F. Anthony, S. Hamon, A. Charrier, and P. Lashermes. 1998. Phylogenetic analysis of chloroplast DNA variation in *Coffea* L. *Mol. Phyl. and Evol., 9*, 109–117.

DaMatta, F.M., and J.D.C. Ramalho. 2006. Impacts of drought and temperature stress on coffee physiology and production: A review. *Braz. J. Plant. Physiol., 18*: 55–81.

DaMatta, F.M., E. Rahn, P. Läderach, R. Ghini, and J.C. Ramalho. 2019. Why could the coffee crop endure climate change and global warming to a greater extent than previously estimated? *Clim. Change, 152*: 167–178.

DaMatta, F.M., R.T. Avila, A.A. Cardoso, S.C. Martins, and J.C. Ramalho. 2018. Physiological and agronomic performance of the coffee crop in the context of climate change and global warming: A review. *J. Agric. Food Chem., 66*(21): 5264–5274.

Davis, A.P. 2010. Six species of *Psilanthus* transferred to *Coffea* (Coffeeae, Rubiaceae). *Phytotaxa, 10*: 41–45.

Davis, A.P. 2011. *Psilanthus mannii*, the type species of *Psilanthus*, transferred to *Coffea*. *Nord. J. Bot., 29*: 471–472.

Davis, A.P., H. Chadburn, J. Moat, R. O'Sullivan, S. Hargreaves, and E.N. Lughadha. 2019. High extinction risk for wild coffee species and implications for coffee sector sustainability. *Sci. Adv., 5*: eaav3473.

Davis, A.P., R. Govaerts, D.M. Bridson, and P. Stoffelen. 2006. An Annotated Taxonomic Conspectus of the Genus *Coffea* (Rubiaceae). *Bot. J. Linn. Soc., 152*: 465–512.

Davis, A.P., T.W. Gole, S. Baena, and J. Moat. 2012. The impact of climate change on indigenous Arabica coffee (*Coffea arabica*): Predicting future trends and identifying priorities. *PLoS One, 7*: e47981

Dussert, S., and F. Engelmann. 2006. New determinants for tolerance of coffee (*Coffea arabica* L.) seeds to liquid nitrogen exposure. *CryoLett., 27*: 169–178.

Dussert, S., N. Chabrillange, F. Engelmann, F. Anthony, and S. Hamon. 1997. Cryopreservation of coffee (*Coffea arabica* L.) seeds: Importance of precooling temperature. *CryoLett., 18*: 269–276.

Dussert, S., N. Chabrillange, F. Engelmann, F. Anthony, J. Louarn, and S. Hamon. 1998. Cryopreservation of seeds of four coffee species (*Coffea arabica, C. costatifructa, C. racemosa* and *C. sessiliflora*): Importance of water content and cooling rate. *Seed Science Research, 8*: 9–15.

Dussert, S., N. Chabrillange, N. Vasquez, F. Engelmann, F. Anthony, A. Guyot, and S. Hamon. 2000. Beneficial effect of post-thawing osmo-conditioning on the recovery of cryopreserved coffee (*Coffea arabica* L.) seeds. *CryoLett., 21*: 47–52.

Eira, M.T.S., C. Walters, and L.C. Caldas. 1999b. Water sorption isotherms in *Coffea* spp. seeds and embryos. *Seed Science Research, 9*: 321–330.

Eira, M.T.S., C. Walters, L.C. Caldas, L.C. Fazuoli, J.B. Sampaio, and M.C.L.L. Dias. 1999a. Tolerance of Coffea spp. seeds to desiccation and low temperature. *Revista Brasileira de Fisiologia Vegetal, 11*: 97–105.

Eira, M.T.S., L.C. Fazuoli, O. Guerreiro Filho, M.B. Silvarolla, M.A.G. Ferrão, A.F.A. Fonseca, R.G. Ferrão, T. Sera, A.A. Pereira, N.S. Sakiyama, L. Zambolim, C.H.S. Carvalho, L. Padilha, and F.F. Souza. 2007. Bancos de germoplasma de café no Brasil: base do melhoramento para produtividade e qualidade. In: Anais do V Simpósio de Pesquisa dos Cafés do Brasil, Águas de Lindóia, Brazil.

Eira, M.T.S., R.B. Reis, F.N.S. Ribeiro, and V.S. Ribeiro. 2005. *Banco de sementes de café em criopreservação: Experiência inédita no Brasil.* Embrapa Recursos Genéticos e Biotecnologia (Circular Técnica, 42), Brasília, Brazil.

Engelmann, F., and M.E. Dulloo. 2007. Introduction. *In*: F. Engelmann, M.E. Dulloo, C. Astorga, S. Dussert, and F. Anthony [Eds.], *Complementary Strategies for Ex situ Conservation of Coffee* (*Coffea arabica* L.) *Genetic Resources. A Case Study in CATIE, Costa Rica*, 1–11. *Topical Reviews in Agricultural Biodiversity*, 66 pp. Biodiversity International, Rome, Italy.

Etienne, H., F. Anthony, S. Dussert, D. Fernandez, P. Lashermes, and B. Bertrand. 2002. Biotechnological applications for the improvement of coffee (*Coffea arabica* L.). *In Vitro Cell. Dev. Biol.: Plant, 38*: 129–138.

Favoretto, P., C.C. Silva, A.G. Tavares, G. Giatti, P.F. Moraes, M.T.V. Lobato, M.B. Silvarolla, O. Guerreiro Filho, and M.P. Maluf. 2017. Assisted-selection of naturally caffeine-free coffee cultivars: Characterization of SNPs from a methyltransferase gene. *Mol. Breed., 37*: 31.

Fazuoli, L.C., H.P. Medina-Filho, W. Gonçalves, O. Guerreiro Filho, and M.B. Silvarolla. 2002. Melhoramento do cafeeiro: Variedades do tipo arábica obtidas no Instituto Agronômico de Campinas. *In*: Zambolim L. [Eds.], *O estado da arte de tecnologias na produção de café*, 163–251. Viçosa, Brazil.

Fazuoli, L.C., M.T. Braghini, M.B. Silvarolla, W. Gonçalves, J.C. Mistro, P.B. Gallo, and O. Guerreiro Filho. 2018. IAC 125

RN: A dwarf coffee cultivar resistant to leaf rust and root-knot nematode. *Crop Breed. Appl. Biotechnol., 18*: 237–240.

Fernandes, I., I. Marques, O.S. Paulo, D. Batista, F.L. Partelli, F.C. Lidon, F.M. DaMatta, J.C. Ramalho, and A.I. Ribeiro-Barros. 2021. Understanding the impact of drought in *Coffea* genotypes: Transcriptomic analysis supports a common high resilience to moderate water deficit but a genotype dependent sensitivity to severe water deficit. *Agronomy, 11*: 2255.

Friis, I., F.N. Rasmussen, and K. Vollesen, 1982. Studies in the flora and vegetation of South-West Ethiopia. *Opera Bot., 63*: 1–70.

Friis, I., S. Demissew, and P.V. Breugel. 2010. Atlas of the potential vegetation of Ethiopia. *Biol. Skrif., 58*: 1–307.

Ge, Y., B. Wang, X. Shi, Z. Zhang, M. Qi, H. Du, P. Qu, K. Jiang, Z. Chen and X. Li. 2024. Multi-Omics Analyses Unravel Genetic Relationship of Chinese Coffee Germplasm Resources. *Forests, 15*(1): 163.

Geeraert, L., E. Hulsmans, K. Helsen, G. Barecha, R. Aerts, and O. Honnay. 2019. Rapid diversity and structure degradation over time through continued coffee cultivation in remnant Ethiopian Afromontane forests. *Biol. Conserv., 236*: 8–16.

Geeraert, L., R. Aerts, G. Berecha, G. Daba, N. De Fruyt, J. D'hollander, K. Helsen, H. Stynen, and O. Honnay. 2020. Effects of landscape composition on bee communities and coffee pollination in *Coffea arabica* production forests in southwestern Ethiopia. *Agric. Ecosyst. Environ., 288*: 106706–106717.

Gomes, L.C., F.J.J.A. Bianchi, I.M. Cardoso, R.B.A. Fernandes, E.I. Fernandes Filho, and R.P.O Schulte. 2020. Agroforestry systems can mitigate the impacts of climate change on coffee production: A spatially explicit assessment in Brazil. *Agric. Ecosyst. Environ., 294*: 106858.

Guerreiro Filho, O., M.B. Silvarolla, and A.B. Eskes. 1999. Expression and mode of inheritance of resistance in coffee to leaf miner *Perileucoptera coffeella. Euphytica, 105*: 7–15.

Hipólito, J., D. Boscolo, and B.F. Viana. 2018. Landscape and crop management strategies to conserve pollination services and increase yields in tropical coffee farms. *Agric. Ecosyst. Environ., 256*: 218–225.

Klein, A.-M., I. Steffan-Dewenter, and T. Tscharntke. 2003a. Fruit set of highland coffee increases with the diversity of pollinating bees. *Proc. R. Soc. B, 270*: 955–961.

Klein, A.-M., I. Steffan-Dewenter, and T. Tscharntke. 2003b. Bee pollination and fruit set of *Coffea arabica* and *C. canephora* (Rubiaceae). *Am. J. Bot., 90*, 153–157.

Koh, I., R. Garrett, A. Janetos, and N.D. Mueller. 2020. Climate risks to Brazilian coffee production. *Environ. Res. Lett., 15*: 104015.

Koutouleas, A., T. Sarzynski, M. Bordeaux, A.S. Bosselmann, C. Campa, H. Etienne, N. Turreira-García, C. Rigal, P. Vaast, J.C. Ramalho, P. Marraccini, and A. Ræbild. 2022b. Shaded-coffee: A nature-based strategy for coffee production under climate change? A review. *Front. Sustain. Food Syst., 6*: 877476.

Krishnan, S., S. Pruvot-Woehl, A.P. Davis, T. Schilling, J. Moat, W. Solano, A. Al Hakimi, and C. Montagnon. 2021. Validating South Sudan as a Center of Origin for *Coffea arabica*: Implications for Conservation and Coffee Crop Improvement. *Front. Sustain. Food Syst., 5*: 445.

Krug, C.A., and A. Carvalho. 1951. The genetics of *Coffea*. *In*: Demerec, M. [Ed.] Advances in Genetics (Vol. 4), 127–158. Academic Press, New York.

Ky, C-L, J. Louarn, S. Dussert, B. Guyot, S. Hamon, and M. Noirot. 2001. Caffeine, trigonelline, chlorogenic acids, and sucrose diversity in wild *Coffea arabica* L. and *C. canephora* P. accessions. *Food Chem., 75*: 223–230.

Labouisse, J.P., B. Bellachew, S. Kotecha, and B. Bertrand. 2008. Current status of coffee (*Coffea arabica* L.) genetic resources in Ethiopia: Implications for conservation. *Genet. Resour. Crop Evol., 55*: 1079–1093.

Läderach, P., J. Ramirez–Villegas, C. Navarro-Racines, C. Zelaya, A. Martinez–Valle, and A. Jarvis. 2017. Climate change adaptation of coffee production in space and time. *Clim. Change, 141*: 47–62.

Lasco, R.D., R.J.P. Delfino, and M.L.O. Espaldon. 2014. Agroforestry systems: helping smallholders adapt to climate risks while mitigating climate change. *WIREs Clim. Change, 5*(6): 825–833.

Lashermes P., M.C. Combes, J. Robert, P. Trouslot, A. D'Hont., F. Anthony, and A. Charrier. 1999. Molecular characterization and origin of the *Coffea arabica* L. genome. *Mol. Gen. Genet., 261*: 259–266.

Lashermes, P., M.C. Combes, P. Trouslot, and A. Charrier. 1997. Phylogenetic relationships of coffee-tree species (*Coffea* L.) as inferred from ITS sequences of nuclear ribosomal DNA. *Theor. Appl. Genet., 94*: 947–955.

Lashermes, P., P. Trouslot, F. Anthony, M.C. Combes, and A. Charrier. 1996. Genetic diversity for RAPD markers between cultivated and wild accessions of *Coffea arabica. Euphytica, 87*: 59–64.

Maluf, M.P., M. Silvestrini, L.M.C. Ruggiero, O. Guerreiro Filho, and C. Colombo. 2005. Genetic diversity of cultivated *Coffea arabica* inbred lines assessed by RAPD, AFLP and SSR marker systems. *Sci. Agric., 62*: 366–373.

Mekbib, Y., K. Tesfaye, X. Dong, J.K. Saina, G.W. Hu, and Q.F. Wang. 2022. Whole-genome resequencing of *Coffea arabica* L. (Rubiaceae) genotypes identify SNP and unravels distinct groups showing a strong geographical pattern. *BMC Plant Biol., 22*: 69.

Melese, Y.Y., and S.A. Kolech. 2021. Coffee (*Coffea arabica* L.): Methods, Objectives, and Future Strategies of Breeding in Ethiopia: Review. *Sustainability, 13*(19): 10814.

Mendonça, A.P., J.V.A. Nonato, V.T. Andrade, B.J.R. Fatobene, M.T. Braghini, A. Prela- Pantano, and O. Guerreiro Filho. 2016. *Coffee arabica* clones resistant to coffee leaf miner. *Crop Breed. Appl. Biotechnol., 16*: 42–47.

Merot-L'anthoene, V., R. Tournebize, O. Darracq, V. Rattina, M. Lepelley, L. Bellanger, C. Tranchant-Dubreuil, M. Coulee,

M. Pegard, S. Metairon, F. Coralie, P. Stoffelen, S.B. Janssens, C. Kiwuka, P. Musoli, U. Sumirat, H. Legnate, J-L. Kambale, J.F. Costa Neto JF, C. Revel, A. Kochko, P. Descombes, D. Crouzillat, and V. Poncet. 2019. Development and evaluation of a genome-wide Coffee 8.5K SNP array and its application for high-density genetic mapping and for investigating the origin of *Coffea arabica* L. *Plant Biotechnol. J., 17*(7): 1418–1430.

Moat, J., J. Williams, S. Baena, T. Wilkinson, T.W. Gole, Z.K. Challa, S. Demissew, and A.P. Davis. 2017. Resilience potential of the Ethiopian coffee sector under climate change. *Nat. Plants, 3*: 17081.

Moat, J., T.W. Gole, and A.P. Davis. 2019. Least Concern to Endangered: Applying climate change projections profoundly influences the extinction risk assessment for wild Arabica coffee. *Glob. Chang. Biol., 25*(2): 390–403.

Molino, J.V.D. M. Silvestrini, J.C. Martinati, O. Guerreiro Filho, and M.P. Maluf. 2010. Functional markers in assisted selection: Study of an expressed SSR locus associated with leaf-miner resistance. *In*: *Proceedings of 23rd International Conference on Coffee Science, 2010*, 387–391. Bali, Indonesia, Curran Associates, Inc.

Moncada, P., and S. McCouch. 2004. Simple sequence repeat diversity in diploid and tetraploid *Coffea* species. *Genome, 47*: 501–509.

Montagnon, C., A. Mahyoub, W. Solano, and F. Sheibani. 2021. Unveiling a unique genetic diversity of cultivated *Coffea arabica* L. in its main domestication center: Yemen. *Genet. Resour. Crop. Evol., 68*: 2411–2422.

Montagnon, C., and P. Bouharmont. 1996. Multivariate analysis of phenotypic diversity of *Coffea arabica. Genet. Res. Crop. Evol., 43*: 221–227.

Montagnon, C., F. Sheibani, and B. Bertrand. 2024. The history and genetic diversity of cultivated Coffea arabica. Advances in Botanical Research. Available online: https://doi.org/10.1016/bs.abr.2024.04.003 (accessed on 2 July 2024).

Montagnon, C., F. Sheibani, T. Benti, D. Daniel, and A.D. Bote. 2022. Deciphering early movements and domestication of *Coffea arabica* through a comprehensive genetic diversity study covering Ethiopia and Yemen. *Agronomy, 12*: 3203.

Moreira, S., C. Pires, G. Marcatti, R. Santos, H. Imbuzeiro, and R. Fernandes. 2018. Intercropping of coffee with the palm tree, macauba, can mitigate climate change effects. *Agric. For. Meteorol., 256/257*: 379–390.

Oliveira, R.R., T.H.C. Ribeiro, C.H. Cardon, L. Fedenia, V.A. Maia, B.C.F. Barbosa, C.F. Caldeira, P.E. Klein, and A. Chalfun-Junior. 2020. Elevated temperatures impose transcriptional constraints and elicit intraspecific differences between coffee genotypes. *Front. Plant Sci., 11*: 1113.

Orozco-Castillo, C., K.J. Chalmers., R. Waugh, and W. Powel. 1994. Detection of genetic diversity and selective gene introgression in coffee using RAPD markers. *Theor. Appl. Genet., 93*: 934–940.

Padilha, L., J.C. Martinati, O. Guerreiro Filho, and M.P. Maluf. 2022. Checagem da pureza genética de amostras de café bourbon por meio de marcador microssatélite. *Embrapa Café (Boletim de Pesquisa e Desenvolvimento, 6), Brasília.* Brazil.

Pereira, L.F.P., J.C.M. Schenk, M.P. Maluf, G.C. Sant'ana, L. Nogueira, L.A. Pereira, B.S.R. da Silva, P.S. Guimarães, O. Guerreiro Filho, and L. Padilha. 2019. Marcadores SNP avaliados em cultivares de café Arábica. *In*: *Anais do X Simpósio de Pesquisa dos Cafés do Brasil*. Vitória, Brazil.

Pinto-Maglio, C.A. 2006. Cytogenetics of coffee. *Braz. J. Plant Physiol.*, 18: 37–44.

Poncet, V., P. van Asten, C.P. Millet, P. Vaast, and C. Allinne. 2024. Which diversification trajectories make coffee farming more sustainable? *Curr. Opin. Environ. Sustain., 68*: 101432.

Pruvot-Woehl, S., S. Krishnan, W. Solano, T. Schilling, L. Toniutti, B. Bertrand, and C. Montagnon. 2020. Authentication of *Coffea arabica* varieties through DNA fingerprinting and its significance for the coffee sector. *J. AOAC Int., 103*: 325–334.

Salojärvi, J., A. Rambani, Z. Yu, R. Guyot, S. Strickler, M. Lepelley, C. Wang, S. Rajaraman, P. Rastas, C. Zheng, D.S. Muñoz, J. Meidanis, A.R. Paschoal, Y. Bawin, T.J. Krabbenhoft, Z.Q. Wang, S.J. Fleck, R. Aussel, L. Bellanger, A. Charpagne, C. Fournier, M. Kassam, G. Lefebvre, S. Métairon, D. Moine, M. Rigoreau, J. Stolte, P. Hamon, E. Couturon, C. Tranchant-Dubreuil, M. Mukherjee, T. Lan, J. Engelhardt, P. Stadler, S.M.C. Lemos, S.I. Suzuki, U. Sumirat, C.M. Wai, N. Dauchot, S. Orozco-Arias, A. Garavito, C. Kiwuka, P. Musoli, A. Nalukenge, E. Guichoux, H. Reinout, M. Smit, L. Carretero-Paulet, O. Guerreiro Filho, M.T. Braghini, L. Padilha, G.H. Sera, T. Ruttink, R. Henry, P. Marraccini, Y. Van de Peer, A. Andrade, D. Domingues, G. Giuliano, L. Mueller, L.F. Pereira, S. Plaisance, V. Poncet, S. Rombauts, D. Sankoff, V.A. Albert, D. Crouzillat, A. Kochko, and P. Descombes. 2024. The genome and population genomics of allopolyploid Coffea arabica reveal the diversification history of modern coffee cultivars. *Nat. Genet., 56*: 721–731.

Sánchez, E., W. Solano, A. Gatica-Arias, M. Chavarria, and E. Araya-Valverde. 2020. Microsatellite DNA fingerprinting of *Coffea* sp. germplasm conserved in Costa Rica through singleplex and multiplex PCR. *Crop. Breed. Appl. Biot., 20*: e27812013.

Saturni, F.T., R. Jaffé, and J.P. Metzger. 2016. Landscape structure influences bee community and coffee pollination at different spatial scales. *Agric. Ecosyst. Environ., 235*: 1–12.

Scalabrin, S., G. Magris, M. Liva, N. Vitulo, M. Vidotto, D. Scaglione, L. Del Terra, M.R. Ruosi, L. Navarini, G. Pellegrino, J.C.B. Mier y Teran, L. Toniutti, F.S. Liverani, M. Cerutti, G. Di Gaspero, and M. Morgante. 2024. A chromosome-scale assembly reveals chromosomal aberrations and exchanges generating genetic diversity in *Coffea arabica* germplasm. *Nat. Commun., 15*: 463.

Scalabrin, S., L. Toniutti, G. Di Gaspero. D. Scaglione, G. Magris, M. Vidotto, S. Pinosio, F. Cattonaro, F. Magni, I. Jurman, M. Cerutti, F.S. Liverani, L. Navarini, L. Del Terra, G. Pellegrino, M.R. Ruosi, N. Vitulo, G. Valle, A. Pallavicini, G. Graziosi, P.E. Klein, N. Bentley, S. Murray, W. Solano, A. Al Hakimi, T. Schilling, C. Montagnon, M. Morgante, and

B. Bertrand. 2020. A single polyploidization event at the origin of the tetraploid genome of *Coffea arabica* is responsible for the extremely low genetic variation in wild and cultivated germplasm. *Sci. Rep., 10*: 4642.

Scott, H. 1953. La vegetation de la hante Ethiopie centrale et meridionale. In: Symposium de l'Association pour l'Etude Taxonomique de la Flore de l'Afrique Tropicale, Bruxelles,1951/10/24-27. *Lejeunier, Revue Botanique, T., 16*: 67–80.

Setotaw, T.A., E.T. Caixeta, G.F. Pena, E.M. Zambolim, A.A. Pereira, and N.S. Sakiyama. 2010. Breeding potential and genetic diversity of "Híbrido do Timor" coffee evaluated by molecular markers. *Crop. Breed. Appl. Biotechnol., 10*: 298–304.

Silvarolla, M.B., P. Mazzafera, and L.C. Fazuoli. 2004. A naturally decaffeinated Arabica coffee. *Nature, 429*: 826–826.

Silvarolla, M.B., P. Mazzafera, and M.M.A. Lima. 2000. Caffeine content of Ethiopian *Coffea arabica* beans. *Genet. Mol. Biol., 23*: 213–215.

Silvestrini, M., M.G. Junqueira, A.C. Favarin, O. Guerreiro Filho, M.P. Maluf, M.B. Silvarolla, and C.A. Colombo. 2007. Genetic diversity and structure of Ethiopian, Yemen and Brazilian *Coffea arabica* L. accessions using microsatellites markers. *Genet. Resour. Crop Evol., 54*: 1367–1379.

Silvestrini, M., M.P. Maluf, M.B. Silvarolla, O. Guerreiro Filho, H.P. Medina, M.M.T. Vanini, A.S. Oliveira, C. Gaspari-Pezzopane, and L.C. Fazuoli. 2008. Genetic diversity of a *Coffea* germplasm collection assessed by RAPD markers. *Genet. Resour. Crop Evol., 55*: 901–910.

Sousa, T.V., E.T. Caixeta, E.R. Alkimim, A.C.B. Oliveira, A.A. Pereira, L. Zambolim, and N.S. Sakiyama. 2017a. Molecular markers useful to discriminate *Coffea arabica* cultivars with high genetic similarity. *Euphytica, 213*: 75.

Sousa, T.V., E.T. Caixeta, E.R. Alkimim, A.C.B. Oliveira, A.A. Pereira, N.S. Sakiyama, M.F.R. Resende-Júnior, and L. Zambolim. 2017b. Population structure and genetic diversity of coffee progenies derived from Catuaí and Híbrido de Timor revealed by genome-wide SNP marker. *Tree. Genet. Gen., 13*: 124.

Steiger, D.L., C. Nagai, P.H. Moore, C.W. Morden, R.V. Osgood, and R. Ming. 2002. AFLP analysis of genetic diversity within and among Coffea arabica cultivars. *Theor. Appl. Genet., 105*(2–3): 209–215.

Sylvain, P.G. 1955. Some observations on *Coffea arabica* L. in Ethiopia. *Turrialba, 5*: 37–53.

Tadesse, M., and Nigatu, L. 1996. An ecological ethnobotanical study of wild spontaneous coffee *Coffea arabica* in Ethiopia. *In: The Biodiversity of African Plants. Proceedings of 14th AETFAT Congress, 22–27 August 1994, Wageningen*, 277–294. Kluwer Academic Publisher, Wageningen,

Tapaça, I.P.E., L. Mavuque, R. Corti, S. Pedrazzani, I.S.A. Maquia, C. Tongai, F.L. Partelli, J.C. Ramalho, I. Marques, and A.I. Ribeiro-Barros. 2023. Genomic evaluation of *Coffea arabica* and its wild relative *Coffea racemosa* in Mozambique: Settling resilience keys for the coffee crop in the context of climate change. *Plants, 12*(10): 2044.

Tesfaye, K., K. Govers, E. Bekele, and T. Borsch. 2014. ISSR fingerprinting of *Coffea arabica* throughout Ethiopia reveals high variability in wild populations and distinguishes them from landraces. *Plant Syst. Evol., 300*: 881–897.

Tesfaye, K., T. Borsch, K. Govers, and E. Bekele. 2007. Characterization of *Coffea* chloroplast microsatellites and evidence for the recent divergence of *C. arabica* and *C. eugenioides* chloroplast genomes. *Genome, 50*: 1112–1129.

Thomas, A.S. 1942. The wild Arabica coffee on the Boma Plateau, Anglo-Egyptian Sudan. *Emp. J. Exp. Agric., 10*: 207–212.

Tran, H.T.M., A. Furtado, C. Alberto, C. Vargas, H. Smyth, L.S. Lee, and R. Henry. 2018. SNP in the *Coffea arabica* genome associated with coffee quality. *Tree. Genet Gen., 14*: 72. https:// doi. org/ 10. 1007/ s11295-.018-.1282-9.

Van der Vossen, H., B. Bertrand, and A. Charrier. 2015. Next generation variety development for sustainable production of arabica coffee (*Coffea arabica* L.): A review. *Euphytica, 204*: 243–256.

Vega, F.E., E. Rosenquist, and W. Collins. 2003. Global project needed to tackle coffee crisis. *Nature, 425*: 343.

Yirga, M.W. Gebreselassie, and A. Tesfaye. 2021. Genetic diversity of coffee (*Coffea arabica* L.) collections for morpho-agronomic traits in southwestern Ethiopia. *East Afri. J. Sci., 15*: 41–154.

Zewdie, B., Y. Bawin, A.J.M. Tack, S. Nemomissa, K. Tesfaye, S.B. Janssens, S. Van Glabeke, I. Roldán-Ruiz, T. Ruttink, O. Honnay, and K. Hylander. 2022. Genetic composition and diversity of Arabica coffee in the crop's centre of origin and its impact on four major fungal diseases. *Molecular Ecology, 32*: 2484–2503.

Zhang, D., F.E. Vega, F. Infante, W. Solano, E.S. Johnson, and L.W. Meinhardt. 2020. Accurate differentiation of green beans of Arabica and Robusta coffee using nanofluidic array of single nucleotide polymorphism (SNP) markers. *Journal of AOAC International, 103*: 315–324.

Zhang, D., F.E. Vega, W. Solano, F. Su, F. Infante, and L.W. Meinhardt. 2021. Selecting a core set of nuclear SNP markers for molecular characterization of Arabica coffee (*Coffea arabica* L.) genetic resources. *Conserv. Genet. Resour., 13*: 329–335.

Zhou, L., F.E. Vega, H. Tan, A.E.R. Lluch, L.W. Meinhardt, W. Fang, S. Mischke, B. Irish, and D. Zhang. 2016. Developing single nucleotide polymorphism (SNP) markers for the identification of Coffee germplasm. *Trop. Plant Biol., 9*: 82–95.

11

Genetic Diversity of Date Palm (*Phoenix dactylifera* L.) and Sustainable Utilization

Akshatha Banadka, Kokila Srinivas, Rashi Bhavsar, S. Sanjay, Akash Shaikh, Nandhini Murali, Praveen Nagella and *Jameel M. Al-Khayri**

Introduction

Phoenix dactylifera L., (date palm), a flowering plant in the palm (Arecaceae or Palmae) family, has held significant value and has served as a fundamental dietary staple and a primary generator of prosperity within arid, irrigated desert regions since ancient times (Wakil et al., 2015). The date palm has been prized for its nutritional qualities, role as a basic food source, and importance as a key source of wealth in arid, irrigated desert regions since antiquity (Petruzzello, 2023).

The tallest of the *Phoenix* species, the date palm grows to 25–30 meters with a stem diameter of 40–50 cm, producing offshoots during early growth. Its upright, pinnate, glossy leaves bear sharp spines on the rachis and form a crown with around 100 leaves, each 3–6 meters long (Zaid and de Wet, 2002a). These perennial plants are dioecious in nature, with separate fruit-bearing female plants and pollen-producing male plants (Wakil et al., 2015). The palm, which typically reproduces through seeds or offshoots, matures sexually in 5–7 years and produces flowers annually. Plantations remain viable for around 50 years, reaching full production at 10–15 years, after which there is a decline in fruit yield (Johnson et al., 2015; Petruzzello 2023). Fruits range in size and shape from 4–7 by 2–3 cm and mature in about 200 days. Kimri (80% moisture), Khalal (60% moisture), Rutab (40% moisture), and Tamar (20% moisture) are the four major stages of the fruit's growth. As they ripen, they turn from green to hues of yellow, orange, red, or brown (Al-Shahib and Marshall, 2002). Figure 1 illustrates various ripening stages of date palms. The seeds feature a raphe that runs lengthwise and is deeply grooved. It has sizes ranging from 7–30 mm (Barrow, 1998). A mature tree yields 20–100 kg based on factors like cultivar and climatic conditions. Commercially, palms are replaced earlier due to declining fruit production (Johnson et al., 2015; Petruzzello, 2023). Over the years, the date production has consistently risen, growing from 2,535 in 1980 to 7,549 metric tons in 2012, as indicated by the FAOSTAT (Al-Khayri et al., 2015a).

Department of Agricultural Biotechnology, College of Agriculture and Food Sciences, King Faisal University, Al-Ahsa 31982, Saudi Arabia.
Email: jkhayri@kfu.edu.sa

Figure 1. Ripening stages of date palm fruit. (A) Khalal, (B) Bisir, (C) Rutab, (D) Tamer.

For operations including pollination, bunch maintenance, and trimming in towering date palm crowns, novel approaches are required. Despite wind pollination, artificial pollination using gathered pollen is essential for success. Bags shield from sunburn and birds, thinning promotes fruit, and bundling secures a crop. Although it is tolerant of droughts, water is essential for various irrigation techniques. Traditional techniques incorporate manure, cover crops, and employ nitrogen as the major nutrient. In the realm of propagation, three main avenues are utilized: seed propagation, offshoot propagation, and tissue culture (Chao and Krueger, 2007a). Although mostly used for breeding, seed propagation has problems with fruit quality and gender diversity. Although offshoot propagation gives consistent quality and earlier fruiting, it is time-consuming and scale-limited. Tissue culture permits homogenous plants, scalable replication, and the reproduction of desired features. Despite pros and cons, each method's selection depends on speed, uniformity, and scalability factors for date palms (Zaid and de Wet, 2002b).

Date palms offer a promising solution for the current food supply uncertainties and anticipated demand growth. Their nutrient-rich fruits are highly prized for their dietary benefits. Date fruits are used in various forms, including whole, pitted or unpitted, dehydrated, and diced. They are used in making cereal, pudding, bread, cakes, cookies, candy bars, ice cream, and date shakes. It is also used to make beer, honey, juice, sugar syrup, vinegar, wine, and other condiments. Beyond food, the date palm tree's trunk, wood, and fiber are used in bags, baskets, cords, fuel, mats, roofs, shades, timber, walls, and boats. Date seeds can be used as feed or decorative beads, and their oil can be used to make soap, palm hearts can be eaten raw or cooked. Dates have medicinal uses for everything from intestinal troubles to sore throats. Notably, date palms are regarded for their aesthetic appeal and landscaping purposes (Chao and Krueger, 2007a).

Taxonomic Background

Phoenix dactylifera, commonly known as the date palm, belongs to the plant family Arecaceae. The systematic classification of *Phoenix dactylifera* L. is presented in Table 1.

Date palms were originally native to Mesopotamia (present-day Iraq) and were dispersed from there over various places with favorable water and climatic conditions for fruit cultivation (Johnson et al., 2013). There were two main routes through which the date palms spread historically. One was from Mesopotamia to places like Arabia and India. The other started in Egypt and went across North Africa to Morocco. It has been transported into regions like the Sahel and Sudan through trade caravans and waterways like the Nile. The Moorish people took date palms from North Africa to Spain in the 8th century. The fruits failed to ripen in Spain's climate. Hence, Spanish folks brought date seeds to the Mexican and Peru regions of America (Johnson et al., 2015).

In the past century, people transported young date palm trees to distant places, establishing vast date farms with unique varieties in countries such as India, Israel, Jordan, Kuwait, Mexico, Namibia, Oman, UAE, USA, Australia, and parts of Africa (Figure 2). Around 150 million or more date palm trees are cultivated globally, ranging from 10°–39°N latitude. However, these trees flourish abundantly between 24°–34°N latitude in the Middle East and North Africa (Johnson et al., 2015). Date palms prefer temperatures ranging between 12.7–27.5°C, while they can withstand temperatures up to 50°C and brief periods of frost as low as –5°C. The ideal temperature ranges between 21–27°C on average, from pollination to fruit maturity (Chao and Krueger, 2007a).

The remarkable genetic diversity within date palms has led to various unique cultivars, each with its special characteristics. This diversity is evident in the numerous date palm cultivars cultivated globally. Amir Hajj, Barhi, Deglet Noor, Dayri, Fardh, Halawy, Hayani, Khadrawy, Maktoom, Medjool, Saidy, Thoory, and Zahidi are some of the date palm cultivars. These inherent genetic variations in these cultivars hold crucial significance for the survival and adaptability of the species (Al-Khayri et al., 2015a).

Modern agricultural practices have indeed increased date palm production, but they have also led to the extensive use of land and resources. Thus, growing date palms optimally requires striking a balance between production and sustainability (Jaradat, 2011). In this regard, the study discusses the need for sustainable approaches to date palm cultivation that consider both agricultural productivity and environmental conservation through careful consideration of research findings and the practical experiences of farmers. The chapter enlists a range of date palm varieties and their genetic variations, assisting in selecting the best cultivars for cultivation. It investigates methods for identifying and choosing optimal varieties, underscoring the importance of adopting cultivation practices that maximize output while reducing negative environmental effects. Furthermore, it explores various scientific techniques to enhance date palm propagation, including tissue culture. Moreover, the chapter emphasizes the significance of modern breeding strategies for developing high-quality date palms with exceptional traits. This comprehensive chapter is a resource for practitioners globally involved in date palm cultivation. It underscores the importance of striking a delicate equilibrium between productivity, genetic diversity, and environmental conservation.

Table 1. Systematic classification of *Phoenix dactylifera* L.

Kingdom	Plantae
Phylum	Magnoliophyta
Class	Liliopsida
Subclass	Arecidae
Order	Arecales
Family	Arecaceae
Genus	*Phoenix*
Species	*dactylifera*

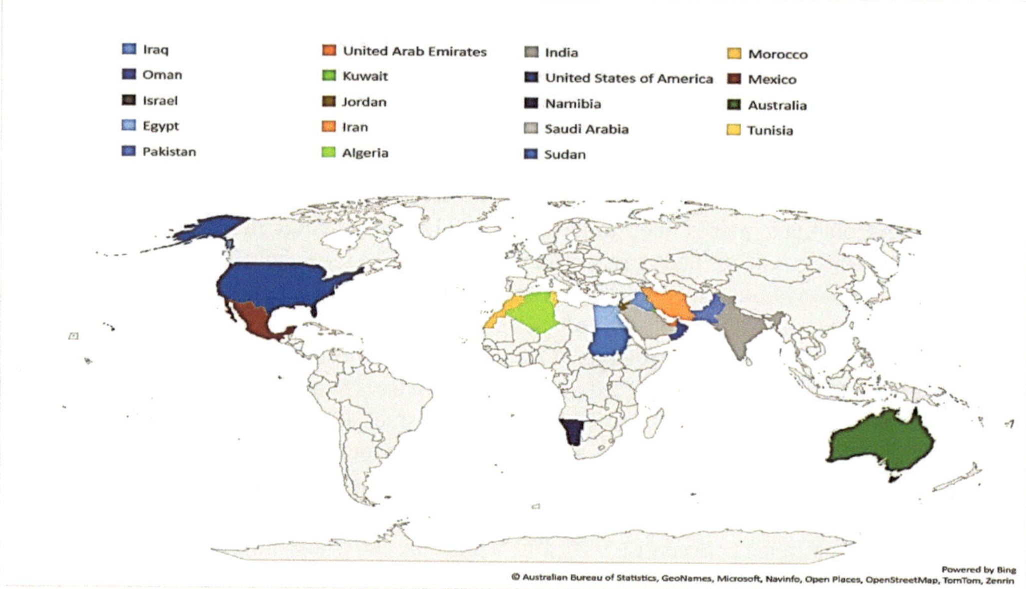

Figure 2. Geographic distribution of *Phoenix* species.

Genetic Diversity in Date Palm: Assessment and Characterization of Germplasm and Genetic Resources

Genetic diversity in date palms is crucial for sustaining and enhancing this economically and culturally noteworthy crop. Being a perennial fruit tree reproducing through vegetative means, the date palm stands out due to its makeup of genetically distinct clones that showcase remarkably heterozygous cultivars (Al-Khayri et al., 2012). Date palm genetic diversity is not distributed randomly or consistently throughout time and space. However, it varies between populations, oases across local areas, and different regions (Chaluvadi et al., 2019) (Table 2). The date palm can be propagated through methods such as planting seeds, utilizing offshoots, and employing techniques like tissue culture or other forms of micro-propagation for purposes like selective breeding and genetic modification (Krueger, 2021). These techniques also help to preserve genetic diversity in groves and minimize the reliance on monovarietal cultivation, which can result in the depletion of the genetic reservoir and

Table 2. Comparison of Date Palm Varieties and Yields Across Geographic Regions.

Geographical Region/ Country	Total Number of Varieties	Yield	References
Iraq	600	0.65 million tons	(Hamza and Al-Hashimi, 2023)
Tunisia	250	0.41 million tons	(Karim et al., 2023)
Oman	250	0.23 million tons	(Al-Hinai et al., 2022)
Pakistan	300	0.57 million tons	(Rehman et al., 2022)
Libya	400	0.17 million tons	(Racchi and Camussi, 2018)
Sudan	400	0.43 million tons	(Elshibli and Korpelainen, 2008)
Egypt	255	1.52 million tons	(Al-Khayri et al., 2015a)
Saudi Arabia	400	0.89 million tons	(Nasser, 2014)
Iran	400	0.98 million tons	(Hajian and Hamidi-Esfahani, 2015)
United Arab Emirates	200	0.53 million tons	(Al-Muaini et al., 2019)
Algeria	600	0.48 million tons	(Bouguedoura et al., 2015)

also develop resilience to biotic and abiotic stresses (Hadrami and Hadrami, 2009). Several significant historical, geographical, ecological, and human-related factors play a crucial role in determining how this genetic diversity is distributed over space and time.

Factors Influencing the Genetic Diversity of Date Palm

The genetic diversity of the date palm is influenced by various interconnected natural, biological, and anthropogenic factors (Figure 3). Depending on the circumstances, these factors can impact regulation favorably or negatively (Jaradat, 2016). For example, in earlier times, the commercial adoption of date palm varieties positively contributed to the enhancement of genetic diversity by introducing various genotypes to different geographical locations. However, more recently, commercial adoption has become limited to only a few varieties of date palms, resulting in a reduction of genetic diversity (Hancock, 2012). Positive regulatory factors serve to decrease genetic drift, while negative regulatory factors serve to increase genetic drift, thus contributing to a reduction in genetic diversity (Messer, 2016).

The phenomenon of genetic drift, which is more pronounced in smaller populations, can lead to random genetic variations within a population over time. Date palms in specific regions or isolated populations can have genetic diversity influenced by genetic drift. The reduction in population size and increase in population isolation can cause habitat loss and fragmentation, and isolation can further restrict population dispersal, ultimately decreasing gene flow and population connectivity (Templeton, 2019). It is important to note that alleles lost to random genetic drift may not be recovered by gene flow, and this could disrupt the migration-drift equilibrium. Because of the decline in genetic diversity, individual fitness and the ability to adapt to environmental changes may ultimately decline (Hays and Fagan, 2016).

The genetic variation among date palms is significantly influenced by their geographic distribution and habitat variation. These palms are native to certain regions of South Asia, the Middle East, and North Africa. Their broad geographic distribution has resulted in genetic variations that illustrate adaptive capacity to diverse environmental circumstances (Zehdi-Azouzi et al., 2015). Empirical

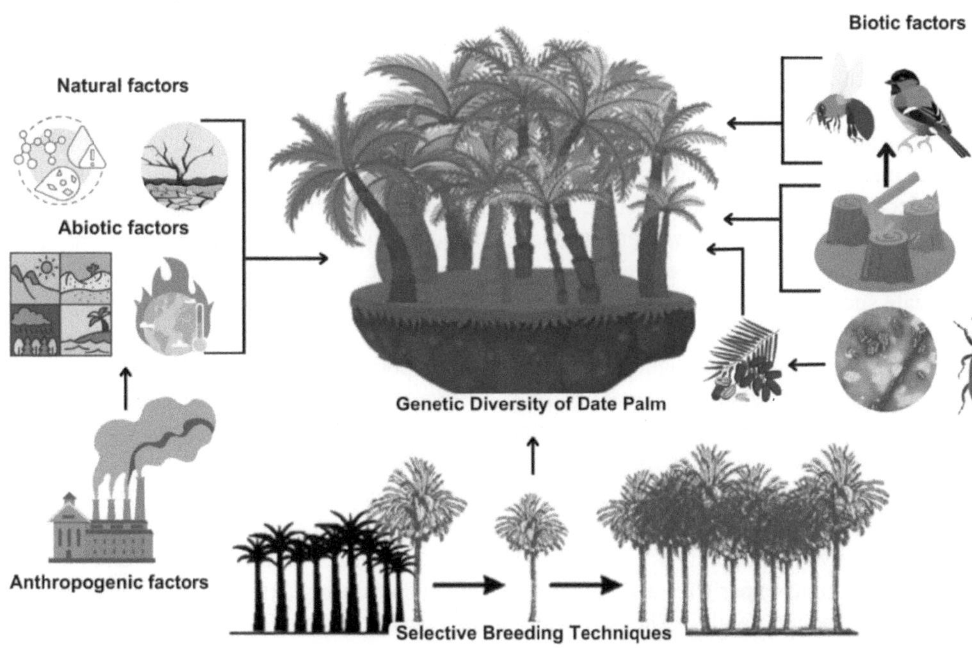

Figure 3. Factors affecting the genetic diversity of date palm.

research supports the influence of geographic factors on the date palm genetic diversity. For instance, Chaluvadi et al. (2019) conducted a study using microsatellite markers to examine the genetic diversity of various date palm varieties. The study conducted by the researchers revealed that the genetic profiles of distinct date palm varieties sourced from various regions were uniquely dissimilar. This finding underscores the favorable impact of geographic distribution on genetic differentiation. In addition, the variation in habitat within these regions plays a major contribution to increasing date palm genetic diversity (Chaluvadi et al., 2019).

The date palm population exhibits genetic adaptations resulting from abiotic factors, such as temperature variations, soil types, moisture availability, and other environmental conditions that display variation across habitats. Furthermore, date palms have evolved to thrive optimally in their specific habitat, further boosting the genetic variation within the species as a whole (Chen et al., 2013).

The translocation of date palms from one geographic area to another, known as date palm migration, can significantly impact their genetic diversity. Introducing new date palm genotypes into different regions through trade and human-mediated migration has been a significant contributing factor (Debnath, 2016).

Adding new genetic material within the local populations can enhance genetic diversity and facilitate the acquisition of novel traits. Due to migration, gene flow allows for the transfer of alleles and genetic traits between populations. This gene flow preserves connectivity and genetic variability among populations of date palms. The introduction of various date palm variants to new regions by means of migration and expeditions has played an essential function in positively influencing genetic diversity by increasing genetic variation (Changtragoon et al., 2017). The promotion of genetic variability in date palms can be attributed to the significant and positive impact of pollination mechanisms, which are among the most influential natural factors. Being dioecious in nature, date palms bear male and female inflorescences on separate trees, thus promoting cross-pollination between different individuals and resulting in genetic material mixing and recombination (Ksiazek-Mikenas et al., 2019). The role of various pollinators, including birds, insects, and wind, is crucial in the transfer of pollen and the promotion of genetic variability within the populations of date palms. Moreover, pollination facilitates genetic interchange and gene flow between populations of date palms, as pollen movement between different geographic locations promotes the transfer of genetic material that contributes to genetic variability both within and across populations. This gene flow maintains genetic connectivity and prevents populations' isolation (Manickavasagan et al., 2012).

The genetically diverse variety of date palm populations is subject to the influence of natural selection, which bestows favor upon individuals who exhibit characteristics that enhance their fitness and probability of survival within a given environment. Assorted biotic and abiotic elements include competition, disease, soil composition, temperature, and the availability of water can all serve as selection pressures (Patankar et al., 2019). Date palms have the capacity to manifest regional acclimation via natural selection, wherein diverse populations acquire specific genetic traits that enable them to thrive in their distinctive ecological niche. Due to these adaptations, date palms can subsist and procreate successfully within their indigenous environments (Kremer et al., 2014).

Biotic stresses refer to the impact of living organisms, such as pathogens, pests, and competitors, on the genetic composition of populations. These stresses can potentially favor individuals with specific genetic traits that enhance their survival and reproductive abilities in the face of these challenges. The presence of pests and pathogens can significantly affect the genetic variation among date palm populations, representing a major form of biotic stress (Singla and Krattinger, 2016). Date palm plantations are prone to a variety of pests and pathogens, which have the capacity to modify the genetic composition of populations. The negative impact of biotic stresses, namely pests and pathogens, on the genetic diversity of date palm populations is a well-documented phenomenon (Chao and Krueger, 2007). The populations experience a significant decline in size during extensive infections, leading to a decline in genetic variability. The only way for the plant to overcome this challenge is through adaptation and accumulating genetic variation over a prolonged period.

However, severe infections may result in the complete loss of the population. Thus, it is imperative to take adequate measures to ensure this factor does not become an extensive negative influence (Ezebilo et al., 2013). The red palm weevil disease, which is frequently encountered in the date palm plantations of the Gulf regions and Egypt, is the result of *Rhynchophorus ferrugineus*, a pernicious pest that consumes date palm tissues and results in significant damage (Kurdi et al., 2021). Fusarium wilt, also called Bayoud disease, is observed in North African date palm plantations and brought on by the *Fusarium oxysporum f.*sp. *albedinis* fungus, a devastating ailment that negatively impacts the date palm. The aforementioned pathogen induces vascular wilt, leading to stunted growth, yellowing of fronds, and eventual plant mortality (El Hassni et al., 2007). Another significant fungal infection brought on by *Fusarium proliferatum* is prevalent in date palm plantations in Iraq, which induces rot in inflorescence, thereby adversely affecting growth and yield (Muhammed, 2012). The selection process may result in changes to the allele frequency within said populations over time. It is worth noting that some pests and pathogens may display genetic diversity or regional variation independently. In light of co-evolutionary dynamics, interactions between populations of date palms and pest or pathogen populations could potentially give rise to genetic differentiation among the date palm populations (Abdel-Baky et al., 2023).

Abiotic stresses, including but not limited to temperature fluctuations, water availability, and salinity, can significantly impact the genetic diversity of the date palm. Specifically, temperature stress can lead to decreased pollen viability, a decline in fruit quality, and overall plant health. Owing to the considerable variation in temperature, salinity of the soil and water, as well as water availability, the ability of the date palm to tolerate such stressors has not kept pace with the mechanisms of adaptation, resulting in the loss of substantial genetic variance (Hazzouri et al., 2020). Moreover, the high concentration of sodium chloride and other salts can harm plant growth and pose a substantial risk to the rapidly expanding date palm sector in a number of North African nations and the Middle East (Jaradat, 2011). Due to climate change and rising global temperatures, particularly in areas that are semi-arid and arid, there has been a marked increase in the loss of date palm plantations due to drought. Additionally, limited access to water and highly salinized soil can curtail the range of date palms and reduce genetic diversity in affected areas (Alhammadi and Edward, 2009). Date palms with narrow physiological tolerances may prove incapable of withstanding severe circumstances like heat waves, cold snaps, flooding, drought, mineral toxicity, salinity, or other abiotic stressors. Conversely, those with broad physiological optima are more likely to be well-adapted or resistant to such stresses, promoting genetic diversity (Atkinson and Urwin, 2012).

Founder effects occur when a small proportion of individuals form a new population. The genetic makeup of the original population may not be entirely reflected in the genetic composition of the new population. Consequently, the genetic variability may be reduced, and common genetic variations may be more prevalent in the founder population. This diminished genetic diversity and smaller population size can result in evolutionary bottlenecks (Choudhuri, 2014).

It is widely acknowledged that date palms have undergone a bottleneck due to their extensive domestication. Moreover, several studies suggest that a secondary bottleneck would be far more severe if it were to arise. This second bottleneck could potentially be attributed to anthropogenic factors or any significant environmental changes caused by anthropogenic factors might be involved (van de Wouw et al., 2010). Anthropogenic factors exert an enormous impact on the genetic variability of date palms. During the early stages of domestication and cultivation, introducing diverse cultivars to various populations and regions positively impacted genetic diversity by augmenting genetic variation (Gros-Balthazard et al., 2013). Nonetheless, the date palm's genetic diversity has come under threat during the past few decades as the domestication, mass cultivation, and mass propagation of the date palm have diminished to a handful of varieties with distinct genotypes and little to no potential variation. Cloned plants, being exact replicas of their parent plants, ensure the continuation of desired traits and characteristics through uniformity. However, this uniformity also reduces the genetic diversity within the cultivated population as no new genetic variations are introduced (Diaz et al., 2003).

The date palm population's environment has encountered repercussions beyond normal growth conditions, as a result of various anthropogenic factors, including pollution and the exploitation of available natural resources. This decline in population size has led to a reduction in genetic variability, resulting in negative influence (Lovich and Bainbridge, 1999). For instance, the discharge of heavy metals into waterways and the atmosphere due to industrial activities is a prime example of the cause of date palm population decline (Al-Khashman et al., 2011). Anthropogenic factors are major contributors to climate change, which is a significant source of concern. The modified climatic conditions profoundly affect the propagation of date palms, causing a reduction in the population of cultivars and increasing genetic vulnerability (Kittel, 2013). Overall, anthropogenic factors have a considerable negative impact on the genetic diversity of date palms, raising the risk of genetic susceptibility and potentially paving the path to genetic erosion (Saaroni et al., 2004; Kittel, 2013). A more comprehensive study into the factors that affect the date palm's genetic diversity may enable us to anticipate susceptibilities and thwart genetic erosion (Abul-Soad et al., 2017). To preserve date palm genetic diversity, a multitude of progressive simulation studies are presently underway, and a more detailed comprehension of the factors that impact it may facilitate our goals (Shabani and Kumar, 2013).

Characterization Methods for Evaluation of Date Palm Varieties

It is crucial to evaluate and understand the genetic diversity of date palm for its effective conservation and utilization. This comprehensive exploration includes various characterizations methods such as morphological characterization, cytological characterization, biochemical characterization, and molecular characterization that collectively provide deep insights into the variations in the date palm populations. Morphological characterization deals with the observable traits, providing a visual understanding of the plant's phenotype (Allam et al., 2021). Cytological characterization investigates cellular structures and chromosomal arrangement, providing insights at the microscopic level (Mohamed and Abd El-Kawy, 2010). Biochemical characterization examines the biochemical composition, offering insights into date palm's metabolic profile (Allam et al., 2021). Lastly, molecular characterization employs advanced genetic techniques, unraveling the complexities of date palm DNA for a high-resolution perspective on its genetic diversity (Elhoumaizi et al., 2002). The combination of these diverse characterization methods ensures a thorough assessment of date palm varieties for effective conservation and breeding strategies.

Morphological Characterization

There exist more than 5,000 date varieties worldwide, each exhibiting differences in the shape, size, and characteristics of their fruit pulp (Table 3). The quality attributes such as appearance, color, chewiness, elasticity, flesh thickness, fruit size, mouthfeel, mouth shear, pit size, sweetness, solubility, and texture have led to the development of diversity in date palms (Ismail et al., 2001). Date varieties are categorized into three classifications: soft, semi-dry, or dry, which is determined by the texture of the fruit as it naturally ripens. Additionally, these cultivars are sorted based on how long it takes for them to yield mature fruit, falling into categories like early, midseason, and late (Chao and Krueger, 2007a). Each cultivar can be a result of sexual reproduction, or it can be derived from offshoots. The sexually reproduced ones exhibit both parent's traits whereas the ones derived from offshoots are true-to-type (Johnson, 2011).

The texture of date fruits can vary, and this classification can help in identifying and categorizing different date varieties. Classifications based on date fruit are soft, semi-soft, semi-dry, dry, sugary, and crunchy dates. The most commonly employed molecular markers including AFLP, ISSR, RAPD, RFLP, and SSR, come with certain constraints with regard to cost, ease of use, robustness, dominance/co-dominance, and polymorphism level. Among these markers, nuclear microsatellites, also known as

Table 3. Traits of different cultivars of date palm across different geographical locations.

Cultivar	Geographical Location	Characteristics of Fruit	Trait of Importance	References
Amari	Tunisia	Dried date, soft, sweet, medium size	Fruit quality	(Sharoba et al., 2022)
Amhat	Egypt	Soft and moist date	Fruit texture	(Mustafa et al., 2014)
Anbarah	Saudi Arabia	Dark colored, chewy, very aromatic, flesh tender, and slightly crispy.	Fruit size	(Aleid et al., 2014)
Aseel	Pakistan	Soft, large in size, oval in shape, and brown in color	Fruit quality	(Abul-Soad et al., 2015)
Barhi	Iraq	Sweet and juicy, yellow color	Superior quality fruits and long stalk	(Alqaisi et al., 2022)
Dayri	Iraq	Semi-dry and strong flavor	Tolerance to moisture	(Siahpoosh et al., 2011)
Deglet Noor	Tunisia and Algeria	Dried date, Semi-soft, medium to dark brown, semi-dry dates	Fruit texture	(Othmani et al., 2020)
Fard	Oman	Semi-moist texture with a soft, mild sweetness	Fruit quality	(Rambabu et al., 2020)
Hadrawi	Iraq	Dried date, dark brown color, fleshy, and sweet date.	Fruit quality	(Al Shaibani, 2021)
Hallawy	Iraq	Yellow or light gold in color, firm, chewy texture, less sweet	Fruit sweetness	(Gupta et al., 2017)
Hayany	Egypt	Long, soft fruit, black and shiny color with less sweet	Fruit texture	(Abdel Salam et al., 2022)
Iteema	Algeria	Amber coloration and an oblong shape	Fruit sweetness	(Thind et al., 2007)
Kabkab	Iran	Light brown to dark black, oval, soft and juicy, Sweet & has a caramel-like & honey flavor	Fruit length	(Salamati et al., 2019)
Khadrawi	Iraq	Soft, melting, caramel-like, flesh not fibrous, pleasant and delicious taste	Fruit texture	(Muhammad Munir, 2020)
Khalas	Oman	Average-sized and reddish dark brown in color, hearty flavor, and a butter-caramel taste	Semi dry cultivar	(Munir et al., 2021)
Khasab	Iraq	Delicate, white flesh with a tint of yellow, thick, and pleasant taste	Fruit quality	(Alam et al., 2023)
Khastawi	Iraq	Small, syrupy, juicy texture with mild sweetness	Resistance to humidity	(Al-Hamdani, 2016)
Lulu	Pakistan, Oman	Less fibrous flesh, thick, soft, and sweet	Salt tolerance	(Montazar et al., 2020)
Maktoom	Iraq	Tender, high sugar, and moisture content	Tolerance to moisture	(Khierallah et al., 2015)
Mazafati	Iran	Cylindrical fruit, dark brown to black fruit, soft flesh, tasty flavor	Fruit sweetness	(Alikhani-Koupaei et al., 2018)
Medjool	Morocco	Dried date, large fruit, soft and sweet, light brown to dark-brown color	Tolerance to moisture	(Al-hajjaj and Y. Ayad, 2018)

Table 3. contd. ...

Table 3. contd. ...

Cultivar	Geographical Location	Characteristics of Fruit	Trait of Importance	References
Piarom	Iran	Dark brown to black, elongated, sweet, wrinkled texture, long ratio length, skin sticky to meat, rubbery flesh, easy to chew	High quality fruits	(Khierallah et al., 2015)
Rabbi	Iran	Long, slender, semi-dry, red to dark brown, fleshy fruit	Fruit texture	(Karizaki, 2017)
Sayer	Iraq	Soft, dark orange-brown, medium size, dry and syrupy	Salt tolerance	(Abdulwahid, 2017)
Thoori	Algeria	Dry, firm in texture semi-sweet, peanut-like nutty flavor	Fruit texture	(Hasnaoui et al., 2011)
Zaghloul	Egypt	Crunchy, deep red or burgundy skin, with crisp flesh	Fruit texture	(Farag et al., 2012)
Zahidi	Iraq	Round, medium sized, dried date, golden color, less sweet	Fruit texture	(Gupta et al,. 2017)

SSR markers, appear to meet the majority of the criteria necessary for achieving precise analysis of date palm fingerprints and phylogenetic relationships (Khanam et al., 2012). ISSR markers' studies performed by Hamza et al. (2012) revealed the clear differentiation of dry and soft cultivars which had different fruit consistency. RAPD, SSR, and AFLP markers were used to distinguish between varietal kinds (texture) in date palm cultivars of the New Valley governorate (Ibrahim et al., 2014). Date fruits require a consistent duration of warm weather from pollination to full ripening, and the duration of this warmth is a key factor in determining the suitability of a particular variety for a specific geographical area (Sirisena et al., 2015). Hammadi et al. (2009) grouped 26 Tunisian date palm cultivars based on fruit characteristics through principal coordinate analysis (PCA) and the Pearson coefficient also showed a correlation between the harvesting season and the percentage of the date palm's midrib that has spun out can be used as a gauge of maturity of date palm. PCA analysis by Al-Faifi et al. (2016) using high polymorphic SSR markers also showed that mid-season varieties of Saudi Arabia grouped apart from early and late varieties.

Biochemical (Nutritional) Characterization of Date Palm

The biochemical (nutritional) characterization of date palm involves a detailed analysis of its nutritional content and composition. It involves a thorough examination of essential components present in date palms such as carbohydrates, fats, proteins, vitamins, and minerals. This nutritional profiling is important for assessing the dietary value and potential health benefits of date palm products. Analysis not only aids in determining the nutritional content for human consumption but also plays a vital role in optimizing agricultural practices and crop management strategies for enhanced date palm productivity. This biochemical characterization is pivotal in deciding the effective utilization of date palms for both nutritional and agricultural purposes (Sobhy and Hafez, 2010). Table 4 lists the nutritional breakdown of various cultivars of dates.

Cytological Characterization

The chromosome number 2n = 36 was reported in date palm (Beal, 1937), though 2n = 28, 32, 34, 36, and 64 were reported in different populations/cultivars (Al-Sahil and Al-Rawi, 1987). Cytological characterization like differences in chromosome numbers leading to phenomena like polysomy, polyploidy, chromosome breakage, or nucleolar heterochromatin aggregation can be used to classify

Table 4. Nutritional composition of various date palm cultivars.

Nutritional component	**Medjool** (Johnson et al., 2015)	**Deglet Noor** (Johnson et al., 2015)	**Ajwa** (Assirey, 2015)	**Gondaila** (Mohamed et al., 2014)	**Barakawi** (Mohamed et al., 2014)	**Jaw** (Mohamed et al., 2014)	**Mishrig** (Mohamed et al., 2014)	**Bittamoda** (Mohamed et al., 2014)	**Madini** (Mohamed et al., 2014)	**Khodari** (Assirey, 2015)
Total Sugars (g/100 g)	66.47	63.35	74.3	80.41	79.55	78.73	80.27	80.11	79.04	79.4
Protein (g/100 g)	1.81	2.45	2.91	4.09	4.03	3.69	3.92	3.72	3.94	3.42
Total lipid (g/100 g)	0.15	0.39	0.47	2.00	1.87	1.79	2.06	1.81	1.71	0.18
Potassium (mg/100 g)	696	656	476.3	135.36	1088.4	974.86	733.52	722.56	691.67	16
Phosphorus (mg/100 g)	62	62	27	208.79	232.04	212.29	150.19	166.67	170.33	289.6
Calcium (mg/100 g)	64	39	187	238.10	276.24	223.46	293.04	277.78	222.20	133
Iron (mg/100 g)	0.90	1.02	-	6.48	4.06	7.06	5.50	5.09	6.91	-
Magnesium (mg/100 g)	54	43	150	120.88	66.30	111.73	109.89	100.00	66.67	150
Dietary fibers (g/100 g)	6.7	8.0	-	2.53	2.67	2.74	2.78	2.37	3.14	-
Moisture			22.8	8.78	9.38	10.68	8.81	10.03	9.90	19.5

date palm variants. The cytological approach is used to analyze genetic variation or instability hence differentiating them based on the origin of the date palm variants and classifying between and within the species. Despite the progress in the cytological methods this is considered insufficient to study varieties of the date palm. This is because characterization using a cytological approach is time consuming and takes a lot of effort making it unreliable (Naqvi et al., 2021).

Molecular Characterization Techniques for Analysis of Genetic Diversity of Date Palm

To maintain the identity and homogeneity of the date palm cultivars, clonal propagation is a process that is usually performed. Hence, DNA fingerprinting technology has instigated a growing necessity nowadays to classify closely related cultivars (Elhoumaizi et al., 2002). Selection of commercial varieties and resulting nomination of desirable genotypes is feasible by linking a gene of interest from molecular markers. Frequent use of phenotypic and molecular markers has been observed in identifying plant genetic diversity. Phenotypic markers are classified based on the 270 characteristics of the fruit, leaves, and spines, which differ according to the variety that is being studied. Specifically, differentiation of the date palm cultivars can be done based on the shape, weight, color, size, aspects of fruit skin, texture, consistency, etc., and biochemical markers like proteins and isozymes. However, the use of morphological traits independently for the characterization of date palm cultivars is challenging since morphological markers are greatly affected by the environment. Due to the influence of specific geographical areas, environmental conditions, and the variation of plant development stage, these serve as inconsiderable and inexplicit indicators since they cannot be estimated accurately (Elhoumaizi et al., 2002; Elmeer and Mattat, 2015). DNA fingerprinting, either in combination with or independently, improves the ability of morphological markers. Molecular markers are an indispensable tool for the efficacious selection of preferred agronomic traits because they are solely based on the genotype and, hence, are considered an irreplaceable resource for crop genetic studies. Analysis of genetic diversity, and gene characterization, and sex determination by using these techniques in the germplasm of date palm has been represented in Table 5. Few breeding programs and studies of crop genetic linkages have generated promising results using DNA markers (Salomon-Torres et al., 2017).

***RAPD (Random amplified polymorphic DNAs)*:** Random amplified polymorphic DNA (RAPD), one of the most widely used molecular techniques, identifies variation in the non-specific regions of the genome. Polymerase Chain Reaction (PCR) is a base for the RAPD technique. It is used to evaluate phylogenetic relationships among the particular genomes of the different date palm cultivars. It is also employed to create linkage maps to aid in the discovery of molecular markers associated with economically significant features. A study proposed differentiation of male and female genotypes of date palm in which 30 RAPD primers were initially screened with highly polymorphic primers found in OPB-20, OPB-15, and OPA, and they discovered three primers that were capable of efficiently differentiating the male genotypes from the females of date palm [(OPB 20), (OPB 15), and (OPA 5)] (Patil and Daspute, 2019). Another study that used RAPD for sex determination was to compare genetic material from four female Egyptian date palm trees and four unknown male plants. The genetic resemblance of the four female date palms (Zaghloul, Amhat, Samany, and Siwi) varied between 87.5–98.9%. According to the banding profiles, males 3 and 4 are genetically similar to the four female cultivars (Soliman et al., 2003). Hence, RAPD is considered to be the most commonly used technique for sex determination in date palms. In another study, pre-screening of 123 primers on four genotypes, 43 accessions, including 37 from Morocco and six from Iraq and Tunisia date palm, resulted in the selection of 19 primers for polymorphism and provided satisfactory results. Hence, RAPD is shown to provide stunning results in the differentiation and identification of different cultivars of date palms (Sedra et al., 1998). Moreover, RAPD has also been extensively used in combination with other techniques like SSR, AFLP, RAFLP, etc.

Table 5. Molecular markers are used for assessing the genetic variation in date palm.

Variety Name or Accession Number	Region	Marker and Method Used	References
Boustammi noire, Boustammi blanche, Tademainte, Hafs, Bouslikhene, Bouzeggar, Aîssa-Iyoub, Belhazit, Azigzao, Bouittob, Ademou, Bouijou, Otokdime, Aguelid, Houa, Ahardane, Mejhoul, Boufeggouss, Jihel, Bouskri, Bourar, Deglet Nour Tunisia, Ftimi Tunisia, Besser Lahlou Tunisia, Barhi Iraq, Zahdi Iraq, Hallaoui Iraq, SH-1001, SH-1002, SH-1003, SH-1004, SH-1005, SH-1006, SH-1007 A, SH-1008, SH-9 SH-1009, SH-1010, NP3, NP4	Morocco	RAPD	(Sedra et al,. 1998)
Deglet Nour, Deglet Bey, Boufegous, Ftimi, Kenta, Kentichi, Ghars Mtig, Zehdi	Tunisia	RAPD	(Trifi et al., 2000)
Ajwa, Bareem, Barney, Nabtet Saif, Nabtet Sultan, Om-Hammam, Om-Kobar, Rabeeha, Shehel, Shishi, Sugai, Sukkary Asfar, Sukkary Hamra	Saudi Arabia	RAPD	(Al-Khalifah et al., 2012)
Zaghloul, Amhat, Samany, Siwi	Egypt	RAPD	(Soliman et al., 2003)
Asada, Barakawi, Brair Bitamoda, Gargoda, Gondaila, Laggai, Khateeb, Medina, Kulma, Shidda Tonisi, Zaglool Galisoog, Sulatani	Sudan	SSR	(Elshibli and Korpelainen, 2008)
Khalas, Shaishi, Barhi, Hilali, Sukari, Khnaizi, Shahil, Khasab, Rzaiz, Lulu, and Phahel.	Qatar	SSR	(Elmeer et al., 2011)
Zahidi Hatary Helaly Khalas, Succary, Anbara, Abu main, Sheshy Barhee Sultana Naboot saif Khadrawy Khush Zabad Khanezy Thuri	Qatar	SSR	(Ahmed and Al-Qaradaw, 2008)
EF015858, EF015859, EF015860 EF015862, EF015863, EF015864 EF015866, EF015868, EF015869 EF015870, EFO15871, EF015873 EF015875, EF015876, EF015878	Algeria and California	SSR	(Akkak et al., 2009)
Usta Umran, Tebarzal Um Al-Dihen, Guntar, Khestawi, Bream Ashrasi MaktomAsfar, Buliani, Leelwi Shwethi Ahmer, Qul Husaini Helawi, Qitaz, Chipchab, Zahdi, Shwethi Asfar, Khadrawi, Baw Adem, Deari, Barhi, Meer Haj Um Al-Blaliz, Ghanami Akhder, Ghnami Ahmer, Khekri, Smeasmi, Ghulami, Greatli	Iraq	SSR	(Khierallah et al., 2011)
Assiane, Boufeggous Tgharas, Aziza bouzid, Aziza manzou, Boufeggous gharas, Taabdount, Afroukh n'tijent, Mäles, Mejhoul, Admam, Tadmant	Morocco	SSR	(Bodian et al., 2012)

Table 5. contd. ...

Table 5. contd. ...

Variety Name or Accession Number	Region	Marker and Method Used	References
Agglany, Amhat, Ammry Baldy unknown, Bartmouda, Bint Aisha, Chammeia, Deglet Noor, Degna, Feryhy, Gargouda, Gondailah, Halawy, Hayany, Hommera, Malakaby, Medjool, Nasser Dein, Oreibby, Red, Sakkoutty, Sammany, Shakngobil, Siwi Taktakt, Acinat, Aggwa, Yellow Balady, Yellow unknown, Zaghloul	Egypt	AFLP	(El-Assar et al., 2005)
Deglet Nour Boufagous, Flimi Kenta, Kintichi, Deglet Bey, Ghars Mttig, Zchdi, Arichti, Khouftimi, Horra Okhet Degla, Hamra Kharroubi, Angou, Bejjou, Goundi, Halwaya, Tazerzit Safra, Tazerzit Soda, Ammari, Besser Hlo, Mahmoudia, Chodak, Rhaimya, Om lal, Om essayed, Bidh Hmam, Gasbi, CIV T124, T138, T158, T169, DFI, DG9, T159, Rochdi, Lemsi, Bouhattam, Smiti, Denga, Aguiwa, Deglet Nour K, Kechdou ahmar, Fhal Ksebba, Rakli Fermla, Tamri	Tunisia	SSR	(Zehdi et al., 2004)
Abada, Amir Hajj, Ashrasi, Badrayah, Barhee, Bentamoda, Boyer, Dayri, Deglet Beida, Deglet Noor, Halawy, Hayany, Haziz, Horra, Javis, Khadrawy, Khir, Medjool, Sayer, Thoory, Zahidi,	California Iraq Sudan Algeria Egypt Morocco	AFLP	(Cao and Chao, 2002)

Amplified fragment length polymorphism (AFLP)
Inter-simple sequence repeat (ISSR)
Microcrystalline cellulose (MCC)
Random amplified polymorphic DNA (RAPD)
Restriction fragment length polymorphism (RFLP)
Simple sequence repeat (SSR)

***RAPD-ISSR (Inter-simple sequence repeat)*:** Three examples demonstrate the combined use of these markers in date palm genetic diversity. The genetic diversity of 10 cultivars of all semi-dry date palm was determined in Egypt. Five primers were utilized for RAPD analysis and five for ISSR analysis to distinguish among the cultivars, out of which eight primers showed high polymorphism (Soliman et al., 2019). Additionally, four cultivars of date palms were studied in Egypt. All of the primers examined showed both intravarietal polymorphism and intervarietal variance. The 10 primers identified 12 distinct polymorphic RAPD markers among the four cultivars. In addition, seven ISSR primers were employed using the four bulked samples. Six primers showed 24 distinct markers that characterize the various cultivars (Hussein et al., 2005) Similarly, Al-Khalifah et al. (2012) employed RAPD markers to examine the genetic diversity of 13 distinct cultivars. The screening of 140 RAPD primers resulted in the selection of 37 primers that exhibited polymorphism, with consistent findings. The distinct banding patterns created by 37 carefully chosen primers distinguished all 13 genotypes.

***RAPD-RFLP (Restriction fragment length polymorphism)*:** On offshoot leaves surrounding the shoots used to commence tissue culture, an RFLP study of five elite date palm cultivars was undertaken. With cDNA probe 1, which was highly polymorphic, differentiation between the five cultivars was simple. A variation across cultivars was also discovered using random primer amplification of whole DNA isolated from branch leaves (Corniquel and Mercier, 1994).

***RAPD- AFLP (Amplified Fragment Length Polymorphism)*:** RAPD banding patterns analysis obtained by PCR amplification with 37 random primers revealed no evidence for somaclonal variants, and the proportion of polymorphic bands was zero out of 259 scored bands. Meanwhile, AFLP banding pattern analysis utilizing 13 primer combinations revealed small genetic differences in the AFLP banding patterns. According to AFLP analysis, the proportion of genetic variants (polymorphism) in tissue culture-derived date palm offshoots of cultivars Sakkoty, Gandila, and Bertamoda was 2.6, 0.79, and 1%, respectively (Saker et al., 2006).

***RAPD and isoenzyme analysis*:** Somaclonal differences in tissue culture-derived date palm plants were studied using isoenzyme analyses and activities of peroxidase (PER), polyphenol oxidase (POD), and glutamate oxaloacetate (GOT), as well as randomly amplified polymorphic DNA (RAPD) fingerprints. It was discovered that the frequency of somaclonal variants varies with age. All of the plants studied had similar isoenzyme patterns for PER and GOT. Variations in the activity of the three enzymes investigated, as well as POD isoenzymes, were found. RAPD analysis revealed genetic differences in around 4% of the date palm plants examined (70 regenerates). Only cultures aged 6 and 12 months were found to have genetic differences. All morphologically aberrant shoots were shown to have genetic differences at the molecular level (Saker et al., 2000).

***SSR (Microsatellites)*:** Tandem repeats of 2–10 base pairs which can be found throughout the date palm genome are called microsatellites, also known as simple sequence repeats (SSRs) (Ahmed et al., 2012). Microsatellites are extensively used in genetic analysis due to their highly polymorphic nature, locus-specificity, codominance, and high reproducibility (Akkak et al., 2009). The increase or decrease in the number of sequence tandem repeats results in the occurrence of polymorphism in microsatellites. It is widely employed to analyze sex, heterozygosity, fixation index, and population identification. According to a study that investigated the performance of 253 SSR microsatellites in which patterns of genetic variation were studied among 1,066 date palms from 411 cultivars in 12 countries (Al-Faifi et al., 2016).

Furthermore, 19 SSR markers with the greatest performance were identified for future genetic diversity investigation in the date palm (Salomon-Torres et al., 2017). Moreover, date palm genetic diversity using 16 loci of microsatellite (SSR) primers and palm germplasm from Sudan representing 37 female and 23 male accessions were studied. As reference material, eight Moroccan female accessions were used. The polymorphism in the SSR markers analyzed was high. At the 16 loci, 343 alleles were discovered. The number of alleles per locus ranged from 14–44, with an average of 21.4. Sudan cultivars (0.841), Moroccan cultivars (0.820), and male accessions (0.799) all had a high degree of anticipated heterozygosity. The mean anticipated and observed heterozygosity identified for Tunisian date palms were 0.70 and 0.61, respectively, but the values obtained in this study were 0.853 and 0.912 (Elshibli and Korpelainen, 2009). In 2009, an analysis of genetic diversity among 15 different cultivars of date palms was conducted in Qatar. Ten of the 16 primer pairs examined for their potential to create predicted SSR banding patterns in Qatari date-palm genotypes exhibited distinct single bands in the majority of the genotypes analyzed (Ahmed and Al-Qaradaw, 2008). In another study, following screening, 17 microsatellite loci were identified and tested on 31 cultivars and clones from Algerian and Californian date palm germplasm. All primer pairs generated an amplification product of the predicted size and revealed a high polymorphism level in the tested samples (Akkak et al., 2009). Another analysis was conducted using 22 microsatellite SSR primers; researchers evaluated genetic diversity in 30 date palm cultivars from Iraq, including 24 female and six male cultivars. The polymorphism in the SSR markers analyzed was high (Khierallah et al., 2011).

The genetic diversity of 128 date palms collected from the Figuig oasis in Morocco was examined. There were 121 female palms, 11 cultivars, and seven males among these palms. For the investigation of their genetic diversity, eighteen primers were utilized. Only 15 primers successfully amplified all of the samples (Bodian et al., 2012). SSR markers were used for analyzing genetic connections among 32 date palms representing typical cultivars growing in diverse geographical locations in Saudi Arabia. Ninety-three unique simple sequence repeat markers for date palm polymorphism detection were created and tested. The average polymorphism information content was 0.595, and the primer discrimination power was 0.662. The heterozygosities anticipated and observed were 0.676 and 0.763, respectively (Al-Faifi et al., 2016). In addition, 14 microsatellite loci were selected to investigate genetic diversity in Tunisian date palm germplasm. A significant amount of polymorphism in 49 accessions from three major oases with minimal regional organization within Tunisia was observed. 100% of local date palm accessions were strongly fingerprinted and easily recognized using only three loci (Zehdi et al., 2004). Another study focused on determining the genetic linkages between 21 endangered date palm varieties from 11 oases in Nefzaoua, southwestern Tunisia. The eight pairs of primers amplified 48 alleles in the accessions tested, indicating a high level of polymorphism with minimal geographical pattern among the evaluated oases. The observed heterozygosity (Ho) average varied from 0.38 (mPdCIR078) to 0.80 (mPdCIR010), with a mean of 0.603. The estimated heterozygosity (He) average varied from 0.54 (mPdCIR035) to 0.78 (mPdCIR025), with a mean of 0.682 (Ben Abdallah et al., 2020). Similarly, an additional study found the genetic diversity among 32 date palm cultivars by using 14 microsatellite primers in Qatar. The results showed that the 14 microsatellite markers produced 98 alleles, and the cluster analysis revealed four significant clusters related to geographical locations (Elmeer et al., 2019). With the aim of determining the genetic and geographical patterns of Nigeria and Saudi Arabia date palms, a study was conducted using six microsatellite markers on 14 cultivars of date palms, 10 from Nigeria and four from Saudi Arabia. The analysis revealed 83.3% polymorphism and hence directed toward high diversity in the cultivars (Yusuf et al., 2015).

***AFLP (Amplified fragment length polymorphism)*:** AFLP is a PCR-based amplification molecular technique, where a selected set out of all the fragments obtained by the restriction digestion of the entire genome is amplified. AFLP is an efficient method to ascertain even minimal amounts of variation, which helps differentiate between highly related cultivars (Diaz et al., 2003). In a study, four sets of amplified fragment length polymorphism (AFLP) markers with near-infrared fluorescence-labeled primers were used to study 47 date palm samples collected from eight sites in Egypt. A total of 350 bands were evaluated, with 233 (66.6%) being polymorphic (El-Assar et al.,

2005). Another AFLP analysis was conducted in Egypt, where 14 date palm cultivars yielded 657 amplicons directing toward a polymorphism level of 45.8% (Adawy et al., 2004). Polymorphisms in 21 California date cultivars were identified using AFLP analysis using near-infrared fluorescence-tagged primers. These findings demonstrated that AFLP markers may be used to identify particular date cultivars accurately. The information will be useful for gathering future data and helping select various parents for cross-hybridization in a breeding endeavor (Cao and Chao, 2002). AFLP markers were used to validate the hypothesis that 'Medjool' is a landrace variety in Morocco, 66 'Medjool' date palm accessions from Morocco, six types of dates from Egypt, and four 'Medjool' accessions and one "Deglet Noor" accession from California were analyzed. Four groups of fluorescent-labeled amplified fragment length polymorphism (AFLP) markers were used to evaluate these accessions; 402 bands were created, with 217 of them being polymorphic (54.0%). The 66 Moroccan 'Medjool' accessions had at least 79% genetic similarity. From the results, the hypothesis was supported and proved right (Elhoumaizi et al., 2005).

mt DNA (Mitochondrial DNA): A study revealed a comprehensive sequencing and transcriptome analysis of the mitochondrial genome in three date palm varieties. It was found that the mitochondrial genome encodes 38 proteins, 30 tRNAs, and three rRNAs (6.5% of gene content). The bulk (93.5%) of the genome sequence was composed of chloroplast-derived coding (10.3%) and non-coding elements. The study found clear tissue-specific gene regulatory patterns, particularly in female and male flowers (Fang et al., 2012).

cp DNA (Chloroplast DNA): An assessment of the molecular signature from eight date palm cultivars from Kingdom of Saudi Arabia using cpDNA sequences rpoB and psbA-trnH was carried out and in conclusion, the analysis clustered these cultivars into three groups with bootstrap values of 58 and 64% and interpreted that the sequences they obtained could be used as molecular signatures for potential date palm under trading and selection of genuine cultivars at the seed stage (Al-Qurainy et al., 2011).

In vitro hermaphroditism: Induction of *in vitro* hermaphroditism in date palm female flowers was performed by Masmoudi-Allouche (2009). The researchers then examined the genetic variability associated with *in vitro* floral hermaphroditism induction in the Tunisian cultivar Deglet Noor. They uncovered two probable candidate gene markers that might assist researchers in better understanding flower development and date palm sex identification (Masmoudi-Allouche et al., 2009).

Utilization of Genetic Diversity for Breeding Improved Cultivars

The efforts being made to preserve the genetic diversity of date palm across the world are aimed not only at conserving the species but also at providing future scientists, researchers, and breeders enough genetic resources to help in breeding improved cultivars for desirable traits like higher yield, better stress tolerance abilities, and so on. The need for this is increasingly becoming evident as wild-type and existing cultivars are facing threats due to climate change, with a significant reduction in the availability of regions suitable for date palm cultivation in the next few decades in major countries like Saudi Arabia (Allbed et al., 2017) and Tunisia (Haj-Amor et al., 2020). These risks are presented in various forms, including predicted lack of soil stability and fertility and lack of irrigation water due to increased salinity. Besides these factors, the date palm also faces competition for resources from invasive species like *Fusarium oxysporum* (Shabani and Kumar, 2013). In light of such threats, breeding programs focusing on producing cultivars resistant to this pathogen have shown some early success and hold immense potential in the future (Johnson et al., 2015). Traditional breeding methods have also yielded cultivars with good fruit quality in the F_1 generation in Morocco, but the program is projected to take at least 30 years from its initiation date to yield suitable F_2 generations for agricultural use (Rivas et al., 2012).

These instances of traditional cross-breeding hold promise but also highlight the major drawback of such programs, which is the time taken to yield commercially relevant results (Bouguedoura et al., 2015). As such, efforts have already been made to shift to new techniques for breeding improved cultivars, mainly via biotechnological approaches. In this context, notable achievements have been made in several instances. *In vitro* pollination of *Phoenix dactylifera* L. with the pollen from dwarf date palm *Phoenix pusilla* resulted in seedless fruit in three cultivars, named 'Barhi', 'Madjhool',''and "Sultana" (Kordrostami et al., 2022). Induced *in vitro* adaptation to high salt tolerance was achieved in the cultivar 'Khalas' through callus cultures (Al-Khateeb et al., 2020). Future prospects for such alternative breeding techniques have also been identified in CRISPR-Cas-based precision breeding, which can help in single-base editing, and result in knock-out or knock-down varieties of date palm, essentially creating designer cultivars to suit growing needs for better yield and higher adaptability (Sattar et al., 2021).

Sustainable Utilization of Date Palm Genetic Resources

The use of diverse date palm types for improving date palm cultivars involves a complex approach that combines traditional and modern breeding techniques. Date palms showcase a diverse genetic makeup, which can be utilized to enhance desirable traits such as yield, disease resistance, adaptability, and fruit quality (Chao and Krueger, 2007). Traditional breeding techniques have been vital in exploiting this range by conducting controlled cross-pollination between selected parent plants. This approach has been used for generations and is crucial for selecting improved cultivars (El Hadrami et al., 2011). Genetic studies have also been influential in understanding the genetic composition of various date palm varieties using advanced tools like DNA sequencing and marker-assisted selection. These studies allow researchers to explore the genetic composition of various date palm varieties. Understanding genetic variety aids in the identification of useful genes or alleles linked with desired qualities, which may subsequently be used to improve cultivars (Mathew et al., 2014). Agronomic trait evaluation entails thoroughly examining several date palm accessions for features like yield, fruit quality, pest and disease resistance, tolerance to environmental stresses (e.g., salt or drought), and overall adaptability. Varieties with excellent features are chosen as possible candidates for future breeding attempts (Bouhouch et al., 2021). Tissue culture techniques are used to conserve and multiply elite date palm genotypes. This allows for the rapid reproduction of superior cultivars by producing a high number of true-to-type plants with desirable features (Al-Khayri et al., 2018). Furthermore, the use of biotechnology, such as genetic transformation and genome editing, provides an avenue for focused improvement of date palm properties. This can include introducing certain genes to confer resistance to diseases or pests, improving fruit quality, or improving the plant's ability to flourish in specific environmental conditions (Johnson, 2011). Collaborations between agricultural research institutions, botanical gardens, and gene banks promote the interchange of various germplasm and information, promoting the preservation and effective utilization of date palm genotype variety (Krueger, 2021). Date palm wild relatives have distinct genetic features that can be used in breeding programs to increase the genetic variety of cultivated cultivars (Zhang et al., 2012). Conservation and genetic exploration of these wild relatives are critical to date palm enhancement. In conclusion, utilizing date palm's extensive diversity has provided a robust pathway for the sustained enhancement of date palm cultivars, ensuring adaptability, high productivity, and the retention of high-quality fruit traits, thereby contributing to the long-term viability of cultivation of the date palm.

Traditional Methods

There exist numerous date palm cultivars that have been created through traditional techniques such as the crossing of breeds and the controlled pollination process. These procedures are targeted at enhancing and advancing the date palm cultivars in terms of their productivity, fruit characteristics, shelf life, and resistance to pests. The research endeavors involve the examination of the impact of pollination timing on the fruit set and the development of date palms. Studies indicate that pollination carried out in the early morning yields higher fruit characteristics and production compared to pollination conducted later in the morning (Munir et al., 2021). The combination of pollen grains with a carrier and SSPG (Stimulative Solution of Pollen Germination) serves to minimize the costs associated with pollination, ensure a good yield of high-quality fruit, and facilitate mechanical pollination (El-Dengawy, 2017). Moreover, the effectiveness of various pollination methods, such as manually pollinating female date palm blossoms in the field using sponge pieces charged with gathered pollen, is compared to the traditional practice employed by farmers involving male inflorescences. The manual pollination method proves to be an effective alternative to the traditional method, particularly in situations where stored pollen needs to be used due to the absence of fresh pollen grains (Abdallah et al., 2014). Date palm cultivars such as 'Khalas' and 'Barhee' are produced through crossbreeding techniques and possess improved disease resistance and adaptability to climatic conditions (El-Beltagi et al., 2021). Additionally, several research studies demonstrate the nutritional enhancement in date palm cultivars resulting from crossbreeding techniques (Rani et al., 2022). Crossbreeding techniques have also contributed to decreased yield among date palm cultivars such as 'Dhakki', 'Kupra', and 'Chohara' (Ahmad et al., 2023). The utilization of traditional methods such as pollination and crossbreeding is essential for the development of improved date palm cultivars with enhanced traits, adaptability, and yield. Research indicates that the implementation of controlled pollination and targeted crossbreeding has significantly enhanced the date palm cultivars, resulting in the availability of a wide range of commercially valuable varieties.

Due to many difficulties inherent in conventional breeding procedures, the discovery of modern breeding techniques boomed, resulting in speedier and more efficient outcomes. With recent advancements in biotechnology, modern breeding has been revolutionized with the introduction of tissue culture, in vitro mutagenesis, cryopreservation, genetic engineering etc. (Al-Khayri et al., 2018).

Tissue Culture Technology

Somatic Embryogenesis

Somatic embryogenesis (also known as asexual embryogenesis) is a highly effective approach for micropropagation of date palms. Date palm embryo-like structures formed from somatic embryogenesis can grow into a whole plant in the same manner that zygotic embryos do. Date palm plantlets can be produced in two phases by somatic embryogenesis. The start of vegetative buds in date palm offshoot hearts was first reported by Poulain et al. (1979). Reynolds and Murashige and Tisserat used different tissues as explants to induce and regenerate somatic embryos from the same cultures (Reynolds and Murashige et al., 1979; Tisserat et al., 1979). To yield acclimatized plants, somatic embryogenesis takes a long time, usually around 2–3 years. To address this, breeders have used several ways to optimize the developmental phases of somatic embryos. Several attempts were made to do the same however, one efficient way of doing so was by altering the culture medium ingredients. Amino acids, auxins, cytokinins, abscisic acid, N-phenyl N-1,2,3-thiadiazol-5-yl urea (TDZ) for the direct somatic embryo, sucrose, silver nitrate as an inhibitor of ethylene, organic additives, biotin and thiamine were added into the culture medium over the course of few years by different scientists for its enhancement (Al- Khayri et al., 2018).

Organogenesis

Organogenesis in plant tissue culture infers initiation and development of organ tissue, resulting in plant regeneration without the callus stage in cultured explants. Organogenesis ensures uniformity in in vitro cultivated plants while reducing genetic variation. There are four major steps involved in organogenesis: (i) meristem induction; (ii) shoot multiplication; (iii) shoot elongation; and (iv) acclimatization. Several studies have examined date palm organogenesis (Bekheet, 2013; Jazinizadeh et al., 2015; Khan and Bi, 2012; Khierallah and Bader, 2007). Aaouine (2003) studied direct-shoot organogenesis and plant regeneration across 30 date palm genotypes. Bekheet (2013) accomplished direct organogenesis shoot tips in the Zaghloul cv. Jazinizadeh et al. (2015) studied the direct organogenesis of cv. Barhee uses various mediums, sucrose levels, and hormone concentrations.

In vitro Mutagenesis

Mutation induction is an established method for introducing desirable features into crop cultivars that are not naturally present or have been lost through evolution (Hasan et al., 2022). The first attempt to conduct mutation induction in date palm was performed by the FAO/IAEA Coordinated Research Project on the development of Bayoud-disease-resistant date palm mutant cultivars in North Africa (Jain, 2005, 2006). Gamma radiation was utilized to select mutants resistant to Bayoud disease in *Fusarium oxysporum* f. sp. albedinis culture via *in vitro* selection and toxin extraction (Sedra and Lazerk, 2011). Al-Enezi and Al-Khayri (2012 a,b) analyzed the impact of X-rays on date palm seedlings, including DNA content, ions, photosynthetic pigments, proline content, growth, and water levels.

Cryopreservation

Cryopreservation is the method of storing *in vitro* produced genetic resources such as shoot tips, callus, cell suspension, microspores, and somatic embryos at extremely low temperatures, often at –196°C in liquid nitrogen (Al-Khayri et al., 2018). Date palm was first cryopreserved in 1979 by Finkle and Ulrich (Hadrami and Hadrami, 2008). Tisserat et al. (1985) found that 10-year-old Deglet Noor cv. date palms produced the same amount of fruit with cryopreserved pollen. Researchers used cold hardening and various sugars to promote cryopreservation of date palm explants. Myriad cryo protocols for date palm, including normal vitrification, droplet vitrification, and encapsulation vitrification are being used (Bekheet et al., 2006; Fki et al., 2017; Alansi et al., 2019).

While modern breeding techniques are still developing, it holds tremendous potential and value, and demonstration of more such methods are upcoming trends to keep up with in the field.

Traditional Uses of Date Palms and Their Cultural Significance

Date palm (*Phoenix dactylifera* Linn.) has extensively spread to Australia, India, Pakistan, Mexico, southern Africa, South America, and the United States since it originated in areas of the Arabian Peninsula, North Africa, and the Middle East. In the Arabic region, date palms survive the temperature changes and grow well. For the past 5,000 years, dates have been cultivated in the Middle East. The documented reports from Iraq are as close to 3000 BC. In Iraq, the date palm was used to construct the moon temple (Jain et al., 2011). In Quran Scripts (Marwat et al., 2016) and Christian traditions, numerous date palm parts have been mentioned. Historical artifacts from Egypt have referenced date palms as symbols, in temples, in roof construction (Azzam, 1960), as life-supporting plantations (Bekheet and El-Sharabasy, 2015), and in the making of traditional craft (Darwish et al., 2019).

There are several uses for every part of the date palm, including the fruit, leaf, flower, seed, and gum. The primary product is the date fruit consumed as food by humans and animals. Timely pruning of date palm yields leaves, midrib, leaf lets, spadix stems, leaf sheath fiber, petiole, kernel, and trunk. The leaves of date palms were used in the construction of walls by wickerwork technique or by Sedda, where a mat is made by sewing together the leaves of the date palm. Each mat is then fastened by threads to create dense sheathing, which provides extremely effective heat conservation (Darwish et al., 2019b). The midrib obtained by either feeding animals the leaflet or by manually removing the leaflets are used for various purposes (Eldeeb, 2017) like breadboards, crates, and bird coops (Zaid and de Wet, 2002), handmade furniture (El-Batraoui and Baṭrāwī, 2016), rural wall and roof sheathing (Darwish et al., 2019), fencing, doors (Piesik, 2012; Wendrich, 2009), and boats (Johnson, 2016), baking bats, discs (Wendrich, 2009), rods for fishing, and as scaffolding to support grapevine growth (Barreveld, 1993). The various applications of date palms are illustrated in Figure 4.

Date palm leaflets, the leftover after preparation of midrib, are a great resource for a range of products. A variety of traditional goods, including sandals (Wendrich, 2009), hats, baskets, and mats (Barreveld, 1993) are made using the Egyptian art of weaving. Krena, which are threads prepared from low-quality leaflets, are utilized as fillers in mattresses, cushions, and bedding (Barreveld, 1993). Leaflets also make heavy-duty ropes and hand brooms (Barreveld, 1993). The spadix of date palm with a good tensile strength is also used for making several handcrafts like household accessories like tissue boxes, tablecloths, heavy-duty mats, decorative trays, bags, and lamp shades (Wendrich, 2009). Leaf-sheath fibers are used in making ropes, which are widely utilized in the UAE for sailing (Piesik, 2012), or ropes made into the net that is capable of transporting large loads over camels in rural areas (Barreveld, 1993). The fibers from date palm leaves can also be used as fuel, to make fishnets, basket handles, brushes, bedding, and to shade living plants and their offshoots (Barreveld, 1993).

They were also used for roofing in Egypt's rural houses (El-Tawil, 1989; Helal, 1989). An analysis of the reed architecture in the Manzala Lake region reveals the existence of spontaneous architecture in Egypt. The Faculty of Fine Arts at Helwan University in Cairo highlights the benefit of improved thermal insulation when applied to residential roofing. Leaf sheath fibers were also used as wicks obtained from remains of the Middle Kingdom and wigs obtained from remains of the Roman Period preserved at the Egyptian Museum in Cairo. Ancient Egyptians also used leaf sheath fibers in making bags and agricultural plows. The same method is still employed today to make bird traps, donkey saddles, supporting belts for palm climbers, and camel bags.

Date palm kernels are used to make jewelery, necklaces, and prayer beads (Ray and Ray, 2003), roasted to make coffee, and heated to make charcoal (Johnson, 2016; Barreveld, 1993). In many rural areas of Egypt and the Arab world, date palm trunks are used as strong and long-lasting columns and beams in traditional housing (El-Tawil, 1989). The trunks of date palms are fashioned into blocks

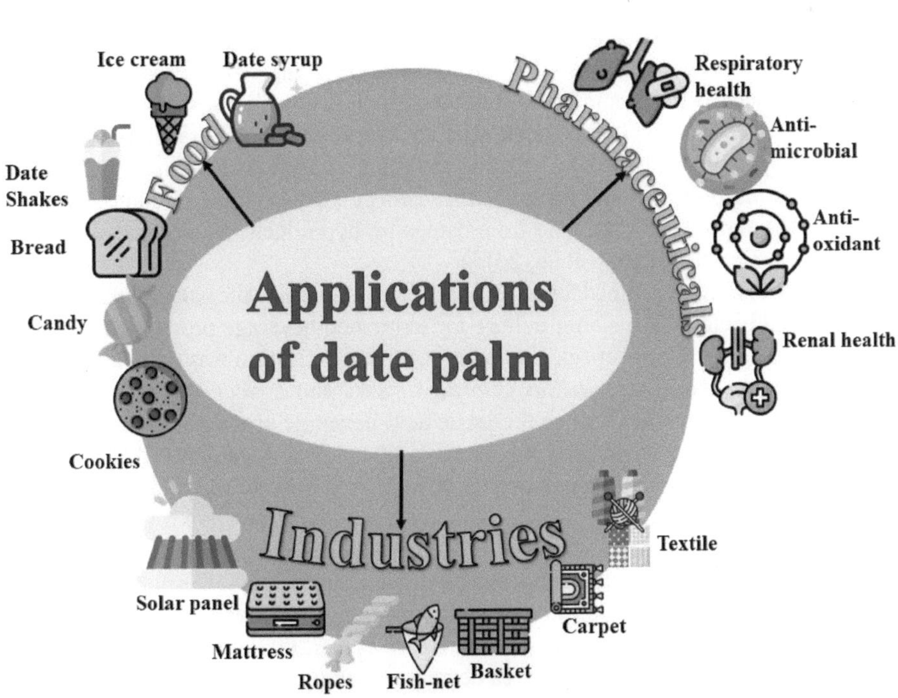

Figure 4. Diverse Applications of Date Palm.

that support and serve as the armrests on tables and chairs. As the improvement in the processing of date palm materials is advancing the applications are also improving. Date palm use is the best example of sustainable material application.

Diverse Uses of Date Palm beyond Its Traditional Roles

Application of Date Palms in Food

Enzymes are biocatalysts that lower the activation energy of a reaction. They have a significant impact on a number of industries, including the manufacturing of paper, pulp, laundry, textiles and pharmaceuticals. They are isolated from microorganisms, animals, and plants. The extent of their uses is determined based on the enzymes' properties. Peroxidase is an important enzyme purified from date palm leaflets (Al-Senaidy and Ismael, 2011), showing strong activity and stability across a broad pH range, relative thermos-stability, broad substrate selectivity, and stability in the presence of high concentrations of ionic metals makes it a potentially useful tool in a variety of industrial and medicinal applications.

In Arab countries, date palms are the most common fruit. The pulp is utilized for various applications or as food. The seeds contain about 5–7% of protein by weight (Aldhaheri et al., 2004). Deglet Nour dates seed proteins have been isolated and characterized. Their emulsifying properties were comparable to soy protein isolate. This protein isolate can be a sustainable application in the food industry. Date palm leaflets are used in microbial protein production (Abou-Zeid, 1991) and successfully applied in the growth of *Trichoderma viride.* Date syrup is used for the production of Carotenoids. *Lactobacillus plantarum* was used in a study by Elsanhoty et al. (2012) to obtain 16–21 mg/kg of dry carotenoid from 5% date syrup. With additional media and growth condition optimization, the yield of dry cell carotenoid has increased to 54.89 mg/kg.

Researchers (EL-Mously et al., 2023a) studied the use of wheat straw and date palm leaflets as a fungus's growth substrate. They were able to cultivate *P. ostreatus T, Pleurotus florida, P. ostreatus M, and Pleurotus ostreatus A*. Date palm leaflets showed higher growth compared to wheat straw. Additionally, date waste has been utilized as a substrate for the growth of *Aspergillus niger* ATCC 16404 and *Saccharomyces cerevisiae* ATCC 1102 (Acourene et al., 2011). Products made from date palms, including fronds, petioles, and leaf sheaths, have high lignocellulose content. These byproducts have the potential to be used to produce lactic acid, an organic acid that is essential for the fabrication of biodegradable plastics and used in a variety of food, pharmaceutical, and leather industries. *Geobacillus stearothermophilus* and *Lactobacillus delbrueskii* subsp. are used in a successful study (Alrumman, 2016) to produce lactic acid from date palm byproducts. Additionally, waste dates are utilized to make lactic acid (Nancib et al., 2015).

Ziadi et al. (2011) have examined for microflora in fresh sap from date palms and for the possibility of using lactic acid bacteria as a starter culture for fermented beverage production. There were ten predominant lactic acid bacterial species. A quick rate of acidification was obtained by using lactic acid strains (autochthonous *Lactobacillus delbrueskii KH3* and *Leuconostoc mesenteriodes 5B4*). Utilizing native isolation could enhance the lactic acid beverage production process. A study was conducted on extracting and characterizing insoluble fiber from waste dates (Ahmed et al., 2013). Fiber was found to have low sodium content and to be rich in potassium, calcium, and magnesium. Date fibers showed a high capacity to hold oil and water. Date fibers are a great food ingredient because of their high mineral content.

Studies have verified that date palm kernel extract contains phytohormones. Its composition includes protein, fat, carbohydrates, 10 distinct amino acids, numerous fatty acids (C12–C20), and numerous minerals like calcium and magnesium. Bauza et al. (2002) have shown that applying 5% date palm kernel twice a week to the eye area decreases aging. Date palm seed oil from the Zahidi, Medjool, Halawy, and Dekel Noor cultivars (Devshony et al., 1992) was processed into methyl esters, and the methyl esters containing fatty acids were examined. In certain cosmetic creams, shampoos, and shaving bar soaps, date seed oil has been effectively substituted for other vegetable oils.

The wide range of applications of natural wax, including cosmetics, coating, and lubricants, has increased the demand for optimizing natural wax extraction. Al Bulushi et al. (2018) conducted experiments to extract natural wax from date palm leaflets. With a value of € 3.78/kg, they were able to economically extract date wax, as opposed to € 7.8/kg, € 2.4/kg, and € 6.2/kg for candelilla, carnauba, and beeswax, respectively. In tilapia feed, waste dates have been used in place of dietary starch (Belal and Al-Jasser, 1997), date palm seeds as a diet for African catfish (Sotolu et al., 2014), dairy cows fed date palm leaflets as roughage (Bahman et al., 1997), date palm leaves and cardboard ensilage to improve the quality of Omani sheep's meat (Mahgoub et al., 2007).

Application of Date Palm in Pharmaceuticals

In Mesopotamia or western India, probable places of origin, dates are also used as part of the medicine. Date palm-based formulations have been mentioned in Ayurvedic texts like Charaka Samhita, Susruta Samhita, Bhavprakash Nighantu, Raj Nighantu, Dhanvantari Nighantu and Kayyadev Nighantu. Date fruit, leaf, flower, seed, and gum use are documented. Fruit has several medicinal uses, including as a tonic for leprosy, thirst, bronchitis, asthma, exhaustion, tuberculosis, fever, vomiting, loss of awareness, and vata disease. Leaves have aphrodisiac and liver-beneficial attributes. Flowers are bitter in taste, and their syrup acts as an expectorant and purgative benefiting the liver and relieving fever and blood complaints (Ahmad et al., 2013). The paste or powder made from seeds is used as a demulcent, expectorant, laxative, nutrient, and treatment for wounds, lesions, inflammation, and gonorrhea, even advised medically for asthmatic patients. Gum is a helpful treatment for genito-urinary system disorders and diarrhea (Ahmad et al., 2013). In Ayurvedic formulations, date, plant parts along with others are used to treat ailments Cough (Vrindamadhav 11.7), Vomiting (Charaka Samhita Chikitsasthan 20.28), Haematuria (Vrindamadhav 35.11), alleviates hoarseness of voice, cough, asthma, and fever (Charaka Samhita Chikitsasthan 8.96), Grahini Roga (Charaka Samhita Chikitsasthan 15.15), abnormal taste (Sushruta Samhita Uttarasthan 39.185), hiccough (Sushruta Samhita Uttarasthan 50.27–28), intrinsic hemorrhage (Sushruta Samhita Uttarasthan 45.20 and Sharma, 2004). In the Middle East and North Africa, including Tunisia, date fruit is a staple food and is also used as a constipation remedy (Trifi et al., 2017).

Exhibiting antimicrobial activity is an inherent property of many plant extracts. Methanol date seed extract has been shown to have antimicrobial activity against bacteria (Jaganathan et al., 2018; El Sohaimy et al., 2015). Ripe date fruit extract has demonstrated hepatoprotective (Saafi et al., 2011). Al-Yahya et al. (2016) have demonstrated the cardioprotective effects of lyophilized aqueous date fruit extract (Ajwa variety) both *in vivo* and *ex vivo*, and they discovered that it inhibits lipid peroxidation, stops endogenous antioxidant consumption, and increases cardiomyoblast cell proliferation by up to 40%. Date fruit acetone/water extract has demonstrated antiatherogenic properties (Borochov-Neori et al., 2015). Date palm extracts have shown a possible promising effect on male infertility (Shehzad et al., 2021).

It was observed that date palm seed oil exhibited antioxidant activity (Boukouada et al., 2014; Mohamed Lemine et al., 2014; Al-Farsi et al., 2005), antimicrobial activity against fungi and bacteria (Golshan Tafti et al., 2017; Trusted Health Products, 2018), UV protective potential (Dammak et al., 2010), diminished oxidative stress caused by H_2O_2 in keratinocytes and epidermal melanocytes (Dammak et al., 2010; Dammak et al., 2007; Ines et al., 2010). Additionally, due to the presence of omega-3 fatty acids, date seed oil is said to be beneficial for hair growth (Perkins, 2016). Women use date palm seed oil in the postpartum period to help restore the uterus's position and balance hormones after giving birth (Lubna, 2020).

Other Applications of Data Palm

Date palm and its byproducts are being tested for wider applications. By-products from date palms have found their way into the textile industry, which has experienced tremendous growth in the last few decades. Modern industry is largely dependent on synthetic fibers, which come from the exhausting resources of oil and cause grave environmental problems. Finding other alternatives as feedstock is a current area of research. Several palm parts, including the midribs, spadix stems, leaflets, mesh, and the trunk at the end of the palm's life, have been used to extract fibers. The above said are rich in cellulose (EL-Mously et al., 2023b). Two types of date palm fibers (DPF) are described by researchers: those that are embedded in the biomass, such as the midrib and spadix stem fibers, and those that are present in a fibrous form, like the leaf sheath. The second type of fibers embedded inside biomass are surrounded by lignin and hemicellulose, posing a challenge to remove the pure fibers without fracturing or harming them. Secondly, the fibers lack the necessary flexibility to be processed into any type of textile because they are hollow and coarse (vascular) (Elseify et al., 2019). Researchers (Elseify et al., 2022) have analyzed the composition of date palm spadix, and midrib (core and skin) in comparison with other plants like Abaca, Banana, and Sisal. Date palm (DP) byproducts showed comparable cellulose, midrib had a higher hemicellulose and lignin content. Thermogravimetric analysis showed DP fibers had better thermal stability and the same crystallinity index. Analysis of a single fiber tensile test revealed that DP fibers' mechanical qualities are on par with, if not superior to, those of other fibers. DP midrib core fibers are a useful reinforcement material for applications requiring a high degree of stiffness. DP Spadix can be used in areas where toughness is required.

Researchers have drawn comparisons between coconut coir and the fibre layer that envelops the date palm trunk in its leaf sheath. DP coir showed higher cellulose and low lignin content (Midani et al., 2022; Nasser et al., 2016; Mekhermeche et al., 2016). DP coir showed good thermal stability and comparable tensile strength (Abdal-hay et al., 2012). Manjula et al. (2017) have prepared a composite fiber comparison showing the need for improvement as dry spots and high void fraction lead to fiber fracture. DP composite showed the least void fraction and bending performance comparable to the other leaf fibers and showed better tensile strength. The DP fibers show lower conductivities. Acoustic absorption analysis showed by comparing with conventional synthetic porous absorbers DP midrib fibers can serve as a viable replacement for synthetic fibers at both high and low frequencies (63–315 Hz and 1000–4000 Hz) (Zunaidi et al., 2017; Gumanová et al., 2022).

For a variety of industries, including cutting-edge ones like natural fiber composites, thermal insulation panels, and biodegradable tableware, DP fiber can serve as a sustainable material basis. Beyond more conventional uses like burlap sacks and ropes, there is a well-established market for things like carpets and backing, gypsum plaster panels, and non-wood paper. The cellulose obtained from DP has been used for the reinforcement of latex (Boufi et al., 2014) showing better mechanical and optical properties.

The obtained cellulose has been used in paper (Hassan et al., 2015) for improvement. Microcrystalline cellulose (MCC) from date palm fruit bunch stalks has been extracted by researchers. Microcrystalline cellulose (MCC) has the benefit of being biodegradable, which has many uses in the food, pharmaceutical, and cosmetic industries as a suspension stabilizer and for water retention (Alotabi et al., 2020). Annual pruning yields date palm leaf sheaths, which are treated as waste, have been used for the preparation of oxidized nanocellulose (Adel et al., 2018) which have good thermal characteristics and water vapor resistance of the bio-nanocomposites. Cellulose whiskers have been extracted from DP rachis and leaflets for the preparation of nanocomposites (Bendahou et al., 2009). Dry tensile strength than those produced from non-treated fibers. In comparison to micro-fibrillated cellulose made from non-treated fibers, date palm fruit stalks treated with an enzyme demonstrated reduced porosity, a more compact and tight structure, and a higher wet and dry tensile strength (Hassan et al., 2014).

Date palm midribs have been used as lumber substituent (Wang and Sun, 2002), furniture (Drake, 1997), eco-friendly wood substitute (Shafie and Zarea-Hosseinabadi, 2019), core blockboards

(Kinawy, 1997) and oriented strand board (Hegazy et al., 2015). Based on the physical properties of the midrib of date palm (El-Mously et al., 1993) they have been used for making chipboard panels and particleboards (Iskanderani, 2008). Particle-boards have also been produced using date palm rachises and trunks (Amirou et al., 2013). Date palm petioles have also been used for the preparation of sandwich cores (Ashori, 2010).

By-products of the date palm are used in construction; for example, leaf sheaths are a key component of mud blocks and are used as a stabilizer, filler, and heat-conservation enhancer (El-Mously and Darwish, 2020). Research has been conducted on the use of midrib for wide-span construction and low-cost solar shading façade (EL-Mously et al., 2023c). Date palm byproducts along with concrete have been tried as alternatives to steel and synthetic fibers (Aljalawi, 2019; Saad et al., 2022). Fibers obtained from the date palm's midribs and petioles (Kareche et al., 2020) led to less cracking of construction material during the drying process. Fibers derived from leaves have been used for the creation of high strength concrete (Saad et al., 2022). Research conducted using fibers from the date palm surface around the trunk was combined with the concrete to improve its ductility and, more significantly, lower its heat conductivity (Kriker et al., 2005).

Because date palm fibers have low thermal conductivity (Boudjemaa et al., 2011) they have been tested for thermal insulation properties. Fibers were obtained from petioles and midrib experiments in mortars (Samir et al., 2014). To create a composite with improved thermal insulation that can be utilized in buildings or solar thermal panels made of date palm fiber and gypsum material, fibers taken from petioles and midribs are being studied. Research is being done on the surface fibers that are taken from the female palm trunks in order to prepare an inexpensive insulation material. Studies have also been carried out in the preparation of polyester composite insulation material, wood cement composite panels, reinforced mortar blocks, and reinforced raw earth bricks (EL-Mously et al., 2023c), which increased the performance. Date palm fibers' inclusion into structural, insulation composites, and brick mixture has great potential. Date palm byproducts have also been used to produce organic fertilizers (Sadik et al., 2012) with soil stabilization properties. A study (Elouaqoudi et al., 2015) has been conducted to understand the microbial composition involved during composting of lignocellulosic residue. Other optimizations (EL-Mously et al., 2023c) concluded that tracking steroidal compounds improved knowledge of the role played by microbial flora in influencing fertilizer quality. Date palm midribs have been used to obtain charcoal (El-Mahrouky et al., 2015) and biochar (Usman et al., 2015).

Date palm byproducts, such as trunk fibers for the removal of $Cd2^{+}$ ions (Al-Ghamdi et al., 2013), date palm seed for the removal of phenol (Suresh Kumar Reddy et al., 2014), and date palm leaflets for the removal of Cu^{2+} and Ag^{+} (El-Shafey and Al-Kindy, 2013), have all been used for the treatment of wastewater. Chemical investigation of date palm fronds indicates they can be good source of cellulose, which has been used for production of lignin and bioethanol (EL-Mously et al., 2023c), ethanol, butanol, and acetone from spoilt date palm fruits (Abd-Alla and Elsadek El-Enany, 2012), biodiesel from date palm seeds (Amani et al., 2013), date pits for the production of jet fuel and green diesel fractions (Al-Muhtaseb et al., 2016), relatively good fuel source for power generation (Sabra et al., 2014) and biogas (Lattieff, 2016). Date palm midribs are also used to produce furfural and side products such as furfural alcohol, 5 Hydroxymethyl furfural (5HMF), and 5 Methyl furfural. These have extensive use in the production of tetrahydro furfuryl alcohol and various synthetic fibers, rubbers, resins resistant to acids and bases, and resins used for strengthening ceramics, as chemical intermediates in the manufacture of lysine, vitamin C, lubricants, and dispersing agents.

Furfural and its byproducts, including furfuryl alcohol, 5 Hydroxymethyl furfural (5HMF), and 5 Methyl furfural, are also produced from date palm midribs (Bamufleh et al., 2013). These are widely used as chemical intermediates in the synthesis of lysine, vitamin C, lubricants, dispersing agents, rubbers, and resins resistant to acids and bases, as well as in the production of tetrahydro furfuryl alcohol and other synthetic fibers, rubbers, and resins used to strengthen ceramics (Huang et al., 2007; Merlo et al., 2009). Hemicellulose has been extracted from date palm leaflets and rachis (Bendahou et al., 2007), while cellulose has been extracted from petioles and rachis (Ramzi et al.,

2010). Additionally, date palm seed powder has been tested by researchers to seal oil well fractures (EL-Mously et al., 2023c). The impact of employing a perforated plate on the date palm leaf sheath fibers' ability to absorb sound has been investigated (Elwaleed et al., 2013).

Conservation of Genetic Diversity of Date Palm

Significance of Conserving the Genetic Diversity

The date palm not only contributes significantly to the agricultural economy of certain parts of the world but also holds commendable socio-cultural importance in such areas, notably in North African countries and the Middle East, where the fruit has been culturally revered and has sustained the livelihood of farmers for thousands of years (El Hadrami and Al-Khayri, 2012). In the past few decades, climate change has negatively impacted the continued existence of several species of flora and fauna has increasingly become evident, with as many as 10,967 species being threatened directly with extinction, as per the latest IUCN records (IUCN, 2021). Unfortunately, the date palm is no exception to this worrying trend. Predictive models have already suggested that many areas in North Africa and certain areas in Saudi Arabia, Iraq, and Iran will become unsuitable for sustained agricultural production of dates in the coming few decades, leading to serious ecological and even economic consequences (Shabani et al., 2012). Moreover, when focusing on the economic aspect, it is to be noted that due to impoverished monoculture as a result of the cultivation of specific high-yielding cultivars, there is now a lack of genetic diversity among farmed varieties, which has placed the perenniality of the crop itself under threat (El Hadrami and Al-Khayri, 2012). Thus, it is a cause for worry that the genetic diversity of the date palm is being threatened not only by climate change but by irresponsible agricultural practices as well, leading to a severe threat to the species. However, it is worth noting that the same genetic diversity which is being threatened can also be leveraged to save the entire species from extinction due to the different varieties capable of showing different levels of adaptability to inhabitable conditions. Furthermore, the genetic variability of the species provides the raw material for breeders to develop improved cultivars with traits like disease resistance, increased yield, and enhanced fruit quality, which are both ecologically and economically desirable (Al-Khayri et al., 2018; Naqvi et al., 2023). Hence, there is considerable impetus for preserving the genetic diversity of date palms, as this would not only help arrest the threat imposed by climate change in the short term but would also provide the foundational work for future breeding programs to address food security issues and the demand for desirable yields. In this regard, adjustment of cultivation practices to enhance resilience to climate change plays a vital role in date palm sustainability (Abul-Soad et al., 2023).

Conservation Methods

Having established the need for conservation of the genetic diversity of the date palm with regards to the immediate threat it faces due to climate change, it is also important to look at the various methods in which this conservation is being carried out presently and the potential for future conservation. At the most basic level, the various conservation methods for any biological species can be categorized under two broad headings: *ex situ* and *in situ*. *Ex situ* refers to conserving the core components of biological diversity of any species outside their natural habitat, usually in controlled laboratory environments. *In situ*, on the other hand, refers to maintaining and recovering viable populations of the concerned species in its natural habitat (Maxted, 2001). *In situ* methods achieve conservation of genetic diversity through the preservation of the ecosystems where the species is found as a whole, to allow the species adequate natural resources to nurture its growth and propagation. There are several different ways in which *ex situ* and *in situ* conservation is practiced, providing a wide range of possibilities for protecting the genetic diversity of date palms.

Ex-situ Conservation Methods

Ex situ conservation methods are increasingly becoming the go-to method for combating genetic erosion in plant species, partly owing to the fact that in such methods, the focus is solely on preserving the genetic diversity itself rather than the preservation of the natural habitat of the plant as a whole (Engels and Ebert, 2021). Moreover, many of these methods utilize efficient methods like seed preservation or DNA preservation, leading to fewer resources than *in situ* conservation. The primary resources used for *ex situ* conservation in date palm currently are:

GeneBanks: Genebanks have been in prevalence for more than half a century, with the explicit aim to preserve the genetic material of plants, and in turn, protect the genetic diversity of both wild and domestic landraces. This is achieved through either the preservation of seeds, genetic material, or the plant as a whole or cuttings in controlled environments (Díez et al., 2018). Collection of samples for gene banks can happen in both random and non-random sampling methods, with the latter helping in the selection of specific cultivars which may have high yield, better stress resistance, and other such desired traits (Hasan et al., 2022). The most useful utility of establishing such genebanks is that not only can the collected samples be conserved to preserve genetic diversity, but they can also serve as initial resources to help populate a fresh geographical area with a selected cultivar (Díez et al., 2018; Hasan et al., 2022). However, gene banks present challenges as well. In field gene banks, the plants are exposed to the outside environment, and are subject to both abiotic and biotic stresses (Panis et al., 2020). In seed gene banks, the ideal lower temperature required for conservation can also lead to desiccation and lowering of germination potential (Jain, 2011). Despite these challenges, most countries with date palms as one of their major economic crops have invested in establishing and maintaining genebanks. The noteworthy ones are Saudi Arabia's field genebank at King Faisal University, Tunisia's genebank at the Centre de Recherches Phoénicicoles, Kuwait's genebank at the Kuwait Institute for Scientific Research, as well as Iraq, Nigeria, and Pakistan (Hasan et al., 2022). In addition to these, efforts have been also made to establish genebanks for date palms in areas far away from the center of origin, in an effort to promote genetic as well as geographical diversity. These include the United States of America (USDA-ARS National Germplasm Repository for Citrus and Dates), and India (Central Institute for Arid Horticulture, CIAH) (Abul-Soad et al., 2017).

Botanical Gardens: Botanical gardens may not seem very different from oasis gardens where date palms are cultivated traditionally, but they serve a completely different purpose. While oasis gardens are maintained for commercial purposes, botanical gardens have evolved from leisure areas for the elite to modern day conservatories, helping to fulfill a growing global need for the preservation of the genetic diversity of plants, particularly overlapping the fields of horticulture, living collections, plant science as well as public education (Westwood et al., 2021). In the context of the date palm, botanical gardens were, in fact, the first documented evidence of an instance of *ex situ* conservation of the plant, when German scientist Engler established a Palm House in the Botanic Garden and Botanical Museum in Berlin-Dahlem in the late 19th century (Griffith et al., 2021). In recent years, the most famous example of date palm conservation via botanical gardens is the Budaiya Botanical Garden in Bahrain, which houses numerous varieties of plants for conservation and research purposes (Almansoori et al., 2015). This example, and several others worldwide, also highlight the research potential that ex situ conservation in botanical gardens provides, including the possibility to produce various hybrids specifically bred for desirable qualities like high yield (Perez et al., 2018).

Cryopreservation: Cryopreservation is arguably the most refined method of ex-situ preservation of plant species available currently, thanks to path-breaking advances in in-vitro conservation techniques in the past few years. As the name suggests, cryopreservation encompasses preserving the vegetal structure of plants, like tissues, cells, or even organs, in liquid nitrogen at a temperature of –196°C, ensuring almost indefinite preservation under controlled conditions and viability after thawing (Benelli, 2021). Particularly in the case of date palm, previous attempts at germplasm preservation via conventional means have proved tedious and futile, and vegetative preservation, like in field gene banks and botanical gardens, presents challenges in the way of environmental exposure, pests, and pathogen infection (Al-Bahrany and Al-Khayri, 2012). In fact, research has already been carried out to optimize cryopreservation techniques to suit the conservation of calli of date palm and successful regeneration after thawing in countries like Saudi Arabia, with a maximum reported survival rate of 86.7% following treatment with a novel concoction named PVS2 for vitrification (rapid freezing) by Alansi et al. (2019). This is a breakthrough because since the date palm possesses recalcitrant seeds, the seeds are unsuitable for cryopreservation, and saving tissue and callus samples is the only way forward (Nguyen et al., 2023). As this is a fledgling field, research is being conducted to further optimize cryopreservation techniques in various countries, with successful cryopreserved samples being maintained in Tunisia (Hamza et al., 2015) and foundational work for future conservation also beginning in India at the National Bureau of Plant Genetic Resources (NBPGR), New Delhi (Pareek, 2015).

In situ Conservation Methods

In situ conservation entails the on-site preservation of genetic resources of a plant species in natural populations. This is achieved via holistic principles of biodiversity and ecological balance and encompasses special "protected areas" like biosphere reserves and national parks (Ajayi, 2019). However, such conservation methods are usually not preferred over *ex situ* methods due to logistical constraints and the threat of exposure to environmental stress. Largely due to such hurdles and the ease of *ex situ* conservation methods, efforts to maintain date palm genetic resources via *in situ* methods have not received much attention and have historically not been included in reviews of the current trends and status of *in situ* conservation in plant species around the world (Jaradat, 2015). There are still certain benefits of practicing in-situ conservation, especially from an ecological perspective. In the context of date palms, the primary resources of *in situ* conservation being utilized include protected areas, and traditional agroforestry systems.

Protected Areas: Protected areas for the conservation of any species, both plants and animals, have become extremely important in the past few decades, not only as a means of conserving the genetic resources and diversity of the natural population but also to protect the natural habitat as well (Heywood, 2019). This is because climate change affects the flora and fauna as well as the ecological balance of the world and can only be reversed through the preservation of natural ecosystems as a whole. Ideally speaking, this means that designated protected areas, such as reserves and national parks, can help conserve the genetic diversity of date palms and their associated ecosystems because these protected areas provide a haven for wild date palm populations, allowing them to evolve and adapt in response to natural selection pressures. In reality, very little efforts have been made to designate protected areas for date palms, and the onus for current conservation efforts mostly falls on the oases and gardens where date palms are cultivated commercially, for example, the Barhi and Medjool cultivars in Egypt and their involvement in crop improvement programs (Bekheet and El-Sharabasy, 2015). In the same vein, active efforts have been made to preserve the already existing oases from environmental exposure and pests, leading to an indirect method of ensuring *in situ* conservation (Issa et al., 2021). Furthermore, orchards and gardens in countries where date palms are not native have also contributed to the *in situ* conservation efforts, like the Elche Palm Grove in Spain, (Rivera et al., 2015), and Big Bend National Park, Texas, USA (Johnson and Macknight, 2023), both being listed as UNESCO World Heritage Sites.

Traditional Agroforestry Systems: Traditional agroforestry systems incorporating date palms play a significant role in conserving genetic diversity. These systems often involve intercropping date palms with other crops, creating dynamic agricultural landscapes that support various plant species in higher densities and maintain local genetic variability, setting them apart from other basic agricultural practices (Bourou et al., 2015). In Egypt, the potential for date palm as ideal candidates for agroforestry has been identified since they can withstand harsh climatic conditions and provide shade and protection to other small plants via intercropping (Bekheet and El-Sharabasy, 2015). In Spain, the early 20th-century practice of agroforestry with date palm proved particularly useful in forest restoration and stabilization of the dunes of Guardamar del Segura, Alicante (Rivera et al., 2015). Unfortunately, these remain as the only few examples of proper agroforestry conservation efforts involving date palms. Despite having thousands of years of natural history and agricultural and cultural presence in many Middle Eastern and North African countries, the agroforestry potential of date palms is yet to be fully realized, and nascent stages of research and development currently provide the hope and prospects of future utilization across the world (Al-Khayri et al., 2015b).

Conclusion and Prospects

In conclusion, this present study provides an extensive overview of various features of date palm genetic resources. It serves as a valuable guide for practitioners worldwide involved in date palm cultivation. The study catalogs various date palm varieties and their genetic variations to aid cultivar selection. It discussed various methodologies for identifying optimal varieties and emphasized the need for cultivation practices that optimize yield while minimizing environmental impact. The study comprehensively examines many factors influencing genetic diversity and enlists various techniques employed in genetic analysis to effectively utilize the fullest potential of date palm genetic resources for its sustainable utilization. Additionally, the study explores alternative propagation methods, such as plant tissue culture and modern breeding strategies to enhance date palm propagation and develop high-quality date palms with exceptional traits.

In the near future, studies on date palm genetics hold great potential. With an understanding of genetic diversity and genetic cultivars and incorporating wild date palm varieties into breeding programs, new genetic traits and diversity in cultivated populations with high-productivity varieties such as drought-resistant, pest-resistant, and drought-resistant varieties can be introduced. Omics technologies, genome sequencing techniques, and gene editing techniques like CRISPR-Cas9 can serve as potential tools for high-yielding, disease-resistant, and climate-adaptive date palm varieties with superior quality cultivars by providing information on date palm genetics and enable precise genetic modifications for desirable traits. Date palm cultivation can be made more environmentally friendly using sustainable agricultural techniques like intercropping and organic farming.

Acknowledgments

The authors express their appreciation for the support provided by the Deanship of Scientific Research, Vice Presidency for Graduate Studies and Scientific Research, King Faisal University, Saudi Arabia [Project No. KFU241297].

References

Alrumman, Sulaiman A. 2016. Enzymatic saccharification and fermentation of cellulosic date palm wastes to glucose and lactic acid. *Braz. J. Microbiol.*, *47*: 110–119.

Abdallah, A.B., N.A. Al-Wusaibai, and Y. Al-Fehaid, 2014. Assessing the efficiency of sponge and traditional methods of pollination in Date palm. *J. Agr. Sci. Technol.*, B.4: 267–271.

Abdal-hay, A., N.P.G. Suardana, D.Y. Jung, K.-S. Choi, and J.K. Lim. 2012 Effect of diameters and alkali treatment on the tensile properties of date palm fiber reinforced epoxy composites. *Int. J. Precis. Eng. Manuf.*, *13*: 1199–1206.

Abd-Alla, M.H., and A.-W. Elsadek El-Enany. 2012. Production of acetone-butanol-ethanol from spoilage date palm (*Phoenix dactylifera* L.) fruits by mixed culture of *Clostridium acetobutylicum* and *Bacillus subtilis*. *Biomass Bioenergy*, *42*: 172–178.

Abdel-Baky, N.F., M.I. Motawei, A.A.S. Al-Nujiban, M.A. Aldeghairi, L.A.M. Al-Shuraym, M.T.M. Alharbi, A.S. Alsohim, and M. Rehan. 2023. Detection of adaptive genetic diversity and chemical composition in date palm cultivars and their implications in controlling red palm weevil, *Rhynchophorus ferrugineus* Oliver. *Braz. J. Biol., 83*: e270940.

Abdel Salam, M., T.K.R. El-Mahdy, and A.M. El-sisi. 2022. Effect of some growth regulators on yield and fruit quality of sewy and hayany date palm cultivars. *Assiut. J. Agric. Sci., 53*: 65–76.

Abdulwahid, A.H. 2017. Mechanisms of date palm *Phoenix dactylifera* salt tolerance, effect of water salinity on some cell wall enzymes activity during development of Sayer date palm fruit. *Bioflux., 9*: 154–216.

Abou-Zeid, A.-Z.A. 1991. Increasing the protein content of palm by-products. *Bioresour. Technol., 37*: 239–242.

Abul-Soad, A.A., S.M. Jain, and M.A. Jatoi. 2017. Biodiversity and conservation of date palm. *In*: *Sustainable Development and Biodiversity*, 313–353. Springer International Publishing, Cham.

Abul-Soad, A.A., S.M. Mahdi, and G.S. Markhand. 2015. Date palm status and perspective in Pakistan. *In*: Al-Khayri, J.M., Jain, S.M., and Johnson, D.V. [Eds.], *Date PGlm genetic Resources and Utilization: Vol. 2: Asia and Europe*, 153–205. Springer Netherlands, Dordrecht.

Abul-Soad A.A., N.H. Mohamed Salomón-Torres R., J.M., Al-Khayri, 2023. Cultivation of date palm for enhanced resilience to climate change. *In*: Abul-Soad, A.A., Al-Khayri, J.M. [Eds.], *Cultivation for Climate Change Resilience, Vol 2. Temperate Fruit Trees*, 138–169. CRC Press, Taylor & Francis Group, LLC. Boca Baton, FL, USA.

Acourene, S., K. Djafri, A. Ammouche, L. Amourache, A. Djidda, M. Tama, and B. Taleb. 2011. Utilisation of the date wastes as substrate for producing baker's yeast and citric acid. *BioTechnology, 10*: 488–497.

Adawy, S., M.M., Saker, A.M.I.N.A Mohamed, H.A. Elitriby. 2004. Genetic relationships of upper Egypt date palm cultivars (*Phoenix dactylifera* L.): II. as revealed by AFLP markers. *In*: *Proceed. Int. Conf. Genet. Eng. Appl.*, 181–195. Sharm El-Sheikh, South Sinai, Egypt.

Adel, A., A. El-Shafei, A. Ibrahim, and M. Al-Shemy. 2018. Extraction of oxidized nanocellulose from date palm (*Phoenix dactylifera* L.) sheath fibers: Influence of CI and CII polymorphs on the properties of chitosan/bionanocomposite films. *Ind. Crops Prod., 124*: 155–165.

Ahmad, A., S. Soni Dutta, K. Varun Singh, and M.K. Santosh. 2013. *Phoenix dactylifera* Linn. (Pind Kharjura): A Review. *Int. J. High Risk Behav. Addict., 4*: 447–451.

Ahmad, R., H.M., Ali, A. Lisek, W.F.A. Mosa, S. Ercisli, and M.A. Anjum. 2023. Correlation among some phenological and biochemical traits in date palm (*Phoenix dactylifera* L.) germplasm. *Front Plant Sci., 14*: 1118069.

Ahmed, J., A. Almusallam, and S.N. Al-Hooti. 2013. Isolation and characterization of insoluble date (*Phoenix dactylifera* L.) fibers. *LWT - Food Sc. Technol., 50*: 414–419.

Ahmed, T.A., and A.Y. Al-Qaradaw. 2008. Molecular phylogeny of Qatari date palm genotypes using simple sequence repeats markers. *Biotechnol., 8*: 126–131.

Ahmed, T., S. Al-Hadidi, and A. Al-Qaradawi. 2012. Inter- and intra-specific genetic variations among Qatari date palm cultivars using inter simple sequence repeat (ISSR) markers. *In*: *Qatar Foundation Annual Research Forum, 2012*(1): EEP87. Hamad bin Khalifa University Press (HBKU Press).,

Ajayi, S.S. 2019. Principles for the management of protected areas. *In*: Ajayi, S.S. [Ed.]. *Wildlife Conservation in Africa*, 85–93. Academic Press.

Akkak, A., V. Scariot, D. Torello Marinoni, P. Boccacci, C. Beltramo, and R. Botta. (2009) Development and evaluation of microsatellite markers in *Phoenix dactylifera* L. and their transferability to other Phoenix species. *Biol. Plant, 53*: 164–166.

Alam, M.Z., S. Al-Hamimi, M. Ayyash, C.T. Rosa, E.M. Yahia, S. Haris, A.H. Al-Marzouqi, and A. Kamal-Eldin. 2023. Contributing factors to quality of date (*Phoenix dactylifera L.)* fruit. *Sci. Hortic., 321*: 112256.

Alansi, S., F. Al-Qurainy, M. Nadeem, S. Khan, M. Tarroum, A. Alshameri, and A.-R.Z. Gaafar. 2019. Cryopreservation: A tool to conserve date palm in Saudi Arabia. *Saudi J. Biol. Sci., 26*: 1896–1902.

Al-Bahrany, A.M., and J.M. Al-Khayri. 2012. Optimizing in vitro cryopreservation of date palm (*Phoenix dactylifera* L.). *Biotechnol., 11*: 59–66.

Al Bulushi, K., T.M. Attard, M. North, and A.J. Hunt. 2018. Optimisation and economic evaluation of the supercritical carbon dioxide extraction of waxes from waste date palm (*Phoenix dactylifera*) leaves. *J. Clean Prod., 186*: 988–996.

Al-Enezi, N.A., and J.M. Al-Khayri, 2012a. Alterations of DNA, ions and photosynthetic pigments content in date palm seedlings induced by X-irradiation. *Int. J. Agric. Biol., 14*: 329–336.

Al-Enezi, N.A. and J.M. Al-Khayri, 2012b. Effect of X-irradiation on proline accumulation, growth and water content of date palm (*Phoenix dactylifera* L.) seedlings. *J. Biol. Sci., 2*: 146–153.

Aldhaheri, A., G. Alhadrami, N. Aboalnaga, I. Wasfi, and M. Elridi. 2004. Chemical composition of date pits and reproductive hormonal status of rats fed date pits. *Food Chem., 86*: 93–97.

Aleid, S.M., B.H. Hassan, S.A. Almaiman, S.H. Al-Kahtani, and S.M. Ismail. 2014. Microbial loads and physicochemical characteristics of fruits from four Saudi date palm tree cultivars: Conformity with applicable date standards. *Food Nutr. Sci., 05*: 316–327.

Al-Ghamdi, A., H. Altaher, and W. Omar (2013) Application of date palm trunk fibers as adsorbents for removal of Cd+2 ions from aqueous solutions. *J. Water Reuse Desalin., 3*: 47–54. https:// doi.org/10.2166/wrd.2013.031.

Al-Faifi, S.A., H.M. Migdadi, S.S. Algamdi, M.A. Khan, M.H. Ammar, R.S. Al-Obeed, M.I. Al-Thamra, E.H. EL-Harty, and J. Jakse. 2016. Development, characterization, and use of genomic SSR markers for assessment of genetic diversity in some Saudi date palm (*Phoenix dactylifera* L.) cultivars. *Electron. J. Biotechnol.*, *21*: 18–25.

Al-Farsi, M., C. Alasalvar, A. Morris, M. Baron, and F. Shahidi. 2005. Comparison of antioxidant activity, anthocyanins, carotenoids, and phenolics of three native fresh and sun-dried date (*Phoenix dactylifera* L.) varieties grown in Oman. *J. Agric. Food Chem.*, *53*: 7592–7599.

Al-hajjaj, H.S., and J.Y. Ayad. 2018. Effect of foliar boron applications on yield and quality of Medjool date palm. *J. Appl. Hortic.*, *20*: 182–189.

Al-Hamdani, K.A. 2016. Effects of Humic acid and k-Humate treatment on vegetative growth, fruits quantitative characters and nutrients content of date palm cv. Khastawi grown in Gypsifrious Soil. *DASJ*, *8*: 218–231.

Alhammadi, M.S., and G.P. Edward. Effect of salinity on growth of twelve cultivars of the United Arab Emirates date palm. Commun. *Soil Sci. Plant Anal.*, *40*: 2372–2388.

Al-Hinai, M., A. Al-Lawati, A. Al-Mamari, A. Al-Jabri, I. Al-Harrasi, A. Al-Amri, Y. Al-Raisi, and M. Ben Salah. 2022. Fingerprinting of Omani date palm cultivars. *In*: *VII International Date Palm Conference 1371*: 209–216. actahort. org, .

Alikhani-Koupaei, M., R. Fatahi, Z. Zamani, and S. Salimi. 2018. Effects of deficit irrigation on some physiological traits, production, and fruit quality of "Mazafati" date palm and the fruit wilting and dropping disorder. *Agric. Water Manage.*, *209*: 219–227.

Aljalawi, N.M.F. 2019. Effect of sustainable palm fiber on high strength concrete properties. *IOP Conf. Ser. Mater. Sci. Eng.*, *518*: 022004.

Al-Khalifah, N.S., E. Askari, and A.E. Shanavas Khan. 2012. Molecular and morphological identification of some elite varieties of date palms grown in Saudi Arabia. *Emir J. Food Agric.*, 456–461.

Al-Khashman, O.A., A.H. Al-Muhtaseb, and K.A. Ibrahim. 2011. Date palm (*Phoenix dactylifera* L.) leaves as biomonitors of atmospheric metal pollution in arid and semi-arid environments. *Environ. Pollut.*, *159*: 1635–1640.

Al-Khateeb, S.A., A.A. Al-Khateeb, M.N. Sattar, and A.S. Mohmand. 2020. Induced *in vitro* adaptation for salt tolerance in date palm (*Phoenix dactylifera* L.) cultivar Khalas. *Biol. Res.*, *53*: 37.

Al-Khayri, J.M., S.M. Jain, and D.V. Johnson. 2015a. *Date Palm Genetic Resources and Utilization: Volume 2: Asia and Europe*. Springer Netherlands.

Al-Khayri, J.M., S.M. Jain, and D.V. Johnson. 2015b. *Date Palm Genetic Resources and Utilization: Volume 1: Africa and the Americas*. Springer Netherlands.

Al-Khayri, J.M., D.V. Johnson, N.S. Al-Khalifah, and S.M. Jain. 2012. Special issue on date palm papers presented at the Arab Palm Conference 2011. *Emir J. Food Agric.*, *24*: 1.

Al-Khayri J.M., P.M. Naik, S.M. Jain, D.V. Johnson. 2018. Advances in date palm (*Phoenix dactylifera* L.) breeding. *In*: Al-Khayri, J.M., Jain, S.M., and Johnson, D.V. [Eds.], *Advances in Plant Breeding Strategies (Vol. 3) Fruits*. 727–771. Springer, Switzerland

Allam, A., K. Djafri, M. Bergouia, E.-H. Khemissat, M. Tama, and B. Taleb. 2021. Morphological and physicochemical characterisation of date palm cultivars from Ghardaïa (southeast Algeria). *Spring*, *185*: 12–24.

Allbed, A., L. Kumar, and F. Shabani. 2017. Climate change impacts on date palm cultivation in Saudi Arabia. *J. Agric. Sci.*, *155*: 1203–1218.

Almansoori, T.A., M.A. Al-Khalifa, and A.M.A. Mohamed. 2015. Date Palm Status and Perspective in Bahrain. *In*: Al-Khayri J.M.,, Jain, S.M., and Johnson, D.V. [Eds.], *Date Palm Genetic Resources and Utilization (Vol. 2) Asia and Europe*, 353–386. Springer Netherlands, Dordrecht.

Al-Muaini, A., S. Green, A. Dakheel, A.-H. Abdullah, W.A. Abou Dahr, S. Dixon, P. Kemp, and B. Clothier. 2019. Irrigation management with saline groundwater of a date palm cultivar in the hyper-arid United Arab Emirates. *Agric. Water Manage.*, *211*: 123–131.

Al-Muhtaseb, A.H., F. Jamil, L. Al-Haj, M.A. Al-Hinai, M. Baawain, M.T.Z. Myint, and D. Rooney. 2016. Efficient utilization of waste date pits for the synthesis of green diesel and jet fuel fractions. *Energy Convers. Manage.*, *127*: 226–232.

Alotabi, M.D., B.A. Alshammari, N. Saba, O.Y. Alothman, L.K. Kian, A. Khan, and M. Jawaid. 2020. Microcrystalline cellulose from fruit bunch stalk of date palm: Isolation and characterization. *J. Polym. Environ.*, *28*: 1766–1775.

Alqaisi, M.R.M., A.S. Al-Jumaili, and H.M. Khlafa. 2022. Molecular detection of date palm (*Phoenix dactylifera* L.) at the vegetative stage by using DNA ISSR marker for cultivar identification. *AIP Conf. Proc.*

Al-Qurainy, F., S. Khan, F.M. Al-Hemaid, M.A. Ali, M. Tarroum, and M. Ashraf. 2011. Assessing molecular signature for some potential date (*Phoenix dactylifera* L.) cultivars from Saudi Arabia, based on chloroplast DNA sequences rpoB and psbA-trnH. *Int. J. Mol. Sci.*, *12*: 6871–6880.

Alrumman, S.A. 2016. Enzymatic saccharification and fermentation of cellulosic date palm wastes to glucose and lactic acid. *Braz. J. Microbiol.*, *47*(1): 110–9. doi: 10.1016/j.bjm.2015.11.015

Al-Sahil, A.A. and A.M.A. Al-Rawi.1987. A study of the cytology of two female cultivars of date palm. *Date Palm J.*, *15*: 123–142.

Al-Senaidy, A.M., and M.A. Ismael. 2011. Purification and characterization of membrane-bound peroxidase from date palm leaves (*Phoenix dactylifera* L.). *Saudi J. Biol. Sci.*, *18*: 293–298.

Al-Shahib, W., and R.J. Marshall. 2002. Dietary fibre content of dates from 13 varieties of date palm *Phoenix dactylifera* L. *Int. J. Food Sci. Technol.*, *37*: 719–721.

Al Shaibani, F.Y.Y. 2021 *The Effect of Natural Elicitors and Cold Storage Period on Quality Improvement of UAE Date Palm Fruits* (*Phoenix dactylifera* cv. Barhi). United Arab Emirates University.

Al-Yahya, M., M. Raish, M.S. AlSaid, A. Ahmad, R.A. Mothana, M. Al-Sohaibani, M.S. Al-Dosari, M.K. Parvez, and S. Rafatullah. 2016. 'Ajwa' dates (*Phoenix dactylifera* L.) extract ameliorates isoproterenol-induced cardiomyopathy through downregulation of oxidative, inflammatory and apoptotic molecules in rodent model. *Phytomedicine*, *23*: 1240–1248.

Amani, M.A., M.S. Davoudi, K. Tahvildari, S.M. Nabavi, and M.S. Davoudi. 2013. Biodiesel production from *Phoenix dactylifera* as a new feedstock. *Ind. Crops Prod.*, *43*: 40–43.

Amirou, S., A. Zerizer, A. Pizzi, I. Haddadou, and X. Zhou. 2013. Particleboards production from date palm biomass. *Eur. J. Wood Prod.*, *71*: 717–723.

Aaouine, M. 2003. Date palm large-scale propagation through tissue culture techniques. *In*: *The Date Palm from Traditional Resource to Green Wealth*, 79–86. Emirates Centre for Strategic Studies and Research, Abu Dhabi, United Arab Emirates.

Ashhab, A.O.A. 2012 Composting mulch of date palm trees through microbial activator in Saudi Arabia. *5*(1): 156–161. https://doi.org/10.13140/RG.2.2.21664.92169.

Ashori, A. 2010. Hybrid Composites from Waste Materials. *J. Polym. Environ.*, *18*: 65–70.

Assirey, E.A.R. 2015. Nutritional composition of fruit of 10 date palm (*Phoenix dactylifera* L.) cultivars grown in Saudi Arabia. *J. Taibah Univ. Sci.*, *9*: 75–79.

Atkinson, N.J., and P.E. Urwin. 2012. The interaction of plant biotic and abiotic stresses: From genes to the field. *J. Exp. Bot.*, *63*: 3523–3543.

Azzam, O.A. 1960. *The Development of Urban and Rural Housing in Egypt*. Ph.D., Swiss Federal Institute of Technology, Zurich.

Bael, J.M. 1937. Cytological studies on the genus *Phoenix*. *Bot. Gaz.*, *15*: 400–407.

Bahman, A.M., J.H. Topps, and J.A. Rooke. 1997. Use of date palm leaves in high concentrate diets for lactating Friesian and Holstein cows. *J. Arid Environ.*, 35: 141–146.

Bamufleh, H.S., Y.A. Alhamed, and M.A. Daous. 2013. Furfural from midribs of date-palm trees by sulfuric acid hydrolysis. *Ind. Crops Prod.*, *42*: 421–428.

Barreveld, W.H. 1993. *Date Palm Products*. Food & Agriculture Organization of the UN (FAO).

Barrow, S.C. 1998. A Monograph of *Phoenix* L. (Palmae: Coryphoideae). *Kew Bull.*, *53*: 513.

Bauza, E., C. Dal Farra, A. Berghi, G. Oberto, D. Peyronel, and N. Domloge. 2002. Date palm kernel extract exhibits antiaging properties and significantly reduces skin wrinkles. *Int. J. Tissue React.*, *24*: 131–136.

Bekheet, S. 2013. Direct organogenesis of date palm (*Phoenix dactylifera* L.) for propagation of true-to-type plants. *Sci. Agric.*, *4*: 85–92.

Bekheet, S.A., H.S. Taha, M.E. Solliman, and N.A. Hassan. 2006. Cryopreservation of date palm (*Phoenix dactylifera* L.) cultured *in vitro*. 283–291.*In*: *III International Date Palm Conference*, 736 pp.

Bekheet, S.A., and S.F. El-Sharabasy. 2015. Date Palm Status and Perspective in Egypt. In: Al-Khayri, J.M., Jain, S.M., and Johnson D.V. [Eds.], *Date Palm Genetic Resources and Utilization* (Vol. 1): *Africa and the Americas*, 75–123. Springer Netherlands, Dordrecht.

Belal, I.E.H., and M.S. Al-Jasser. 1997. Replacing dietary starch with pitted date fruit in Nile tilapia *Oreochromis niloticus* (L.) feed. *Aquac. Res.*, *28*: 385–389.

Ben Abdallah H., A. Laajimi, et al. 2020 Genetic diversity of endangered date palm (*Phoenix dactylifera* L.) in the oases of Nefzaoua, Tunisia, using SSR markers. *Fruits*, *75*: 84–91.

Bendahou, A., A. Dufresne, H. Kaddami, and Y. Habibi. 2007. Isolation and structural characterization of hemicelluloses from palm of *Phoenix dactylifera* L. *Carbohydr. Polym.*, *68*: 601–608.

Bendahou, A., Y. Habibi, H. Kaddami, and A. Dufresne. 2009. Physico-chemical characterization of palm from *Phoenix dactylifera* L, preparation of cellulose whiskers and natural rubber based nanocomposites. *J. Biobased Materials Bioenergy*, *3*: 81–90.

Benelli, C. 2021. Plant Cryopreservation: A look at the present and the future. *Plants*, *10*: 2744. https://doi.org/10.3390/plants10122744.

Bodian, A., N. Khadidiatou, N. Khadidiatou, P. Madialacké, D. Ibrahima, and S. Maurice. 2012. Analysis of genetic diversity of date palm (*Phoenix dactylifera* L.) cultivars from Mauritania using microsatellite markers. *International J. Sci. Advanced Techno.*, *2*(3): 96–104.

Borochov-Neori, H., S. Judeinstein, A. Greenberg, N. Volkova, M. Rosenblat, and M. Aviram. 2015. Antioxidant and antiatherogenic properties of phenolic acid and flavonol fractions of fruits of 'Amari' and 'Hallawi' date (*Phoenix dactylifera* L.) varieties. *J. Agric. Food Chem.*, *63*: 3189–3195.

Boudjemaa, A., B. Adel, B. Abderrahim, I. Laurent, and F. Magali. 2011. Renewable materials to reduce building heat loss: Characterization of date palm wood. *Energy Build.*, *43*: 491–497.

Boufi, S., H. Kaddami, and A. Dufresne. 2014. Mechanical performance and transparency of nanocellulose reinforced polymer nanocomposites. *Macromol. Mater. Eng.*, *299*: 560–568.

Bouguedoura, N., M. Bennaceur, S. Babahani, and S.E. Benziouche. 2015. Date Palm status and perspective in Algeria. *In*: Al-Khayri, J.M., Jain, S.M., and Johnson, D.V. [Eds.], *Date Palm Genetic Resources and Utilization (Vol. 1): Africa and the* Americas,125–*168*. Springer Netherlands, Dordrecht.

Boukouada, M., Z. Ghiaba, N. Gourine, I. Bombarda, M. Saidi, and M. Yousfi. 2014. Chemical composition and antioxidant activity of seed oil of two Algerian date palm cultivars (*Phoenix dactylifera*). *Nat. Prod. Commun., 9*: 1777–1780.

Bourou, S., N. Woin, and M.A. Elhoumaizi. 2015. Date palm status and perspective in Cameroon. *In*: Al-Khayri, J.M., Jain, S.M., and Johnson, D.V. [Eds.], *Date Palm Genetic Resources and Utilization: (Vol. 1): Africa and the Americas*, 411–427. Springer Netherlands, Dordrecht.

Cao, B.R., and C.-C.T. Chao. 2002. Identification of date cultivars in California using AFLP Markers. *HortScience, 37*: 966–968.

Chaluvadi, S.R., P. Young, K. Thompson, B.A. Bahri, B. Gajera, S. Narayanan, R. Krueger, and J.L. Bennetzen. 2019. Phylogeny, and analysis of genetic variation in a diverse collection of date palm (*Phoenix dactylifera*) and related species. *Plant Divers, 41*: 330–339.

Changtragoon, S., R. Jalonen, and A.J. Lowe. 2017. Use of molecular markers in the conservation management of tropical trees. *In*: *Sustainable Development and Biodiversity*, 155–195. Springer International Publishing, Cham.

Chao, C.T., and R.R. Krueger. 2007a. The Date Palm (*Phoenix dactylifera* L.): Overview of biology, uses, and cultivation. *HortScience, 42*: 1077–1082.

Chao, C.T., and R.R. Krueger. 2007b. The date palm (*Phoenix dactylifera* L.): Overview of biology, uses, and cultivation. *HortScience, 42*: 1077–1082. This is duplicated?

Chen, Y., H. Zhou, and Y. Chen. 2013. Adaptation strategies of desert riparian forest vegetation in response to drought stress. *Ecohydrol., 6*: 956–973.

Choudhuri, S. 2014. Fundamentals of molecular evolution. *In*: *Bioinformatics for Beginners*, 27–53. Elsevier.

Corniquel, B., and L. Mercier. 1994. Date palm (*Phoenix dactylifera* L.) cultivar identification by RFLP and RAPD. *Plant Sci., 101*:163–172.

Dammak, I., F. Ben Abdallah, S. Boudaya, L. Keskes, S. Besbes, A. El Gaied, H. Attia, H. Turki, and B. Hentati. 2007. Effects of date seed oil on normal human skin *in vitro*. *Eur. J. Dermatol., 17*: 516–519.

Dammak, I., S. Boudaya, F. Ben Abdallah, H. Turki, and H. Attia. 2010. Effect of date seed oil on p53 expression in normal human skin. *Connect Tissue Res., 51*: 55–58.

Darwish, E.A., H. El-Mously, and A. Abdelrahman. 2019. The technical heritage of date palm leaves utilization in traditional handicrafts and architecture in Egypt and the Middle East. *In*: *By-products of Palm Trees and Their Applications*. Materials Research Forum LLC.

Debnath, S.C. 2016. Genetic diversity and erosion in berries. *In*: *Sustainable Development and Biodiversity*, 75–129. Springer International Publishing, Cham.

Devshony, S., E. Eteshola, and A. Shani. 1992. Characteristics and some potential applications of date palm (*Phoenix dactylifera* L.) seeds and seed oil. *J. Am. Oil Chem. Soc., 69*: 595–597.

Diaz, S., C. Pire, J. Ferrer, and M.J. Bonete. 2003. Identification of *Phoenix dactylifera* L. varieties based on amplified fragment length polymorphism (AFLP) markers. *Cell Mol. Biol. Lett., 8*: 891–899.

Díez, M.J., L. De la Rosa, I. Martín, L. Guasch, M.E. Cartea, C. Mallor, J. Casals, et al. 2018. Plant Genebanks: Present situation and proposals for their improvement. The Case of the Spanish Network. *Front. Plant Sci., 9*: 1794.

Drake, P.A. 1997. The composite panel industry: A global market assessment. *In*: *Proceedings of 31st International Particleboard/Composite Materials Symposium*, 29–39. Washington State University, Pullman.

El-Assar, A.M., R.R. Krueger, P.S. Devanand, and C.-C.T. Chao. 2005. Genetic analysis of Egyptian date (*Phoenix dactylifera* L.) accessions using AFLP markers. *Genet. Resour. Crop Evol., 52*: 601–607.

El-Batraoui, M., Baṭrāwī, M. 2016. *The Traditional Crafts of Egypt*. The American University in Cairo Press.

Eldeeb, A. 2017. Recycling agricultural waste as a part of interior design and architectural history in Egypt. *In*: *The International Conference: Cities' Identity Through Architecture and Arts* (*CITA*). The Academic Research Community Publication

El-Beltagi, H.S., H. Ghazzawy, E.M.B. Mahdy, and H.M. Ali-Dinar.2021. Impact of geographical distribution on genetic variation of two date palm cultivars in arid regions. *Fresenius Environ. Bull., 30*: 11513–11523.

El-Dengawy, E.F.A. 2017. Improvement of the pollination technique in Date Palm. *J. Plant Production, 8*: 307–314.

El Hadrami, A., and J.M. Al-Khayri. 2012. Socioeconomic and traditional importance of date palm. *Emir J. Food Agri., 24*: 371–385.

El Hassni, M., A. El Hadrami, F. Daayf, M. Chérif, E.A. Barka, and I. El Hadrami. 2007. Biological control of bayoud disease in date palm: Selection of microorganisms inhibiting the causal agent and inducing defense reactions. *Environ. Exp. Bot., 59*: 224–234.

Elhoumaizi, M.A., P.S. Devanand, J. Fang, and C.C.T. Chao. 2005. Confirmation of 'Medjool' date palm (*Phoenix dactylifera* L.) as a landrace variety by a study of 'Medjool' accessions from Morocco Using AFLP Markers. *HortScience, 40*: 1121A–1121.

Elhoumaizi, M.A., M. Saaidi, A. Oihabi, and C. Cilas. 2002. Phenotypic diversity of date-palm cultivars (*Phoenix dactylifera* L.) from Morocco. *Genet. Resour. Crop Evol., 49*: 483–490.

El-Mahrouky, M., A.H. El-Naggar, A.R. Usman, and M. Al-WABEL. 2015. Dynamics of CO_2 emission and biochemical properties of a sandy calcareous soil amended with Conocarpus waste and biochar. *Pedosphere, 25*: 46–56.

Elmeer, K., and I. Mattat. 2015.Genetic diversity of Qatari date palm using SSR markers. *Genet. Mol. Res. 14*: 1624–1635.

Elmeer, K., I. Mattat, A. Al-Malki, A.-G. Al-Mamari, K. BoJulaia, A. Hamwieh, and M. Baum. 2019. Assessing genetic diversity of Shishi date palm cultivars in Saudi Arabia and Qatar using microsatellite markers. *Int. J. Hortic. Sci. Technol., 6*: 1–9.

El-Mously, H., and E.A. Darwish. 2020. Date palm byproducts: History of utilization and technical heritage. *In*: Midani, M., Saba, N., and Alothman, O.Y. [Eds.], *Date Palm Fiber Composites: Processing, Properties, and Applications*, 3–71. Springer Singapore, Singapore.

El-Mously, H., M.M. El-Morshedy, and M.M. Megahed. 2012 (November). Evaluation of particleboard made of palm leaves midribs as compared with flax board. *In*: *Proceedings of the 4th International Conference on Production Engineering and Design for Development*, 27–29. Cairo, Egypt.

EL-Mously, H., M. Midani, and E.A. Darwish. 2023a. *Date Palm Byproducts: A Springboard for Circular Bio Economy*, 355–366. Springer.

EL-Mously, H., M. Midani, and E.A. Darwish. 2023b. Date Palm Byproducts in Fibers, Textiles, and Composites. *In*: *Date Palm Byproducts: A Springboard for Circular Bio Economy*, 103–128. Springer Nature Singapore.

EL-Mously, H., M. Midani, and E.A. Darwish. 2023c. Date Palm Byproducts in Enzymes, Food, Beverage, Pharmaceuticals, Cosmetics, and Natural Wax. *In*: *Date Palm Byproducts: A Springboard for Circular Bio Economy*, 93–102. Springer Nature Singapore.

Elouaqoudi, F.Z., L. El Fels, S. Amir, G. Merlina, A. Meddich, L. Lemee, A. Ambles, and M. Hafidi. 2015. Lipid signature of the microbial community structure during composting of date palm waste alone or mixed with couch grass clippings. *Int. Biodeterior. Biodegradation, 97*: 75–84.

Elsanhoty, R.M., I.A. Al-Turki, and M.F. Ramadan. 2012. Screening of medium components by Plackett–Burman design for carotenoid production using date (*Phoenix dactylifera*) wastes. *Ind. Crops Prod., 36*: 313–320.

Elseify, L.A., M. Midani, A.A. El-Badawy, S. Awad, and M. Jawaid. 2022. Comparative study of long date palm (*Phoenix dactylifera* L.) midrib and spadix fibers with other commercial leaf fibers. *Cellulose, 30*: 1927–1942.

Elseify, L.A., M. Midani, L.A. Shihata, and H. El-Mously. 2019. Review on cellulosic fibers extracted from date palms (*Phoenix dactylifera* L.) and their applications. *Cellulose, 26*: 2209–2232.

El-Shafey, E.-S.I., and S.M.Z. Al-Kindy. 2013. Removal of Cu^{2+} and Ag^{+} from aqueous solution on a chemically carbonized sorbent from date palm leaflets. *Environ. Technol., 34*: 395–406.

Elshibli, S., and H. Korpelainen. 2008. Microsatellite markers reveal high genetic diversity in date palm (*Phoenix dactylifera* L.) germplasm from Sudan. *Genetica, 134*: 251–260.

Elshibli, S., and H. Korpelainen. 2009. Biodiversity of date palms (*Phoenix dactylifera* L.) in Sudan: Chemical, morphological and DNA polymorphisms of selected cultivars. *Plant Genetic Resources, 7*: 194–203. 10.1017/S1479262108197489.

El-Tawil H. 1989. *Environment and Architecture in Siwa*. Ph.D. thesis, College of Fine Arts, Alexandria University, Egypt.

Elwaleed, A.K., N. Nikabdullah., M.J.M. Nor, M.F.M. Tahir, R. Zulkifli 2013. Experimental investigation of sound absorption properties of perforated date palm fibers panel. *IOP Conference Series: Materials Science and Engineering, 46*: 012027. https://doi.org/10.1088/1757-899X/46/ 1/012027.

Engels, J.M.M., and A.W. Ebert. 2021. A critical review of the current global ex situ conservation system for plant agrobiodiversity. I. History of the development of the global system in the context of the political/legal framework and its major conservation components. *Plants, 10*: 1557. https://doi.org/10.3390/plants10081557.

Ezebilo, E., and M. Elsafi, L. 2013. On-farm diversity of date palm (*Phoenix dactylifera* L) in Sudan: A potential genetic resources conservation strategy. *Sustain. Sci. Pract. Policy, 5*: 338–356.

Fang, Y., H. Wu, T. Zhang, M. Yang, Y. Yin, L. Pan, X. Yu, et al. 2012. A complete sequence and transcriptomic analyses of date palm (*Phoenix dactylifera* L.) mitochondrial genome. *PLoS One, 7*: e37164.

Farag, K.M., A.S. Elsabagh, H.A. El-Ashry, et al. 2012. Fruit characteristics of 'Zaghloul' date palm in relation to metaxenic influences of used pollinator. *Am. Eurasian J. Agric. Environ. Sci., 12*: 842–855.

Fki, L., O. Chkir, W. Kriaa, A. Nasri, E. Baklouti, R.B. Masmoudi, A. Rival, N. Drira, and B. Panis. 2017. *In vitro* cryopreservation of date palm caulogenic meristems. *Methods Mol. Biol., 1638*: 39–48.

Finkle, B.J., J.M., Ulrich, D.W., Rains, B.B., Tisserat, and G.W. Schaefer. 1979. Survival of alfalfa, *Medicago sativa*, rice *Oryza sativa* and date palm *Phoenix dactylifera*, callus after liquid nitrogen freezing. *Cryobiology, 16*: 583.

Golshan Tafti, A., N. Solaimani Dahdivan, and S.A. Yasini Ardakani. 2017. Physicochemical properties and applications of date seed and its oil. *Food Chem. Toxicol., 24*: 1399–1406.

Griffith, M.P., A. Meyer, and A. Grinage. 2021. Global *ex situ* conservation of palms: Living treasures for research and education. *Front.fFor Glob. Chang., 4*. https://doi.org/10.3389/ffgc.2021.711414.

Gros-Balthazard, M., C. Newton, S. Ivorra, M. Tengberg, J.-C. Pintaud, and J.-F. Terral. 2013. Origins and domestication of date palm (*Phoenix dactylifera* L.). The state of the art and the study perspectives. *Rev. ethnoécologie*. https://doi.org/10.4000/ethnoecologie.1524.

Gumanová, V., L. Sobotová, T. Dzuro, M. Badida, and M. Moravec. 2022. Experimental survey of the sound absorption performance of natural fibers in comparison with conventional insulating materials. *Sustain. Sci. Pract. Policy, 14*: 4258.

Gupta, A., R.K. Godara, V.K. Sharma, and A.K. Panda. 2017. Artificial Pollination: A tool for improving fruiting traits in date palm (*Phoenix dactylifera* L.). *Chem. Sci. Rev. Lett.* *6*(22): 1312–1320.

Hadrami, I.E., and A.E. Hadrami. 2009. Breeding Date Palm. *In*: Jain, S.M., and Priyadarshan, P.M. [Eds.], *Breeding Plantation Tree Crops: Tropical Species*, 191–216. Springer New York, New York, NY.

Haj-Amor, Z., T. Kumar Acharjee, L. Dhaouadi, and S. Bouri. 2020. Impacts of climate change on irrigation water requirement of date palms under future salinity trend in coastal aquifer of Tunisian oasis. *Agric. Water Manage.*, *228*: 105843.

Hajian, S., and Z. Hamidi-Esfahani. 2015. Date Palm Status and Perspective in Iran. *In*: Al-Khayri, J.M., Jain, S.M., Johnson, D.V. [Eds.],*Date Palm Genetic Resources and Utilization* (*Vol. 2): Asia and Europe*, 19–47. Springer Netherlands, Dordrecht.

Hammadi, H., R. Mokhtar, E. Mokhtar, and F. Ali. 2009. New approach for the morphological identification of date palm (*Phoenix dactylifera* l.) cultivars from Tunisia. https://www.pakbs.org/pjbot/PDFs/41(6)/PJB41(6)2671.pdf. Accessed 13 Sep 2023.

Hamza, H., M.A. Benabderrahim, M. Elbekkay, G. Ferdaous, T. Triki, and A. Ferchichi. 2012. Investigation of genetic variation in Tunisian date palm (*Phoenix dactylifera* L.) cultivars using ISSR marker systems and their relation with fruit characteristics. *Turk. J. Biol.* *36*: 449–458.

Hamza, H., M. Jemni, M.A. Benabderrahim, A. Mrabet, S. Touil, A. Othmani, and M.B. Salah. 2015. Date Palm Status and Perspective in Tunisia. *In*: Al-Khayri, J.M., Jain, S.M., and Johnson, D.V. [Eds.], *Date Palm Genetic Resources and Utilization (Vol. 1)*: *Africa and the Americas*, 193–221. Springer Netherlands, Dordrecht.

Hamza, R.S., and A.G. Al-Hashimi. 2023. Effect of different drying methods on some mineral elements, amino acids and fatty acids of the fruits of some Iraqi palm cultivars. *Biodiversitas Journal of Biological Diversity*, *24*(1): 359–369. https://doi.org/10.13057/biodiv/d240143.

Hancock, J.F. 2012. *Plant Evolution and the Origin of Crop Species*. CABI

Hasan, M., H.M. Abdullah, A.S. Hasibuzzaman, and M.A. Ali. 2022. Date Palm Genetic Resources for Breeding. *In*: Priyadarshan, P.M., and Jain, S.M. Eds.], Cash Crops: Genetic Diversity, Erosion, Conservation and Utilization, 479–503. Springer International Publishing, Cham.

Hasnaoui, A., A. Elhoumaizi, A. Hakkou, B. Wathelet, and M. Sindic. 2011. Physico-chemical characterization, classification, and quality evaluation of date palm fruits of some Moroccan cultivars. *J. Sci. Res.*, *3*: 139–149. https://doi.org/10.3329/jsr.v3i1.6062.

Hasan M., H.M. Abdullah, A.S.M. Hasibuzzaman, and M.A. Ali. 2022. Date palm genetic resources for breeding. *In*: Priyadarshan, P., and Jain, S.M. [Eds.], *Cash Crops*, 479–503. Springer International Publishing, Cham.

Hassan, M.L., J. Bras, E.A. Hassan, C. Silard, and E. Mauret. 2014. Enzyme-assisted isolation of microfibrillated cellulose from date palm fruit stalks. *Ind. Crops Prod.*, *55*: 102–108.

Hassan, M.L., J. Bras, E. Mauret, S.M. Fadel, E.A. Hassan, and N.A. El-Wakil. 2015. Palm rachis microfibrillated cellulose and oxidized-microfibrillated cellulose for improving paper sheets properties of unbeaten softwood and bagasse pulps. *Ind. Crops Prod.*, *64*: 9–15.

Hays, C.G., and C.M. Fagan. 2016. *Conservation Biology, Evolution, and Encyclopedia of Evolutionary Biology*, 347–353. Elsevier. doi:10.1016/b978-0-12-800049-6.00301-2

Hazzouri, K.M., J.M. Flowers, D. Nelson, A. Lemansour, K. Masmoudi, and K.M.A. Amiri. 2020. Prospects for the study and improvement of abiotic stress tolerance in date palms in the post-genomics era. *Front. Plant Sci.*, *11*: 293.

Hegazy, S., K. Ahmed, and S. Hiziroglu. 2015. Oriented strand board production from water-treated date palm fronds. *BioResources*, *10*: 448–456.

Helal A. 1989. *Spontaneous Architecture in Egypt: Analytical Study of Reed Architecture in Manzala Lake Region*. Faculty of Fine Arts, Helwan University, Cairo.

Heywood, V.H. 2019. Conserving plants within and beyond protected areas: Still problematic and future uncertain. *Plant Divers*, *41*: 36–49.

Huang, W., H. Li, B. Zhu, Y. Feng, S. Wang, and S. Zhang. 2007. Selective hydrogenation of furfural to furfuryl alcohol over catalysts prepared via sonochemistry. *Ultrason. Sonochem.*, *14*: 67–74.

Hussein E., S. Adawy, S. Ismail, and H El-Itriby. 2005. Molecular characterization of some Egyptian date palm germplasm using RAPD and ISSR markers. *Arab J. Biotech.*, *8*: 83–98.

Ibrahim, I.A., M.H. Hashem, A.A. Hemeida, M.M. Hassan, and A.I. Maksoud. 2014. Characterization of genetic diversity of Date palm (*Phoenix dactylifera* L.) cultivars collected from New Valley governorate (El-Kharga and Dakhleh) based on morphological variability and molecular markers. *Life Sci. J.*, *11*: 879–889.

Ines, D., B. Sonia, B.A. Fatma, B. Souhail, A. Hamadi, T. Hamida, and H. Basma. 2010. Date seed oil inhibits hydrogen peroxide-induced oxidative stress in human epidermal keratinocytes. *Int. J. Dermatol.*, *49*: 262–268.

Iskanderani, F.I. 2008. Physical properties of particleboard panels manufactured from *Phoenix dactylifera*-L (date palm) mid-rib chips using urea formaldehyde binder. *Int. J. Polym. Biomater.*, *57*: 979–995.

Ismail, B., I. Haffar, R. Baalbaki, and J. Henry. 2001. Development of a total quality scoring system based on consumer preference weightings and sensory profiles: Application to fruit dates (Tamr). *Food Qual. Prefer.*, *12*: 499–506.

Issa, M., S. Abd Elhamed, and H. Alam. 2021. A new strategy for controlling three devastating pests attacking date palm plantations in El Bahariya and Siwa Oases, Egypt. *Egyptian International Journal of Palms*, *1*: 39–56.

IUCN. 2021. Issues Brief: Species and climate change. *In*: *International Union for Conservation of Nature*. https://www.iucn.org/resources/issues-brief/species-and-climate-change. Accessed 17 Aug 2023.

Jaganathan, V., M. Shanmugavadivu, and S. Ganesh. 2018. Preliminary phytochemical screening and anti-bacterial activity of date seed methanolic extract. *Int. J. Adv. Res. Biol. Sci.*, *5*(2): 209–215.

Jain S.M. 2005. Major mutation-assisted plant breeding programs supported by FAO/IAEA. Plant *Cell Tissue Organ Cult.*, *82*: 113–123. https://doi.org/10.1007/s11240-004-7095-6.

Jain, S. 2011. Prospects of *in vitro* conservation of date palm genetic diversity for sustainable production. *Emir. J. Food Agric.*, *23*: 110.

Jain, S.M. 2006. Radiation-induced mutations for developing Bayoud disease resistant date palm in North Africa. *In*: *Proceeding of the International Workshop on True-to-typeness of Date Palm Tissue Culture-derived Plants*, *16*: 47–62.

Jain, S.M., J.M. Al-Khayri, and D.V. Johnson. 2011. *Date Palm Biotechnology*, 1–743. Springer Science & Business Media. Springer Dordrecht.

Jaradat, A.A. 2011. Biodiversity of date palm. *In*: *Encyclopedia of Life Support Systems: Land Use, Land Coverand Soil Sciences*, 1–31. Eolss Publishers, Oxford, UK.

Jaradat, A.A. 2016. Genetic Erosion of *Phoenix dactylifera* L.: Perceptible, Probable, or Possible. *In*: *Sustainable Development and Biodiversity*, 131–213. Springer International Publishing, Cham.

Jaradat, A.A. 2015. Biodiversity, genetic diversity, and genetic resources of date palm. *In*: Al-Khayri, J.M., Jain, S.M., and Johnson, D.V. [Eds.], *Date Palm Genetic Resources and Utilization (Vol. 1): Africa and the Americas*, 19–74. Springer.

Jazinizadeh, E., Zarghami, R., Majd, A., Iranbakhsh, A., and G. Tajaddod. 2015. *In vitro* production of date palm (*Phoenix dactylifera* L.) cv. 'Barhee' plantlets through direct organogenesis. *Biol. Forum*, *7*: 566–572.

Johnson, D.V. 2016. Unusual date palm products: Prayer beads, walking sticks, and fishing boats. *Emir. J. Food Agric.*, *28*: 12. https://doi.org/10.9755/ejfa.2015-11-1021.

Johnson, D.V. 2011. Introduction: Date palm biotechnology from theory to practice. In: Jain SM, Al-Khayri, J.M., and Johnson, D.V. [Eds.], *Date Palm Biotechnology*, 1–11. Springer Netherlands, Dordrecht.

Johnson, D.V., J.M. Al-Khayri, and S.M. Jain. 2015. Introduction: Date Production Status and Prospects in Africa and the Americas. *In*: Al-Khayri, J.M., Jain, S.M., and Johnson, D.V. [Eds.], *Date Palm Genetic Resources and Utilization (Vol. 1): Africa and the Americas*, 3–18. Springer Netherlands, Dordrecht.

Johnson, D.V., J.M. Al-Khayri, and S.M. Jain. 2013. Seedling date palms (*Phoenix dactylifera* l.) as genetic resources. *Emir. J. Food Agric.*, 809–830.

Johnson, D.V., and J.C. Macknight. 2023. Date palm (*Phoenix dactylifera*) and other exotic orchard Fruit Crops in Big Bend, Texas. *PalmArbor*, 1–15.

Kareche, A., B. Agoudjil, B. Haba, and A. Boudenne. 2020. Study on the durability of new construction materials based on mortar reinforced with date palm fibers wastes. *Waste and Biomass Valorization*, *11*: 3801–3809.

Karim, K., A.A. Malik, H. Hamza, E. Mohamed, and S.S. Sharma. 2023. Genetic diversity of Tunisian and Indian date palm cultivars and genotypes revealed by AFLP markers. *Research Square*.

Karizaki, V.M. 2017. Iranian dates and ethnic date-based products. *J. Ethn. Foods*, *4*: 204–209.

Khan, S., and T.B. Bi. 2012. Direct shoot regeneration system for date palm (*Phoenix dactylifera* L.) cv. Dhakki as a means of micropropagation. *Pak. J. Bot.*, *44*: 1965–1971.

Khanam, S., A. Sham, J.L. Bennetzen, and M.A.M. Aly. 2012. Analysis of molecular marker-based characterization and genetic variation in date palm (*Phoenix dactylifera* L.). *Aust. J. Crop Sci.*, *6*: 1236–1244.

Khierallah, H.S., and S.M. Bader. 2006. Micropropagation of date palm (*Phoenix dactylifera* L.) var. Maktoom through direct organogenesis, 213–224. *In*: *III International Date Palm Conference*, 736 pp.

Khierallah, H.S.M., S.M. Bader, M. Baum, and A. Hamwieh. 2011. Genetic diversity of Iraqi date palms revealed by microsatellite polymorphism. *J. Am. Soc. Hortic. Sci.*, *136*: 282–287.

Khierallah, H.S.M., S.M. Bader, K.M. Ibrahim, and I.J. Al-Jboory. 2015. Date Palm Status and Perspective in Iraq. *In*: Al-Khayri, J.M., Jain, S.M., and Johnson D.V. [Eds).], *Date Palm Genetic Resources and Utilization (Vol. 2): Asia and Europe*, 97–152. Springer Netherlands, Dordrecht.

Kinawy, M.S.A. 1997. Study of the appropriate conditions of the press-cycle for the manufacture of palm midrib-core blockboards [Master]. The Faculty of Engineering, Ain Shams University

Kittel, T.G.F. 2013. The vulnerability of biodiversity to rapid climate change. *In*: *Climate Vulnerability*, 185–201. Elsevier.

Kordrostami, M., M. Mafakheri, and J.M. Al-Khayri. 2022. Date palm (*Phoenix dactylifera* L.) genetic improvement via biotechnological approaches. *Tree Genet. Genomes*, *18*: 26.

Kremer, A., B.M. Potts, and S. Delzon. 2014. Genetic divergence in forest trees: understanding the consequences of climate change. *Funct. Ecol.*, *28*: 22–36.

Kriker, A., G. Debicki, A. Bali, M.M. Khenfer, and M. Chabannet. 2005.Mechanical properties of date palm fibres and concrete reinforced with date palm fibres in hot-dry climate. *Cement and Concrete Composites*, *27*: 554–564.

Krueger, R.R. 2021. Date Palm (*Phoenix dactylifera* L.) Biology and Utilization. *In*: Al-Khayri, J.M., Jain, S.M., and Johnson, D.V. [Eds.], *The Date Palm Genome, (Vol. 1): Phylogeny, Biodiversity, and Mapping*, 3–28. Springer International Publishing, Cham.

Ksiazek-Mikenas, K., J.B. Fant, and K.A. Skogen. 2019. Pollinator-mediated gene flow connects green roof populations across the urban matrix: A paternity analysis of the self-compatible forb *Penstemon hirsutus*. *Front. Ecol. Evol.*, *7*. https://doi.org/10.3389/fevo.2019.00299.

Kurdi, H., A. Al-Aldawsari, I. Al-Turaiki, and A.S. Aldawood. 2021. Early detection of red palm weevil, (Olivier), infestation using data mining. *Plants*, *10*: 95. https://doi.org/10.3390/plants10010095.

Lattieff, F.A. 2016. A study of biogas production from date palm fruit wastes. *J. Clean Prod.*, *139*: 1191–1195.

Lisa, S.M., M. Spannagl, A. Al-Malki, B. George, M. F. Torres, E. K. Al-Dous, E. K. Al-Azwani, E. Hussein, S. Mathew, K.F.X. Mayer, Y. A. Mohamoud, K. Suhre, and J. A. Malek. 2014. A first genetic map of date palm (Phoenix dactylifera) reveals long-range genome structure conservation in the palms. *BMC Genomics.*, *15*: 285.

Lovich, J.E., and D. Bainbridge. 1999. Anthropogenic degradation of the southern California desert ecosystem and prospects for natural recovery and restoration. *Environ. Manage.*, *24*: 309–326.

Lubna, M. 2020. Date kernels, a free treatment for poor breast milk production, strengthening eyelashes, maintaining the uterus, fighting cancer, cirrhosis of the liver, and kidney failure. In: Health and Environment. https://www.elaard.com/73977. Accessed 26 Sep 2023

Mahgoub, O., I.T. Kadim, M.H. Al-Busaidi, K. Annamalai, and N.M. Al-Saqri. 2007. Effects of feeding ensiled date palm fronds and a by-product concentrate on performance and meat quality of Omani sheep. *Anim. Feed Sci. Technol.*, *135*: 210–221.

Manickavasagan, A., M. Mohamed Essa, and E. Sukumar. 2012. *Dates: Production, Processing, Food, and Medicinal Values*. CRC Press

Manjula, R., N.V. Raju, R.P.S. Chakradhar, E. Kalkornsurapranee, and J. Johns. 2017. Influence of chemical treatment on thermal decomposition and crystallite size of coir fiber. *Int. J. Thermophys.*, *39*: 1–11.

Marwat, S.K., W. Bakhsh, M. Shoaib, E. Khan, and S. Javaria. 2016. Ethnobotanical and religious importance of the Quranic Plant-Nakhl (*Phoenix dactylifera* L.) review. *Am. Eurasian J. Agric. Environ. Sci.*, *16*: 842–851.

Masmoudi-Allouche, F., A. Châari-Rkhis, W. Kriaâ, R. Gargouri-Bouzid, S.M. Jain, and N. Drira. 2009. *In vitro* hermaphrodism induction in date palm female flower. *Plant Cell Rep.*, *28*: 1–10.

Maxted, N. 2001. *Ex Situ*, *In Situ* Conservation. *In*: Levin, S.A. [Ed.], *Encyclopedia of Biodiversity*, 683–695. Elsevier, New York.

Mekhermeche, A., A. Kriker, and S. Dahmani. 2016. Contribution to the study of thermal properties of clay bricks reinforced by date palm fiber. *AIP Conf. Proc.*, *1758*: 030004.

Merlo, A.B., V. Vetere, J.F. Ruggera, and M.L. Casella. 2009. Bimetallic PtSn catalyst for the selective hydrogenation of furfural to furfuryl alcohol in liquid-phase. *Catal. Commun.*, *10*: 1665–1669.

Messer, P.W. 2016. Neutral models of genetic drift and mutation. *In*: *Encyclopedia of Evolutionary Biology*, 119–123. Elsevier.

Midani, M., Elseify, LA, T. Hamouda, and A. Hassanin. 2022. Comparison of coconut coir and date palm coir (sheath fiber) and their composites. *In*: Jawaid, M. [Ed.], *Coir Fiber and its Composites*, 331–345. Woodhead Publishing.

Mohamed Lemine, F.M., M.V.O. Mohamed Ahmed, L. Ben Mohamed Maoulainine, Z.E.A.O. Bouna, A. Samb, and A.O.M.S. O Boukhary. 2014. Antioxidant activity of various Mauritanian date palm (*Phoenix dactylifera* L.) fruits at two edible ripening stages. *Food Sci. Nutr.*, *2*: 700–705.

Mohamed, M.A., and A.M. Abd El-Kawy. 2010. Cytological Studies for Date Palm (*Phoenix Dactylifera* L) Tissue Culture Derived Plants. *Report and Opinion*, *2*: 17–21,

Mohamed, R.M.A., A.S.M. Fageer, M.M. Eltayeb, and I.A. Mohamed Ahmed. 2014. Chemical composition, antioxidant capacity, and mineral extractability of Sudanese date palm (*Phoenix dactylifera* L.) fruits. *Food Sci. Nutr.*, *2*: 478–489.

Montazar, A., R. Krueger, D. Corwin, A. Pourreza, C. Little, S. Rios, and R.L. Snyder. 2020. Determination of actual evapotranspiration and crop coefficients of California date palms using the residual of energy balance approach. *Water*, *12*: 2253.

Muhammed, A.H. 2012. Inflorescence rot disease of date palm caused by *Fusarium proliferatum* in Southern Iraq. *Afr. J. Biotechnol.*, *11*: 8616–8621.

Munir, M., M.R. Alhajhoj, and A.K.M. Sallam. 2021. Impact of pollination time of the day on the fruit, yield, and quality traits of date palm cultivar Khalas. *J. Appl. Hortic.*, *23*(3): 348–352.

Mustafa, E.A.M., S.A.A. Heiba, M.M.S. Saleh, N.E. Ashour, D.A. Mohamed, M. El-Migeed, et al. 2014. Effect of different pollinizer sources on yield, fruit characteristics and phylogenetic relationships with Amhat cv. date palm (*Phoenix dactylifera* L.) in Egypt using RAPD markers. *Int. J. Agric. Res.*, *9*: 331–343.

Nancib, A., N. Nancib, A. Boubendir, and J. Boudrant. 2015. The use of date waste for lactic acid production by a fed-batch culture using *Lactobacillus casei* subsp. rhamnosus. *Braz J. Microbiol.*, *46*: 893–902.

Nasser, R.A.-S. 2014. An evaluation of the use of midribs from common date palm cultivars grown in Saudi Arabia for energy production. *BioResources*, *9*: 4343–4357.

Nasser, R.A., M.Z.M. Salem, S. Hiziroglu, H.A. Al-Mefarrej, A.S. Mohareb, M. Alam, and I.M. Aref. 2016. Chemical analysis of different parts of date palm (*phoenix dactylifera* l.) using ultimate, proximate, and thermo-gravimetric techniques for energy production. *Energies*, *9*: 374.

Naqvi, S.A., M. Waseem, T.A. Naqvi, M.J. Jaskani, and I.A. Khan. 2023. Diversity, conservation, and utilization of date palm germplasm. *In*: *Date Palm*, 65–89. CABI, GB.

Naqvi S.A., W. Shafqat, M.S. Haider, et al. 2021. Gender determination of date palm. *In*: *The Date Palm Genome*, 161–177, (Vol. 1). Springer International Publishing, Cham.

Nguyen, V.-A., P.T. Nguyen, M.-A. Le, A. Bazrafshan, S. Sisunandar, S. Kalaipandian, S.W. Adkins, and Q.T. Nguyen. 2023. A practical framework for the cryopreservation of palm species. *In vitro Cellular & Developmental Biology – Plant*, *59*: 425–445.

Othmani, A., M. Jemni, K. Kadri, S. Amel, F. Artés, and J.M. Al-Khayri. 2020. Preharvest fruit drop of date palm (*Phoenix dactylifera* L.) CV. Deglet Nour at kimri stage: Development, physico-chemical characterization, and functional properties. *Int. J. Fruit Sci.*, 20: 414–432.

Panis, B., M. Nagel, and I. Van den Houwe. 2020. Challenges and prospects for the conservation of crop genetic resources in field genebanks, in *in vitro* collections and/or in liquid nitrogen. *Plants*, *9*. https://doi.org/10.3390/plants9121634.

Pareek, S. 2015. Date Palm Status and Perspective in India. *In*: Al-Khayri, J.M., Jain, S.M., and Johnson, D.V. [Eds.], *Date Palm Genetic Resources and Utilization (Vol. 2): Asia and Europe*, 441–485. Springer Netherlands, Dordrecht.

Patankar, H.V., I. Al-Harrasi, L. Al Kharusi, G.A. Jana, R. Al-Yahyai, R. Sunkar, and M.W. Yaish. 2019. Overexpression of a Metallothionein 2A gene from date palm confers abiotic stress tolerance to yeast and *Arabidopsis thaliana*. *Int. J. Mol. Sci.*, *20*: 2871.

Patil, A.M., and A.A. Daspute. 2019. Identification and validation of sex specific DNA Markers in Date Palm (*Phoenix Dactylifera* L.) using RAPD Markers. *J. Adv. Sci. Technol.*, *16*: 220–228.

Perez, T.M., O. Valverde-Barrantes, C. Bravo, T.C. Taylor, B. Fadrique, J.A. Hogan, C.J. Pardo, J.T. Stroud, C. Baraloto, and K.J. Feeley. 2018. Botanic gardens are an untapped resource for studying the functional ecology of tropical plants. *Philos. Trans. R. Soc. Lond. B Biol. Sci.*, *374*: https://doi.org/10.1098/rstb.2017.0390.

Perkins, S. 2016. The benefits of date seed oil for hair. *In*: *NaturallyCurly.com*. https://www.naturallycurly.com/ curlreading/curl-products/the-benefits-of-date-seed-oil-for-hair-si. (Accessed 26 Sep 2023.)

Petruzzello, M. 2023. Date palm. *In*: *Encyclopedia Britannica*. https://www.britannica.com/plant/date-palm. (Accessed 1 Sep 2023.)

Piesik, S. 2012. *Arish: Palm-leaf Architecture*. Thames & Hudson, USA.

Poulain, C., A. Rhiss, and G. Beauchesne. 1979. Multiplication vegetative en culture *in vitro* du palmier-dattier (*Phoenix dactylifera* L.). *J. Plant Physiol.*, *119*: 227–235.

Racchi, M.L., and A. Camussi. 2018. The date palms of Al Jufrah - Libya: A survey on genetic diversity of local varieties. *JAEID*, *112*: 161–184.

Rambabu, K., G. Bharath, A. Hai, F. Banat, S.W. Hasan, H. Taher, and H.F. Mohd Zaid. 2020. Nutritional quality and physico-chemical characteristics of selected date fruit varieties of the United Arab Emirates. *Processes*, *8*: 256.

Ramzi, K., F. Ikhlass, and M.M. Farouk. 2010. Study of liquids absorption and retention capacities of new cellulosic materials and sodium cellulose carboxylmethylate prepared from *Posidonia*. *Fibers Polym.*, *11*: 593–597.

Rani, V.G., P. Thangamathi, H. Balasubramanian, et al. 2023. Evaluation and assessment of nutritional composition for quality profiling of *Phoenix dactylifera* cultivars using multivariate analytical tools. *Ecol. Genet. Genom.*, *27*: 100173.

Ray, H.P., and Ray. 2003. *The Archaeology of Seafaring in Ancient South Asia*. Cambridge University Press.

Reddy S.K.K., P. Kannan, A. Al Shoaibi, C. Srinivasakannan, 2014. KOH-based porous carbon from date palm seed: Preparation, characterization, and application to phenol adsorption. *Water Sci. Technol.*, *70*(10): 1633–1640. https://doi.org/10.2166/wst.2014.419.

Rehman, M., A. Ahmed, A. Tareen, M. Mazahir, and A. Mariam. 2022. *Physico-chemical Properties of DatePalm Cultivars Grown in Punjab*. Available at SSRN.

Reynolds J.F., and T. Murashige. 1979. Asexual embryogenesis in callus cultures of palms. *In Vitro*, *15*: 383–387. https://doi.org/10.1007/bf02616146.

Rivas, M., R.L. Barbieri, and L.C. da Maia. 2012. Plant breeding and *in situ* utilization of palm trees. *Cienc. Rural*, *42*: 261–269.

Rivera, D., C. Obón, F. Alcaraz, E. Carreño, E. Laguna, A. Amorós, D.V. Johnson, G. Díaz, and A. Morte. 2015. Date palm status and perspective in Spain. *In*: Al-Khayri, J.M., Jain, S.M., and Johnson, D.V. [Eds.], *Date Palm Genetic Resources and Utilization (Vol. 2): Asia and Europe*, 489–526. Springer Netherlands, Dordrecht.

Saad, M., I.S. Agwa, B. Abdelsalam, and M. Amin. 2022. Improving the brittle behavior of high strength concrete using banana and palm leaf sheath fibers. *Mech. Adv. Mater. Struct.*, *29*: 564–573.

Saafi, E.B., M. Louedi, A. Elfeki, A. Zakhama, M.F. Najjar, M. Hammami, and L. Achour. 2011.Protective effect of date palm fruit extract (*Phoenix dactylifera* L.) on dimethoate induced-oxidative stress in rat liver. *Exp. Toxicol. Pathol.*, *63*: 433–441.

Saaroni, H., A. Bitan, E.B. Dor, and N. Feller. 2004.The mixed results concerning the "oasis effect" in a rural settlement in the Negev Desert, Israel. *J. Arid Environ.*, *58*: 235–248.

Sabra, W., C. Groeger, P.N. Sharma, and A.-P. Zeng. 2014. Improved n-butanol production by a non-acetone producing *Clostridium pasteurianum* DSMZ 525 in mixed substrate fermentation. *Appl. Microbiol. Biotechnol.*, *98*: 4267–4276.

Sadik, M.W., A.O. Al Ashhab, M.K. Zahran, and F.M. Alsaqan. 2012. Composting mulch of date palm trees through microbial activator in Saudi Arabia. *Int. J. Biochem.Biotechnol.*, *1*(5): 156–161.

Saker, M.M., S.S. Adawy, A.A. Mohamed, and H.A. El-Itriby. 2006. Monitoring of cultivar identity in tissue culture-derived date palms using RAPD and AFLP analysis. *Biol. Plant*, *50*: 198–204.

Saker, M.M., S.A. Bekheet, H.S. Taha, A.S. Fahmy, and H.A. Moursy. 2000. Detection of somaclonal variations in tissue culture-derived date palm plants using isoenzyme analysis and RAPD fingerprints. *Biol. Plant, 43*: 347–351.

Salamati, N., H. Dehghanisanij, and P. Bayat. 2019. The effect of surface and subsurface drip irrigation system on qualitative and quantitative characteristics of dates of Kabkab cultivar. *Irrigation Sci. Engg., 42*: 153–165.

Salomon,T.R, Ortiz-U.N, Villa-A.C, and Yaurima, V. 2017. Assessment SSR markers used in analysis of genetic diversity of date palm (*Phoenix dactylifera* L.). *Plant Cell Biotechnol. Mol. Biol., 18*: 269–280.

Samir, B., D. Zied, D. Mokhtar, R. Elhadj, and B. Rabah. 2014. Thermal and mechanical performance of natural mortar reinforced with date palm fibers for use as insulating materials in building. *Energy Build,, 81*: 98–104.

Sattar, M.N., Z. Iqbal, and J.M. Al-Khayri. 2021. CRISPR-Cas based precision breeding in date palm: future applications. *In*: Al-Khayri, J.M., Jain, S.M., and Johnson, D.V. [Eds.], The *Date Palm Genome (Vol. 2): Omics and Molecular Breeding*, 169–199. Springer International Publishing, Cham.

Sedra, M., and B.H. Lazrek, 2011. *Fusarium oxysporum* f. sp. albedinis toxin characterization and use for selection of resistant date palm to Bayoud Disease. *In*: Jain, S., Al-Khayri, J., and Johnson, D. [Eds.], *Date Palm Biotechnology*, 253–270. Springer, Dordrecht.

Sedra, M.H., P. Lashermes, P. Trouslot, and M.-C. Combes. 1998. Identification and genetic diversity analysis of date palm (*Phoenix dactylifera* L.) varieties from Morocco using RAPD markers. *Euphytica, 103*: 75–82.

Shabani, F., and L. Kumar. 2013. Risk levels of invasive *Fusarium oxysporum* f. sp. in areas suitable for date palm (*Phoenix dactylifera*) cultivation under various climate change projections. *PLoS One, 8*: e83404.

Shabani, F., L. Kumar, and S. Taylor. 2012. Climate change impacts on the future distribution of date palms: A modeling exercise using CLIMEX. *PLoS One, 7*: e48021.

Shafie, M., and H. Zarea-Hosseinabadi. 2019. Eco-friendly laminated strand lumber from date palm rachis. *Drv Ind.*, 70: 359–367.

Sharoba, A., G. Elsharnouby, and Z. Ibrahim. 2022. Physicochemical properties and microbiological quality of dates syrup prepared from some Egyptian and Iraqi dates palm (*Phoenix dactylifera* L.) fruits. *Egypt J. Chem., 0*: 0–0.

Shehzad, M., H. Rasheed, S.A. Naqvi, J.M. Al-Khayri, J.M. Lorenzo, M.A. Alaghbari, M.F. Manzoor, and R.M. Aadil. 2021. Therapeutic potential of date palm against human infertility: A review. *Metabolites, 11*(6): 408. https://doi.org/10.3390/metabo11060408.

Siahpoosh, A., F. Gol Fakhrabadi, and F. Jorkesh. 2011. Determination and comparison of antioxidant activity of aqueous and methanol extracts of date palm (*Phoenix dactylifera* L.var dayri). *Research in Medicine, 35*: 81–68.

Singla, J., and S.G. Krattinger. 2016. Biotic stress resistance genes in wheat. *In*: *Encyclopedia of Food Grains*, 388–392. Elsevier.

Sirisena, S., K. Ng, and S. Ajlouni. 2015. The emerging Australian date palm industry: Date fruit nutritional and bioactive compounds and valuable processing by-products. *Compr. Rev. Food Sci. Food Saf., 14*: 813–823.

Sobhy, A.E.-S., and E.E. Hafez. 2010. Biochemical and nutritional characterizations of date palm fruits (*Phoenix dactylifera* L.). *J. Appl. Sci. Res., 6*: 1060–1067.

Soliman, S.S., B.A. Ali, and M.M.M. Ahmed. 2003.Genetic comparisons of Egyptian date palm cultivars (*Phoenix dactylifera* L.) by RAPD-PCR. *Afr. J. Biotechnol., 2*: 86–87.

Soliman, K.H., Reda R., and Sherif, El-S. 2019. Genetic diversity of semi-dry date palm (*Phoenix dactylifera* L.) Cultivars in Egypt, 1–14.

Sotolu, A.O., A.A. Kigbu, and A.J. Oshinow. 2014. Supplementation of date palm (*Phoenix dactylifera*) seed as feed additive in the diets of juvenile African catfish (Burchell, 1822). *J. Fish Aquat. Sci., 9*: 359–365.

Suresh Kumar Reddy, K., P. Kannan, A. Al Shoaibi, and C. Srinivasakannan. 2014. KOH-based porous carbon from date palm seed: Preparation, characterization, and application to phenol adsorption. *Water Sci. Technol., 70*: 1633–1640.

Templeton, A.R. 2019. Genetic Drift. *In*: Human Population Genetics and Genomics, 101–127. Elsevier.

Thind, S.K., N. Kaur, and S.K. Thatai. 2007. Sori as potential indices for evaluating date palm germplasm against Graphiola leaf spot. *Plant Disease Research, 22*: 45–46.

Tisserat, B., M.F., Gabr, and M.T. Sabour. 1985. Viability of cryogenically treated date palm pollen. *Date Palm J., 4*: 25–32.

Trifi, M., A.O.B.O.M. Salem, S. Zehdi, E. Cherif, S. Rhouma, A. Othmani, S. Dkhil-Dakhlaoui, and M. Marrakchi. 2017. Les ressources génétiques du palmier dattier en Tunisie. *In*: *Biotechnologies du palmier dattier*, 23–32. IRD Éditions.

Trusted health products. 2018. What's so great about date seed oil. https://www.trustedhealthproducts.com/blogs/food-health/whats-great-date-seed-oil. Accessed 17 Aug 2023

Usman, A.R.A., A. Abduljabbar, M. Vithanage, Y.S. Ok, M. Ahmad, M. Ahmad, J. Elfaki, S.S. Abdulazeem, and M.I. Al-Wabel. 2015. Biochar production from date palm waste: Charring temperature induced changes in composition and surface chemistry. *J. Anal. Appl. Pyrolysis, 115*: 392–400.

van de Wouw, M., C. Kik, T. van Hintum, R. van Treuren, and B. Visser. 2010. Genetic erosion in crops: Concept, research results, and challenges. *Plant Genet. Resour., 8*: 1–15.

Wakil, W., J.R. Faleiro, T.A. Miller, G.O. Bedford, and R.R. Krueger. 2015. Date palm production and pest management challenges. *In*: Wakil, W., Romeno Faleiro, J., Miller, T.A. -Eds.], *Sustainable Pest Management in Date Palm: Current Status and Emerging Challenges*, 1–11. Springer International Publishing, Cham.

Wang, D., and X.S. Sun. 2002. Low density particleboard from wheat straw and corn pith. *Ind. Crops Prod., 15*: 43–50.

Wendrich W.Z. 2009. Basketry. *In*: Selin, H. [Ed.], *Ancient Egyptian Materials and Technology. Encyclopaedia of the History of Science, Technology, and Medicine in Non-Western Cultures*, 392–394. Springer, Dordrecht.https://doi.org/10.1007/978-1-4020-4425-0_8470.

Westwood, M., N. Cavender, A. Meyer, and P. Smith. 2021. Botanic garden solutions to the plant extinction crisis. *Plants, People, Planet*, *3*: 22–32.

Yusuf A., A. Culham, W. Aljuhani, et al. 2015. Genetic diversity of Nigerian date palm (*Phoenix dactylifera*) germplasm based on microsatellite markers. *Int. J. Bio-Sci. Biotechnol.*, *7*: 121–132. 10.14257/ijbsbt.2015.7.1.12.

Zhang, G., L. Pan, Y. Yin, W. Liu, D. Huang, T. Zhang, L. Wang, C. Xin, Q. Lin, G. Sun, M.B. Abdullah, X. Zhang, S. Hu, I. S. Al-Mssallem, and J. Yu. 2012. Large-scale collection and annotation of gene models for date palm (*Phoenix dactylifera* L.). *Plant Mol. Biol.* *79*: 521–536.

Zaid, A., and P.F. de Wet. 2002a. Botanical and systematic description of the date palm. *In*: Zaid, A. [Ed.], *Date Palm Cultivation*. Food and Agricultural Organization of the United Nations, Rome.

Zaid, A., and P.F. de Wet. 2002b. Date palm propagation. *In*: Zaid, A. [Ed.], *Date Palm Cultivation*. Food and Agricultural Organization of the United Nations, Rome.

Zehdi-Azouzi, S., E. Cherif, S. Moussouni, M. Gros-Balthazard, S. Abbas Naqvi, B. Ludeña, K. Castillo, et al. 2015. Genetic structure of the date palm (*Phoenix dactylifera*) in the Old World reveals a strong differentiation between eastern and western populations. *Ann. Bot.*, *116*: 101–112.

Zehdi, S., M. Trifi, N. Billotte, M. Marrakchi, and J.C. Pintaud. 2004. Genetic diversity of Tunisian date palms (*Phoenix dactylifera* L.) revealed by nuclear microsatellite polymprohism. *Hereditas*, *141*: 278–287.

Ziadi, M., S. Mrsquo hir, N. Kbaier, M. Hamdi, and A. Ferchichi. 2011. Microbiological analysis and screening of lactic acid bacteria from Tunisian date palm sap. *Afr. J. Microbiol. Res.*, *5*: 2929–2935.

Zunaidi, N.H., W.H. Tan, M.S.A. Majid, and E.A. Lim. 2017. Effect of physical properties of natural fibre on the sound absorption coefficient. *J. Phys. Conf. Ser.*, *908*: 012023.

12

Genetic Diversity of *Balanitis roxburghii* Planch., and Sustainable Utilization

Guggalada Govardhana Yadav[1,2] and *Hosakatte Niranjana Murthy*[1,3,*]

Introduction

The genus *Balanites* Delile, classified within the family Zygophyllaceae, encompasses nine species: *B. aegyptiaca* (L.) Delile, *B. angolensis* (Welw.) Welw., *B. glabra* Mildbr. & Schltr., *B. maughamii* Sprague, *B. pedicillaris* Mildbr. and Schltr., *B. rotundifolia* (Tiegh.) Blatt., *B. roxburghii* Planch., *B. triflora* Tiegh., *B. wilsoniana* Dawe and Sprague, and 11 infraspecific taxa (Sands, 2001; POWO, 2024a). The *Balanites* sps. are small to large-sized trees that thrive mainly in drier parts of Africa, western Asia, India, and Myanmar. They produce drupes with a single large seed that stores a high amount of oil. Renowned for their numerous medicinal properties attributed to diverse bioactive compounds, predominantly saponins, the species also exhibit phenolics, flavonoids, coumarins, steroids, and alkaloids (Sands, 2001; Murthy et al., 2021). These plants serve as a nutritional resource in various regions of Africa and India, with *B. aegyptiaca*'s fruit pulp, leaves, seeds, and seed oil being widely utilized for their rich content of sugars, proteins, fiber, minerals, and fats (Murthy et al., 2021). Noteworthy is the fact that *Balanites* sps. host a substantial amount of seed oil and an appreciable quantity of diosgenin in their parts. *B. aegyptiaca*, for instance, contains up to 50.22% of seed oil (Chapagain and Wiesman, 2005), *B. orbicularis* 31% (Radunz et al., 1985) and *B. wilsoniana* 30% (Burkill, 1985), acknowledged for their edibility in various parts of Africa. The seed oil of *B. aegyptiaca* emerges as a rich source of nutraceuticals and essential fatty acids, exhibiting proven anti-diabetic activity, thereby offering potential as a new edible oil source (Chapagain et al., 2009; Al Ashaal et al., 2010). Additionally, biodiesel prepared from this seed oil meets international biodiesel standards (EN-14214), presenting an opportunity for commercial seed oil production through systematic cultivation (Chapagain et al., 2009). Coming to the diosgenin (a phytosteroid sapogenin), *B. aegyptiaca* comprises 4.08% sapogenin in its fruit pericarp (Hardman and Sofowora, 1972), 3% in the mesocarp, and 2.74% in seed kernels (Chapagain and Wiesman, 2005); also, *B. rotundifolia*

[1] Department of Botany, Karnatak University, Dharwad – 580003, India.
[2] Department of Botany, A.D.B. First Grade College, Harapanahalli – 583131, India.
[3] Department of Horticulture Science, Chungbuk National University, Cheongju 28644, Republic of South Korea.
* Corresponding author: hnmurthy60@gmail.com

and *B. wilsoniana* exhibit impressive diosgenin quantity in their parts (Hardman and Wood, 1972; Sofowora and Hardman, 1973). These facts suggest the potential for commercial diosgenin production through systematically cultivating these species.

Genus *Balanites* is represented by only one species in India, i.e., *B. roxburghii* (Figure 1A). Being native to India, this species is distributed throughout the dry and arid lands of India, including the Peninsular region (Yadav and Murthy, 2022). Regionally, it is known as desert date in English, ingala or ingaleru in Kannada, nanjunda marm in Tamil, gaara or garrah pandu in Telugu, ingudi in Sanskrit and hingora in Hindi. parallel to its allied species, *B. roxburghii* also possesses numerous applications in traditional medicine systems; its parts—leaves, fruit pulp, seed, and seed oil—are used as famine foods. It also accommodates high quantities of seed oil and diosgenin in its parts. In this regard, the systematic cultivation of this plant is essential to exploit its full potential to be a source of high-value products mentioned above in India.

In this chapter, we commence by exploring the distribution and morphology of *Balanites roxburghii*. Subsequently, we delve into its diverse benefits, taxonomic background, and the collection and assessment of germplasm through morphological, biochemical and cytological characteristics. We also discuss superior genotypes and utilization and conservation of germplasm, encompassing a comprehensive discussion of its various aspects.

Figure 1. Morphology of *Balanites roxburghii*. A – Habit; B – Twig showing leaves and thorns; C – Flowers; D – Fruits; E – Seed kernels.

Distribution and Morphology of *Balanites roxburghii*

B. roxburghii is native to India and distributed throughout the drier regions of India. It is generally found in scrubby and deciduous forests and also along roadsides, banks of streams, hill bases, and seasonal water-logged flood plains, where the water availability is comparatively more, at least in the form of groundwater. When it is present in thick forests, it prefers open areas. It grows rich in sandy, black cotton, red, and alluvium soils; avoids growing in rocky and stony areas. In India, it is majorly found in South India except for Kerala state, Maharashtra, Madhya Pradesh, Gujarat, Rajasthan, Uttar Pradesh, West Bengal, Chhattisgarh, Punjab and Haryana (Sands, 2001; Yadav and Murthy, 2022).

B. roxburghii is a semi-evergreen shrub or small tree reaching up to 9 meters (Figure 1A). The stem reaches a diameter of 20–30 cm with bark up to 6 mm thick. The thorns are dull green, 0.8–3.5 cm long, axillary, occasionally branched and bearing leaves and flowers (Figure 1B). Leaves are elliptic or obovate, bifoliate, subsessile, 2–3.8 × 0.6–4 cm, with a small third leaflet arising between the two major ones (Figure 1B). Flowers are arranged in a cymose pattern, 5-merous, actinomorphic and bisexual (Figure 1C). Sepals have tomentellous outside and villous within and a glabrous margin. Petals glabrous outside and silky villous within (Figure 1C). Stamens 10, disc cupular with 10-lobed glandular margins (Figure 1C). Ovary hemispherical, silky villous outside. The fruit is a drupe, 3–7 cm long, 1.8–3.5 cm wide, with five significant grooves present, often with five minor grooves (Figure 1D). The exocarp is yellowish and leathery, the mesocarp is brownish and pulpy, and there is a hard endocarp inside in which the seed kernel lies and that weighs up to 4 g (Figure 1E). Flowering and fruiting happens between March and February (Cooke, 1902; Sands, 2001).

Traditional Medicinal Values of *B. roxburghii*

B. roxburghii is well-known among India's rural population for its medicinal applications where it is distributed. The whole plant, mainly fruit, is used for snake bites (Chopra et al., 1956; Nadakarni, 1954). The bark is used as antihelmintic and spasmolytic and to treat cough and skin diseases (Virendra et al., 2009). The seeds are used as expectorants and colics and to treat burns and skin diseases (Chopra et al., 1956; Chatterjee and Parkrashi, 1997). The fruit is used as a detergent to clean clothes and deworm children, bark to poison the fish, and seed oil to cure burns and freckles (Nadakarni, 1954). The pulp and leaves are used in curing jaundice in South India. In rural areas of South India, the peeled ripe fruits are kept in the cooking salt, which facilitates the adhering of pulp to salt crystals and subsequent obvious ingredients of the food they prepare.

Nutritional Value

Balanites roxburghii's leaves and fruit pulp are used for edible purposes during the food scarcity period in some regions of South India. The cattle, sheep and goats graze the leaves and ripened fruits, especially during the hot summer when the greenery becomes rare, except for a few species, such as *Acacia* and *Prosopis*. The seed kernels are good sources of oil, protein and fiber, with a composition of 43.2–61.0, 10.96, and 15.12%, respectively (Yadav et al., 2022, 2023; Table 1). The seed cake is rich in minerals, such as phosphorus (7.62 mg/g DW), potassium (14 mg/g DW), calcium (6.05 mg/g DW), magnesium (6 mg/g DW), iron (1419 µg/g DW), zinc (431 µg/g DW), and copper (36.80 µg/g DW) (Table 1). The seed cake has a 2.72 mg GAE (gallic acid equivalent)/g DW of phenolics and showed impressive antioxidant activity (Table 1). The seed oil is rich in essential fatty acids and other nutraceuticals (Yadav et al., 2022, 2023; Table 1). Linoleic acid is one of the significant essential fatty acids and it is found in the range of 18.24–40.80%, whereas oleic (28.12–56.46%) and palmitic acid (16.75–27.25%) are two prominent fatty acids. The seed oil is also rich in carotenoids, tocopherols, and phytosterols compared to commercial edible oils (Table 1). The cytotoxicity assay against 3T3-L1 preadipocytes showed no cell death up to 100 ppm concentration and projected this as a new edible oil source (Yadav et al., 2023).

Other Important Products (Diosgenin and Seed Oil)

Apart from its edible quality, seed oil can also be used to produce various industrial products. For instance, Khanvilkar et al. (2016) prepared biodiesel from the *B. roxburghii* seed oil. They studied its fuel properties: density, kinematic viscosity, acid value, heat capacity, lubricity, iodine value, and bromine number. The results showed that the biodiesel was as good as that produced from *B. aegyptiaca* and soybean oil and met the specification of bio-diesel standards ASTM-D6751. Zang et al. (2018) prepared soap and lotion using seed oil of its allied species, *B. aegyptiaca*, and reported impressive characteristics. Similarly, the seed oil of *B. roxburghii* could also be explored for its cosmetic applications.

Table 1. Nutritional composition of seed kernels of *Balanites roxburghii* (KA-10A accession; Yadav et al., 2022).

Components	**Composition**
Proximate composition (%)	
Moisture	12.87 ± 1.23
Fat	43.20 ± 0.46
Protein	10.96 ± 0.09
Carbohydrate	5.74 ± 0.03
Ash	4.60 ± 0.12
Fiber	15.12 ± 1.21
Energy (Kcal/100 g)	422.81
Macroelements (mg/g DW)	
Nitrogen	28.40
Phosphorous	7.62
Potassium	14.00
Sulphur	5.60
Sodium	0.32
Calcium	6.05
Magnesium	6.00
Microelements (μg/g DW)	
Boron	28.96
Zinc	431.00
Iron	1419.00
Manganese	44.50
Copper	36.80
Phytochemicals composition	
Total phenolic content (mg GAE/g)	2.72 ± 0.03
Tannins (mg TAE/g DW)	1.99 ± 0.02
Physicochemical properties of seed oil	
Colour	Yellow
State at room temperature	Liquid
Density (g/cm^3)	0.915
Refractive index	1.33
Free fatty acid content (%)	0.31 ± 0.01
Peroxide value (meq O_2/kg)	69.98 ± 5.18
Ioddine value (I_2/100 g)	81.19 ± 0.72
Unsaponification value (%)	1.93 ± 0.14
Carotenoids (mg/kg)	5.36 ± 0.39
Tocopherol composition of seed oil (mg/100 g oil)	
α	7.29 ± 0.03
β+γ	1.79 ± 0.01
δ	0.72 ± 0.01

Table 1. contd. ...

Table 1. contd. ...

Components	**Composition**
Phytosterol composition of seed oil (mg/100 g oil)	
Ergosterol	5.16 ± 0.53
Stigmasterol + campesterol	40.78 ± 1.60
β-Sitosterol	126.90 ± 1.08
Squalene	17.45 ± 0.03
Fatty acid composition of seed oil (% composition)	
Palmitic acid (16:0)	17.79 ± 0.15
Stearic acid (18:0)	6.31 ± 0.01
Oleic acid (18:1)	56.38 ± 0.28
Linoleic acid (18:2)	18.77 ± 0.44
Linolenic acid (18:3)	0.38 ± 0.03
1,9,17-Docosatriene (22:3)	0.23 ± 0.03
Total saturated fatty acids	24.10
Total monounsaturated fatty acids	56.38
Total polyunsaturated fatty acids	19.38

DW – Dry weight; TAE – Tannic acid equivalent; GAE – Gallic acid equivalent

Diosgenin is a triterpenoid steroidal sapogenin in high demand in the drug industry as a precursor for steroidal hormones and corticosteroid production (Jesus et al., 2016)). The diosgenin content of *B. roxburghii* is impressive. Varshney and Vyas (1982) reported 0.97% in root wood, 0.36% in stem wood, 0.62% in seed kernel, 0.32% in fruit pulp, and 0.15% in epicarp. In another study, Amalraj and Shankarnarayan (1987) reported up to 5.6% diosgenin content in the fruits. These studies put *B. roxburghii* to be a good source of diosgenin in India.

Taxonomy

The taxonomic treatment of *Balanites roxburghii* is very interesting and comprehensively discussed by Sands (2001). The close resemblance of *B. roxburghii* with one of the African species, *B. aegyptiaca*, led many workers to consider both species the same. However, Voigt (1845) for the first time recognized this plant as *B. aegyptiaca* var. *indica* based on the description of Roxburgh's *Xymenia aegyptiaca* and Graham's *B. aegyptiaca*. Planchon (1854) raised this Indian plant to species level, as *Balanites roxburghii*, based on two distinct characteristics; the first is that the Indian plant bears leaves and flowers on the thorns (Figure 1B), whereas the African plant does not and the second is the inner surface of the petals are villous in the Indian plant (Figure 1C) and it is glabrous in the African plant. Duthie (1903) identifies one more distinguishing characteristic of the Indian plants, i.e., no elongation of the ovary during the development, unlike in the African plant. Even though many taxonomists clarified that *B. roxburghii* is a distinct species from *B. aegyptiaca*, many authors prefer using the name *B. aegyptiaca* for Indian specimens (Sands, 2001). In the present study, we considered only the articles named *Balanites roxburghii* to avoid confusion.

Germplasm Collection and Diversity of Genetic Resources

Plant genetic diversity is a broad term that encompasses all the variation existing among different genotypes or populations of a species concerning the total genetic makeup of genotypes (Bhandari et al., 2017; Swarup et al., 2021). Evolutionary forces, such as selection, mutation, migration, and genetic drift lead to the creation of genetic diversity within a species and are apparent in landraces, cultivars, traditional varieties, and crop wild relatives (Bhandari et al., 2017). This variability helps a species withstand or adapt to varied environmental conditions; it also acts as a storehouse of genes responsible for valuable traits with commercial value. The first essential step in assessing the variability within a species is collecting genotypes, and populations from different geographical locations. Subsequently, assessment is done by three major methods, viz., 1) morphological analysis, 2) biochemical analysis, and 3) molecular marker analysis (Govindaraj et al., 2015; Bhandari et al., 2017). Alongside this, cytological characteristics are also helpful in some species.

Thus, as said above, it is necessary to collect the germplasm and assess the diversity of *Balanites roxburghii*, with as many methods as possible, to bring into cultivation and subsequent commercial exploitation. The following sections discuss the diversity of *B. roxburghii* as analyzed by various methods.

Morphological Variability

Morphological or phenotypic characteristics have traditionally been used to analyze the genetic diversity of a species, and farmers used to select individuals with the best-looking seeds, fruits, or other desirable traits for the following year's plantation (Acquaah, 2012). The morphological variability studies in *Balanites roxburghii* are majorly concentrated on fruits, including their parts, such as epicarp, mesocarp and seed kernels. Amalraj and Shankarnarayan (1987), assessed the fruit and leaf-related traits in 50 accessions collected from four districts of Rajasthan state of India, viz., Sirohi, Pali, Jodhpur, and Baramer. The fruit-related traits included length, width, circumference, weight, and volume. The fruit length was 2.6–8.2 cm, width 2.2–6.3 cm, circumference 7.0–19.4 cm, weight 6.0–80.0 g, and volume 5–106 cm^3. Among these traits, fruit volume was highly variable with a coefficient of variation (CV) of 54.09%, followed by weight, length, and width with CVs of 44.56, 21.48, and 18.03%, respectively. The leaf characters, length and breadth were also analyzed; average length and breadth were 3.45 and 1.67 cm, respectively, with the CVs of 18.93 and 26.84%, respectively.

Yadav and Murthy (2022) assessed the phenotypic diversity of 45 *B. roxburghii* accessions collected from South Indian states, viz., Andhra Pradesh, Karnataka, Telangana, and Tamil Nadu (Figure 2). In the study, 12 fruit-related traits were considered; among them, three were qualitative, viz., fruit shape, fruit apex, and ridges on fruit, whereas nine were quantitative, viz., weight, length and width of fruits, weight of pulp (mesocarp), rind (epicarp), seed kernel and endocarp, and percentage composition of pulp and seed kernel. Among the qualitative characters, fruit shape was highly variable with six types; elliptic shape was the prominent type found in 46.67% (n = 21) of accessions, followed by obovate (17.78%, n = 8), oblong (13.33%, n = 6), round (11.11%, n = 5), ovate (8.89%, n = 4), and pyriform (2.22%, n = 1). Fruit apex trait had three variables; flattened type was prominent and found in 55.56% of the accessions, followed by round and acute types. Ridges on the fruit is one important character because in fruit with pale or no ridges it is easy to remove the epicarp, whereas the deep, prominent ridges make the epicarp deeply inserted in that region and makes removing the epicarp very difficult. This trait might be helpful while processing the fruits; 68.89% (n = 31) of the accessions had pale lines, and 22.22% (n = 10) only had prominent ridges; the remaining had no ridges.

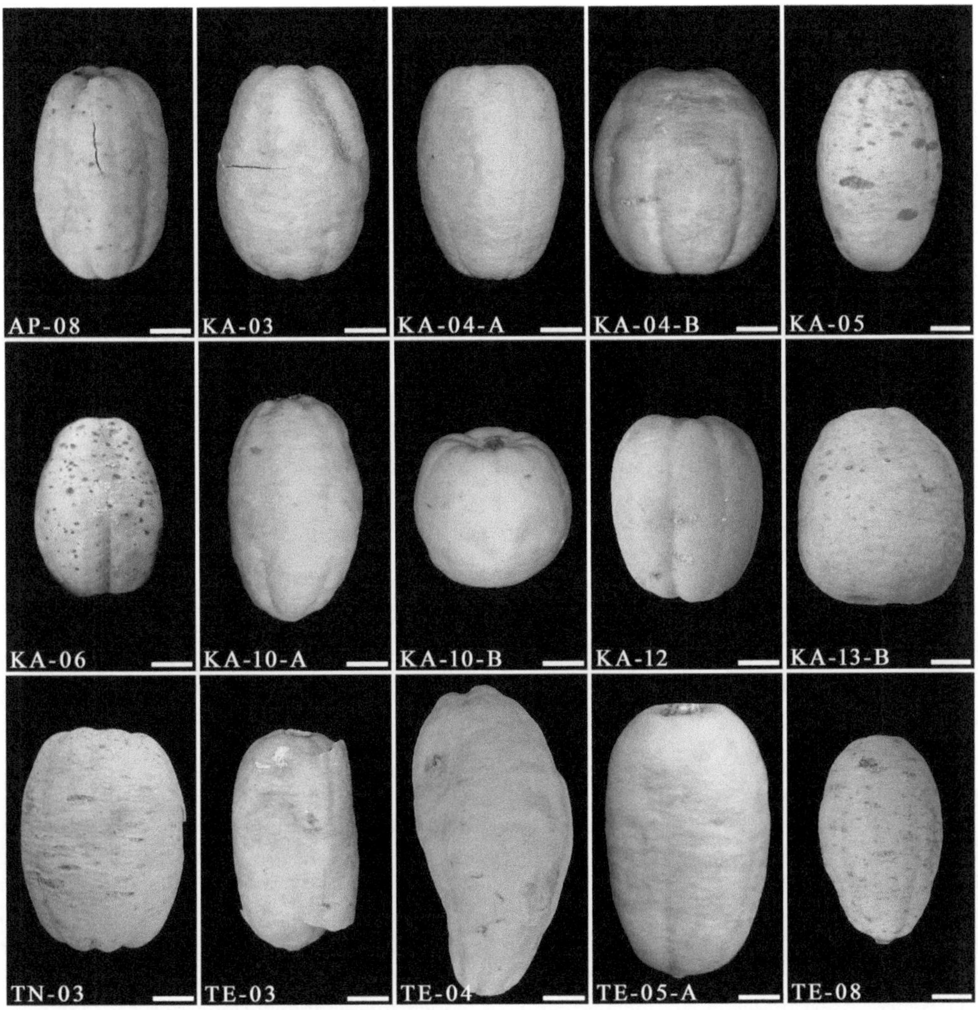

Figure 2. Morphological variability in fruits of *Balanites roxburghii* superior accessions. Scale: 1 Bar = 1 cm

Fruit dimensions are the significant traits to be considered in the variability analysis as they directly influence the composition of all other parts. Fruit weight ranged from 10.23–37.69 g, fruit length from 3.56–7.66 cm, and fruit width from 2.72–4.37 cm. Compared to the fruits of southern Indian accessions, the fruits of Rajasthan state (Amalraj and Shankarnarayan, 1987) are larger. Pulp weight ranged between 1.71–10.24 g and seed kernel weight between 0.77–3.87 g. The percentage composition of pulp was up to 35.62% and that of seed kernels was up to 14.37%. Among the quantitative traits studied, pulp weight was highly variable with a CV of 35.46%, followed by endocarp weight (CV = 31.78%), seed kernel weight (CV = 31.16%), rind weight (cv = 30.63%), and fruit weight (CV = 28.59%).

Correlation between these traits showed that fruits with greater length tend to have heavy endocarp (r = 0.78) and large seed kernel (r = 0.56), whereas fruits that are wider tend to have more pulp (r = 0.81) and epicarp (r = 0.82). Thus, the appearance of fruit could be a primary screening method in identifying desirable genotypes for cultivation purposes.

Cytological Characterization

Cytological characterization is one of the important methods of plant genetic diversity analysis. It includes the analysis of cytological features, such as chromosome size, secondary constriction, centromere position, arm ratio, ploidy level, DNA content, and total chromosome length (Albert et al., 2010; Das et al., 2010). Research on the cytological characteristics of *Balanites roxburghii* is scant and primarily focuses on identifying chromosome numbers. Sanjappa (Love, 1979) reported the chromosome number of *B. roxburghii* as n = 18. Nevertheless, a more comprehensive study by Ram et al. (2008) involved the collection of *Balanites* specimens from four locations in the arid regions of Rajasthan, India. Their observations revealed chromosomal mosaicism among different accessions and within individual plants of the same accession. Among the four accessions, two consistently exhibited gametic chromosome numbers of n = 18, while one accession displayed a combination of n = 18 and n = 16. Remarkably, another accession showed a highly variable range of gametic numbers, specifically n = 9, 11, 13, 14, and 16, but never 18. Despite Ram et al. (2008) referring to the species as *B. aegyptiaca*, Sands (2001) and POWO (2024b) clearly state that *B. aegyptiaca* is not found in India. In India, the genus *Balanites* is exclusively represented by *B. roxburghii*. Therefore, it is evident that when the authors mention *B. aegyptiaca*, they are referring to *B. roxburghii*. It seems that *B. aegyptiaca* also exhibits chromosome mosaicism as evidenced by various reports, such as Hall and Walker (1991) and Goldblatt (1979) report n = 9 for *B. aegyptiaca*, whereas CABI (2024) mentions the existence of n = 8. Chromosome mosaicism is a common phenomenon reported in many species, which is often associated with gene mutations, endopolyploidy, genome reduction, erratic cell division, and chromosome irregularities (Ram et al., 2008). However, Hall and Walker (1991) and Goldblatt (1979) suggest the basic chromosome number for the genus *Balanites* as x = 9.

Biochemical Characterization

Biochemical characterization involves the assessment of various metabolomics, such as proteins, enzymes, isozymes and secondary metabolites, such as phenolics, flavonoids, alkaloids, and terpenoids (Razzaq et al., 2019). The variability of biochemicals depends on many factors, such as edaphic, stress, environmental, and genotypic. However, constant exposure of plants to these environmental conditions leads to permanent changes in the genetic constituent (Nino et al., 2007; Dalawai et al., 2021). Hence, assessing secondary metabolites leads to identifying the superior genotypes to introduce into cultivation. Biochemical characterization in *Balanites roxburghii* is majorly concentrated on diosgenin content and seed oil nutraceuticals.

Diosgenin is distributed in all the parts of *B. roxburghii*, and its content was already mentioned earlier in this article. Diosgenin variability between 0.1–5.6% was found in the fruits of 50 accessions collected from Jodhpur, Baramer, Pali, and Sirohi districts of Rajasthan state, India (Amalraj and Shamkarnarayan, 1987). Recently, we have analyzed diosgenin variability in the epicarp, pulp, and seed kernels of southern Indian populations and found enormous variability with CVs of 61.93, 49.98, and 61.03%, respectively (unpublished data).

The seed oil content and nutraceuticals vary significantly among the 19 accessions of *B. roxburghii* collected in southern India (Yadav et al., 2023). The seed oil content ranged from 43.18–60.99%, with a CV of 7.99%. The nutraceuticals, such as carotenoids, ranged between 2.94–22.58 mg/kg oil (CV = 54.95%), α-tocopherols between 1.21–10.55 mg/100 g oil (CV = 33.92%), β and γ tocopherols together between 0.89–4.98 mg/100 g oil (CV = 49.75%) and δ tocopherol between 0.0 (not detected) to 3.44 mg/100 g oil (CV = 116.90%). Phytosterols, such as ergosterol, were found in the range of 0.0 (not detected) to 20.61 mg/100 g oil (CV = 78.32%), stigmasterol and campesterol together in the range of 12.36–60.69 mg/100 g oil (CV = 40.25%), β-sitosterol in the range of 52.50–147.28 mg/100 g oil (CV = 24.97%), and squalene in the range of 3.98–65.45 mg/100 g oil (CV = 113.53%). The total phenolic content of the oil was found at 2.22–10.18 mg gallic acid equivalent/100 g oil concentrations. In contrast, individual phenolics, such as gallic acid, protocatechuic acid, epicatechin, *trans*-cinnamic acid, quercetin, and kaempferol, varied considerably among the accessions. The fatty acids are prominent components of the oil, and *B. roxburghii* seed oil was majorly constituted by oleic, linoleic, palmitic, and stearic acids. They showed much variability with the CVs of 19.08, 21.28, 14.12, and 13.14%, respectively.

Utilization of Germplasm

Balanites roxburghii should be systematically cultivated to harness its complete potential as a dependable source of diosgenin and edible oil. Despite this imperative, information on its breeding and improvement process is scarce. Nevertheless, certain studies have successfully identified genotypes through morphological and biochemical characterization. These findings could prove valuable in upcoming breeding and improvement programs for *B. roxburghii*.

Yadav and Murthy (2022) categorized 45 accessions of *B. roxburghii* from southern India based on fruit morphological traits, forming four clusters. Two clusters exhibited normal-sized fruits, while the third and fourth clusters featured larger-sized fruits. The morphological traits of superior genotypes have been summarized in Table 2 and their variability is presented in Figure 2. Cluster 3 comprised two accessions (KA-04A and TE-04) demonstrating superiority in seed kernels, with weights ranging between 3.21–3.84 g and fruit weights between 31.17–33.41 g. Cluster 4 included six accessions (KA-03, KA-04B, KA-13A, KA-13B, TN-03, and TE-05A) excelling in pulp weight, ranging between 6.98–10.24 g, and fruit weight between 29.57–37.69 g. The authors emphasize the preference for accessions within Cluster 3 for seed kernel-related traits and those within Cluster 4 for pulp-related traits.

In the subsequent phase of their research, Yadav et al. (2023) meticulously selected nineteen morphologically superior accessions from South India and performed a comprehensive seed oil characterization. Employing k-means clustering methods for accession grouping resulted in the identification of five clusters based on seed oil content and nutraceuticals. Each cluster showcased accessions excelling in specific traits. Seed oil quantity and some of its nutraceuticals of superior genotypes have been summarized in Table 2. Cluster 2 comprised nine accessions (AP-02A, AP-07, KA-04B, KA-05, KA-12, TN-03, TE-03, TE-05A, and TE-07B) demonstrating superiority in oil quantity and carotenoid content. Clusters 3 and 4 each encompassed a single accession. The accession in Cluster 3 (KA-06) exhibited the highest levels of squalene and β-sitosterol. At the same time, the one in Cluster 4 (AP-08) displayed elevated amounts of β, γ, and δ tocopherol, ergosterol, stigmasterol, and campesterol. Cluster 5 comprised five accessions (KA-03, KA-10A, KA-10B, KA-13B, and TE-08) featuring the highest α-tocopherol content and stigmasterol, campesterol, β-sitosterol, and squalene. Given the superior fruit morphology and seed oil-related traits exhibited by these accessions, it is recommended to prioritize them in future breeding, improvement, and domestication programs focused on seed oil-related traits.

As outlined in the preceding sections, *Balanites roxburghii* is a versatile fruit-bearing tree species prevalent in the arid regions of India. India's *B. roxburghii* distributed regions experience minimal rainfall, limited irrigation resources and favourable soil conditions. These areas present an opportunity for commercial cultivation of *B. roxburghii*, offering valuable products, such as seed oil and diosgenin and contributing to the economic well-being of local communities. However, no information is available on the systematic cultivation practices and release of new varieties of *B. roxburghii*.

Table 2. Some major morphological and seed oil traits of morphologically and biochemically superior accessions of *Balanites roxburghii* (Yadav and Murthy, 2022; Yadav et al., 2023).

Sl. No.	Accession	State	Fruit weight (g)	Pulp weight (g)	Seed kernel weight (g)	Seed oil content (%)	Carotenoids (mg/kg oil)	Tocopherols (mg/100 g oil)			**Phytosterols (mg/100 g oil)**			**Squalene (mg/100 g oil)**
								α	β+γ	δ	**Ergosterol**	**Stigmasterol + campesterol**	**β-Sitosterol**	
1.	AP-08	Andhra Pradesh	21.60±0.73	3.89±0.25	2.24±0.13	45.87 ± 1.23	5.57 ± 0.48	5.78 ± 0.02	4.98 ± 0.02	3.44 ± 0.04	20.61 ± 10.90	60.69 ± 2.15	85.19 ± 4.54	14.67 ± 0.06
2.	KA-03	Karnataka	29.57±0.71	7.33±0.29	2.51±0.14	48.85 ± 1.08	5.01 ± 0.05	5.69 ± 0.00	1.03 ± 0.00	0.19 ± 0.01	5.42 ±0.49	32.68 ± 0.52	124.77 ± 1.13	8.19 ± 0.08
3.	KA-04A	Karnataka	31.17±1.02	6.34±0.19	3.21±0.17	51.13 ± 0.80	5.11 ± 0.71	5.80 ± 0.01	1.80 ± 0.00	0.31 ± 0.02	ND	17.40 ± 0.14	61.87 ± 0.50	9.58 ± 0.03
4.	KA-04B	Karnataka	34.98±1.58	8.39±0.67	2.65±0.19	51.38 ± 0.75	5.69 ± 1.30	1.21 ± 0.00	0.90 ± 0.00	0.24 ± 0.01	2.18 ± 0.04	31.24 ± 0.13	95.70 ± 0.81	8.42 ± 0.39
5.	KA-05	Karnataka	23.26±1.00	5.17±0.2	2.51±0.16	48.43 ± 0.65	6.05 ± 0.84	4.79 ± 0.00	1.58 ± 0.00	0.47 ± 0.00	3.93 ± 0.53	25.73 ± 0.06	98.03 ± 0.79	7.95 ± 0.24
6.	KA-06	Karnataka	20.55±0.69	6.11±0.26	1.99±0.10	45.03 ± 0.62	7.76 ± 0.07	5.06 ± 0.00	2.75 ± 0.00	0.67 ± 0.04	4.54 ± 0.03	47.09 ± 0.22	147.28 ± 5.36	65.45 ± 0.06
7.	KA-10A	Karnataka	22.94±0.67	6.31±0.27	1.61±0.07	43.18 ± 0.45	5.34 ± 0.37	7.27 ± 0.02	1.77 ± 0.00	0.70 ± 0.01	5.14 ± 0.55	40.80 ± 1.62	127.04 ± 1.06	17.43 ± 0.02
8.	KA-10B	Karnataka	20.32±0.90	7.24±0.37	2.39±0.18	47.71 ± 0.30	6.39 ± 0.60	7.22 ± 0.01	2.26 ± 0.00	0.38 ± 0.06	4.77 ± 0.75	33.72 ± 0.10	121.06 ± 1.29	14.16 ± 0.40
9.	KA-12	Karnataka	25.38±0.86	4.93±0.26	2.24±0.12	60.99 ± 1.25	6.03 ± 0.10	3.80 ± 0.00	0.89 ± 0.00	0.58 ± 0.03	3.92 ± 0.02	28.66 ± 0.09	108.35 ± 1.12	6.42 ± 0.07
10.	KA-13B	Karnataka	30.91±2.88	7.07±0.72	2.03±0.26	43.92 ± 0.80	12.02 ± 0.32	10.34 ± 0.01	1.63 ± 0.01	ND	9.90 ± 0.07	57.32 ± 0.09	139.22 ± 1.11	8.07 ± 0.01
11.	TN-03	Tamil Nadu	37.69±0.94	10.24±0.52	2.47±0.13	49.72 ± 0.73	12.68 ± 0.01	6.89 ± 0.01	2.48 ± 0.01	1.09 ± 0.01	10.22 ± 0.05	38.96 ± 0.70	103.85 ± 0.83	9.04 ± 0.01
12.	TE-03	Telangana	28.11±0.92	4.72±0.48	2.34±0.29	47.66 ± 0.50	22.58 ± 0.25	9.75 ± 0.09	1.40 ± 0.00	ND	6.74 ± 0.84	19.25 ± 0.12	90.09 ± 0.73	5.51 ± 0.21
13.	TE-04	Telangana	33.41±0.92	5.05±0.25	3.87±0.15	48.23 ± 0.60	9.15 ± 0.02	8.58 ± 0.02	3.46 ± 0.01	0.73 ± 0.07	11.79 ± 0.24	29.02 ± 0.26	64.86 ± 0.52	3.98 ± 0.00
14.	TE-05A	Telangana	36.38±0.91	8.58±0.24	2.98±0.19	46.17 ±0.43	11.52 ± 0.69	8.98 ± 0.05	1.08 ± 0.00	ND	4.34 ± 0.03	22.69 ± 0.34	102.62 ± 0.83	7.69 ± 0.01
15.	TE-07B	Telangana	21.19±0.69	6.48±0.34	2.36±0.18	49.49 ± 0.29	6.99 ± 0.01	7.12 ± 0.02	3.25 ± 0.00	1.06 ± 0.01	3.97 ± 1.08	36.58 ± 0.39	88.33 ± 0.81	7.94 ± 0.12
16.	TE-08	Telangana	23.99±1.09	5.65±0.38	2.62±0.10	45.16 ± 0.33	8.24 ± 0.46	10.55 ± 0.06	2.04 ± 0.00	0.73 ± 0.01	5.67 ± 0.23	38.68 ± 0.34	113.46 ± 0.90	8.57 ± 0.01

ND – Not detected

Conservation of Genetic Resources

Both *in situ* and *ex situ* conservation methods complement each other in protecting plant genetic resources, which in turn is crucial in protecting agrobiodiversity (Valkoun and Damania, 1993). However, *in situ* conservation offers many advantages over *ex situ* conservation strategies such as gene banks and gardens (Ledig, 1986). One such advantage is, *in situ* conservation, the populations are free to evolve in the natural conditions and keep the germplasm in tune with the changing environment, whereas in *ex situ*, it is not possible, and germplasm is static (Valkoun and Damania, 1993; Ledig, 1986). Concerning the *Balanites roxburghii*, the populations, including the superior one identified by Yadav and Murthy (2022) and Yadav et al. (2023), are all situated in forests that are under protection by the respective state Governments. The protected areas include reserved forests, wildlife sanctuaries, and national parks. Thus, the populations are being conserved naturally. Apart from this, there need to be some special strategies to ensure the efficient conservation of *B. roxburghii* in both *in situ* and *ex situ*.

Conclusion

Balanites roxburghii exhibits widespread distribution across the arid regions of India, showcasing significant variability in morphological, biochemical, and cytological characteristics. Numerous studies have pinpointed superior genotypes with the potential for integration into future breeding, crop improvement and cultivation programs for *B. roxburghii*. The substantial topographic, edaphic and climatic diversity in India likely plays a pivotal role in fostering variability among *B. roxburghii* populations. Nevertheless, prolonged exposure to these environmental conditions induces genetic changes, as evidenced by numerous studies. Thus, molecular marker-based variability analysis is essential to validate these observations in *B. roxburghii*. Furthermore, comprehensive analyses are required for variables, such as fruit yield, tree height, canopy area, and additional morphological and biochemical traits. These investigations are crucial for geneticists and breeders, providing insights to identify, improve and select potential genotypes for future cultivation programs.

References

Acquaah, G. 2012. History of plant breeding. *In:* G. Acquaah. [Ed.], *Principles of Plant Genetics and Breeding*, 2nd Edn., 22–39. John Wiley & Sons, Ltd., West Sussex, UK. https://doi.org/10.1002/9781118313718.

Al Ashaal, H.A., A.A. Farghaly, M.M.A. El Aziz, and M.A. Ali. 2010. Phytochemical investigation and medicinal evaluation of fixed oil of *Balanites aegyptiaca* fruits (Balantiaceae). *Journal of Ethnopharmacology*, *127*: 495–501. https://doi.org/10.1016/j.jep.2009.10.007.

Albert, P.S., Z. Gao, T.V. Danilova, and J.A. Birchler. 2010. Diversity of chromosomal karyotypes in maize and its relatives. *Cytogenetic and Genome Research*, *129*: 6–16. https://doi.org/10.1159/000314342.

Amalraj, V.A., and K.A. Shankarnarayan. 1987. Variability in *Balanites roxburghii* Pl. in the Indian desert. *Journal of the Bombay Natural History Society*, *84*: 476–478.

Bhandari, H.R., A.N. Bhanu, K. Srivastava, M.N. Singh, Shreya, and A. Hemantaranjan. 2017. Assessment of genetic diversity in crop plants: An overview. *Advances in Plants & Agriculture Research*, *7*: 279–286. https://doi.org/10.15406/apar.2017.07.00255.

Burkill, H.M. 1985. *The Useful Plants of West Tropical Africa*, (Vol. 1), *Families A–D*. Royal Botanic Gardens, Kew, U.K.Chapagain, B., and Z. Wiesman. 2005. Variation in diosgenin level in seed kernels among different provenances of *Balanites aegyptiaca* Del (Zygophyllaceae) and its correlation with oil content. *African Journal of Biotechnology*, *4*: 1209–1213.

Chapagain, B.P., Y. Yehoshua, and Z. Wiesman. 2009. Desert date (*Balanites aegyptiaca*) as an arid lands' sustainable bioresource for biodiesel. *Bioresource Technology*, *100*: 1221–1226. https://doi.org/10.1016/j.biortech.2008.09.005.

Chatterjee, A., and S.C. Parkrashi. 1997. *The Treatise of Indian Medicinal Plants*, (Vol. 3), Publication and Information Directed, New Delhi, India.

Chopra, R.N., S.L. Nayar, and I.C. Chopra. 1956. *Glossary of Indian Medicinal Plants*. CSIR, New Delhi, India.

Cooke, T. 1902. *The Flora of the Presidency of Bombay* (Vol. 1). Taylor and Francis, London. UK.

Dalawai, D., C. Aware, J.P. Jadhav, and H.N. Murthy. 2021. RP-HPLC analysis of diterpene lactones in leaves and stem of different species of *Andrographis*. *Natural Product Research, 35*: 2239–2242. https://doi.org/10.1080/14786419.2019.1662004.

Das, A.B., I.C. Mohanty, D. Mahapatra, S. Mohanty, and A. Ray. 2010. Genetic variation of Indian potato (*Solanum tuberosum* L.) genotypes using chromosomal and RAPD markers. *Crop Breeding and Applied Biotechnology, 10*: 238–246. https://doi.org/10.1590/S1984-70332010000300009.

Duthie, J.F. 1903. *Flora of the Upper Gangetic Plain 1*. Superintendent of Government Printing, Calcutta, India.

Goldblatt, P. 1979. Miscellaneous chromosome counts in angiosperms, II. Including new family and generic records. *Annals of the Missouri Botanical Garden, 66*: 856–861. https://doi.org/10.2307/2398925.

Govindaraj, M., M. Vetriventhan, and M. Srinivasan. 2015. Importance of genetic diversity assessment in crop plants and its recent advances: An overview of its analytical perspectives. *Genetiocs Research International*, Article ID 431487. http://dx.doi.org/10.1155/2015/431487.

Hall, J.B., and D.H. Walker. 1991. *Balanites aegyptiaca*: A monograph. School of Agricultural and Forest Sciences, University of Wales, Bangor, UK.

Hardman, R., and E.A. Sofowora. 1972. A reinvestigation of *Balanites aegyptiaca* as a source of steroidal sapogenins. *Economic Botany, 26*: 169–173. https://doi.org/10.1007/BF02860779.

Hardman, R., and C.N. Wood 1972. Effect of 2,4,5-trichlorophenoxyacetic acid on the yield of diosgenin and yamogenin from seeds of *Balanites orbicularis* during germination. *Phytochemistry, 11*: 1067–1071. https://doi.org/10.1016/S0031-9422(00)88454-X.

Jesus, M., A.P.J. Martins, E. Gallardo, and S. Silvestre. 2016. Diosgenin: recent highlights on pharmacology and analytical methodology. *Journal of Analytical Methods in Chemistry*, Article ID 4156293. http://dx.doi.org/10.1155/2016/4156293.

Khanvilkar, P., G. Patel, P.S. Nagar, and S.N. Shah. 2016. *Balanites roxburghii* Plant oil as potential non-edible feedstock for biodiesel production. *Energy and Power, 6*: 21–27.

Ledig, F.T. 1986. Conservation strategies for forest gene resources. *Forest Ecology and Management, 14*: 77–90. https://doi.org/10.1016/0378-1127(86)90093-9.

Love, A. 1979. IOPB chromosome number reports LXIII. *Taxon., 28*: 265–279. https://doi.org/10.1002/j.1996-8175.1979.tb00499.x.

Murthy, H.N., G.G. Yadav, Y.H. Dewir, and A. Ibrahim. 2021. Phytochemicals and biological activity of desert date (*Balanites aegyptiaca* (L.) Delile). *Plants, 10*: 32. https://doi.org/10.3390/plants10010032.

Nadakarni, K.M. 1954. *Indian Materia Medica*, 3rd Edn. Bombay Popular Prakashan, Mumbai, India.

Nino, J., D.A. Jiménez, O.M. Mosquera, and Y.M. Correa. 2007. Diosgenin quantification by HPLC in a *Dioscorea polygonoides* tuber collection from Colombian flora. *Journal of the Brazilian Chemical Society, 18*: 1073–1076. https://doi.org/10.1590/S0103-50532007000500030.

Planchon, J.E. 1854. Affinitis et synonymie de quelques genres nouveaux ou peu connus. *Annales des Sciences Naturelles; Botanique Ser., 4* (2): 256–266.

POWO. 2024a. *Balanites*. Available on https://powo.science.kew.org/taxon/urn:lsid:ipni.org:names:331380-2. (Accessed on 20 Jan. 2024.)

POWO. 2024b. *Balanites aegyptiaca*. Available on https://powo.science.kew.org/taxon/urn:lsid:ipni.org:names:813589-1. (Accessed on 20 Jan. 2024.)

Radunz, A., W. Grosse, and J. Mewi-Sutz. 1985. Seeds of the East African savannah bush *Balanites orbicularis* as a possible new source of lipids for commercial use. *Journal of the American Oil Chemists' Society, 62*: 1251–1252. https://doi.org/10.1007/BF02541837.

Ram, H., K. Arun, S.K. Sharma, and S.R. Rao. 2008. Genetic variation and chromosomal mosaicism in the natural populations of *Balanites aegyptiaca* (L.) Del. from Rajasthan, India. *Journal of Sustainable Forestry, 26*: 328–343. https://doi.org/10.1080/10549810701879776.

Razzaq, A., B. Sadia. A. Raza, M. Khalid Hameed, and F. Saleem. 2019. Metabolomics: A way forward for crop improvement. *Metabolites, 9*: 303. https://doi.org/10.3390/metabo9120303.

Sands, M.J.S. 2001. The desert date and its relatives: A revision of the genus *Balanites*. *Kew Bulletin, 56*: 1–128. https://doi.org/10.2307/4119431.

Sofowora, E.A., and R. Hardman. 1973. Steroids, phthalyl esters, and hydrocarbons from *Balanites wilsoniana* stem bark. *Phytochemistry, 12*: 403–406. https://doi.org/10.1016/0031-9422(73)80028-7.

Swarup, S., E.J. Cargill, K. Crosby, L. Flagel, J. Kniskern, and K.C. Glenn. 2021. Genetic diversity is indispensable for plant breeding to improve crops. *Crop Science, 61*: 839–852. https://doi.org/10.1002/csc2.20377.

Valkoun, J., and A.B. Damania. 1993. Methodologies for *in situ* conservation of plant genetic Resources. *JIRCAS International Symposium Series, 2*: 168–173.

Varshney, I.P., and P. Vyas. 1982. Saponin and sapogenin contents of *Balanites roxburghii*. *International Journal of Crude Drug Research, 20*: 3–7.

Virendra, S., T. Priyanka, J.R. Patel, M.L. Kori, and V.K. Dixit. 2009. Preliminary phytochemical and antiasthmatic studies on stem bark of *Balanites roxburghii* Planch. *International Journal of Pharmaceutical and Clinical Research, 1*: 40–42.

Voigt, J.O. 1845. *Hortus Suburbanus Calcuttensis*. Bishop's College Press, Calcutta, India.

Yadav, G.G., and H.N. Murthy. 2022. Analysis of phenotypic variation and selection of superior genotypes of *Balanites roxburghii* Planch. from South India. *Genetic Resources and Crop Evolution, 69*: 1993–2009. https://doi.org/10.1007/s10722-022-01366-x.

Yadav, G.G., V. Manasa, H.N. Murthy, and A.W. Tumaney. 2023. Chemical composition and nutraceutical characterization of *Balanites roxburghii* seed oil. *Journal of Food Composition and Analysis, 115*: 104952. https://doi.org/10.1016/j.jfca.2022.104952.

Yadav, G.G., H.N. Murthy, and Y.H. Dewir. 2022. Nutritional composition and in vitro antioxidant activities of seed kernel and seed oil of *Balanites roxburghii*: An underutilized species. *Horticulturae, 8*: 798. https://doi.org/10.3390/horticulturae8090798.

Zang, C.U., A.A. Jock, H.I. Garba, and Y.I. Chindo. 2018. Application of desert date (*Balanites aegyptiaca*) seed oil as potential raw material in the formulation of soap and lotion. *American Journal of Analytical Chemistry, 9*: 423–437. https://doi.org/10.4236/ajac.2018.99033.

13

Genetic Diversity of Pequi Fruits (*Caryocar brasiliense*) and Sustainable Utilization

Bruno Henrique Gomes,[1,*] *Mariana Gonçalves Mendes,*[2] *Ana Maria Bonetti*[2] and *Ana Paula Oliveira Nogueira*[1]

Introduction

The Cerrado (tropical savanna, Figure 1) is one of the most biodiverse Brazilian biomes, and is considered a world hotspot for biodiversity conservation and preservation (Myers et al., 2000; Peyre, 2021). It's the second largest biome in Brazil, with a territorial extension of approximately 21% of the national territory. It has a seasonal climate, with a rainy period (October to March) followed by a dry period (April to September) with diverse fauna and flora. The diverse phytophysionomic composition of the Cerrado creates an ideal environment for studying the environment and plant population variations.

The process of fragmentation due to activities linked to population growth and the development of agribusiness puts the maintenance, and environmental balance of Cerrado at risk (Ribeiro and Rodrigues, 2006). The direct implications of fragmentation on biodiversity are indiscriminate reduction of areas, extinction, and evolutionary compromise due to loss of genetic variability and may lead to reduced adaptation of species to environmental changes (Melo Jr et al., 2004).

The species that has emerged is the *Caryocar brasiliense* Cambes (Malpighiales: Caryocaraceae) (Figure 2). The species is a fruitful of great importance, especially for extraction, since they intended its fruits for food, as well as contributing as an important source of income for the needy populations of the regions of occurrence (Araujo, 1995; Kerr et al., 2007). The extensive period of supply of the fruits of the plumber, combined with the heterogeneity of the producing regions provide the existence of differences between their physical attributes, as a result, there is a wide genetic variability (Vera et al., 2005, 2007).

The *C. brasiliense* is a species of extractive economic base that feeds various families and serves as an alternative source of income, mainly for the rural environment (Santos et al., 2013). The

[1] Genetics and Plant Breeding Laboratory, Biotechnology Institute, Federal University of Uberlândia, Uberlândia, Minas Gerais, Brazil.
[2] Genetics Laboratory, Institute of Biotechnology, Federal University of Uberlândia, Uberlândia, Minas Gerais, Brazil.
Email: b.hgomes@hotmail.com
* Corresponding author

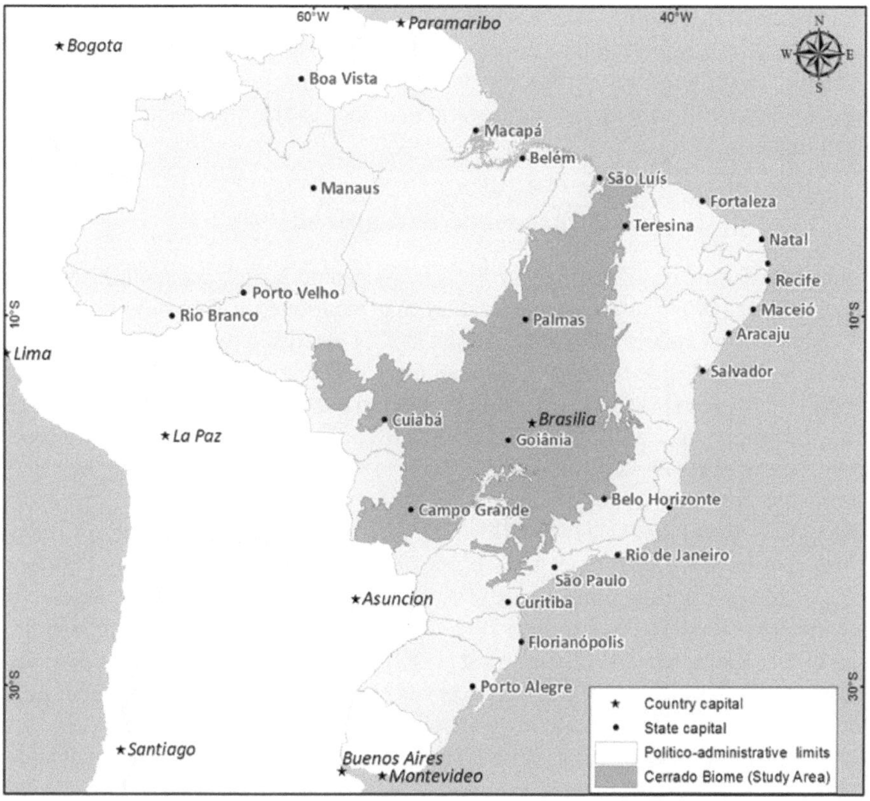

Figure 1. Location of the Cerrado biome in the Brazilian territory.

Figure 2. Morphology of the tree, inflorescence, flower, and fruit of the pequi tree grown at the Fazenda Água Limpa Experimental Station - UFU, Uberlândia–MG. A) pequi tree; B) side view of the floral buds; C) top view of the floral buds; D) pequi tree flower; E) fruits; F) arrangement of the pyrenes in the fruit, on the left fruit without thorns and the right fruit with thorns at the endocarp; G) details of the pyrenes, on the left fruit without thorns (yellowish color) and the right fruit with thorns endocarp (orange color); H) details of the endocarp with (right) and without thorns (left).

species offers various applications as a source of food, energy matrix, prospecting of molecules with pharmacological effects, reforestation, and woodworking among other uses. Something widely used the fruits (pequi) and leaves for both feeding and therapeutic purposes (Vieira and Martins, 2000). In cooking, they used fruit pulp in various preparations, such as rice with pequi, chicken with pequi, paçocas, and liqueurs.

Taxonomic Background

The Caryocaraceae family is difficult to describe and has a complicated taxonomic history. According to Prance and Silva (1973), the first publication that mentioned a Caryocaraceae member was possibly Clusius' book *Exoticorum* (1605). Aublet (1775) identified two new taxa that he classified in *Polyandria tetragynia*, subsequently known as Caryocar. Caryocar was categorized as "Genera incertae sedis" by Jussieu (1789) in his Genera Plantarum, which is still used to signify the inability to determine the proper place of a genus. Meyer's (1818) publication *Anthodiscus* described the only other Caryocaraceae genus known today. De Candolle (1824) was the first to segregate Caryocar into its own family, Rhizoboleae, in his *Prodromus*.

Bentham and Hooker (1862) classified Rhizoboleae as a Ternstroemiaceae tribe and Caryocaraceae as a distinct family. Caryocaraceae were classified as Ternstroemiaceae by Baillon (1872). Szyszylowicz (1895) recognized Caryocaraceae as a family in his book, which Hutchinson (1926) later moved into the order Theales. Phylogenetic studies by Wettstein (1935) placed Caryocaraceae in the Guttiferales order. Gundersen (1950), Cronquist (1968), and Takhtajan (1969),, however, reclassified Caryocaraceae in the Theales order. Caryocaraceae is now classified as a Malpighiales order. When completing the Bayesian phylogenetic analysis on the genome of chloroplasts of *C. brasiliense*, Nunes et al. (2020) were able to grasp previously unresolved connections in Malpighiales, putting Caryocaraceae as a distinct group in order.

The species commonly known as *Caryocar* spp. belongs to the family Caryocaraceae, of the order Malpighiales, and has 26 species distributed in two genera: Anthodiscus G.F.W. Meyer (10 species) and *Caryocar* L. (16 species) (Prance, 1990). In Brazil, there are 10 species of the genus *Caryocar* and three of the genus *Anthodiscus* (Souza, 2005). The most prevalent in the Brazilian territory is *C. brasiliense*, but *C. villosum*, *C. coriaceum*, *C. barbinerve*, *C. crenatum*, and *C. edulis* are found in the country. The *C. brasiliense* occurs throughout Cerrado (Figure 1), with wide distribution in Bahia, Ceará, Pará, Maranhão, Piauí, Tocantins, Mato Grosso, Matos Grosso do Sul, Minas Gerais, and São Paulo.

Characterization of Evaluation

Morphological Characterization

Studies carried out with *C. brasiliense* and *C. villosum*, indicate that the species allogamy. Prothandria (male sexual organs ripen before females) and hercogamy (spatial separation of anthers and stigmas) function as major mechanisms against autogamy; however, auto-pollination can occur in small proportions (Gribel and Hay, 1993; Martins and Gribel, 2007). The species has a chromosomal number of $2n = 46$ (Ehrendorfer et al., 1984; Prance, 1990).

The plants that characterize the family Caryocaraceae are dicotyledonous (Magnoliopsidas) and in their majority of medium to large size, can reach more than 10 meters in height, has a sturdy trunk with a diameter of up to 2 meters, coated by a dark shell and scratched. The branches are wide and the branches extend across the side inserted in opposite forms, are long, thick, and slightly inclined with basal branching. The leaves are opposite, long, petiolate, trifoliolata, and have serrated, crewed, dented, or rarely whole edges (Prance, 1990). About the inflorescence, they are characterized by being paniculated and inserted into the terminal racemes (Joly, 1985). The pedicels are articulated near the flower, which is hermaphrodite, cyclic, large, and actinomorphic (Prance, 1990). The calyx is characterized by being penta-lobed, and embroidered and the flower has five petals, being welded at the base, which fall after anthesis together with the stems. The anthers are bilocular, oblong, and longitudinally dehiscent (Joly, 1985).

The flowers of the squid have a yellow or white color, large and visible, with terminal inflorescence paniculate, hermaphrodite, cyclical, and actinomorphic, with 5–6 sepals and many petals, welded at the base or free and extrafloral nectar in the calyx (Oliveira and Pie, 1998). The Androecium are formed by numerous red stems welded at the base or forming groups with glandular filets and ovary containing between 4–20 carpels, and many loci each with only one egg (Dickison, 1990).

The fruit, known as pequi, is heterogeneous in size. It is composed of the exocarp or pericarp (external shell, green, or brown-green). The external mesocarp (white pulp), and the internal mesocarp, are oily and of color ranging from light yellow to orange, due to the presence of carotenoids. Under the inner mesocarp is the hard and woody endocarp, inside which is housed a white, oily, and edible almond. The fruit has 1–4 pyrenes. The pyrene is the unit of dispersion that consists of the seed enveloped by the endocarp (Gomes et al., 2022) and the internal mesocarp that covers the endocarps, constituting the edible part (Almeida et al., 1998; Melo Jr et al., 2004).

Thorns may be present in the endocarp or absent (Kerr et al., 2007). When the presents are thin, black or red, they are arranged in groups on the endocarp, pointed towards the pulp (Melo Jr et al., 2004; Vilas Boas, 2013). They are thin and rigid, 2–5 mm in length, and serve as a protection for the seed (Melo Jr et al., 2004). Seeds have a reniform shape, and oily of white in color (Almeida et al., 1998). The embryo exposes a curved root emerging from a seed.

Biochemical Characterization

C. brasiliense is an important source of nitrogen, phosphorus, potassium, calcium, and magnesium, mainly in their leaves (Medeiros and Haridasan, 1985). In almonds, there are large amounts of manganese, zinc, copper, magnesium, phosphorus, sodium, iron, and calcium (Mariano-da-Silva et al., 2009). Substances such as friedelin, friedelanol, oleic acid, β-sitosterol, stigmasterol, ellagic acid, and β-amyrine are present in the leaves (Ascari et al., 2013; Oliveira et al., 1968). Anthocyanic heterosides, saponins, tannins, flavonoids, steroids, triterpenes, and reducing glycosides are identified in leaves (Magalhães et al., 1988) and flower buds. Pequi has considerable carbohydrate values (19.60%) and titrated acidity of 0.9–2.0 (Vilas-Boas, 2004; Gomes et al., 2019).

The protein and lipid content of the pequi stands out about other species native to Cerrado such as araticum (Annonaceae), baru (*Dipteryx alata*), buriti (*Mauritia flexuosa*), cagaita (*Tenocalyx dysentericus*), jatobá (*Hymenaea courbaril*), and mangaba (*Hancornia speciosa*) (Almeida, 1998) presenting 2.64% protein, value less than only the jatobá (6%) and baru (3.87%). The lipid content of pequi is higher (20%) compared to other cerrado species (Almeida, 1998; Gomes et al., 2019). The amount of lipids present in the pulp and seed of its fruits provides high-energy values.

The pulp and almonds are predominantly unsaturated fatty acids. Fruit oil contains miristic acid, palmitic, stearic, oleic, and linoleic acid, vitamin A, and carotenoids (Guedes et al., 2017). The pulp contains 36% to 66% oil (Faria-Machado et al., 2015) with a high content of carotenoids (Aquino et al., 2009) and minerals (Silva et al., 2009). Pequi pulp oil contains about 54–60% oleic acid (Miranda-Vilela et al., 2009) which has important purposes as food, cosmetic and oleochemical use due to its high oxidative stability (Guedes et al., 2017). In the oil there is about 35–42% palmitic acid (Miranda-Vilela et al., 2009), which makes it suitable for margarine and fat applications (Guedes et al., 2017). In smaller quantities of palmitoleic acid (0,67%), linoleic (1,36%), linolenic (0,51%), stearic (1,28%), and arachidonic (0,12%) (Miranda-Vilela et al., 2009). Pulp oil has a high content of carotenoids with concentrations of up to 246 mg g^{-1} and triacyl glycerols (Aquino et al., 2009). The main triacyl glycerols present in pulp oil are 3-dipalmitoil-2-oleyl-glycerol, 1,2-dioleyl-3-palmitoyl-glyserol, and palmitoil-oleil-estearoyl-glucerol (Facioli and Gonçalves, 1998; Guedes et al., 2017). Among the main functions of fatty acids are the conformation of cell membranes, being also precursors of substances such as prostaglandins, thromboxanes, and leukotrienes.

Phenolic compounds make up the large class of food phytochemicals, which have various functions, and can act as antioxidants, blockers of the action of specific enzymes that cause inflammation, inhibit platelet agglomeration and the activation of carcinogens (Manach et al., 2004). Studies have shown that *C. brasiliense* has a high concentration of phenols, such as flavonoid, quercetin, and quercetin 3-O-arabinoside, in addition to acidic components such as gallic acid and quinic acid in fruit and peel mainly (Khouri et al., 2007; Roesler et al., 2008; Miranda-Vilela et al., 2009). Concerning the presence of carotenoids in the fruit, the main carotenoids present in the internal mesocarp, before and after conventional cooking, have been identified and quantified. The fruits presented, before cooking, β-carotene (9.35 mg g^{-1}), z-carotene (17.79 mg g^{-1}, cryptophlavine (12.14 mg.g^{-1}), β-cryptoxanthin (93.67 mg.g^{-1}, anteraxanthin (79.14 mg.g^{-1}), zeaxanthin (4.32 mg.g^{-1}), and mutatoxanthine (14.68 mg.g^{-1}). After cooking, the presence of the carotenoids β-carotene (12.17 mg g^{-1}), z-carotene (8.17 mg.g^{-1}), cryptophlavine (12.65 mg.g^{-1}, β-cryptoxanthin (7.99 mg.g^{-1}, anteraxanthin (31.28 mg.g^{-1}), zeaxanthin (79.73 mg.g^{-1}), and mutatoxanthine (2. mg.g^{-1}). The α and β carotenes together account for 10% of the total carotenoid pigments in the pequi pulp (Lima et al., 2007).

In the volatile aromas of pequi there are high terrors of ethyl hexanoate (52.9%) followed by small amounts of ethyl octanoate (4.6%), tetrahydrofurfuryl alcohol (4.3%), ethyl butanoate (4.1%), butyl palmitate (3.7%), isobutyl stearate (2.6%), and 3-methylvaleric acid (2.6%) (Maia et al., 2008). More than 95% of the constituents of essential oils in pulps were identified in studies (Geőcze et al., 2013). Among these chemical components, hydrocarbons, fatty acids, and terpenoids have been identified, and mostly ethyl hexanoate, ethyl octanoate, acyclic monoterpene β-ocymene, and hexanoic acid, as well as several other components in smaller quantities (Geőcze et al., 2013).

The pulp of pequi has a large amount of fiber (Lima et al., 2007). Dietary fibers are a set of plant-derived substances that are resistant to digestion and absorption in the human small intestine. The content of raw fiber contained in the pulp is high, approximately 13%. Regular consumption of fiber in the diet was related to the reduction of the risk of many pathological conditions, including a reduction in post-prandial glucose levels (Lima et al., 2007). In addition to the fibers, the considerable presence of pectin content in the pulp makes the fruit viable for processing. The function of pectin is to confer firmness, flavor retention, and aroma (Gancz et al., 2006), ensuring the stability of the pulp of the pequi.

The peculiar characteristics of the peach allow the species to present not only nutritional importance but also pharmacological action. Secondary metabolites are found in various parts of the plant, such as leaves, flowers, shells, and fruits, among others. These chemical constituents produced by secondary plant metabolism, such as compounds terpenes, alkaloids, and glycosides, present biological activities whose mechanism of action should be investigated (Miranda-Vilela, 2016; Palmeira et al., 2016; Vale et al., 2018; Torres et al., 2016; Traesel et al., 2016; Pires de Oliveira et al., 2021).

Molecular Characterization

Molecular markers are characteristics of DNA (deoxyribonucleic acid) that differentiate two or more individuals and are genetically inherited. The different molecular markers differ by the technology used to reveal DNA polymorphism variability and thus vary in the ability to detect differences between individuals, cost, ease of use, consistency, and repetitiveness. The main types of molecular markers can be classified into two groups, according to the methodology used to identify them: hybridization or DNA amplification via the polymerase chain reaction (PCR). The use of molecular markers is useful in plant genetic improvement, as it provides an unlimited number of DNA polymorphisms, independent of the environmental effects and the physiological stage of the plant. It sallows early and accurate identification of individuals with the best combination of favorable alleles, which can be used in genetic improvement (Londe, 2010).

Collevatti et al. (1999) developed the first microsatellite markers for molecular studies in *C. brasiliense.* In the author's research, it was noted that Mendelian inheritance and segregation were confirmed for all 10 loci in open pollination half-brother families, as well as the transferability of these 10 Loci to five other species of the same genus. The number of alleles per locus ranged from 10–22, with an average of 16 and expected heterozygosity ranging from 0.84–0.94.

In Londe (2010), molecular markers RAPD (Random Amplification of Polymorphic DNA) were used to characterize genotypes of squid with and without thorns in the endocarp. Since pequi is an allogamous plant, the use of codominant markers allows better analysis of the polymorphic pattern of the species, which helps to elucidate the genetic diversity between different genotypes.

Utilization of Germplasm

Utilization in Breeding and Improvement Programs

As it is a native species, one of the first stages of pre-improvement of *C. brasiliense* is the collection and identification of the best materials with a view to forming a base population. This population will serve as access to genetic material for collecting seeds and installing provenance and progeny tests. These tests are important due to the heterogeneity of the edaphoclimatic environment of its natural populations, and the great genetic variability observed among *C. brasiliense* genotypes (Gomes et al., 2022).

C. brasiliense has great variability for characters such as flowering and fruiting time, fruit production, dehiscence and shape of fruits, physical-chemical characteristics of fruits, nutritional factors of fruits, presence or absence of thorns, propagation (germination and emergence of seedlings), survival and rate of vegetative development (Vera et al., 2005; Vera et al., 2007; Moura et al., 2013; Hernández et al., 2020). Estimates of genetic parameters help in defining conservation and genetic improvement strategies in perennial species since it is possible to separate the environmental and genetic factors that interfere with phenotypic expression through the efficient use of quantitative genetic techniques.

C. brasiliense is a native species in the process of domestication. Its maintenance is concentrated in areas of the closet and germplasm banks where various studies are carried out such as plant production, phytosanitary, and genetic diversity both from phenotypic and molecular characteristics, to establish strategies of genetic improvement, domestication, and conservation of the species. Most of the pre-improvement works of the species are carried out with fruit-producing plants with thorns at the endocarps, from the analysis of data coming from the physical characteristics of the fruits. The biometric analysis of the fruits allows the researcher to obtain useful information for the conservation and exploitation of natural resources in a rational way (Gusmão et al., 2006). The phenotypic analysis contributes important information for the detection of genetic variability within and between populations of the same species and allows us to relate the genetic variability with environmental factors (Carvalho et al., 2003).

The differences found between and within populations are influenced by environmental factors and, when a species has a large geographical distribution, these differences occur due to adaptation to the climate of these regions (Rodrigues et al., 2006), as is the case with the *C. brasiliense* species. With the advance of anthropic actions, there is a need for the preservation and sustainable use of various fruit plants, and the use of biometric data makes it possible to estimate the productivity and yield of fruits (Rivas and Barilani, 2004), allowing the selection of the most productive and adapted to certain environments

When observing the allelic frequencies in *C. brasiliense*, it behaves in a varied way between and within populations. However, much of the variability observed in cuttings is found within populations and no allele fixation occurs (Melo Jr et al., 2004). As a result, the high variability benefits the existence of selection in be benefic the heterozygotes leading to low endogamy in populations (Melo Jr et al.,

2004). The distinct morphological characteristics are common in native plants of the Cerrado even between fruits of the same area, indicating that the region of occurrence of the fruits is an important factor in the variability of the character.

The characterization of the fruits to their physical aspects allows the obtaining of information that can be used in the determination of genetic variability, quality parameters, yield, and above all, in genetic breeding programs (Oliveira et al., 2009). Information such as mass, size of fruits and seeds, and number of seeds per fruit, for example, are characteristics of each species, with high influence on the environment (Alves et al., 2005). Due to the variability of morphological characters in the *C. brasiliense*, the analysis of fruits utilizing physical and chemical characters allows one to differentiate individuals from a population (Melo et al., 2001; Cruz and Carvalho, 2003) and species of the same genus (Cruz et al., 2001). In addition to the classification of ecological groups, characterization of germplasm (Alves et al., 2003), and various other applications.

The individual variation possibility of selection of offspring aimed at improving some characteristic of commercial interest. The seeds, because they are reproductive organs, present great stability in the vegetative characteristics, due to the greater genetic control and the pressure of selection exercised during the evolutionary process (Alves et al., 2010). The patterns of genetic variation that occur between and within peat populations allow for appropriate selection, conservation, and handling techniques, mainly concerning populations of natural occurrence. Studies of genetic diversity in natural populations of *C. brasiliense* have been developed in an attempt to measure the level of inheritable information in individuals, species, and populations (Gomes et al., 2022). Furthermore, knowledge of diversity allows the study of the extent and distribution of genetic variation between species and of taxonomic and evolutionary investigations.

Conservation of Useful Germplasm/Genetic Resources

The increasingly frequent dissemination of the qualities of pequi as a food has considerably increased the consumption of this fruit in the country. This has been causing intense pressure on the native population, causing genetic erosion and killing plants through nutritional depletion, when extractivism is carried out without any management plan. Therefore, the conservation of accessions and selected matrices in *ex situ* germplasm banks, followed by a process of domestication and encouragement of the cultivation of superior genetic materials is the main measure to save this species from extinction.

Although there are populations at the beginning of domestication *C. brasiliense* is a species formed by typically wild populations. *In situ* conservation should be the preferred germplasm conservation strategy, with *ex situ* conservation being a complementary form. Among the risk factors that can cause genetic erosion in the species, two deserve special attention due to their high potential for negative impact on the genetic variability of populations: habitat alteration and extractivism.

Habitat change results mainly from the occupation of areas of the Cerrado for agricultural activities. With deforestation to implement pastures and crops, local populations are eliminated or fragmented, which can lead to the loss of alleles or changes in the reproductive dynamics of the remaining subpopulations, with risks to long-term survival. The cutting and sale of wood's *C. brasiliense* is prohibited in the South, Southeast, Northeast, and Center-West regions of Brazil. In areas converted to pasture, plants remaining from the original vegetation can remain productive and in good condition for long periods. Even so, the contribution of these individuals to the conservation of germplasm is quite limited, since the seeds produced will hardly find appropriate conditions for germination and establishment. The contribution to gene flow via pollen may occur if other individuals occur in reserve areas at a distance below the reach of pollinators. In areas converted into annual crops, the remaining *C. brasiliense* plants find it difficult to survive for long periods due to management practices that greatly alter the local environment.

Another factor with an important impact on the conservation of *C. brasiliense* is the extractive exploitation of fruits for local consumption or commercialization. As it is highly appreciated by the region's population and even in urban centers outside the Cerrado region, there is intense harvesting activity, concentrating precisely on the most productive plants whose fruits have the most appreciated characteristics. This process, if carried out unplanned, can cause negative selection on populations, favoring the reproduction of agronomically inferior individuals. The design of an *in situ* conservation plan for *C. brasiliense* must integrate different strategies in a complementary way, from large conservation units to legal reserves and permanent preservation areas of private properties.

The *ex situ* conservation of *C. brasiliense* is still incipient. The few existing collections are maintained *in vivo* and are characterized more as working collections than collections with a sufficient sampling of the species diversity. Pioneering initiatives include the collections at Embrapa Cerrados, in the Federal District; the Federal University of Goiás, in Goiânia; the Federal University of Minas Gerais, in Montes Claros; the Federal University of Vales do Jequitinhonha and Mucuri, in Diamantina, Minas Gerais, from Emater Goiás, and the Federal University of Uberlândia, in Uberlândia Minas Gerais.

Concluding Remarks

The species *Caryocar brasiliense* Cambessedes, commonly known as the pequi tree, emerges as a crucial element within the Cerrado's ecosystem. Beyond its ecological importance, pequi serves as a valuable economic resource, contributing significantly to the livelihoods of local populations. The species is prized for its fruits, which are not only used for sustenance but also represent a source of income, particularly for rural communities. The review delves into the taxonomic background of *C. brasiliense*, providing a historical overview of its classification within the Caryocaraceae family and the Malpighiales order. The molecular characterization employing markers like microsatellites and RAPD sheds light on the genetic diversity within pequi populations. Such insights are essential for conservation, breeding programs, and domestication efforts to ensure the resilience of this species under increasing anthropogenic pressures. Furthermore, the morphological and biochemical characterizations provide a comprehensive understanding of *C. brasiliense.* The intricate details of its physical attributes, reproductive mechanisms, and biochemical composition, including the presence of essential nutrients, fatty acids, and bioactive compounds, highlight its multifaceted significance. The pequi's potential applications in various domains, such as food, energy, pharmacology, and woodworking, further underscore its versatility. The study emphasizes the need for conservation strategies, both *in situ* and *ex situ*, to safeguard the genetic diversity of pequi. As the species faces challenges from habitat alteration and extractive exploitation, careful planning and integrated conservation efforts become imperative. In situ conservation plans, ranging from large conservation units to private property reserves, should be devised to address the diverse threats to pequi populations. *Ex situ* conservation initiatives, while still in their early stages, hold promise in complementing in situ efforts. In summary, the research not only contributes valuable insights into the biology, ecology, and genetic diversity of *C. brasiliense* but also advocates for sustainable conservation and management strategies. As pequi continues to play a vital role in the lives of local communities and the ecological balance of the Cerrado, understanding and preserving this species is crucial for the long-term health and resilience of the savanna biome.

Acknowledgments

The author thanks the Federal University of Uberlândia (UFU), the Research Support Foundation of the State of Minas Gerais (FAPEMIG), National Council for Scientific and Technological Development (CNPq) e a Coordination for the Improvement of Higher Education Personnel (CAPES).

References

Almeida, S.D., T.S.A. Costa, and J.A. Silva. 1998. Frutas nativas do cerrado: caracterização físico-química e fonte potencial de nutrientes. *Cerrado: ambiente e flora*, 247–285.

Alves, E.U., R.D.L.A Bruno, A.P.D. Oliveira, A.U. Alves, A.U. Alves, and R.C.D. Paula. 2005. Influence of size and provenance on the germination and vigor of *Mimosa caesalpiniifolia* Benth. *Seeds. Rev. Árvore, 29*: 877–885.

Alves, R.M., M.D.V.D. Resende, B.D.S. Bandeira, T.M. Pinheiro, and D.C.R. Farias. 2010. Evaluation and selection of *Theobroma grandiflorum* parents, in Belem, Para. *Rev. Bras. Frutic., 32*: 204–212. https://doi.org/10.1590/S0100-29452010005000010.

Alves, R.M., A.A.F. Garcia, E.D. Cruz, and A. Figueira. 2003. Selection of morpho-agronomic descriptors for cupuaçuzeiro germplasm characterization. *Pesq. Agropec. Bras. 38*: 807–818. https://doi.org/10.1590/S0100-204X2003000700004.

Aquino, L.P., F.Q. Ferrua, S.V. Borges, R. Antoniassi, J.L.G. Correa, and M.A. Cirillo. 2009. Influence of pequi drying (*Caryocar brasiliense* Camb.) on the quality of the oil extracted. *Food Science and Technology, 29*: 354–357. https://doi.org/10.1590/S0101-20612009000200018

Araujo, F.D. 1995. A review of *Caryocar brasiliense* (Caryocaraceae): an economically valuable species of the central Brazilian cerrados. *Econ. Bot.*, 40–48. https://doi.org/10.1007/BF02862276.

Aublet, J.B.C.F. 1775. *Histoire des plantes de la Guiane francaise*. London, Paris: Pierre-François Didot Jeune.

Baillon, H. 1872. *Histoire des plantes: Ménispermacées, berbéridacées, nympheacées.... Tome troisième*. Librairire de L. Hachette et Cie.

Bentham, G., and J.D. Hooker. 1862. Ternstroemiaceae tribus Rhizoboleae. *In*: *Genera Plantarum 1* (Vol. 1): 180–181. London: Reeve.

Carvalho, J.E.U.D., R.F.R.D. Nazaré, and W.M.O.D. Nascimento. 2003. Physical and physicochemical characteristics of a bacuri (*Platonia insignis* Mart.) type with better industrial yield. *Rev. Bras. Fruticultura, 25*: 326–328. https://doi.org/10.1590/S0100-29452003000200036.

Collevatti, R.G., R.V. Brondani, and D. Grattapaglia. 1999. Development and characterization of microsatellite markers for genetic analysis of a Brazilian endangered tree species *Caryocar brasiliense. Heredity*, 83(6): 748–756. https://doi.org/10.1046/j.1365-2540.1999.00638.x

Cronquist, A. 1968. *The Evolution and Classification of Flowering Plants*. Houghton Miffin.

Cruz, E.D., and J.E.U.D. Carvalho. 2003. Fruit and seed biometry and germination of Micropholis cf. venulosa Mart. & Eichler (Sapotaceae). *Acta Amazonica, 33*: 389–398. https://doi.org/10.1590/S0044-59672003000300005.

Cruz, E.D., F.D.O. Martins, and J.E.U.D. Carvalho. 2001. Fruit and seed biometry and germination of jatoba-curuba (*Hymenaea intermedia* Ducke, Leguminosae-Caesalpinioideae). *Rev. Bras. Bot., 24*: 161–165. https://doi.org/10.1590/S0100-84042001000200005.

De Candolle, A.P. 1824. *Rhizoboleae*. Paris: Prodromus systematis naturalis regni vegetabilis.

Dickison, W.C. 1990. A study of the floral morphology and anatomy of the Caryocaraceae. *Bull. Torrey Bot. Club.*, 123–137. https://doi.org/10.2307/2997051.

Ehrendorfer, F., W. Morawetz, and J. Dawe. 1984. The neotropical angiosperm families Brunelliaceae and Caryocaraceae: first karyosystematical data and affinities. *Pl. Syst. Evol., 145*: 183–191. http://www.jstor.org/stable/23669986.

Facioli, N.L., and Gonçalves, L.A. 1998. Piqui (*Caryocar brasiliense* Camb) oil triglyceride composition modification by enzymatic way. *Quím. Nova*: *21*: 16–19. https://doi.org/10.1590/S0100-40421998000100003.

Faria-Machado, A.F., A. Tres, S.M. van Ruth., R. Antoniassi, N.T. Junqueira, P.S.N. Lopes, and H.R. Bizzo. 2015. Discrimination of pulp oil and kernel oil from pequi (*Caryocar brasiliense*) by fatty acid methyl esters fingerprinting, using GC-FID and multivariate analysis. *J. Agric. Food Chem., 63*(45). https://doi.org/10.1021/acs.jafc.5b03699.

Gancz, K., M. Alexander, and M. Corredig. 2006. *In situ* study of flocculation of whey protein-stabilized emulsions caused by addition of high methoxyl pectin. *Food Hydrocolloids, 20*(2–3): 293–298. https://doi.org/10.1016/j.foodhyd.2005.02.022.

Geőcze, K.C., L.C.A. Barbosa, P.H. Fidêncio, F.O. Silvério, C.F. Lima, M.C.A. Barbosa, and F.M. Ismail. 2013. Essential oils from pequi fruits from the Brazilian Cerrado ecosystem. *Food Research International, 54*(1): 1–8. https://doi.org/10.1016/j.foodres.2013.06.005.

Gomes, B.H., C.F. Lima, M.G. Mendes, A.M Bonetti, C.R. Furstenau, M.V. Faria, and A.P.O. Nogueira. 2019. Chemical composition and nutritional of pequi fruits (*Caryocar brasiliense* Camb.) with and without thorns at the endocarp. *Int. J. of Food Sci. and Nuts, 4*: 48–55.

Gomes, B.H., M.V. Faria, M.G. Mendes, A.M. Bonetti, R.J. Oliveira Júnior, and A.P.O. Nogueira. 2022. Genetic diversity and correlation between morphological traits of pequi fruits (*Caryocar brasiliense* Camb.) with and without thorns at the endocarp. *An. Acad. Bras. Ciênc, 94*. https://doi.org/10.1590/0001-3765202220210016.

Gribel, R., and J.D. Hay. 1993. Pollination ecology of *Caryocar brasiliense* (Caryocaraceae) in Central Brazil cerrado vegetation. *J. Trop. Ecol.*, 9(2): 199–211. https://doi.org/10.1017/S0266467400007173.

Guedes, A.M.M., R. Antoniassi, M.C. Galdeano, R. Grimaldi, M.G. de Carvalho, A.E. Wilhelm, and A.G. Marangoni. 2017. Length-scale specific crystalline structural changes induced by molecular randomization of pequi oil. *J. Oleo Sci., 66*(5): 469–478. https://doi.org/10.5650/jos.ess16192.

Gundersen, A. 1950. *Families of Dicotyledons*. Waltham: Chronica Botanica Co, 1950.

Gusmão, E., F. de Almeida Vieira, and É.M. da Fonseca Júnior. 2006. Fruits and endocarps biometry of murici (*Byrsonima verbascifolia* Rich. ex A. Juss). *Cerne*, *12*(1): 084–091.

Hernández, L.H., F.A.D. Santos, E.C. Palomino, E.V. Tambarussi, and C.B. Moraes. 2020. Genetic Variation of Trees of *Caryocar brasiliense* for Fruit Morphometric Traits. *Floresta Ambient.*, *28*. https://doi.org/10.1590/2179-8087-FLORAM-2018-0097.

Hutchinson, J. (1926). *Families of Flowering Plants*. London: Oxford Clarendon Press.

Joly, A.B. 1998 *Botânica–introdução à taxonomia vegetal*, 324–329. 12 Edn São Paulo, Nacionalp.

Jussieu, A.L. 1789. *Genera Plantarum*. Paris: Herissant.

Kerr, W.E., F.R.D. Silva, and B. Tchucarramae. 2007. Pequi (*Caryocar Brasiliense* Camb.): informações preliminares sobre um pequi sem espinhos no caroço. *Rev. Bras. Frutic.*, *29*: 169–171. https://doi.org/10.1590/S0100-29452007000100035 .

Khouri, J., I.S. Resck, M. Poças-Fonseca, T.M. Sousa, L.O. Pereira, A.B. Oliveira, and C.K. Grisolia. 2007. Anticlastogenic potential and antioxidant effects of an aqueous extract of pulp from the pequi tree (*Caryocar brasiliense* Camb). *Genet. Mol. Biol.*, *30*: 442–448. https://doi.org/10.1590/s1415-47572007000300024.

Lima, A.D., A.M.D.O. Silva, R.A. Trindade, R.P. Torres, and J.O.R.G.E. Mancini-Filho. 2007. Chemical composition and bioactive compounds in the pulp and almond of pequi fruit. *Rev. Bras. Frutic.*, *29*: 695–698. https://doi.org/10.1590/S0100-29452007000300052.

Londe, L.N., C. Ueira-Vieira, W.E. Kerr, and A.M. Bonetti. 2010. Characterization of DNA polymorphisms in *Caryocar brasiliense* (Camb.) in populations with and without thorn at the endocarp by RAPD markers. *An. Acad. Bras. Ciênc.*, *82*: 779–789. https://doi.org/10.1590/S0001-37652010000300024.

Magalhães, H.G., H. Monteiro Neto, M.H. Lagrota, M.D. Wigg, L.A.S. Guimarães, M.A.S.O. Loja, and R.R. Araújo.1988. Estudo estrutural do pequizeiro *Caryocar brasiliense* Camb. Caryocaraceae, sob o aspecto farmacoquímico e botânico. *Rev. Bras. Farm.*, *69*(1–3): 31–41.

Maia, J.G.S., Andrade, E.H.A., and da Silva, M.H.L. 2008. Aroma volatiles of pequi fruit (*Caryocar brasiliense* Camb.). *J. Food Compos. Anal.*, *21*(7): 574–576. https://doi.org/10.1016/j.jfca.2008.05.006.

Manach, C., A. Scalbert, C. Morand, C. Rémésy, and L. Jiménez. 2004. Polyphenols: food sources and bioavailability. *Am. J. Clin. Nutr.*, *79*(5): 727–747. https://doi.org/10.1016/j.jfca.2008.05.006.

Mariano-da-Silva, S., J.D.D.A. Brait, F.P.D. Faria, S.M.D. Silva, S.L.D. Oliveira, P.F. Braga, and F.M.D.S. Mariano-da-Silva. 2009. Chemical characteristics of pequi fruits (*Caryocar brasiliense* Camb.) native of three municipalities in the State of Goiás-Brazil. *Food Sci. Technol.*, *29*: 771–777. https://doi.org/10.1590/S0101-20612009000400011.

Martins, R.L., and R. Gribel. 2007. Pollination of *Caryocar villosum* (Aubl.) Pers. (Caryocaraceae), an emergent tree in central Amazônia. *Rev. Bras. Bot.*, *30*: 37–45. https://doi.org/10.1590/S0100-84042007000100005.

Medeiros, R.A., and M. Haridasan. 1985. Seasonal variations in the foliar concentrations of nutrients in some aluminium accumulating and non-accumulating species of the cerrado region of central Brazil. *Plant and Soil*, 433–436.

Melo, W.M.C., R. Pinho, J. Santos, and D.F. Ferreira. 2001. Use of morphoagronomical characters and molecular markers to evaluate genetic divergence among maize hybrids. *Ceres*, *48*(276): 195–207.

Melo Jr, A.F., D. Carvalho, J.S.R. Póvoa, and E. Bearzoti. 2004. Genetic structure of natural populations of pequizeiro (*Caryocar brasiliense* Camb.). *Scientia Forestalis*, *66*: 55–65.

Meyer, G.F.W. 1818. *Primitiae florae essequeboensis*. Gottingen: Sumptibus H. Dieterich.

Miranda-Vilela, A.L., C.K. Grisolia, I.S. Resck, and M.A. Mendonça. 2009. Characterization of the major nutritional components of *Caryocar brasiliense* fruit pulp by NMR spectroscopy. *Quimi. Nova*, 32: 2310–2313. https://doi.org/10.1590/S0100-40422009000900013.

Miranda-Vilela, A.L., I.F. Ribeiro, and C.K. Grisolia. 2016. Association between interleukin 6–174 G/C promoter gene polymorphism and runners' responses to the dietary ingestion of antioxidant supplementation based on pequi (*Caryocar brasiliense* Camb.) oil: a before-after study. *Genet. Mol. Biol.*, 39: 554–566. https://doi.org/10.1590/1678-4685-GMB-2015-0299.

Moura, L.R., S.R. Orpinelli, J.H. Sousa, M.B Faleiro, E.C. Conceição, D.M. Sugita, and V.M. Moura. 2017. *Caryocar brasiliense* shell ethanol extract in chronic cardiotoxicity induced by doxorubicin in rats. *Pesquisa Pesq. Vet. Bras.*, *37*: 713–724. https://doi.org/10.1590/s0100-736x2017000700011.

Myers, N., R.A. Mittermeier, C.G. Mittermeier, G.A. Da Fonseca, and J. Kent. 2000. Biodiversity hotspots for conservation priorities. *Nature*, *403*(6772): 853–858. https://doi.org/10.1038/35002501.

Nunes, R., U.J.B.D. Souza, C.P. Targueta, R.B. Pinto, T.N. Soares, J.A.F. Diniz-Filho, and M.P.D.C. Telles. 2020. Complete chloroplast genome sequence of *Caryocar brasiliense* Camb.(Caryocaraceae) and comparative analysis brings new insights into the plastome evolution of Malpighiales. *Genet. Mol. Biol.*, *43*. https://doi.org/10.1590/1678-4685-GMB-2019-0161.

Oliveira, M.E.B.D., N.B. Guerra, A.D.H.N. Maia, R.E. Alves, D.D.S Xavier, and N.M.D.S Matos. 2009. Physical characteristics of pequi fruits from chapada do Araripe-CE Rev. *Bras. Frutic.*, 31: 1196–1201. https://doi.org/10.1590/S0100-29452009000400038.

Oliveira, P.S., and M.R. Pie. 1998. Interaction between ants and plants bearing extrafloral nectaries in cerrado vegetation. *An. Soc. Entomol. Bras.*, *27*: 161–176. https://doi.org/10.1590/S0301-80591998000200001.

Palmeira, S.M., P.R. Silva, J.S. Ferrão, A.A. Ladd, M.L. Dagli, C.K. Grisolia, and F.J Hernandez-Blazquez. 2016. Chemopreventive effects of pequi oil (*Caryocar brasiliense* Camb.) on preneoplastic lesions in a mouse model of hepatocarcinogenesis. *Eur. J. Cancer Prev.*, *25*(4): 299–305. https://doi.org/10.1097/CEJ.0000000000000187.

Pires de Oliveira, I., F. Freitas de Lima, L.M. Spadoti, and C.H. Lescano. 2021. *Caryocar brasiliense* Camb., *C. villosum* (Aubl.) and *C. coriaceum* Wittm. *In*: *Fruits of the Brazilian Cerrado: Composition and Functional Benefits*, 53–77.

Peyre, G. 2021. Terrestrial Biodiversity Hotspots: Challenges and Opportunities. *In*: *Encyclopaedia of the UN Development Goals*. Springer Nature. https://doi. org/10.1007/978-3-319-71065-5.

Prance, G.T. 1990. The genus Caryocar L. (Caryocaraceae): An underexploited tropical resource. *Advances in Economic Botany*, 177–188.

Ribeiro, R.A., and F.M. Rodrigues. 2006. Genética da conservação em espécies vegetais do cerrado. *Revista de Ciências Médicas e Biológicas*, *5*(3): 253–260.

Rivas, M., and A. Barilani. 2004. Diversidad, potencial productivo y reproductivo de los palmares de *Butia capitata* (Mart.) Becc. de Uruguay. *Agrociencia Uruguay*, *8*(1): 11–20.

Rodrigues, A.D.C., J.T.A. Osuna, S.R.O.D. Queiroz, and A.P.S. Rios. 2006. Biometria de frutos e sementes e grau de umidade de sementes de angico (*Anadenanthera colubrina* (vell.) brenan var. cebil (griseb.) altschul) procedentes de duas áreas distintas. *Revista Científica Eletrônica de Engenharia Florestal*, *4*(8): 1–15.

Roesler, R.., R.R. Catharino, L.G. Malta, M.N. Eberlin, and G. Pastore. 2008. Antioxidant activity of *Caryocar brasiliense* (pequi) and characterization of components by electrospray ionization mass spectrometry. *Food Chemistry*, *110*(3): 711–717. https://doi.org/10.1016/j.foodchem.2008.02.048.

Santos, F.S., R.F. Santos, P.P. Dias, L.A .Zanão Jr, and F. Tomassoni. 2013. A cultura do Pequi (*Caryocar brasiliense* Camb.). *Acta Iguazu*, *2*(3): 46–57.

Souza, V.C., and Lorenzi, H. 2005.Botânica Sistemática: Guia Ilustrado Para Identificação das Famílias de Angiospermas da Flora Brasileira, Baseado em APG II; Nova Odessa Instituto Plantarum: Nova Odessa, Brazil, 2005; p. 640.

Szyszylowicz, I. 1895. Caryocaraceae. *In*: Engler, H.G.A. and Prantl, K.A.E. *Die natürlichen Pflanzenfamilien*. *3*(6): 153–157.

Takhtajan, A. 1969. *Flowering Plants Origin and Dispersal*. Washington, D.C.: Smithsonian Institution Press.

Torres, L.R.D.O., F.C. de Santana, F.L. Torres-Leal, I.L. de Melo, L.T. Yoshime, E.M. Matos-Neto, and J. Mancini-Filho. 2016. Pequi (*Caryocar brasiliense* Camb.) almond oil attenuates carbon tetrachloride-induced acute hepatic injury in rats: Antioxidant and anti-inflammatory effects. *Food and Chemical Toxicology*, *97*: 205–216. https://doi. org/10.1016/j. fct.2016.09.009.

Traesel, G.K., S.E.L.T. Menegati, A.C. Dos Santos, R.I.C. Souza, G.R.V. Boas, P.N. Justi, and S.A. Oesterreich. 2016. Oral acute and subchronic toxicity studies of the oil extracted from pequi (*Caryocar brasiliense,* Camb.) pulp in rats. *Food and Chemical Toxicology*, *97*: 224–231. https://doi.org/10.1016/j.fct.2016.09.018.

Vale, A.F., H.H. Ferreira, E.J. Benetti, A.C.S. Rebelo, A.C.R. Figueiredo, E.C. Barbosa, and K. Simões. 2018. Antioxidant effect of the pequi oil (*Caryocar brasiliense*) on the hepatic tissue of rats trained by exhaustive swimming exercises. *Braz. J. Biol.*, *79*: 257–262. https://doi.org/10.1590/1519-6984.180015.

Vera, R., R.V. Naves, J.L. do Nascimento, L.J. Chaves, W.M. Leandro, and de E.R.B. Souza. 2005. Caracterização física de frutos do pequizeiro (*Caryocar brasiliense* Camb.) no Estado de Goiás. Pesqui. *Agropecu. Trop.*, *35*(2): 71–79.

Vera, R., E.R.B. de Souza, E.P. Fernandes, R.V. Naves, M.S.S. Júnior, M. Caliari, and P.A. Ximenes. 2007. Caracterização física e química de frutos do pequizeiro (*Caryocar brasiliense* Camb.) oriundos de duas regiões no estado de Goiás, Brasil. Pesqui. *Agropecu. Trop.*, *37*(2): 93–99.

Vieira, R.F., and M.V. Martins, 2000. Recursos genéticos de plantas medicinais do cerrado: uma compilação de dados. *Rev. Bras. Pl. Med.*, *3*(1): 13–36.

Vilas-Boas, B.M.V., A.D.P. Alves, J.A. Alves, L.J. Rodrigues, T.C. Alves, and E.V.D. Boas. 2013. Caracterização física, química e bioquímica do mesocarpo interno de frutos do pequizeiro colhidos em diferentes estádios de desenvolvimento. *Ciência Rural*, *43*: 2285–2290. https://doi.org/10.1590/S0103-84782013001200027.

Wettstein, R. 1935. Caryocaraceae. *In*: *Handbuch der systematischen Botanik*. Leipzig.

14

Genetic Diversity of Wild Apple and Sustainable Utilization

Shivali Sharma,[1] *Aatifa Rasool*,[2,*] *Gaurav Sharma*,[1] *Javid I Mir*[3] and *Sheikh Mansoor*[4,*]

Introduction

One of the most significant fruit trees in the world is the apple (*Malus* × *domestica*). In addition to giving people delicious fruit, apples are used as a raw material for a variety of foods and beverages, including jams, juices, and fruit wines (Bhargava and Bansal, 2021). Apples also play a significant role in the agricultural economy. In several European nations, consumption is more than twice as high. But apples may also be used for a plethora of other things, such as preparation of beauty treatments, fermented goods, distilled alcohol, and seed oil (which is a by-product of making juice and cider). Numerous serious illnesses have been associated with a lower risk of consuming apples (Boyer and Liu, 2004; Hyson, 2011). Apples have few calories: 100 grams of fresh fruit slices have just 50 calories and no cholesterol or saturated fats. Their tart flavour is also a result of their high malic acid content. Antioxidants, phytonutrients, flavonoids, and polyphenolics are abundant in apples. Together, these substances support the development of balanced gut flora and help the body defend against the damaging effects of free radicals. Apple fruits are rich in beta-carotene and vitamin C. They are also a great source of B-complex vitamins, including pyridoxine (vitamin B-6), thiamine, and riboflavin. Additionally, apples have trace amounts of calcium, phosphorus, and potassium in them. The peel of the fruit contains significant quantities of antioxidants and two-thirds of the fruit's nutritional fibre. Apples are grown in various temperate parts of the planet and exported all over the world due to their easy shipping and excellent retention after storage. According to FAO (2022), China leads global apple production, contributing 49% of 46% of the global harvested land. Europe and Asia are primary importers, while the Americas and Europe dominate exports. This interdependence emphasizes apples' international significance, prompting substantial breeding investments. Apples serve not only as a popular fruit but also as a crucial genetic resource. Their genetic diversity, vital for

[1] Department of Fruit Science, Rani Laxmi Bai, Central Agricultural University Jhansi Uttar Pradesh, 284003.
[2] Department of Fruit Science, Sher e Kashmir University of Agricultural Sciences and Technology of Kashmir, 190025.
[3] Department of Plant Biotechnology, Central Institute of Temperate Horticulture Srinagar, J&K India, 191132.
[4] Department of Plant Resources and Environment, Jeju National University, Republic of Korea 63243.
* Corresponding email: aatifaw74@gmail.com; mansoorshafi21@gmail.com

breeding new cultivars, attracts considerable attention and research. As a global fruit, apples satisfy diverse preferences and exhibit adaptability, doubling as an agricultural genetic resource treasure trove. The extensive genetic traits within apples, encompassing color, shape, and quality, impact plant responses to the environment. Ecosystem stability, crop development, and new variety breeding rely heavily on apple germplasm resources. The genetic diversity of apples enhances their resilience to diseases, pests, climatic variations, and stress, ultimately elevating production and quality. In essence, apples stand as not just a common fruit but a cornerstone in global agriculture, embodying genetic wealth crucial for sustainable cultivation and innovation (Basannagari and Kala, 2013). Additionally, genetic diversity allows for the development of new, more competitive varieties through breeding. The objective of this study is to investigate the genetic diversity of apple germplasm resources, analyse the value of these resources to the growth of the apple industry from various angles, and assess the efficiency and drawbacks of current conservation strategies. In addition, the study will assess current conservation strategies and provide helpful references for enhancing and developing more scientific conservation strategies in the future.

Genetic Diversity of Apple Germplasm Resources

The Origin and Evolution of Apple

The apple (*Malus* x *domestica*), a valuable and age-old fruit, has a lengthy evolutionary history that includes several evidences of human and cultural activity. Apples originated in Asia and Europe. The oldest apple gene pool was formed by wild apple trees that thrived in Asia's high mountains (Gao et al., 2015). In addition to progressively acquiring diversity in their natural habitat, these wild apple trees also played a significant role in the development of agricultural civilization among humans. As early human cultures realized that wild apples could be eaten, they progressively created the first artificial types using selective breeding and farming techniques. This process is the result of interactions between man and nature as well as the domestication and development of plants. As commerce and cultural contacts increased, apples eventually made their way throughout the globe. Apple's growing traditions in many places originated from the historical commerce and warfare that facilitated the transfer of apples from Asia to Europe. People go through this process of constant selection, breeding, and reproduction to produce a distinct apple variety. Not only is the origin and evolution of the apple a topic of botany and genetics, but it is also a narrative spanning human civilization. The wisdom of man and nature's mutual domestication, as well as the diversity created by cross-cultural interaction, are woven throughout the narrative. Comprehending the apple's origins in detail will not only help people appreciate the fruit's significance to human civilization, but it will also give scientists a solid foundation on which to debate apple genetic diversity and conservation tactics in the future.

Methods for Determining Genetic Diversity

The process of determining genetic diversity is crucial for the preservation and application of apple germplasm resources. Scientists have used a variety of approaches to investigate the genetic diversity of apple germplasm resources.

Technology Utilizing Molecular Markers

Analyzing DNA or RNA sequences, such as random amplified polymorphic DNA (RAPD), simple repeat sequence (SSR), single nucleotide polymorphism (SNP), etc., is one of the most widely utilized methods for assessing the genetic variety of apples. It can highlight genetic variations and diversity levels across various apple cultivars (Ahmad et al., 2021).

Based on Morphological Characteristics

This is another crucial approach for assessing the genetic diversity of apples. Researchers use quantitative and qualitative analysis of fruit morphology, leaf characteristics, tree structure, and other factors to identify the variations and differences between various varieties.

Evaluation of Biological Traits

This is essential for ascertaining genetic diversity. It is possible to determine the genetic diversity of a variety by observing and documenting its growing patterns, blooming duration, fruiting traits, and other features.

Classification and Characteristics of Germplasm Resources of Apple

The apple tree is a significant fruit tree with a wide variety of germplasm resources that exhibit remarkable variation in form, biological traits, and genetic traits.

Geographical classification: It is the process of grouping apple varieties based on where they came from and where they are currently found. Globally, apple cultivars can differ greatly in terms of quality, disease resistance, and growing conditions. Apples from various nations and areas may also vary in fruit form and development patterns. For instance, the Maras apple from Central Asia is thought to represent the original wild species of apple (Marconi et al., 2018).

Genetic classification: This method relies on the genotypes and genetic traits of different apple cultivars. The qualities of disease resistance, fruit size, flavor, and maturity are determined by the genetic heritage of the many apple types. Higher genetic diversity in older cultivars is typically beneficial for breeding and conservation efforts.

Fruit characteristics: Fruit qualities including size, colour, shape, and flavor provide the basis for another kind of categorization. This categorization is frequently used to identify between several apple varieties that are used for cooking, processing (producing juices, jams, etc.), and fresh meals.

Classification by use: This method divides apple types into groups based on how they are used—for example, as fresh food, in processing, or cooking.

Vulnerability of Apple Genetic Resources

Volk et al. (2015) evaluated Malus genetic resources for susceptibility. They concluded that the few cultivars being planted, the age of orchards, the expense of replanting with new cultivars, and the dearth of breeding programs all pose threats to the apple harvest. New cultivars with enhanced tolerance to biotic and abiotic stress are required, as is a steady supply of premium products for customers. Cultivars that require less inputs and are simpler to harvest are needed since producers are experiencing several obstacles that are driving up production costs. Volk et al. (2015) similarly concluded that in order to make greater use of the variety maintained in *ex situ* collections, including wild species, apple breeding efforts in the future will need to incorporate advancements in genomic technology together with enhanced phenotyping. Although the USA was the study's primary emphasis, many of its conclusions apply to Apple's vulnerabilities in other nations and areas. According to Vincent et al. (2013), less than 50 *ex situ* accessions comprise 55% of the wild relatives of *M.* × *domestica*.

Furthermore, samples from wild species are frequently transferred among collections, which reduces the amount of variety that has been researched. Presently, the International Union for Conservation of Nature (IUCN, 2017) lists 29 species of *Malus* on its red list of vulnerable species. Because there is insufficient data on the majority of these species, it is impossible to determine their conservation status with accuracy. After they are assessed in further detail, a few of these may be classified as threatened. Three species—*M. baccata*, *M. fusca*, and *M. hupehensis*—are categorized as of least concern, which indicates that they are not at risk in the wild at this time. While *M. komarovii* and *M. niedzwetzkyana* are categorized as endangered, *M. sieversii* is classed as vulnerable (mostly because of habitat destruction) (IUCN, 2016, 2007a, 2007b). The representation of *Malus* species in the IUCN categorization system has probably been impacted by the paucity of knowledge regarding the taxonomy, genetic diversity, distribution, and conservation status of these species. A hub of diversification for several fruit and nut species, including *M. sieversii*, are the ancient forests of China and Central Asia. While there may be some protection for trees located inside park or forest preserve boundaries, wild apple tree-rich natural areas are vulnerable to lose from agricultural encroachment, urbanization, hybridization, and other alternative uses (Forsline et al., 2002). Challenges faced by genebanks that manage *ex situ* collections include rising maintenance costs, decreased support for regular operations, more regulations and costs associated with acquiring and sharing accessions, increased threats *in situ* to valuable local cultivars and wild relatives that represent significant gaps in collections, and rising user demand for genotyping, evaluation, and other accessions data (Table 1).

Conservation Status of Germplasm Resources of Apple

Ex situ Conservation

Apple genetic resource collections are comparatively long-term investments created for a range of objectives, including public interest, breeding, conservation, research, and dissemination. Significant apple collections are maintained by commercial firms, NGOs, botanical gardens, arboreta, national, subnational, or local genebanks. Wøldikes Frugthave in Denmark, which had 278 apple varieties in the 1870s, and the Alnarp collection in Sweden, which served as the inspiration for Dahl's pomology, published in 1905–1929, are two instances of defunct (or newly reconstructed) collections. Certain collections that are still in existence were founded as early as the 1920s; they include the UK National Fruit Collection at Brogdale, which is presently run by the University of Reading, and the collection of the N.I. Vavilov Research Institute of Plant Industry in Russia (Loskutov, 1999).

The German Pillnitz collection dates to around 1930, while the Swedish Balsgård collection was founded in the late 1940s. The Chinese Academy of Agricultural Sciences established its collection in the 1950s, whereas the French collection was started by the National Institute for Agricultural Research (INRA) in 1946. Most other apple genebanks were established in a more contemporary period. As part of a broader national network of clonal repositories, the National Plant Germplasm System apple collection was established in the United States in the 1980s (Postman et al., 2006).

Conservation in Clonally Propagated Crops

Since apples are clonally reproduced, genebank conservation approaches for them are very different from those for annual seed crops (Volk and Walters, 2003). Since apple varieties are clonally propagated—typically by grafting—every accession tree is an identical duplicate of the mother tree. It is not feasible to propagate particular cultivars as seeds since their seeds were produced during a cross-pollination event and are not genetically identical to the mother tree.

M. × *domestica* field collections are mainly kept as grafted trees in an orchard, allowing for the genotype or cultivar conservation. This makes it easier to employ collecting materials for breeding, phenotypic assessments, and bud dispersal. There are apple genebanks with plantations of grafted trees or seedlings that mimic wild species populations. With the help of these accessions, the public can now have access to wild species' fruits and flowers for phenotypic and genotypic analyses as well as breeding initiatives.

Generally, apple varieties are grafted onto seedstocks that have been enhanced. In the past, open-pollinated apple trees' seeds were used to generate rootstocks. Gene banks, like the Agricultural

Table 1. Primary wild relatives of *Malus* × *domestica* and reported resistance/tolerance to abiotic and biotic stresses.

	Common Name	**Taxon**	***M. sieversii***	***M. sylvestris***	***M. orientalis***	***M. prunifolia***	**References**
1.	Apple scab	*Venturia inaequalis* (Cooke) G. Winter		Resistant			Bassett et al. (2011)
2	Bot canker	*Botryosphaeria* (Sawada) Shang		Resistant	Resistant		Bus et al. (2005)
3	European canker	*Neonectria ditissima*	Resistant		Resistant	Resistant	Forsline and Aldwinckle, (2004)
4	Fire blight	*Erwinia amylovora* (Burrill) Winslow et al.		Resistant			Harshman et al. (2017)
5	Powdery mildew	*Podosphaera leucotricha* (Ellis & Everh.) E.S. Salmon		Resistant	Resistant	Resistant	Hokanson et al. (2001)
6	Wooly apple aphid	*Eriosoma lanigerum* Hausmann		Resistant	Resistant	Resistant	Isutsa and Merwin (2000)
7	Light brown apple moth	*Epiphyas postvittana* Walker		Resistant	Resistant	Resistant	Khan et al. (2014)
8	Apple leaf curling midge	*Dasineura mali* Kieff.		Resistant	Resistant	Resistant	Kumar et al. (2010)
9	Phytophthora	*Phytophthora*		Resistant	Resistant	Resistant	Luby et al. (2002)
10	Western white root rot	*Rosellinia nacatrix* Prill.		Resistant	Resistant	Resistant	Luby et al. (2001)
11	Violet root rot	*Helicobasidium mompa* Tanaka		Resistant	Resistant	Resistant	Maguylo and Bassett (2014)
12	Apple maggot	*Rhagoletis pomonella* Walsh		Resistant	Resistant	Resistant	Norelli et al. (2017)
13	Blue mold	*Penicillium expansum* Link			Resistant	Resistant	Norelli et al. (2017)
14	Apple replant disease	Various		Resistant	Resistant	Resistant	Norelli et al. (2017)

Research Service (USDA-ARS) collection located in Geneva, New York, of the United States Department of Agriculture, include specific collections of accessions of apple rootstocks. Although the process of breeding apple rootstocks is time-consuming and resource-intensive, contemporary rootstock cultivars are very valuable as nursery stock (Fazio et al., 2015; Volk et al., 2015).

Disadvantages

Apple cultivar and wild species conservation in field collections can be labor- and space-intensive, and they are vulnerable to pests, diseases, and environmental hazards. Pruning and trimming trees inside orchards are necessary to prevent crop rotation. After 25–50 years, trees may require frequent repropagation, depending on the rootstock and orchard circumstances. Field collections can also be kept in part in different formats, such as potted bonsai trees in greenhouses.

In vitro Conservation Techniques

In vitro culture of apples is typically feasible; methods for introduction, multiplication, and rooting are known and may be used for a wide variety of *M.* × *domestica* cultivars as well as other species of *Malus* (Kovalchuk et al., 2009; Sedlak et al., 2006, 2004). Apple genetic transformation technologies have been made possible by biotechnological procedures that employ *in vitro* techniques (Malnoy et al., 2010). Using *in vitro* bioreactor systems, several apple genetic resources (especially novel cultivars and rootstocks) have been mass-produced (Brischia et al., 2002; Chakrabarty et al., 2007, 2003).

Disadvantages

Due to the necessity of routine culture transfers, *in vitro* propagation is labor-intensive. It is concerning when soma clonal variation occurs, especially when cells are kept *in vitro* for extended periods (Oh et al., 2007). Facilities that have alarm systems and protocols in place to stop culture loss from system failures (such as overheating or excessive cooling) are required to house *in vitro* collections.

Cryopreservation Techniques

Wang et al. (2018) suggest that apple genetic resources can be efficiently preserved through cryopreservation using dormant buds and shoot tips. The initial genebanks employing apple cryopreservation techniques utilized dormant buds as the explant, as observed in studies by Forsline et al. (1998) and Towill et al. (2004). The typical process involves collecting dormant bud scions during winter, desiccating them until they reach around 30% moisture content, gradually cooling them to –30°C to –35°C, and then placing them in liquid nitrogen vapor for long-term preservation, as outlined by Forsline et al. (1998), Towill et al. (2004), and Höfer (2015). Viability is most effectively evaluated by grafting the buds into rootstocks after warming and rehydration. Notably, viable dormant buds have been successfully recovered even after ten years in cryopreserved conditions (Volk et al., 2008a). Additionally, apple shoot tips can undergo cryopreservation according to the methods described by Wang et al. (2018). These shoot tips, approximately 1 mm in size, are excised from in vitro plants, subjected to necessary pre-treatments, immersed in cryoprotectants (or encapsulated and dehydrated), and then rapidly frozen in liquid nitrogen. Various cryopreservation procedures for Malus, the genus to which apple belongs, have been documented in previous studies (Li et al., 2015; Feng et al., 2013; Niino et al., 1992; Paul et al., 2000; Zamecnik et al., 2007).

Disadvantages

Apple genetic resources can be preserved as particular genotypes in secondary sites that might not face the same environmental risks as the field harvest by using cryopreservation techniques. Equipment specific to cryopreserved collections, dependable liquid nitrogen supplies, and sophisticated monitoring systems are needed. While many gene banks have used the dormant bud cryopreservation technique, field collections may not be able to fully utilize it if the trees are not placed in a way that allows them to achieve adequate dormancy (Bilavčik et al., 2015; Guyader et al., 2012; Volk et al., 2017; Vogiatzi et al., 2012, 2011).

Conservation of Pollen

Breeding programs may access the alleles present in an accession that pollen has captured (Zhang et al., 2009). Breeders who want to cross trees that are not present at the same site or whose blooming dates are not coordinated may find pollen to be a very helpful conservation goal (Sharafi, 2011).

According to Zhang et al. (2009), pollen from the Rosaceae family is desiccation resistant and may be kept in liquid nitrogen for long-term preservation. There are techniques for gathering, desiccating, and storing pollen (or anther) (Ganeshan et al., 2008; Xu et al., 2014). Certain gene banks retain accessions in many complementary forms, such as grafted trees, seedlings, assessment orchards, seeds (wild species), and frozen explants Volk et al. (2015). One such gene bank is the USDA-ARS National Plant Germplasm System. Field-grown grafted trees are accessible for breeding, budwood dispersion, and phenotypic and genetic assessments.

Conservation of Seeds

Long-term seed storage effectively preserves the variety of wild species populations, while cryogenic storage offers a safe fallback if field trees are destroyed (Kushnarenko et al., 2010; Volk et al., 2015). As of right now, no plans for apple cultivar seed conservation have been made public. Since apples are self-incompatible, only the maternal parent is known for seed samples. As a result, just 50% of the variety that is being preserved comes from the first accession. Of the 91 institutions from 24 countries that have contributed collection information in the Genesys database, 70 of the 91 institutions solely keep Apple accessions in the field, and 92% of the accessions are preserved as field collections. Additionally, around 5% of the accessions are kept as seeds. The Vavilov Institute in Russia is the custodian of the majority of these seed collections (Genesys, 2017).

In situ Conservation

The preservation of wild apple species in their natural environments has been a goal. Ancient *M. sieversii* woodlands in the highlands of eastern Kazakhstan, western China, and adjacent Kyrgyzstan, Uzbekistan, and Tajikistan are the source of the domesticated apple. These woods' vestiges may still be seen today, most notably in eastern Kazakhstan's Djungarskii and Zailiiski Mountains (Dzhangaliev, 2003). In Kazakhstan and Kyrgyzstan's natural preserves, efforts have been undertaken to conserve old apple woods (Dzhangaliev, 2003).

In 2003, Dzhangaliev AD. According to Hokanson et al. (1997), «about 90% of the wild apples that existed near Almaty in 1935 are now gone», indicating that human encroachment was harming Kazakhstan's apple forest. Although field collections can preserve variety over an extended period of time, they did not consider this to be feasible due to the associated expenses and the small amount of variability that could be preserved, other than as seed. They concluded that it would be more economical to keep populations in nature reserves, which would also permit ongoing evolution and adaptation.

Diversity Maintenance and Introduction of New Germplasm

New germplasm introduction and variety preservation are essential to the expansion and conservation of apple genetic resources. Two components of diversity conservation include preserving the present apple varieties and preserving genetic variation throughout time. This means preserving the genetic variety of apples and protecting wild and traditional varieties from the pressures of industrial cultivation. Furthermore, it involves maintaining and expanding the range of currently accessible cultivars by innovative and scientific methods to enhance their disease resistance and adaptability, ensuring their growth and yield in diverse environments

(Papp et al., 2020). Enhancing apple genetic resources is the aim of introducing new germplasm. This might entail searching for new wild apple types with exceptional disease resistance, flavor, or adaptability by exploring the area and gathering them. Scientific breeding and genetic improvement research are also included in the introduction of novel germplasm. Through mating or genetic improvement, new genotypes are incorporated into current varieties to enhance their qualities and features.

A Genomics-based Approach to Conservation

The management and conservation of apple genetic resources can benefit from new opportunities presented by the genomics-based conservation strategy. This strategy contributes to a more thorough and in-depth understanding of the genetic linkages and diversity among apple varieties by making use of cutting-edge genomics methodologies and molecular biology tools (Buiteveld et al., 2021).

High throughput genetic data may be obtained via genomics technology, assisting in the identification and confirmation of genetic variations and traits across various breeds. We can precisely identify and characterize the genetic diversity, genetic linkages, and genomic differences between various types of apples by using genomic analysis of apple genetic resources. This will aid in a better understanding of the composition and traits of apple genetic resources as well as the breeding and conservation programs' choice of appropriate varieties. Apple genetic resources may be digitized and informative with the use of the genomics-based conservation approach. To support the long-term conservation and use of resources, researchers may compile and manage a significant quantity of genetic data by creating the genomic database of apple germplasm resources. In addition to facilitating resource exchange and sharing, this digital approach to management fosters international collaboration.

Continuous Monitoring and Management

The preservation of apple genetic resources is greatly aided by ongoing maintenance and observation. To guarantee prompt conservation and use of these resources, this strategy entails long-term monitoring and management of apple varieties and their diversity. To determine the amount, distribution, and quality of apple germplasm resources across various locations, continuous monitoring entails conducting routine surveys and evaluations of these resources (Gepts, 2006). The ongoing management of apple resources also includes the collection, preservation, classification, and digital management of genetic resources to ensure the integrity and sustainability of resources. This monitoring can assist in identifying threatened varieties and in taking prompt action for protection and conservation. Finding new types and perhaps valuable resources is aided by ongoing monitoring and management techniques.

A Global Strategy for the Conservation and Use of Apple Genetic Resources

In addition to human encroachment and climate change, hybridization with cultivated apples poses a serious danger to the conservation of the closely related *Malus* progenitor species in protected areas. Growing risks to the genetic integrity of *M. sylvestris* and *M. sieversii* in protected forests were noted by Feurtey et al. (2017) as a result of gene flow with *M.* × *domestica* growing nearby. The Piedmont-Mountain border zone and other low-mountain areas have had the greatest influence on gene flow. This is due to the cultivation of apple cultivars in gardens and the past use of wild rootstocks for grafting *M.* × *domestica*. This has increased the risk of gene flow and increased the level of admixture in the adjacent wild populations.

As proposed by Feurtey et al. (2017), sizable and well-chosen *in situ* reserves may be able to preserve the genetic variety seen in wild populations of Malus species if they are managed successfully indefinitely. Even while regular access to these reserves could be restricted, protected places might provide gene banks with a variety of sources. Site selection and geographic considerations for protection should ideally be guided by both genetic and phenotypic data gathered from these natural populations (Hokanson et al., 1997). Additionally, there are records of apple varieties that were once cultivated at certain historic places. When site managers want to transplant historic cultivars at their locations, access to these cultivars is essential. With respect to local, landrace, and historic cultivars, it is critical to ensure that key significant cultivars are conserved and secured, likely through a partnership amongst *ex situ* gene banks, local organizations, and the international community.

Challenges and Problems: At present, a series of challenges and problems facing Apple genetic resources will directly affect its long-term protection, utilization, and sustainable development.

Loss and Reduction of Genetic Diversity

The decline in genetic diversity is one of the main problems. A decrease in genetic diversity may result in a decline in the overall genetic performance of apple germplasm, weakening disease resistance, adaptability, and other crucial characteristics. Commercial cultivation, which favours a small number of superior varieties, has caused the marginalization and loss of many traditional or wild apple varieties (Wang et al., 2019).

Ecological Pressure is Also An Important Issue

The apple-producing environment is under threat from several factors, including disease, pest insects, climate change, and human activity. As a result, there is a chance that apple genetic resources may become endangered or perhaps extinct in some locations. The health and diversity of the apple population may suffer long-term harm as a result of the continuous effects of these variables. Numerous management and conservation problems exist as well. Long-term conservation and utilization are not given enough attention, and there are deficiencies in the management, conservation, and gathering of genetic resources as well as in the comprehensive monitoring and assessment systems that are lacking. The tendency toward commercial farming has resulted in the disregard for a variety of cultivars, endangering the genetic diversity and cultural legacy of apples.

Erosion of Genetic Diversity Due to Population Size

An Idea of a Bottleneck: The most frequent occurrence in cross-pollinated crops is inbreeding, which is widely recognized to have negative consequences and reduce population fitness in small outcross populations by recombining undesirable genes (recessive identical alleles). Severe population decreases, often known as the "genetic bottleneck", in natural populations result in a loss of genetic diversity as well as an increased vulnerability to infectious pests and diseases, which in turn increases the likelihood that a particular crop would go extinct. The percentage of initial heterozygosity preserved every generation is predicted by genetic models to be $[1 - (1/2Ne)]$. where G is the actual population size and E is the effective population size, which is often smaller than G. Thus, a population of $Ne = 10$ individuals loses 5% of its heterozygosity per generation. This indicates that severe bottlenecks degrade heterozygosity and genetic diversity (Pimm et al., 1989). Therefore, plant breeders have been advised to maintain the optimum population

Climate Change and Its Impact on Plant Genetic Resources

Within the extinction vortex, a crucial stage is decreased genetic variety. To meet the new and growing demands on germplasm for climate change adaptation in agriculture, gene banks need to be improved. Food baskets may be significantly impacted by gene banks' inclusion of various traits in their screening procedures and thorough collections that include what are now regarded as minor crops.

Importance of Genetic Diversity

The evaluation of genetic diversity in farmed crops has significant implications for breeding and genetic resource conservation. Numerous molecular markers have been extensively employed to identify cultivars in numerous species and analyse genetic variability. Molecular markers have demonstrated success in distinguishing between cultivars, categorizing synonyms, detecting cultivars that were incorrectly labeled, determining genetic linkages, and providing insights into the process of domestication (Wunsch and Hormaza, 2002). Because SSR markers have a high polymorphism level, are abundant, co-dominantly inherited, are relatively easy to analyze, and are reproducible, they are the ideal DNA markers for analysing genetic connections and diversity among crop species (Schlotterer, 2004). Apple researchers have created hundreds of microsatellite markers, some of which have been plotted on genetic linkage maps.

Significance of Genetic Conservation of Crop Plants

The strain from an expanding population, the urbanization of agricultural areas, and the quick industrialization of many aspects of our daily lives are all contributing to the direct and indirect erosion of biodiversity. For instance, a number of factors, including urbanization, coastal development, deforestation, land degradation, and environmental stress, are causing a substantial number of plant species to become extinct, particularly vital food crops for agriculture. However, large-scale production of genetically homogeneous maize and potato types has resulted in system-driven famines, such as the Southern corn leaf blight pandemic and the Irish potato famine in the United States. Even after these historical occurrences, the significance of PGR was not widely acknowledged until the expansion of the green revolution among farmed crops endangered the preservation of landraces. It was anticipated that the enhanced crop types brought forth by green revolution technology would boost farmers' incomes. To preserve PGR in the majority of stable food crops worldwide, the Consultative Group of International Agricultural Researches (CIGAR) established gene banks and domestication research centres. Centre for domestication: potatoes (Peru), rice (North China), wheat and barley (Middle/ Near East and North Africa), and maize (Mexico); visit http://www.cigar.org/centre/index.html for further details.) The UN and the Food and Agriculture Organization (FAO) backed the Convention on Biological Variety (CBD) and the International Treaty on Plant Genetic Resources (ITPGR), two international accords that acknowledge the critical role that genetic variety conservation plays. Such a pact remains one of the key supremos in food production, both now and in the future. According to CBD, Article 2, which was established at the Rio de Janeiro Earth Summit, genetic diversity is the fundamental component of biodiversity and variety within and between species as well as within ecosystems (Evenson et al., 2003). Research methods developed from IPRI's global project «strengthening the scientific basis of *in situ* conservation of agricultural biodiversity» were used by M. Smale, M. Istvan, I. Devra, and Jarvis in "The economics of conserving agricultural biodiversity on the farm: *Proceedings of a Workshop Hosted by the Institute for Agro Botany (IA) and the International Plant Genetic Resource Institute (IPGRI '02),* Godollo,(Run on)Hungary, May 2002."

The issue, though, is that contemporary crop types in particular were created largely to produce large amounts of food within well-managed production environments. Due to the very fluctuating stress conditions, they face, these types are frequently unsuitable for low-income farmers operating in marginal production contexts (Evenson et al., 2003]. According to research by Ceccarelli et al. (2002), landraces and traditional varieties have been shown to exhibit stronger stability (adaptation over time) in low-input agriculture in marginal situations. As a result, their cultivation may increase farm-level resilience in the face of shocks to food supply. This is particularly true in many regions of Ethiopia with difficult agroclimatic conditions, delayed technical advancement, and underdeveloped and inadequately equipped market institutions (Bruinsma et al., 2003).

Assessment of Genetic Diversity in Crop Plants

Numerous methods, including (i) morphological, (ii) biochemical characterization/evaluation (allozyme) in the progenomic era, and (iii) DNA (or molecular) marker analysis, particularly single nucleotide polymorphism (SNPs) in the postgenomic era, are regularly used to assess genetic diversity within and between plant populations. Markers may show comparable inheritance patterns to any other characteristic, such as codominant, recessive, or dominant. A marker is considered codominant if it is possible to identify the genetic pattern of homozygotes from that of heterozygotes. Codominant markers typically provide greater information than dominant markers.

Assessment of Genetic Diversity by Morphological Markers

Large tracts of land are frequently needed for these field experiments, which makes it potentially more expensive than molecular assessment in western (developed) countries and equally expensive in Asian and Middle Eastern (developing) countries when taking labor costs and availability into account. These markers are based on visually accessible traits like flower colour, seed shape, growth habits, and pigmentation. They also do not require expensive technology. These marker sizes are used for crop development and the conservation of any characteristic for a particular purpose. Since traits are frequently subject to phenotypic plasticity, it is imperative to determine the ideal population size and its representatives prior to quantifying genetic diversity. This ensures that diversity assessment is free from bias and allows for the assessment of diversity in the presence of environmental variation, which cannot be ignored in addition to genotypic variation. These kinds of markers—such as prickly seeds, bristling panicles, and variations in the color of flowers or leaves—are still advantageous and necessary for differentiating mature plants from their genetic contamination in the field.

Assessment of Genetic Diversity by Biochemical Markers

Allelic variations of enzymes known as isozymes, which are identified by electrophoresis and particular staining, are the second class of genetic markers known as biochemical markers. In nature, isozyme markers are codominant. They have straightforward inheritance and detect variety at the functional gene level. Small quantities of plant material are all that are needed to detect it. However, there are only a few enzyme markers that are known to exist, and even these have unique structural issues and are not the sole enzymes. As a result, there is not much room to investigate the resolution of genetic variety.

Assessment of Genetic Diversity by Molecular Markers

Molecular markers, which include a wide range of DNA molecular markers and may be used for both genetic and molecular variation analysis, are the third and most used form of genetic markers. These markers can identify variations resulting from chromosomal duplication, inversion, deletion, and/or insertion. Because these markers are only found near or associated with the genes that govern the characteristics, they do not affect the phenotypic of the traits of interest. Both dominant and codominant inheritance patterns are present for these markers. Genetic properties of various markers vary (e.g., they might be codominant or dominant, amplify defined or anonymous locations, include expressed or non-expressed sequences, etc.). A genomic locus that is identified using a probe or particular starter (primer) and that, by its existence, clearly separates the chromosomal feature it represents from its bordering areas at the extremities is known as a molecular marker (Barcaccia et al., 2000]. Molecular markers and phenotypic manifestation of a genetic characteristic may or may not correlate. Regardless of the growth, differentiation, development, or defensive status of the cell, they are stable and detectable in all tissues, which gives them several benefits over traditional, phenotype-based alternatives. Furthermore, environmental, pleiotropic, and epistatic influences do not confuse them. Since the focus of our work is on genomic advancements and their use in crop genetic diversity assessment, we won't be going into great detail on pre-genomic period techniques.

Use of Genetic Resources in Apple Breeding and for Sustainable Fruit Production

Recurrent Breeding Strategies

The Horticulture and Food Research Institute of New Zealand Ltd. (Hort Research) has developed an Apple Genetics Population with the primary goals of preserving biodiversity for cultivar development (Noiton and Alspach, 1996) and offering genetic information related to significant apple characteristics (Oraguzie et al., 2000, 2001). Noiton and Alspach (1996) stress the importance of this population in maintaining biodiversity for future cultivar development, while Oraguzie et al. (2000, 2001) emphasize its role in providing crucial genetic information on important apple traits. The established population is positioned to introduce novel fruit characteristics, making it a vital resource for implementing a long-term breeding strategy based on recurrent selection principles. This initiative underscores the institute's dedication to sustaining genetic diversity and supporting ongoing cultivar improvement efforts in apple breeding (Noiton and Alspach, 1996; Oraguzie et al., 2000).

Increasing Genetic Diversity

A project at Agroscope FAW Wädenswil aims to enhance the genetic basis in apple breeding by employing adapted recurrent selection strategies. Families were derived from open-pollinated seeds sourced from various apple cultivars and crab apples from global repositories (Notion et al., 1999). The chosen strategy, inspired by successful approaches in New Zealand, focuses on maintaining and augmenting genetic variability. Diallele crosses were conducted, involving cultivars like 'Zitronenreinette', 'Oberrieder Glanzreinette', 'Discovery', 'Delbard Jubilé', 'Champagner Reinette', 'Pilot', 'Angold', 'Reglindis', A 810-390 (Pl2), 'Schweizer Orangenapfel', HL 75-27 ('Goldspur' x 'Antonovka'), HL 10-04 ('Goldspur' x 'Glockenapfel'), and 'Nabella.' Notably, no major resistance gene against apple scab was intentionally included in the crosses. The primary objective of these crosses is to enhance the genetic diversity concerning both fruit quality and disease resistance. The initial seedling selected for advanced testing is FAW 13959, derived from the cross between 'Zitronenreinette' and 'Discovery'. This ongoing initiative represents a proactive effort to broaden the genetic foundation in apple breeding, emphasizing traits related to fruit quality and disease resistance.

Strategies for Durable Disease Resistance

Developing apple varieties resistant to diseases poses a significant challenge, and molecular techniques present a promising solution. Liebhard et al. (2003) advocate the use of molecular markers and genetic linkage maps to identify major resistance genes and quantitative trait loci (QTL) associated with resistance. Combining different functional resistances in a cultivar can mitigate the risk of resistance breakdown. Molecular selection, targeting stacked genetic resistance against apple scab (*Venturia inaequalis*) and powdery mildew (*Podosphaera leucotricha*), has shown potential, with various resistances detected using available markers (Tartarini et al., 1999; Hemmat et al., 2002; Gygax et al., 2004; Markussen et al., 1995; Seglias and Gessler, 1997; James and Evans, 2004).

At Agroscope FAW Wädenswil, advancements in marker-assisted selection, including mass selection and cost reduction, have been achieved (Frey et al., 2004). Multiplex reactions can analyze up to eight markers simultaneously, and systems have been developed for automated switching between SCAR-based and microsatellite-based screenings. Given the transient nature of major resistances like Vf, Vm, and Vr, alternative strategies incorporating major genes and partial resistance from genetic resources show promise. Developing fire-blight-resistant apple and pear cultivars involves exploiting genetic variation in germplasm and using QTL markers. Forsline and Aldwickle (2002) screened the USDA Apple Collection for natural occurrences of fire blight, finding no major resistance genes, suggesting a rational approach utilizing QTLs for fire blight resistance caused by *Erwinia amylovora*. These strategies collectively contribute to the development of durable disease-resistant apple varieties, underscoring the significance of molecular tools in modern breeding practices.

Sources for Increased Genetic Diversity

Recent efforts to broaden the genetic diversity of apples have focused on collecting material from their original centers. Forsline et al. (1995) successfully gathered 65 accessions (18,000 seeds) representing three endemic species during an expedition to Kazakhstan and Kyrgyzstan. U.S. quarantine restrictions limited the collection of cuttings from elite materials. Twelve expeditions across seven ecosystems obtained unique germplasm from unexplored areas, currently undergoing selection for various fruit and tree traits. The Fruit Genebank in Dresden-Pillnitz (Germany) collected 7,000 seeds and scions from 55 accessions and 28 Malus species during a 2001 expedition to Chinese provinces. Switzerland conducted a national inventory of fruit genetic resources from 2000–2005, contributing to the safe conservation of diverse genetics. The study also examined the nutritional value and health implications of apples. Stushnoff et al. (2003) analyzed phenolic and antioxidant diversity in 321 Malus species, selections, and cultivars. The data aids breeders in parental selection and supports postharvest physiology studies. The comprehensive approach involves future collections from core collections, wild apples in Central Asia, and the University of Minnesota germplasm collection, contributing to the understanding, preservation, and utilization of apple genetic resources for breeding programs and public health.

Issues for Access and Distribution

The global system could have a big impact by making Apple genetic resources more accessible in the long term. International apple varieties are important for breeding and production, and better access to these resources could benefit local economies, research, and breeding programs. Exploring local varieties is crucial for understanding how apples adapt to the environment or resist pests and diseases. These local varieties can also be useful for specific traits like robustness, tolerance to stresses, and cider quality.

Untapped genetic resources are found in wild apple species, prompting global exploration of policy issues related to apple genetic material exchange through the SMTA. Halewood et al. (2013)

delved into fundamental concerns regarding the implementation of the ITPGRFA, applicable to Malus germplasm exchange. Addressing the need to enhance access to non-publicly held genetic resources and distinguish between research, breeding, and commercialization access is crucial. The international distribution of Malus genetic resources poses challenges, particularly for clonally propagated cultivars. Documentation of existing in vitro cultivar collections can be valuable for distribution purposes. Looking ahead, the creation of a pathogen-free set of apple genetic resources may facilitate their global distribution, as suggested by Halewood et al. (2013).

Prospects

The global *ex situ* conservation system for Malus faces challenges in efficiency, cost-effectiveness, rationality, and security. To enhance Malus genetic resource conservation and utilization, priority actions are proposed. Strengthening the long-term conservation of international cultivars involves optimizing redundancies across gene banks. For wild species, connecting *ex situ* seed conservation with *in situ* seed gardens is proposed to bolster restoration efforts. While responsibility for local cultivars may rest at the national or local level, the global system is envisioned to play a role in monitoring, characterizing, genotyping, and sharing information for their enduring conservation. Creating linkages among diverse conservers is crucial to addressing urgent challenges in Malus genetic resource conservation. Collaboration is proposed among various users involved in conservation efforts, fostering global actions through comprehensive accession-level information systems and cooperative initiatives. The highest priority action involves establishing the Global Apple Diversity Platform to safeguard Apple diversity. This platform aims to facilitate cooperation among *ex situ* apple collections, *in situ* conservation efforts, and user communities for long-term sustainability and enhanced utilization. Implementing this global strategy begins with an international meeting on Malus genetic resources, bringing together key gene banks, governmental bodies, NGOs, and researchers. This meeting aims to establish the Global Apple Diversity Platform, promoting knowledge sharing and action planning. Participation involves registration as a collection holder or an individual committed to contributing to apple diversity conservation and use. The platform's management includes a secretariat with a coordinator, an apple/temperate fruit tree genetic resources expert, and a database manager overseen by a steering group. The secretariat could be housed at an international organization, national gene bank, or an existing regional initiative, with key national gene banks taking leadership roles. Governance specifics would be defined through stakeholder consultations. Long-term sustainability requires resources for the secretariat, meetings, database development, collective activities, addressing collection threats, and short-term projects. The Global Apple Diversity Platform aims to build a diverse community, establish an international information platform, form working groups, monitor apple genetic resources, coordinate a global genotyping effort, verify the identity of local cultivars, develop core collections, implement communication and advocacy, and provide training, to those identified through studies, surveys, workshops, and genebank visits.

References

Ahmad, R., M.A. Anjum, S. Naz, and R.M. Balal. 2021. Applications of molecular markers in fruit crops for breeding programs: A Review. *Phyton.*, *90*(1): 17–34.

Barcaccia, G., E. Albertini, D. Rosellini, S. Tavoletti, and F. Veronesi. 2000. Inheritance and mapping of 2n-egg production in diploid alfalfa. *Genome*, *43*: 528–537.

Basannagari, B., and C.P. Kala. 2013, Climate change and apple farming in Indian Himalayas: A study of local perceptions and responses. *PLoS One*, *8*(10): e77976.

Bassett, C.L., D.M. Glenn, P.L. Forsline, M.E. Misniewski, and R.E. Farrell. 2011. Characterizing water use efficiency and water deficit responses in apple (*Malus* × *domestica* Borkh. and *Malus sieversii* Ledeb) M. Roem. *HortScience*, *46*: 1079–1084.

Bhargava, A., and A. Bansal. 2021. Classification and grading of multiple varieties of apple fruit. *Food Anal. Methods,14*: 1359—1368.

Bilavčik, A., J. Zamecnik, and M. Faltus. 2015. Cryotolerance of apple tree bud is independent of endodormancy. *Frontiers Plant Sci.*, *6*: 695.

Boyer, J., and R.H. Liu. 2004. Apple phytochemicals and their health benefits. *Nutrition J.*, *3*: 5.

Brischia, R., E. Piccioni, and A. Standardi. 2002. Micropropagation and synthetic seed in M.26 apple rootstock (II): A new protocol for production of encapsulated differentiating propagules. *Plant Cell Tissue Organ Cult.*, *68*: 137–141.

Buiteveld, J., H.J.K. Putten, L. Kodde, I. Laros, G. Tumino. N.P. Howard, E. van de Weg, and M.J.M. Smulders. 2021. Advanced gene bank management of genetic resources of European wild apple, Malus sylvestris, using genome-wide SNP array data. *Tree Genet. Genomes*, *17*: 32.

Bus, V.G.M., F.N.D. Laurens, W.E. Van de Weg, R.L. Rusholme, E.H.A. Rikkerink, S.E. Gardiner, H.C.M. Bassett, L.P. Kodde, and K.M. Plummer. 2005. The Vh8 locus of a new gene-for-gene interaction between *Venturia inaequalis* and the wild apple *Malus sieversii* is closely linked to the Vh2 locus in Malus pumlia R12740-7. *New Phytol.*, *166*: 1035–1049.

Ceccarelli, S., and S. Grando. 2002. Plant breeding with farmers requires treating the assumptions of conventional plant breeding: Lesson from the ICCRDA Barely Program *In*: D.A. Cleveland, and D. Soleri [Eds.], *Farmers, Scientists, and Plant Breeding Integrating Knowledge and Practice*. CABI International, New York, NY, USA.

Chakrabarty, D., E.J. Hahn, Y.J. Yoon, and K.Y. Paek. 2003. Micropropagation of apple rootstock M.9 EMLA using bioreactor. *J. Hort. Sci., Biotechnol.*, *78*: 605–609.

Chakrabarty, D., Y.H. Dewir, E.J. Hahn, S.K. Datta, and K.Y. Paek. 2007. The dynamics of nutrient utilization and growth of apple rootstock 'M9 EMLA' in temporary versus continuous immersion bioreactors. *J. Plant Growth Regul.*, *51*: 11–19.

Evenso, R.E., and D. Gollin. 2003. Assessing the impact of the Green Revolution, 1960 to 2000. *Science*, *300*: 758–762.

Falco, D.S., J.P. Chavas, and M. Smale. 2007. Farmer management of production risk on degraded lands: The role of wheat variety diversity in the Tigray region, Ethiopia. *Agric. Econ.*, *36*: 147–156.

FAO. 2022. *The State of the World's Genetic Resources for Food and Agriculture*. FAO, Rome, Italy.

Fazio, G., T.L. Robinson, and H.S. Aldwinckle. 2015. The Geneva Apple Rootstock Breeding Program. *Plant Breed. Rev.*, *39*: 379–447.

Feng, C.H., Z.H. Cui, H. Cui, B.Q. Li, L. Chen, Y.L. Ma, Y.H. Zhao, and Q.C. Wang. 2013. Duration of sucrose pre-culture is critical for shoot regrowth of *in vitro*-grown apple shoot-tips cryopreserved by encapsulation-dehydration. *Plant. Cell. Tiss. Organ. Cult.*, *112*: 369–378.

Feurtey, A., A. Cornille, J.A. Shykoff, A. Snirc, and T. Giraud. 2017. Crop to-wild gene flow and its fitness consequences for a wild fruit tree: Towards a comprehensive conservation strategy of the wild apple in Europe. *Evol. Appl.*, *10*: 180–188.

Forsline, P.L., and H.S. Aldwinckle. 2002. Natural Occurrence of Fire Blight in USDA Apple Germplasm Collection after 10 Years of Observation. *Acta Hort.*, *590*: 351–357.

Forsline, P.L. 1995. Adding diversity to the national apple germplasm collection: Collecting wild apple in Kazakhstan. *New York Fruit Quart.* *3*: 3–6.

Forsline, P.L., and H.S Aldwinckle. 2004. Evaluation of *Malus sieversii* seedling populations for disease resistance and horticultural traits. *Acta Hort.*, *663*: 529–534.

Forsline, P.L., L.E. Towill, J.W. Waddell, C. Stushnoff, W.F. Lamboy, and J.R. McFerson. 1998. Recovery and longevity of cryopreserved dormant apple buds. *J. Amer. Soc. Hort. Sci.*, *123*: 365–370.

Frey, J.E., B. Frey, C. Sauer, and M. Kellerhals. 2004. Efficient low-cost DNA-extraction and multiplex fluorescent PCR method for high-throughput marker-assisted selection (MAS) in apple breeding. *Plant Breed.*, *123*: 554–557.

Ganeshan, S., P.E. Rajasekharan, S. Shashikumar, and W. Decruze. 2008. Cryopreservation of pollen. *In*: Reed, B.M. [Ed.], *Plant Cryopreservation*, 443–464. Springer, New York, NY.

Gao, Y., F. Liu, K. Wang, D. Wang, X. Gong, L. Liu, C.M. Richards, A.D. Henk, and G.M. Volk. 2015. Genetic diversity of Malus cultivars and wild relatives in the Chinese National Repository of Apple Germplasm Resources. *Tree Genet. Genomes*, *11*: 106.

Genesys. 2019. *Accession Browser*. The Crop Trust, Bonn, Germany.

Gepts, P. 2006. Plant Genetic Resources Conservation and Utilization: The Accomplishments and Future of a Societal Insurance Policy. *Crop Science*, *46*: 2278–2292.

Guyader A., P. Chatelet, and F. Engelmann. 2012. First results on cryopreservation by dormant bud technique of a set of Malus and Pyrus cultivars from the INRA Biological Resources Centre. *In*: Grapin, A., Keller, E.R.J., Lynch, P.T., Panis, B., Revilla Bahillo, A., and Florent, E. [Eds.], *Cryopreservation of Crop Species in Europe. COST Action 871: CryoPlanet. Proceedings of the Final Meeting*, 141–144. European Cooperation in Science and Technology (COST), Brussels, Belgium,

Gygax, M., L. Gianfranceschi, R. Liebhard, M. Kellerhals, C. Gessler, and A. Patocchi. 2004. Molecular markers linked to the apple scab resistance gene Vbj derived from Malus baccata jackii. *Theor. Appl. Genet.*, *109*: 1702–1709.

Halewood, M., E. Andrieux, L. Crisson, J. Rwihaniza, J. Wasswa, M. Edmond, K. Koffi, T.Y. Dorji, M.R. Bhatta, and D. Balma. 2013. Implementing "mutually supportive" access and benefit sharing mechanisms under the Plant Treat, Convention on Biological Diversity, and Nagoya Protocol. *Law Dev.*, *9*: 68–96.

Harshman, J., K. Evans, H. Allen, R. Potts, J. Flamenco, and H. Aldwinckle, M. Wisniewski, and J. Norelli. 2017. Fire blight resistance in wild accessions of *Malus sieversii. Plant Dis.*, *101*(10): 1738–1745.

Hemmat, M., S.K. Brown, and N.F. Weeden. 2002. Tagging and mapping scab resistance genes from R12740-7A apple. *J. Amer. Hort. Sci.*, *127*: 365–370.

Höfer, M. 2015. Cryopreservation of winter-dormant apple buds: Establishment of a duplicate collection of Malus germplasm. *Plant Cell Tissue Organ. Cult.*, *121*: 647–656.

Hokanson, S.C., W.F. Lamboy, A.K. Szewc-McFadden, and J.R. McFerson. 2001. Microsatellite (SSR) variation in a collection of *Malus* (apple) species and hybrids. *Euphytica*, *118*: 281–294.

Hokanson, S.C., J.R. McFerson, P.L. Forsline, PL, W.F. Lamboy, J.J. Luby, A.D. Djangaliev, and H.S. Aldwinckle. 1997. Collecting and managing wild Malus germplasm in its center of diversity. *HortScience*, *32*: 173–176.

Hyson, D.A. 2011. A comprehensive review of apples and apple components and their relationship to human health. *Adv. Nutr.*, *2*: 408–420.

Isutsa, D., and I.A. Merwin. 2000. Malus germplasm varies in resistance or tolerance to apple replant disease in a mixture of New York orchard soils. *HortScience*, *35*: 262–268.

IUCN. 2007a. *Malus niedzwetzkyana.* Participants of the FFI/IUCN SSC Central Asian regional tree Red Listing workshop, Bishkek, Kyrgyzstan (11–13 July, 2006). The IUCN Red List of Threatened Species, 2007.

IUCN. 2007b. *Malus sieversii.* Participants of the FFI/IUCN SSC Central Asian regional tree Red Listing workshop, Bishkek, Kyrgyzstan (11–13 July, 2006).

IUCN. 2016. *Malus komarovii.* Rhodes, L., Maxted, N., Kim, H., Son, S.-W., Kim, Y.-S. The IUCN Red List of Threatened Species 2016.

IUCN. 2017. *The IUCN Red List of Threatened Species 2017-1.* International Union for Conservation of Nature and Natural Resources, Cambridge, UK.

James, C.M., and K. Evans. 2004. Identification of Molecular Markers linked to the Mildew Resistance Genes Pl-d and Pl-w in Apple. *Acta Hort.*, *663*: 123–128.

Khan, M.A., K.M, Olsen, V. Sovero, M.M. Kushad, and S.S. Korban. 2014. Fruit quality traits have played critical roles in domestication of the apple. *Plant Genome*, *7*(3): 1–18.

Kovalchuk, I., Y. Lyudvikova, M. Volgina, and B.M. Reed. 2009. Medium, container, and genotype all influence *in vitro* cold storage of apple germplasm. *Plant Cell Tissue Organ Cult.*, *96*: 27–136.

Kumar, S., R.K. Volz, P.A. Alspach, and V.G.M. Bus. 2010. Development of a recurrent apple breeding programme in New Zealand: A synthesis of results, and a proposed revised breeding strategy. *Euphytica*, *173*: 207–222.

Kushnarenko, S., E. Salnikov, M. Nurtazin, Z. Mukhitdinova, I. Rakhimbaev, and B.M. Reed. 2010. Characterization and cryopreservation of *Malus sieversii* seeds. *Asian and Australasian J. Plant. Sci. Biotechnol.*, *4*: 5–9.

Li, B.Q., C.H. Feng, L.Y. Hu, M.R. Wang, G. Volk, and Q.C. Wang. 2015. Recovery patterns and genetic integrity in *Malus* shoot tips cryopreserved using droplet-vitrification and encapsulation-dehydration techniques. *J. Biotechnol.*, *214*: 182–191.

Liebhard, R., B. Koller, A. Patocchi, M. Kellerhals, W. Pfammatter, M. Jermini, and C.Gessler. 2003. Mapping Quantitative Field Resistance Against Apple Scab in a Fiesta x Discovery' Progeny. *Phytopathology*, *93*: 493–501.

Loskutov, I.G. 1999. *Vavilov and His Institute: A History of the World Collection of Plant Genetic Resources in Russia*, 188pp. Run on. International Plant Genetic Resources Institute, Rome, Italy.

Luby, J., P. Forsline, H. Aldwinckle, V. Bus, M. and Geibel, M. 2001. Silk Road apples: Collection, evaluation, and utilization of *Malus sieversii* from Central Asia. *HortScience*, *36*: 225–231.

Maguylo, K., and C. Bassett. 2014. Phenotyping *M. sieversii* collections from Kazakhstan for leaf traits and tree architecture. *Acta Hort.*, *1058*: 335–341.

Malnoy, M., E.E. Boresjza-Wysocka, J.J. Norelli, M.A. Flaishman, D. Gidoni, and H.S. Aldwinckle. 2010. Genetic transformation of apple (*Malus* × *domestica*) without use of a selectable marker gene. *Tree Genet. Genomes*, *6*: 423–433.

Marconi, G., N. Ferradini, L. Russi, L. Concezzi, F. Veronesi, and E. Albertini. 2018. Genetic characterization of the apple germplasm collection in central Italy: The value of local varieties, *Front. Plant Sci.*, *9*: 1460.

Markussen, T., J. Krüger, H. Schmidt, and F. Dunemann. 1995. Identification of PCR-based markers linked to the powdery mildew resistance gene Pl1 from Malus robusta in cultivated apple. *Plant Breed.*, *114*: 530–534.

Niino, T., A. Sakai, H. Yakuwa, and K. Nojiri. 1992. Cryopreservation of *in vitro*-grown shoot tips of apple and pear by vitrification. *Plant Cell Tissue Organ Cult.*, *28*: 261–266.

Noiton, D.A.M., and P.A. Alspach. 1996. Founding clones, inbreeding, coancestry, and status number of modern apple cultivars. *J. Amer. Society. Hort. Sci.*, *121*: 773–782.

Norelli, J., M. Wisniewski, G. Fazio, E. Burchard, B. Gutierrez, E. Levin, and S. Droby. 2017. Genotyping-by-sequencing markers facilitate the identification of quantitative trait loci controlling resistance to *Penicillium expansum* in *Malus sieversii. PLoS One*, *12*(3): e0172949.

Oh, T.J., M.A. Cullis, K. Kunert, I. Engelborghs, R. Swennen, and C.A. Cullis. 2007. Genomic changes associated with somaclonal variation in banana (*Musa* spp.). *Physiol. Plant*, *129*: 766–774.

Oraguzie, N.C., M.E. Hofstee, P.A. Alspach, L.R. Brewer, and C. Howard. 2000. Quantitative genetic studies in a recurrent selection program in apple. *Acta Hort.*, *538*: 191–195.

Papp, D., L Gao, R. Thapa, D. Olmstead, and Khan, A., 2020. Field apple scab susceptibility of a diverse Malus germplam collection identifiers potential sources of resistance for apple breeding. *CABI Agriculture and Bioscience 1*: 16.

Paul, H., G. Daigny, and B.S. Sangwan-Norreel. 2000. Cryopreservation of apple (*Malus × domestica* Borkh.) shoot tips following encapsulation-dehydration or encapsulation-vitrification. *Plant Cell Rep.*, *19*: 768–774.

People's Trust for Endangered Species. 2019. *Fruit Finder*. People's Trust for Endangered Species, London, UK.

Pimm, S.L., J.L. Gittlaman, G.F. McCracken, and M. Gilpin. 1989. Genetic bottlenecks: Alternative explanations for low genetic variability. *Trends in Ecology and Evolution*, *4*: 176–177.

Postman, J., K. Hummer, E. Stover, R. Krueger, P. Forsline, L.J. Granke, F. Lee, T. Ayala-Silva, and B. Irish. 2006. Fruit and nut genebanks in the U.S. National Plant Germplasm System. *HortScience*, *41*: 1188–1194.

Sedlak, J., F. Paprštein, A. Bilavčik, and J. Zámečnik. 2006. Proliferation and cold hardening of in vitro grown apple shoot tips. *Acta Hort.*, *725*: 467–470.

Sedlak, J., F. Paprštein, A. Bilavčik, and J. Zámečnik. 2004. *In vitro* cultures and cryopreservation as a tool for conserving of fruit species. *Bulletin of Botanical Gardens*, *13*: 65–67.

Seglias, N.P., and C. Gessler. 1997. Genetics of apple powdery mildew resistance derived from *Malus zumi* (Pl2). *IOBC/WPRS BULL.*, *20*: 195–208.

Sharafi, Y. 2011. Investigation on pollen viability and longevity in *Malus pumila* L., *Pyrus commonis* L., and *Cydonia oblonga* L., *in vitro*. *J. Medicinal Plant Res.*, *5*(11): 2232–2236.

Stushnoff, C., A.E. McSay, J.J. Luby, and P.L. Forsline. 2003. Diversity of phenolic antioxidant scavenging capacity in the apple germplasm collection. *Acta Hort.*, *623*: 305–312.

Tartarini, S., L. Gianfranceschi, S. Sansavini, and C. Gessler. 1999. Development of reliable PCR markers for the selection of the Vf gene conferring scab resistance in apple. *Plant Breed.*, *118*: 183–186.

The IUCN Red List of Threatened Species 2007. http://dx.doi.org/10.2305/ IUCN.UK.2007.RLTS.T32363A9693009.en (accessed October 28, 2016).

Towill, L.E., P.L. Forsline, C. Walters, J.W. Waddell, and J. Laufmann. 2004. Cryopreservation of Malus germplasm using a winter vegetative bud method: Results from 1915 accessions. *CryoLett.*, *25*: 323–334.

Vincent, H., J. Wiersema, S. Kell, H. Fielder, S. Dobbie, N.P. Castañeda-Álvarez, L. Guarino, R. Eastwood, B. León, and N. Maxted. 2013. A prioritized crop wild relative inventory to help underpin global food security. *Biol. Conserv.*, *167*: 265–275.

Vogiatzi, C., B.W. Grout, A. Wetten, and T.B. Toldam-Andersen. 2011. Cryopreservation of winter-dormant apple buds: I Variation in recovery with cultivar and winter conditions. *CryoLett.*, *32*(4): 358–366.

Vogiatzi, C., B.W.W. Grout, and A. Wetten. 2012. Cryopreservation of winter-dormant apple: III Bud water status and survival after cooling to –30°C and during recovery from cryopreservation. *CryoLett.*, *33*: 160–168.

Volk, G.M., C.T. Chao, J.L. Norelli, S. Brown, G. Fazio, C. Peace, J. McFerson, G.Y. Zhong, and P. Bretting. 2015. The vulnerability of U.S. apple (*Malus*) genetic resources. *Genetic Resour. Crop Evol.*, *62*: 765–794.

Volk, G.M., M. Jenderek, and C.T. Chao. 2017. Prioritization of Malus accessions for collection cryopreservation at the USDA-ARS National Center for Genetic Resources Preservation. *Acta Hort.*, *1172*: 267–272.

Volk, G.M., and C. Walters. 2003. Preservation of genetic resources in the national plant germplasm clonal collections. *Plant Breed. Rev.*, *23*: 291–344.

Volk, G.M., J. Waddell, R. Bonnart, L. Towill, D. Ellis, and M. Luffman. 2008a. High viability of dormant Malus buds after 10 years of storage in liquid nitrogen vapour. *CryoLett.*, *29*: 89–94.

Wang, M.R., L. Chen, J.A. Teixeira da Silva, G.M. Volk, and Q.C. Wang. 2018. Cryobiotechnology of apple (*Malus* spp.): Development, progress, and future prospects. *Plant Cell Rep.*, *37*: 689–709.

Xu, J., L. Li, Q. Liu, Y. Shi, J.G. Peng, M.X. Jia, and Y. Liu. 2014. Wide-scale pollen banking of ornamental plants through cryopreservation. *CryoLett.*, *35*: 312–319.

Zamecnik, J., M. Faltus, and A. Bilavcik. 2007. Cryoprotocols used for cryopreservation of vegetatively propagated plants in the Czech cryobank. *Adv. Hort. Sci.*, *21*: 247–250.

Zhang, Y.L., R.D. Chen, C.J. Huang, and Y. Liu. 2009. Cryobanking of *Prunus mume* pollen and its application in cross-breeding. *CryoLett.*, *30*: 165–170.

15

Genetic Diversity of Peach (*Prunus persica* L. Batsch.) and Sustainable Utilization

Ikra Manzoor,[1,*] *Syed Zainab Kashani,*[2] *Gazala Gulzar,*[3] *S.H. Wani,*[4]
Neha Sharma[5] and *M.A. Mir*[1]

Introduction

Among the stone fruits found in temperate areas, peach (*Prunus persica* L. Batsch.) had a global production of 17 MMT with China as the leading producer (FAO, 2023). Peach is a member of the *Rosaceae* family and based on genomic and phenotypic evidence, it is believed to be native to southwestern China (Wang et al., 2012; Cao et al., 2014).

Domestication

China which is considered the place of origin of peaches accounts for the large wild germplasm (landraces and wild types) contributing to the heterogeneity and multiplicity as compared to the remaining parts of the world where peaches can be found. Through the ancient Silk Road, there was propagation of peach throughout the entire China and to the whole world after it was domesticated. Currently, almost 200 countries of the world grow peach fruit crops (FAO, 2018).

In China, the heterozygosity of the wild and the cultivated types in peach have been understood by doing tremendous studies in the field and collecting germplasm (Yu et al., 2010, 2019). Besides this, from various continents of the world which includes places like North America, Europe, and Oceania, there is the import of nearly 400 exotic varieties (Chinese Crop Germplasm Resources

[1] Division of Fruit Science, Faculty of Horticulture, Sher-e-Kashmir University of Agricultural Sciences & Technology of Kashmir, Shalimar, Srinagar-190025
[2] Division of Vegetable Science, Faculty of Horticulture, Sher-e-Kashmir University of Agricultural Sciences & Technology of Kashmir, Shalimar, Srinagar-190025
[3] Division of Plant Pathology, Faculty of Horticulture, Sher-e-Kashmir University of Agricultural Sciences & Technology of Kashmir, Shalimar, Srinagar-190025
[4] Mountain Research Centre for Field Crops, Sher-e-Kashmir University of Agricultural Sciences & Technology of Kashmir, Khudwani-192101
[5] Division of Vegetable Science, IARI, Pusa, New Delhi-110012
* Corresponding author: manzoorikra@gmail.com

Information System, 2020). In the Mediterranean basin, peach got entry through China and from Persia (Iran); early writers of Greece and Rome reiterated it. Throughout the European Mediterranean, peach was propagated by Romans. The secondary centers of diversity of peaches are Iran, Central Asia, Caucasus, Italy, Spain, and California (Zeven and Zhokovsky, 1975).

In India, chilling peaches are cultivated in the Himalayan regions of Jammu & Kashmir, Himachal Pradesh, and Uttarakhand while those with less chilling requirements are cultivated in Punjab, Haryana, Uttar Pradesh, and Nilgiri hills (Thakur et al., 2023).

Genetic Resources

The peach is a predominant temperate stone fruit which is a member of the genus *Prunus* in the family Rosaceae which is a fruit of choice for human beings. *P. mira* Koehne, *P. davidiana* (Carr.) Franch., and *P. kansuensis* Rehd.,also originated in China and are the three common wild species of peach (Wang et al., 2012)

P. mira Koehne is the ancestor of the domesticated peaches found in the plateau of Tibetas depicted by whole genome sequencing due to their wide genetic connections within (Cao et al., 2014).

*Prunus davidiana*is is mainly found in the valley of the Yellow River and areas of Mt. Taihang (Wang et al., 2012), with certain features such as an early date of bloom, a mesocarp with thin tissues, and great height of trees as compared to other peach species and cultivars making it a superior source of germplasm for breeding experiments for ornamental cultivars. *Prunus davidiana* is a great material as it is resistant to the pest *Myzus persicae* (peach aphid), by knowing about the stages of development of the size of fruit based on infection under an artificially directed environment (Sauge et al., 1998) and under uncontrolled conditions (Wang et al. 2001). Because of its high resilience to RKN (*Meloidogyne incognita),* it is a great rootstock source for domesticated peaches (Cao et al., 2014).

In China, the germplasm of peach follows a predominant effect for breeding experiments in other countries of the world. Through the import of the cultivar, namely, Shanghai Shui Mi, which served as the parental source during the beginning of the 20th century, 'The Hakuto' cultivar was selected by Japan (Xu et al., 2006; Yamamoto et al., 2003) along with the famous cultivar 'Elberta' released by the USA. The two cultivars, viz., 'Hakuto' and 'Elberta' used as parental lines, paved the way for the breeding of new varieties of peach (Aranzana et al., 2012; Scorza et al,. 1985). In the USA, South Africa, Brazil, Argentina, Australia, China, Spain, Italy, France, and Japan, a lot of effort has been put into peach breeding over the years (Martinez-Gomez et al., 2003), with a production of almost 2,000 new cultivars; half of these have been registered in and come from the USA while only 5% are from China (Hancock et al., 2008; Sansavini et al., 2006).

The varieties in China were developed by pioneering the cultivars, viz., 'Shanghai Shui Mi' ('Chinese Cling') and 'Bai Hua Shui Mi'. In Nanjing and Beijing germplasms, the two cultivars, viz., 'Okubo' and 'Hakuto' were interbred which were sourced from Japan for developing peaches with white color and less acidic nature.'NJN76', two other cultivars, viz., 'Mayfire'and 'Legrant' produced nectarines by interbreeding.The propagation of indigenous cultivars in China was generally dependent on the sexual mode of reproduction, i.e., seed (Wang and Zhuang, 2001). The majority of peaches in Japan were either selections or mutants of two cultivars, viz., 'Hakuto' and 'Hakuho' (Xu et al., 2006; Yamamoto et al., 2003). Novel genetic backgrounds should be identified to reduce the low genetic fidelity in peach breeding programs from the use of few founders (Aranzana et al., 2010; Aranzana et al., 2003; Bouhadidaet al., 2010; Xie et al., 2010; Cheng and Huang, 2009; Wunsch et al., 2006). The various peach species are shown in Figure 1.

Cytogenetics

The cultivated peach is autogamous following a diploid nature with ($2n = 2x = 16$) and the size of the genome is 5.9×10^9 bp/2x nucleus (Baird et al., 1994).

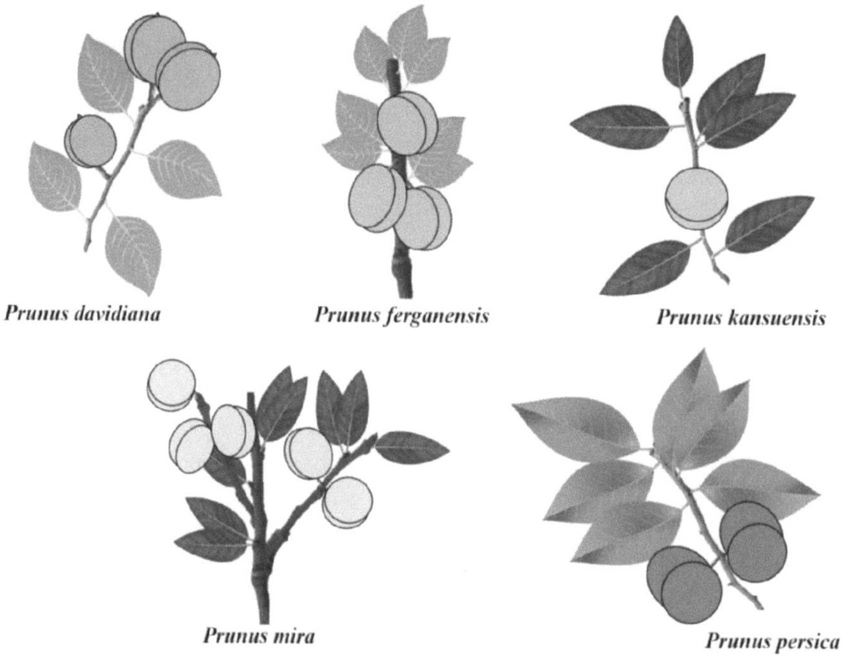

Figure 1. Different species of peach (*Prunus* spp.) exploring genetic diversity.

Table 1. Native peach species with diploid ploidy levels (adapted from Scorza and Okie,1990).

Species	Chromosome Number (2*n*)
P. davidiana (Carr.) Franch	16
P. ferganensis (Kost. & Rjab) Kov. & Kost.	16
P. kansuensis Rehd.	16
P. mira Koehne	16
P. persica (L.) Batsch	16

Problems of Breeding

1. Germplasm exchange has become difficult due to the patenting of new peach varieties.
2. Only one seed per fruit needs to perform a large number of crosses for a sizable population.
3. The replant problems and nematodes do not allow breeders to use the same land again for growing progenies.
4. In low-chill peaches, embryo rescue is to be done and poor *ex vitro* survival is a problem.

Objectives

1. Development of cultivars with high yield and good quality to have better returns through high-density plantations.
2. Development of cultivars suitable for processing.
3. Development of varieties resistant to fruit borer, fruit fly leaf spot, etc.
4. Less chilling hour requirement and resistance to high temperatures during summers.
5. To broaden the genetic base to have diverse gene sources for improvement.
6. Development of early bloomers which flower between 60 and 70 DAF.
7. Resilience to RKN and flooding problems.
8. Dwarf nature of the scion and the stock.

9. Nutrition problems such as Iron (Fe) chlorosis and lack of zinc (Zn) in soil as well as plants.
10. Reduction in tree vigor to aid in mechanization.

Chemotaxonomy

In peach tissues, the amount of secondary metabolites differ to a great extent, and the concentration varies with respect to the cultivar and the environment where they are grown. In peaches and nectarines, cyanidin is the predominant red color compound in general, the main anthocyanin compound in the peach fruit is cyanidine 3-glucoside. The derivatives of hydroxycarboxylic acid, anthocyanins, flavonols, and flavan3-ols are the general phenolic compounds found in peaches and nectarines. Peaches and nectarines contain various compounds, viz., cyanidin-3-O-glucoside, cyanidin-3-O-rutinoside, flavonols (i.e., quercetin-3-O-glucoside and quercetin-3-O-rutinoside), flavan 3-ols (catechin, epicatechin, and proanthocyanidins, including procyanidin B1), and others as reported by Ravagliaet al. (2013), Clifford (2003), and Hong et al. (2004).

The major phenols in peaches include chlorogenic acids, caffeic acid, catechin, and procyanidin B3 (i.e., catechin- (4β-8) -catechin). The important hydroxycinnamic acid derivatives found in peach are chlorogenic acid and non-chlorogenic acid whereas compounds like procyanidin B1 (epicatechin-(4β-8) -catechin), catechin, and epicatechin, which is the main type of flavan 3-ols, and flavonols in the skin of peach fruit in comparison to the flesh of the peach fruit. Peach and nectarine skin areas are rich sources of anthocyanins with few concentrations of pigments present near the stone tissues of the fruit. The main color compounds for nectarines and peaches include cyanidin3-glucoside and cyanidin 3-rutinoside. A few varieties are also rich in cyanidine 3-acetyl glucoside and cyanidin 3-galactoside. Compounds, such as quercetin 3-glucoside and quercetin 3-rutinoside include important flavonol compounds in peaches and nectarines and are found mainly inside the skin (Shahidi and Naczk, 2004).

Cytomorphology

As compared to many herbaceous species (e.g., corn, tomato, or *Arabidopsis*), peach is not very well defined genetically, and very little research has been done about its cytology (Jelenkovic and Harrington, 1972). Among the tree fruit species, peach is best-characterized (Mowrey et al., 1990). There are almost 40 Mendelian traits which have been described: phenotypic (Hansche, 1988; Monet, 1988; Ramming, 1991) and biochemical (Arulsekaret al., 1986).

Besides this, for 20 other traits, heritability has been estimated (Monet, 1988). A molecular genetic linkage map, from intraspecific F2 populations, has been developed as a permanent resource for geneticists and breeders (Abbott et al,. 1992; Chapparro et al., 1992; Eldredge et al., 1992).

Various genetic and molecular biological studies show that genomic size depicts the basic genetic estimation analysis. Information regarding (n) amount of DNA in the nucleus (C value) is crucial involving fundamental and applied research for organizing the size of the genome, relationships among species, study of the expression of genes, and improvement of germplasm (Bennett, 1984).

For analysis of genetic variation and hybridization programs, linkage maps have been created and for depicting the length of recombination of nuclear genomes and correlating this genetic distance with physical distance, genome size is of paramount importance (Meagher et al., 1988).

Dickson et al. (1992) conducted a flow cytometric analysis of members of the Rosaceae in a survey of the four subfamilies, which included six *Prunus* species and two peach cultivars. Baird et al. (1994) employed flow cytometry studies with laser technique, and studied the DNA content within the nucleus for peach lines having polyploidy, and developed it artificially comprising the following, viz., x, 2x, 3x, and 4x, which are chimeras (cytoplasmic) of the periclinal type. DNA amount of nucleus in peachfruit varies from 0.30 pg with (x) type of nuclei to 1.23 pg with (4x) type of nuclei. The peach genome of (2x) type is quite compact and was found to have 0.60 ± 0.03 pg, i.e., 5.8×10^8 nucleotide bp. There is a series of arithmetic depicted by polyploid types, due to the size of the genome directly related to the status of ploidy with the amount of DNA directly related to several chromosomes.

Genetic Tansformation through Tissue Culture and Transgenics

Many regeneration protocols have been developed (Table 2) as studied by Hammerschlag et al. (1985) employing diverse plant parts in peaches explants, viz., developing calli from embryos (immature), cotyledonary part (both mature and immature), and explants from leaf tissue. In this study, using immature embryonic tissues, plants of peach were regenerated without using callus (unorganized cell mass) obtained through mature embryonic tissues. After transfer from a medium containing 4.5 μM 2, 4-D, and 0.45 μM BA to a medium having 0.27μM NAA and 2.2 μM, friable callus cells obtained from embryonic tissue (immature), a white nodular regenerative callus was acquired. For three subcultures, this callus regenerated its morphogenetic ability. Using embryonic tissue (mature), the callus which was green and had nodules (tumor-like growth) without regeneration was obtained. After putting unorganized cell mass into a medium with a 5-fold NAA concentration and a 2-fold BA concentration, shoots were found to be regenerated and induced rooting under dark conditions, having 28.5 μM IAA concentration in the medium. Pooler et al. (1995) reported that in three rootstocks of peach, viz., 'Flordaguard', 'Nemared', and 'Medaguard', using cotyledonary tissue of stored seed which is fully mature, regeneration of shoots was found to have maximum success by using Murashige and Skoog (MS) medium for the culture of cotyledonary tissues, under a dark period for about 3 weeks, containing 2.5% of sucrose as a carbon source with an amalgamation of IBA (1.25 or 2.5 mM) and TDZ (6.25 or 12.5 mM) concentrations. The three rootstocks, viz., 'Flordaguard', 'Nemared', and 'Nemaguard' had high regeneration recovery, viz., 60%, 33%, and 6%, respectively. The rate of regeneration is not affected by the time of storage of seed, i.e., 1–3 years. The roots are produced by shoots with a three-fourth of regeneration rate. With this protocol of regeneration, which is efficient and easy, seeds that are in storage might be utilized around the year; along with that it might be a novel method of regeneration for sexually raised rootstocks of peach in introgression of genes. Gentile et al. (2002) developed a protocol by using the leaf tissue of tissue-cultured plants of peach to produce adventitious shoots through the regeneration step. A sexually produced line of peach, i.e., P16C15, which is immature along with the lines, viz., Babygold 6, 842 Standard, San Giorgio, and Yumyeong which are mature, excision of the apical part was done which were then inoculated under a dark period for 21 days onto a 6-benzyladenine (BA) and α-naphthaleneacetic acid (NAA) containing a medium which was then put into a medium without auxins for culturing it for 21 days under light conditions. Using the apex of the plant, there was excision of the first four leaves in the apex which

Table 2. Genetic Transformation and tissue culture in peach (*Prunus persica* L. Batsch).

Technique	Explant Used	References
Shoot tip culture	Cotyledonary tissues	Pooler et al. (1995)
Adventitious shoot culture	Leaf tissue	Gentile et al. (2002)
Agrobacterium tumefaciens-mediated genetic transformation	Mature seed	Perez-Clemente et al. (2004)
Agrobacterium tumefaciens-mediated genetic transformation	cotyledons, the axis of the embryo, and pieces of hypocotyl	Padilla et al. (2006)
Organogenesis	meristematic bulks (MBs)	Sabbadiniet al. (2019)
Shoot regeneration	Leaf explants	Zong et al. (2019)
Agrobacterium tumefaciens-mediated genetic transformation	parts of hypocotyls in peach, leaf tissues, and shoots	Xu et al. (2020)
In vitro culture	Leaf explants	Zheng et al. (2022)
Somatic embryogenesis	Expanding leaves and anthers	Ricci et al. (2022)
Callus culture	Adventitious roots	Gao et al. (2023)

had been inoculated similarly. The callus was produced by leaves, in the dark conditions at the base of the petiole which put forth adventitious shoots after putting it to a medium without auxins inoculated under conditions of light. For several months, the maintenance of the callus for morphogenesis was done. As Perez-Clemente et al. (2004) proved that by employing a marker (*in vivo*), namely (GFP), i.e., green fluorescent protein, engineered plants of peach (*Prunus persica* L.) via modification of genetic constitution in the pieces of the embryo from mature seed as explants' material and to accomplish it, an efficient, fast, and reproducible alteration method along with regeneration was given with an added benefit as this type of explants' material is available round the year. In this research, the vector employed was strain C58 (pMP90) of *Agrobacterium tumefaciens* to alter the genetic material as it has pBin19 binary plasmid. This study has used a selectable marker, viz., Nospro-*npt*II-Noster cassette, along with a vital reporter gene (CaMV35 Spro-*sgfp*-CaMV35Ster cassette) which encodes for the modified version of (sGFP), i.e., green fluorescent protein. This study used sections of embryo co-cultured under *in vitro* conditions with the vector *Agrobacterium tumefaciens* following a detection process, and there was regeneration of transgenic shoots. Shoots show that the young leaves at the apical portion had more expression of sGFP which sustained. During the detection of GFP expression under *in vivo* conditions, a quick, efficient, and convenient distinction for a modified expression, along with escape buds, might have been obtained. The discarding of escapes was followed by rhizogenesis during which engineered shootlets put forth roots under tissue culture conditions; thereafter, amplification via the Polymerase chain reaction technique was done to detect the recovered plantlets. By the southern blotting technique, transgene (*sgfp*) was noticed for fixed genetic amalgamation. In the second generation, for transgenic peach plants, expression of green fluorescent protein had been kept at a high range. Padilla et al. (2006) developed a protocol to produce an environment suitable for transferring a desirable gene via *Agrobacterium tumefaciens* as a vector for which the starting material (explants) such as cotyledons, the axis of the embryo, and pieces of the hypocotyl using seeds that had not germinated along with slices from the internode of epicotyls obtained using seeds which had germinated by employing the modification method with *Agrobacterium tumefaciens*. By employing two strains of *Agrobacterium tumefaciens*, three types of plasmids, and four types of promoters, viz., CaMV35s, (Aocs) 3AmasPmas ('super-promoter'), mas-CaMV35s, and CAB, the GUS (uidA) marker gene has been studied. Using six *Agrobacterium tumefaciens* lines, one plasmid, i.e., pLC101, and the doubleCaMV35s promoter, i.e., dCaMV35s, the green fluorescent protein was tested. As compared to the CAB promoter, more GUS expression was produced by the CaMV35s promoter. Even though two strains of *Agrobacterium tumefaciens*, viz., EHA105 and LBA4404 have been used containing a similar type of plasmid (pBIN19), they showed variable expression of GUS depicting association among the two, i.e., *Agrobacterium tumefaciens* line and the plasmid used. Fourteen days after co-cultivation an amalgamation with the three, i.e., *Agrobacterium tumefaciens* strain (EHA105), plasmid type (pBIN19), and the promoter (CaMV35s) had a maximum success of genetic modification using internodal part of epicotyl (56.8%), cotyledonary region (52.7%), leaf tissues (20%), and axis of the embryo (46.7%) as determined through rate of GUS expression by explants. Under the promoter type (dCaMV35s), the expression for green fluorescent protein was maximum with the internodal part when used as an explant, having only an 18–19% rate. When there was a combination containing plasmid (pCL101) having green fluorescent protein along with five lines of a vector, i.e., *Agrobacterium tumefaciens*, maximum transformation rates have been 20–21% (for the internodal part as well as the cotyledonary one, respectively). Since there was co-cultivation of nine peach genotypes to the strain of *Agrobacterium tumefaciens* (EHA105) along with plasmid (pCL101) having green fluorescent protein, with maximum transformation rates as 26–28% for the cotyledonary region and internodal pieces, respectively. Green fluorescent protein depicts a viable marker for genetic alteration, permitting a method of genetic engineering that is quite non-destructive. It has been found that in this study, rates of GFP transformation were low. So, in peach, it is vital to standardize an expression of this marker gene. Sabbadini et al. (2019), through genetic alteration employing the vector *Agrobacterium tumefaciens* with a tissue culture technique of organogenesis involving callus formation, i.e., indirectly described an *in vitro* regeneration protocol for the rootstock of peach, a rootstock, viz., Hansen 536, a cross of *Prunus persica* and *Prunus amygdalus* by using meristematic bulks (MBs) as explants. In this study, it was found that an effective regeneration process of shoots of adventitious nature has been done by culturing meristematic bulks (MBs) of the Hansen 536 rootstock upon a suitable growth medium with an altered Woody Plant medium (WPM) having 6-benzyladenine at the concentration of 4.4 μM, NAA, i.e., 1-naphthaleneacetic acid (0.1 μM) and plant agar S1000 (B&V) (6.0 g/l). Meristematic bulks (MBs) were utilized as explants for genetic engineering via *Agrobacterium tumefaciens* to incorporate 'ihp35S-PPV194' an RNAi assembly to tackle the Plum pox virus. Using certain concentrations, i.e., 0, 17, or 42 μM, for both detection via green fluorescent protein (GFP) and selection via kanamycin, transgenic events were detected. Using a molecular approach, strains of callus proliferation with green fluorescent protein were identified for stable expression of the ihp35S-PPV194: eGFP gene assembly. To create a successful shoot regeneration system from leaf explants of 'Hansen 536' (*Prunus dulcis* × *Prunus persica*), Zong et al. (2019) evaluated 10 culture media, including Murashige and Skoog (MS) or woody plant medium (WPM) supplemented with different plant growth regulators. 0%–36.1% were in the range of accidental shoot regeneration frequencies, and 32.3–36.1% were the best regeneration rates on WPM medium containing 8.88 μm 6-benzyl amino-purine (BAP) and 0.98–3.94 μm indole-3-butyric acid (IBA); 80% of the regenerated shoots that were directly transplanted into the substrate were able to root and survive, demonstrating their high rooting ability. Using all three of the *Agrobacterium tumefaciens* strains tested, transient transformation showed effective delivery of the β-glucuronidase (GUS) reporter gene (gusA). OD600 0.5–1.0 was used during the four days of cocultivation. Recent research conducted by Xu et al. (2020) describes a transformation protocol using explants for producing genetically modified hair-like roots for a compound plant having shoots (wild type) along with roots (engineered) with the help of a vector *Agrobacterium rhizogenes* with explants such as parts of hypocotyls in peach, leaf tissues, and shoots. In this study, different types of starting material, viz., leaf tissues, pieces of hypocotyl, and shoots, could produce engineered hair-like root systems, having almost 50% modification rate, in the case of parts of hypocotyls used. When shoots are infected with *Agrobacterium rhizogenes*, composite plants are generated for developing engineered hair-like roots of adventitious type. To study the role of *PpMYB10.1*, regulatory gene for anthocyanin in the genetically modified hair-like root system, a compound plant system here has been in order along with *PpUFGT* and *PpGST*, the two downstream genes, have been under a firm activation process. This study has provided a rapid and reproducible genetic alteration system via *Agrobacterium rhizogenes* vectors to know the functioning of genes, and the transgenic system, and detect the linkage between microorganisms in the root-rhizosphere for the peach fruit crop.

Zheng et al. (2022) provided evidence of the functions of H3K27me3 modification and DNA methylation in regulating chromatin status during Peach callus development. The findings showed that mCG was predominant in genic regions and that mCHH and mCG were predominant throughout the genome. DNA hypomethylation and H3K27me3 demethylation were observed during the *in vitro* culture of leaf explants, and these changes may have the effect of activating auxin- and cytokinin-related regulators to cause the formation of calluses. While the H3K27me3 demethylase inhibitor GSK-J4 dramatically decreased callus development, the DNA methylation inhibitor 5-azacytidine could significantly increase callus development.

In a research by Ricci et al. (2022), somatic embryogenesis (SE) in the peach hybrid rootstock 'GF677' and various peach cultivars using expanding leaves and anthers as explants was done. Cultured in different induction media with N6-benzylaminopurine (BAP) and 2,4-dichloro phenoxy acetic acid (2,4-D), the somatic explants did not yield somatic embryos. However, cream-colored calli resembling pre-embryogenic structures were observed when anthers from 'Big Top'. 'Tardibelle', and 'Big Bang' were cultured on SE induction media supplemented with 4.4 μM BAP along with 4.5 or 9 μM 2,4-D. Similarly, Gao et al. (2023) conducted an extensive large-scale analysis of the callus induction rate (CIR) in peach trees, revealing substantial variability within the peach germplasm. Late-maturing cultivars, notably, displayed significantly elevated CIR levels. Furthermore, cultivars with high CIR demonstrated a propensity for adventitious root (AR) development during callus induction, with a positive correlation noted between CIR and the capacity for AR regeneration. Additionally, PpLBD1 emerged as a promising candidate gene implicated in the regulation of callus adventitious rooting in peach trees.

Genetic Assessment through Molecular Marker Approach

Genetic Diversity

Peach (*Prunus persica* L. Batsch) is a commercially viable stonefruit grown in temperate regions of the world as the fruits are edible. The current scenario demands the development of cultivars/ varieties that are resilient to the increasing trend of climate change; it is necessary to search for germplasm suited to warmer climatic conditions. In the last century, it was a predominant fruit crop in the La Palma islands of Spain, but in the recent past, peach cultivation has been forsaken and was found under wild conditions (Perez et al., 2020). Research by Perez et al. (2020) has been done to estimate genetic diversity studies in peaches for assessment of local germplasm along with its prevention of degradation. Along La Palma islands (Spain), peach trees were sampled. In this study, there was a prospection of 89 locally available accessions of peach and with the help of 10 single-sequence repeat (SSR) loci, 28 different genotypes were analyzed. By comparing these genotypes with 95 landraces of peach (Spanish type) preserved under off-site conditions along with 26 extra specimens that had been brought from eight other countries, this study revealed the fact that the La Palma island germplasm of peach is less diverse. Besides, it is found that there exists a link between the germplasm of La Palma islands and other peaches of Spanish type, depicting that the germplasm in the Peninsula of Iberia and simultaneous interbreeding along with identification of the locally available accessions is congenial to the climatic conditions of the island (i.e., subtropical). Based on fruit type and geographic origin, the structure of the population shows a grouping of the samples. Also, Aranzana et al. (2010) estimated the genetic diversity, structure of the population, and LD, i.e., linkage disequilibrium studies for commercial peach cultivars and based upon SSRs, genetic distance was found, and by studying features of fruit, viz., melting type of fleshing peaches, melting type of flesh in nectarines and non-melting types, peach cultivars had been grouped into three classes. As compared to melting peaches and nectarines, non-melting types possess more allelic combinations along with greater homozygosity. Except, ("Admiral Dewey", "Early Crawford", and "Chinese Cling"), pioneer varieties of the United States clubbed with melting types of commercial value, depict that their genetic material is well represented within present-day varieties.

By analyzing the structure of a population, the samples were subdivided into sub-populations in the same manner. Doing linkage disequilibrium (LD) studies among three sub-populations which were either non-structured or scarcely structured depicting high estimates of (LD) preservation in peach germplasm ranging from 13–15 cM. By utilizing a single-sequence repeat (SSR) set, this study confirmed earlier research for estimating diversity in peaches along with population structure providing extra avenues to enforce the rights of breeders and the breeding programs. For determining the mutation frequency of markers along with a study on pedigree patterns, especially the pioneer genetic material of modern varieties, that helps to know the peach evolution aspects, SSR data were used.

There can be a precise determination of linkage disequilibrium (LD) preservation via the autogamous behavior of the domesticated genetic material of peach and an impediment that started during the initiation of current fruit breeding programs. High LD paves the way for sequencing of the whole genome of the peach, which is important for genotypic analysis of horticulturally prominent characters.

To determine the alterations in genetic structures and gene selections linked to the differentiation of peaches (*Prunus persica*) and their wild relatives, Akagi et al. (2016) examined the accessions of these plants. Numerous differences in the genetic structures and delineations between domesticated peaches and their wild relatives, as well as between peach landraces and modern fruit (F) and modern ornamental (O-A) cultivars, were found by analyzing genome-wide informative single-nucleotide polymorphism loci.

Koohi et al. (2020) employed 67 EST-SSR markers to detect genetic variation in peach germplasm and generated 302 bands. The higher SI index value (1.903) indicated increased genetic variability in the genotypes. The dendrogram, based on Jaccard's similarity coefficients, delineated four distinct clusters in the entire germplasm. Furthermore, the analysis of molecular variance (AMOVA) revealed that 70.68% of the total variation originated from within-population variation, while 29.32% stemmed from variation among populations. Similarly, Trifonova et al. (2021) examined a set of 85 peach cultivars by employing 12 SSR markers to evaluate their genetic diversity and relationships. The detected degree of polymorphism was found to be comparable to earlier evaluations of genetic variability in peach cultivars. With an average PIC value of 0.49, the mean number of alleles per locus was 5.67.

Mas-Gómez et al. (2022) used the new Peach 18K SNP v2 array to genotype 287 peach accessions to investigate the genetic diversity and single nucleotide polymorphism (SNP)-trait associations in the Spanish germplasm. The new SNP array's high density made it possible to identify 30 synonymy groups that were previously unknown when using low-density markers. Furthermore, a potential large-scale molecular event in 'Starcrest', a Springcrest sport, was discovered that may indicate the replacement of a 13.5 Mb chromosome.

Demirel et al. (2023) evaluated the 52 local peach genotypes and two commercial peach cultivars from the Iğdır province of Türkiye, using 32 ISSR markers to assess genetic diversity and population structure. There was a total of 213 scored alleles, 154 of which were polymorphic. With an average of 0.60, the Jaccard similarity coefficient between the genotypes of peaches varied from 0.40–0.89. It was verified by the results of PCoA and population structure analysis that 54 genotypes were divided into four groups. Also, Li et al. (2023) conducted molecular characterization on 39 late-mature peaches (*Prunus persica* L.) accessions, employing ISSR and SRAP markers to gauge genetic divergence. ISSR primers showed 70.09% polymorphism, while SRAP markers exhibited 73.33% polymorphism. The similarity coefficients, based on ISSR and SRAP markers, ranged from 0.65–0.89, suggesting a diverse genetic peach base. The various studies are depicted in Table 3. The genetic diversity for resistant genes in peaches is shown in Figure 2.

Table 3. Genetic diversity assessment in Peach (*Prunus persica* L.) via molecular markers approach.

Species	Marker Used	References
Prunus persica L.	SNP	Akagi et al. (2016)
Prunus persica L.	SSR	Perez et al. (2020)
Prunus persica L.	EST-SSR	Koohi et al. (2020)
Prunus persica (L.) Batsch	SSR	Trifonova et al. (2021)
Prunus persica (L.) Batsch	SNP	Mas-Gómezet al. (2022)
Prunus persica L.	ISSR	Demirel et al. (2023)

Functional Genomic Studies and Genome Sequencing Initiatives in Peach

One of the impediments to breeding programs in peaches is the lack of availability of molecular markers. In peach, with the help of high-throughput sequencing programs, a genome-scale uncovering via single-nucleotide polymorphism (SNP) was employed for creating and assessing next-generation IlluminaInfinium SNP genotyping array and for that The International Peach SNP Consortium (IPSC) was formed in 2012. With the help of Illumina and Roche/454 sequencing platforms on 56

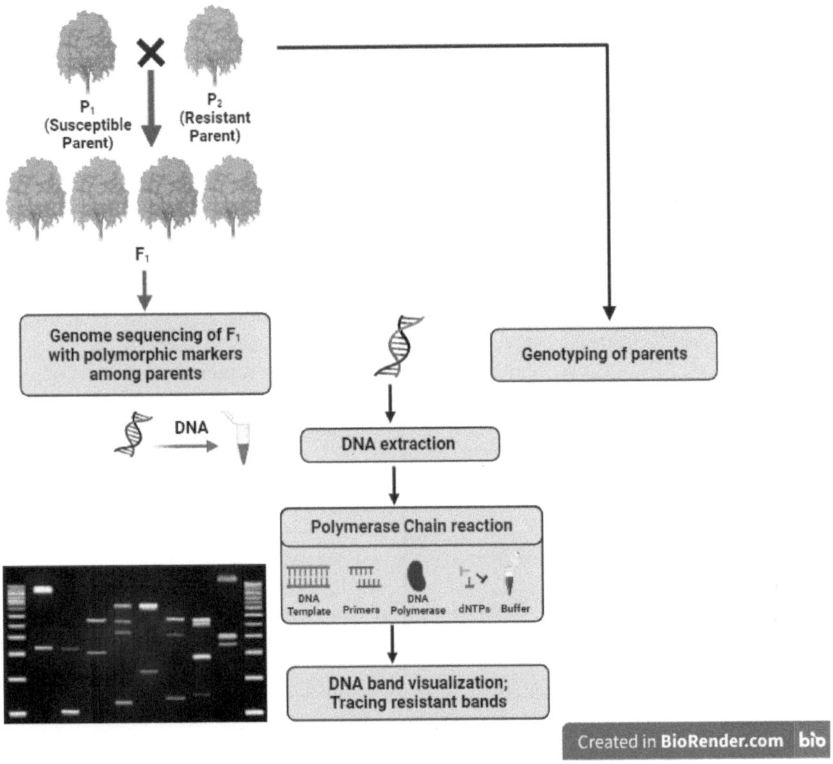

Figure 2. The genetic diversity for resistant genes in Peach (*Prunus persica* L.) via molecular marker approach.

accessions of peach, whole genome re-sequencing of peach was done. In the IPSC SNP array v1 of peach, almost eight, 144 SNPs set has been counted covering all chromosomes of peach, i.e., eight in number, spanning 26.7 kb among SNPs as reported by Verde et al. (2012). The sequencing of the peach genome was done additionally by Verde et al. (2013) which describes that with the help of a high-throughput shortgun array for whole genome sequencing in double haploid, each genotype cv. 'Lovell' (PLov2-2N; 2n = 2x = 16), 265 Mb genome size has been revealed with the prediction of 27, 852 genes coding for proteins as well as antisense type of RNAs. Also, the sequencing data of six distantly related peach accessions was used to perform genome-wide DNA marker detection. A total of 114,227~178,968 small insertion/deletions (InDels), 8386~12,298 structure variants (SVs), 2111~2581 copy number variants (CNVs), and 229,357~346,940 simple sequence repeats (SSRs) were found and annotated. There were 650,693~1,053,547 single nucleotide polymorphisms (SNPs) recorded (Guan et al., 2019). Peach genome sequencing helps breeders and researchers to differentiate among various species by conducting comparative genomic analysis between them. In the Rosaceae family, there exists a macro-synteny among them at the level of the genome which facilitated the detection and assessment of pathways leading to metabolic studies found in conserved germplasm along with deciphering of genetic markers to determine various quality-related traits.

Conclusions and Future Perspective

Peach being a stone fruit of perennial nature faces certain restrictions during its improvization at the genetic level through classical approaches of breeding with crosses from related species having desired traits which take time. To overcome this, the methods of recent biotechnology have been developed to widen the breeding of fruit crops. Recently, genome sequencing involving high-throughput is finding ways to regulate the genetic makeup of fruit crops focusing on lacunas and targeting desired genes.

Since information about complex genetic pathways is not clear and frequent. Therefore, advanced approaches of RNAi and CRISPR-*Cas* help identify the vast genetic complexes of breeding with valuable knowledge. Also, the Genome-wide association study (GWAS) will resolve issues of mapping and reducing large-scale population assessment opening avenues of breeding programs for a better and sustainable future in peach breeding.

References

Abbott, A., L. Belthoff, S. Rajapakse, R. Ballard, W.V. Baird, R. Monet, R. Scorza, P. Morgans, and A. Callahan. 1992. Development of a genetic linkage map of peach. *Plant Genome, 1*: 22.

Akagi, T., H. Toshio, Y. Hideaki, M.G. Thomas, and T. Ryutaro. 2016. Genome-wide view of genetic diversity reveals paths of selection and cultivar differentiation in peach domestication. *DNA Research, 23*(3): 271–282.

Aranzana, M.J., E.K. Abbassi, W. Howad, and P. Arús. 2010. Genetic variation, population structure and linkage disequilibrium in peach commercial varieties. *BMC Genetics, 11*: 1–11.

Aranzana, M.J., E. Illa, W. Howad, and P. Arús. 2012. A first insight into peach [*Prunus persica* (L.) Batsch] SNP variability. *Tree Genetics and Genomes, 8*: 1359–1369.

Aranzana, M., J. Carbó, and P. Arús. 2003. Microsatellite variability in peach [*Prunus persica* (L.) Batsch]: Cultivar identification, marker mutation, pedigree inferences and population structure. *Theoretical and Applied Genetics, 106*: 1341–1352.

Arulsekar, S., D E. Parfitt, and D.E. Kester. 1986. Comparison of isozyme variability in peach and almond cultivars. *Journal of Heredity, 77*(4): 272–274.

Baird, W.V., A.S. Estager, and J.K. Wells. 1994. Estimating nuclear DNA content in peach and related diploid species using laser flow cytometry and DNA hybridization. *Journal of the American Society for Horticultural Science, 119*(6): 1312–1316.

Bennett, M.D. 1984. The genome, the natural karyotype and biosystematics. *Plant Biosystematics*, 41–66.

Bouhadida, M., M.A. Moreno, M.J. Gonzalo, J.M. Alonso, and Y. Gogorcena, Y. 2011. Genetic variability of introduced and local Spanish peach cultivars determined by SSR markers. *Tree Genetics &Genomes, 7*: 257–270.

Cao, K., Z. Zheng, L. Wang, X. Liu, G. Zhu, W. Fang, S. Cheng, P. Zeng, C. Chen, X. Wang, and M. Xie. 2014. Comparative population genomics reveals the domestication history of the peach, *Prunus persica*, and human influences on perennial fruit crops. *Genome Biology, 15*: 1–15.

Chapparro, C., D. Werner, D.O.Malley, and R. Sederoff. 1992. Targeted mapping and linkage analysis in peach. *Plant Genome, 21*: 9–11.

Cheng, Z., and H. Huang. 2009. SSR fingerprinting Chinese peach cultivars and landraces (*Prunus persica*) and analysis of their genetic relationships. *Scientia Horticulturae, 120*(2): 188–193.

Chinese Crop Germplasm Resources Information System. 2020. https://www.cgris.net/cgris_english.html. (Accessed on 04 February 2024).

Clifford, M.N., K.L. Johnston, S. Knight, and N. Kuhnert. 2003. Hierarchical scheme for LC-MS n identification of chlorogenic acids. *Journal of Agricultural and Food Chemistry, 51*(10): 2900–2911.

Demirel, S., M. Pehluvan, and R. Aslantaş. 2023. Evaluation of genetic diversity and population structure of peach (*Prunus persica* L.) genotypes using inter-simple sequence repeat (ISSR) markers. *Genetic Resources and Crop Evolution*, 1–12.

Dickson, E.E., K. Arumuganathan, S. Kresovich, and J.J. Doyle. 1992. Nuclear DNA content variation within the Rosaceae. *American Journal of Botany, 79*(9): 1081–1086.

Eldredge, L., R. Ballard, W.V. Baird, A. Abbott, P. Morgens, A. Callahan, R. Scorza, and R. Monet. 1992. Application of RFLP analysis to genetic linkage mapping in peaches. *HortScience, 27*(2): 160–163.

Food and Agriculture Organization of the United Nations. 2018 .https://www.fao.org/news/archive/news-by-date/2018/en/ (Accessed on 09 February 2024).

Food and Agriculture Organization of the United Nations. 2023. https://www.fao.org/documents/card/en?details=cc3017en. (Accessed on 09 February 2024).

Gao, L., J. Liu, L. Liao, A.Gao, B.N. Njuguna, C. Zhao, B. Zheng, and Y. Han. 2023. Callus induction and adventitious root regeneration of cotyledon explants in peach trees. *Horticulturae, 9*(8): 850.

Gentile, A., S. Monticelli, and C. Damiano. 2002. Adventitious shoot regeneration in peach [*Prunus persica (L.)*Batsch]. *Plant Cell Reports, 20*: 1011–1016.

Guan, L., K. Cao, Y. Li, J. Guo, Q. Xu, and L. Wang. 2019. Detection and application of genome-wide variations in peach for association and genetic relationship analysis. *BMC Genetics, 20*(1): 1–13.

Hammerschlag, F.A., G. Bauchan, and R. Scorza. 1985. Regeneration of peach plants from callus derived from immature embryos. *Theoretical and Applied Genetics, 70*: 24–251.

Hancock, J.F., T.M. Sjulin, and G.A. Lobos. 2008. Strawberries. In *Temperate fruit crop breeding: Germplasm to Genomics*, 393–437. Dordrecht: Springer Netherlands.

Hansche, P.E. 1988. Two genes that induce brachytic dwarfism in peach. *HortScience, 23*(3): 604–606.

Hong, N.H., T.D. Xuan, E. Tsuzuki, H. Terao, M. Matsuo, and T.D. Khanh. 2004. Weed control of four higher plant species in paddy rice fields in Southeast Asia. *Journal of Agronomy and Crop Science, 190*(1): 59–64.

Jelenkovic, G., and E. Harrington. 1972. Morphology of the pachytene chromosomes in *Prunus persica. Canadian Journal of Genetics and Cytology, 14*(2): 317–324.

Koohi Dehkordi, M., T. Beigzadeh, and K. Sorkheh. 2020. Novel in silico EST-SSR markers and bioinformatic approaches to detect genetic variation among peach (*Prunus persica* L.) germplasm. *Journal of Forestry Research, 31*: 1359–1370.

Li, W., Y. Ma, Y. Kou, Z. Zeng, D. Qiu, and C. Ma. 2023. Analysis of genetic diversity and relationships between late-mature peach (*Prunus persica* L.) varieties assessed with ISSR and SRAP markers. *Fruit Research, 3*(1).

Martínez-Gómez, P., S. Arulsekar, D. Potter, and T.M. Gradziel. 2003. An extended interspecific gene pool available to peach and almond breeding as characterized using simple sequence repeat (SSR) markers. *Euphytica, 131*: 313–322.

Mas-Gómez, J., C.M. Cantín, M.A. Moreno, and P. J. Martínez-García. 2022. Genetic Diversity and Genome-Wide Association Study of Morphological and Quality Traits in Peach Using Two Spanish Peach Germplasm Collections. *Frontiers in Plant Science, 13*: 854770.

Meagher, R.B., M.D. McLean, and J. Arnold. 1988. Recombination within a subclass of restriction fragment length polymorphisms may help link classical and molecular genetics. *Genetics, 120*(3): 809–818.

Monet, R. 1988. Peach genetics: Past, present, and future. *In*: *II International Peach Symposium, 254*: 49–58.

Mowrey, B.D., D.J. Werner, and D.H. Byrne. 1990. Inheritance of isocitrate dehydrogenase, malate dehydrogenase, and shikimate dehydrogenase in peach and peach × almond hybrids. *Journal of the American Society for Horticultural Science, 115*(2): 312–319.

Padilla, I.M., A. Golis, A. Gentile, C. Damiano, and R. Scorza. 2006. Evaluation of transformation in peach *Prunus persica* explants using green fluorescent protein (GFP) and beta-glucuronidase (GUS) reporter genes. *Plant Cell, Tissue, and Organ Culture, 84*: 309–314.

Pérez, V., N. Larrañaga, D. Abdallah, A. Wünsch, and J.I. Hormaza. 2020. Genetic diversity of local peach (*Prunus persica*) accessions from La Palma island (Canary Islands, Spain). *Agronomy, 10*(4): 457.

Pérez-Clemente, R.M., A. Pérez-Sanjuán, L. García-Férriz, J.P. Beltrán, and L.A. Cañas. 2005. Transgenic peach plants (*Prunus persica* L.) produced by genetic transformation of embryo sections using the green fluorescent protein (GFP) as an *in vivo* marker. *Molecular Breeding, 14*: 419–427.

Pooler, M.R., and R. Scorza. 1995. Regeneration of peach [*Prunus persica* (L.) Batsch] rootstock cultivars from cotyledons of mature stored seed. *HortScience, 30*(2): 355–356.

Ramming, D.W. 1991. Genetic control of a slow-ripening fruit trait in nectarine. *Canadian Journal of Plant Science, 71*(2): 601–603.

Ravaglia, D., R.V. Espley, R.A. Henry-Kirk, C. Andreotti, V. Ziosi, R.P.Hellens, G. Costa, and A.C. Allan. 2013. Transcriptional regulation of flavonoid biosynthesis in nectarine (*Prunus persica*) by a set of R2R3 MYB transcription factors. *BMC Plant Biology, 13*(1): 1–14.

Ricci, A., S. Sabbadini, O. Navacchi, L. Burgos, and B. Mezzetti. 2022. Approaches applied to induce regeneration, via somatic embryogenesis, in the peach rootstock GF677 (*P. persica× P. amygdalus*) and in different peach cultivars using leaves and anthers as starting explants. *In*: *XXXI International Horticultural Congress (IHC2022): International Symposium on In Vitro Technology and Micropropagated Plants, 1359*: 241–248.

Sabbadini, S., A. Ricci, C. Limera, D. Baldoni, L. Capriotti, and B. Mezzetti. 2019. Factors affecting the regeneration, via organogenesis, and the selection of transgenic calli in the peach rootstock Hansen 536 (*Prunus persica× Prunus amygdalus*) to express an RNAi construct against PPV virus. *Plants, 8*(6): 178.

Sauge, M.H., J. Kervella, and T. Pascal. 1998. Settling behaviour and reproductive potential of the green peach aphid *Myzuspersicae* on peach varieties and a related wild Prunus. *Entomologiaexperimentalisetapplicata, 89*(3): 233–242.

Sansavini, S., D. Bassi, and A. Gamberini. 2006. Peach breeding, genetic and new cultivar trends. *Acta Horticulturae, 713*: 23–52.

Scorza, R., and W.R. Okie. 1990. Peaches. *In*: J.N. Moore and J.R. Ballington [Eds.], *Genetic Resources of Temperate Fruit and Nut Crops*, 177231. ISHS: Wageningen.

Scorza, R., S.A. Mehlenbacher, and G.W. Lightner. 1985. Inbreeding and coancestry of freestone peach cultivars of the eastern United States and implications for peach germplasm improvement. *Journal of the American Society for Horticultural Science, 110*(4): 547–552.

Shadidi, F., and M. Naczk. 2004. *Phenolics in Foods and Nutraceuticals*. Routledge. Taylor and Francis Group.

Thakur, D., K. Thakur, M. Thakur, A. Negi, and V. Kumar. 2023. Performance of some Japanese Plum (*Prunussalicina*) Cultivars in Mid-hill Condition of Kullu Valley, Himachal Pradesh. *International Journal of Bioresource and Stress Management, 14*: 1495–1501.

Trifonova, A.A., K.V. Boris, N.V. Mesyats, V.A. Tsiupka, A.V. Smykov, and I.V. Mitrofanova. 2021. Genetic diversity of peach cultivars from the collection of the Nikita botanical garden based on SSR markers. *Plants, 10*(12): 2609.

Verde, I., A.G. Abbott, S. Scalabrin, S. Jung, S. Shu, F. Marroni, T. Zhenentyayeva, M. T. Dettori, J. Grimwood, F. Cattonaro, A. Zuccolo, L. Rossini, J. Jenkins, E. Vendramin, L.A. Meisel, V. Decroocq, B. Sosinski, S. Prochnik, T. Mitros, A. Policriti, G. Cipriani, L. Dondini, S. Ficklin, D.M. Goodstein, P. Xuan, C. Del Fabbro, V. Aramini, D. Copetti, S. Gonzalez, D.S. Horner, R. Falchi, S. Lucas, E. Mica, J. Maldonado, B. Lazzari, D. Bielenberg, R. Pirona, M. Miculan,

A. Barakat, R. Testolin, A. Stella, S. Tartarini, P. Tonutti, P. Arus, A. Orellana, C. Wells, D. Main, G. Vizzotto, H. Silva, F. Salamini, J. Schmutz, M. Morgante, and D.S. Rokhsar. 2013. The high quality draft genome of peach (*Prunus persica*) identifies unique patterns of genetic diversity, domestication and genome evolution. *Nature Genetics, 45*: 487–496.

Verde, I., N. Bassil, S. Scalabrin, B. Gilmore, C.T.Lawley, K. Gasic, D. Micheletti, U.R. Rosyara, F. Cattonaro, E. Vendramin, and D. Main. 2012. Development and evaluation of a 9K SNP array for peach by internationally coordinated SNP detection and validation in breeding germplasm. *PloS One, 7*(4): e35668.

Wang, L.R., G.R. Zhu, W.C. Fang, K. Cao, X. Wang, C. Chen, P. Zhao, and X. Wang. 2012. Peach genetic resource in China. *Beijing: China Agriculture Press, 755*: 756.

Wang, L., G. Zhu, W. Fang, Q. Zuo, and L. Han. 2001. Study on the resistance to peach aphid (*MyzuspersicaeSulzer*) of peach germplasm. *Journal of Fruit Science, 18*(3): 145–147.

Wang, Z.H., and E.J. Zhuang. 2001. *China Fruit Monograph:pPeach Flora.*

Wünsch, A., M. Carrera, and J.I. Hormaza. 2006. Molecular characterization of local Spanish peach [*Prunus persica* (L.) Batsch] germplasm. *Genetic Resources and Crop Evolution, 53*: 925–932.

Xie, R., X. Li, M. Chai, L. Song, H. Jia, D. Wu, M. Chen, K. Chen, M.J. Aranzana, and Z. Gao. 2010. Evaluation of the genetic diversity of Asian peach accessions using a selected set of SSR markers. *ScientiaHorticulturae, 125*(4): 622Zhuang. 2001. *China Fruit Monograph*629.

Xu, D.H., S. Wahyuni, Y. Sato, M. Yamaguchi, H. Tsunematsu, and T. Ban. 2006. Genetic diversity and relationships of Japanese peach (*Prunus persica* L.) cultivars revealed by AFLP and pedigree tracing. *Genetic Resources and Crop Evolution, 53*: 883–889.

Xu, S., E. Lai, L. Zhao, Y. Cai, C. Ogutu, S. Cherono, Y. Han, and B. Zheng. 2020. Development of a fast and efficient root transgenic system for functional genomics and genetic engineering in peach. *Scientific Reports, 10*(1): 1–9.

Yamamoto, T., K. Mochida, and T. Hayashi. 2003. Shanhai Suimitsuto, one of the origins of Japanese peach cultivars. *Journal of the Japanese Society for Horticultural Science, 72*(2): 116–121.

Yu, M., R. Ma, Z.Shen, and Z. Cai. 2010. Research advances in peach germplasm in China. *Jiangsu Journal of Agricultural Sciences, 26*(6): 1418–1423.

Zeven, A.C., and P.M. Zhokovsky. 1975. *Dictionary of Cultivated Plants and Their Centers of Diversity*. Centre for Agricultural Publishing and Documentation Wageningen.

Zheng, B., J. Liu, A. Gao, X. Chen, L.Gao, L. Liao, B. Luo, C.O. Ogutu, and Y. Han. 2022. Epigenetic reprogramming of H3K27me3 and DNA methylation during leaf-to-callus transition in peach. *Horticulture Research, 9*: uhac132.

Zong, X., B.J. Denler, G.H. Danial, Y. Chang, and G.Q. Song. 2019. Adventitious shoot regeneration and *Agrobacterium tumefaciens*-mediated transient transformation of almond × peach hybrid rootstock "Hansen 536". *HortScience, 54*(5): 936–940.

16

Genetic Diversity of Jamun (*Syzygium cumini* Skeels) and Its Sustainable Utilization

Anuradha Sane,[1] *Rekha, A.*[2,*] and *P.C. Tripathi*[3]

Introduction

Jamun, botanically called *Syzygium cumini* Skeels, belongs to the family Myrtaceae (Chase and Reveal, 2009). The Jamun is also known by common names such as Indian blackberry, Java plum, Jambu, black plum, Jambul, Kalajam, Phalinda, Rajamun, damson plum, and duhat plum (Nadkarni, 1976). It is native to India, Burma, Ceylon, and the Andaman Islands (Zeven and de Wet, 1982) and naturalized throughout Myanmar, Sri Lanka, East Africa, South East Asia, and Eastern Australia (USDA-ARS, 2008; Figure 1). Besides, the trees can be found throughout South America, and Madagascar, and have also naturalized in the warmer sections of the United States of America (Florida and Hawaii). It is an economically and ecologically important native tree of tropical moist deciduous and evergreen forests of the Western Ghats, a biodiversity hotspot (Kadavul and Parthasarathy, 1999). It is also cultivated in many countries as an avenue tree and as a source of fruit for human consumption (Corlett, 2005). It is found grown in wild and semi-wild tropical and subtropical parts of India, viz., Punjab, Haryana, Uttar Pradesh, Maharashtra, Rajasthan, Gujarat, Madhya Pradesh, Bihar, Chhattisgarh, Jharkhand, Karnataka, Kerala, Tamil Nadu, Andhra Pradesh (Singh et al., 2011) and Himachal Pradesh (Misra et al., 1975). It is available throughout the Indian plains and in the Kumaon hills up to 1,600 m. In India, it is grown in the Indo-Gangetic plains, lower ranges of the Himalayas, and Kumaon hills in the north to Tamil Nadu in the south. The world production of Jamun is 13.5 million tons per annum and India's share is 15.4% (Kshirsagar et al., 2019). It has gained tremendous importance and recognition in the recent past not only because of its hardy nature but also for its medicinal and nutritional properties.

It possesses great potential for commercial use in fallow and dryland horticulture. Additionally, the timber derived from this tree is utilized in construction and agricultural implements, and it displays resistance to the effects of water (Chundawat, 1990). The Jamun tree has the ability to withstand prolonged periods of drought and flooding, making it well-suited for windbreaks and

[1] ICAR – Division of Fruit Crops, Indian Institute of Horticultural Research, Bengaluru 560089, India
[2] ICAR – Indian Institute of Horticultural Research, Bengaluru 560089, India
[3] ICAR – Horticulture Science division, ICAR, Krishi Bhavan, New Delhi -110012, India
* Corresponding author: arekha2001@yahoo.com

Figure 1. Distribution of *Syzygiumcumini* Skeels (Orwa et al., 2009).

roadside plantations. It can be found growing in forests and along roadsides, as well as in backyards and bunds, making it a significant component in agroforestry systems. Considering the increasing relevance and demand for this fruit crop, the management and effective utilization of the vast genetic resources available is crucial in expanding organized jamun orcharding to meet market demands (Anushma et al., 2023).

Nutritional Value

Jamun fruits exhibit a delectable taste, and a pleasantly flavored profile, and are much liked by the general populace. They are predominantly utilized for desserts and the preparation of delightful beverages, such as jellies, jam, squash, wine, vinegar, and pickles (Ochse et al., 1961). Jamun is renowned for its high nutritional value, as it contains carbohydrates, free amino acids, and a plethora of minerals, including sodium, potassium, calcium, phosphorous, iron, and zinc, as well as water-soluble vitamins such as ascorbic acid, thiamine, and niacin (Table 1). The bark, leaves, fruits, and seeds of this fruit have been employed since ancient times to alleviate a myriad of ailments, with

Table 1. Nutritional value of Jamun fruit (100 g) (Sarvade et al., 2016).

Particulars	Values
Energy	60 Kcal
Carbohydrates	15.56
Dietary fiber	0.3–0.9 g
Fat	0.15–0.30 g
Protein	0.72 g
Vitamins	
Thiamine (B 1)	0.006 mg
Riboflavin (B2)	0.012 mg
Niacin (B3)	0.260 mg
Pantothenic acid (B5)	0.160 mg
Vitamin (B6)	0.038 mg
Vitamin C	14.3 mg
Calcium	19.0 mg
Magnesium	15.0 mg
Phosphorous	17.0 mg
Potassium	79 mg
Sodium	4 mg

a particular focus on diabetes. Considering its myriad health benefits, the demand for this fruit has witnessed significant growth in recent times. Despite having diverse medicinal and nutraceutical properties, limited information is available on documentation of ethnobotanical uses and available diversity and the pattern of variation in genetic resources and variability in its anti-diabetic compound in different genotypes.

Medicinal Properties

According to the World Health Organization (WHO), about 65–80% of the world's population in developing countries depend essentially on plants for their primary healthcare due to ancient practices, poverty, and lack of access to modern medicine. Different parts of the plant are used in folk medicine in India and elsewhere (Jain, 2000). It finds applications in almost all types of medicines like Ayurveda, Unani, Siddha, Sri Lankan medicine, Tibetan traditional medicine, and also in Homeopathy systems. It is an ancient medicinal plant with an illustrious medical history and has been the subject of classical reviews for over 100 years. Various traditional practitioners in India use different parts of the plant in the treatment of diabetes, blisters in the mouth, cancer, colic, diarrhea, digestive complaints, dysentery, piles, pimples, and stomachaches (Jain, 1991). During the last four decades, numerous folk medicinal reports on the anti-diabetic effects of this plant have been cited in the literature. In Unani medicine, various parts of Jamun are used as a liver tonic, to enrich blood, strengthen teeth and gums, and form lotions for removing ringworm infection of the head (Damasceno et al., 2002). It is known to have several other medicinal properties such as hypoglycemic, cardio-protective, antioxidant, anti-allergic, anticancer, hepatoprotective, radioprotective, chemoprotective, and gastro-protective properties that have been proven on animal models by different researchers. The leaves and bark are used for controlling blood pressure and gingivitis, astringent, antiscorbutic, diuretic, antidiabetic, and chronic diarrhea and enlargement of the spleen (Achrekar et al., 1991). Various parts of the Jamun have been claimed to have hypolipidemic, ulcer onset prevention, nitric oxide scavenging, and free radical scavenging properties (Rizvi et al., 2022). In their study, Sane et al. (2022) recorded the ethnobotanical uses of jamun fruits in the tribal regions of Karnataka (Biligirirangana hills) and Odisha (Mayurbhanj). They emphasized that documenting indigenous knowledge about jamun would contribute to enriching and conserving the genetic resources of this fruit in the specific areas where it naturally occurs and would support sustainable utilization of this fruit tree.

Studies indicate that jamun seeds and pulp offer potential benefits for diabetic patients, including blood glucose reduction and the prevention of complications like neuropathy and cataracts (Sagrawat et al., 2006; Helmstadter, 2008). Jamun seeds are a valuable source of ellagitannins (ETs), including compounds like corilagin, 3,6-hexahydroxy diphenoyl glucose, and its isomer 4,6-hexahydroxy diphenoyl glucose, as well as gallic acid and ellagic acid (EA). Additionally, the jamun plant contains anthocyanins, glucoside, isoquercetin, kaempferol, and myricetin. The seeds are also believed to contain alkaloids such as jambosine and glycoside jambolin (or antimellin), which stop the conversion of starch into sugar (Swami et al., 2012).

Different extracts of Jamun were proven to have a variety of pharmacological activities. The reported health benefits can be attributed to the presence of various phytochemicals like polyphenols, terpenes, anthocyanins, and flavonoids. The presence of anthocyanin and ellagitannins in the pulp is important in overcoming oxidative stress-related illnesses (Baliga et al., 2011). A detailed review by Chakraborty et al. (2023) highlighted the significant benefits of fruit and seeds in managing metabolic syndrome such as diabetes, hypertension, hyperlipidemia, and obesity. These positive effects contribute to overall health protection against metabolic syndromes (Rizvi et al., 2022). Interestingly, even before the discovery of insulin, jamun was used for diabetes treatment, either alone or in combination with other hypoglycemic plants, even in Europe (Helmstadter, 2008).

Other Uses of Jamun Tree

Fruits are highly perishable and utilized in the preparation of value-added products such as jam, jelly, juice, wine, and fruit bars. Wood serves as a valuable fuel source. The reddish-grey or brownish heartwood is utilized in carpentry. The wood is hard and also finds use in making boats, agricultural implements, cartwheels, bridge construction, and even crafting musical instruments, particularly the guitar. Tannin extracted from the bark yields a brown dye.

Taxonomic Background

The Jamun has a wide distribution in tropical and warm temperate regions across the globe, particularly in biodiversity hotspots. The genus *Syzygium* Gaertn. is considered one of the largest groups of flowering plants. It occurs worldwide in tropical and subtropical areas, including South Africa, South America, Southeast Asia, and Australia. *Syzygium* species are characterized by deciduous petals, terminal or occasionally lateral inflorescences in the form of cymes, and independent cotyledons in fruit, which are reported as drupes (Akoegninou et al., 2006).

The extensive diversity within the *Syzygium* genus poses challenges for species classification. Consequently, numerous taxonomists have conducted research to categorize these species. The most recent study employed a phylogenetic approach (Craven and Biffin, 2010). These *Syzygium* species inhabit diverse ecosystems, often as medium to large-sized sub-canopy trees, significantly influencing a wide range of organisms. Several other species within the *Syzygium* genus, including *S. caryophyllatum, S. aromaticum, S. aqueum, S. grande*, and *S. myrtifolium*, are used as spices or fruits in the pharmacology and horticulture industry (Low et al., 2022). The genus *Syzygium* has many species, of which *Syzygium cumini* is the main species of commercial interest distributed across India. The other economically important *Syzygium* species are as follows.

Syzygium aromaticum (Linn.) Merrill & Perry. syn. *Caryophyllus aromaticus* Linn. is a conical evergreen tree, 9–12 m in height with grayish smooth bark, lanceolate leaves, and flowers borne in terminal clusters. The flower bud is an important source of spice the ‘Clove’.

Syzygium jambos (Linn.) Alston. (syn. *Eugenia jambos* Linn.) is also known as rose apple. It is a small to medium-sized tree having narrow lanceolate leaves with glabrous bark. The flowers with numerous 5–8 cm long stamens are attractive. The fruits are light yellow-white and rose-scented with persistent calyx. The seed is polyembryonic and it is being grown in Assam, Bihar, Andhra Pradesh, Tamil Nadu, West Bengal, coastal areas of Maharashtra, and Gujarat.

Syzygium malaccense (Linn.) Merrill and Perry. (syn. *Eugenia malaccensis* Linn.) is known as the Malayan apple or watery rose apple. It is a large, shrub-sized tree with a nearly straight trunk and densely foliaged crown, attractive red flower sub-globose fruits found chiefly in the gardens in Bengal and South India.

Syzygium samarangense (Blume.) Merrill & Perry. syn. *Eugenia javanica* Lam., is a 5–15 m tall tree, with a low branched trunk and widely branched crown, found in Andaman and Nicobar Islands. The fruits are waxy, bell-shaped resembling cashew apples also known as water apples.

Syzygium alternifolium (Wight) Walp. syn. *Eugenia alternifolia* Wight, is a moderate-sized or large tree, with flowering during the dry season, found in Assam, Andhra Pradesh, and Tamil Nadu.

Syzygium aqueum (Burm. f.) Alston syn. *Eugenia aquea* Burm. f is a much-branched tree, grown to an altitude of 1,350 m in Assam and Meghalaya and cultivated in Bengal.

Syzygium bracteatum (Willd.) Raizada syn. *Eugenia bracteata* Roxb., a shrub or small tree, is found in Arunachal Pradesh, Assam, Meghalaya, and Tamil Nadu.

Syzygium fruitecosum (Roxb.) DC is a medium-sized and evergreen tree found in Bangladesh; fruits are edible and small in size. Its seeds are utilized in preventing chronic diseases.

Syzygium densiflorum, trees are evergreen, native to the Western Ghats in India, and suitable rootstock for jamun as it is resistant to termite attack.

Syzygium uniflora Linn. (Surinam cherry or Pitanga cherry), is a small tree found in South India and bears small-sized fruits having bright red colour and aromatic flavor (Mishra et al., 2014).

Syzygium polyanthum (Wight.) Walp*:* A tree with a single trunk and clear, dense canopy shape, and can reach 25 m. high. It has a dark brown, rough grooved bark. Single leaves are arranged opposite elliptic-round shaped or obovate (obovate) with a pointed tip. Compact white flowers are fragrant and reddish. Sweet fruit is a round, red-colored, or dark red fruit. Its natural habitat is the forest area at an altitude of 5–1,000 MSL. This species is often planted in home gardens for the leaves and fruit.

Syzygium pycnanthum Merr. and L.M. Perry is a small tree, up to 15 m high, and a trunk with no buttresses. Single leaves are oppositely arranged, dark green on the upper surface and pale green on the lower surface. The leaf shape is ovate-oblong-lanceolate acute-acuminate leaf tip (pointy-tapered). Compact flowers with short flowers, crown is purplish-white flowers, white-colored greenish-purple petals, have many stamens. The fruit is a berry that is round, light green, dark purplish-red, or reddish-green.

Syzygium racemosum (Bl.) DC. It is a small tree with dense branching. The bark is light gray-brown. leaf is elliptical-oblong leaf shape. Young leaves are reddish-copper. The inflorescence is terminal or axillary panicles with yellowish-white flowers, with a crown like a small calyptra. The fruit is yellowish-green, rounded bell-shaped with a diameter, the leaves are thin copper-colored young leaves; it has small round white flowers.

Syzygium littorale (Bl.) Amshoff: The tall stature of trees can reach 10–20 m. Leaves are lanceolate-oblong-shaped arranged in opposite directions. The inflorescence terminal has white flowers. The fruit is round, campanulate (bell-shaped), green, and yellow.

Mudiana (2016) reported six species of *Syzygium*, namely, *S. cumini, S. littorale, S. pycnanthum, S. polyanthum, S. racemosum, and S. samarangense. S. pycnanthum* is the most frequently found in Gunung Baung, East Java, Indonesia. *S. polyanthum and S. samarangense* are the only species that are known to be cultivated. Four other species are wild and have not been explored for their potential utilization.

Botany

Jamun is a slow-growing, evergreen large tree with a grayish-brown thick bark with woody scales. The tree can reach heights of up to 30 m, is long-lived, and can survive more than 100 years. The leaves are oblong-ovate to elliptical, leathery, and 6–12 cm long and have an aroma similar to turpentine. The inflorescence (panicle) is mostly borne below the leaves on the branchlets, measuring 4–6 cm in length. The trees start flowering from March to April and the fruits mature by May or June. Flowers are fragrant, small about 5 mm in diameter greenish-white produced in clusters, bisexual, and occur in dichotomous paniculate cymes. The calyx is funnel-shaped, about 4 mm long and toothed. The petals adhere together as a small disk. The stamens are many and measure up to the calyx. The fruit is a fleshy drupe, oblong to ovoid, measures 1.5–3.5 cm long, dark-purple or nearly black, with one large seed. The fruit exhibits a blend of sweetness, mild sourness, and astringency, often resulting in a purple-stained tongue (Singh et al., 2019; Bajpai et al., 2012). The seed is recalcitrant; seeds germinate well when fresh, but viability is lost within two weeks of open storage at room temperature.

Germplasm Collection

The Indian subcontinent is home to a significant population of jamun trees, earning it the epithet 'Jambudweep'. These trees are widely distributed across India, spanning from the north to south and east to west. They thrive in various regions, including the Gangetic plains, the Western peninsular region, and the Western Ghats. Jamun trees grow both in cultivated lands and wild conditions, and their hardy nature and high-yielding potential have made them important as a dry-land horticultural crop. The advantage of variations in these traits can be taken to evolve a superior seedling selection/strain. Identification of elite or plus trees for total yield and quality of fruits is the starting point of improvement of any tree. The selection of a large number of elite trees serves as an immediate seed source with fairly high genetic gain. In Jamun, significant variations were reported in plant growth, fruit shape, fruit size, fruit color, and pulp content, as well as total soluble solids (TSS), acidity, and early fruit bearing. These diverse traits offer an opportunity to develop superior seedling selections. Jamun is cross-pollinated and plants are seed propagated leading to wide variability in nature. Closely related species are found in distinct ecological zones and are not exploited for their unique characteristics. The species is under threat because of neglect and is prone to destruction due to developmental activities. As a result of urbanization and intensive cultivation, the jamun crop is experiencing significant genetic erosion and the extinction of multiple species. Some of the edible species of *Syzygium* are planted throughout the tropics worldwide. Assessment of the genetic diversity of different populations is important to form a basis for conservation, genetic tree improvement, and promotion or domestication of populations with desirable traits. In the search for sources of resistance to biotic and abiotic challenges, the genetic diversity of related wild species is very valuable (Singh et al., 2011). In Iran, promising accessions of Jamun were identified and selected based on pomological characterizations such as fruit weight, color, and taste (Khadivi et al., 2023) which can be cultivated directly in orchards or used as parents in breeding programs. Principal component analysis showed that fruit-related characters are very important in differentiating selections. An extensive investigation was conducted to determine the natural variability among jamun seedling trees and to discover superior genotypes with high fruit quality. A wide range of seedling progenies concerning fruit shape and size, pulp colour, TSS, acidity, and earliness are available, particularly in Uttar Pradesh, Gujarat, and Maharashtra, providing good scope for selection. A wide range of physicochemical characteristics of fruits are available offering the possibility of selecting a variety suitable for fresh market and processing. Small seed size, high pulp content, and better biochemical properties are considered to be ideal characteristics (Sanjay and Singh, 2012; Mishra et al,. 2014). An intensive survey of the existing natural variability among the jamun seedling trees resulted in the identification of superior genotypes in Maharashtra (Keskar et al., 1989), West Bengal (Kundu et al., 2001), Goa (Priya Devi et al., 2002), Karnataka (Ghojage et al., 2011), Uttar Pradesh and Jharkhand (Patel et al., 2005; Singh et al., 2019), Tamil Nadu (Latheef and Kavino, 2018), and subtropical areas of Jammu (Plathia et al., 2018). A significant positive correlation was observed for fruit weight with pulp weight, fruit volume, pulp-to-seed ratio, and titratable acidity, whereas seed percentage had a significant negative correlation with fruit weight (Srivastava et al., 2012).

At ICAR-Indian Institute of Horticulture Research (IIHR), Bengaluru, characterization, and evaluation of germplasm in the field gene bank led to the identification of superior genotypes (IIHRJ-3, IIHRJ-14, and IIHRJ-12) with lesser plant height, canopy spread, early initiation of flowering, bold fruits, moderate TSS and high pulp content (Anushma and Sane, 2018). Likewise, Dharwad-6 differed from other genotypes in terms of vegetative characteristics as shown in cluster analysis (Ajay Kumar et al., 2021). A few unique selections, including seedless and year-round fruiting and genotypes with high percentages of polyembryony, have been identified (Tripathi et al., 2021; Rekha et al., 2020). Based on diversity analysis of 96 jamun accessions maintained in the field gene bank at IIHR, Bengaluru, J-6, and IC0621981 were found superior with respect to fruit parameters. Thirteen accessions had a good pulp-to-seed ratio, the highest fruit weight (16g) was observed in accession I-6. Proximity analysis of these 96 jamun accessions based on 15 various traits (fruit, leaf, seed, pulp related) resulted in 10 distinct clusters. Contribution of various traits to genetic diversity showed that fruit girth followed by seed girth together accounted for one-fourth of variability enabling the identification of lines with bold fruits while, seed length alone contributed to 90.9% variability (ICAR-IIHR, 2019, 2020). Maximum genetic distance (based on all traits) was observed, in particular, between the pairs (Collection 12 and Collection 2; MP-5 and Collection 2; KHA-1 and Konkan Bahadoli), which could be used for further hybridization trials in jamun (Kumar et al., 2020).

Systematic Germplasm Collections in India

Over the past three decades, systematic germplasm collections have been carried out across various regions in India. These valuable collections are meticulously maintained in the field gene banks under the Indian Council of Agricultural Research (ICAR) (Laishram and Singh, 2011; Annual Report, 2017). The elite trees selected for total yield, size and quality of fruit, etc., are the starting point of any fruit tree improvement activity. From the physicochemical evaluation of jamun fruits collected from different locations, it is concluded that the genotypes exhibited tremendous variability for either one or more characteristics. It offers more avenues for establishing desirable clones by simple seedling selection. Minimal descriptors for the characterization and evaluation of jamun and other agri-horticultural crops have been published and are available in the public domain (Mahajan et al., 2002). It helps in adopting uniform definitions of descriptors and descriptor states and, thereby enables them to generate standardized, authentic, and uniform data so that a sound database networking may be developed for the exchange and utilization of information. A modified three-digit extended Biologische Bundesanstalt Bundessortenamt and Chemische Industrie (BBCH) scale was developed and 3-digit codes were assigned to describe the critical phenological stages of Jamun. Eight primary stages and 36 secondary growth stages have been described with illustrations that occur sequentially in Jamun under subtropics (Singh et al., 2021). To augment the germplasm wealth, it is essential to collect and evaluate the variability present in *Syzygium cumini* and related wild and semi-wild species. The unexplored regions in India for jamun collection include the North-eastern states, Western Ghats (Kerala, Tamil Nadu, and Maharashtra), Chhattisgarh, and the Andaman and Nicobar Islands (Anushma et al., 2023).

Cytological Characterization

The genus *Syzygium* has a basic chromosome number of 11 ($x = 11$). In the Myrtaceae family, diploidy ($2n = 22$) is the most common ploidy level. In addition to diploids, polyploids have been observed, including tetraploids in species like S. jambos and *S. paniculatum* ($2n = 44$) (Oginuma et al., 1993). However, Jena et al. (2003) discovered a somatic chromosome number of $2n=66$ in *Syzygium cumini,* $2n = 44$ in *S. aqueum,* and $2n = 22$ in *S. malaccensis*. Sharma et al. (2020) found three cytotypes with $2n = 22$, 44, and 66 in the meiotic studies on five different populations from north India. A reproductive ecological study of a *Syzygium* species indicated that both self-compatibility and self-incompatibility exist in this genus with self-compatibility being the most common (Sanewski, 2010).

Biochemical Characterization

Jamun contains various bioactive compounds, including terpenes, flavonoids, anthocyanin, lipids, alkane, phenols, and phenylpropanoid, and a rich source of nutrition and medicine (Arya et al., 2019). The seeds are reported to contain an alkaloid 'jambosine' and glycoside 'jambolin' or 'antimellin', which arrests the conversion of starch into sugar in the body (Morton, 1987). The seed powder of *S. cumini* has increasingly been used for diabetes management in recent years (Proma et al., 2018). The scientific evidence for the antidiabetic potential of jamun tree parts was investigated by elucidating the mechanism of action of bioactive principles. Some of the phytochemical constituents such as friedelin, epifriedelanol, betulinic acid, beta-sitosterol, petunidin, kaempferol, quercetin, myricetin, and begin can bind more effectively than the synthetic drug acarbose and inhibit the activity of α-amylase enzyme and thus slows down the glucose release in the bloodstream (Smruthi et al., 2016). Seeds are a good source to control hyperglycemic conditions among diabetic patients and their top phytochemicals, gallic acid and kaempferol, can be used in combination with other synthetic anti-diabetic agents to stimulate drug activity with minimal dose and side effects (Rashid et al., 2022). There are significant differences in the concentrations of primary and secondary metabolites within a genus, as well as between species obtained from distinct geographic and climatic zones (Singh et al., 2019). In a separate study, Sharma et al. (2015) profiled the metabolites in Jamun fruit pulp using mass spectrometry techniques such as NMR (Nuclear Magnetic Resonance), HPLC (High-Performance Liquid Chromatography), MS (Mass Spectrometry), GC-MS (Gas Chromatography-Mass Spectrometry), MALDI-TOF (Matrix-Assisted Laser Desorption/Ionization Time-of-Flight). Sixty-eight chemically distinct fruit pulp metabolites, such as anthocyanins, anthocyanidins, sugars, phenolics, and volatile chemicals, were identified. Among phytochemicals, anthocyanin is the most abundant flavonoid in jamun due to its purple color. In a study at IIHR, biochemical evaluation of jamun germplasm revealed that the collection-5 and IC-0621981 (from Madhya Pradesh) had the highest anthocyanin content in terms of cyaniding-3- glucosides (308.24/100g FW), DPPH (2,2-Diphenyl-1-picrylhydrazyl, 24 mg/g), FRAP (Ferric Reduction Activity Potential, 20.41 mg/g) and total phenols (11.14 mg/g) (ICAR-IIHR 17; 18). The biochemical profiling of jamun collected from different locations in India (Karnataka and Odisha) has provided ample evidence for the diversity of secondary metabolites such as phenols, flavonoids, and anti-oxidants (Sane et al., 2017, 2022). Significant variation in alpha-glucosidase inhibitory activity was observed among the 24 tested jamun genotypes, the highest inhibitory activity was observed in selection-45, followed by collection-2 and BRH-1 (ICAR-IIHR 2019). A study on the variability of jamun genotypes for available minerals in the seed identified PGR-14, MP-5, Collection-9, and Selection-45 as the rich source of iron, manganese, zinc, and copper, respectively. Identification of the principal functional component responsible for nutritional and therapeutic functions would help to append a silver lining to the jamun-based food sector (Vani et al., 2023). The alpha-glucosidase inhibitory activity (IC_{50}) of seeds was 88.3±9.7 (ICAR-IIHR, 2021). The genes related to alkaloid and glucoside biosynthesis showed adaptive evolution in *S. cumini* that emphasizes the genomic basis for its pharmacological properties (Chakraborty et al., 2023).

Molecular Characterization

DNA markers are used to assess genetic variation in gene banks and play an important role in the development of conservation strategies, such as identifying overlaps or inconsistencies in the regulation of *ex situ* germplasm collections. Molecular markers have several advantages over traditional morphological markers as they provide data that can be analyzed objectively (Joshi et al., 1999). With the advent of molecular assisted breeding where DNA markers are used extensively, it is now possible to reduce the duration for developing suitable genotypes. Over the years, several types of molecular markers have been used in jamun, including random amplified polymorphic DNA (RAPD), inter simple sequence repeats (ISSR), and microsatellites or simple sequence repeats (SSR). Molecular markers such as ISSR and RAPD primers have been used for the differentiation of diverse genetic bases of Indian accessions (Shakya et al., 2010), wild jamun genotypes of Andaman and Nicobar Islands (Ahmad et al., 2012), phylogenetic relationships among jamun genotypes (Singh et al., 2014) and nucellar and zygotic seedlings identification (Talang et al., 2022). In a study to develop microsatellite markers in *Syzygium*, Rai et al. (2013) found high transferability of SSR of *Psidium guajava* for *Syzygium aromaticum*. The transferability of SSR to related species is advantageous for the characterization, conservation, and breeding of these species. Similarly, Nogueira et al. (2016) reported variation in the transferability of SSR among species within the genera. They also concluded that *Campomanesia* and *Myrciaria* genera showed higher transferability than the *Syzygium*. At IIHR, SSR primers for Jamun were developed through successful cross-species amplification using Guava primers (ICAR-IIHR, 2018). A total of 17,457 SSR primers were designed using the partial genome sequence data in Jamun, and these primers were used to assess the genetic diversity of different genotypes of Jamun and related species including those collected from northeast region of India. The availability of genome sequences and markers will accelerate the utilization of Jamun germplasm resources in breeding (ICAR-IIHR, 2021).

Huma Asif et al. (2013) gave a complete sequence and bioinformatics analysis of the chloroplast genome of *Syzygium* and studied structural organization, gene order, gene content, location, and distribution of repetitive sequences. They also compared data with the partial *S. cumini* chloroplast DNA sequences available in GenBank which provide an impetus for further studies related to intra- and inter-specific genetic variation in *S. cumini*. Furthermore, they reported that the variation intergenic spacer located between the IRA (Inverted Repeat A)/LSC (Large Single-Copy) border and the first gene of the LSC region can be useful for phylogeographic studies in *Syzygium* and other members of Myrtaceae. In another study, the chloroplast genome assembly of *S. cumini* showed a circular genome of 158,509 bases with 83 protein-coding genes (Chakraborty et al., 2023). They also reported the identification of evolutionary signatures and gene family evolution of all the key *S. cumini* genes involved in the phenylpropanoid-flavonoid biosynthesis pathway that have been recognized as critical mechanisms showing natural variation for species adaptation (Guo, 2013). The assembled genome size was comparable to the *S. malaccense* chloroplast genome (Tao et al., 2020).

The transcriptome of the draft genome of *S. cumini* was sequenced and assembled and it has a total size of 709.9 Mbp with 61,195 coding genes (Chakraborty et al., 2023). They reported that the genome contained 3.25% heterozygosity and was inferred as a tetraploid genome. Moreover, the phylogenetic position of *S. cumini* was established using a comprehensive genome-wide analysis including species from 18 Eudicot plant orders. The adaptive evolution of secondary metabolism pathways is associated with specifically the anti-diabetic property, of this species conferred by the bioactive compounds that act as nutraceutical agents in modern medicine. They also identified several SNPs that could be useful as intra-specific markers to address lineage sorting from polymorphic ancestry within the Myrtaceae family.

Utilization of Germplasm

In a perennial crop like jamun, an important way to develop a new high-yielding variety with specific desirable characteristics is initially by selecting superior genotypes based on the phenotypic features from among the existing variants and using them directly in future breeding programs. For this, it is essential to identify and effectively conserve the existing genetic resources with the utmost certainty. Traditionally, jamun is propagated by seeds, hence there is considerable variation among the trees, owing to its cross-pollination in nature (Purseglove, 1981). Thus, a great deal of variation is observed in trees and fruits within jamun populations. Crop improvement work in Jamun is through a selection of seedling trees based on promising horticultural characteristics. Varieties are being selected and naming is based on origin, fruit shape, color. Some selections are made from the available diversity by the local farmers, and researchers for large fruit size, high pulp content, and small seed size.

Because of allogamous in nature and predominance of seed propagation, enormous variability exists with respect to morphology, floral, and physio-chemical characteristics (Prakash et al., 2010). In this regard, an attempt has been made to bring out high-yielding clones through clonal selection among available germplasm resources in the region (Mishra et al., 2016). To initiate any crop improvement program, selection, and hybridization are important methods. Notably, Percy and Bose (1965) observed that an F1 hybrid resulting from a cross between the *Alba* variety of water apple (*S. javanicum*) and rose apple (*S. jambos*) exhibited prolific bearing with larger fruits than those of the parents. *Syzygium* species can be distinguished based on fruit size and color, and leaf and flower characteristics. In the Indian subcontinent, there are two main morphotypes of Jamun that have been recognized (Morton, 1987). Kattha jamun, which is small and acidic to taste, and the Ras Jaman, which is oblong, dark-purple, or bluish, with a pink, sweet fleshy pulp, and small seeds. Additionally, there are "Adaka jamun" that have bold seeds and a pink-colored pulp, and "seedless types" with chaffy seeds and a pulp content of 80–85%.

The "Ra jamun" variety is widely cultivated in north India whereas 'Paras', a large-sized fruit is cultivated in Gujarat, whereas another type found near Varanasi lacks seeds. Apart from these, some local selections growing in Haryana and Uttar Pradesh are Kaatha, Badama, Jathi, Ashada, and Bhado. These selections are based on the size, shape, taste, fruiting period, and maturity of fruits. The jamun fruit demand is increasing day-by-day because of its nutritive value and that will require selected plants of superior quality with high yield potential.

Jamun Breeding

The presence of variability opens the scope for further breeding applications and selection. The selected superior genotypes based on the physico-chemical characteristics of jamun can be utilized for commercial plantations in India. The initial step in tree improvement involves identifying elite or high-performing trees based on total yield and fruit quality. Additionally, these elite trees provide valuable scion material for vegetative propagation, which can directly contribute to improved planting materials. To enhance the utilization and conservation of existing genetic resources, it is essential to study aspects such as the center of origin, diversity, characterization, taxonomy, systematic phenology, floral biology, and pollination (Singh et al., 2019).

Breeding Objectives

1. To develop varieties with early bearing, and high-quality fruits with early synchronized ripening. Quality fruits are characterized by a high pulp percentage, large size, small seed, and an attractive shape (either oval or oblong). Additionally, they exhibit a deep purple or bluish-black coloration and contain ample juice.
2. To develop a variety with rapid juvenile growth, a dwarf tree stature suitable for high-density planting and ease of harvesting.
3. To develop a variety having resistance to pests such as fruit borer and fruit fly and diseases like anthracnose and leaf spot (Mishra et al., 2016).

Varieties Released

Rajendra Jamun: Identified from Bihar Agricultural College, Bhagalpur, Bihar for big-sized fruits, early maturing, and high-yielding.

Narendra Jamun: Selection made at Acharya Narendra Deva University of Agriculture and Technology (ANDUAT), Ayodhya. It is an improved selection with better fruit qualities. The fruit is oblong type, with higher fruit weight and higher pulp: seed ratio.

Konkan Bahadoli: The variety was released by Dr. Balasaheb Sawant Konkan Krishi Vidyapeeth (DBSKKV), Dapoli. The variety is a selection from local Jamun trees that grow in Bahadoli. It has heavy and cluster-bearing habits with bold fruits (25–28 g), small seeds, high pulp content (83.3%), and better table and processing qualities.

Dhupdal: The variety is a selection from local jamun trees that grow in Belgaum. The selection was identified by Kittur Rani Channamma College of Horticulture (KRCCH), Arabhavi, for its large-sized sweet fruits.

Selection 1: Selection made by Banaras Hindu University, Varanasi. It is promising about fruit weight (14.55 g), minimum seed weight (1.73 g), higher pulp content (90.05), higher total soluble solid (21.23%), and total sugar (20.24%) (Prakash et al., 2010).

Goma Priyanka: Released from Central Horticultural Experiment Station (CHES), ICAR-CIAH (Central Institute of Arid Horticulture), Godhra, Gujarat. It is a semi-dwarf, spreading growth habit, precocious bearer (starts flowering in 4th year), and suitable for high-density planting. Fruits are good in taste, bold (19.86 g), 85.06% pulp, 16.87°Brix TSS, vitamin C (45.44 mg/100 g), and yield potential of 30 kg/plant

Chintamani selection-1: It is a selection from a farmer's field in Chintamani. It has regular bearing and an average yield of 150 kg/tree. Fruit parameters include fruit weight of 17.9 g, pulp recovery of 88.9%, TSS 15°B, and acidity of 0.25% (Venkataramana et al., 2020).

Thar Kranti: It is semi-dwarf, matures in 75 days, and good yield (65 kgs/pl). Fruits are oblong with 20.10 g fruit weight, 85.57% pulp, 17.10°Brix TSS, and 0.40% acidity. The Total sugar content is 12.50% with 6.2% reducing sugar and 48.45 mg/100 g vitamin C, and it is suitable for high-density planting.

Jamwant: The variety was released from ICAR-CISH (Central Institute of Subtropical Horticulture), Lucknow. An improved variety, of bold, large-sized bunches with attractive deep color fruits. High pulp:seed ratio (90–92%), TSS (16–17°B), fruits are bold with an average weight of 24.05 g, oblong fruits (3.9 x 3.03 cm), high ascorbic acid (49.88 g/100 g), anti-oxidant (38.03 mg AEAC), and yield 80–90 kg /pl (Singh et al., 2012).

CISH J-42: Released from ICAR-CISH, Lucknow. Fruits are unique, seedless with pulp content of 97–98% and TSS 14–15° Brix.

Seedless Jamun (IC0635379; INGR21168): Identified by IIHR, Bengaluru from the collections made from the Western Ghats. The collection does not have even rudimentary seeds and the whole fruit is edible. Pink spots are visible in the transverse section of the fruit. The selection can be used for table and processing purposes.

Round-the-year Fruiting Genotype: identified from farmer's field in Tumkur, fruits twice a year (June-July and Nov.-Dec.). The average fruit weight is 8.6 g, and oblong in shape with a projected base. The fruits are dark purple in color with pinkish-white pulp, acidic sweet to taste with 13.86 °B TSS (Tripathi et al., 2021).

At ICAR-IIHR, standardization of gamma irradiation dose in seeds for mutation breeding indicated that seeds treated with doses ranging from 20–1000 Gy experienced reduced germination percentage and slower germination speed. A 100 Gy dose was found to be lethal, resulting in the lowest seedling height and the highest number of distorted leaves. When evaluating growth characters in the mutant population of jamun, significant variation was observed in morphological traits. The mutant line, II-1-19 exhibited the highest average fruit weight of 13.71 g while II-4-20 had the highest TSS content of 12.67 °Brix. These findings contribute to our understanding of mutation breeding and its impact on plant characteristics.

Conservation of Germplasm

Jamun germplasm is mainly conserved as field genebanks (FGBs) in research institutes and agricultural universities in India. The germplasm collections are planted at a distance of 8 m x 8 m and are maintained by adopting a recommended package of practices for Jamun. Efforts were made by Biodiversity International, different ICAR institutes, and universities in this regard. ICAR-Indian Institute of Horticultural Research has collected 108 collections including closely related species like *S. alternifolium, S. samarangense*, and *S. operculatum* from states such as Karnataka, Tamil Nadu, Andhra Pradesh, Gujarat, Rajasthan, Odisha, Jharkhand, Uttar Pradesh, Madhya Pradesh, and Himachal Pradesh and are being maintained in the field gene bank at IIHR, Bengaluru. ICAR-National Bureau of Plant Genetic Resources (NBPGR) maintains 20 collections; ICAR-Central

Institute for Subtropical Horticulture (CISH) has 54 collections, and ICAR-Central Institute for Arid Horticulture (CIAH) has 69 collections in the field gene banks. Additionally, several state agriculture universities, including Bihar Agriculture College, ANDUAT (Acharya Narendra Deva University of Agriculture and Technology), Ayodhya, and Mahatma Phule Krishi Vidyapeeth (MPKV), Rahuri, contribute to this wealth of germplasm. These efforts play a crucial role in preserving and enhancing the genetic diversity for future breeding programs. Jamun trees are not only essential for economic prosperity but also contribute significantly to ecological conservation by providing habitat, and shelter, and promoting biodiversity. Preserving wild Jamun demands habitat protection, restoration, and sustainable agriculture through agroforestry research, community engagement, and supportive policies. Integrating traditional knowledge with modern conservation ensures a comprehensive approach, actively preserving genetic diversity, ecological balance, and cultural relevance through multidimensional conservation efforts for Jamun.

Conclusion

The abundant antioxidants and phenolic compounds found in jamun fruit have contributed to its growing popularity due to its potential health benefits. Despite this, commercial cultivation of jamun remains challenging for several reasons. To overcome these obstacles, it is essential to focus on developing cultivars that exhibit excellent quality, tolerance to abiotic stresses, and regular bearing. Additionally, promoting value-added products derived from jamun as nutrient-rich foods and implementing effective postharvest management techniques will be crucial for establishing jamun as a viable commercial fruit. Clinical trials are essential to validate the therapeutic properties of a plant before it can be promoted as a medicinal resource. In the case of jamun, there exists significant potential for research to deepen our understanding of this fruit and leverage its health benefits. Further exploration of nutrigenetic and nutrigenomic aspects related to jamun can contribute to medical science and enhance natural approaches to combating illnesses. Despite being underutilized and underexploited, jamun holds agricultural significance. Raising awareness about its health advantages is crucial for encouraging regular consumption. To achieve this, promoting agricultural and industrial development and launching campaigns to highlight its benefits are essential. By fostering innovative consumption ideas, we can maximize the intake of important bioactive compounds from jamun, aiding in the fight against diseases.

References

Achrekar, S., G.S. Kaklij, M.S. Pote, and S.M. Kelkar. 1991. Hypoglycemic activity of Eugenia jambolana and Ficus bengalensis: Mechanism of action. *In Vivo (Athens, Greece)*, *5*(2): 143–147.

Ahmad, Israr, S. Bhagat, P. Simachalam, and R.C. Srivastava. 2012. Molecular characterization of *Syzygium cumini* (wild jamun) from A&N Islands. *Indian Journal of Horticulture*, *69*(3): 306–311.

Ajay Kumar, K., A. Rekha, R. Venugopalan, M.K. Honnabhyraiah, S. Mohankumar, and K.S. Shivashankara. 2023. Variability in morphological parameters of Jamun (*Syzygium cumini* Skeels) genotypes. *International Journal of Minor Fruits, Medicinal, and Aromatic Plants*, *9*(2): 01–13.

Ajay Kumar, K., K. Mounika, M.M. Shulee Ariina, and C.S. Maiti. 2021. Study of Jamun (*Syzygium cumini* Skeels.) Genotypes Based on Shelf life and Interspecific Crossability. *International Journal of Plant & Soil Science*, *33*(24): 606–609.

Akoègninou, Akpovi, W.J. Van der Burg, and L. Jos G. Van der Maesen. 2006. *Flore analytique du Bénin*. No. 06.2. Backhuys Publishers.

Anushma, P.L., and Anuradha Sane, 2018. Assessing variability in morphological traits of jamun (Run on) (*Syzygium cumini* (L.) Skeels) genotypes. *Journal of Plant Development Sciences*, *10*(11): 629–632.

Anushma, P.L., Linta Vincent, and Anuradha Sane. 2023. Jamun. *In*: *Fruit and Nut Crops*, 1–(Run on)21. Springer Nature, Singapore. APG IV 2016. An update of the Angiosperm Phylogeny Group classification for the orders and families of flowering plants: APG IV. *Botanical Journal of the Linnean Society*, *181*(1): 1–20.

Arya, S.S., K. Pegu, and P.D. Sadawarte. 2019. Bioactive Compounds and Health Benefits of Jamun. *In*: *Bioactive Molecules in Food*, 2297–2315. Springer, Cham.

Huma, Asif, Asifullah Khan, Asif Iqbal, Ishtiaq Ahmad Khan, Berthold Heinze, and M. Kamran Azim. 2013. The chloroplast genome sequence of *Syzygium cumini* (L.) and its relationship with other angiosperms. *Tree Genetics and Genomes, 9*: 867–877.

Bajpai, A., A.K. Singh, and H. Ravishankar. 2012. Reproductive phenology, flower biology, and pollination in jamun (*Syzygium cuminii* L.). *Indian Journal of Horticulture, 69*(3): 416–419.

Baliga, Manjeshwar Shrinath, Harshith P. Bhat, Bantwal Raghavendra Vittaldas Baliga, Rajesh Wilson, and Princy Louis Palatty. 2011. Phytochemistry, traditional uses, and pharmacology of *Eugenia jambolana* Lam. (black plum): A review. *Food Research International, 44*(7): 1776–1789.

Chakraborty, Abhisek, Shruti Mahajan, Manohar S. Bisht, and Vineet K. Sharma. 2023. Genome sequencing of *Syzygium cumini* (Jamun) reveals adaptive evolution in secondary metabolism pathways associated with its medicinal properties. *Frontiers in Plant Science, 14.*

Chase, Mark W., and James L. Reveal. 2009. A phylogenetic classification of the land plants to accompany APG III. *Botanical Journal of the Linnean Society, 161*(2): 122–127.

Chundawat, B.S. 1990. *Arid Fruit Culture*, 165–171. Oxford and IBH Co. Pvt. Ltd. New Delhi. Corlett, R.T. 2005. Interactions between birds, fruit bats, and exotic plants in urban Hong Kong, South China. *Urban Ecosystems, 8*: 275–283.

Craven, Lyn A., and Ed Biffin. 2010. An infrageneric classification of *Syzygium* (Myrtaceae). *Blumea-Biodiversity, Evolution, and Biogeography of Plants, 55*(1): 94–99.

Damasceno, D.C., G.T. Volpato, I. de MP. Calderon, and M.V.C. Rudge. 2002. Study of *Averrhoa carambola* and *Eugenia jambolana* extracts purchased from manipulation drugstore on the experimental diabetes. *Revista Brasileira de Toxicologia, 15*: 09–14.

Ghojage, A.H., G.S.K. Swamy, V.C. Kanamadi, R.C. Jagdeesh, P. Kumar, C.P. Patil, and B.S. Reddy. 2011. Studies on variability among best selected genotypes of Jamun (*Syzygium cumini* Skeels.). *Acta Horticulturae, 890*(2): 255–260.

Guo, Ya-Long. 2013. Gene family evolution in green plants with emphasis on the origination and(Run on)evolution of *Arabidopsis thaliana* genes. *The Plant Journal, 73*(6): 941–951.

Helmstädter, A. 2008. *Syzygium cumini* (L.) Skeels (Myrtaceae) against diabetes: 125 years of research. *Die Pharmazie-An International Journal of Pharmaceutical Sciences, 63*(2): 91–101.

ICAR-IIHR. 2016. *Annual Report 2015–2016*. ICAR-Indian Institute of Horticultural Research, 216pp. Bengaluru 560089.

ICAR-CIAH. 2017. *Annual Report 2017–2018*. ICAR Central Institute of Arid Horticulture, 262pp. Bikaner 334006,

ICAR-IIHR. 2018. *Annual Report 2018–2019*. ICAR- Indian Institute of Horticultural Research, 212pp. Bengaluru 560089.

ICAR-IIHR. 2021. *Annual Report. 2020*. ICAR-Indian Institute of Horticultural Research, 187pp. Bengaluru 560089.

ICAR-NBPGR. 2020. *ICAR-National Bureau of Plant Genetic Resources*, 266pp. New Delhi.

Jain, S.K. 1991. *Dictionary of Indian Folk Medicine and Ethnobotany*. Deep Publications.

Jain S.K. 2000. Human aspects of plant diversity. *Economic Botany, 54*: 459–470.

Jena, S., P. Sahoo, and A.B. Das. 2003. New reports of chromosome number and genome size in eight mangroves from coastal Orissa. *Caryologia, 56*(3): 353–358.

Joshi, S.P., P.K. Ranjekar, and V.S. Gupta. 1999. Molecular markers in plant genome analysis. *Current Science, 77*(2): 230–240.

Kadavul, K., and N. Parthasarathy 1999. Plant biodiversity and conservation of tropical semi-evergreen forest in the Shervarayan hills of Eastern Ghats, India. *Biodiversity and Conservation, 8*: 419–437.

Keskar, B.G., A.R. Karale, B.C. Dhawale, and K.G. Chouhari. 1989. Improvement of jamun by (Run on)selection. *Maharashtra J. Hort., 4*: 117–120.

Khadivi, A., F. Mirheidari, A. Saeidifar, and Y. Moradi. 2023. Selection of the promising accessions of jamun (*Syzygium cumini* (L.) Skeels) based on pomological characterizations. *Food Science and Nutrition, 11*(1): 470–480.

Khan, S., Vaishali, and V. Sharma. 2010. Genetic differentiation and diversity analysis of medicinal tree *Syzygium cumini* (Myrtaceae) from ecologically different regions of India. *Physiology and Molecular Biology of Plants, 16*: 149–158.

Kshirsagar, R.B., G.B. Desai, A.R. Sawate, and N.M. Deshmukh. 2019. Physico-chemical and nutritional properties of Jamun (*Syzygium cumini*) seed. *Journal of Pharmacognosy and Phytochemistry, 8*(2): 211–213.

Kumar, K.A., A. Rekha, R. Venugopalan, K.S. Shivashankara, S. Mohan Kumar, M.K.

Honnabyraiah, and N. Jyothi. 2020. Grouping of jamun genotypes based on fruit quality parameters. *Indian Academy of Horticultural Sciences* (*IAHS*), *77*(3): 412–417.

Kundu, S. 2001. Evaluation of some local types of jamun (*Syzygium cuminii* (L.) Skeels) of West Bengal. *Environment and Ecology, 19*(4): 872–874.

Laishram, M., and Y.S. Singh. 2011. Germplasm status of minor and underutilized fruits of India. *In*: *Proceedings of the International Symposium on Minor Fruits and Medicinal Plants for Health and Ecological Security* (*ISMF & MP*), 19–22. West Bengal, India.

Latheef, A., and M. Kavino. 2018. Variability in Genotypes of Jamun (*Syzygium cumini* Skeels) for Morphological and Yield Characters. *Madras Agricultural Journal, 105*(10–12): 601.

Low, Y.W., S. Rajaraman, C.M. Tomlin, J.A. Ahmad, W.H. Ardi, and K.A. Parusuraman. 2022. Genomic insights into rapid speciation within the world's largest tree genus *Syzygium*. *Nature Communications, 13*(1): 5031.

Mahajan, R.K., K.K. Gangopadhyay, Gunjeet Kumar, V.K. Dobhal, Umesh Srivastava, P.N. Gupta, and S.K. Pareek. 2022. Minimal descriptors of agri-horticultural crops. *In*: *Part ll J. Fruit Crops-National Bureau of Plant Genetic Resources*, 223–228. New Delhi.

Mishra D.S., A.S., R. Kumar, S. Singh, A.K. Singh, and G.S.K. Swamy. 2014. Jamun. *In*: Ghosh, S.N. [Ed.], *Tropical and Subtropical Fruit Crops: Crop Improvement and Varietal Wealth*, 376–399. JAYA Publishing House.

Mishra, D.S., S. Singh, A.K. Singh, and V. Yadav. 2016. Future Fruit Crops for Semi-arid Conditions of Western India. *In*: *Compendium Winter School on Exploitation of Underutilized Fruits of Arid and Semi-Arid Region* (S.S. Lakhawat), 187–192. Udaipur.

Misra R.S., and P.N. Bajpai. 1975. Studies on floral biology of Jamun (Java Plum) *Syzygium cuminii* (L) Skeels. *Indian Journal of Horticulture*, *32*(1, 2): 15–24.

Morton, J. 1987. *Fruits of Warm Climates, Miami*. Julia Morton Winterville, North Carolina.

Mudiana, D. 2016. Syzygium diversity in Gunung Baung, East Java, Indonesia. *Biodiversitas Journal of Biological Diversity*, *17*(2): 733–740.

Nadkarni, K.M. 1976. *Indian Materia Medica*. Bombay: Popular Prakashan Ltd.

Nogueira, A.M., A. Ferreira, and M.F. da Silva Ferreira. 2016. Transferability of microsatellites from *Psidium guajava* to *Eugenia*, *Myrciaria*, *Campomanesia*, and *Syzygium* species (Myrtaceae). *Plant Molecular Biology Reporter*, *34*: 249–256.

Ochse, J.J., M.J. Soule, M.J. Jr. Dij Kman, and C. Wehlberg. 1961. *Tropical and Subtropical, Agriculture*. Mac. Millan, New York.

Oginuma, K., A. Kato, H. Tobe, S.G. Mathenge, and F.D. Juma. 1993. Chromosomes of some woody plants in Kenya. *Acta Phytotaxonomica et Geobotanica*, *44*(1): 53–58.

Orwa, C., A. Mutua, R. Kindt, R. Jamnadass, and A. Simons. 2009. *Agroforestry Database: A Tree Reference and Selection Guide. Version 4*. Research Gate.

Patel, V.B., S.N. Pandey, S.K. Singh, and B. Das. 2005. Variability in jamun (*Syzygium cuminii* Skeels) accessions from Uttar Pradesh and Jharkhand. *Indian Journal of Horticulture*, *62*(3): 244–247.

Percy, L.S., and T.K. Bose. 1965. Studies on hybrid of *Syzygium cumini* Skeels. *Ind. J. Hortic.*, *22*: 87–88.

Mishra, D.S., S. Singh, A.K. Singh, and V. Yadav. 2016. Future Fruit Crops for Semi-arid Conditions of Western India. *In*: *Compendium Winter School on Exploitation of Underutilized Fruits of Arid and Semi-Arid Region* (S.S. Lakhawat), 187–192. Udaipur. Plathia, M., V.K. Wali, P. Bakshi, and N. Sharma. 2018. Studies on variability of seedling origin Jamun (*Syzygium cumini* (L.) Skeels) germplasm growing in subtropical region of Jammu, India. *Int. J. Curr. Microbiol. App. Sci.*, *7*(2): 3363–3368.

Prakash, J., A.N. Maurya, and S.P. Singh. 2010. Studies on variability in fruit characters of Jamun. *Indian Journal of Horticulture*, *67*(4): 63–66.

Priya Devi, S., M. Thangam, A.R. Desai, and P.G. Adsule. 2002. Studies on variability in physico-chemical characters of different Jamun (*Syzygium cumini*) accessions from Goa. *Indian Journal of Horticulture*, *59*(2): 153–156.

Proma, N.M., J. Naima, M.R. Islam, J.A. Papel, M.M. Rahman, and M.K. Hossain. 2018. Phytochemical constituents and anti-diabetic properties of *Syzygium cumini* Linn. Seed. *Int. J. Pharm. Sci. Res.*, *9*: 1806–1814.

Purseglove, J.W., E.G. Brown, C.L. Green, and S.R.J. Robbins. 1981. *Spices*, *2*: 447–813. Longman Group Ltd.

Rai, M.K., M. Phulwaria, and N.S. Shekhawat. 2013. Transferability of simple sequence repeat (SSR) markers developed in guava (*Psidium guajava* L.) to four Myrtaceae species. *Molecular Biology Reports*, *40*: 5067–5071.

Rashid, F., A Javaid, U.A. Ashfaq, M., Sufyan, A. Alshammari, M Alharbi, M.A. Nisar, and M. Khurshid. 2022. Integrating pharmacological and computational approaches for the phytochemical analysis of *Syzygium cumini* and its anti-diabetic potential. *Molecules*, *27*(17): 5734.

Rekha, A., H.D. Talang, and Anuradha Sane. 2020. Assessing polyembryony among jamun (*Syzygium cumini* Skeels.) collections. *International Journal of Current Microbiology and Applied Sciences*, *9*(2): 658–663.

Rizvi, M.K., R. Rabail, S. Munir, M. Inam-Ur-Raheem, M.M.N. Qayyum, M. Kieliszek, A. Hassoun, and R.M. Aadil. 2022. Astounding Health Benefits of Jamun (*Syzygium cumini*) toward Metabolic Syndrome. *Molecules*, *27*(21): 7184.

Sagrawat, H., A.S. Mann, and M.D. Kharya. 2006. Pharmacological potential of *Eugenia jambolana*: A review. *Pharmacognosy Magazine*, *2*(6): 96–105.

Sane, A., A. Rekha, A. Shamina, and M.R. Jasmin. 2017. Ethno-botanical and Phyto-chemical analysis of Jamun (*Syzigium cumini*): A case study in tribal belt of Karnataka, India. *In*: *Proceeding of International Symposium on Horticulture Priorities and Emerging Trends*, Bengaluru, India

Sane, A., A. Rekha, A. Shamina, M.R. Jasmin, and P. Srinivas. 2022. Ethno-botanical and biochemical evaluation of Jamun (*Syzigium cumini*) in tribal belts of Odisha and Karnataka. *In*: *Proceedings in the National Conference on Tribal Horticulture*, Andhra Pradesh, India.

Sanewski, G. 2010. *Understanding the Cropping Behaviour of Riberry* (*Syzygium leuhmannii*). Union offset printing. Canberra RIRDC publication No. 10/194.

Sanjay, S., and A.K. Singh. 2012. Studies on variability in jamun (*Syzygium cuminii* Skeels) from Gujarat. *Asian Journal of Horticulture*, *7*(1): 186–189.

Sarvade, S., D.S. Gautam, S. Bhalawe, and P.K Bisen. 2016. An overview of potential multipurpose agroforestry tree species, *Syzygium cuminii* (L.) Skeels in India. *Journal of Applied and Natural Science*, *8*(3): 1714–1719.

Shakya, R., S.A Siddiqui, N. Srivastava, and A. Bajpai. 2010. Molecular characterization of jamun (*Syzygium cumini* L. Skeels) genetic resources. *International Journal of Fruit Science*, *10*(1): 29–39.

Sharma, N., B. Singh, M.S Wani, R.C. Gupta, and T.H. Habeeb. 2020. Morphological, Cytological, and Chemotypic Variation of Essential Oils in *Syzygium cumini* (L.) Skeels. *Analytical Chemistry Letters*, *10*(5): 609–619.

Sharma, R.J., R.C. Gupta, A.K. Bansal, and I.P. Singh. 2015. Metabolite Fingerprinting of *Eugenia jambolana* Fruit Pulp Extracts using NMR, HPLC-PDA-MS, GC-MS, MALDI-TOF-MS and ESI-MS/MS Spectrometry. *Natural Product Communications*, *10*(6): p.1934578X1501000644.

Singh, Y., P. Bhatnagar, and S. Kumar. 2019. A review on bio-active compounds and medicinal strength of Jamun (*Syzygium cumini* Skeels). *Int. J. Chem. Stud.*, *7*(4): 3112–3117.

Singh, A.K., A. Bajpai, and H. Ravishankar. 2012. CISH J-37 (IC0587715; INGR11025), a Jamun (Run on)(*Syzygium cuminii*) Germplasm, with Bold Fruit and High Pulp Quality (Pulp Content 90–92% and TSS 16–17° Brix). *Indian Journal of Plant Genetic Resources*, *25*(2): 197–221.

Singh, J.P., A.K. Singh, A. Bajpai, and I.Z. Ahmad. 2014. Comparative analysis of DNA polymorphisms and phylogenetic relationships among *Syzygium cumini* Skeels based on phenotypic characters and RAPD technique. *Bioinformation*, *10*(4): 201–208.

Singh, A.K., A. Bajpai, S. Rajan, S.S. Das, and K.K. Mishra. 2021. Modified BBCH codification and correlation of phenological characteristics with climatic variables in jamun (*Syzigium cuminii* Skeels). *Scientia Horticulturae*, *283*: 110081.

Singh, S., A.K. Singh, H.P. Singh, B.G. Bagle, and T.A. More. 2011. Jamun. *Technical Bulletin*, 46. ICAR, New Delhi. Smruthi, G., V. Mahadevan, V. Vadivel, and P. Brindha. 2016. Docking studies on antidiabetic molecular targets of phytochemical compounds of *Syzygium cumini* (L.) Skeels. *Asian J. Pharm. Clin. Res.*, *9*(3): 287–293.

Srivastava, V., P. Kumar, and P.N. Rai. 2012. Correlation study for physicochemical characters in Jamun. *HortFlora Research Spectrum*, *1*: 83–85.

Swami S.B., N.J. Thakor, M.M. Patil, and P.M. Haldankar. 2012. Jamun (*Syzygium cumini* (L.): A Review of Its Food and Medicinal Uses. *Journal of Food and Nutrition Sciences*, *3*(8): 1100–1117.

Talang H.D., A. Rekha, M.R. Dinesh, and K.S. Shivashankar. 2017. Physicochemical Variability of Selected Genotypes of Jamun (*Syzygium cumini* Skeels). *Indian J. Hortic.*, *7*(1): 64–66.

Tao, L., Z. G. Shi, and Q. Y. Long 2020. Complete chloroplast genome sequence and phylogenetic analysis of *Syzygium malaccense*. *Mitochondrial DNA Part B.*, *5*(3): 3567–3568.

Tripathi, P.C., A. Rekha, A. Sane, V.K. Rao, and M. Arivalagan. 2021. Exploring jamun diversity: Few unique selections. *Current Horticulture*, *9*(1): 61–62.

USDA-ARS. 2008. *Germplasm Resources Information Network* (*GRIN*). Online Database. Beltsville, Maryland, USA: National Germplasm Resources Laboratory. https://npgsweb.ars-grin.gov/gringlobal/taxon/taxonomysearch.aspx.

Vani, N.U., P.C Tripathi, T.R. Rupa, A.P. Singh, and V.K. Rao. 2023. Evaluation of Jamun genotypes for seed mineral content. *Fruits*, *78*(2): 1–8.

Venkataramana, P., Sivappa, S.K. Priyadarshini, and M. Mahesh. 2020. A new Jamun (*Syzygium cumini* L.) variety Chintamani selection-1. *International Journal of Chemical Studies*, *8*(1): 1874–1877.

Zeven A.C., and De J.M. Wet. 1982. *Dictionary of Cultivated Plants and Their Regions of Diversity: Excluding Most Ornamentals, Forest Trees, and Lower Plants*, 1–11. Pudoc. Wageningen.

17

Genetic Diversity of *Citrus* Species and Sustainable Utilization

T.K. Hazarika

Introduction

Within the family Rutaceae, citrus is one of the most widely grown fruits worldwide, cultivated in over 130 countries, primarily in tropical and subtropical climates. The major economically important species of the family Rutaceae are *Citrus sinensis* (L.) Osbeck (Sweet orange), *C. reticulata* Blanco (mandarin), *C. grandis* (L.) Osbeck (pummelo), *C. paradisi* Mac. f. (grapefruit), *C. limon* (L.) Burm. f. (lemon), *C. aurantifolia* (Christm) Swing. (lime), and *C. aurantium* (L.) (sour orange). Globally sweet oranges accounted for approximately 68% of citrus production, mandarin (27%), Lemon and Limes (13%), and others (7%). Citrus is believed to have its primary center of origin in South and Southeast Asia, particularly in the region extending from northeast India, eastward through the Malayan Archipelago to China and Japan, and southward to Australia (Mabberley, 2004). Citrus trees are found between latitude 40°N and 40°S where conditions are not too dry, not too moist, and not too cold.

As per FAOSTAT-2011 global Citrus production was 127 million tons which is the top-ranked among all the fruit crops, followed by Banana (98 million tons), Apple (70 million tons), and Grape (68 million tons). The highest citrus-producing countries are China (22.94 million tons), Brazil (22.70 million tons), the USA (10.44 million tons), and India (8.26 million tons) as per FAOSTAT 2012. In the last three decades, the highest citrus production per year in the world was reported from Brazil. But now China is producing the maximum number of citrus fruits in the world as per the FAO report 2012. In the last decade, India's position improved from the sixth rank to the fourth.

Citrus fruits are well known for their specific taste, aroma, and oils. Citrus fruits are abundant in phenolic acid, flavonoids, carotenoids, limonoids, dietary fiber, vitamins, minerals, and essential oils. The role of citrus species in the prevention of several communicable diseases has been examined and it has been reported that citrus fruits exhibit numerous biological properties owing to their bioactive profile. The phenolic composition of many *Citrus* species from different parts of the world has been reported (Xi et al., 2014).

Department of Horticulture, Aromatic and Medicinal Plants, School of Earth Sciences and Natural Resources Management, Mizoram University, Aizawl, Mizoram.
Email: tridip28@gmail.com

Taxonomic Background

The taxonomy of *Citrus* L. and allies has been a challenge for botanists. Most of its complexity derives both from the biology of the species and its cultivation history. The species have great ability to cross and to produce hybrids (both intra and intergeneric). Some of these hybrids are fertile by zygote-derived embryos or can become fertile by spontaneous formation of adventitious nucellar embryos, which contributes to the maintenance of genetic stability and the perpetuation of the hybrids as apomictic clones (Chapot, 1975; Scora, 1975). Associated with these biological properties, species of *Citrus* were transferred between areas in the Far East (Chapot, 1975; Scora,. 1975), where crossing between species became possible, blurring the taxonomic limits previously maintained by geographic isolation. This resulted in a warm of morphological forms whose number ranges indifferent interpretations from as few as 11 (Engler, 1931) to as many as 145 species (Tanaka, 1977).

The taxonomic situation is further complicated by shifting boundaries between *Citrus* and allied genera. The history of the subfamily Aurantioideae was marked by the proposal of new genera separated from *Citrus*, such as *Poncirus*, *Fortunella*, and *Microcitrus*. Engler initially treated the allied species of *Citrus* in just one genus, but later on, he accepted this subdivision (Engler, 1931). However, there are no strong morphological characters to separate these genera, and many of the supposedly diagnostic characters of one genus can be found in the species of another. Burkill (1931) treated Swingle's genera as subgenera of *Citrus*, but his proposal was largely overlooked. More recently, Mabberley (1998), when revising *Citrus* and related genera in Australia, merged with *Citrus* the genera *Fortunella*, *Eremocitrus* Swingle, and *Microcitrus*.

One of the most usually accepted classifications for the subtribe Citrinae (Swingle and Reece, 1967) is a progressive series: "Primitive Citrus Fruit Trees", "Near-Citrus Fruit Trees", and "True Citrus Fruit Trees".

The "true citrus fruit trees" group of the subtribe *Citrinae* (Rutaceae) includes six genera, i.e., *Fortunella*, *Eremocitrus*, *Poncirus*, *Clymenia*, *Microcitrus,* and *Citrus*. The most commercial genus of this group is *Citrus* with several species, 16 and 162 according to Swingle and Reece (1967) and Tanaka (1977), respectively. Citrus are the most widely cultivated fruit trees in the world. Species within the other five genera may have desirable traits such as resistance to pests and pathogens when used as rootstocks, or they may serve as a source of germplasm for genetic improvement of citrus (Spiegel-Roy and Goldschmidt, 1996). Possibly excluding *Clymenia*. *Citrus* is cross-compatible with the other four genera of this group and intergeneric hybrids have been established through sexual hybridization for citrus breeding improvement (Iwamasa and Nito, 1988).

Currently, two different classification systems of *Citrus* are commonly accepted for the *Citrus* taxonomy, the Swingle system (Swingle, 1943; Swingle and Reece, 1967) and that of Tanaka (1954, 1961). Swingle (1943) recognized only 16 valid species under two subgenera, i.e., *Citrus* and *Papeda*, 10 under subgenus *Citrus,* and 6 under subgenus *Papeda* and did not accord species status to budsports or natural hybrids. The subgenus *Citrus* is characterized by pulp vesicles that are nearly free from oil droplets and never contain acrid oil. Petioles with narrow wings or wingless or if broadly winged, are subcordate and never 3/4th as broad as the leaf blade, flowers large and fragrant, stamens cohering in bundles. The subgenus *Citrus* comprises 10 species, viz., *Citrus tachibana, C. medica, C. limon, C. reticulata, C. indica, C. grandis, C. paradisi, C. aurantifolia, C. sinensis*, and *C. aurantium*. On the other hand, the sub-genus *Papeda* is characterized by pulp vesicles containing numerous droplets of acrid oil. Petioles are long and very broadly winged but not cordate, often nearly as broad as the leaf blade. Stamens are usually free. Six species are included under the subgenus *Papeda*: *C. ichangensis, C. latipes, C. micrantha, C. celebica, C. macroptera,* and *C. hystrix*. On the other hand, Tanaka proposed his system of classification of *Citrus* more elaborately. He included almost all sorts of geographic and morphological variability of various *Citrus* plants and proposed an exhaustive list of species (Naqvi, 2001). In 1954, in *Species Problems in Citrus*, Tanaka published a classification of the genus *Citrus* in which two subgenera, *ArchiCitrus* and *MetaCitrus,* eight sections, 13 subsections, eight groups, two subgroups, two micro-groups, and 145 species were distinguished. In 1961, he added two new subsections, another group, and 12 new species to his system, taking the total to 157 species (Nicolosi, 2007). In 1977, he again increased the number of species up to 162 that mostly represented variants and hybrids. (Table 1)

Tanaka's classification is more complex as compared to that of Swingle's because of the far greater number of species included in each subgenus. Tanaka very often has given the rank of species to cultigens, chance seedlings, apomicts, etc. On the other hand, Swingle's classification was easier to understand although it does not provide an exhaustive description of *Citrus* systematic. It was very concise, but not free from such inclusions by admitting *C. paradise* and *C. limon* as species and ignoring the most distinctive forms which are probably more valid than his *C. indica* or *C. tachibana.* Even so, Swingle's system is the most used, with some additions and deletions and adoption from Tanaka's system (Nicolosi, 2007).

Subsequent classifications of *Citrus* proposed by various other botanists (Bhattacharya and Dutta, 1956; Singh and Nath, 1969) are mere modifications of either Swingle's or Tanaka's system. Hooker (1875), Bonavia (1880–1890), Lushington (1910), and Singh (1967) attempted to study and classify Indian *Citrus* from botanical as well as horticultural perspectives. Scora (1975) and Barrett and Rhodes (1976) suggested that there are only three 'basic' true species of *Citrus* within the subgenus *Citrus* as follows: citron (*C. medica* L.), mandarin (*C. reticulata* Blanco), and pummelo (*C. maxima* L. Osbeck). Later, Scora (1988) added *C. halimi* as another true species. Other species within this subgenus are hybrids derived from these true species, species of subgenus Papeda, or closely related genera. Taxonomic characterization leading to unambiguous identification of *Citrus* species and their genetic resources are essential requisites for Citrus breeding, Citriculture, and Citrus industry. In a systematic account of Indian Citrus, Nair and Nayar (1997) followed primarily the scheme of Swingle and Reece (1967) and partly that of Tanaka (1977) including 18 taxa, comprising eight species under subgenus *Citrus*, three under subgenus Papeda, and seven other indigenous Citrus varieties with a suspected hybrid origin and uncertain taxonomic affinities. This idea has recently been supported by data derived from molecular markers (Federici et al., 1998; Nicolosi et al., 2000; Barkley et al., 2006). Recently Mabberley (1998, 2004) circumscribed Citrus in a broader sense by merging three of the closely related genera, viz., Fortunella Swingle, Eremo Citrus Swingle, and Micro Citrus Swingle within it. Systematics of Citrus thus continues to be an intriguing topic for further research and discussion.

Table 1. The genus *Citrus*: A summary.

Species	Common Name	Year Named	ProbableOrigin	Probable Native Habitat	Seed Reproduction
C. medica	Citron	1753	True Species	India	Sexual
C. aurantium	Sour orange	1753	Hybrid	China	Nucellar
C. sinensis	Sweet orange	1757	Hybrid	China	Nucellar
C. maxima	Pummelo	1765	True species	China	Sexual
C. limon	Lemon	1766	Hybrid	India	Partly sexual
C. reticulata	Mandarin	1837	True species	China	Variable
C. aurantifolia	Lime	1913	Hybrid	Malaya	Partly sexual
C. paradisi	Grapefruit	1930	Hybrid	Barbados	Nucellar
C. tachibana	Tachibana	1924	Unknown	Japan	Sexual
C. indica	Indian wild orange	1931	Unknown	India	Sexual
C. hystrix	Mauritiuspapeda	1813	Unknown	S.E. Asia	Sexual
C. macroptera	Malesianpapeda	1860	Unknown	S.E. Asia	Sexual
C. celebica	Celebes papeda	1898	Unknown	Celebes	Sexual
C. ichangensis	Ichangpapeda	1913	Unknown	China	Sexual
C. micrantha	Papeda	1915	Unknown	Philippines	Sexual
C. latipes	Khasi papeda	1928	Unknown	N.E. India	Sexual

Source: Krueger and Navarro, 2007.

Collection and Assessment of Germplasm/Genetic Resources

Although Citrus originated in Southeast Asia and there is a lot of diversity in this region; however, in this region of now, there is a loss of genetic materials for Citrus. Considering the importance of Citrus genetic resources, efforts have been made for *ex situ* conservation of Citrus germplasm in Southeast Asian countries. Still, large *ex situ* citrus collections of Citrus are found in Argentina, Australia, Brazil, Corsica, Morocco, New Zealand, South Africa, Spain, Turkey, and the United States. Some of the larger collections contain many selections of the same variety, so the genetic diversity is less than might be expected from the number of accessions (Malik et al., 2012).

In southern China, a wide range of genetic diversity of *Citrus* and related genera is still present *in situ*. In China, the exploration and collection of indigenous citrus genetic resources began in the 1950s and 1960s but was interrupted by the Cultural Revolution of 1967–1972. Governmental surveys resumed during the 1970s and 1980s and uncovered a number of new putative species, including *C. honghensis*, *C. mangshanensis*, *C. daoxianensis*, and *Poncirus polyandra*. Beginning in the early 1960s, a National Citrus Germplasm Repository was established at Beibei, Chongqing, Sichuan province, and regional citrus germplasm repositories in Huangyan, Zhejiang province; Guiling, Guangxi province; Zhangsa, Hunan province; Guangzhou, Guangdong province; Jiangjin, Sichuan province; Wuzhung province; and Hubei province. As of 1996, the National Citrus Germplasm Repository had 1,041 accessions (decreased from 1,200), of which indigenous, bud mutations, and nucellars accounted for 58%, 5%, and 37%, respectively. The Huangyan, Guiling, Zhangsa, and Guangzhou regional repositories had 128, 216, 40, and 140 accessions, respectively. Regional Citrus Research Institutes in Shantou, Guangdong; Ichang, Hubei; Thouyang, Hunan; Ganzhou, Jianxi; Yuchi, Yunnan; and Wu, Jiangsu also maintain small amounts of citrus germplasm, as so botanic gardens such as Xithanbanna and Guanzhon. As of 2000, the regional and national citrus germplasm repositories established in the 1980s–1990s were mostly no longer functional. The only one that remained open was Beibei, which served as the national collection and maintains around 900 accessions (https://www.ars.usda.gov/ARSUserFiles/20361000/Crop_Vulnerability_2004.pdf).

The national citrus germplasm repository of China was established in 2007 at Beibei, Chongqing, Sichuan province. Besides the national citrus gene bank there are many other sites found in China where wild citrus germplasm has been maintained. Among them Huazhong Agricultural University (HZAU), Huangyan, Zhejiang province; Guiling, Guangxi province; Changsha, Hunan province; Guangzhou, Guangdong province; Jiangjin, Chongqing municipality; Wuzhung province; and Hubei province are notable (Krueger and Navarro, 2007). HZAU has its citrus germplasm collection center where more than 400 accessions are maintained, along with 100 embryo callus lines. The HZAU collection includes trifoliate oranges, tangerines, kumquat, lemon, lime, pummelo, sweet oranges, papeda, citron, and grapefruit (Biswas et al., 2020).

Australia has several *ex situ* collections. It is the center of origin for several related genera (most notably *Eremocitrus* and *Microcitrus*), that are included in the collections, as well as in certain botanic gardens (e.g., Royal Botanic Garden, Sydney, and Brisbane Botanic Gardens) and arboreta (e.g., Waite Research Institute, University of Adelaide Arboretum) (https://www.ars.usda.gov/ARSUserFiles/20361000/Crop_Vulnerability_2004.pdf).

Citrus germplasm collection is maintained in Australia by the State Government, Departments of Primary Industries, and the Commonwealth Scientific and Industrial Research Organization (CSIRO). Apart from these, there are several other citrus germplasm banks (Royal Botanic Garden, Sydney; Brisbane Botanic Gardens; Waite Research Institute, University of Adelaide). Australia mainly follows *ex situ* conservation for wild and cultivated citrus and related genera (Biswas et al., 2020).

Brazil initiated citrus germplasm collection in the year 1948. Initially, about 470 citrus accessions were collected from Riverside California, USA, and maintained in the Brazilian citrus gene bank; new local accessions were then added to the gene bank and now the total collection reached about 1,546 accessions. These collections include nucellar clones, newly introduced cultivars, and new hybrid collections for breeding and rootstocks (Biswas et al., 2020).

Southeast Asia (including Malaysia) is rich in indigenous germplasm. There are four collections in Malaysia, the main one being the University of Malaya (Rimba Ilmu) Botanical Garden (over 140 accessions representing 25 genera and 53 species); the others (Jerangau Statio, Trengganu; Kuala Kangsar, and Cameron Highlands) are maintained by MARDI. There are also some *in situ* conservation efforts, such as at the Taman Negara National Park in Pahang and the Danum Valley in Sabah. There are three collections in Thailand with 585 total accessions. Among them, Phichit Horticultural Research Center has a collection of mostly native pummelos; and Nan Horticultural Research Station has approximately 70 accessions of mandarins, sweet oranges, and citrus relatives, of which approximately 25% are native. In the Philippines, the main collection is maintained by the National Plant Genetic Resources Laboratory of the Institute for Plant Breeding in Los Baños, and consists mostly of commercial and imported varieties; reportedly there are two other collections with slightly over 100 accessions. There are also three collections in Indonesia (498 accessions) and at least two in Vietnam (National Institute of Agricultural Science and Technology and Phu Ho Fruit Research Center), which contain materials collected by an IBPGR-sponsored program in 1992 (https://www.ars.usda.gov/ARSUserFiles/20361000/Crop_Vulnerability_2004.pdf).

National Institute for Agricultural Research (INRA) and (the French Agricultural Research Centre for International Development) (CIRAD), both national institutes located at San Giuliano, Corsica, France maintained a total of 928 genotypes in the field; among these 340 were mandarins and about 220 sweet oranges (Roose et al., 2015). France uses cryopreservation of the citrus seeds collection for security backup. The collections are often evaluated by molecular markers Simple Sequence Repeat (SSR), Single Nucleotide Polymorphism (SNP), Indel markers, and phenotyping. Phenotyping is performed according to International Plant Genetic Research Institute (IPGRI) descriptors. Biochemical traits and abiotic stress tolerance are also evaluated. Every year in INRA and CIRAD about 25,000 citrus buds are distributed in 22 locations in France and other Mediterranean countries (Biswas et al., 2020).

Asia's largest collections are in Japan. The Fruit Tree Research Stations in Tsukuba, Okitsu, and Kuchinotsu have large collections that have several citrus relatives. There are also three other collections of citrus germplasm in Nagasaki, Kagoshima, and Okinawa. Japan has been active in collecting in Southeast Asia, Nepal (1983–1985), and Vietnam (1996) through IBPGR-coordinated cooperative programs. Accessions collected from these ventures are maintained in Japan (https://www.ars.usda.gov/ARSUserFiles/20361000/Crop_Vulnerability_2004.pdf).

Around 1,500 indigenous and exotic citrus accessions are conserved and maintained in 20 field gene banks at different locations in India. Among them, a large collection of about 600 accessions including Citrus spp., *Poncirus trifoliata*, and *Severinia species* are maintained at Central Citrus Research Institute (CCRI), Nagpur. For long-term preservation, Indian collections are also maintained using cryopreservation (Singh et al., 2001; Sharma et al., 2004).

In India, the Northeast region is the center of origin/diversity. A wide range of genetic diversity undoubtedly exists in these areas. There is an *in situ* gene sanctuary for citrus in the Garo Hills in the northeast, which is a field gene bank with 627 accessions. Other regions of diversity include the central and northwest Himalayas, Maharashtra, and the southern peninsula. In India, *ex situ* conservation of citrus germplasm began in the 1950s. *Ex situ* collections consist of around 500 accessions at eight sites (Chetalli, Bangalore (Bengaluru), Rahuri, Tirupati, Abohar, Bhatinda, Yercaud, New Delhi). There are smaller collections at 14 additional sites (Akola, Barapani, Birouli, Hessaraghatta, Katol, Ludhiana, Nurpur, Parbhani, Pantnagar, Pedong, Periyakulam, Sirmour, Srirampur, Tinsukia). The *ex situ* collections in India are mostly of rootstock varieties and a few local cultivars, with not much diversity represented.

The Spanish Citrus Germplasm gene bank maintains around 620 accessions, including 425 accessions from 51 Citrus species, 53 from 20 Citrus relatives, and 142 intra- and inter-specific hybrids. Spanish germplasm collection is maintained at fields, screenhouses, and cryopreservation (Biswas et al., 2020).

The University of California, Riverside (UCR) maintains over 1,000 accessions in UCR gene banks. UCR collection includes commercial citrus varieties, breeding lines, other citrus germplasm, and wild relatives, particularly important for research and extension activities. Apart from the UCR collection, the United State Department of Agriculture (USDA) Agricultural Research Service (ARS) National Clonal Germplasm Repository for Citrus and Dates (NCGRCD) also maintain and preserve 400 accessions of citrus genetic resources (Biswas et al., 2020).

Characterization of Evaluation

Morphological Characterization

Characterization and assessment of diversity are essential for the identification of distinct genotypes, for deciphering genetic relationships including parentages, and for efficient management and utilization of germplasm. Morphological characterization is still the basic and initial step for diversity assessment before employing any other advanced methods. The morphological marker is known for its coverage in the study of agronomic traits in addition to convenience. Further, the technique is relatively cheaper and easier to conduct. Morphological characterization is a way for the description and classification of germplasm and flower color, growth habits are routinely used, morphological characters include both qualitative and quantitative characteristics, which are the strongest determinants of the agronomic value and systematic classification of fruit crops and are considered as an initial step for cultivar identification and diversity assessment. Morphological characterization of plants can be done by observing quantitative and qualitative characters that can be seen from the leaf color, stem color, leaf length (cm), leaf width (cm), leaf petiole length (cm), leaf petiole color, flower color, type of flowering, and fruit color. The study of floral morphology is important for understanding self-compatibility and pollen sterility, which have a major role in fruit breeding and fruit sets are essential prerequisites for initiating any breeding program. Besides, such information would also be useful in taxonomical studies. Morphological analyses are still needed not only for identifying and evaluating the genetic diversity and genetic relationship among the citrus group but also for validating the desired characteristics of varieties. In Citrus, morphological characterization is very useful for describing the differences in the citrus group under different agroecological conditions. Several authors have investigated and morphologically characterized different selections of Citrus plants, to increase the number of genotypes with the potential to be used in breeding programs or to be released as new varieties (Koehler-Santos et al., 2003). Morphological properties have been the main character used for the recognition and description of plant taxa (Duminil and Michele, 2009; Dwari and Mondal, 2011). The important role of morphology is attributable to its advantageous properties such as ease of examination, high variation, established descriptive terminology, and accessibility to herbarium specimens (Szczepaniak and Cieslak, 2011). Moreover, morphology applies to all levels in the taxonomic hierarchy. Morphology has always been the first taxonomic evidence for the recognition and delimitation of infraspecific categories. The key role of morphology in defining infraspecific taxa is shown by the fact that no formal taxonomic status would be assigned when there are no morphological differences among populations of particular plant species (Cires et al., 2009; Lohwasser et al., 2010). At present, the morphological study is still considered important and has been deployed as an initial step for cultivar identification and diversity assessment at the field level (Elameen et al., 2010).

Taxonomic interactions within citrus groups are highly complicated. This complexity in relations might be due to the recurrent occurrence of spontaneous mutants and interspecific hybrids in nature. Although the *Citrus* genus is very complex involving a multitude of species and their hybrids, but still the morphological characters especially the fruit and leaf characters are useful for visual scoring and distinguishing accessions within a species. It allows simple grouping of accessions, development of core collections, identification of gaps in the collection, identifying specific germplasm for breeding programs, etc. The crop improvement programs in citrus are mainly based on the depiction of morphological, cytological, and genetic characters of various cultivars (Fatima et al., 2015; Ahmad et al., 2018). The morphological flexibility is considered a major hindrance in the validation of phenotypic diversity. However, the inheritance of agronomic traits of citrus was reported to be governed by multiple genes and evaluated only by their morphological characterization (Liu and Deng, 2007). Morphological descriptions are the basic passport data of the genotype under the given environmental conditions and that is why they are being used widely (Khan et al., 2008; Kinley and Chinawat, 2011; Baswal et al., 2017; Gupta et al., 2018; Kaur et al., 2022). The identification of superior genotypes within the collected germplasm is valuable for genetic up-gradation and crop conservation (Gaikward et al., 2018).

Several authors have investigated and characterized morphologically different selections of citrus plants, to increase the number of genotypes with the potential to be used in breeding programs or to be released as new varieties. In lime and lemon, Dubey et al. (2015) opined that traits like fruit weight, fruit length, juice volume, and seed content had significant contributions in lime and lemon diversity, and could be used in differentiation among the accessions of citrus. Characters related to plants, flowers, fruits, and leaves have been used to describe and characterize distinct Mandarin varieties and their hybrids (Domingues et al., 1999). Several studies have been conducted for morphological characterization of different species of Citrus which includes Hossain and Rabbani (2009) in lemon (*Citrus limon*), Singh et al. (2010) in rangpur lime (*Citrus limonia* Osbeck.), Dorji and Yapwattanaphun (2011) in mandarin orange (*Citrus reticulata*), Hazarika et al. (2013, 2017) in hatkora (*Citrus macroptera* Mont.), Hazarika et al. (2013, 2016) in pummelo (*Citrus grandis* (L.) Osbeck), Marboh et al. (2015) in *Citrus* spp. Kayesh et al. (2017) in lemon (*Citrus limon*) Gupta et al. (2018, 2020) in Sweet Orange (*Citrus sinensis* Osbeck), Dubey et al. (2019) in Indian pummelo (*Citrus maxima*), Rohini et al. (2020) in rough lemon (*Citrus × jambhiri* Lush.), Kumari et al. (2021) in acid lime (*Citrus aurantifolia* Swingle), Khiavi and Golein (2021) in acid lime (*Citrus aurantifolia* Swingle), Sudo et al. (2021) in Satsuma mandarin (*Citrus unshiu* Marcow.), Chan et al. (2022) in Gunung Omeh Citrus (*Citrus nobilis* Lour), Sangma and Suresh (2022) in *Hatkora (Citrus macroptera* Mont.), Yin et al. (2023) in 'Hongjiangcheng' sweet orange, Muda et al. (2023) in sweet orange (*Citrus sinensis*)

Cytological Characterization

Phenotypic variations among the species are quite interesting as the genotype determines their characters (Raghuvanshi, 1962) and it may further be modified due to a variety of climatic and topographic conditions prevailing in the state. Now it has been established that nine ($n = 9$) is the haploid number (Krug, 1943) and polyploidy is a common phenomenon for this genus (Iwamasa and Nito, 1988). On the other hand, confusion and disagreement have prevailed in the interpretation of species in Citrus. The earlier pomologist, Swingle (1943) classified *Citrus* into 16 species, Tanaka (1977) divided it into 162 species, and recent molecular taxonomists Deng et al. (1995) have considered only 3 genuine botanical species, namely *C. grandis, C. reticulata,* and *C. medica*. Hence, cytological investigation is very helpful to confirm the systematic position of important species. The divergent views of the classification of *Citrus* are due to the ambiguous morphological characters that have been responsible for the confusion existing in their characterization. The karyomorphological data can be used as supporting evidence in solving problems in taxonomy and the delineation of *Citrus* species (Herrero et al., 1996). Iwamasa and Nito (1988) reported that diploidy has been generally widespread in the genus, *Citrus*, and 9 is the basic chromosome number without any contradiction. Kumar and Subramaniam (1986) reported that 18 is the diploid number in *C. reticulata Blanco, C. aurantifolia* Swin., *C. grandis* (L.) Osbeck, *C. aurantium* Linn., *C. limon* Burm f., and *C. medica* Linn. The probable progenitor of mandarin orange, *C. indica* Tan, has also been shown to have

2n = 18 (Singh, 1981) and similarity of the morphological features of the chromosome with other species has so far not been reported.

Most species of the genus *Citrus* are characterized by polyembryony, which consists of the production of 1–40 adventive embryos by the nucellus (Fusurato, 1957) so that two or more embryos develop in a single seed. The trait of adventive nucellar embryony in *Citrus* has long been a subject of interest to taxonomists. Among the three true species in sub-genus *Citrus*, i.e., *C. medica* (citron), *C. maxima* (pummelo), and *C. reticulata* (mandarin), the first two species are strictly monoembryonic with only sexual offspring whereas *C. reticulata* is polyembryonic.

Meiotic events in *Citrus* and its inter-specific and inter-generic hybrids are quite interesting. However, the meiotic behavior in *Citrus* is mainly regular; irregularities are infrequent (Agarwal, 1989). Analysis of meiotic chromosome pairing in hybrids not only is a classical and authentic approach to understanding species relationships but it also helps in the genetic stability of polyploids (Yan et al., 1997). Cultivated *Citrus* species have been hybridized with some wild relatives such as *Murraya*, *Severinia*, *Atalantia*, and *Swinglea*, in order to introduce desirable traits, mainly resistance to pests and pathogens (Barrett, 1977). The identification of chromosomes of different genomes could be a simple method of identifying *Citrus* hybrids and is thus important for future work (Cameron and Frost, 1968). Therefore, the present study was taken up to resolve the cryptic hybridity of *Citrus* species present in their natural habitat through cytogenetic tools.

Most Citrus genomes are highly heterozygous, and it is practically impossible to develop homozygous lines through conventional hybridization, due to sexual incompatibility, nucellar embryony, severe inbreeding depression, and long juvenility. Gametic embryogenesis is a single-step approach to produce homozygous clones from heterozygous parents, from which most Citrus haploids were generated. *In situ,* parthenogenesis induced by irradiated pollen, followed by *in vitro* embryo culture has been reported in Citrus (Chiancone et al., 2006). Haploid plantlets have been recovered by anther culture from *Poncirus trifoliata* L. Raf. (Hidaka et al., 1979) and *C. madurensis* Lour. (Chen et al., 1980). A doubled haploid plantlet has been obtained from the hybrid No. 14 of *C. ichangensis* × *C. reticulata* (Deng et al., 1992). Haploid plantlets and highly embryogenic haploid calli were recovered from *C. clementina* Hort. ex Tan. (Germanà et al., 2005). Haploid, but albino embryos, arose from cultures of 'Mapo' tangelo (*C. deliciosa* × *C. paradisi*) (Germanà et al., 1997). In other reports, haploid and diploid calli, embryos, and leafy structures, but no green plants, were produced.

Gogoi et al. (2005) conducted cytological studies of the seven species, namely, *Citrus reticulate* Blanco (var. Khasi mandarin), *C. aurantifolia* Swin. (var. Round fruited), *C. indica* Tan., *C. aurantium* Linn. (var.Chakala), *C. grandis* (L.) Osbeck (var. Pathalipahar), *C. limon* Burmf. (var. Assam lemon), and *C. medica* Linn. (var. Birajora) have revealed diploidy, 2n = 18. It was found that seven species studied possess small, rod-shaped, and morphologically more or less similar chromosomes. No numerical variation of the chromosome was observed, but gross morphological features of the chromosome complements were variable from one species to another and also distinct for each species. Thus, similarities in chromosome number indicate that the seven species must have originated from a common ancestral stock, and the inclusion of seven species under the same genus is justified. Moreover, consideration of their absolute chromosome length, TF%, and karyotype formulae indicates their individuality and a strong genetic relationship. It is, therefore, recommended that the seven species must be treated as separate species.

Germana et al. (2013) did the Cytological and molecular characterization of three gametoclones of *Citrus clementina* designated *Enterococcal surface protein* (ESP), Fos-related Antigen (FRA), and Inhibitor of *T-Cell Apoptosis* (ITA) in order to elucidate genomic rearrangements provoked by haploidization. Their results revealed that ESP and FRA were haploid, but ITA was tri-haploid. Homozygous patterns, represented by a single peak (allele), were observed among the three plants at almost all SSR loci distributed across the entire diploid donor genome.

Yasuda et al. (2016) studied the cytogenetics and evolution of the genus *Fortunella.* Among the 6 *Fortunella* species examined, *Fortunella hindsii* var. *chintou* Swing. (Hongkong kumquat) showed the simplest Chromosomal Microarray Analysis (CMA) karyotype composition. On the other hand, close relationships were found among the 3 species *F. margarita* (Lour.) Swing. (Oval kumquat), *F. japonica* (Thunb.) Swing. (Round kumquat), *F. crassifolia* Swing. (Meiwa kumquat). *Fortunella polyandra* (Ridl.) Tan. (Malayan kumquat) and *F. obovata* hort. ex Tan. (Changshou kumquat) had type E chromosomes, which is an elemental chromosome type for Citrus. We concluded that there are only two true species for the genus *Fortunella, F. hindsii* and *F. margarita* complex, which includes *F. margarita, F. japonica,* and *F. crassifolia,* and that *F. polyandra* and *F. obovata* should be classified as natural or horticultural hybrids.

Biochemical Characterization

Isozymes have been used as powerful tools both in the characterization of cultivars and in genetic and phylogenetic studies for many plant species, including citrus (Tanksley and Orton, 1983). Isozyme variation and inheritance have been studied extensively in citrus species (Torres et al., 1978; Soost and Torres, 1981; Hirai et al., 1986; Hirai and Kajiura, 1987). Torres et al. (1978) used this technique to characterize distinct Citrus biotypes and to distinguish nucellar from zygotic seedlings. Furthermore, isozymes have been used to develop fingerprints for cultivar identification and to reveal genetic relationships among cultivars (Ashari et al., 1989), to study the phylogenic origins of several biotypes (Hirai et al., 1986), and to identify intergeneric hybrids (Rahman and Nito, 1994) and the products of somatic hybridization (Grosser et al., 1994) in Citrus. To date, however, genetic control of isozymes in the 'true citrus fruit trees' has been studied only in *Citrus* and *Poncirus* (Torres et al., 1985). Isozymes have been used as biochemical markers to study taxonomic relationships between citrus species (Rahman and Nito, 1994; Protopapadakis and Papanikolaus, 1999). This method is useful in distinguishing cultivars derived by sexual reproduction but does not generally distinguish between cultivars differentiated by mutations (Herrero et al., 1996; Rahman et al., 2001).

Isozyme analysis is a useful technique for determining the genetic variability, identification of nucellar and zygotic seedlings, and identification of somatic hybrids and studies in phylogeny. In *Citrus*, more than 20 isozyme loci have been genetically characterized. Isozyme analysis is increasingly being used to distinguish nucellar and zygotic seedlings (Bhat et al., 2010). The identification of intergeneric and interspecific somatic hybrids obtained by protoplast fusion has also been done using isozyme analysis (Grosser and Gmitter, 1990; Tusa et al., 1990). In Citrus, cultivars within the same species could have arisen through mutation, and given species status and isozyme profiles have shown that sweet orange and grapefruit have similarities (Hirai and Kajiura, 1987; Roose, 1988).

Fang De-qui et al. (1993) analyzed the leaf or bark tissue extracts of 108 biotypes of Citrus and its five related genera by polyacrylamide gel electrophoresis for isozymes of peroxidase, glutamate oxaloacetate transaminase, phosphoglucose isomerase, phosphoglucose mutase, superoxide dismutase, tetrazolium oxidase, NADP+- dependent malate dehydrogenase and esterase. Based on the cluster analysis, the total Citrus biotypes are classified into seven groups, i.e., Honghe papeda, Mauritius papeda, Pummelo, Ichang papeda, Citron, Lemon-lime, and Mandarin-orange. Ichang papeda is suggested to be the third subgenus of the genus Citrus. The Fuming trifoliata orange is accepted as a new species of *Poncirus-P. polyandra* S. Q. Ding. Microacrumen (small-fruited mandarin) is the primitive form of the mandarins. It seems better to consider "Horse-nose Mandarin" or "Xipigoushigan mandarin" as the typic mandarin originated in China. Euacrumen (large-fruited mandarin) might have arisen in two ways. One is the hybridization of Microacrumen and *C. sinensis*, and the other is just the variation of Microacrumen. Fortunella obovata Tan. may be a hybrid of Fortunella and Citrus. *C. unshiu* Marc. arose from the nucellar mutation of "Huangyanbendiguangju mandarin" (*C. nobilis* Lour.) in Huangyan of Zhejiang Province. *C. junos* Sieb. came from the hybridization of Ichang papeda and mandarin. *C. grandis* Osbeck and mandarin may have been involved in the origin of *C. aurantium* L. *C. limonia* Osbeck may be a hybrid between *C. medica* L. and mandarin.

Rahman et al. (1994) investigated the Isozyme inheritance and variation of phosphoglucoisomerase (PGI) for six genera of 'true citrus fruit trees' (Rutaceae) and their intergeneric hybrids. The progeny segregated showed Mendelian inheritance at locus Pgi-1 in Fortunella, Eremocitrus, and Microcitrus. Some common isozymes were detected among the six genera. Their results suggested that a genetic relationship exists among all the genera of the "true citrus fruit trees" group.

King et al. (1996) investigated the isozymes for triploid Citrus, their diploid siblings, and diploid progenitors of four enzymes; malate dehydrogenase, 6-phosphogluconate dehydrogenase, shikimate dehydrogenase, and phosphoglucose isomerase, with polyacrylamide gel electrophoresis. Of the four enzymes investigated, only allozymes for shikimate dehydrogenase exhibited consistent differences over a wide range of Citrus cultivars. Greater band density was evident in the allozyme contributed by the diploid gamete. The band density ratio between allozymes for triploid Citrus was close to 0.5, while for diploid Citrus band density ratios were close to 1.0. This effect is due to the extra protein coded by the additional gene dose and was not observed in diploids. Shikimate dehydrogenase proved to be an accurate molecular marker for distinguishing between diploid and triploid Citrus for heterozygous progeny.

Elisiario et al. (1999) conducted the isozyme analysis of 'Carvalhais' mandarin and their results revealed that all the analyzed trees of 'Carvalhais' showed similar monomorphic patterns for each isozyme system, indicating that this mandarin cultivar originated as a single initial clone.

Cabrita et al. (2001) analyzed 22 Citrus varieties (sweet oranges, lemons, grapefruits, clementines, and several other mandarin biotypes) by isozyme in order to assess their genetic relationships. Four isozyme systems (GOT, IDH, PGM, and PGI) revealed seven loci, four of them polymorphic. Sweet oranges, lemons, grapefruits, and clementines were discriminated at the species level but individual varieties within each species remained undistinguishable by the analyzed isozyme systems. Nevertheless, a specific isozyme pattern was found for almost each one of the mandarin varieties except Carvalhais and Fremont, which shared an identical isozyme pattern.

Hossain and Rabbani (2010) investigated the genetic diversity of 73 lemon genotypes using three isozymes, viz., glutamate oxaloacetate transaminase (GOT), malate dehydrogenase (MDH), and peroxidase (PER) isozymes. A total of 25 different electrophoretic zymotypes were observed for three isozymes studied. Eight zymotypes with GOT, 10 with MDH, and seven with PER were formed by 22, 39, and 12 bands at different Rf values, respectively. The result revealed that isozymes are useful genetic markers for variability analysis of lemon and other species of *Citrus*.

Yamamoto et al. (2011) studied the genetic relationships among local Citrus accessions of Japan using isozyme. Four loci were detected for three enzymes examined in isozyme analysis. Four and three kinds of genotypes were detected in glutamate oxaloacetate transaminase (GOT)-1 and GOT-2, respectively. At least six genotypes were observed in peroxidase (PX). All accessions showed the same genotype except for one accession of superoxide dismutase (SOD)-3. In general, accessions that belong to the same species or type showed identical genotypes; however, some diversity of genotypes was observed within the same species and type.

Molecular Characterization

Markers have been used over the years for the classification of plants. Markers are any trait of an organism that can be identified with confidence and relative ease and can be followed in a mapping population (Bhat et al., 2010). They can be defined as heritable entities associated with economically important traits under the control of polygenes (Beckman and Soller, 1986). Molecular markers are any kind of molecule indicating the existence of a chemical or a physical process. Molecular markers include biochemical constituents (e.g., secondary metabolites in plants) and macromolecules (e.g., proteins and deoxyribonucleic acid) (Joshi et al., 1999). These macromolecules show easily detectable differences among different strains of a species or different species. Strauss et al. (1992) distinguished the molecular markers into two classes. Biochemical molecular markers derived from the chemical products of gene expression, i.e., protein-based markers and molecular genetic markers derived from direct analysis of polymorphism in DNA sequences, i.e., DNA-based markers.

DNA markers-based polymerase chain reactions (PCR) are becoming increasingly popular for various uses for which isozymes are used besides the construction of linkage maps and marker-assisted selection. RAPD has been successfully used to measure genetic diversity (Sawazaki et al., 1997; Coletta et al., 1998) and to understand the genetic origin of cultivars (Russo et al., 1998). Fang et al. (1997) found that the acid lesser trait in pummelo 2240 is controlled by a single recessive gene called acitric.

Several molecular markers, particularly polymerase chain reaction (PCR) based DNA markers have now been utilized in inferring systematic and phylogenetic relationships and identification of cultivars and crop relatives in several economically important plant groups (Weising et al., 2005). In Citrus, a wide variety of DNA based markers has been used in order to study their genetic variation as well as phylogenic and taxonomic relationship among different genera, and some of the important examples are: Microsatellite (Pang et al., 2003), RAPD (Nicolosi et al., 2000; Federici et al., 1998), RFLP (Federici et al., 1998), ISSR (Gulsen and Roose, 2001a, b; Fang et al., 1998), organelle genome analysis (Yamamoto et al. 1993), PCR-RFLP analysis of non-coding regions of chloroplast DNA (cpDNA) (Nicolosi et al., 2000; Gulsen and Roose, 2001a, b), RAPD and PCR-RFLP (Federici et al.,1998; Abkenar et al., 2004), RAPD, SCAR, and PCR–RFLP (Nicolosi et al., 2000; Jena et al., 2009), AFLP (Liang et al., 2007; Pang et al., 2007), SSR (Barkley et al., 2006), ISSR (Fang et al., 1998; Shahsavar et al., 2007) and sequence data analysis of ITS region of nrDNA (Xu et al., 2006; Kyndt et al., 2010; Pessina et al., 2011) and non-coding chloroplast DNA (cpDNA) regions (Chase et al., 1999; Araujo et al., 2003; Morton et al., 2003; Lu et al., 2011). These molecular studies have provided some insight into Citrus phylogeny and the species concept was generally supported. The most important advance was that molecular evidence supported the hybrid origin of many so-called species, for example, sweet orange, grapefruit, lemon, etc., and identified their putative parental species (Nicolosi et al., 2000).

RAPD markers have been used for analysis of genetic diversity in Citrus (Coletta Filho et al., 1998; Das et al., 2004), for characterization of Citrus hybrids in breeding programs (Bastianel et al., 1998), cultivar identification (Deng et al., 1996; Mariniello et al., 2004) and for phylogenetic analysis (Machado et al., 1996; Nicolosi et al., 2000). RAPDs have been employed most widely, since this technique is simpler and less expensive than RFLPs (Dos Santos et al., 1994). RAPD analysis has been also used in Citrus to build genomic maps (Cai et al., 1994), to identify markers linked to relevant agronomic traits (Cheng and Roose, 1995; Gmitter et al., 1996), and for taxonomy studies (Luro et al., 1992). In Citrus, a number of examples are there where RAPD markers have also been used for genetic diversity analysis (Shaaban et al., 2006; Shahsavar et al., 2007; Hvarleva et al., 2008), chimeras (Sugawara et al., 2002), and phylogenetic analysis (Nicolosi et al., 2000) of Citrus. Moreover, in Citrus several traits of horticultural importance, including resistance to Citrus tristeza virus (Gmitter et al., 1996), nematode resistance (Ling et al., 1994), and dwarfing (Cheng and Roose, 1995)

SSRs have proven to be the marker of choice in Citrus breeding research, because of their variability, ease of use, accessibility of detection, and reproducibility (Jannati et al., 2009; Fu et al., 2010). SSR was extensively exploited for phylogenetic studies in Citrus (Pang et al., 2003; Barkley et al., 2006), for the assessment of genetic variability (Corazza-Nunes et al., 2002; Golein et al., 2005), and for the construction of genetic maps (Cristofani et al., 2003).

ISSR markers have successfully been used in Citrus to identify closely related varieties (Fang and Roose, 1997), to determine genetic diversity, characterization, and assess phylogenetic relationships among the Citrus and related genera (Fang et al., 1998; Gulsen and Roose, 2001; Shahsavar et al., 2007; Uzun et al., 2009; Marak and Laskar, 2010); and to fingerprint and group trifoliate accessions (Fang et al., 1997), ISSR has been previously used to fingerprint trifoliate orange germplasm accessions (Fang et al., 1997) and other closely related Citrus cultivars (Fang and Roose, 1997).

Utilization in Breeding and Improvement Programs

Collection of germplasm either from exotic or indigenous sources is an important method of citrus breeding. In the introduction, the introduced material can either be used directly as varieties or it can be used in future breeding programmes for the improvement of local varieties or the development of entirely new varieties. Through introductions, all the present-day commercial citrus fruits were dispersed throughout the world. Blood Red Malta, Valencia, Hamlin, Washington Naval, Mosambi, Kinnow, Pineapple, Cleopatra mandarin, Clementine mandarins, Lisbon, Eureka, Villafranca lemons, Marsh seedless, Duncan, Foster grapefruits, were introduced from the USA. Kinnow mandarin, a hybrid between King x Willow leaf mandarin was introduced to South India, Punjab, and Himachal Pradesh. Nagpur Mandarin was introduced from Aurangabad by Raghojee Bhonsle II. The introduced material can either be used directly as varieties or it can be used in future breeding programs for the improvement of local varieties or the development of entirely new varieties. Introduced citrus types like trifoliate orange, severinia, eremocitrus, etc. can be used in breeding programs for improving disease resistance or to overcome soil and water-related problems like salinity. Cleopatra mandarin is highly salt tolerant, whereas, trifoliate orange is resistant to phytophthora, citrus nematode, and citrus tristeza virus and induces dwarfing effects.

Selection

Most of the present-day cultivars of citrus are not the result of planned hybridization programs, but they have come through the selection of natural bud sport mutants or chance seedlings. In any type of breeding program, the selection of superior seedlings is of immense importance. In selection, superior cultivars are selected from natural variability. This involves the survey of different areas, collection of germplasm and their evaluation in the target environment and then selecting the desirable ones. Selection of highly heritable characters such as plant height, days to flowering, the color of fruits, disease resistance, is easy and effective even based on individual plants, but for characters like yield, evaluation of replicated trials is a must which is to be conducted in the target environment. The method of clonal selection resulted in the evolution of a number of high-yielding varieties like Yuvraj Blood Red, an early maturing and seedless variety from Blood Red, and Mudkhed Seedless, a seedless bud mutant of Nagpur mandarin. Similarly, in acid lime, Pramalini, and Vikram, were selected at Marathwada Agricultural University, Parbhani which yields 30–35% more yield than existing Kagzi lime varieties. Sai Sarbaty Rahuri (Maharashtra), Jai Devi (Tamil Nadu), and Tenali (Andhra Pradesh) are also important selections of acid lime. Chakradhar lime, a seedless and thornless selection of Kagzi lime was also the result of selection. PKM-1 was developed at Periyakulam. A seedless selection of acid lime was also made in Himachal Pradesh. Selections namely Dabwali Baramasi and PAU selection in lemon were developed due to selection and they produced the heaviest crop. Kiyomi tangor and the triploid grapefruit Oroblanco have also evolved by selection (Hazarika, 2023).

Hybridization

Hybridization in citrus has been under practice for over 90 years with the objective to breed superior cultivars for specific characteristics like pest and disease resistance, cold hardiness, better fruit quality, dwarfness, thornlessness, precocity, and salt tolerance. Hybridization involves the controlled crossing of two genetically different parents to obtain a desirable variety. Reshuffling the genes in the existing germplasm creates new varieties. For any planned program of hybridization, a collection of germplasm is a must. Different varieties with desirable characteristics are utilized for hybridization to achieve specific objectives like producing early maturing varieties with high yield and quality of fruits, developing rootstocks having resistance to biotic and abiotic stresses wider adaptability. Kinnow mandarin is a well-known hybrid between King Orange and willow leaf mandarin which is a commercial variety of Punjab. Trifoliate orange and Kumquats, two related genera of citrus are extensively used in crossing with citrus. Melgold is a hybrid between pummelo and grapefruit which can be grown on Troyer citrange or Rough lemon rootstocks. In lime, Hybrid 2, Hybrid 4, and N 52 are resistant to citrus canker) (Hazarika, 2023) (Table 2).

Table 2. Hybrids of citrus.

Name of Hybrid	Parents
Intergeneric hybrids	
Citranges	*Poncirus trifoliata* (Citrange) x *Citrus sinensis*
Citrangequats	*Poncirus trifoliata* x Kumquat
Citrumelos	*Poncirus trifoliata* x *C. paradise*
Citremon	*Poncirus trifoliata* x *Fortunella japonica*
Limequat	*C. aurantifolia* x *Fortunella japonica*
Orangequat	*C. reticulata* x *Fortunella japonica*
Intrageneric hybrids	
Lemonage	*C. limon* x *C. sinensis*
Lemonimes	*C. limon* x *C. aurantifolia*
Lemandarins	*C. limon* x *C. reticulate*
Tangors	*C. sinensis* x *C. reticulate*
Tangelos	*C. reticulata* x *C. paradise*

Mutation Breeding

Mutation refers to any heritable change in the genetic makeup, but the breeder is generally interested only in those mutations that alter a phenotype. At the molecular level, mutations can alter DNA by base substitution, insertion, deletion, or subsequence rearrangement. Any of these types of changes may cause a phenotypic change. Mutation breeding refers to the development of new cultivars or germplasm by the deliberate production and selection of new mutations. In some cases, the objective of the breeder is to create a mutation in an existing cultivar that improves a defect such as seediness. After finding and evaluating a selection with a desirable trait, the original cultivar containing the mutation would be released. The breeder may also utilize mutation breeding to obtain selection with desired mutations, and then use this germplasm as a parent in a hybridization program.

Mutations result in the form of mutated buds or sports, part of the trees, or the entire tree. Citrus and its related genera are extremely prone to mutations. Most of the present-day scion and rootstock cultivars are the results of these natural mutations selected from commercial plantations (Table 3). In mutation breeding, seeds or budwood are irradiated with gamma rays, and thus a new variety can be developed. Mutation breeding can be successfully employed in citrus. Natural somatic mutations are common in citrus. "Star Ruby" is an excellent example of the development of new varieties by using mutation in citrus. Hence, evaluation and selection are possible from natural mutants, e.g., Iyo tangor for seedlessness, Satsuma and Naval orange for early season, and Ruby grapefruit for good color. In citrus, artificial mutations were also induced with the help of chemical and physical mutagens. Induction of mutation through physical and chemical mutagens are successful way of improvement in citrus (Hazarika, 2023).

Polyploidy Breeding

Diploidy is the general rule in citrus, but the occurrence of polyploidy has been reported in many cultivars. Hong Kong wild Kumquat and Fortunella Hindsii are tetraploids. Hybrid tetraploid seedlings

Table 3. Citrus cultivars developed through mutation.

Species	Original Cultivar	Mutant Cultivar	Type of Mutation
Naval orange	Bahia	Baianinha	Limb sport
	Washington	Navelina	Limb sport
	Washington	Navelate	Limb sport
	Washington	Marrs	Limb sport
	Washington	Leng	Limb sport
Mandarin	Owari	Clausellina	Bud sport
Grapefruit	Walters (White fleshed)	Foster (Pink fleshed)	Limb sport
	Foster (Pink fleshed)	Hudson (deep red fleshed)	Bud sport
	Hudson (deep red fleshed)	Star Ruby (Deepest red peeled)	Limb sport
	Thompson (light pink-fleshed)	Redblush (Deep red fleshed)	Limb sport
	Henderson	Flame	Bud sport

seldom occur from crossing diploids, but spontaneous tetraploid in citrus occurred as nucellar seedlings. Therefore, nucellar tetraploids can be derived from cultivars that produce a moderately high percentage of nucellar seedlings. On the other hand, triploids have favorable characteristics and yield well, but they are sterile. Triploids are generally more vigorous than tetraploids. All have few or very few seeds. Triploids show promise for commercial use when selection for high yield is successful (Hazarika, 2023).

Polyploidy breeding is very useful in the production of larger and seedless fruits. The triploids, which are generally seedless can be produced by crossing tetraploid and diploid species and the resulting triploids are seedless due to improper meiosis. The development of Melogold and Oroblanco grapefruit hybrids are good examples of the triploid varieties

Tissue Culture

Tissue culture has been exploited to create genetic variability from which crop plants can be improved, to improve the state of health of the planted material, and to increase the number of desirable germplasms available to the plant breeder. Tissue culture and micropropagation protocols have been described for a number of Citrus species and explant sources (Edriss and Burger, 1984; Duran-Vila et al., 1989). The environmental conditions and composition of culture media are known to be crucial for the growth of tissue cultures (Duran-Vila et al., 1992; Randall, 1994). Impact of MS and Murashige and Tucker (MT) media with various kinds and concentrations of cytokinins on the tissue culture of *Citrus reticulata* revealed that MT medium supplemented with 0.5 mg L^{-1} BA gave the greatest percentage of shoots and roots numbers (Te- Chato and Nudoung, 1998). Regeneration of different species of Citrus has been already investigated using MS medium supplemented with BA 3 mg L^{-1} or with BA 1 mg L^{-1} (Pena et al.,1997; Cervera et al., 2000; Dominguez et al., 2000).

Pérez-Tornerov et al. (2010) investigated the influence of the basal medium and different plant growth regulators on micropropagation of nodal explants from mature trees of lemon cultivars and the best results were obtained with 2 mg l^{-1} BA and 1 or 2 mg l^{-1} GA. Similarly, Jajoo (2010) developed an efficient and highly reproducible plant regeneration protocol from the nucellar embryo of *Citrus limonia* by using an MS medium. BA 2.22 mM induced the highest number of multiple shoots 18.26 shoots per explant. On transfer of individual shoots to root-inducing MS medium supplied with auxins, IBA (2.46 μM) and BA (1.11 μM) proved to be the best combination. This media resulted in 78.80 % rooting and produced plantlets with an average of 5.53 roots/shoot.

Species/Cultivars Released

Hybrids of Citrus

Poncirus hybrids

Citrange: It is a hybrid of the sweet orange and the trifoliate orange. Fruit is usually much larger and more orange-like in appearance. In general, the citranges exhibit some degree of intermediacy between the parental species. The influence of the trifoliate orange is strongly marked in the citranges by the trifoliate nature of their leaves, the acidity and bitterness of their fruits, and the cold hardiness of the trees. The influence of the sweet orange is shown, in the evergreen nature of the trees, though a few are semi-deciduous and in their greater vigour. Of great horticultural importance in connection with their use as rootstocks is the fact that with few exceptions they come remarkably true from seed. Since they exhibit some of the most desirable features of the trifoliate orange combined with the greater vigor and wider range of soil adaptation of the sweet orange, some of them have achieved importance as rootstocks.

Citrangequat: It is a trigeneric hybrid of a citrange and kumquat. Combining the cold-hardiness of the kumquat and trifoliate orange, the citrangequats appear to be more cold-resistant than the citranges or the calamondin and kumquat, for they are said to succeed in regions too cold for these fruits. The tree is vigorous, erect, thorny, or thornless, with mostly trifoliate leaves; highly cold-resistant. Fruits are bitter but are considered edible by some at the peak of their maturity. The fruit resembles the oval kumquat, mostly very acidic.

Citrangedin: A hybrid between citrange (*Poncirus trifoliata*) and calamandin.

Citrangor: A hybrid between Trifoliate orange (*Poncirus Trifoliata)* and sweet orange (*C. sinensis*). The Crown is compact or dense, not weeping. Petiole glabrous, long, wings narrow, adjoining the blade. Leaflets one or three, margin bluntly toothed, rachis wings absent, fruit broader than long or as broad as long or longer than broad, rind light green to orange or red-orange, rind texture rough, firmness leathery, navel absent, flesh orange or yellow, taste sour.

Cicitrange: A back cross hybrid between citrange and *Poncirus trifoliatea.*

Citrumelo: *Citrus paradisi* Macf. × *Citrus trifoliata* (L.) Raf. It is a cross of grapefruit and trifoliate orange. Swingle Citrumeolo is a hybrid of Duncan grapefruit and *Poncirus trifoliata.* Phytophthora foot rot and nematode tolerance of Swingle citrumelo make it suitable for replanting sites. Trees are vigorous, and large and produce intermediate to high yields. Swingle citrumelo rootstock produces fruit with high juice and soluble solids content and mid-range acidity. It is a superior rootstock for grapefruit producing high yields of large, excellent quality fruit with high juice content.

Citrandarin: Citrandarin is a cold-hardy intergeneric hybrid of *Citrus trifoliata* (Trifoliate Orange) and *Citrus reticulata* (Mandarin). It can tolerate down to +10° F (–12°C) for short durations depending upon cultivar and individual seedling variation.

Citremon: *Citrus limon* (L.) Burm. × *Poncirus trifoliata* L. C-1499 is one popular cultivar of Citremon.

Citradia: A hybrid between sour orange (*Citrus aurantium*) and Trifoliate orange (*Poncirus trifoliata*). Crown compact or dense, not weeping. Petiole glabrous, length short or long, wings medium, adjoining the blade. Leaflets one, margin crenate/crenulate, Fruit as broad as long or longer than broad, rind yellow to orange, rind texture slightly rough, firmness leathery, navel absent, flesh orange, taste sour. These hybrids greatly resemble citranges but seem to be even more vigorous and cold-resistant.

Citrumquat: A hybrid between Trifoliate orange (*P. trifoliata*) and Kumquat (*Fortunella japonica* or *F. margarita*)

Fortunella Hybrids

Hybrids of Fortunella include the following:

Limequat: Limequats are Mexican lime x kumquat hybrids. Trees are vigorous, evergreen, the single leaflets having narrowly-winged petioles; nearly spineless or with a few short thorns; more cold-tolerant than the lime but not as hardy as the kumquat; very resistant to wither tip. Fruit much like the Mexican lime. There are three named cultivars: Eustis, Lakeland, and Tavares. All limequats are more cold-resistant than the lime parent but much less than kumquats. None of the limequats has achieved commercial importance for the fruit, but Eustis and Lakeland are grown to some extent as ornamentals

Indio Mandarinquat: Mandarin x kumquat hybrid (*Citrus reticulata* Blanco X *Citrus japonica* Thunb. sp.). Orange bell-shaped fruit is much larger than a typical kumquat. The sweet peel is eaten along with the tart flesh for a unique flavor combination. This variety usually blooms during the summer months and produces abundant crops of fruit that stay on the tree during the winter months. Indio looks like a giant kumquat, with similar tangy-sweet flesh and an edible rind.

Nippon Mandarinquat: Satsuma mandarin x kumquat hybrid. The name 'Orangequat' originally given to this class of hybrid is misleading since its parentage involves a mandarin rather than an orange. This variety originated from a cross between the Satsuma mandarin (*C. unshiu*) and Meiwa kumquat (*Citrus japonica* Thunb. 'Meiwa'). It is an attractive ornamental and the fruit makes excellent marmalade. The tree is slow-growing, medium-small, spreading; foliage dark green. The fruit is small (smaller than Satsuma but larger than kumquat), broadly oval to obovate; orange-colored; rind relatively thick and spongy; flavor mild and pulp acid. Matures early but holds well on the tree for several months.

Calamondin: Calamondin is a cross between the mandarin (*Citrus reticulata*) var. *austera* Swingle) and oval kumquat (*Citrus japonica* Thunb. 'Nagami'). It is itself a parent in many other hybrids, Faustrimedin among them *(Citrus × microcarpa* **Bunge** × *Citrus australasica* **F. Muell.).** Calamondin has inherited more qualities from the mandarin than kumquat. Whereas the rind of kumquats is usually edible, the rind of Calamondin resembles those of the genus *Citrus* in being often too sour for consumption, especially in the larger fruits (4 - 6 cm in diameter). The inner parts of the flower (pedicel and ovary) are also more like those of the *Citrus*. Calamondin is increasingly popular as a potted house plant.

Citrangequat: Combining the cold-hardiness of the kumquat and trifoliate orange, the citrangequats appear to be more cold-resistant than the citranges or the Calamondin and kumquat. The tree is vigorous, erect, thorny, or thornless, with mostly trifoliate leaves; highly cold-resistant. The fruit resembles the oval kumquat, mostly very acidic.

Procimequat: A hybrid between limequat and Hong Kong kumquat (*Fortunella hindsii*). Crown open, not weeping. Petiole glabrous, length short, wings narrow, adjoining the blade. Leaflets one, margin crenate/crenulate, Fruit is broader than long, rind yellow, yellow-orange, orange or red-orange, rind texture slightly rough, flesh yellow, taste sour.

Sunquat: A hybrid of lemon and kumquat. *Citrus japonica* Thunb. sp. × *Citrus limon.* The fruit usually known as Lemonquat is a Sunquat, having an edible rind. It is most likely a type of mandarinquat, possibly a hybrid of clementine and a Meiwa kumquat. The whole fruit with its peel is edible and has no off-flavors. Sunquat is said to be excellent for marmalade. The cultivated variety 'Rio Grande Valley' is sold as a lemonquat.

Yuzuquat: A trigeneric hybrid between Yuzu lemon and 'Nagami' kumquat. The fruit often is used as a lemon substitute and is very seedy.

Intrageneric Hybrids

Tangor: The tangor (*C. reticulata* × *C. sinensis*) is a hybrid of the mandarin orange (tangerine, *Citrus reticulata*) and the sweet orange (*Citrus sinensis*). The name 'tangor' is a portmanteau using the 'tang' of tangerine and the 'or' of "orange". The popular varieties of Tangor are King ("King of Siam"; formerly *Citrus nobili*), Murcott ("honey Murcott"; "Murcott honey orange"; 'red'; "big red"), Ortanique (originally found in Jamaica, the name comes from the words 'orange', 'tangerine', and 'unique'), Temple ('Magnet' of Japan) and Umatilla (misnomer "Umatilla tangelo").

Tangelo: The tangelo (*C. reticulata* × *C. maxima* or *C. paradisi*), is a hybrid of tangerine and pummelo or grapefruit. Widely known as honeybells, tangelos are the size of an adult fist, have a tangerine taste, and are juicy. They generally have a loose skin and are easier to peel than oranges, readily distinguished from them by a characteristic 'nipple' at the stem. The Orlando and minneola are two popular varieties of Tangelo. The early maturing Orlando tangelo is noted for its juiciness, mild and sweet flavor, large size, and flat-round shape with a characteristic knob. Most Minneola tangelos are characterized by a stem-end neck, which tends to make the fruit appear bell-shaped. The fruit is usually fairly large.

Lemonime: The lemonime (*Citrus limon*) × (*Citrus aurantiifolia*) fruit is larger than a regular lime. The plant is reported to be resistant to wither tip and *Citrus* scab. The most important variety is the Perrine of Florida. Other varieties include: 'Fourny' (Corsica), 'Oscar' (Italy), 'La Valette' (Malta), 'Mohtasseb' (Morocco), 'Khangi' (Nepal), and 'AK' (Turkey).

Lemmonnage: *Citrus limon* x *Citrus sinensis,* A hybrid of lemon and orange.

Lemandarin: *Citrus limon* × *Citrus reticulata.* Lemandarin or Mandarin lemon or Mandarin lime belongs to a group of several closely related types of *Citrus*. They resemble the mandarin in appearance but taste more like limes. It is thought to have originated in China, where it is called the Canton lime or Canton lemon. It has three well-known varieties: Rangpur, Otaheite, and Kusaie.

Cultivars of Citrus

Mandarins

Mandarins "the loose-skinned orange" include a group of *Citrus* fruits that are easy-peeling with a nice rind color. Mandarin, its hybrids, and relatives together form the largest group in the genus *Citrus*. The number of commercially important mandarin varieties almost equals the number of all other horticultural *Citrus* types combined. The grouping of all mandarins into 3 species by Swingle earlier seemed too arbitrary and the division into 36 mandarin species and their subtypes by the Japanese author Tanaka seemed too elaborate. The system followed by Robert Willard Hodgson in his book, *Horticultural Varieties of Citrus*, divides mandarins into four species. *Deliciosa* (Mediterranean mandarin), *nobilis* (King type), and *unshiu* (satsumas) formed species of their own because of their origin, distribution, and characteristics, and all other mandarin types (*common mandarin, clementines, tangerines, tangors, tangelos, and their hybrids*) were grouped under *reticulata*.

Satsuma mandarin group: This mandarin group is one of the most famous mandarin in the world and is commonly grown in Japan. Some of the important cultivars include: Miyagawa, Okitsu, Seto, Miho, Kuno, Matsuyama, Obawase, and Iseki are a few early ripening varieties while Owari, Silverhill, "Dobashi Beni", and Kimbrough are late ripening varieties.

Willow leaf mandarin group: Also known as Mediterranean mandarin. In comparison with the other mandarins, the most distinctive characteristics of this species are the small size and narrow-lanceolate form of leaves; the mild and pleasantly aromatic flavor of the juice; the distinctive nature and fragrance of the rind oil and the plump and almost spherical seeds. Another distinctive characteristic not confined to this mandarin is the spreading-drooping habit of growth. A few of the cultivars under this group are Avana, Chios, Emperor, Montegrina, Natal, and Willowleaf.

King mandarin group: Fruits are the largest of the mandarins and oblate to depressed globose. Rind thick, moderately adherent but peelable; surface moderately smooth to rough and warty. Deep yellowish-orange to orange at maturity. King of USA, Kunembo of Japan, Japon, Guanxi Shagan, Huang Yen Man Chieh, Huangyan Bendi-guangju, Shagan, and Som-Chuck are some common cultivars.

Common mandarin: There are numerous cultivars in this group and some of the important cultivars are described below:

Khasi mandarin: It is a commercial cultivar of mandarin in northeastern India. Trees are medium to large with erect habit, dense foliage, both thorny and thornless. Fruits depressed, globose, oblate to ovate, orange-yellow to bright orange, surface smooth, glossy, rind thick to moderate, adherence slight, base even, occasionally short-necked, segments moderate in number, generally 8–10; abundantly juicy, juice orange colored with a peculiar sour sweet blend; seeds 9–25, polyembryonic. Possesses a very good post-harvest life.

Kinnow mandarin: It is believed to be a first-generation hybrid between King mandarin (*C. nobilis*) and willow leaf mandarin (*C. deliciosa*). First introduced in Punjab, it has performed well and now gained commercial significance in the Kinnow region of Punjab. Plants are medium to large, erect, symmetrical, dense foliage with a few scattered spines; leaves broadly lanceolate. Fruits medium, globose to oblate base flattened, deep orange-yellow coloured on ripening; rind medium to thin, adherent to segments, number of segments 9–10, easily peelable, very juicy; seeds 12–25; fruit maturity in mid-January; somewhat irregular bearer.

Coorg mandarin: The most important commercial variety in south India, it is particularly grown on a large scale in Coorg and Wynad tracts and the Chikmagalur district of Karnataka. Trees are very vigorous and upright with compact foliage, sparingly spinous. Fruits medium to large, bright orange in color, oblate to globose in shape, base necked or depressed, rind thin to medium thick, easily peeled, segments 8–12; juice abundant with deep chrome color and attractive flavour; seeds 14–30, matures during February–March.

Nagpur mandarin: Among all commercial mandarin types of India, It is the most important cultivar and its commercial cultivation is confined in Satpura hills (Vidarbha region), Nagpur of central India. Trees are vigorous and spineless with compact foliage; leaves are narrowly lanceolate. Fruits medium, sub globose, cadmium coloured, surface smooth, base slightly drawn out with glandular furrows; rind thin, loosely adherent; easily peelable, segments 10–12, juice abundant, saffron coloured; seeds 4–10. It matures during January–February

Darjeeling and Sikkim Mandarin: Its cultivation is mainly concentrated in the Darjeeling district of West Bengal and Sikkim of Northeast India. Fruits globose to oblate, orange-yellow in color, juice moderate, orange in color.

Ponkan: Ponkan is probably the most widely grown mandarin in the world, being heavily grown in China, India, and Brazil. Rind medium-thick, fairly loosely adherent; surface relatively smooth but pebbled, with prominent, sunken oil glands; orange-colored at maturity. Flesh color orange; tender and melting, juicy; flavor mild, pleasant, and aromatic.

Clementine: Clementine is a hybrid of the Mediterranean mandarin and a sweet orange. It is an early maturing monoembryonic cultivar with good-quality fruits. Fruit size is variable, ranging from medium to medium. Well-known clementine cultivars include Clementines Marisol, Clemenules, Oroval, Clementines Algerian, Nour, Sidi Aissa, Monreal and Carte Noir, Corsica, Fina and Caffin, Bruno, Esbal, Fina Sodea, Orogrande, and Ragheb.

Dancy: Fruits are oblate to pear-shaped and of medium size. Peel is deep orange-red to red, smooth, glossy at first but lumpy and fluted later, thin, leathery, and tough. Pulp dark orange, of fine quality, richly flavored. This cultivar is grown in many parts of the world and is also known as Bijou, Christmas, Lady, Moragne, Obeni-mikan, Trimble, Welschart, and Weshart.

Beauty: The tree is vigorous, medium to large, upright-spreading, virtually thornless, with dense foliage of medium-sized, broadly lanceolate leaves. Fruits are medium in size, and oblate. The rind is thin, and firm, but easily removed; the surface is smooth and glossy; the color is orange-red at maturity. The flesh is orange-colored, tender, juicy, and sprightly flavored.

Sweet Orange

This group is commonly known as tight-skinned orange. The increased demand for fresh juice and the development of very early and very late maturing varieties have contributed to its present popularity and its year-round availability. Some popular cultivars under this group are described below:

Ambersweet: It is a complex *Citrus* hybrid. A mid-season sweet orange was crossed with a clementine x Orlando tangelo hybrid. It can be peeled more easily than other oranges. The fruit resembles those of navel oranges in size and appearance more than other types and has good orange rind and juice color at maturity. Trees are moderately vigorous and upright in shape. Foliage is usually fairly dense. Young shoots have small thorns.

Berna: Tree slow-growing, compact, medium-small, very productive. Fruits medium-small, seeds few or none. Rind medium-thick, firm; surface finely pebbled. The flesh is well-colored; moderately juicy; and flavor sweet. Late in maturity (end of March).

Jaffa: Also known as "Florida Jaffa" or "affa Blood Orange" was introduced to Florida in 1883 from Palestine. Fruits are small to medium, commercially seedless (0–6 seeds), with a thin, smooth peel. The Peel color is yellow. The flesh is melting in texture and of very high quality, producing a thick, nectar-like juice. It usually achieves maturity by Christmas time. Flecks of pigment sometimes occur in the fruit during cooler winter conditions.

Hamlin: Hamlin is an early variety; the first fruit reaches maturity in October and is seedless, Juicy, and very productive. Hamlin is also known as Norris.

Pineapple: The fruit is medium-large, somewhat flattened on both ends, with a moderately thick, smooth peel that develops a good orange color under cool night conditions. The juice color and quality are very good. It usually contains 15–25 seeds. It succeeds on sour orange and rough lemon rootstocks. If the crop is allowed to remain too long on the tree, it may induce alternate bearing. It is the favorite midseason orange, sweet with a flavor that resembles a pineapple, and good for processing.

Shamouti: The peel is thick but comes off easily in segments. The fruit is juicy, with a distinct flavor, and of good quality. Fruits are large, oval to ellipsoid with rounded apex, deep orange in color, nearly seedless, and sweet.

Valencia: The fruits are medium to large and commercially seedless (0–6 seeds). The peel is thin and the pulp is tender and very juicy. Valencia is a late variety and keeps well on the tree. It can be picked until late spring, or even early summer, which prolongs the season and increases productivity. Fruit is of high quality but in the heat and without cool nights colour break does not occur and the fruit remains greenish in the tropics even when fully ripe. Campbell nucellar, Cutter, Delta, Dom Joao, Frost nucellar, Harward Late, Midknight, Olinda, Rhode Red, Cotidian, and Valencia Late are a few of the popular varieties of Valencia.

Navel oranges: Navel oranges syn. Bahia oranges get their name from the navel-like protrusion at the lower end of the fruit. It is a secondary embryo inside the same fruit. Navel oranges have no functioning pollen. They do not cross-pollinate with other citruses and produce seedless fruit. Navel oranges are propagated from budwood and recently by cloning. Cloning is the most cost-effective way of propagation, especially in developing countries. Many varieties grow somewhat slower and smaller than other orange varieties.

Washington Naval: It is the second most important orange variety in the world after Valencia. It is the leading variety in Brazil, California, Paraguay, South Africa, Australia, and Japan. Fruit is medium to large, yellow, exceptionally delicious, round and seedless, rich in flavor, and has a slightly pebbled orange rind, easily peeled, segments are seedless, loose, very juicy, very sweet, and texture is firm but melting. The sugar-to-acid balance is on the sugary side but has enough acids to make a pleasing *Citrus* flavor. It is early in maturity.

Mosambi: Fruits are medium to large, slightly oblate to globose, color light yellow to pale orange, rind medium, tight, difficult to peel; flesh firm, juicy with straw yellow color, seeds 15–25.

Malta Blood Red: Fruits medium to large, ellipsoid to spherical, cadmium yellow in color with scarlet blush, apex broad, rind thin to medium, tight, flesh blood red when fully ripe, seeds many.

Sathgudi: Fruits spherical to globose, attractive orange-yellow, medium to large, segments 10–12, juicy, sweet with acid tinge, moderate flavor, seeds 10–20.

Lemon

Eureka: Tree is spreading, moderately vigorous, sparsely foliaged, less thorny. More prone to wind blemish, sunburn, and frost damage than other varieties, less cold-resistant than other varieties. Flowers are tinged with pink. Fruit have a smooth thin rind, high juice content, high acid level, low number of seeds, and good flavor.

Lisbon: In comparison with Eureka, Lisbon is more cold-resistant, more productive, and vigorous. It has denser foliage and the fruit inside the canopy are better protected from sun, wind, and cold. Trees are also much thornier. Fruits have a less pronounced nipple and slightly rougher rind texture. In suitable conditions, Lisbon outyields Eureka by 20–25% due to the bigger size of trees and fruit.

Volkamer: It was first thought of as a variant of mandarin lime. More recently it has been identified as a cross of lemon and sour orange. The attractively dense foliage makes it an excellent ornamental tree. Fruits are lemon-shaped, wide, and with a rough, bright reddish rind. Flesh and juice are yellow-reddish. The fruit has few or no seeds, tastes slightly bitter, and has a pleasantly fresh taste and aroma. It can be used in cooking instead of as a lemon. It is used as rootstock because of its resistance to many diseases.

Galgal: It is an important local item of trade in the Himachal Pradesh, Jammu Kashmir, and Punjab states of India. Tree vigorous, upright or spreading but irregular and open, with stout branches, numerous thick spines; leaves large and dull-green. Flowers are large, purple-tinged, and produced in spring only. New shoot growth purple-tinted. Fruit medium-large to large, oblong to ellipsoid, usually with short blunt-pointed nipple, sometimes depressed and flat. Rind medium-thick; surface usually smooth, tightly adherent; color pale to golden yellow. Segments about 10, flesh color pale yellow, moderately juicy; flavor very sour and with a trace of bitterness. Seeds are numerous and large.

Assam lemon: Fruits elliptic to oblong-obovate, lemon yellow, medium to large, long, apex nippled and base rounded, rind medium thick, axis hollow, segments 11–13, pulp light yellow or whitish, juicy, acidic, normally seedless or occasionally few seeds.

Baramasi: Fruits ovoid to globose in shape, medium-sized, light yellow in color, rind medium to thick, flesh firm, acidic with good flavor, seedless or occasionally few seeds.

Lucknow seedless: Fruits oblong, lemon yellow, smooth, apex nippled, base rounded, rind thin, axis hollow, segments 10–13, pulp light yellow and coarse, juicy, flavor good and sour, seeds absent to few, Fruits ripen from November to January.

Kagzi kalan: Fruits medium, spherical, yellow with apex slightly nippled, base rounded, rind thin, smooth, flesh acidic, light yellow, juicy, 8–13 seeds. Susceptible to tristeza, canker, and die-back diseases.

Pant lemon 1: Fruits medium, 80–100 g, round and smooth, rind thin, juicy, tolerant to tristeza, canker and die back.

Limes

West Indian lime: West Indian lime (Key lime, Mexican lime, Common lime) is the common round lime grown especially in Mexico, Florida, and West Indies. In India, it is known as Kagzi lime. Color dark green when immature to yellowish green when ripe. Fruits small, round obovate or short elliptical, round base, sometimes with a short neck, apex round, rind thin, tight, flesh color greenish yellow, highly acidic, seeds small, moderate in number.

Pramalini: Bears fruits in clusters of 3–7 and yields 30% more than normal kagzi lime. Fruits have 57% juice.

Vikram: Fruits bear in clusters of 5–10 and produce some off-season fruits during September, May, and June. It gives 30–32% higher yields than local strains.

Selection 49: Prolific bearer, fruits larger sized, juice abundant, fruits have a tendency to bear summer crops and show tolerance to canker, tristeza, and leaf miner.

Seedless lime: It is a selection from kagzi lime. Fruits oblong, skin thin, primrose colored, prolific bearer, yields double than normal but late.

Tahiti Lime (*C. latifolia*)**:** Tahiti lime (Bearss lime, Persian lime) is the elongated or ovate seedless lime, often with beautiful spring green flesh and juice. It is less acidic and often juicier than the West Indian lime. Trees are thornless and bigger than Mexican lime. Leaves are bigger and darker. Fruits are often more oval. It is slightly more tolerant to cold than the Mexican lime but still prefers warmer areas than lemon. The extremely rare seeds are strongly monoembryonic. Most seeds are flat, have a shriveled look, and rarely germinate.

Indian lime: Indian lime (Palestine sweet lime) is a sweet lime of Indian origin often thought to be a hybrid of limetta and lime. It has less sugar than sweet orange but almost no citric acid and therefore tastes sweeter. The tree has big shiny leaves that are often cupped or rolled. Fruit is large and usually has a pronounced nipple, color turns orange-yellow at maturity. It is one of the few limes that can be enjoyed fresh. In India, the fruit is also cooked whole and eaten as a dessert or preserved. The following two are the varieties of Sweet lime.

Mitha Chickna: Fruits spherical, globose, yellow, glossy surface with oil glands, rind thin, leathery, flesh medium coarse, juicy, sweet, seeds few.

Mitha Tora: Fruits large with depressed apex and necked back, lemon yellow colored, rind tough and thick with oil glands, flesh yellowish-white, juicy, sweet, and well flavored.

Rongpur lime: It is indigenous to India. Trees are evergreen, spreading in growth habit, highly productive, rind and pulp orange colored, thin, and readily separate from the pulp. It is commonly used as Rootstock.

Pummelo

Tahitian: A typical pummelo but with a thin peel of greenish-yellow color and the flesh has a greenish shade as well but may become amber in full maturity. Fruit is juicy and taste is sweet and has a flavor of melon or lime. Tahitian pummelo is often used in the development of new cultivars of both pummelo and grapefruit.

Siamese sweet: It is an acidless, sweet-tasting pummelo from Thailand. Fruit is oblate to broad ovoid; pulp white, with large, crisp, non-juicy sacs easily separating from each other; mild-flavored but faintly bitter. The tree is dwarf with drooping branches and hairy new growth.

Sweetie: Sweetie is one of the newest pummelo hybrids. Its parents are Siamese Sweet and a variant of Marsh grapefruit. The flesh is pale yellow and the taste is mild and sweet. This is one of the pummelos that retain some green on its rind even when ripe.

Chandler: It is a cross between the Siamese Sweet (white, acidless) and Siamese Pink (acid) pummelos. The fruit is an almost perfect globe, medium size to very big with a smooth peel that sometimes has a pinkish tinge. The pulp is pink to medium red, fine-grained, tender, and fairly juicy. Segment walls are thin, flavor very good; subacid, about 12% sugar, seedy.

Honey: Fruits are seedless and sweet, flesh semi-transparent, color light green to lemon yellow, Fruit weight (½–2 kg). The main harvest is from mid-September to mid-October, but the fruit is available from August to February. 100 ml juice contains 9.2–9.9 g sugar, 0.73–1.01 g acid, 49–52 mg Vitamin C. 100g fruit contains 0.6 g fat, 0.7g protein, and 57 Kcal energy.

Cocktail: It is a cross of the Siamese Sweet pummelo and Frua mandarin. Trees are large and vigorous. Fruit sizes vary from the size of an orange to the size of a grapefruit, have thinner peels than most pummelos; the fruit has seeds, and is very juicy; peel, flesh, and juice are mandarin-colored; and the flavor is pleasant and sub-acid. Matures in early winter and the fruits hold well on the tree.

Conservation of Useful Germplasm

In situ Conservation

The conservation of germplasm in their natural environment is known as *in situ* conservation. This approach is particularly useful for the preservation of land plants in a near natural habitat along with several wild relatives with genetic diversity. *In situ* conservation efforts are required to safeguard existing Citrus genetic diversity.

India established the world's only "Citrus Gene Sanctuary" at Nokrek Biosphere Reserve in Garo hills of Meghalaya for *in situ* conservation in 1981, covering about 10,266 ha (Singh, 1981). It is a unique sanctuary, probably the first of its kind in the world. It is endowed with a highly specified microclimate with a combination of tropical and mild temperate seasons and very high humidity and rainfall. The gene sanctuary is a part of the buffer zone of Nokrek Biosphere Reserve and spreads over the east, west, and south Garo hills of Meghalaya. Citrus species growing inside the gene sanctuary are still safe, however, the slow rate of regeneration of these species and increasing human intervention around the gene sanctuary area are causes of serious concern. This "Citrus Gene Sanctuary" deserves the status of "Heritage Site" under the provisions existing in the "Biological Diversity Act, 2002" of India (Anonymous. 2004). Sharma et al. (2004) and Malik et al. (2013) suggested the creation of more nature reserves, gene sanctuaries, and gene parks, and the inclusion of Citrus species in the social forestry system to safeguard the genetic resources and to allow their further evolution under the natural stresses. Species-specific conservation strategies should be adopted depending on biological status, population size, diversity, and endemism (Malik et al., 2012).

In situ On-farm Conservation

On-farm conservation is an important component of genetic resources conservation especially in marginal and diverse agro-ecosystems. The vast indigenous genetic diversity of Citrus in Southeast Asia is presently being maintained on farms, especially in small orchards, backyards, and farmers' fields. In northeastern India and the foothills of northwestern and central Himalayan tracts, specific species of *Citrus* have been maintained by local people for generations.

In-situ on-farm conservation is a common practice in *Citrus* diversity-rich areas. Conservation of Citrus genetic resources needs urgent attention to protect the existing genetic diversity and to promote the cultivation of rare and endangered species that are of great relevance in the socio-economical structure of tribal populations of this area. In *Citrus*, "local selections" or "farmers' varieties" have been selected or identified since time immemorial in India. Due to their local socio-economic importance and the preference for fruits as a food, in cultural or religious rituals, on-farm conservation for these semi-domesticated, natural selections/cultivars/farmer's varieties has been attempted. These local cultivars are being grown as isolated plants or in small numbers in homestead gardens, farmers' fields, and backyards. Such selections need urgent attention for further characterization, evaluation, taxonomical identity, and on-farm conservation. *In situ* on-farm conservation of Kachai lemon in Kachai village of Okhrul district of Manipur, India is a model of exclusive farmer participation in on-farm conservation and sustainable horticulture providing livelihood support to tribal farmers (Malik et al., 2013). *In situ*, conservation areas for wild and semiwild species and several local Citrus types have been proposed based on a detailed survey in Northeast India. This recent concept and role of custodian farmers in on-farm conservation may be further strengthened, especially since the Citrus genetic resources of India are highly diverse (Sthapit et al., 2013, Malik et al. 2012).

Ex situ Conservation

Field Genebank

Field genebanks provide easy and ready access to the plant genetic resources of Citrus, for characterization, evaluation, or utilization. The routine maintenance of trees in the field genebank is most expensive when compared to other conservation methods (Engelmann, 2010). In addition, distribution and exchange from field genebanks is difficult because of the vegetative nature of the material and the greater risks of disease transfer from country to country (Engelmann, 1997). In the field of genebanks, trees are highly subjected to risks of losses caused by several biological and climatic factors like pests and pathogens attacks like dieback, mineral deficiency, drought, weather damage, and human error (Withers and Engels, 1990; Malik et al., 2012).

Table 4. Important Citrus growing countries and Field Genebank.

Country	Field Genebank
USA	Citrus Research and Education Center, University of Florida, Florida; National Clonal Germplasm Repository at Riverside, California
Japan	Fruit Tree Research Stations, Okitsu, and Kuchnotsu
China	National Citrus Germplasm Repository, Beibei Chongqing, Sichuan Province; Citrus Research Institute of the Chinese Academy of Agriculture Sciences, Beibei, Chongqing
Australia	Queensland Horticulture Institute, Queensland
Malaysia	MARDI, Cameron Highlands; University of Malaya, UM; Botanical Gardens, Rimba Ilmu
Thailand	Nan Horticultural Research Station
Indonesia	National Plant Genetic Resources Laboratory, Institute of Plant Breeding, Los Banos, Laguna
Vietnam	National Institute of Agricultural Sciences and Technology

Source: Malik et al. 2012.

In vitro Conservation

Shoot tips, stem sections, root sections, leaf sections, stem internodes, and epicotyl segments have been used as explants for micropropagation followed by *in vitro* conservation and cryopreservation using various techniques (Marin and Duran-Vila, 1991; De Carlo and Lambardi, 2005; Liu and Deng, 2007).

Cryopreservation

Cryopreservation is one of the safe storage for *Citrus* species. In *Citrus,* to obtain genetically stable plants with high recovery percentages, cryopreservation of seeds and embryonic axes is the only suitable method for the long-term conservation of citrus germplasm. However, seeds of many *Citrus* species show non-orthodox (recalcitrant or intermediate) seed storage behavior and therefore, cannot be stored using conventional storage methods.

Conclusion

Conservation of *Citrus* genetic resources needs urgent attention in order to protect the existing genetic diversity and to promote the cultivation of rare and endangered species that are of great relevance in the socio-economical structure of the tribal population of this area. *Ex situ* and *in situ,* conservation of rare and endangered species should urgently be undertaken to ensure safe storage of germplasm for dynamic conservation and sustainable use of genetic resources of all available *Citrus* germplasm. Cryopreservation and *in vitro* storage of seeds can also be attempted to safeguard the germplasm. Creation of more natural reserves, gene sanctuaries, and gene perks and inclusion of *Citrus* species in the social forestry system to safeguard the genetic resources and to allow their further evolution under natural stresses. It is high time to establish a field genebank at different sites as a revival policy and thereafter proper care needs to be taken for maintenance for better conservation of the *Citrus* diversity. To reduce the risk of germplasm loss, in addition to field genebanks, *in vitro* genebanks have to be developed.

Strategies for future genomics research in *Citrus* include the development of molecular markers, construction of genetic and physical maps, identification of genes and gene expression, and whole genome sequencing. Genomic technologies such as analysis and sequencing of the genome, identification of its genes and their expression, recombination, and diversity can be applied for the genetic improvement of the crop.

References

Abkenar, A.A., S. Isshiki, and Y. Tashiro. 2004. Phylogenetic relationships in the true Citrus fruit trees revealed by PCR-RFLP analysis of cp DNA. *Sci. Hort., 102*: 233–242.

Agarwal, P.K. 1989. Cytogenetical investigations in Rutaceae V. Cytomorphology of the three intergeneric hybrids of *Citrus* and *Poncirus*. *Cytologia, 54*: 705–708.

Ahmed, S., H.S. Rattanpal, and G. Singh. 2018. Diversity assessment of grapefruit (*Citrus* × *paradisi*) and tangelo (*Citrus* × *tangelo*) under Indian conditions using physico-chemical parameters and SSR markers. *Appl. Ecol. Env. Res., 16*: 5343–5358.

Anonymous, 2004. *The Biological Diversity Act, 2002 and Biological Diversity Rules, 2004,* 57pp. National Biodiversity Authority 2004. Frontline Offset Printers, Triplicane, Chennai-India.

Araujo, E.F., L.P. Queiroz, and M.A. Machado. 2003. What is Citrus? Taxonomic implications from a study of cp-DNA evolution in the tribe Citreae (Rutaceae subfamily Aurantioideae). *Org. Divers Evol., 3*: 55–62.

Ashari, S., D. Aspinall, and M. Sedgley. 1989. Identification and investigation of relationships of mandarin types using isozyme analysis. *Sci. Hort., 40*: 305–315.

Barkley, N.A., M.L. Roose, R.R. Krueger, and C.T. Federici. 2006. Assessing genetic diversity and population structure in a Citrus germplasm collection utilizing simple sequence repeat markers (SSRs). *Theor. Appl. Genet., 112*: 1519–1531.

Barrett, H.C. 1977. Intergeneric hybridization of *Citrus* and other genera in *Citrus* cultivar improvement. *Proc. Int. Soc. Citriculture, 2*: 586–589.

Barrett, H.C., and A.M. Rhodes. 1976. A numerical taxonomic study of affinity relationships in cultivated Citrus and its close relatives. *Syst. Bot., 1*: 105–136.

Bartholomew, E.T., and W.B. Sinclair. 1952. *The Lemon Fruit*. Berkeley, CA: University of California, Berkeley.

Bastianel, M., S.F. Schwarz, H.D. Coletta Filho, L.L. Lee, M.A. Machado, and O.C. Koller. 1998. Identification of zygotic and nucellar tangerine seedlings (*Citrus* spp.) using RAPD. *Genet. Mol. Biol., 21*: 123–127.

Baswal, A.K., H.S. Rattanpal, G.S. Uppal, and K.S. Gill. 2017. Assessment on performance and variability in different sweet orange (*Citrus sinensis* Obseck) cultivars under Punjab conditions. *J. Appl. Nat. Sci., 9*: 780–783.

Beckman, J.S., and M. Soller. 1986. Restriction fragment length polymorphism and genetic improvement of agricultural species. *Euphytica, 3*: 111–124.

Bhat, Z.A., W.S. Dhillon, R. Rashid, J.A. Bhat, W.A. Dar, and M.Y. Ganaie. 2010. The role of Molecular Markers in improvement of fruit crops. *Not. Sci. Biol., 2*: 22–30.

Bhattacharya, S.C., and S. Dutta. 1956. *Classification of Citrus Fruits of Assam*, 110pp. Sc. Monogr. 20. ICAR, New Delhi,

Biswas M.K., Bagchi M., Deng X., and Chai Lijun. 2020. Genetic Resources of Citrus and Related Genera. *In*: Gentile, A., La Malfa, S., and Deng, Z. [Eds.], *The Citrus Genome. Compendium of Plant Genomes*, 23–32. Springer International Publishing. Springer Nature Switzerland AG.

Burkill, I.H. 1931. An enumeration of the species of Paramignya, Atalantia, and Citrus found in Malaya. Gard. *Bull. Straits Settlem., 5*: 212–220.

Cabrita, L., P. Elisiário, J. Leitão, and A. Guerreiro. 2001. Assessment of the genetic relationships among citrus species and varieties by isozyme and RAPD markers. *Acta Hort., 546*: 177–181.

Cai, Q., C.L. Guy, and G.A. Moore. 1994. Extension of the linkage map in Citrus using random amplified polymorphic DNA (RAPD) markers and RFLP mapping of cold-acclimation-responsive loci. *Theor. Appl. Genet., 89*: 604–614.

Cameron, J.W., and H.B. Frost. 1968. Genetics, breeding and nucellar embryony. *In*: Reuther, W., Batchelor, L.D., and Webber H.J. [Eds.], *The Citrus Industry* (Vol. II), 325–370. University of California Press; California, USA.

Cervera, M., C. Ortega, A. Navarro, L. Navarro, and L. Pena. 2000. Generation of Transgenic Citrus plants with tolerance-to-salinity gene HAL2 from yeast. *J. Hort. Sci. Biotechnol., 75*: 26–30.

Chan, S.R.O.S., B.S. Achmad, and Ferdinant. 2022. Morphological Characterization of Gunung Omeh Citrus (*Citrus nobilis* Lour) in Guguak District, Lima Puluh Kota Regency. *IOP Conf. Series: Earth and Env. Sci., 1097*: 012032. doi:10.1088/1755-1315/1097/1/012032.

Chapot, H. 1975. The Citrus plant. *In*: *Citrus Technical Monograph. No. 4*, 6–13.

Ciba-Geigy Agrochemicals.

Chase, W., C.M. Morton, and J.A. Kallunki. 1999. Phylogenetic relationships of Rutaceae: a cladistic analysis of the subfamilies using evidence from rbcL and atpB sequence variation. *Amer. J. Bot., 8*: 1191–1199.

Chen, Z., H. Wang, and H. Liao. 1980. The induction of Citrus pollen plants in artificial media. *Acta Genet. Sinica, 7*: 189–192.

Chen, X.M., A.R. Tait, and D.D. Kitts. 2017. Flavonoid composition of orange peel and its association with antioxidant and anti-inflammatory activities. *Food Chem., 218*: 15–21.

Cheng, F.S., and M.L. Roose. 1995. Origin and inheritance of dwarfing by the Citrus rootstock and Poncirus trifoliata flying dragon. *J. Amer. Soc. Hort. Sci., 120*: 286–291.

Chiancone, B., A. Tassoni, N. Bagni, and M.A. Germanà. 2006. Effect of polyamines on in vitro anther culture of *Citrus clementina* Hort ex Tan. *Plant Cell Tissue and Organ Culture, 87*: 145–153.

Cires, E., C. Cuesta, E.L. Peredo, M.A. Revilla, and J.A.F. Prieto. 2009. Genome size variation and morphological differentiation within Ranunculus parnassifolius group (Ranunculaceae) from calcareous screes in the northwest of Spain. *J. Syst. Evol., 281*: 193–208.

Coletta Filho, H.D., M.A. Machado, M.L.P.N. Targon, M.C.P.Q.D.G. Moreira, and J. Pompeu Jr. 1998. Analysis of the genetic diversity among mandarins (*Citrus* spp.) using RAPD markers. *Euphyt., 102*: 133–139.

Corazza-Nunes, M.J., M.A. Machado, W.M.C. Nunes, M. Cristofani, and M.L.P.N. Targon. 2002. Assessment of genetic variability in grapefruits (*Citrus paradisi* Macf.) and pummelos (*C. maxima* Burm. Merr.) using RAPD and SSR markers. *Euphyt., 126*: 169–176.

Cristofani, M., M. Machado, V. Novelli, A. Souza, and M. Targon. 2003. Construction of linkage maps of *Poncirus trifoliata* and *Citrus sunki* based on microsatellite markers. *In*: *Proc. Int. Soc. Citricul., 1*: 175–178.

Das, A., Mondal, B., Sarkar, J., and Chaudhuri, S. 2004. Genetic diversity analysis of Citrus cultivars and rootstocks of the Northeastern India by RAPD markers. *J. Hort. Sci. Biotech., 79*: 850–854.

De Carlo, A., and Lambardi, M. 2005. *Cryopreservation of Citrus Germplasm*. Villa Gualino, Turin, Italy.

Deng, Z.N., A. Gentile, E. Nicolosi, G. Continella and E. Tribulato. 1996. Parentage determination of some Citrus hybrids by molecular markers. *Proc. Int. Soc. Citricul.* 2: 849–854.

Deng ZN, Gentile A, Nicolosi E, Continella G, Tribulato E (1996) Parentage determination of some
Citrus hybrids by molecular markers. Proc Int Soc Citricul 2:849–854

Deng, J.N., A. Gentile, E. Nicolosi, F. Domina, A. Vardi, and E. Tribulato. 1995. Identification of *in vivo* and *in vitro* lemon mutants by RAPD markers. *J. Hort. Sci., 70*: 117–125.

Deng, X.X., Z.A. Deng, S.Y. Xiao, and W.C. Zhang. 1992. Pollen derived plantlets from anther culture of Ichang papeda hybrids No.14 and Trifoliate orange. *Proceedings of the International Society of Citriculture*, 190–192. Acireale. Italy.

Di Majo, D., M. Giammanco, M. La Guardia, E. Tripoli, S. Giammanco, and E. Finotti. 2005. Flavanones in citrus fruit: Structure-antioxidant activity relationships. *Food Res. Int.*, *38*: 1161–1166.

Domingues, E.T., V.C. Souza, C.M. Sakuragui, J. Pompeu Jr, R.M. Pio, J. Teofilo Sobrinho, and J.P. Souza. 1999. Caracterizacaomorfologica de tangerinas do banco ativo de germoplasma de citros do centro de citricultura Sylvio Moreira/IAc. *Scientia Agri., Piracicaba*, *56*: 197–206.

Dominguez, A., J. Guerri, M. Cambra, L. Navarro, P. Moreno, and L. Peña. 2000. Efficient Production of Transgenic Citrus Plants Expressing the Coat Protein Gene of Citrus Tristeza Virus. *Plant Cell Reports*, *19*: 427–433.

Dorji, K., and Y. Chinawat. 2011. Morphological Identification of Mandarin (*Citrus reticulata* Blanco) in Bhutan. *Kasetsart J. Nat. Sci.*, *45*: 793–802.

Dos Santos, J.B., J. Nienhuis, P. Skroch, J. Tivang, and M.K. Slocum. 1994. Comparison of RAPD and RFLP genetic markers in determining genetic similarity among Brassica oleracea L. genotypes. *Theor. Appl. Genet.*, *87*: 909–915.

Dubey, A.K., R.M. Sharma, O.P. Awasthi, N. Sharma, and A. Kholia. 2019. Morphological characterization of Indian pummelo (*Citrus maxima*). *Ind. J. Agril. Sci.*, *89*(9): 1523–1528.

Dubey, A.K., R.M. Sharma, O.P. Awasthi, M. Srivastava, and N. Sharma. 2015. Genetic diversity in lime (*Citrus aurantifolia* Swing.) and lemon (*Citrus limon* (L.) Burm.) based on quantitative traits in India. *Agrofor. Syst.* doi: 10.1007/s10457-015-9867-x.

Duminil, J., and M.D. Michele. 2009. Plant species delimitation: A comparison of morphological and molecular markers. *Plant Biosystem.*, *143*: 528–542.

Duran-Vila, N., Y. Gogorcenna, V. Ortega, J. Ortiz, and L. Navarro. 1992. Morphogenesis and tissue culture of sweet orange (*Citrus sinensis* (L.) Osb. Effect of temperature and photosynthetic radiations. *Plant Cell Tis. Org. Cult.*, *29*: 11–18.

Duran-Vila, N., V. Ortega, and L. Navarro. 1989. Morphogenesis and tissue cultures of three Citrus species. *Plant Cell Tiss. Org. Cult.*, *16*: 123–133.

Dwari, S., and A.K. Mondal. 2011. Systematic studies (morphology, anatomy and palynology) of economically viable grass *Brachiaria mutica* (Forsskil) Stapf in eastern India. *Afr. J. Pl. Sci.*, *5*: 296–304.

Edriss, M.H., and D.W. Burger. 1984. *In vitro* propagation of troyer citrange from epicotyl segments. *Sci. Hort.*, *23*: 159–162.

Elameen, A., A. Larsen, S.K. Sonja, S. Fjellheium, S. Msolla, and E. Masumba. 2010. Phenotypic diversity of plant morphological and root descriptor characters within a sweet potato, *Ipomoea batatas* (L.) Lam., germplasm collection from Tanzania. *Genet. Res. Crop Evol.*, *58*: 397–407.

Elisiario, P.J., G.G. Santos, A.R. Guerreiro, F. Ollitrault Patrick, Luro, and J. LeitaÄo. 1999. Isozyme analysis revealed that the Portuguese mandarin 'Carvalhais' originated as a single clone. *Sci. Hort.*, *82*: 145–152.

Engelmann, F. 2010. Use of biotechnologies for the conservation of plant biodiversity. *In Vitro Cell. Dev. Biol. –Plant.* doi: 10.1007/s11627-010- 9327-2.

Engler, A. 1931. Rutaceae. *In*: Engler, A., and Prantl, K. [Eds.], *Die naturlichen Pflanzenfamilien* (2nd Edn.), *19*, 187 –359. Engelmann, Leipzig.

Fang, De-qiu, Zhang, Wen-cai, and Xiao, Shun-Yuan. 1993. Studies on Taxonomy and Evolution of Citrus and Its Related Genera by Isozyme Analysis. *J. Syst. Evol.*, *31*(4): 329–352.

Fang, D.Q., R.R. Krueger, and M.L. Roose. 1998. Phylogenetic relationships among selected Citrus germplasm accessions revealed by inter-simple sequence repeat (ISSR) markers. *J. Amer. Soc. Hort.*, *123*: 612–617.

Fang, D.Q., M.L. Roose, R.R. Krueger, and C.T. Federici. 1997. Fingerprinting trifoliate orange germplasm accession with isozymes, RFLPs, and inter-simple sequence repeat markers. *Theor. Appl. Genet.*, *95*: 211– 219.

Fang, D.Q., and M.L. Roose. 1997. Identification of closely related Citrus cultivars with inter-simple sequence repeat markers. *Theor. Appl. Genet.*, *95*: 408– 417.

Fatima, B.M., S. Usman, M.S. Khan, I.A. Khan, and M.M. Khan. 2015. Identification of citrus polyploids using chromosome counts, morphological and SSR markers. *Pak. J. of Agril. Sci.*, *52*: 107–114.

Federici, C.T., D.Q. Fang, R.W. Scora, and M.L. Roose. 1998. Phylogenetic relationships within the genus *Citrus* (Rutaceae) and related genera as revealed by RFLP and RAPD analysis. *Theor. Appl. Genet.*, *96*: 812–822.

Fu, J., Z.J. Peng, X.D. Cai, and W.W. Guo. 2010. Regeneration and molecular characterization of interspecific somatic hybrids between Satsuma mandarin and two seedy sweet oranges for scion improvement. *Plant Breed.*, 01773. doi:10.1111/j.1439- 0523.

Fusurato, K. 1957. Studies on polyembryony in *Citrus*. *Rep. Kihara Inst. Biol. Res.*, *8*: 40–48.

Gaikwad, K.A., S.R. Patil, P.K. Nagree, and N.R. Potdukhe. 2018. Morphological characterization of citrus rootstock genotypes. *Int. J. Chem. Stud.*, *6*: 516–529.

Germanà, M.A., B. Chiancone, O. Lain, and R. Testolin. 2005. Anther culture in *Citrus clementina*: A way to regenerate tri-haploids. *Aust. J. Agr. Res.*, *56*: 839–845. ISSN: 0004–9409.

Germanà, M.A., and G. Reforgiato. 1997. Haploid embryos regeneration from anther culture of 'Mapo' tangelo (*Citrus deliciosa* x *C. paradisi*). *Adv. Hort. Sci.*, *11*: 147–152.

Giorgetti, L., P. Aleza, E. Carrera, C. Chen, B. Chiancone, G. Costantino, D. Dambier, X. Deng, C.T. Federici, Y. Froelicher, W. Guo, V. Ibáñez, J. Juárez, K. Kwok, F. Luro, M.A. Machado, M. Angel Naranjo, L. Navarro, P. Ollitrault, G. Ríos, M.L. Roose, M Talon, Q., Xu, and G. Gmitter Jr Fred. 2013. Cytological and molecular characterization of three gametoclones of *Citrus clementina*. *BMC Plant Biology*, *13*: 129.

Gmitter, F.G., S.Y. Xiao, S. Huang, X.L. Hu, S.M. Garnsey, and Z. Deng. 1996. A localized linkage map of the Citrus tristeza virus resistance gene region. *Theor. Appl. Genet.*, *92*: 688–695.

Gogoi, M., B. Singh, P. Rethy, and A.K. Mishra. 2005. Cytological studies in some Citrus species of Arunachal Pradesh. *Ind. J. Genet. Plant Breed.*, *65*(01): 70–72.

Golein, B., Talaie, A., Z. Zamani, A. Ebadi, and A. Behjatnia. 2005. Assessment of Genetic Variability in Some Iranian Sweet Oranges (*Citrus sinensis* [L.] Osbeck) and Mandarins (*Citrus reticulata* Blanco) Using SSR Markers. *Int. l. J. Agril. Biol.*, 1560–8530/2005/07–2–167–170.

Grosser, J.W., and G.G. Gmitter, Jr. 1990. *Plant Breeding Rev.*, *8*: 339–397.

Grosser, J.W., E.S. Louzada, F.G. Gmitter Jr., and J.L. Chandler. 1994. Somatic hybridization of complementary Citrus rootstocks: Five new hybrids. *Hort Sci.*, *29*(7): 812–813.

Gulsen, O., and M.L. Roose. 2001a. Chloroplast and nuclear genome analysis of the parentage of lemons. *J. Amer. Soc. Hort. Sci.*, *126*: 210–215.

Gulsen, O., and M.L. Roose. 2001b. Lemons: Diversity and relationships with selected Citrus genotypes as measured with nuclear genome markers. *J. Amer. Soc. Hort. Sci.*, *126*: 309–317.

Gupta, S., H. Monika, Rattanpal, G. Sidhu, and G. Singh. 2018. Genetic diversity in sweet orange (*Citrus sinensis* Osbeck) germplasm based on fruit morphological and physiological traits. *Res. on Crops*, *19*(3): 425–429.

Gupta, S., Monika, H. Rattanpal, G. Sidhu, and G. Singh. 2020. Morphological characterization of sweet orange (*Citrus sinensis* osbeck) germplasm under subtropical conditions. *Ind. J. Pl. Genet. Res.*, *33*: 224–229.

Hazarika, T.K. 2023. Citrus. *In*: Rajasekharan, P.E., and Rao, V.R. [Eds.], *Fruit and Nut Crops. Handbooks of Crop Diversity: Conservation and Use of Plant Genetic Resources*. Springer, Singapore. https://doi.org/10.1007/978-981-99-1586-6_15-1.

Hazarika, T.K., J. Lalchhanmawia, A.C. Lalrinfeli, Lalthlamuani, Shukla, and B.P. Nautiyal. 2017. Assessment of genetic variability in the endangered *Citrus macroptera* Mont. ('hatkora') from Mizoram, Northeast India. *Genet. Res. Crop Evol.*, *64*: 321–330.

Hazarika, T.K., M. Lalbiakngheti, and B.P. Nautiyal. 2013. Genetic variability in physico-chemical characteristics of some pummelo [*Citrus grandis* (L.) Osbeck] accessions from Mizoram. *Ind. J. Hort.*, *70*: 431–434.

Hazarika, T.K., B. Lalawmpuii, and B.P. Nautiyal. 2013. Studies on variability in physico-chemical characters of hatkora (*Citrus macroptera* Mont.) collections of Mizoram. *Ind. J. Hort.*, *70*(4): 480–484.

Hazarika, T.K., Lalthlamuani, Lalchhanmawia, J, Lalrinfeli, and Nautiyal, B.P. 2016. Variability in physico-chemical characteristics and selection of superior types among local pummelo (*Citrus grandis* (L.) Osbeck) germplasms from Mizoram, Northeast India. *Cur. Sci.*, *111*(8): 25.

Herrero, R., M.J. Asins, E.A. Carbonell, and L. Navarro. 1996. Genetic diversity in the orange subfamily Aurantioideae. I. Intraspecies and intragenus genetic variability. *Theor. Appl. Genet.*, *92*, 599–609.

Hidaka, T., Y. Yamada, and T. Shichijo. 1979. *In vitro* differentiation of haploid plants by anther culture in *Poncirus trifoliata* (L.) Raf. *Jap. J. Breed.*, *29*: 248–254.

Hirai, M., and I. Kajiura. 1987. Genetic analysis ofleafisozymes in Citrus. *Japan. J. Breed.*, *37*: 377–388.

Hirai, M., I. Kozaki, and I. Kajiura. 1986. Isozyme analysis and phylogenic relationship of Citrus. *Japan Journal of Breeding*, *36*: 377–389.

Hossain, M.I., and M.G. Rabbani. 2009. Morphological characterization of lemon germplasms. *J. Sher-e-Bangia Agric. Univ.*, *3*(2): 45–50.

Hossain, M.I., and M.G. Rabbani. 2010. Genetic diversity of lemon genotypes assessed by isozymes. *Sher-e-Bangla Agric. Univ.*, *4*(1): 6–11.

https://www.ars.usda.gov/ARSUserFiles/20361000/Crop_Vulnerability_2004.pdf

Hvarleva, T., T. Kapari-Isaia, L. Papayiannis, A. Atanassov, A. Hadjinicoli, and A. Kyriakou. 2008. Characterization of *Citrus* cultivars and clones in Cyprus through microsatellite and RAPD analysis. *Biotechnol & Biotechnol Eq.*, *22*: 787–794.

Iwamasa M., and N. Nito. 1988. Cytogenetics and the Evolution of Modern Cultivated Citrus. *In*: *Proceeding of the Sixth International Citrus Congress*, 265–275. [Eds.], Goren, R., and Mendel, K.]. Tel Aviv, Israel.

Jajoo, A. 2010. *In vitro* Propagation of Citrus limonia Osbeck through Nucellar Embryo Culture. *Curr. Res. J. Biol. Sci.*, *2*: 6–8.

Jannati, M., R. Fotouhi, A. Pourjan, and A.Z. Salehi. 2009. Genetic diversity analysis of Iranian Citrus varieties using microsatellite (SSR) based markers. *J. Hort. For.*, *1*: 120–125.

Jena, S.N., S. Kumar, and K.N. Nair. 2009. Molecular phylogeny in Indian Citrus L. (Rutaceae) inferred through PCR-RFLP and trnL-trnF sequence data of chloroplast DNA. *Sci. Hort.*, *199*: 403–416.

Joshi, S.P., P.K. Ranjekar, and V.S. Gupta. 1999. Molecular markers in plant genome analysis. *Curr. Sci.*, *77*: 230–240.

Kaur, H., G.S. Sidhu, N.K. Sarao, R. Singh, and G. Singh. 2022. Assessment of genetic diversity of mandarin cultivars grown in major citrus regions of world using morphological and microsatellite markers. *Hort. Env. and Biotech*. https://doi.org/10.1007/s13580-021-00404-4.

Kayesh E., S. Das, M.M.R. Rajib, and M.M. Islam. 2017. Morphological characteristics and fruit quality of selected lemon genotypes. *Ann. Bangladesh Agric.*, *21*(1, 2): 111–116.

Khan, M.M., S. Mumtaz, S. Ahmad, M. Abbas, and I.A. Khan. 2008. Some studies on the morphology of Kinnow mandarin and Feutrell's Early. *Pak. J. Agril. Sci.*, *45*: 424–431.

Khiavi, S.J., and B. Golein. 2021. Morphological Markers as a Tool for Investigating Genetic Diversity of Some Lime Genotypes. *J. Hort.Sci.*, *34*(4): 645–662.

King, B., L. Lee, and P. Scott. 1996. Identification of triploid Citrus by isozyme analysis. *Euphytica*, *90*: 223–231. doi:10.1007/BF00023862.

Kinley, D., and Y. Chinawat. 2011. Morphological Identification of Mandarin (Citrus reticulate Blanco) in Bhutan. *Kasetsart J.*, *45*: 793–802.

Koehler-Santos, P., A.L.C. Dornelles, and L. Freitas, Bd. 2003. Characterization of mandarin Citrus germplasm from southern Brazil by morphological and molecular analyses. *Pesq. Agro. Brasil*, *38*: 797–806.

Krueger, R.R., and L. Navarro. 2007. Citrus germplasm resources. *In*: I.A. Khan [Ed.],. *Citrus Genetics, Breeding, and Biotechnology*, 45–140. CAB International, Wallingford.

Krug, C.A. 1943. Chromosome numbers in the subfamily Aurantioideae, with special reference to the genus Citrus. *The Bot. Gaz.*, *104*: 602–611.

Kumar, V., and B. Subramaniam. 1986. *Chromosome Atlas of Flowering Plants of the Indian Subcontinent* (Vol. 1), 356–360. BSI, Calcutta.

Kumari, S., A. Sharma, R. Salgotra, M. Sharma, and P. Singh. 2021. Morphological studies of Acid Lime (*Citrus aurantifolia* Swingle) genotypes of northwestern regions of India. *Biol. Forum*, *13*: 1020–1032.

Kyndt, T., T.N. Dung, P. Goetghebeur, H.T. Toan, and G. Gheysen. 2010. Analysis of ITS of the rDNA to infer phylogenetic relationship among Vietnamese *Citrus* accessions. *Genet. Resour. Crop Evol.*, *57*: 183–192.

Liang, G., G. Xiong, Q. Guo, Q. He, and X. Li. 2007. AFLP analysis and the taxonomy of Citrus. *Acta Hort.*, *760*: 137–142.

Ling, D., F.G. Gmitter, L.W. Duncan, and S.Y. Xiao. 1994. Inheritance of Citrus nematode resistance and its linkage with RAPD markers in Citrus. *HortSci.*, *29*: 1235–1241.

Liu, Y.Z., and X.X. Deng. 2007. Citrus Breeding and Genetics in China. *The Asian and Austr. J. Pl. Sci. Biotech.*,*1*: 23–28.

Lohwasser, U., A. Dittbrenner, H. Budahn, F. Marthe, and A. Borner. 2010. Taxonomy of plant genetic resources-use of morphological, molecular and phytochemical data in order to verify existing classifications. *Agril. Consp. Scien.*, *75*: 175–178.

Lu, Z., Z. Zhou, and R. Xie. 2011. Molecular phylogeny of the true Citrus fruit trees group (Aurantioideae,
Rutaceae) as inferred from chloroplast DNA sequences. *Agric. Sci. China* 10: 49–57.

Luro, F., F. Laigret, J.M. Bove, and P. Ollitrault. 1992. Application of random amplified polymorphic DNA (RAPD) to Citrus genetics taxonomy. *Proc. Int Citri.*, *1*: 225–228.

Mabberley, D.J. 1998. Australian Citrerae with notes on other Aurantoideae (Rutaceae). *Telopea*, *7*: 333–344.

Mabberley, D.J. 2004. Citrus (Rutaceae): A review of recent advances in etymology, systematics and medical applications. *Blumea*, *49*: 481–498.

Machado, M.A., H.D. Coletta Filho, M.L.P.N. Targon, and J. Pompeu, Jr. 1996. Genetic relationship of Mediterranean madarins *Citrus deliciosa* Tenore using RAPD markers. *Euphyt.*, *92*: 321–326.

Malik, S.K., R. Chaudhury, S. Kumar, O.P. Dhariwal, and D.C. Bhandari. 2012. *Citrus Genetic Resources in India: Present Status and Management*, 184pp. National Bureau of Plant Genetic Resources, New Delhi, India.

Malik, S.K., S. Kumar, I.P. Singh, O.P. Dhariwal, and R. Chaudhury. 2013. Socio-economic importance, domestication trends, and *in situ* conservation of wild Citrus species of Northeast India. *Genet. Resour. Crop Evol.*, *60*: 1655–1671.

Marak, C.K., and M.A. Laskar. 2010. Analysis of phenetic relationship between *Citrus indica* Tanaka and a few commercially important Citrus species by ISSR markers. *Sci. Hort.*, *124*: 345–348.

Marboh, E.S., A.K. Singh, A.K. Dubey, and J. Prakash. 2015. Analysis of genetic variability among citrus (Citrus spp) genotypes using morphological traits. *Ind. J. Agril. Sci.*, *85*: 203–211.

Marin, M.L., and N. Duran-Vila. 1991. Conservation of Citrus Germplasm *in vitro*. *J. Amer. Soc. Hort. Sci.*, *116*: 740–746.

Mariniello, L., M.G. Sommella, A. Cozzolino, P. Di Pierro, D. Ercolini, and R. Porta. 2004. Identification of campania *Citrus limon* by Random Amplification DNA markers. *Food Biotech.*, *18*: 289–297.

Morton, C.M., M. Grant, and S. Blackmore. 2003. Phylogenetic relationships of the Aurantioideae inferred from cpDNA sequenced data. *Am. J. Bot.*, *90*: 1463–1469.

Nair, K.N. and M.P. Nayar. 1997. Rutaceae. *In*: Hajra, P.K., Nair, V.J., and Daniel, P. [Eds.], *Flora of India* (Vol. IV), 259–408. Botanical Survey of India, Calcutta.

Naqvi, N.S. 2001. Morphology and taxonomy of Citrus. *In*: [Singh, S., and Naqvi, S.A.M.H.], *Citrus*, 69–80. International Book Distributing Co. Lucknow, India.

Nicolosi, E., Z.N. Deng, A. Gentil, S. Lamalfa, G. Continella, and E. Tribulato. 2000. Citrus phylogeny and genetic origin of important species as investigated by molecular markers. *Theor. Appl. Genet.*, *100*: 1155–1166.

Nicolosi, E. 2007. Origin and taxonomy. *In*: [Ed. Khan, I.], *Citrus Genetics, Breeding and Biotechnology*. CAB International, UK.

Pang, X., C. Hu, and X.X. Deng. 2003. Phylogenetic relationships among Citrus and its relatives as revealed by SSR markers. *Acta. Genet. Sin.*, *30*: 81–87.

Pang, X.M., C.G. Hu, and X.X. Deng. 2007. Phylogenetic relationship within Citrus and related genera as inferred from AFLP markers. *Genet. Resour. Crop Evol.*, *54*: 429–436.

Pena, L., M. Cervera, J. Juarez, A. Navarro, J.A. Pina, and L. Navarro. 1997. Genetic transformation of lime (*Citrus aurantifolia* Swing.): Factors affecting transformation and regeneration. *Plant Cell Rep.*, *16*: 731–737.

Pérez-Tornero, O., C.I. Tallon, and I. Porras. 2010. An efficient protocol for micropropagation of lemon (*Citrus limon*) from mature nodal segments. *Plant Cell Tiss. Org. Cult.*, *100*: 263–271.

Pessina, D., R. Gentili, G. Barcaccia, S. Nicole, G. Rossi, S. Barbestia, and S. Sgorbatia. 2011. DNA content, morphometric, and molecular marker analyses of *Citrus limonimedica*, *C. limon* and *C. medica* for the determination of their variability and genetic relationships within the genus Citrus. *Sci. Hort.*, *129*: 663–673.

Peterson, J.J., J.T. Dwyer, G.R. Beecher, S.A. Bhagwat, S.E. Gebhardt, D.B. Haytowitz, and J.M. Holden. 2006. Flavanones in oranges, tangerines (mandarins), tangors, and tangelos: A compilation and review of the data from the analytical literature. *J. Food Comp. Anal.*, *19*: S66–S73.

Protopapadakis, E., and Y. Papanikolaus. 1999. Use of four isozymatic systems in lemon and lemon-like citrus cultivars to detect their genetic diversity. *J. Hort. Sci. Biotech.*, *74*: 26–29.

Raghuvanshi, S.S. 1962. Cytological studies in genus Citrus IV. Evolution in genus *Citrus*. *Cytologia*, *27*: 172–188.

Rahman, M.M., N. Nito, and S. Isshiki. 1994. Genetic analysis of phosphoglucoisomerase isozymes in true citrus fruit trees. *Sci. Hort.*, *60*(1–2): 17–22.

Rahman, M.M., and N. Nito. 1994. Phylogenetic relationships among the "true citrus fruit trees" by glutamate oxaloacetate transaminase isozymes analysis. *J. Jpn. Soc. Hort. Sci.*, *62*: 755–760.

Rahman, M.M., N. Nito, and S. Isshiki. 2001. Cultivar identification of 'Yuzu' (Citrus junos Sieb. Ex Tanaka) and related acid citrus by leaf isozymes. *Sci. Hort.*, *87*: 191–198.

Randall, P. Niedz. 1994. Growth of embryogenic sweet orange callus on media varying in the ratio of nitrate to ammonium nitrogen. *Plant Cell Tiss. Org. Cult.*, *39*: 1–5.

Roose, M.L. 1988. Isozymes and DNA restriction fragment length polymorphisms in Citrus breeding and systematics *In*: Goren, R. and K. Mendel (Eds.). Proceedings of 6th international Citrus Congr, Balaban Publishers, Rehovot, Israel, pp 155–165.

Roose, M.L., F.G. Gmitter Jr, R. Lee, K. Hummer, M. Machado, S. Ashmore, X. Deng, G. Ancillo, M.C. Vives, M. Volk (2015) Development of a global conservation strategy for citrus genetic resources. Adv. Hortic. 1065: 75–84.

Russo, M.P., M.L. Rosa, A. Astuto, and G.R. Reupero. 1998. *Adv. Horti. Sci.*, *12*: 85–88.

Sangma, P.M., and C.P. Suresh. 2022. Study on morphological characterization of *Citrus macroptera* Mont collected from Garo hills district Meghalaya, India. *Int. J. Sci. Res. Archive*, *7*: 344–356.

Sawazaki, H.E., G.W. Muller, and L. Sodek. 1997. *Revista Brasileira de Biologia.*, *57*: 337–342.

Scora, R.W. 1975. On the history and origin of Citrus. *Bull. Torrey Bot. Club*, *102*: 369–375.

Scora R.W. 1988. Biochemistry, Taxonomy and evolution of modern cultivated Citrus. R.K. Goren, K. Mendel (Eds.), Proc. 6th Int. Citrus Cong., vol. 1, Margraf Publishers, Weikersheim, Germany (1988), pp. 277–289

Shaaban, E.A., S.K.H. Abd-EL-Aal, N.S. Zaied, and A.A. Rizkalla. 2006. Assessment of genetic variability on Rerence orange accessions using RAPD-DNA markers. *Res. J. Agric. Biol. Sci.*, *2*: 564–570.

Shahsavar. A.R., K. Izadpanah, E. Tafazoli, and B.E. Tabatabaei Sayed. 2007. Characterization of Citrus germplasm including unknown variants by inter-simple sequence repeat (ISSR) markers. *Sci. Hort.*, *112*: 310–314.

Sharma, B.D., D.K. Hore, and S.G. Gupta. 2004. Genetic resources of Citrus of northeastern India and their potential use. *Genet. Resour. Crop Evol.*, *51*: 411–418.

Singh, B. 1981. *Establishment of First Gene Sanctuary in India for Citrus in Garo Hills*. Concept. Pub. Co., New Delhi.

Singh. H., H.S. Rattanpal, G.S. Sidhu, and T.S. Chahal. 2010. Study on Physio-morphological characteristics among Six Rangpur Lime (*Citrus limonia* osbeck.) strains. *J. Tree Sci.*, *29*(1, 2): 48–56.

Singh, I., S. Singh, R. Srivastava, and K. Singh. 2001. Exploration and collection of citrus germplasm from NEH region (Meghalaya) of India. *Ind. J. Plant Genet. Resour.*, *14*(1): 70–73.

Singh, R., and N. Nath. 1969. Practical approach to the classification of Citrus. *In*: *Proc. Intl Citrus Symp.*, *1*: 435–440. (Run on) Champman, H.D. [Ed.].

Soost, R.K., and A.M. Tortes. 1981. Leaf isozymes as genetic markers in Citrus. *In*: *Proc. Int. Soc. Citriculture*, *1*: 7–10.

Spiegel-Roy, P., and E.E. Goldschmidt. 1996. *Biology of Citrus: Biology of Horticultural Crops*. Cambridge University Press, Cambridge.

Sthapit, B., Lamers, H., and Rao, R. 2013. Custodian farmers of agricultural biodiversity: Selected profiles from South and Southeast Asia. Paper presented at Workshop on Custodian Farmers of Agricultural Biodiversity. New Delhi, India. *Bioversity International*.

Strauss, S.H., J. Bonsquet, V.D. Hipkins, and Y.P. Hong. 1992. Biochemical and molecular genetic markers in biosystematic studies of forest trees. *New Forests*, 6: 125–158.

Sudo, M., K. Yasuda, M. Yahata, M. Sato, A. Tominaga, H. Mukai, G. Ma, M. Kato, and H. Kunitake. 2021. Morphological Characteristics, Fruit Qualities and Evaluation of Reproductive Functions in Autotetraploid Satsuma Mandarin (*Citrus unshiu* Marcow.). *Agronomy, 11*: 2441.

Sugawara, K., T. Wakizuka, A. Oowada, T. Moriguchi, and M. Omura. 2002. Histogenic identification by RAPD analysis of leaves and fruit of newly synthesized chimeric Citrus. *J. Amer. Soc. Hort. Sci.*, *127*: 104–107.

Swingle, W.T. 1943. The botany of Citrus and its wild relatives of the orange Sub-family. *In*: Webber, H.J., and Batchelor, L.D. [Eds.], *The Citrus Industry* (Vol. 1),129–474. University of California, USA.

Swingle W.T., and P.C. Reece. 1967. The botany of Citrus and its wild relatives. *In*: Reuther, W., Batchelor, L.D., and Webber, H.J. [Eds.], *The Citrus Industry*, 190–340. University of California Press, Berkley.

Szczepaniak, M., and E. Cieslak. 2011. Genetic and morphological differentiation between *Melica ciliata* L. and *M. transsilvanica* Schur (Poaceae) in Europe reveals the non-presence of M. ciliata in the Polish flora. *Acta Soc. Bot. Poloniae, 80*: 301–313.

Tanaka, T. 1977. Fundamental discussion of citrus classification. *Studia Citrologica, 14*: 1–6.

Tanaka, T. 1954. *Species Problem in Citrus* (*Revisioaurantiacearum, IX*). Japan Soc. Prom. Sci., Veno, Tokyo.

Tanaka, T. 1961. Contribution to the knowledge of Citrus classification. *Reports Citrologia*, 107–114.

Tanksley, S.D., and T.J. Orton. 1983. *Isozymes in Plant Genetics and Breeding,* Part B, 472 pp. Elsevier, Amsterdam.

Te-Chato, S., and S. Nudoung. 1998. Tissue culture of Citrus reticulata Blanco cv. Shogun and gene transformation by Agrobacteria. *Proc. Abstracts IMT-GT UNITE Conf.*, Songkxhla.

Torres, A.M., T.M. Lastovicka, T.E. Williams, and R.K. Soost. 1985. Segregation distortion and linkage of *Citrus* and *Poncirus* isozymes. *J. Hered., 76*: 289–294.

Torres, A.M., R.K. Soost, and U. Diedenhofen. 1978. Leaf isozymes as genetic markers in Citrus. *Am. J. Bot., 65*: 869–881.

Tripoli, E., M. La Guardia, S. Giammanco, D. Di Majo, and M. Giammanco. 2007. Citrus flavonoids: Molecular structure, biological activity, and nutritional properties: A review. *Food Chem., 104*: 466–479.

Tusa, N., J.W. Grosser, and G.G. Gmitter, Jr. 1990. *J. Amer. Soc. Horti. Sci., 115*: 1043–1046.

Uzun, A., O. Gulsen, G. Kafa, and U. Seday. 2009. Field performance and molecular diversification of lemon selections. *Sci. Hort., 120*: 473–478.

Weising, K., H. Nybom, K. Wolff, and G. Kahl. 2005. *DNA Fingerprinting in Plants: Principles, Methods, and Applications* (2nd Edn.). Taylor and Francis Group, Boca Raton, FL.

Withers, L.A., and J.M.M. Engels. 1999. The test tube genebank: A safe alternative to field conservation. *IBPGR Newsl. Asia and the Pacific, 3*: 1–2.

Xi, W., B. Fang, Q. Zhao, B. Jiao, and Z. Zhou. 2014. Flavonoid composition and antioxidant activities of Chinese local pummelo (*Citrus grandis* Osbeck.) varieties. *Food Chem., 161*: 230–238.

Xu, C.J., L. Bao, B. Zhang, Z.M. Bei, X.Y. Ye, S.L. Zhang, and K.S. Chen. 2006. Parentage analysis of huyou (*Citrus changshanensis*) based on internal transcribed spacer sequences. *Plant Breeding, 125*: 519–522.

Yamamoto, M., S. Kobayashi, Y. Nakamuram, and Y. Yamada. 1993. Phylogenic relationship of Citrus revealed by diversity of cytoplasmic genomes. *In*: Hayashi, T., Omura, M., and Scott, N.S. [Eds.], *Technique on Gene Diagnosis and Breeding in Fruit Trees*, 39–46. FIRS Japan.

Yamamoto, M., R. Kouno, T. Nakagawa, T. Usui, T. Kubo, and S. Tominaga. 2011. Isozyme and DNA Analyses of local Citrus germplasm on Amami Islands Japan. *J. Japan. Soc. Hort. Sci., 80*(3): 268–275.

Yan, G., A.R. Ferguson, M.A. McNeilage, and B.G. Murray. 1997. Numerically unreduced (2n) gametes and sexual polyploidization in *Actinidia. Euphyt., 96*: 267–272.

Yasuda, K., M. Yahata, and H. Kunitake. 2016. Phylogeny and Classification of Kumquats (*Fortunella* spp.) Inferred from CMA Karyotype Composition. *The Hort. J., 85*(2): 115–121.

Yin, P., W. Ding, H. Zhang, X. Liu, H. Zhang, J. Zeng, and J. Xu. 2023. Morphological, physiological and molecular characteristics of the seedless 'Hongjiangcheng' sweet orange. *Hort. Pl. J., 9*(3): 437–449.

II. Genetic Diversity and Utilization of Nut Plants

18

Genetic Diversity of Walnut (*Juglans regia* L.) and Its Sustainable Utilization

Rafiq Ahmad Shah,[1,*] *Hamidullah Itoo,*[1] *Parshant Bakshi,*[2] *Amit Jasrotia,*[2] *Khurshid Ahmad Sheikh,*[1] *Ajit Pal Singh*[3] and *Bilal Ahmad Padder*[4]

Introduction

Persian walnut is the only species in the Juglandaceae family typically cultivated for edible nuts (Zaini et al., 2020; Hussain et al., 2021; Pollegioni et al., 2017). The *Juglans* genus has around 21 species, with *Juglans regia* being the most significant in terms of commerce (Mapelli et al., 2013; Arzani et al., 2008). Walnuts were first domesticated in Central Asia and then spread to other temperate areas of the world, such as Europe, Italy, Spain, France, and China (Pollegioni et al., 2014). Around 4.5 million tons of walnuts are produced worldwide each year across 59 different nations (Candemir, 2021). China is the largest producer of walnuts, accounting for over half of the world's walnut production (Cui et al., 2023). The USA and Turkey are also significant producers of walnuts, with these two nations together and China producing 81% of the world's walnuts from an area of 69.22% of the world's walnut acreage (Wang et al., 2022a). Walnuts are nutritionally dense and therefore are considered the "bread of the future" (Jacimovic et al., 2020; Itoo et al., 2023). The edible portion of walnut (40–60 % kernel) is a rich source of fats and fatty acids of which the majority percentage (70%) are polyunsaturated fatty acids such as linoleic (18:2) and linolenic acid (18:3) (Hussain et al., 2021; Goodarzi et al., 2023; Nguyen and Vu, 2023). The other components found are proteins (16–24%), carbohydrates (12–21%), fiber (1.5–2%), minerals (1.7–2%), folic acid, vitamin E, phytosterols, flavonoids, melatonin, and polyphenols (Persic et al., 2018; Behyan et al., 2018). Walnut has the most diverse phenolic profile and highest phenolic content among tree nuts. In recognition of their

[1] Ambri Apple Research Centre, Pahnoo Shopian, Sheri-e-Kashmir University of Agricultural Sciences and Technology of Kashmir -192303, Srinagar (J&K).

[2] Division of Fruit Science, Sher-e-Kashmir University of Agricultural Sciences and Technology of Jammu, Faculty of Horticulture and Forestry, SKUAST-Jammu, Chatha, Jammu, J&K-180009 (J&K).

[3] Division of Fruit Science, Sant Baba Bhag Singh University, Khiala, Jalandhar, India-144030.

[4] Plant Virology and Molecular Plant Pathology Laboratory, Division of Plant Pathology, Sher-e-Kashmir University of Agricultural Sciences and Technology of Kashmir, Shalimar, Kashmir-190 025, Srinagar (J&K).

* Corresponding author: rafiq_masoodi@rediffmail.com

several health benefits, they are incorporated into the diet for a stronger and healthier body. Phenolics are powerful antioxidants with anti-carcinogenic (Fan et al., 2023), anti-inflammatory (Papoutsi et al., 2008), antibacterial (Vieira et al., 2019), anti-aging (Malekjani and Jafari, 2022), anti-fungal (Sytykiewicz et al., 2015), and neuroprotective properties (Orhan et al., 2011). The consumption of walnut kernels has increased in recent years because of their inclusion in common food diets and evidence connecting their consumption in different forms to a wide range of health benefits (Ros, 2010; Zec et al., 2020). This has skyrocketed production in just three decades from one million tons to over three million tons in 2020 (Goodarzi et al., 2023).

Taxonomic Background

In the *Juglans* genus, there are approximately 22 species all with 2n = 32 (Zhang et al., 2020) and their genome size varied from 571–594 Mb (Chen et al., 2019). The different walnut species are taxonomically grouped into four sections based on phylogenetic analysis and fruit morphology. The first three sections are *Rhysocaryon*, which includes black walnuts of America, *Cardiocaryon*, which includes Japanese, Manchurian, and Chinese walnuts, and *Trachycaryon*, which includes butternut of Eastern North America (Aradhya et al., 2007; Britton et al., 2007; Dandekar et al., 2005). The fourth section, *Juglans*, only has a single species called *Juglans regia*, commonly grown in temperate zones. The *Juglans regia* L. and related walnut species nut pictures are presented in Figure 1. All the 22 *Juglans* genus species L. are diploid (2n = 32) with 16 chromosomes and a genome size of approximately (600–6153 Mbp), which is roughly one-fourth of the human genome size (Vahdati et al., 2019; Sideli et al., 2020). Across Central Asia, different forms of walnut exist with distinct morphological and biological characteristics. The seven wild and cultivated forms of walnut are: (1) large fruit cultivars (*Juglans regia* L. var. *macrocarpa* DC); (2) thin-shelled, dessert cultivars (*Juglans regia* L. var. *tenera* DC); (3) Almond shaped cultivars (*Juglans regia* L. var. *bartheriana* DC); (4) cluster type cultivars (*Juglans regia* L. var. *racemose*); (5) Late blooming cultivars (*Juglans regia* L. var. *serotina* DC); (6) hard-shelled cultivars (*Juglans regia* L. var. *dura* DC; and (7) kapa gall producing cultivars (*Juglans regia* L. var. *caulioma mihi Sczepotiev* DC) with each group containing dozens of cultivars. *Juglans regia* L. has a wide distribution from eastern Europe to China, including western Asia, the Himalayan regions of Pakistan, India, Nepal, Bhutan, and more (Khan et al., 2010; Aradhya et al., 2017). The natural range of *J. regia* is quite extensive, spanning from China in the east to Turkey in the west. It also extends from Kazakhstan in the north to the temperate regions of India and Nepal in the south (McGranahan and Leslie, 2012; Martinez-Garcia et al., 2016). Because of the uncertainty surrounding the number and locations of refugia for *J. regia*, it is unclear whether current populations are descended from refugia in southern Europe, the Balkans, the Carpathian mountains, Anatolia, the Zagros mountains, or even China (Martinez-Garcia et al., 2016; Aradhya et al., 2017).

The determination of the native geographic range of walnuts has been a challenge due to its long history of cultivation and widespread use across Eurasia, from China to Western Europe (Pollegioni et al., 2017). As a result, the origin and evolutionary history of *J. regia* is a topic of debate among scholars. Many researchers believe Central Asia, which comprises a large and diverse region, is the center of origin and diversity (Dandekar et al., 2005; Mamadjanov, 2005). In Asia, the walnut survived and grew in isolated patches after the coldest Last Glacial period from western China through Central Asia to the Caucasus (Aradhya et al., 2017). The fragmentation and differentiation of walnut populations in these isolated patches were promoted by barriers to gene flow and desertification in the Holocene period (Pollegioni et al., 2014, 2015). The Asian walnut landscape was modified by the exchange and trading of walnut through green corridors such as the Silk Road, which facilitated the walnut extension across and beyond Asia (Vahdati, 2014). Recent phylogenomic studies have revealed that *J. regia* and its landrace *J. sigillata* have a hybrid origin (Zhang et al., 2019). *J. sigillata* is a sympatric species indigenous to southwestern China (Wang et al., 2015; Zhao et al., 2018) and is found to freely hybridize with the closely related species *J. regia*.

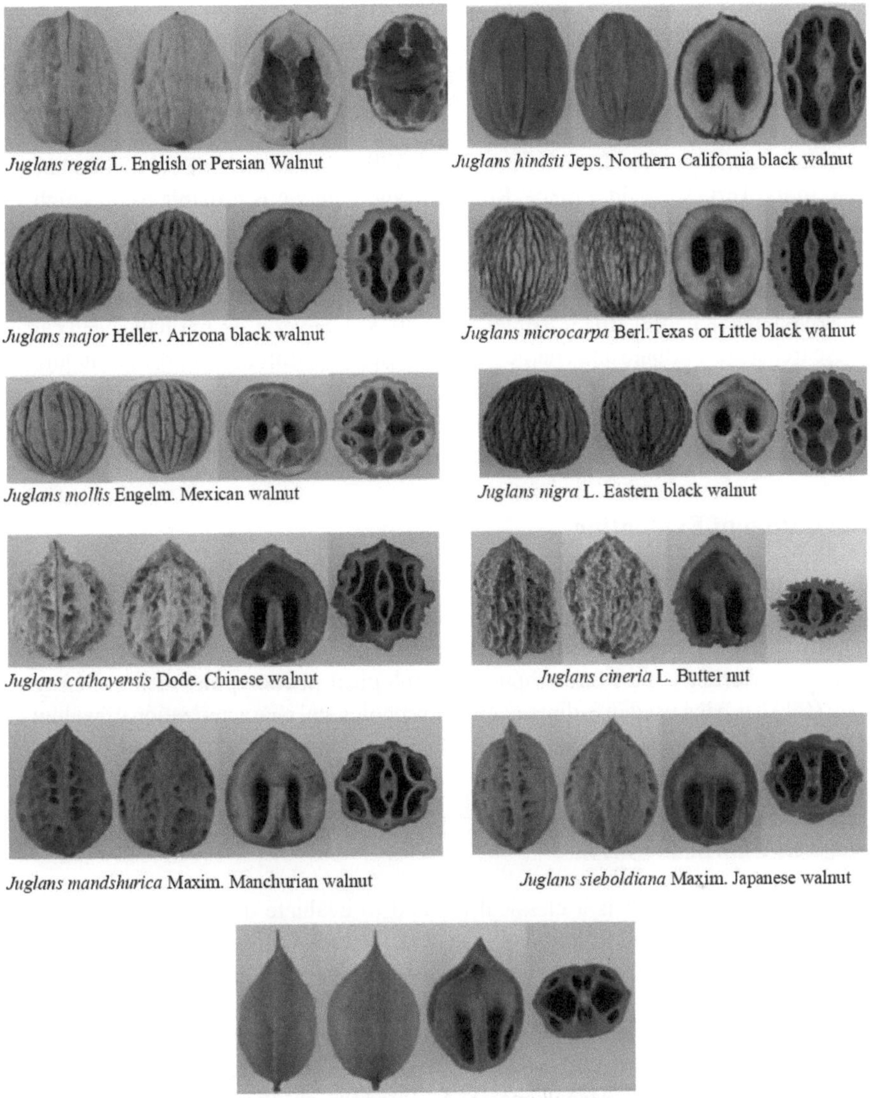

Figure 1. Variability in walnut depicting the phenotypic characteristics of different species of walnut.
Source: Different species of walnut available in INRA walnut germplasm (Bemard et al., 2018).

In contrast to the Asian walnut, there is no molecular evidence that shows that the walnut survived in the Last Glacial Maximum in Europe despite the presence of fossil records (Fornari et al., 2001). The Isozyme analysis revealed the genetic differentiation within the European population was similar to the Asiatic population but between population stands was less than Asian stands. This shows there was probably a frequent transfer of walnut selections in different regions across Europe such as Italy (Pollegioni et al., 2011b). This affected the genetic structure of *J. regia*, reducing genetic diversity by selection and increasing homogeneity by dispersal.

Collection and Assessment of Germplasm/Genetic Resources

Walnut genetic diversity is a key material for any crop breeding program (Vahdati et al., 2019). This diversity is found in both wild and cultivated populations, especially when farmers haven't strongly selected any genotype for specific traits. However, selecting walnuts for breeding is challenging for a couple of reasons. Firstly, its allogamous nature makes it difficult to control pollination (Topcu, 2021). Secondly, the long juvenile phase means that it takes a long time to evaluate the genotype, unlike other fruit crops (Laurens et al., 2003). Additionally, the agamic propagation of walnuts through grafting is also quite challenging (Poggetti et al., 2017). These problems have helped in maintaining the walnut genetic diversity across its cultivation areas. In Eurasian countries such as Iran, China, India, Turkey, Spain, France, Romania, and others it is possible to find high variability in nut and physiological traits. A large population of wild plants and wide variability in traits can be observed in these areas. The most desirable characteristics for diversity analysis and characterization are nut weight, kernel weight, late leafing, kernel color, shell thickness, and resistance to diseases and pests (Soveili and Khadivi, 2023; Rezaei et al., 2018a). Several studies have been carried out by different researchers to evaluate and characterize the walnut germplasm for different purposes. Most of the walnut plantations have not been subjected to intensive plant breeding and many of these tree populations exhibit a rich genetic diversity, which can only be fully exploited once it is accurately recognized and assessed.

Characterization of Evaluation

Morphological markers are extensively used for the evaluation and characterization of walnut germplasm (Shah et al., 2023a). In addition to these markers, cytological and biochemical markers have been utilized. However, with the availability of better molecular tools their use in characterization has been limited. Recently high-throughput morphological phenotyping such as X-ray computed tomography (CT) has been used in 3-dimensional morphological characterization of walnut genotypes (Bernard et al., 2020c). The sequencing of the Chandler walnut cultivar genome has opened a new era of rapid characterization of walnut diversity through SNP genotyping.

Morphological Characterization

Morphological characterization is a classical method to evaluate diversity based on tree growth, phenological, inflorescence, and fruit traits (Shah et al., 2023a). Different studies on walnut diversity utilized a combination of traits or a single trait to evaluate the germplasm. Evaluation of morphological characteristics is very useful for the direct identification of superior genotypes, collection, and conservation of germplasm, and utilization of genotypes as parents for breeding programs (Fatahi et al., 2010; Bernard et al., 2018b). However, this traditional method requires skilled professionals, which can be costly and prone to inaccuracies. Additionally, this approach is time-consuming and laborious, and its results are affected by the environment. Recently Bernard et al. (2020c) developed a robust method that gives accurate and complete morphological measurements of walnut using X-ray computed tomography (CT) and can help in quantifying traits that were previously difficult to measure such as rugosity, sphericity, and shape indexes. In addition, the filling ratio without cracking opens the walnut. Although the morphological characterization may produce ambiguous results, they are still essential for the management of germplasm collections

Different morphological traits of plants and nuts' morphometric properties have been utilized by different researchers to select superior plants from different regions (Mikdat et al., 2010; Rezaei et al., 2018b; Khadivi et al., 2019a; Shah et al., 2021; Poggetti et al., 2017; Akca et al., 2015; KilicogLu and Yasar, 2018; Yavuz and Pirlak, 2023; Manthos et al., 2023; Soveili and Khadivi, 2023; Zhang et al., 2023, Sarikhani et al., 2023). High variability in nut traits (nut shape, nut size, shell thickness, kernel color, kernel fill, kernel color, and kernel percentage) has been reported in walnut trees in different regions (Cosmulescu and Botu, 2012; Diaz and Fernandez-Lopez, 2005). The nut phenotypes are the most stable characters to distinguish different genotypes (Krska et al., 2009) and have been approved by the International Union for the Protection of New Varieties of Plants (UPOV) for examining the distinctiveness, uniformity, and stability of the cultivar (Shah et al., 2023a). However, these are not adequate to distinguish all the cultivars particularly those that have been selected from the intrapopulation which showed sometimes similar phenotypic traits. Therefore, in addition to morphological characters we need an unambiguous characterization such as genetic fingerprinting to supplement the morphological characterization (Chen et al., 2014).

Cytological Characterization

In addition to examining the morphological features, scientists have also studied the genetic diversity of plants through cytology, such as cell karyotype. The variation in chromosome and genome size can also be an important tool in studying genetic diversity. Flow cytometry was used to determine the ploidy level of walnut samples in *in vitro* conditions (Grouh et al., 2011). Cell karyotyping in haploid root tip cells revealed that among the 16 chromosomes, 10 of them were metacentric, 5 metacentric, and 1 sub-telocentric (Vahdati et al., 2012). These chromosomal studies are very important for fruit genetics and breeding and are used to study taxonomy, phylogeny, evolution, hybridization, cytogenetics, and genetic variation (Nathewet et al., 2009). The evaluation of the genome size of superior genotypes in Iran through morpho-flow cytometry was very helpful in evaluating the genetic diversity and in predicting the nut weight and nut size of walnut trees (Khorami et al., 2018). Cytological studies have been utilized to find out the production of apomictic seedlings in walnut genotypes (Solhjoo et al., 2023). However, with the development of more advanced molecular techniques, the use of these methods for walnut genotyping has decreased. Cytological diversity analysis has many limitations such as the limited number and low resolution of markers. Additionally, these methods could be sensitive to environmental factors, which could further complicate the true diversity analysis.

Biochemical Characterization

Isozymes and allozymes have been used as biochemical markers to characterize walnut genotypes. Nonetheless, iso-enzyme analysis is considered a quick way to evaluate diversity and only requires a small sample size (Fornari et al., 2001). However, their use is limited due to their low resolution of diversity and small number of markers, as well as their susceptibility to environmental factors. The effect of the environment could vanish if the ecological conditions of cultivation are the same. In biochemical classification studies, the primary biochemical parameters (markers) pertaining to walnut cultivars that are assessed include oil, fatty acid composition, proteins, carbohydrates, and lipids. Other factors that are taken into consideration during the classification process include total phenolic content, total antioxidant activity, and minerals.

In recent years there has been increased focus on studying substances with high biochemical content and functional properties that benefit human health. Walnuts are one of the fruits that fall into this category, and it's being evaluated to select and identify the genotypes with health-beneficial compounds. Walnut kernels have numerous health benefits as they are a rich source of antioxidants, antibacterial and anticancer compounds, and various oxidative compounds such as vitamin E, carotenoids, ellagic acid, melatonin, and polyphenols (Shah et al., 2022; Rahimipanah et al., 2010, Dogan and Akgul, 2005). Walnuts are a highly nutritious food that contains about 60% oil, which varies with the type of cultivar, location, and irrigation rate of cultivation (Rebufa et al., 2022). Additionally, walnuts are an excellent source of protein, making them an essential fruit for human nutrition. As a result, walnuts are classified as a strategic species for human nutrition and are included in the Food and Agriculture Organisation of the United Nations (FAO) list of priority plants (Gandev, 2007). Walnut oil comprises various triacylglycerols such as monoacylglycerols, diacylglycerols, free fatty acids, sterols, sterol esters, and phosphates, and it is known to have one of the highest amounts of polyunsaturated fatty acids (up to 78%) among vegetable oils (Miraliakbari, 2005; Masoodi et al., 2022). Furthermore, walnuts (*Juglans regia* L.) and other species of *Juglans* contain major fatty acids such as oleic, linoleic (omega-6), and linolenic acids (omega-3), which are essential dietary fatty acids (Fatima et al., 2018; Zec et al., 2020; Beyhan et al., 2017; Wang et al., 2020) (Table 1). In addition, walnuts are a great source of dietary minerals such as K, P, Mg, Fe, Ca, Na, Zn, and Cu, flavonoids, vitamins (such as Vitamin A and E), crude fiber, and pectin (Cosmulescu et al., 2009; Aryapak and Ziarati, 2014; Wang et al., 2021; Asgari et al., 2020).

Table 1. Comparative biochemical properties (Fatty acid composition) of different *Juglans species* L.

***Juglans sigillata* L. (Iron walnut)**			***Juglans regia* (Common walnut)**	
Fatty acids	Range (%)	Average (%)	Range (%)	Average (%)
Palmitic acid (C16:0)	5.09–8.73	6.63	4.67–7.33	5.94
Palmitoleic acid (C16:1)	0.00–0.22	0.07	0.00–0.02	0.04
Stearic acid (C18:0)	1.16–3.36	2.27	1.92–3.81	2.70
Oleic acid (C18:1)	12.05–38.68	22.08	13.74–36.30	19.89
Linoleic acid (C18:2)	42.72–73.06	59.45	47.82–67.48	61.30
Linolenic acid (C18:3)	6.35–25.92	9.20	6.34–13.48	9.80
Arachidic acid (C20:0)	0.00–0.08	0.01	0.00–0.10	0.02
Eicosenoic acid (C20:1)	0.00–0.21	0.09	0.00–0.21	0.14
SFA	7.13–11.27	8.91	7.47–9.56	8.67
MUFA	12.21–38.89	22.25	13.92–36.55	20.07
PUFA	53.98–79.66	68.65	55.27–77.43	71.11

SFA: saturated fatty acid, MUFA: monounsaturated fatty acid, PUFA: polyunsaturated fatty acids. Adopted from Wang et al. (2020).

Molecular Characterization

As stated earlier morphological and biochemical markers have limitations in terms of their usefulness and are vulnerable to environmental influences (Shah et al., 2023a). As a result, DNA-based markers have been developed and have become increasingly significant in revealing the true degree of walnut diversity. Unlike previous markers, molecular markers are stable, detectable in all tissues regardless of growth, development, and differentiation, and are not affected by changes in environmental conditions, and pleiotropic and epistatic effects of genes or traits (Mondini et al., 2009). Both dominant markers, random amplified polymorphic DNA (RAPD), amplified fragment length polymorphism (AFLP) and co-dominant markers, restriction fragment length polymorphism (RFLP), SSR and SNP, which may be PCR-based (SSR) and non-PCR-based (RFLP) have been employed in the molecular characterization of walnut (Ahmed et al., 2012; Bernard et al., 2018a, 2020b; Shah et al., 2020; Bayazit et al., 2007).

Markers are useful tools for distinguishing between different genotypes at the genetic level. This is because different genotypes may have variations in amplification band size, which could be a result of nucleotide realignment or mutations in the genome sequence (Collard et al., 2005). With the recent sequencing of the genome of walnut Cv. 'Chandler', researchers can now easily identify distinct genotypes by analyzing differences in the genome sequence alignment (Martinez-García et al., 2016; Stevens et al., 2018). This has opened up new avenues for research and has made the process of identifying and characterizing genotypes much more efficient. Molecular markers have proven to be incredibly useful in providing a wealth of information about genome structure, organization, and evolution (Cervera et al., 2000; Li et al. 2011). From mapping traits to constructing linkage maps, improving germplasm, and identifying parents, to population genetics, molecular markers can do it all (Crespel et al., 2002; Pearl et al., 2003). They can even help identify and position genes associated with important traits, which aid in gene cloning and marker-assisted selection (Thompson

2023; Orman et al., 2020). The use of cutting-edge genomic techniques like genomic selection and association mapping, and the availability of SNP chips has accelerated the process of walnut molecular breeding. Overall, the use of molecular markers has greatly advanced our understanding of genetics and genomics and continues to play a critical role in many areas of biological research.

Analyzing the genome's microsatellite loci is one of the most widely used techniques for determining the genetic diversity of walnuts. SSR markers were first created specifically for *Juglans nigra* as well as *Juglans regia* L., and are now frequently utilized for the investigation of polymorphism in walnuts. Consequently, because of the high degree of cross-reproducibility within the *Juglans* genus, these markers were successfully utilized to investigate the interspecific diversity of *J. regia*. The most comprehensive studies on the genetic diversity of walnut collections (217 walnut accessions and 36 accessions of other *Juglans* species) using 13 SSR markers was conducted by a team of researchers at the INRA (National Research Institute for Agriculture) Research Center (Bernard et al., 2018a). The presence of two main groups, which largely corresponded to the ecological and geographical origin of the varieties showed the importance of SSRs in revealing the population structure and differentiation of walnut. In order to assess the effectiveness of using 13 SSR markers in determining the genetic structure of INRA collections a high throughput 700K SNP genotyping array technology was used (Bernard et al., 2019). Surprisingly, the analysis showed that both approaches had a similar level of information content and were equally effective in assessing the genetic structure of collections. Another comparable study of the use of SSRs in 189 varieties and breeding forms of gene bank collections from 14 nations of the world. Two major groupings were identified by the study, comprising accessions from Greece and the Middle East, as well as from Europe and North Africa. It's intriguing to learn that a variety of studies on gene pool collections and local populations around the globe have made use of microsatellites such as in Europe by (Pollegioni et al., 2011a; Ebrahimi et al., 2017; Vischi et al., 2017), East Asia (Gunn et al., 2010; Wang et al., 2022b; Zhou et al., 2021), Central and South Asia (Shah et al. 2020; Itoo et al., 2023; Pollegioni et al., 2014, 2015; Magige et al., 2022), Middle East (Ebrahimi et al., 2011; Shamlu et al., 2018; Orhan et al., 2020) and North America (Aradhya et al., 2010) (Table 2). Although SSRs are most commonly used for the diversity analysis in walnut with the development and validation of high-throughput genomic tools such as high density 700K SNP genotyping array (Marrano et al., 2019a), SNPs are more appropriate for revealing the population structure (Arab et al., 2019), developing core collections and genome-wide association studies in walnut (Bernard et al., 2020a; Marrano et al., 2019b).

Utilization of Germplasm

In plant breeding operations, breeders can introduce new cultivars by leveraging the genetic diversity of plants (Govindaraj et al., 2015a). However, the breeding cycle for many plants, especially fruit trees, can be quite lengthy. This poses a significant challenge for breeders, particularly in the case of nut crops (van Nocker and Gardiner, 2014). One way to tackle this challenge is by introducing superior genotypes (Roor et al., 2017) that can help shorten the breeding cycle. Another effective strategy is to screen fruit tree germplasm in origin and diversity centers, which can help breeders identify new cultivars (Arzani et al., 2008). The assessment of germplasm is vital for future crop improvement programs. By examining these resources for various characteristics, we can gather useful information to identify and select beneficial genetic resources (Akca et al., 2020; Pop et al., 2013; Shah et al., 2023b) such as identifying spring frost-resistant genotypes (Khadivi et al., 2019b) and or selection of genotypes with superior traits (Panahi et al., 2022; Bartha et al., 2023). Morphological studies have been carried out in different walnut diversity areas to identify the group of individuals or accessions with valuable traits in the first stage and subsequently evaluated at the DNA level through microsatellite markers to evaluate the heterogeneity of the group (Karimi et al., 2014;, Davoodi et al., 2021).

Table 2. Overview SSR of molecular diversity studies of walnut (*Juglans regia* L.) genotypes/cultivars/populations and their mean genetic diversity parameters

Areas of Study	Sample Size of Genotypes/ Populations	Number of Markers Screened and (Name)	Mean Genetic Diversity Parameters (Range)	Main Application	Reference and Year
Italy	456 walnut trees	10 (WGA89,1,202,4,276,118,69,9,321, 331,456)	TNA: 62, Size range: 159-299 Na: 6.2 (3-14), Ne: 2.996 (1.805-4.14) He:0.6444 (0.446-0.759), Ho:0.597 (0.423-0.758) PIC:0.573 (0.382-0.708)	Genetic diversity and its impact by humans	Pollegion et al. (2011)
Italy (Asian range)	Seven phytogeographic regions (39 Persian walnut populations)	14 (WGA1,4,9, 27,32,69, 72, 79,89, 118, 202,276,321, and 331)	Na: (3-6.857), Ne: (2.18-3.941) Ho: (0.464-0.661) He: (0.439-0.699)	Genetic diversity and relationship determination, gene flow	Pollegion et al. (2014)
Italy (Eatern Italian Alps)	two hundred and fifteen wild accessions	20 SSR (WGA1,4,9,27,69,71,72,89,118,2 25, 321,331, 349, 376, Contig40,156,642,721, 1528,1681,1692.	Na: 4 (2-9), He:0.516 (0.059-0.745), Ho:0.462 (0.051-0.712) PIC:0.438 (0.058-0.698)	Genetic diversity and relationship determination	Vischi et al. (2017)
Romania (Romanian National Collection of Walnut)	20 important Romanian and international walnut cultivars	7(WGA1,69,89,118, 276,332, 376)	TNA: 47 bp size range: 158-253. Na: 6.71 (3-13), Ne: 3.66 (2.35-6.34), He:0.718 (0.590-0.86), Ho:0.650 (0.50-0.75) PIC:0.656 (0.512-0.82)	Genetic diversity and relationship determination	Pop et al. (2013)
Iran (Kerman province)	66 genotypes of six populations	17(WGA001,004,009,027,032,069, 071, 089, 118, 202, 225,276,321, 331,332, 349, 376,	TNA: 147, bp size range: 175-330 Na: 8.65 (4-11), Ne: 5.16 (2.28-7.98), He:0.79 (0.65-0.87), Ho:0.23 (0.00-0.85) PIC:0.72 (0.56 -0.82) I:1.79 (1.18-2.22)	Genetic diversity and gene flow	Mohsenipoor et al. (2010)

Iran (Hamedan province)	105 individuals from seven populations	11(WGA1,9,89,202,225,276,321,332,349, 32, 27)	TNA: 63, Size range: 120-282bp Na: 5.73 (3-11), Ne: 2.92 (2.45-6.21), He:0.707 (0.598-0.848), Ho:0.683 (0.596-0.745) PIC:0.685 (0.495 -0.852)	Genetic diversity and relationship determination	Karimi et al. (2010)
Iran (Fars,Chahar-Mahal e (Shahrekord) and Tehran (Karaj), Azarbayejane-sharghi (Mianeh) provinces	31 superior genotypes and 4 foreign cultivars	9 (WGA1,9,27,32,71,69,202,276,376)	TNA: 39, Size range: 120-275 Na (39), Ne (0.80-6.63), DP (0.77 to 0.65), He (0.57-0.93), Ho (0.48-0.86)	Genetic diversity and relationship determination	Ebrahimi et al. (2011)
Iran (Urmia, Iran)	16 accessions and five cultivars collected from (North West Iran)	9(WGA1,5,32,71,89,118,202, 276, 27)	TNA: 34 and Size range: 160-290 Na: 4.25 (34), Ne: 2.92 (1.89-4.06), PIC:0.64 (0.49 -0.75), He:0.63 (0.47-0.75), Ho:0.62 (0.23-0.95)	Genetic diversity and Identity of accessions	Mahmoodi et al. (2013)
Iran (northern and western regions of Iran)	Samples from 12 populations of walnut	10(WGA1,9,3,7,8,2,2, 3, 3, 3)	TNA: 83 bp size range: 160-275 Na: 8.3 (6-11), Ne: 4.94 (3.35-7.54), He:0.788 (0.669-0.870), Ho:0.655 (0.413-0.864)	Genetic diversity estimation	Karimi et al. (2014)
Iran (Kerman province in southeastern Iran)	six walnut populations	17(WGA1,4,9,69,89,118,202,225,276,32 1,331, 332,349,376,27,32)	TNA: 147, bp size range: 175-330 Na: 8.65 (4-11), Ne: 5.16 (2.28-7.98), PIC:0.72 (0.56 -0.82), Ho:0.23 (0.00-0.85), He:0.79 (0.65-0.87)	Genetic diversity and gene flow esti mation.	Vahdati et al. (2015)

Table 2. contd. ...

Table 2. contd. ...

Areas of Study	Sample Size of Genotypes/ Populations	Number of Markers Screened and (Name)	Mean Genetic Diversity Parameters (Range)	Main Application	Reference and Year
Karaj, Iran	93 precocious seedlings and 10 as control	9(WGA1,9,27,32,89,118,202,276,321)	TNA: Na: 6.89 (4-11), Ne: 3.56 (1.73-5.51), PIC:0.65 (0.43 -0.82), Ho:0.61 (0.34-0.74), He:0.69 (0.42-0.83)	Genetic diversity and relationship determination	Ebrahimi et al. (2017)
Iran (Azadshahr province)	39 superior individuals	10(WGA001,005,009,032,054,069,071,089, 202, 276)	Na: (7-9), Ne (2-9.55), I: (0.85 to 2.35), He: (0.5-0.9), Ho: (0.54-0.93)	Genetic diversity and relationship determination and gene flow.	Shamlu et al. (2018)
Iran	39 individual genotypes	17(WGA1,4,9,27,32,69,71,89,118,202,225, 276,321,331,332,	Na: 3.59(2-4), Ne:2.63 (1.19-3.99), Ho:0.56 (0.18-0.80), PIC: 0.29(0.08-0.43)	Genetic diversity and relationship determination	Nickravesh et al. (2021)
India (North-western Himalayas of Kashmir)	82 walnut genotypes	13(WGA1,4,5,9,27,32,69,71,89,118,202, 276, 321).	TNA: 26, Size range: 200-1200bp Na: 2(1-5) PIC:0.39(0.21-0.49) RP:1.34(.51-2.07) MI:0.69 (0.21-0.49)	Genetic diversity and relationship determination	Ahmed et al. (2012)
India(North-western Himalayas of Kashmir)	96 walnut genotypes	19(WGA1,9,27,32,42,69,71,72,76,79,89,11 8, 202, 225, 276,321, 331,332,349)	TNA:29, bp size range: 200-1200 Na: 1.52(1-3) PB:1.36(1-3) PIC:0.168(0.00-0.44) RP:1.91(1.04-2.38) MI: 0.244(0.00-0.88)	Genetic diversity and relationship determination	Shah et al. (2016)
India (North-western Himalayas of Kashmir)	84 walnut genotypes	25 (WGA1,42,376,71,70,276,89,72,79,76 ,202, 4,9,69,11,5,118,45,331,349,321,32,332,27 ,225)	Na:2.34 (2-3) PIC:0.54 (0.36-0.86)	Genetic diversity and relationship determination	Shah et al. (2020)

France (INRA walnut germplasm collection)	253 accessions	15(WGA001,004,009,027,069,072,202,276, 349,376,JR0160,1739,1817,6160, 6439)	TNA:116, Among 217 *J.regia* accessions Na: 8.92 (2-17), Ne: 2.84 (1.18-4.61), PIC:0.52 (0.15 -0.75), Ho:0.47 (0.10-0.68), He:0.56 (0.15-0.78)	Genetic diversity and relationship determination	Bernard et al. (2018)
China	286 (Nine populations)	8(WGA01,42,70,71,72,76,79,89)	TNA:73, Size rage: 153-306 bp Na: 4.60 (3.25-6), Ne: 2.39 (1.75-3.35), Ho:0.414 (0.281-0.492), He:0.525 (0.401-0.68)	Genetic diversity and relationship determination	Wang et al. (2008)
China (foothills of Khawa Karpo, NW Yunnan)	220 walnut trees (159 *J. Regia* and 61 *J. Sigillata*)	14(A089F, A09F, A237N, A332N, A242N, A276N, A276N,A321F,A338N,A004F,A118H, A210N,A202H,A318F,A331N)	For 159 *J. Regia* samples Na: 7.50 (4-15), Ne: 2.943 (1.51-4.088), Ho:0.5684 (0.275-0.8592), He:0.6241 (0.3425-0.7721) I: 1.266 (0.670-1.66)	Genetic diversity estimation among species and human impact and conser vation	Gunn et al. (2010)
China	40 accessions	13(BFU-Jr19,29,49,67,82,161,179,230,286, 319, BJF-AS170,192)	TNA:104 Na: 8.00 (5-11), Ne: 4.887 (2.415-6.751), Ho:0.446 (0.225-0.750), He:0.769 (0.586-0.870) I: 0.742 (0.689-0.834)	Genetic diversity and SSR development	Zhang et al. (2013)
China	35 cultivars	12(WJR007,031,061,069,073,100,115,142, 202,265,279,281)	TNA: 113, Size range: 101-294bp Na: 9.4 (6-16), Ne: 4.67 (2.25-8.39), Ho:0.62 (0.31-0.75), He:0.77 (0.56-0.89) I: 1.73 (1.38-2.39) PIC: 0.73 (0.53-0.87)	SSR fingerprinting Of cultivars based On nut phenotypes	Chen et al. (2014)

Table 2. contd. ...

Table 2. contd. ...

Areas of Study	Sample Size of Genotypes/ Populations	Number of Markers Screened and (Name)	Mean Genetic Diversity Parameters (Range)	Main Application	Reference and Year
China (Tibet)	86 genotypes	12(WGA 1,42,71,72,76,89,4,376,321,276 ,225)	TNA:119, Na: 9.92 (5-13), Ne: 3.34 (1.27-7.21), Ho:0.539 (0.186-0.9302), He:0.588 (0.2139-0.8662)	Genetic diversity and relationship	Wang et al. (2015)
China	144 individuals, from 13 geographically distinct regions in China	10(JH84548,89978, JM61666, 68820, 78331, 5969, JC8125, JR4616,6439, and 3773)	Na: 3.4 (2-6), Ne: 3.4 (1.5-7.9), Ho:0.273 (0.050 -0.577), He:0.381 (0.049-0.624), He:0.337 (0.048-0.542)	Genetic diversity and relationship	Han et al. (2016)
Yunnan, Guizhou, and Sichuan province of southwestern China	15 populations from *J. regia* (n = 190)	25 EST SSR(JC8125,7329,5411,2995, JR4964, 4616,3773,1165,6638,6439,3434,6160,18 17, 3147,JH2096,604484548,91908,86514,42 753, JM5969,61666,78331,68820)	within J. regia population TNA: 180, bp size range: 133-478 Na: 2.261 (1.16-3.06), Ne: 1.674 (1.053-2.139), I:0.518 (0.111 -0.735), Ho:2.291 (0.124-0.467), He:0.306 (0.074-0.419)	Genetic diversity and gene flow of closely related walnut species	Yuan et al. (2018)
China	31 walnut populations	22 (JC8125, 7329, 2995,5411, JM61666,5969,68820,783 31,JH89978,84548, 42753, 86514,91908,2096,6044, JR4616,3773,4964,6439,1165,6160,3147, and 1817)	TNA:95 Na: 6.79 (2-19) He:0.30 (0.08-0.50) Ho:0.23 (0.13-0.40)	Genetic diversity and population studies in China	Feng et al. (2018)
China(Yunnan Province)	14 populations comprising 1122 individuals	20SSR	Na: 17.5(7-27) PIC:0.57(0.26-0.78) RP:1.19(0.55-1.86) I: 1.19(0.55-1.86)	Genetic diversity and relationship	Chen et al. (2020)
China (Qinling mountains	239 seeds and 18 mother plants	12 SSR (WGA05,321,27,32,01,118,04,71, 376,202,276,09)	TNA:239, Size range: 122-294 Na: 16(7-34) Ne: 3.41 (2.15-4.85) PIC:0.774(0.545-0.953) Ar: 6.85 (4.00-9.06) Ho:0.803 (0.548- 0.927) He:0.796 (0.563- 0.957)	Genetic diversity and genetic contributions	Zhou et al. (2020)

Kyrgyzstan	11 natural populations	16(eight nSSR and eight EST SSR)	Na: 3.24(2.94-3.44) Ho:0.538(0.470-0.573) He:0.524(0.435-0.571)	Genetic diversity evaluation and conservation management	Torokeldiev et al. (2018)
Kyrgyzstan	(44 populations, 581 trees)	7(WGA001,004,009,027,079, 202, 276)	Na: 4.11 (5-24), Ho: 0.589 He: 0.560	Combined microsatellite data with metric data	Roor et al. (2017)
Spain	57 cultivars	19 (WGA1,4,5,9,32,54,69,72,79,89,118,2 02, 225,276,321,331,332,349,376)	TNA: 97, bp size range: 105-295 Na: 5 (3-10) Ne: 2.561 (1.256-3.850) PD:0.74 (0.36 -0.90) He:0.578 (0.431-0.740) Ho:0.517 (0.228-0.737)	Genetic diversity and relationship determination	Ruiz-Garcia et al. (2011)
Turkey	59walnut genotypes and cultivars	16(WGA1,4,5,9,69,71,89,118,202,225,276, 321,331,332,349,376)	TNA: 118, Size range: NA Na: 7.4(2-17), PIC:0.841 RP:9.84,	Genetic diversity and relationship determination	Dogan et al. (2014)
Turkey (Central Anatolia region)	91 (74 accessions, 8 commercial cultivars and 9 international cultivars)	45(JRHR214458, CUJRD462, JRHR215674, CUJRA202, CUJRA218, JRHR227254, JRHR219728, JRHR226941, CUJRA212, JRHR223323, JRHR215721, JRHR218769, CUJRB321, JRHR231293, JRHR212270 and others	TNA:390,bp size range: 133-330. Na: 9 (3-19) Ne: 4.06 (1.89-7.92) PIC:0.68 (0.42 -0.86) He:0.73 (0.87-0.47) Ho:0.65 (0.34-0.96) PIC:0.68(0.42-0.86)	Genetic diversity estimation	Guney et al. (2021)
Turkey	32 local genotypes and two 2 cultivars	21(WGA1,4,5,9,27,32,54,69,71,72,79,89 ,118, 202,225,276,321,331,332,349,376)	TNA: 135 Na: 6.43 (3-11) He:0.62 (0.41-0.80) Ho:0.60 (0.39-0.80) PIC:0.68 (0.54 -0.83)	Genetic diversity and its conservation	Orhan et al. (2020)

Table 2. contd. ...

Areas of Study	Sample Size of Genotypes/ Populations	Number of Markers Screened and (Name)	Mean Genetic Diversity Parameters (Range)	Main Application	Reference and Year
Russia (black sea basin countries)	62 samples	11(WGA001, 005, 009, 054,069,089, 202,276,321, 349,376)	TNA: 104, bp size range 105-283 Na: 9.45 (6-16) Ne: 4.46 (2.73-9.50) He:0.67 (0.44-0.86) Ho:0.75 (0.633-0.895) I:1.60 (1.18-2.44)	Genetic diversity and relationship determination	Balapanov et al. (2019)
USA	114 genotypes Midwestern USA (n = 34), Hungary (n = 30), Slovakia (n = 28), and Italy (n = 22)	10 (WGA1, WGA9, WGA27, WGA32, WGA69WGA89, WGA118, WGA202, WGA276, WGA 321.	TNA: 74 alleles, bp size: 120-295 Na: 7.4 (6-13) Ne: 4.52 (2.50-6.70) He:0.75 (0.60-0.85) Ho:0.61 (0.43-0.81) I:1.60 (1.06-2.02)	Genetic diversity and relationship determination	Ebrahimi et al. (2017)
USA	189 genotypes, from 25 different regions in 14 countries on three continents	10 (WGA1, WGA9, WGA27, WGA32, WGA69, WGA89, WGA118, WGA202, WGA276, WGA321.	Na: 11.5 (3-25) Ne: 4.2 (1.9-7.6) PIC:0.69 (0.37 -0.86) He:0.73 (0.48-0.87) Ho:0.62 (0.43-0.77) I:1.59 (0.70-2.34)	Genetic characterization and relationship	Ebrahimi et al. (2016)
USA	48 accessions (47 Cultivars and Paradox)	14(WGA1,4,9,69,89,118,202,225,276,321, 331,332,349,376)	bp size range:143-277 Na: 5.21 (3-8) Ho:(0.41-0.82)	Genetic characterization and cultivar identification	Dangl et al. (2005)
USA	Nine species and five hybrids	10 (WGA4,33,82,90,142,147,148,204,221, 256)	Na:4.43, He:0.610, Ho: 0.401		Ross-Davis and Woeste (2008)
Hungary (Budapest)	Hungarian and introduced walnut genotypes	12 (WGA1, WGA9, WGA27, WGA72, WGA89, WGA118, WGA202, WGA276, WGA321, WGA331, JR1817, JR6160)	TNa: 50 Na: 3.5 (2-6) Ne: 2.30 (1.28-3.84) PIC:0.69 (0.37 -0.86) He:0.515 (0.219-0.740) Ho:0.514 (0.250-0.833) I:0.918 (0.37-1.45)	Genetic background of Hungarian Varieties	Bujdoso and Cseke (2021)

TNa: Total number of alleles **Na**: Number of different alleles, **Ne**: Number of effective alleles, **I**: Shannon's information Index, **Ho**: Observed Heterozygosity, **He**: Expected heterozygosity, **DP**: Discrimination power.

Utilization in Breeding and Improvement Programs

Walnut breeding started in the 20th century and the main goals of walnut breeding are developing high-yielding cultivars with easy-to-shell crack, light kernel color with good taste, well-adapted flowering and harvesting dates, and resistance to major diseases and pests (Cosmulescu and Botu, 2012). At the University of California, several controlled crosses were done between the French cultivars, selected for good quality and late leafing trait, and Californian cultivars such as Payne, chosen for lateral bearing and precocity (Forde, 1975). The results of these crosses helped them in the release of ten cultivars and an additional three promising cultivars such as Chandler, Howard, and Sunland. Tulare was released in the second phase of the breeding improvement program and an additional three varieties, Sexton, Gillet, and Forde were released (Vahdati et al., 2019). Three new cultivars, Ivanhoe, Solano, and Durham were released from 2010–2016 (Vahdati et al., 2019; Bernard et al. 2018b). Other breeding improvement programs in walnuts include scion and rootstock development for biotic and abiotic stresses and introgression of cherry leaf roll virus (CLRV) in commercial cultivars. In the French breeding program, the Cv. Chandler and Pedro's Californian cultivars were used as male parent plants and Franquette, Grosjean, and Marbot as female plants to release lateral-bearing cultivars. The crosses between different parents produced various lateral-bearing cultivars such as Fernor, Fernette, Ferjean, Feradam, and Ferbel. In Iran, a lot of walnut diversity exists which helps the breeders to select the promising genotypes based on traits of interest from the existing population (Hassani et al., 2013, 2012a, b). Some superior genotypes such as Jamal (protandrous), Damavand (protogynous) (Hassani et al., 2012b, a), Persia, Caspian, and Chaldoran (Hassani et al., 2019) (Figure 2). In northwestern Himalayan region of Jammu and Kashmir, India, five promising genotypes such as Hamdan and Sulaiman (Bhat and Ahmed, 2003), Parbat (Shah et al., 2023b), GL0109, which was released as Bhushan (Sharma et al., 2014) have been identified from the existing heterogenous population and released as promising cultivars with good attributes of nut and kernel traits. China has a long history of walnut cultivation and more than 164 seedling landraces have been noted (Chen et al., 2014) but the main cultivars of national interest are Bofeng, Luguang, Zha343, Xiangling, Hanfeng, Zhongling, Jinlong, and others. The main objective of the Chinese walnut improvement is early yielding, good yield quality, and tolerance to anthracnose and other diseases. Global walnut production depends upon intraspecific hybridization such as carried out in America and Iran to create the new commercial cultivars (Li, 2017; Hassani et al., 2019). On the other hand, in southwest China, the formation of interspecific hybrids between *J. regia* and *J. sigillata* have better features compared to their parent species, developed both naturally and artificially (Wu et al., 2009). Late spring frost is a limiting factor for the cultivation of walnuts in different countries (Fallah et al., 2022; Szugyi-Bartha et al., 2021; Crepinsek et al., 2009). This is more pronounced now because of climate change (Hassankhah et al., 2017; Hajinia et al., 2021). To mitigate this problem, identifying and developing suitable spring frost-resistant varieties will be an efficient strategy. The most common and easy method is to identify late leafing genotypes from diverse populations (Akca and Ozongun, 2004; Fallah et al., 2022; Aslantas, 2006). In addition, the natural or controlled cross is utilized to breed cultivars with desired traits such as late leafing and superior nut and kernel traits (Sutyemez et al., 2021).

Species/Cultivars Released

Considering the high genetic diversity in origin and primary diversity centers of walnut areas, walnut explorations in these areas have been utilized for selecting superior genotypes for commercial cultivation or for breeding purposes. In Turkey such as the first superior genotypes to be chosen from seedling populations in various Turkish locations were Yalova-1, Yalova-2, Yalova-3, and Yalova-4, along with Sebin and Bilecik (Orman et al., 2020). Later on, the Maras-18 cultivar was selected from the Kahramanmaraş province because it produced more nuts of superior quality and bloomed in a homogamous manner. Another cultivar called Kaman-1 was chosen from interior Anatolia and is

Figure 2. Few promising varieties released by some countries from seedlings walnut (Juglans regiia L.) germplasm.

known for its high yield and good nut characteristics (Sutyemez, 2016). In Kyrgyzstan, the Institute of Nut and Fruit Species has developed a number of cultivars including Ak-Tersky, Gavinsky, Ostrovershinnii, and Desertnii, and holds a unique collection of 300 cultivars and forms (Molnar et al., 2011). In Central Asia alone more than 350 seedling selections have been reported so far. Eight new superior selections have been selected from the indigenous walnut population for popularization in Kyrgyzstan. These selections have traits of late blooming (2K and AB), can self-pollinate each other (3A, 4A, and 11A), and selections 7A and 15A are precocious and have the ability to bloom twice (Beyeler et al., 2011). In Uzbekistan, breeders at the institute Bostandik branch have developed Bostandisky, Rodina, Gvardeisky, Tonkoskorlupnii, Ideal, Uzbeksky krupnoplodnii, Panfilovets, and Kazahstansky walnut cultivars (Mirzaev et al., 2004; Mavlyanova et al., 2005). Over 20 cultivars have been produced by Tajikistani breeders. The most prominent ones are Muminobadskiy #38, Tadjikskiy #25, Gissarskiy #7, and Muminobadskiy #55. In the USA, the most prominent cultivars released through intervarietal hybrization are Chandler, Hartley, Sunland, and others. In China, the walnut is grown in 21 provinces. Chinese walnut explorers have collected more than 800 forms of varieties but the main important ones are Bofeng, Luguang, Zha343, Xiangling, Hanfeng, Zhongling, Jinlong and Jinboxiang Series, Zhongling Series, Jinlong Series, and Luguo Series. Furthermore, the Shandong Institute of Pomology released 17 varieties including Luguo 2 and Luwen 1 through intraspecific and interspecific hybridization (Zhang et al., 2014). The important cultivars released and grown in different locations are described in (Table 3).

Conservation of Useful Germplasm/Genetic Resources

Plant genetic resources are indeed crucial for the sustainability of agriculture and global food safety. With climate change and the growing human population, it is becoming increasingly important to conserve these resources. In an effort to improve phenotypic traits, breeders always rely on the genetic diversity of the germplasm collections. Perennial genetic resources such as walnut of superior plants are mainly conserved *ex situ* in the field as grafted cultivars but it needs a lot of space, skillful labor for identification, and cost for proper care of the germplasm (Hammer and Teklu, 2008; Van Hintum et al., 2000). Furthermore, it is very difficult to manage and exploit all collections (Govindaraj et al., 2015b). Although large collections will help conserve the whole variability of the population, large collections frequently experience redundancy problems and high accession similarity, which can lower the effectiveness of the usefulness of germplasm for breeding operations. To simplify management problems of large collections, a core collection methodology has been devised (Brown, 1989). Different types of data such as eco-geographical, morphological, and molecular data are used to develop core collections. Among these types, the development of core collection through molecular markers is preferred over others as it reflects the variation at the genetic level. Furthermore, each entry in the core collection is based on variation in DNA sequences, irrespective of phenotypic variation (Mahmoodi et al., 2019; Shah et al., 2021). In *in situ* conservation, an area is delineated and protected to conserve germplasm diversity. The largest naturally occurring walnut forests are found in southern Kyrgyzstan of central Asia (Mamadjanov, 2005) and the second largest walnut forests in central Asia are found in Tajikistan. In these forests, the large walnut area is protected to conserve the indigenous germplasm. Tissue culture may also be used for the conservation of walnut diversity but no major success has been achieved so far in this area.

Conclusion

This chapter highlights the importance of walnut diversity characterization, utilization, and conservation. Although a number of methods have been utilized for the characterization. It is very important to use modern genomic tools to differentiate the closely related genotypes and create of core collection of walnuts. In Central Asia, the presence of large populations of walnut forests has to

Table 3. Overview of walnut cultivars released and their important traits.

Cultivar	Important Traits	Cultivation Location
Gizavezhda	High yield, average nut, tender shell, very light kernel color, cold-hardy, and resistant to pests and diseases	Albania
Leshnica	Medium yield, lateral bearing, large nut, light kernel color, aromatic	Albania
Smokthina	Medium yield, lateral bearing, average nut, light tasty kernel	Albania
Trompito	High yield, early leafing, medium kernel color	Argentina
Ivarto	Low-yield, pollinizer for medium and late varieties	Argentina
Rote Donaunuss	Moderate yield, early to mid-early leafing, medium nut size	Austria
Weinberg 2	Early leafing, large nut, thin shell, light kernel color, good kernel quality	Austria
Axel	Medium yield, late leafing, large nut	Belgium/ Netherland
Dryanovski	Early flowering, medium nut size, light kernel color, high kernel quality	Bulgaria
Izvor 10	Moderate yield, medium nut size, thin shell, light kernel color	Bulgaria
Plovdivski	Large nut, high kernel percentage, high kernel quality, resistance to bacterial disease	Bulgaria
Proslavski	Large nut, good kernel quality, resistance to bacterial disease	Bulgaria
Silistrenski	Late flowering, medium nut size, frost resistant	Bulgaria
Broadview	Early leafing, precocious, homogamous, large nut, good kernel quality	Canada
Zha 343	High yield, lateral bearing, thin shell, light kernel color	China
Zanmei	High yield, lateral bearing, light kernel color	China
Jinlong 1	Moderate yield, terminal bearing, large nut, thin shell, light kernel color	China
Wen 185	High yield, lateral bearing, large nut, thin shell, light kernel color	China
Xiangling	High yield, lateral bearing, thin shell, light kernel color	China
Zhonglin 1	High yield, lateral bearing, thin shell, extra light kernel color	China
Liaoning 1	High yield, lateral bearing, thin shell, extra light kernel color	China
Jinboxiang1	High yield, lateral bearing, thin shell, light kernel color	China
Luguo 2	High yield, lateral bearing, thin shell, light kernel color	China
Luguo 7	High yield, thin shell, light kernel color, lateral bearing	China
Daixiang	High yield, thin shell, light kernel color, lateral bearing, dwarf	China
Xinfeng	High yield, thin shell, light kernel color, lateral bearing	China
Xinxin 2	High yield, thin shell, light kernel color, lateral bearing	China
Mars	High yield, late leafing, thin shell, good kernel quality	Czech Republic
Jupiter	High yield, large nut, thin shell, late leafing,	Czech Republic
Saturn	High yield, large nut, good kernel color	Czech Republic
Apollo	Good yield, terminal bearing, early flowering, large nut, good kernel color	Czech Republic
Sychrov	Medium nut size, red kernel color, thin shell, high kernel quality	Czech Republic
Franquette	Fair yield, terminal bearing, late leafing, good kernel quality, extra light kernel color, used as pollinizer for 'Chandler'	France
Lara	High yield, lateral bearing, medium leafing	France
Fernor	High yield, lateral bearing, late leafing, extra light kernel color	France

Table 3. contd. ...

Table 3. contd. ...

Cultivar	Important Traits	Cultivation Location
Fernette	Good yield, lateral bearing, extra light kernel color, used as pollinizer for 'Chandler'	France
Ferbel	High yield, lateral bearing, large nut, thin shell, good kernel quality	France
Ferouette	High yield, lateral bearing, extra light kernel color, large nut	France
Feradam	High yield, lateral bearing, extra light kernel color, medium leafing	France
Ferjean	High yield, lateral bearing, thin shell, extra light kernel color	France
Meylanaise	Moderate yield, late leafing, good kernel quality, Used as pollinizer	France
Ronde de Montignac	Late leafing, terminal bearing, high kernel quality, used as pollinizer,	France
Rubis	Early leafing, good kernel quality, red kernel color	France
Akura	Moderate yield, moderate lateral bearing, light kernel color	Georgia
Kaspura	High yield, moderate lateral bearing	Georgia
Avenisuri	Moderate yield, terminal bearing, large nut, thin shell	Georgia
Alazani	High yield, lateral bearing	Georgia
Aragvi	Moderate yield, terminal bearing, thin shell	Georgia
Atskuri	High yield, moderate lateral bearing, thin shell, frost resistance	Georgia
Drianovski	High yield, moderate lateral bearing, large nut	Georgia
Aufhausener Baden	Large nut, mid-early flowering, old favorite German cultivar	Germany
Finkenwerder Deichnuss Royal	Large nut, precocious, thin shell, high kernel quality	Germany
Geisenheimer	Moderate yield, medium nut size, medium shellthickness	Germany
Kurmarker	Medium nut size, good kernel quality	Germany
Ledema	Protogynous, large and heavy nut,	Germany
Moselaner	Large nut, light kernel color, good kernel quality	Germany
Ockerwitzer Lange	High yield, large nut, frost resistant.	Germany
Seifersdorfer Runde	Good yield, early flowering, thin shell, light kernel color, frost resistant	Germany
Spreewalder	High yield, early leafing, precocious, light kernel color, good kernel quality	Germany
Weinheimer	Medium nut size, late leafing, high kernel quality, light kernel color	Germany
Weinsberg 1	Moderate to high yield, early flowering, large nut	Germany
Wunder von Monrepos	Medium nut size, late leafing, high kernel quality	Germany
ZP–1, 2, 3, 4, 5	'Chandler' hybrids, lateral bearing, blight resistant, moderate to late leafing, light kernel color.	Greece
Milotai Kései℗	Late leafing and flowering, lateral bearing, nut like Milotai 10, higher tolerance to blight compared to Milotai 10	Hungary
Alsószentiváni 117	Moderate yield, terminal bearing, light kernel color	Hungary
Alsószentiváni 118	High yield, mid-early flowering, large nut	Hungary
Milotai 10	High yield, moderate lateral bearing, light kernel color	Hungary
Tiszacsécsi 83	High yield, moderate lateral bearing	Hungary
Esterhazy II	Early leafing, medium to large nut size, light kernel color, good kernel quality	Hungary
Hamdan	Medium sized nuts with good kernel quality	India

Table 3. contd. ...

Table 3. contd. ...

Cultivar	Important Traits	Cultivation Location
Sulaiman	Medium sized nuts with good kernel quality	India
Parbat	Yield and quality comparatively good than local cultivars	India
Bhushan	Large sized nuts with good kernel recovery percentage	India
Jamal	Moderate yield, terminal bearing, medium light kernel color	Iran
Damavand	Early leafing, used as pollinizer for Jamal	Iran
Sorrento	Moderate yield, terminal bearing	Italy
Malizia	High yield, moderate lateral bearing	Italy
Qingxiang	High yield, terminal bearing, light kernel color	Japan
Kyrgyzskya Bomba	Large nut, great kernel quality, light kernel color	Kyrgyzstan
Ak Terek	Medium nut, great kernel quality, light kernel color	Kyrgyzstan
Oshsky	Medium nut size, great kernel quality	Kyrgyzstan
Uygursky	Large nut, great kernel quality, light kernel color	Kyrgyzstan
Ostrovershinny	Large nut, great kernel quality	Kyrgyzstan
Immuniy	Medium nut, great kernel quality, light kernel color	Kyrgyzstan
Desertniy	Large nut, great kernel quality, light kernel color	Kyrgyzstan
Pescianski	High yield, partial lateral bearing, thin shell, frost resistant, extra light kernel	Moldova
Calarasi	High yield, terminal bearing, frost resistant, light kernel	Moldova
Ovata	Large nut, terminal bearing, very cold resistant	Moldova
Carpatica	Huge nut, terminal bearing	Moldova
Amphyon	High yield, high kernel quality, low susceptibility to disease	Netherland
Dionym	High yield, high kernel quality, low susceptibility to disease	Netherland
Big & Easy	Late flowering, thin shell, good kernel quality	Netherland
Blanco	Large nut, homogamous	Netherland
Coenen	Early flowering, large nut, thin shell	Netherland
Lange van Lod	Large nut, late leafing, high kernel quality, slight tolerance to late spring frost	Netherland
Rex	High yield, late leafing, light color	New Zealand
Shannon	High yield, light kernel color, blight resistant	New Zealand
Meyric	High yield, late leafing, thin shell, high kernel quality	New Zealand
Wilsons Wonder	Large nut, light kernel color	New Zealand
Valcor	High yield, terminal bearing, thin shell, light kernel color	Romania
Valmit (Verisval)	Thin shell, terminal bearing, light kernel color,	Romania
Valrex	High yield, terminal bearing, large nut, thin shell	Romania
Sibişel 44	Moderate yield, terminal bearing, large nut	Romania
Jupâneşti	Precocious, high yield, terminal bearing, thin shell	Romania
Velniţa	Precocious, high yield, terminal bearing	Romania
Valstar	Precocious, high yield, terminal bearing light kernel	Romania
Valcris	Precocious, high yield, terminal bearing light kernel	Romania
Timval	High yield, terminal bearing, large nut	Romania

Table 3. contd. ...

Table 3. contd. ...

Cultivar	Important Traits	Cultivation Location
Miroslava	Terminal bearing, large nut	Romania
Ovidiu	Terminal bearing, large nut	Romania
Anica	Terminal bearing, large nut	Romania
Sibişel 252	Precocious, high yield, terminal bearing	Romania
Ciprian	Precocious, high yield, terminal bearing, large nut	Romania
Claudia	Precocious, high yield, terminal bearing	Romania
Germisara	Moderate yield, terminal bearing, large nut	Romania
Şuşiţa	Precocious, high yield, terminal bearing, light kernel	Romania
Ronutex	High yield, terminal bearing, large nut	Romania
Belbeksky 70	Late leafing, light kernel color, moderate shell thickness	Russia Federation
Vynoslivy	High yield, light kernel color, resistance to low temperature	Russia Federation
Pervomaysky	Late leafing, thin shell, light kernel color	Russia Federation
Krymsky Urozhayany	High yield, thin shell, light kernel color	Russia Federation
Elit	Late leafing, precocious, light kernel color	Slovenia
Krka	Homogenous flowering, high yield, bright kernel	Slovenia
Sava	Intermediate fruit-bearing, late leafing, moderate yield	Slovenia
Fischenthal	Medium nut size, thin shell, cluster bearing, homogamous	Switzerland
Giswill	Lateral bearing, interesting for wood production	Switzerland
Nyffenegger	Terminal bearing, thin shell, red kernel color	Switzerland
Rote Gubler	Medium nut size, light kernel color, forest resistant	Switzerland
Yalova 1	Fair yield, terminal bearing, large nut, thin shell	Turkey
Yalova 3	Fair yield, terminal bearing, thin shell, light kernel color	Turkey
Sebin	High yield, moderate lateral bearing, thin shell, light kernel color	Turkey
Bilecik	Moderate yield, terminal bearing	Turkey
Maras 18	Moderate yield, lateral bearing, light kernel color, high kernel percentage, very early harvest,	Turkey
Sütyemez 1	Moderate yield, lateral bearing, extra-large nut, light kernel color, very early harvest	Turkey
Kaman 1	High yield, lateral bearing, thin shell, light kernel color	Turkey
Maraş 12	High yield, terminal bearing, moderate nut, thin shell, light kernel color, extra high kernel percentage	Turkey
Diriliş	High yield, lateral bearing, late leafing, light kernel color, high kernel percentage, thin shell, early harvest	Turkey
15 Temmuz	High yield, lateral bearing, very late leafing, light kernel color, high kernel percentage thin shell, early harvest	Turkey
Bayrak	High yield, lateral bearing, thin shell, extra light kernel color, high kernel percentage	Turkey
Akça	High yield, lateral bearing, late leafing, good kernel quality, extra light kernel color, used as pollinizer for 'Chandler'	Turkey
Niksar 1	Moderate yield, lateral bearing, late leafing, used as pollinizer for cv. Chandler	Turkey

Table 3. contd. ...

Table 3. contd. ...

Cultivar	Important Traits	Cultivation Location
Eureka	Fair yield, terminal bearing, poor kernel color	USA
Scharsch Franquette	Late leafing, medium thin shell, light kernel color	USA
Hartley	Moderate yield, terminal bearing, light kernel color	USA
Payne	High yield, lateral bearing, precocious, early leafing, early harvest, light kernel color	USA
Vina	High yield, lateral bearing, poor color	USA
Pedro	High yield, lateral bearing	USA
Tehama	Moderate yield, lateral bearing, light kernel color, used as pollinizer for cv. Serr	USA
Serr	Moderate yield, moderate lateral bearing, light kernel color, thin shell, excellent kernel quality, susceptible to pistillate flower abscission (PFA)	USA
Chandler	High yield, lateral bearing, medium leafing, extra light kernel color, thin shell	USA
Howard	High yield, lateral bearing, medium leafing, large nut, thin shell, light kernel color	USA
Sunland	High yield but susceptible to nut drop, lateral bearing, large nut, thin shell, light kernel color	USA
Cisco	Moderate yield, terminal bearing, medium light kernel color, high susceptibility to blight, used as pollinizer for cv. Chandler	USA
Tulare	High yield, lateral bearing, light kernel color, susceptible to winter cold	USA
Robert Livermore	Moderate yield, lateral bearing, red kernel color	USA
Sexton	High yield, lateral bearing, very precocious, light kernel color	USA
Gillet	High yield, lateral bearing, large nut, light kernel color, low susceptibility to blight	USA
Forde	Good yield, lateral bearing, light kernel color, low susceptibility to blight	USA
Ivanhoe	High yield, lateral bearing, very precocious, very early harvest, thin shell, extra light kernel color	USA
Solano	High yield, lateral bearing, extra light kernel color	USA
Durham	Good yield, early harvest, lateral bearing, large nut, light kernel color	USA
Ideal	Small nut from secondary flower, high kernel quality, precocious, cluster bearing habit	Uzbekistan
Hybridiy	Medium size nut	Uzbekistan
Pioner	Large size nut	Uzbekistan
Kazahstansky	High kernel quality	Uzbekistan
Bostonliksky	Large nut	Uzbekistan
Rodina	Large nut, high kernel quality, high frost resistance, low susceptibility to anthracnose	Uzbekistan
Parkent	Large nut	Uzbekistan
Nani	Large nut	Uzbekistan
Gvardiesky	High kernel quality	Uzbekistan
Panfilovets	High kernel quality	Uzbekistan
Tonkoskorlupnii	High kernel quality	Uzbekistan
Ubilini	High yield, high kernel quality, resistance to spring frosts, precocious	Uzbekistan
Bostandik	Large nut, high kernel quality, resistance to spring frosts, precocious	Uzbekistan

Adopted from Vahadati et al. (2019).

be protected to conserve the important allelic resources of traits such as lateral bud, late flowering, and frost resistance plants. For better future gain in the walnut improvement program, more diverse areas should be covered particularly Central Asia to understand the true genetic diversity of different regions. This will also provide an opportunity to identify and conserve breeding material, especially biotic and abiotic stress genotypes which will be major issues in the face of climate change. This will also ensure sustainable production or increased production to meet both national and international demand. Additionally, mapping out patterns of genetic diversity will help to determine the areas that require urgent *in situ* and *ex situ* conservation of walnuts.

References

Ahmed, N., J.I. Mir, R.R Mir, N.A. Rather, R. Rashid, S. Wani, W. Shafi, H. Mir, and M.A. Sheikh. 2012. SSR and RAPD analysis of genetic diversity in walnut (*Juglans regia* L.) genotypes from Jammu and Kashmir, India. *Physiol. Mol. Biol. Plants*, *18*(2): 149–160.

Akca, Y., Y. Bilgen, and S. Ercisli. 2015. Selection of superior Persian walnut (*Juglans regia* L.) from seedling origin in Turkey. *Acta Sci. Pol. Hortorum Cultus.*, *14*(3): 103–114.

Akca, Y. and S. Ozongun. 2004. Selection of late leafing, late flowering, laterally fruitful walnut (*Juglans regia*) types in Turkey. *NZ J. Crop Hortic. Sci.*, *32*(4): 337–342.

Akca, Y., Y.B. Yuldasulu, E. Murad, and K. Vahdati. 2020. Exploring of walnut genetic resources in Kazakhstan and evaluation of promising selections. *Int. J. Hortic. Sci. Technol.*, *7*(2): 93–102.

Arab, M., M.A. Marrano, R. Abdollahi-Arpanahi, C.A. Leslie, H. Askari, D.B. Neale, and K. Vahdati. 2019. Genome-wide patterns of population structure and association mapping of nut-related traits in Persian walnut populations from Iran using the Axiom *J. regia* 700K SNP array. *Sci. Rep.*, *9*(1): 6376–6376. https://doi:10.1038/s41598-019-42940-1.

Aradhya, M., D. Velasco, Z. Ibrahimov, B. Toktoraliev, D. Maghradze, M. Musayev, Z. Bobokashvili, and J.E. Preece. 2017. Genetic and ecological insights into glacial refugia of walnut (*Juglans regia* L.). *PloS One*, *12*(10): e0185974.

Aradhya, M., K. Woeste, and D. Velasco. 2010. Genetic diversity, structure, and differentiation in cultivated Walnut (*Juglans regia* L.). *Acta Hortic.*, *861*: 127–132. https://doi:10.17660/actahortic.2010.861.16.

Aradhya, M.K. D. Potter, F. Gao, and C.J. Simon. 2007. Molecular phylogeny of Juglans (Juglandaceae): A biogeographic perspective. *Tree Genet.*, *3*: 363–378.

Aryapak, S., and P. Ziarati. 2014. Nutritive value of Persian walnut (*Juglans regia* L.) Orchards. *Am. Eurasian J. Agric. Environ. Sci.*, *14*(11): 1228–1235.

Arzani, K., H. Mansouri-Ardakan, A. Vezvaei, and M.R. Roozban. 2008. Morphological variation among Persian walnut (*Juglans regia*) genotypes from central Iran. *NZ. J. Crop Hortic. Sci.*, *36*(3): 159–168.

Asgari, K., M. Labbafi, F. Khodaiyan, M. Kazemi, and S.S. Hosseini. 2020. Valorization of walnut processing waste as a novel resource: Production and characterization of pectin. *J. Food Process. Preserv.*, *44*(12): e14941.

Aslantas, R. 2006. Identification of superior walnut (*Juglans regia*) genotypes in northeastern Anatolia, Turkey. *NZ. J. Crop Hortic. Sci.*, *34*(3): 231–237.

Bartha, K.S., G. Bujdoso, and E. Stefanovits-Banyai. 2023. Biochemical Methods for the Characterisation of Walnut Genotypes with Different Levels of Frost Tolerance. *Erwerbs-Obstbau.*, *65*(2): 293–298.

Bayazit, S., K. Kazan, S. Gulbitti, V. Cevik, H. Ayanoglu, and A. Ergul. 2007. AFLP analysis of genetic diversity in low chill requiring walnut (*Juglans regia* L.) genotypes from Hatay, Turkey. *Sci. Hortic.*, *111*(4): 394–398. https://doi:10.1016/j.scienta.2006.11.006.

Bernard, A., T. Barreneche, A. Donkpegan, L. Fabrice, and E. Dirlewanger. 2019. *Management of a Walnut Germplasm Collection: Which of SSR or SNP Markers Are Most Suitable to Preserve Biodiversity?* Cold Spring Harbor Laboratory. https://doi:10.1101/2019.12.17.879627.

Bernard, A., T. Barreneche, A. Donkpegan, F. Lheureux, and E. Dirlewanger. 2020a. Comparison of structure analyses and core collections for the management of walnut genetic resources. *Tree Genet.*, *16*(5): 1–14.

Bernard, A., T. Barreneche, F. Lheureux, and Dirlewanger, E. 2018a. Analysis of genetic diversity and structure in a worldwide walnut (*Juglans regia* L.) germplasm using SSR markers. *PloS One*, *13* (11): e02080217–e0208021. https://doi:10.1371/journal.pone.0208021.

Bernard, A., T. Barreneche, F. Lheureux, and Dirlewanger, E. 2020b. SSR genetic diversity assessment of the INRAE's walnut (*Juglans* spp.) germplasm collection. *Acta Hortic.*, Article 1297: 3777–3784. https://doi:10.17660/actahortic.2020.1297.50.

Bernard, A., T. Barreneche, F. Lheureux, E. Dirlewanger, and F. Lheureux. 2020c. 3D characterization of walnut morphological traits using X-ray computed tomography. *Plant Methods*, *16*(1): 17–11.

Bernard, A., F. Lheureux, and E. Dirlewanger. 2018b. Walnut: Past and future of genetic improvement. *Tree Genet.*, *14*(1): 1. https://doi.org/10.1007/s11295-017-1214-0.

Beyeler, S., M. Rehnus, G.A. Venglovskaya, and J.P. Sorg. 2011. *Bibliography on the Walnut-fruit Forests of Southern Kyrgyzstan*. Zurich: Bishkek, Kyrgyzstan.

Beyhan, O., A. Ozcan, H. Ozcan, N. Kafkas, S. Kafkas, M. Sutyemez, and S. Ercisli. 2017. Fat, fatty acids and tocopherol content of several walnut genotypes. *Not. Bot. Horti. Agrobo., 45*(2): 4377–4441.10.15835/nbha45210932.

Bhat, A.R., and M.F. Ahmed. 2003. Hamdann and Sulaiman: New walnut varieties for North-western Himalayas. *Indian Hortic., 47*: 33–34.

Britton, M.T., C.H. Leslie, G.H. McGranahan, and A.M. Dandekar. 2007. Walnuts. *In*: Pua, E.C., and Davey, M. [Eds.], *Transgenic Crops V. Biotechnology in Agriculture and Forestry 60*: 349–370. Springer, Berlin, Heidelberg. https://doi.org/10.1007/978-3-540-49161-3_14.

Brown, A.H.D. 1989. Core collections: A practical approach to genetic resources management. *Genome, 31*(2): 818–824.

Candemir, S. 2021.Walnut Production, Trade, and Forecast. *In*: Gubdesli M.A [Eds.], *Walnut*, 7–22. Iksad Publications, Ankara.

Cervera, M., T.D. Remington, J.M. Frigerio, V. Storme, B. Ivens, W. Boerjan, and C. Plomion. 2000. Improved AFLP analysis of tree species. *Can. J. For. Res., 30*(10): 1608–1616. https://doi.org/doi:10.1139/x00-085.

Chen, F., J. Chen, Z. Wang, J. Zhang, X. Li, M. Lin, Y. Song, and L. Zhang. 2019. Genomics: Cracking the mysteries of walnuts. *J. Genet., 98*: 1–3.

Chen, L., Q. Ma, Y. Chen, B. Wang, and D. Pei. 2014. Identification of major walnut cultivars grown in China based on nut phenotypes and SSR markers. *Sci. Hortic., 168*: 240–248.

Collard, B.C.Y., M.Z.Z. Jahufer, J.B. Brouwer, and E.C.K. Pang. 2005. An introduction to markers, quantitative trait loci (QTL) mapping and marker-assisted selection for crop improvement: The basic concepts. *Euphytica, 142*(1–2): 169–196. https://doi:10.1007/s10681-005-1681-5.

Cosmulescu, S., and M. Botu. 2012. Walnut biodiversity in southwestern Romania resource for perspective cultivars. *Pak. J. Bot., 44*(1): 307–311.

Cosmulescu, S.N., A. Baciu, G. Achim, B. Mihai, and I. Trandafir. 2009. Mineral composition of fruits in different walnut (*Juglans regia* L.) cultivars. *Not. Bot. Horti. Agrobo., 37* (2): 156–160.

Crepinsek, Z., M. Solar, F. Stampar, and A. Solar. 2009. Shifts in walnut (*Juglans regia* L.) phenology due to increasing temperatures in Slovenia. *J. Hortic. Sci. Biotechnol., 84*(1): 59–64.

Crespel, L., M. Chirollet, C. Durel, D. Zhang, J. Meynet, and S. Gudin. 2002. Mapping of qualitative and quantitative phenotypic traits in Rosa using AFLP markers. *Theor. Appl. Genet., 105*(8): 1207–1214. https:// doi:10.1007/s00122-002-1102-2.

Cui, M., Q. Li, Z. Ni, Y. Han, Y. Zheng, R. Mo, D. Shen, and Y.J.F. Liu. 2023. The Nutritional Qualities of Walnuts and Their Planted Soils from China: Level and Relationship. *Forests, 14*(7): 1369. https://doi.org/10.3390/f14071369.

Dandekar, A., C. Leslie, and G. McGranahan. 2005. Juglans regia Walnut, 307–323. *In*: Gabrielle J. Persley. [Ed.]. The Doyle Foundation, Glasgow, Scotland. doi:10.1079/9780851996622.0307.

Davoodi, F., M. Rezaei, P. Heidari, H. Hokmabadi, and S. Lawson. 2021. Identification and DNA fingerprinting of some superior Persian walnut genotypes in Iran. *Erwerbs-Obstbau, 63*(4): 393–402.

Diaz, R., and J. Fernandez-Lopez. 2005. Genetic variation at early ages for several traits of interest for timber-production breeding of *Juglans regia*. *Can. J. For. Res., 35*(2): 235–243.

Dogan, M., and A. Akgul. 2005. Fatty acid composition of some walnut (*Juglans regia* L.) cultivars from east Anatolia. *Grasas Aceites., 56*(4). https://doi:10.3989/gya.2005.v56.i4.101.

Ebrahimi, A., R. Fatahi, and Z. Zamani. 2011. Analysis of genetic diversity among some Persian walnut genotypes (*Juglans regia* L.) using morphological traits and SSRs markers. *Sci. Hortic., 130*(1): 146–151. https://doi:10.1016/j.scienta.2011.06.028.

Ebrahimi, A., A. Zarei, J.R. McKenna, G. Bujdoso, and K.E. Woeste. 2017. Genetic diversity of Persian walnut (*Juglans regia*) in the cold-temperate zone of the United States and Europe. *Sci. Hortic., 220*: 36–41. https://doi:10.1016/j.scienta.2017.03.030.

Fallah, M., M. Rasouli, D. Hassani, S.S. Lawson, S. Sarikhani, and K. Vahdati. 2022. Tracing Superior Late-Leafing Genotypes of Persian Walnut for Managing Late-Spring Frost in Walnut Orchards. *Hortic., 8*(11): 1003.

Fan, N., J.L. Fusco, and D.W. Rosenberg. 2023. Antioxidant and Anti-Inflammatory Properties of Walnut Constituents: Focus on Personalized Cancer Prevention and the Microbiome. *Antioxidants, 12*(5): 982.

Fatahi, R., A. Ebrahimi, and Z. Zamani. 2010. Characterization of some Iranians and foreign walnut genotypes using morphological traits and RAPD markers. *Hortic. Environ. Biotechnol., 51*(1): 51–60.

Fatima, T., U. Showkat, and S.Z. Hussain. 2018. Nutritional and health benefits of walnuts. *J. Pharmacogn. Phytochem., 7*(2): 1269–1271.

Forde, H.I. 1975. Pollination and blooming habits of walnuts. 439–455.

Forde, H.I., and W.H. Griggs. 1975. Pollination and blooming habits of walnuts. English, Miscellaneous (No. 2753), Leaflet, Division of Agricultural Sciences.

Fornari, B., M.E. Malvolti, D. Taurchini, S. Fineschi, I. Beritognolo, E. Maccaglia, and F. Cannata. 2001. Isozyme and Organellar DNA analysis of Genetic Diversity in natural/Naturalised European and Asiatic Walnut (*Juglans regia* L.) populations. *Acta Hortic., 544*: 167–178. https://doi:10.17660/actahortic.2001.544.23.

Gandev, S. 2007. Budding and grafting of the walnut (*Juglans regia* L.) and their effectiveness in Bulgaria. *Bulg. J. Agric. Sci.*, *13*(6): 683.

Goodarzi, H., D. Hassani, L. Pourhosseini, S. Kalantari, and A. Lashgari. 2023. Total lipid and fatty acids components of some Persian walnut (*Juglans regia*) cultivars. *Sci. Hortic.*, *321*: 112252. https://doi.org/10.1016/j.scienta.2023.112252.

Govindaraj, M., M.Vetriventhan, and M. Srinivasan. 2015a. Importance of genetic diversity assessment in crop plants and its recent advances: An overview of its analytical perspectives. *Genet. Res. Int.*, *2015*: 1–14.

Govindaraj, M., M. Vetriventhan, and M. Srinivasan. 2015b. Importance of genetic diversity assessment in crop plants and its recent advances: An overview of its analytical perspectives. *Genet. Res. Int.*, *2015*: 431487. https://doi.org/10.1155/2015/431487.

Grouh, M.S.H., K. Vahdati, M. Lotfi, D. Hassani, and N.P. Biranvand. 2011. Production of haploids in Persian walnut through parthenogenesis induced by gamma-irradiated pollen. *J. Am. Soc. Hort. Sci.*, *136*(3): 198–204.

Gunn, B.F., M. Aradhya, J.M. Salick, A.J. Miller, Y. Yongping, L. Lin, and H. Xian. 2010. Genetic variation in walnuts (*Juglans regia* and *J. sigillata*, Juglandaceae): Species distinctions, human impacts, and the conservation of agrobiodiversity in Yunnan, China. *Am. J. Bot.*, *97*(4): 660–671.

Hajinia, Z., S. Sarikhani, and K. Vahdati. 2021. Exploring low-chill genotypes of Persian walnut (*Juglans regia* L.) in west of Iran. *Genet. Resour. Crop Evol.*, *68*(6): 2325–2336. https://doi:10.1007/s10722-021-01131-6.

Hammer, K., and Y. Teklu. 2008. Plant genetic resources: selected issues from genetic erosion to genetic engineering. J. Agric. Rural Dev. Trop. Subtrop., *109*(1): 15–50.

Hassani, D., J. Atefi, R. Haghjooyan, R. Dastjerdi, M. Keshavarzi, M.R. Mozaffari, A. Soleimani, A.R. Rahmanian, F. Nematzadeh, and A. Malmir. 2012a. Damavand, a new Persian walnut cultivar as a pollinizer for Iranian walnut cultivars and genotypes. *Seed and Plant Improvement J.*, *28*(3): 529–531.

Hassani, D., J. Atefi, R. Haghjooyan, R. Dastjerdi, M. Keshavarzi, M.R. Mozaffari, A. Soleimani, A.R. Rahmanian, F. Nematzadeh, and A. Malmir. 2012b. Jamal, a new Persian walnut cultivar for moderate-cold areas of Iran. *Seed and Plant J.*, *28*(3): 525–527.

Hassani, D., R. Haghjooyan, A. Soleimani, J. Atefi, and A. Loni. 2013. Evaluation of some Walnut cultivars and selections in Iran. *Acta Hortic.*, Article *981*: 59–64. doi:10.17660/actahortic.2013.981.4.

Hassani, D., M.R. Mozaffari, R. Dastjerdi, M. Keshavarzi, A. Soleimani, R. Rezaee, and J. Atefi. 2019. Four New Persian Walnut Cultivars of Iran: Persia, Caspian, Chaldoran, and Alvand. *Hortic. Sci. Horts.*, *55*(7): 1162–1163. https://doi.org/10.21273/HORTSCI15044-20.

Hassankhah, A., K. Vahdati, M. Rahemi, D. Hassani, and K.S. Sarikhani. 2017. Persian walnut phenology: Effect of chilling and heat requirements on bud break and flowering date. *Int. J. Hortic. Sci. Technol.*, *4*(2): 259–271.

Hussain, S.Z., B. Naseer, T. Qadri, T. Fatima, and T.A. Bhat. 2021. Walnut (*Juglans regia*)-morphology, taxonomy, composition, and health benefits. *In: Fruits Grown in Highland Regions of the Himalayas: Nutritional and Health Benefits*, 269–281. Springer. doi: 10.1007/978-3-030-75502-7.

Itoo, H., R.A. Shah, S. Qurat, A. Jeelani, S. Khursheed, Z.A. Bhat, M.A. Mir, G.H. Rather, S.M. Zargar, and M.D. Shah. 2023. Genome-wide characterization and development of SSR markers for genetic diversity analysis in northwestern Himalayas Walnut (*Juglans regia* L.). *3 Biotech.*, *13*(5): 136.

Jacimovic, V., M. Adakalic, S. Ercisli, D. Bozovic and G. Bujdoso. 2020. Fruit quality properties of walnut (*Juglans regia* l.) genetic resources in Montenegro. *Sustainability*, *12*(23): 9963.

Karimi, R., A. Ershadi, A. Ehtesham Nia, M. Sharifani, M. Rasouli, A. Ebrahimi, and K. Vahdati. 2014. Morphological and molecular evaluation of Persian walnut populations in northern and western regions of Iran. *J. of Nuts*, *5*(02): 21–31.

Khadivi, A., A. Montazeran, M. Rezaei, and A. Ebrahimi. 2019a. The pomological characterization of walnut (*Juglans regia* L.) to select the superior genotypes: An opportunity for genetic improvement. *Sci. Hortic.*, *248*: 29–33.

Khadivi, A., A. Montazeran, and P. Yadegari. 2019b. Superior spring frost-resistant walnut (*Juglans regia* L.) genotypes identified among mature seedling origin trees. *Sci. Hortic.*, *253*: 147–153.

Khan, M.W., I.A. Khan, H. Ahmad, H. Ali, S. Ghafoor, M. Afzal, F.A. Khan, M. Shah, and S.G. Afridi. 2010. Estimation of genetic diversity in walnut. *Pak. J. Bot.*, *42*(3): 1791–1796.

Khorami, S.S., K. Arzani, G. Karimzadeh, A. Shojaeiyan, and W. Ligterink. 2018. Genome size: A novel predictor of nut weight and nut size of walnut trees. *HortScience.*, *53*(3): 275–282.

KilicogLu, O.C., and A. Yasar. 2018. Selection of superior walnut types with lateral bud fruitfulness and high nut quality. J. *of Agricultural Faculty of Gaziosmanpasa University*, *35*(3): 297–304.

Krska, B., J. Pramukova, and M. Vachun. 2009. Inheritance of some pomological traits in Minaret× Betinka apricot progeny. *Hortic. Sci.*, *36*(3): 85–91.

Laurens, F., N. Moutier, A. Belouin, F. Delort, J. M. Legave, E. Costes, J.L. Regnard, and P. E. Lauri. 2003. Morphological and Architectural Traits on Fruit Trees which Could Be Relevant for Genetic Studies: A Review. *In*: *12th Eucarpia Symposium on Fruit Breeding and Genetics* [Eds.], Laurens, F. and Evans, K., 349–356. https://doi.org/10.13140/2.1.1734.7843.

Li, L. 2017. *Growth and Propagation Characteristics of Juglans Interspecific Hybrid Varieties*. Haidian: Chinese Academy of Forestry.

Li, P., X. Cui, Y. Li, and Z. Wang. 2011. Epitope mapping and identification on a 3D model built for the tartary buckwheat allergic protein TBb. *Acta Biochimica. et Biophysica. Sinica*, *43*(6): 441–447.

Magige, E.A., P.Z. Fan, M.C. Wambulwa, R. Milne, Z.Y. Wu, Y.H. Luo, R. Khan, H.Y. Wu, H.L. Qi, and G.F. Zhu. 2022. Genetic Diversity and Structure of Persian Walnut (*Juglans regia* L.) in Pakistan: Implications for Conservation. *Plants*, *11*(13): 1652.

Mahmoodi, R., M.R. Dadpour, D. Hassani, M. Zeinalabedini, E. Vendramin, S. Micali, and F.Z. Nahandi. 2019. Development of a core collection in Iranian walnut (*Juglans regia* L.) germplasm using the phenotypic diversity. *Sci. Hortic.*, *249*: 439–448. doi:10.1016/j.scienta.2019.02.017.

Malekjani, N., and S.M. Jafari. 2022. Valorization of olive processing by-products via drying technologies: A case study on the recovery of bioactive phenolic compounds from olive leaves, pomace, and wastewater. *Crit. Rev. Food Sci. Nutr.*, *63*(29): 9797–9815. doi: 10.1080/10408398.2022.2068123.

Mamadjanov, D.K. 2005. Walnut Fruit Forests and Diversity of Walnut Tree (*Juglans regia* L.) In Kyrgyzstan. *Acta Hortic.*, 705: 173–176. doi:10.17660/actahortic.2005.705.20.

Manthos, I., D. Rouskas, T. Sotiropoulos, and M. Botu. 2023. Description of Two Promising Walnut (*Juglans regia* L.) Selections with Lateral Bud Fruitfulness and Large Nuts. *Hortic.*, *9*(7): 820.

Mapelli, S., A. Bertani, M. Malvolti, I. Olimpieri, P. Pollegioni, E. Alexandrovski, E. Butkov, and E. Botman. 2013. Resources of *Juglans regia* in Uzbekistan: A valuable step along the silk road. *In*: *I International Symposium on Fruit Culture and Its Traditional Knowledge along Silk Road Countries* , Article 1032, 55–62.

Marrano, A., P.J. Martínez-Garcia, L. Bianco, G.M. Sideli, E.A. Di Pierro, C.A. Leslie, K.A. Stevens, M.W. Crepeau, M. Troggio, C.H. Langley. 2019a. A new genomic tool for walnut (*Juglans regia* L.): development and validation of the high-density Axiom™ *J. regia* 700K SNP genotyping array. *Plant Biotechnol. J.*, *17*(6): 1027–1036.

Marrano, A., G.M. Sideli, C.A. Leslie, H. Cheng, and D.B. Neale. 2019b. Deciphering of the genetic control of phenology, yield, and pellicle color in Persian walnut (*Juglans regia* L.). *Front. Plant Sci.*, *10*: 1140.

Martínez-Garcia, P.J., M.W. Crepeau, D. Puiu, D. Gonzalez-Ibeas, J. Whalen, K.A. Stevens, R, Paul. T.S. Butterfield, M.T. Britton, and R.L. Reagan. 2016. The walnut (*Juglans regia*) genome sequence reveals diversity in genes coding for the biosynthesis of non-structural polyphenols. *The Plant J.*, *87*(5): 507–532.

Masoodi, L., A. Gull, F.A. Masoodi, A. Gani, J. Nissar, T. Ahad, G.A. Nayik, S.A. Mukarram, B. Kovacs, and J. Prokisch. 2022. An Overview on traditional vs. green technology of extraction methods for producing high-quality walnut oil. *Agronomy*, *12*(10): 2258.

Mavlyanova, R.F., F.K. Abdullaev, P. Khodjiev, D.E. Zaurov, T.J. Molnar, J.C. Goffreda, T.J. Orton, C.R. Funk. 2005. Plant genetic resources and scientific activities of the Uzbek Research Institute of Plant Industry. *HortScience*, *40*(1): 10–14.

McGranahan, G., and C. Leslie. 2012. *Walnut Fruit Breeding*, 827–846. https://doi.org/10.1007/978-1-4419-0763-9_22/COVER.

Mikdat, S., U.Y. Kadir, and R.D. Ali. 2010. Selection and determination of some significant properties of superior walnut genotypes. *Sci. Res. Essays*, *5*(19): 2987–2996.

Miraliakbari, H. 2005. *Tree Nut Oils: Chemical Characteristics, Oxidation, and Antioxidants*. Ph.D. dissertation, Memorial University of Newfoundland.

Mirzaev, M.M., U.M. Djavacynce, D.E. Zaurov, J.C. Goffreda, T.J. Orton, E.G. Remmers, and C.R. Funk. 2004. The Schroeder Institute in Uzbekistan: Breeding and germplasm collections. *HortScience*, *39*(5): 917–921.

Molnar, T.J., D.E. Zaurov, J.M. Capik, S.W. Eisenman, T. Ford, L.V. Nikolyi, and C.R. Funk. 2011. Persian walnuts (*Juglans regia* L.) in Central Asia. *Annu. Rep. North Nut Grow. Assoc.*, *101*: 56–69.

Mondini, L., A. Noorani, and M.A. Pagnotta. 2009. Assessing plant genetic diversity by molecular tools. *Diversity*, *1*(1): 19–35.

Nathewet, P., T. Yanagi, Y. Iwastubo, K. Sone T. Takamura, and N. Okuda. 2009. Improvement of staining method for observation of mitotic chromosomes in octoploid strawberry plants. *Sci. Hortic.*, *120*(3): 431–435.

Nguyen, T.H.D., and D.C. Vu. 2023. A review on phytochemical composition and potential health-promoting properties of walnuts. *Food Rev. Int.*, *39*(1): 397–423.

Orhan, E., S.P. Eyduran, D. Poljuha, M. Akin, T. Weber, and S. Ercisli. 2020. Genetic diversity detection of seed-propagated walnut (*Juglans regia* L.) germplasm from Eastern Anatolia using SSR markers. *Folia Hortic.*, *32*(1): 37–46. doi:10.2478/fhort-2020-0004.

Orhan, I.E., I.P. Suntar, and E.K. Akkol. 2011. *In vitro* neuroprotective effects of the leaf and fruit extracts of *Juglans regia* L.(walnut) through enzymes linked to Alzheimer's disease and antioxidant activity. *Int. J. Food Sci. Nutr.*, *62*(8): 781–786.

Orman, E. D. Ates, E. OZkuru, S. Hepaksoy, S. Kafkas, and M.B. Tanyolac. 2020. Association mapping of several nut characters in walnut (*Juglans regia* L.). *Turk. J. Agric. For.*, *44*(2): 208–227. doi:10.3906/tar-1909-35.

Panahi, B.S. Tajaddod, H. Mohammadzadeh Jallali, M.A. Hejazi, and M. Zeinalabedini. 2022. Variability and association among some pomological and physiochemical traits in spring frost tolerant genotypes of Persian walnut (*Juglans regia* L.) and selection of genotypes with superior traits based on machine learning algorithms. *Genet Resour. Crop Evol.*, *69*(1): 1–13.

Papoutsi, Z. E. Kassi, I. Chinou, M. Halabalaki, L.A. Skaltsounis, and P. Moutsatsou. 2008. Walnut extract (*Juglans regia* L.) and its component ellagic acid exhibit anti-inflammatory activity in human aorta endothelial cells and osteoblastic activity in the cell line KS483. *Br. J. Nutr.*, *99*(4): 715–722.

Pearl, H.M., C. Nagai, P.H. Moore, D.L. Steiger, R.V. Osgood, and R. Ming. 2003. Construction of a genetic map for arabica coffee. *Theor. Appl. Genet.*, *108*(5): 829–835. doi:10.1007/s00122-003-1498-3.

Persic, M., M. Mikulic-Petkovsek, A. Slatnar, A. Solar, and R.J.F.C. Veberic. 2018. Changes in phenolic profiles of red-colored pellicle walnut and hazelnut kernel during ripening. *Food Chem.*, *252*: 349–355.

Poggetti, L., P. Ermacora, G. Cipriani, F. Pavan, and R. Testolin. 2017. Morphological and carpological variability of walnut germplasm (*Juglans regia* L.) collected in North-Eastern Italy and selection of superior genotypes. *Sci. Hortic.*, *225*: 615–619. doi:10.1016/j.scienta.2017.07.056.

Pollegioni, P., K. Woeste, F. Chiocchini, S. Del Lungo, M. Ciolfi, I. Olimpieri, V. Tortolano, J. Clark, G.E. Hemery, and S. Mapelli. 2017. Rethinking the history of common walnut (*Juglans regia* L.) in Europe: Its origins and human interactions. Plos One, *12*(3): e0172541.

Pollegioni, P., K. Woeste, I. Olimpieri, D. Marandola, F. Cannata, and M. Emilia Malvolti. 2011a. Long-term human impacts on genetic structure of Italian walnut inferred by SSR markers. *Tree Genet. Genom.*, *7*: 707–723.

Pollegioni, P., K. Woeste, I. Olimpieri, D. Marandola, F. Cannata, and M. Emilia Malvolti 2011b. Long-term human impacts on genetic structure of Italian walnut inferred by SSR markers. *Tree Genet. Genom.*, *7*(4): 707–723.

Pollegioni, P., K.E. Woeste, F. Chiocchini, S. Del Lungo, I. Olimpieri, V. Tortolano, J. Clark, G.E. Hemery, S. Mapelli, and M.E. Malvolti. 2015. Ancient humans influenced the current spatial genetic structure of common walnut populations in Asia. *PloS One*, *10*(9): e0135980.

Pollegioni, P., K.E. Woeste, F. Chiocchini, I. Olimpieri, V. Tortolano, J. Clark, G.E. Hemery, S. Mapelli, and M.E. Malvolti. (2014). Landscape genetics of Persian walnut (*Juglans regia* L.) across its Asian range. *Tree Genet. Genom.*, *10*(4): 1027–1043.

Pop, I.F., A.C. Vicol, M. Botu, P.A. Raica, K. Vahdati, and D. Pamfil. 2013. Relationships of walnut cultivars in a germplasm collection: Comparative analysis of phenotypic and molecular data. *Sci. Hortic.*, *153*: 124–135. doi:10.1016/j.scienta.2013.02.013.

Rahimipanah, M., M. Hamedi, and M.J.A.J.F.S.T Mirzapour. 2010. Antioxidant activity and phenolic contents of Persian walnut (*Juglans regia* L.) green husk extract. *African J. Food Sci. and Tech.*, *1*(4): 105–111.

Rebufa, C., J. Artaud, and Y. Le Dreau. 2022. Walnut (*Juglans regia* L.) oil chemical composition depending on variety, locality, extraction process and storage conditions: A comprehensive review. *J. of Food Composition and Analysis*, *110*: 104534.

Rezaei, Z., A. Khadivi, B. Valizadeh Kaji, and A. Abbasifar. 2018a. The selection of superior walnut (*Juglans regia* L.) genotypes as revealed by morphological characterization. *Euphytica*, *214*: 1–14.

Rezaei, Z., A. Khadivi, B. ValizadehKaji, and A. Abbasifar. 2018b. The selection of superior walnut (*Juglans regia* L.) genotypes as revealed by morphological characterization. *Euphytica*, *214*(4). doi:10.1007/s10681-018-2153-z.

Roor, W., H. Konrad, D. Mamadjanov, and T. Geburek 2017. Population Differentiation in Common Walnut (*Juglans regia* L.) across Major Parts of Its Native Range Insights from Molecular and Morphometric Data. *J. Heredity*, *108*(4): 391–404. doi:10.1093/jhered/esw122.

Ros, E. 2010. Health benefits of nut consumption. *Nutrients*, *2*(7): 652–682.

Sarikhani, S., K. Arzani, M.R. Roozban, and K. Vahdati. 2023. Phenological and pomological evaluation reveals high diversity among walnut populations in southwestern Iran. *J. Nuts*, *14*(2): 129–140.

Shah, R.A., P. Bakshi, A. Jasrotia, H. Itoo, and R. Gupta. 2022. Bio-Chemical Composition of Some Walnut (*Juglans regia* L.) Genotypes of northwestern Himalayan region. *Bangladesh J. Bot.*, *51*(1): 93–101.

Shah, R.A., P. Bakshi, A. Jasrotia, H. Itoo, B.A. Padder, R. Gupta, G. Kour, and D. Dolkar 2023a. Morphological to Molecular markers: Plant genetic diversity studies in walnut (*Juglans regia* L.): A review. *Erwerbs-Obstbau.* doi:https://doi.org/10.1007/s10341-023-00892-x.

Shah, R.A., P. Bakshi, A. Jasrotia, V.K. Wali, S. Sharma, M. Gupta, R.K. Gupta, and M. Jamwal. 2023b. Comparative morpho-molecular characterization of elite walnut variety Parbat (JWSP-06) with local selections of northwestern Himalayan region of Jammu and Kashmir, India. *Sci. Hortic.*, *319*: 112176.

Shah, R. A., P. Bakshi, N. Sharma, A. Jasrotia, H. Itoo, R. Gupta, and A. Singh. 2021. Diversity assessment and selection of superior Persian walnut (*Juglans regia* L.) trees of seedling origin from North-Western Himalayan region. *Resources, Environment, and Sustainability*, *3*: 100015. doi:10.1016/j.resenv.2021.100015.

Shah, R.A., P. Baksi, A. Jasrotia, D.J.I. Bhat, R. Gupta, and M. Bakshi. 2020. Genetic diversity of walnut (*Juglans regia* L.) seedlings through SSR markers in north-western Himalayan region of Jammu. *Bangladesh J. Bot.*, *49*(4): 1003–1012. doi:10.3329/bjb.v49i4.52517.

Shamlu, F., M.S. Rezaei, A. Lawson, A. Ebrahimi, Biabani, and A. Khan-Ahmadi. 2018. Genetic diversity of superior Persian walnut genotypes in Azadshahr, Iran. *Physiology and Molecular Biology of Plants: An International J. of Functional Plant Biology*, *24*(5): 939–949. doi:10.1007/s12298-018-0573-9.

Sharma, R.M., K. Kour, B. Singh, S. Yadav, N. Kotwal, J.C. Rana, and R. Anand. 2014. Selection and characterization of elite walnut (*Juglans regia* L.) clone from seedling origin trees in north western Himalayan region of India. *Australian J. of Crop Science*, *8*(2): 257–262.

Sideli, G., M. P. Mcatee, A. Marrano, B.J. Allen, P.J. Brown, T.S. Butterfield, A.M. Dandekar, C.A. Leslie, and D.B. Neale. 2020. Genetic analysis of walnut (*Juglans regia* L.) pellicle pigment variation through a novel, high-throughput phenotyping platform. *G3: Genes, Genomes, Genetics*, *10*(12): 4411–4424.

Solhjoo, S., R. Fatahi, Z. Zamani, A. Chehregani, F. Palumbo, and G. Barcaccia. 2023. Is apomixis occurring in walnut (*Juglans regia* L.)? New data from progeny molecular tests and cytological investigations shed light on its reproductive system. *Frontiers in Plant Science*, *14*: 1270381.

Soveili, S., and A. Khadivi. 2023. Selecting the superior late-leafing genotypes of Persian walnut (*Juglans regia* L.) using morphological and pomological evaluations. *BMC Plant Biol.*, *23*(1): 379.

Stevens, K.A., K. Woeste, S. Chakraborty, M.W. Crepeau, C.A. Leslie, P.J. Martinez-Garcia, D. Puiu, J. Romero-Severson, M. Coggeshall, and A.M. Dandekar. 2018. Genomic variation among and within six Juglans species. *G3: Genes, Genomes, Genetics*, *8*(7): 2153–2165.

Sutyemez, M. 2016. New Walnut Cultivars: Maras 18, Sutyemez 1, and Kaman 1. *HortScience*, *51*(10): 1301–1303. doi:10.21273/hortsci10972-16.

Sutyemez, M., S.B. Bukucu, and A. Ozcan. 2021. "Helete Gunesi", a new walnut cultivar with late leafing, early harvest date, and superior nut traits. *Agriculture*, *11*(10): 991.

Sytykiewicz, H., G. Chrzanowski, P. Czerniewicz, B. Leszczynski, I. Sprawka, R. Krzyzanowski, and H.J.P.J.O.E.S. Matok. 2015. Antifungal activity of *Juglans regia* (L.) leaf extracts against candida albicans isolates. *Polish J. of Environmental Studies,. 24*(3): 1339–1348.

Szugyi-Bartha, K., G. Bujdoso, V. Froemel-Hajnal, S. Szugyi, and E. Stefanovits-Banyai. 2021. Evaluation of the frost tolerance of Hungarian-bred walnut cultivars. *Acta Biol. Szeged*, *65*(2): 163–170.

Thompson, M.B. 2023. *Intraspecific Hybrid Identification of Black Walnuts Via Marker Assisted Selection*. M.S. Thesis, College of Agriculture, Missouri State University.

Topcu, H. 2021. Conventional and Molecular Breeding in Walnut. *In*: Muhammet Ali Gundesli: *Walnut*, 249. Ankara, Turkey: Iskad Publishing House.

Vahdati, K. 2014. Traditions and Folks for Walnut growing around the Silk Road. *Acta Hortic.*, *1032*: 19–24. doi:10.17660/actahortic.2014.1032.1.

Vahdati, K., M.M. Arab, S. Sarikhani, M. Sadat-Hosseini, C.A. Leslie, and P.J. Brown. 2019. Advances in Persian walnut (*Juglans regia* L.) breeding strategies. *Advances in Plant Breeding Strategies: Nut and Beverage Crops*, *4*: 401–472.

Vahdati, K., M. Lotfi, M. and Sadat Hosseini Grouh. 2014. Karyotype analysis of haploid plants of walnut (*Juglans regia* L.). *In*: *II International Symposium on Biotechnology of Fruit Species, Acta Hort.*, Article No.1048, 225–228.

Van Hintum, T.J., A. Brown, and C.Spillane. 2000. Core collections of plant genetic resources (No. 3). *Bioversity International*. IPGRI.

Van Nocker, S., and S.E. Gardiner. 2014. Breeding better cultivars, faster: Applications of new technologies for the rapid deployment of superior horticultural tree crops. *Horticulture Research*, *1*: 14022. https://doi.org/10.1038/hortres.2014.22.

Vieira, V., C. Pereira, T.C.S.P. Pires, R.C. Calhelha, M.J. Alves, O. Ferreira, L. Barros, and I.C.F.R. Ferreira. 2019. Phenolic profile, antioxidant and antibacterial properties of *Juglans regia* L.(walnut) leaves from the northeast of Portugal. *Ind. Crops Prod.*, *134*: 347–355.

Vischi, M., C. Chiaba, S. Raranciuc, L. Poggetti, R. Messina, P. Ermacora, G. Cipriani, D. Paffetti, C.Vettori, and R. Testolin. 2017. Genetic Diversity of Walnut (*Juglans regia* L.) in the Eastern Italian Alps. *Forests*, *8*(3): 81. doi:10.3390/f8030081.

Wang, G., D. Zhong, H. Liu, T. Yang, Q. Liang, J. Wang, R. Zhang, and Y. Zhang. 2021. Water soluble dietary fiber from walnut meal as a prebiotic in preventing metabolic syndrome. *J. of Functional Foods*, *78*: 104358.

Wang, P., L. Zhong, H. Yang, F. Zhu, X. Hou, C. Wu, R. Zhang, and Y.J.L. Cheng. 2022a. Comparative analysis of antioxidant activities between dried and fresh walnut kernels by metabolomic approaches. *LWT*, *155*: 112875.

Wang, Q., L. Fang, J. Chen, Y. Hu, Z. Si, S. Wang, L. Chang, W. Guo, and T. Zhang, 2015. Genome-wide mining, characterization and development of microsatellite markers in Gossypium species. *Scientific Reports*, *5*(1): 1–10.

Wang, R., D. Zhong, S. Wu, Y. Han, Y. Zheng, F. Tang, Z. Ni, and Y. Liu. 2020. The phytochemical profiles for walnuts (*J. regia and J. sigillata*) from China with protected geographical indications. *Food Sci. Technol.*, *41*: 695–701.

Wang, Z., H. Zhang, B. Tong, B. Han, D. Liu, P. Zhang, and D. Hu. 2022b. The study on genetic diversity and genetic structure of *Juglans mandshurica* in Shandong Province based on EST-SSR. https://doi.org/10.21203/rs.3.rs-1294633/v1 (preprint only, unpublished).

Wu, G.L., H.J. Meng, Y.Y. Hao, Q.L. Liu, D.Wang, and J.B. Tian. 2009. Thirty years of breeding walnut in China. *Acta Hortic.*, *861*: 109–118. https://doi.org/10.17660/ActaHortic.2010.861.14.

Yavuz, M., and L. Pirlak. 2023. Selection of Superior Properties of Walnut Genotypes in Eregli-Halkapınar (Konya) and Ayrancı (Karaman) Walnut Population. *Selcuk J. of Agriculture and Food Sciences*, *37*(3): 457–465.

Zaini, P.A., N.G. Feinberg, F.S. Grilo, H.J. Saxe, M.R. Salemi, B.S. Phinney, C.H. Crisosto, and A.M. Dandekar. 2020. Comparative proteomic analysis of walnut (*Juglans regia* L.) pellicle tissues reveals the regulation of nut quality attributes. *Life*, *10*(12): 314.

Zhang, B.W., L.L. Xu, N. Li, P.C. Yan, X.H. Jiang, K.E. Woeste, K. Lin, S.S. Renner, D.Y. Zhang, and W.N. Bai. 2019. Phylogenomics reveals an ancient hybrid origin of the Persian walnut. *Mol. Biol. Evol.*, *36*(11): 2451–2461.

Zhang, J., W. Zhang, F. Ji, J. Qiu, X. Song, D. Bu, G. Pan, Q. Ma, J. Chen, and R. Huang. 2020. A high-quality walnut genome assembly reveals extensive gene expression divergences after whole-genome duplication. *Plant Biotechnol. J., 18*(9): 1848.

Zec, M.M., I. Krga, M. Takic, J. Debeljak-Martacic, G. Koriccanac, S. Rankovic, T. Popovic, M. Pantelić, and M. Glibetic, 2020. Walnut consumption induces a tissue-specific omega-6/omega-3 decrease in high-fructose-fed Wistar rats. *ACS Omega, 5*(43): 28136–28145.

Zhang, M.Y., Y. Xu, K. Xiang, and X.L. Wang. 2014. Review of walnut breeding research at the shandong institute of pomology. *In*: *2014 International Society for Horticultural Science* (*ISHS*), 55–60. Leuven, Belgium. doi:10.17660/ActaHortic.2014.1050.5.

Zhang, Y. N., Y. Dong, Y. Hao, J. Chen, Z. Zhang, S. Hou, J. Su, and J. Qi Wu. 2023. Nuts phenotypic diversity analysis and character evaluation of 109 high-yield walnut individual trees. *J. Nanjing For Univ., 47*(3): 87.

Zhao, P., H.J. Zhou, D. Potter, Y.H. Hu, X. J. Feng, M. Dang, L. Feng, S. Zulfiqar, W. Z. Liu, and G. F. Zhao. 2018. Population genetics, phylogenomics, and hybrid speciation of *Juglans* in China determined from whole chloroplast genomes, transcriptomes, and genotyping-by-sequencing (GBS). *Mol. Phylogen. Evol., 126*: 250–265.

Zhou, H., P. Zhao, K. Woeste, and S. Zhang. 2021. Gene flow among wild and cultivated common walnut (*Juglans regia*) trees in the Qinling Mountains revealed by microsatellite markers. *J. For. Res., 32*: 2189–2201.

19

Genetic Diversity of Cashew (*Anacardium occidentale* L.) and Sustainable Utilization

Siddanna Savadi, Muralidhara B.M., Manjunatha K., Mohana G.S.* and *Eradasappa E.*

Introduction

Cashew (*Anacardium occidentale* L.) is a tropical tree with its origins in the eastern Brazil region. In global trade, cashew stands second among the edible tree nuts. It is cultivated in > 30 countries over 6 million hectares, and world cashew nut production is > 4.9 million tons (UNCTAD, 2021). Vietnam, Nigeria, India, and Cote d'Ivoire are the major producers of cashew nuts (FAOSTAT, 2019). During the medieval era, the Spanish sailors introduced cashews to Panama and Central America, while the Portuguese introduced them to Mozambique and India (De Castro, 1994). It is believed that cashews spread to other regions of South Asia from India (Berry and Sargent, 2011). The Portuguese purpose of introducing cashews to its colonies was to check soil erosion and recover the wastelands (Bradtke, 2007).

The kernel within the cashew nut is the main economic product of the cashew tree. It is rich in nutrients and possesses about 21% proteins, 23% carbohydrates, and 46% fats. Among the fats, 80% are unsaturated fats, which help in maintaining reduced cholesterol levels in the body. Besides, cashew kernels are rich in minerals such as calcium, iron, magnesium, phosphorus, potassium, sodium, and zinc, and vitamins such as thiamin, riboflavin, niacin, beta-carotene, retinol, and vitamin K (Rico et al., 2016). Kernels are extracted from the raw cashew nuts through industrial processing. The whole kernels are consumed as deserts, and the pieces are used in confectionery and industries. Cashew kernel oil is important in the cosmetic industry. The cashew apple, or the pseudo fruit to which the nut is attached, possesses several important phytochemicals and is another edible part of the cashew. Mature cashew apples are processed into products such as jam, jellies, and beverages (Wickens and Chomsky, 1995). A byproduct of the cashew nut processing industry referred to as cashew nut shell liquid (CNSL), phospholipids present in a nutshell, have immense industrial and green energy potential and are currently used in insulating varnishes, brake linings, etc. (Lubi and Thachil, 2000). In recent decades, cashew cultivation has assumed great priority, especially in developing countries, due to

ICAR-Directorate of Cashew Research (DCR), Puttur, Dakshina Kannada, Karnataka 574202, India.
* Corresponding author: siddannasavadi@gmail.com

its economic and social importance. The global market for cashew nuts is worth about US$7 billion (Straits Research Report, 2022). The western countries, viz., the USA, the European Union, and the Middle East countries, viz., the UAE and Saudi Arabia are the major importers of cashew kernels. Besides, the worth of CNSL in the global market is about US$ 250 million (Straits Research Report, 2022). In recent years, the global consumption of cashew nuts has increased due to its health benefits and increasing earnings in developing countries, especially in India and China (Kahlmann and Kohn, 2018). However, cashew production is too low to meet future demand for cashews (Kahlmann and Kohn, 2018). For instance, the demand for raw cashew nuts in India is projected to be 4–5 million tons by 2050. The major reasons for low cashew nut productivity are the use of low-yielding varieties, losses due to biotic and abiotic stresses, and meager crop management (Saroj and Mohana, 2016). Genetic improvement of cashews to improve yield and quality with scientific interventions began in the 1950s in Brazil and India (Bhat et al., 2010; Queiroz et al., 2012), and it has resulted in many cashew cultivars with enhanced yield.

Taxonomic Background

The cashew tree is a member of the Anacardiaceae family, in which other economically significant fruit and nut trees, viz., mango and pistachio, also exist. Cashew belongs to the genus *Anacardium*, and the word Anacardium is a derivative of the Greek words 'Ana' denoting 'upwards' and 'Cardium' denoting 'heart', which refers to the nut as the heart of the fruit situated outwardly. The *Anacardium* genus comprises more than eight different species, such as *A. nanum* St. Hilaire (a small shrub), *A. subterraneum* Liais (a small shrub with an underground trunk acting as a water reserve), *A. microcarpum* Ducke (a dwarf tree form), and *A. spruceanum* Bth. (Palei et al., 2019).

Although there are several significant characteristics to determine a species within the genus *Anacardium*, Valeriano (1972) proposed that distinguishing these *Anacardium* species into dwarf and giant species is a sure, rational, and practical means of grouping. In the genus, these species have extensive variations in the size and shape of the nuts, color, size, and shape of the peduncle, leaf size, leaf shape, etc. Among these *Anacardium* species, *A. occidentale* is the lone species with commercial importance and is cultivated over a large area. Marking nut (*Semecarpus anacardium*), which has apples and nuts similar to cashew apples and nuts, respectively (Mangesh et al., 2022). However, the sizes of the apple and nut are smaller compared to the cashew.

The cashew tree is a small to medium-sized tree growing about 10–15 m tall with a canopy spread of 10–15 m (Salifou and Masawe, 2018). The cashew tree has a deep taproot system with widespread lateral roots. Leaves are simple, complete, obtuse, and green and are positioned alternately on the shoots. The inflorescence, a panicle with a large number of flowers comprising both perfect (hermaphrodite) flowers and staminate (male) flowers, is produced at the terminal position of the shoots. The fresh flowers are white in color and turn pink a few hours (4–5) after opening. Both perfect (hermaphrodite) flowers and staminate (male) flowers have five sepals, five petals, one large stamen, and 6–14 small stamens. Both large and small stamens produce pollen grains and are sticky by nature. The pistil of a cashew flower has three parts, viz., stigma, style, and ovary (Wunnachit and Sedgley, 1992). Cashew is highly cross-pollinated due to the sticky nature of the pollen, and honeybees are the major pollinators of cashews (Vanitha and Raviprasad, 2019).

Usually, two vegetative and one reproductive flush occur in the annual cycle of cashews. However, two reproductive blooms are also observed in a year in regions with two dry seasons (Ohler, 1979). In cashew, 37 phenological growth stages were identified, codified, and described according to the BBCH (Biologische Bundesantalt, Bundessortenamt and Chemische Industrie) scale, a uniform phenological scale accommodating different growth stages in different species (Adiga et al., 2019).

Cashew fruit or nut is formed after the fertilization of hermaphrodite flowers, and it grows attached to a swollen pedicel or hypocarp called a cashew apple, which is misconceptualized as a fruit (Malhotra et al., 2017). Botanically, drupe is the fruit type of cashew. Cashew nut consists of three distinct layers, viz., exocarp, mesocarp, and endocarp. Cashew nut has a grayish hard coat (nutshell) on the outside and a kernel bounded by a brownish peel called the testa inside the nutshell. Cashew nut or seed has two cotyledons and the embryonic axes. Cashew apple, the pseudofruit, originates from the thalamus, receptacle, or stalk existing external to the ovary. During fruit development, the nut grows fast and attains its full size, and then the cashew apple starts growing and developing at a faster rate to attain its full size and maturity (Adiga et al., 2019).

Collection and Assessment of Germplasm/Genetic Resources

Germplasm is the entire hereditary material, i.e., all the alleles of various genes, existing in a crop species and its close wild relatives. This is also referred to as genetic resources, gene pools, or genetic stock (Singh et al., 2020). Habitat destruction is the main factor responsible for the reduction of biodiversity, and hence there is an increasing concern about the need to preserve important genetic resources for sustained food production and agriculture. Over the years, the collection and conservation of crop germplasm using different technologies has become more compelling due to concern over the loss of biological diversity and the economic importance of accessibility to the germplasm resources required for sustained food production. Germplasm is collected from centers of diversity, genebanks, gene sanctuaries, farmer's fields, markets, seed companies, etc. (Singh et al., 2020).

The main activities in collection and conservation include a collection of germplasm by introduction, exchange, quarantine, and conservation: *in situ* (or in nature or on the farm) and *ex situ* (seeds cold storage; *in vitro* culture; cryopreservation); and characterization of germplasm at morphological, biochemical, cytological, and molecular levels (Mariante et al., 2009). The germplasm collection may include the collection of propagating materials such as seeds, plants, or plant parts valuable in crop breeding, research, and conservation efforts. Germplasm collections comprise different kinds of plant materials used in breeding and research. The germplasm of crops comprises current and obsolete varieties, breeders' lines, superior varieties, intermated populations, primitive landraces, specifically adapted ecotypes, and researchers developed genetic stocks with unique genes, mutations, duplications, translocated chromosomes, tissue culture-derived cybrids, etc. (Singh et al., 2020).

Characterization and Evaluation of Germplasm

Characterization of germplasm refers to the recording of distinctly identifiable traits of accessions in collections, and germplasm evaluation refers to the recording and describing of the genetic material in a gene bank for characters that are important in crop improvement. Usually, a set of traits referred to as 'Descriptors' that provide the basic description of genotype are used for the evaluation of germplasm accessions (Asati et al., 2023). Lists of descriptors comprise the basic description of the characters and their classes of expression according to the criteria used by breeders and other germplasm users. Descriptors include highly heritable characters, viz., morphological features, biochemical characters, as well as molecular markers, which enable easy and quick distinction among genotypes and allow their grouping, as well as to verify the trueness-to-type (Volkova, 2011; Silva et al., 2013).

Morphological Characterization

Conventionally, the genetic diversity of crop species was evaluated based on variations in morphological characters (Schut et al., 1997). Besides, morphological characters have also been utilized as a tool for the classification of cultivars, and the evaluation of such morphological characters has remained the first step in understanding the genetic relationships in the majority of breeding programs as they are easy to record and show potential indications of genetic improvement (Van Beuningen and Busch, 1997). Cashew breeding is mainly based on the selection of useful morphological and agronomic characteristics such as nut size, nut weight, tree size/canopy, panicle size, and yield (Chabi Sika et al., 2015; Sethi et al., 2016; Palei et al., 2019; Matos et al., 2019; Adu-Gyamfi et al., 2020).

Although morphological traits are easy to record and used regularly in breeding programs, they are highly influenced by environmental factors and are insufficient in number to differentiate a large number of accessions or genotypes, so distinguishing cultivars of crops based on morphological traits is sometimes ambiguous, especially for the quantitative traits. Besides, the process takes more time and is subject to individual prejudice (Ikabanga et al., 2017).

Cytological Characterization

Cytogenetic analysis, the study of the chromosomes using tools such as flow cytometry and fluorescence in situ hybridization (FISH) provides information on the number and structure of chromosomes in genomes. Knowledge of the basic cytology (number, structure, and behavior of chromosomes) of accessions is essential for achieving the desired improvement in crop improvement programs. However, cytological investigations of cashews are limited compared to other important tropical industrial crops such as oil palm, coffee, cacao, and tea. Cytological investigations are also useful in resolving ambiguities about the origin and evolution of plant species. Variations in the chromosome structures or numbers would affect the DNA content and result in genetic changes. Therefore, chromosomal or cytological investigations aid in determining the evolutionary path of new species. Cytogenetics has been utilized in breeding for the development of improved cultivars, especially to understand traits like infertility in species (Aliyu and Awopetu, 2007).

In cashew, there was ambiguity about the chromosome number and it was reported to have a polymorphic chromosome number of 2n = 24, 30, 40, 42 (Deckers et al., 2001). Recently, Aliyu and Awopetu (2007) investigated the number, structure, and behavior of chromosomes in Indian and Brazilian cashew populations using microscopy. Cytological investigation of both populations revealed a diploid chromosome number of 2n = 42 and a haploid chromosome number of n = 21, suggesting a lack of polymorphism in the chromosome number. The karyotyping showed that the chromosomes were mostly symmetric, composed mainly of metacentric pairs and several sub-metacentric. The total length of all 21 chromosomes was 51.10 μm and 56.00 μm in the Indian and Brazilian cashew populations, respectively. The range of chromosome lengths was 1.00–4.20 μm and 1.00–4.50 μm for Indian and Brazilian cashew populations, respectively. The karyotyping in the Indian population suggested six long sub-metacentric, one long metacentric, five intermediate metacentric, four intermediate sub-metacentric, and five small metacentric chromosomes, while in the Brazilian population, six long sub-metacentric, one long metacentric, one long sub-telocentric, nine intermediate metacentric, two intermediate sub-metacentric, and two small metacentric chromosomes existed during regular mitotic division (Aliyu and Awopetu, 2007). Karyotyping results showed slight variations in chromosome size and shape, especially the presence of a pair of sub-telocentric chromosomes in the Brazilian population, which suggests intraspecies variation. The slight variations in chromosome sizes and shapes can be explained using Stebbins (1971) gene duplication and multiple-strand hypotheses. A high degree of genomic relationship in populations suggests the exchange of genes among accessions and common origins.

Generally, 21 bivalents were observed in meiotic divisions of both populations, but a few triads in 3 Brazilian accessions suggest irregularities during meiosis (microsporogenesis), which might have an impact on the fertility of their pollens (Aliyu and Awopetu 2007). Further, reports on the genome size of cashew accessions are scanty. Aliyu et al. (2014) examined the nuclei of 43 cashew accessions from diverse regions using flow cytometry with *Solanum lycopersicum* var. Stupicke has a genome size of 1.96 pg/2C as an internal reference standard to determine the absolute nuclear DNA content of accessions. Flow cytometric analysis showed that the absolute genome size of cashew accessions varied from 0.743 pg/2C (363 Mb/1C) to 0.899 pg/2C (440 Mb/1C), with a mean of about 0.857 pg/2C (419 Mb/1C), suggesting differences in the genome sizes of accessions and intra-specific variations in nuclear DNA content, but no endopolyploidy or endoreduplication was observed (Aliyu et al., 2014).

Biochemical Characterization

Leveraging biochemistry to understand useful chemical compounds will facilitate sustainable crop production through enhanced crop quality and stress tolerances, reductions in post-harvest losses, and increased yields (Razzaq et al., 2019). Biochemistry has allowed for an understanding of how metabolisms of different compounds occur and the mechanisms that play a key role in plant growth and development (Klee, 2010).

Traditionally, the quality of cashew nuts was described by morphological descriptors such as kernel size and shape. However, the versatile uses of cashew nuts and other products of the cashew tree necessitate a systematic examination of the chemical composition of kernels and other products. So far, the chemical composition of cashew kernels or other products is barely recognized as a breeding objective in cashew breeding programs. Despite the availability of a large number of cashew cultivars, very limited biochemical data is available to substantiate their high nutritional content and potential uses in industries and food preparation (Aremu et al., 2006; Akinhanmi et al., 2008; Kshirsagar et al., 2011). A thorough analysis of the biochemical compositions of cashew kernels and other products in cashew accessions will strengthen the breeding programs to focus on improving desirable nutritional content in future cashew cultivars.

The germplasm characterization using high-throughput biochemical fingerprinting technologies would facilitate enhanced use of germplasm with better quality and other agronomic traits in crop improvement programs (Razzaq et al., 2019). The consumption of raw cashew apples is inhibited due to the astringency present. Astringency is mainly due to tannin content, and accessions with low tannin content in cashew apples become important in breeding programs. It is demonstrated that apples of *A. microcarpum* have a very low tannin content (0.12%) compared to the clones (CCP 09 and CP 76), indicating the importance of the development of cultivars with low tannin content in cashew apples (Costa et al., 2003).

Molecular Characterization

Traditionally, genetic diversity and genetic studies in crop species were investigated based on variations observed in morphological (quantitative and qualitative) characters. However, morphological characters are often influenced by environmental variations and stages of development, especially the quantitative characters, and are less in number to unequivocally characterize germplasm accessions. Recently, molecular markers have been extensively used for genetic diversity and other genetic studies as they exhibit high polymorphism, are not influenced by environmental variations or stages of development, are unlimited in number, are usually more objective, and are easier to analyze compared to morphological markers. Hence, the use of molecular markers along with morphological markers would allow unequivocal characterization and use of germplasm. Until recently, few molecular markers and genomic resources were available for genetic studies and breeding in cashews. In the beginning, first-generation dominant DNA markers such as RAPD, ISSR, and/or AFLP markers were used to study genetic diversity in the germplasm accessions of different countries (Mneney et al., 2001; Dhanaraj et al., 2002; Archak et al., 2003; Aliyu, 2012; Jena et al., 2016; dos Santos et al., 2019; Borges et al., 2018). Of late, codominant markers, viz., microsatellite or simple sequence markers (SSR), InDels, and single nucleotide polymorphism (SNP) markers, have been developed in cashews. However, until recently, the number of SSRs for genetic studies in cashews was very scarce (Croxford et al., 2006). Recently, the codominant markers, namely, SSRs and InDel markers, have been enhanced in cashews (Savadi et al., 2022; Savadi et al., 2023a, b). SSR markers, and in recent times, InDel markers, have been widely applied for genetic studies and breeding in several crop plants because of their procedural simplicity and low cost. Moreover, SSR and InDel markers are cross-species amplifiable and can be used for genetic studies in related species as well. Cashew (*A. occidentale*) SSR markers were amplifiable in *A. humile*, a shrub growing in the Brazilian savannas (Cota et al., 2012; Soares et al., 2013). Genetic relatedness and gene diversity of germplasm allow classification of germplasm which is useful for breeders to design breeding schemes. Molecular characterization of cashew germplasm accessions conserved in different countries has been carried out to understand genetic relatedness and gene diversity (Table 1).

Table 1. List of molecular characterization studies in cashew germplasm accessions conserved in different countries (adopted from Savadi et al. 2020 and updated).

Study	Country	Genetic Resource Number and Type	Marker Type	Results
Mneney et al. (2000)	Tanzania	Cultivars from different regions of the world and 20 elite Tanzanian varieties	RAPD	A high degree of genetic similarity was found among accessions from Tanzanian, India, and Mozambique whereas accessions from Brazil were most distinct from the other regions of the world
Dhanaraj et al. (2001)	India	90 germplasm accessions	RAPD	Euclidean distance suggested moderate to high genetic diversity
Archak et al. (2003a)	India	19 germplasm accessions	RAPD ISSR, AFLP	AFLP was found to be superior for genetic analysis than RAPD and ISSR
Archak et al. (2003b)	India	24 selections and 11 hybrids	RAPD, ISSR	Low genetic variations suggest the need for the induction of more diverse parental lines in breeding programmes
Archak et al. (2009)	India	91 germplasm accessions	AFLP	No definite population differentiation was found indicating cashew being introduced at a single location in India
Aliyu and Awopetu (2007)	Nigeria	3 populations	Protein-isoenzyme	Moderate genetic diversity among Nigerian collections and clustering indicated the eco-geographical origin of accessions. Further, the closer genetic relationship was observed between Indian and Nigerian populations
Chipojola et al. (2009)	Malawi	4 Malawian populations	Morphological markers	Similarity values ranged from 35–66%, accessions formed four clusters with 14 sub-clusters in clustering
Thimmappaiah et al. (2009)	India	100 germplasm accessions	RAPD, ISSR	Moderate genetic diversity and accessions formed 13 clusters. Correlation was observed between the groupings based on molecular profiling and the morphological characters
Aliyu et al. (2012)	Nigeria	187 germplasm accessions	SSR	Existence of a high level of redundancy (homogenous group) in the Nigerian germplasm which indicated the need for the development of a core collection Revealed the existence of a narrow genetic base of germplasm
Chavan and Raut (2013)	India	11 varieties of Eastern Vidarbha zone	Morphological, RAPD	Varieties were classified into 3 clusters and the similarity coefficient ranged from 0.47–0.94
Ranathunge et al. (2014)	Sri Lanka	15 germplasm accessions	RAPD	The average genetic distance was 0.257 indicating the close genetic relatedness among the genotypes
Dasmohapatra et al. (2014)	India	25 cultivars	RAPD, ISSR	Genotypes are classified into 14 clusters. Varieties BBSR-1 and Vengurla-7 were grouped into a single cluster which are phenotypically similar
Jena et al. (2015)	India	12 promising cultivars	RAPD	Cultivars were grouped into 2 clusters
Chabi Sika et al. (2015)	Benin	60 germplasm accessions	SSR	Germplasm accessions were grouped based on the geographical regions

Table 1. contd. ...

Table 1. contd. ...

Study	Country	Genetic Resource Number and Type	Marker Type	Results
Sethi et al. (2016)	India	20 hybrids and their 8 parents	RAPD	Two hybrids H 6 and D 19 showed a high level of genetic dissimilarity compared to other genotypes. Besides, the genotype-specific bands helpful for fingerprinting of genotypes were identified
Thimmappaiah et al. (2016)	India	40 varieties	RAPD, ISSR, SSR	Genetic similarity between pair of varieties was high among the 40 varieties developed in the Indian cashew breeding program
Borges et al. (2018)	Brazil	25 accessions of 2 populations of cashew and its wild relatives	ISSR	The high degree of genetic polymorphism was observed (91.3%). STRUCTURE analysis showed 2 distinct groups
Hawerroth et al. (2019)	Brazil	77 genotypes including hybrids and their parents	RAPD, ISSR	Genotypes were classified into 21 clusters suggesting high genetic divergence among the breeding material
dos Santos et al. (2019)	Brazil	8 populations (30 individuals/population) of northern Piaui state	ISSR	Genetic diversity was higher in wild than in domesticated groups, significant structure existed in 8 populations, and wild populations showed strong heterogeneity within small regions despite high chances of gene flow
Pereira et al. (2019)	Brazil	542 plants of Bushy Cashew (*A. humile*) from Goiás state	Physical, physicochemical and chemical traits	High heterogeneity among accessions, cashew apples have contained high levels of vitamin C and carotenoids suggesting the high nutritional status of *A. humile*
Kouakou et al. (2020)	Côte d'Ivoire	221 individuals from commercial orchards of different regions	SSR	Total genetic variability at the intra-population level is 98.6% with only 1.4% variability among populations, the average gene flow was high suggesting gene flow within populations maintains high intra-population genetic diversity
Savadi et al. (2021)	India	23 accessions	SSR	Dendrogram analysis of 23 accessions formed two main clusters with two sub-clusters in each. Clustering of the accessions had no relationship with the geographic region of collection.
Savadi et al. (2022)	India	64 accessions	Genic SSRs	A total of 206 alleles were detected and germplasm accessions were clustered into three major clusters. PCoA analysis showed differentiation of accessions according to geographic regions to a certain extent. Structure analysis based on the Bayesian model-based clustering showed the existence of 3 genetic subpopulations (K=3) in studied accessions.
Savadi et al. (2023a)	India	64 accessions	InDel markers	The 64 accessions were clustered into three main clusters and 200 alleles were detected. AMOVA analysis revealed about 6% of genetic variations were attributable to among populations while 94% were attributable to individuals within and among the population

Utilization of Germplasm

Utilization in Breeding and Improvement Programs

Wide variability exists in the germplasm collections of different countries and in the wild in Brazil, which offers great scope for cashew genetic improvement (Cota et al., 2012; Soares et al., 2013). Evaluation of both local and exotic germplasm is needed to identify germplasm with useful alleles or allele combinations contributing to yield stability, food quality, and reduced dependence on external inputs (Dwivedi et al., 2016; Wang et al., 2017). Hence, a comprehensive evaluation of promising germplasm clones across locations with defined ecologies for yield stability and adaptability is vital to cashew breeding programs. The diversity present in the cashew fruit (nut and cashew apple) is shown in Figure 1 and a dwarf genotype conserved at ICAR-DCR, Puttur germplasm collection is shown in Figure 2.

Breeding objectives in cashews comprise the development of new varieties with high yield potential, dwarf and compact canopy type, short flowering duration, a high bisexual flower ratio, improved nut size with a high shelling percentage, tolerances to biotic (tea mosquito bug, diseases), and abiotic stresses (cold and salt tolerance), high nutritional quality, better cashew apple parameters, high and free CNSL type cultivars. Cashew is an export-oriented crop and hence, more emphasis is given to the development of new varieties with high yielding and bold size nuts to get kernels of superior grade. Achieving this objective is a great challenge to breeders as these two traits are in trade-off. A cashew breeder should intend to develop a new variety with stable and high yields over a period of time and geographic regions and minimum yield levels should be at least 10 kg/tree at the potential bearing tree stage under normal density planting. Cashew varieties with dwarf and compact canopy types, i.e., trees with around 4 m in height are needed for high-density planting systems, which allow maximum productivity per unit area especially during the initial years of planting. In Brazil, the majority of the commercially cultivated varieties have been dwarf type since 1980s. Further, dwarf types minimize the cost of cultivation areas as regular pruning is not required like in vigorous varieties. In cashews, there is wide diversity in flowering duration. However, short flowering duration varieties are preferred as they permit harvesting of maximum yields in a short time and also reduce the expenditure for crop management especially on the pests and disease control and also on harvesting of nuts. High flower sex ratio (ratio of hermaphrodite to staminate flowers), i.e., more of hermaphrodite/bisexual flowers is one of the major yields contributing character in cashew.

Figure 1. A view of diversity in the fruit size and color variations in cashew germplasm.

Figure 2. A view of dwarf cashew germplasm accessions at ICAR-DCR, Puttur.

Therefore, breeders should select trees with a higher sex ratio as parents in breeding programs. It will enhance the proportion of progenies with a high sex ratio and culminate in a higher fruit set. In cashew, Tea Mosquito Bug (*Helopeltis anotnii*) is a major sucking insect pest causing yield losses of up to 100% if control measures are not taken. However, there is no resistance source in existing germplasm. Hence, developing varieties having field tolerance and escaping type varieties, i.e., mid-season flowering types to escape from the attack of this pest. Further, attempts should made to generate novel germplasm providing resistance to this insect pest through mutations and genetic engineering for utilization in breeding resistant varieties. At ICAR-DCR, Puttur, mutation breeding is being used as a method for developing genetic stocks for resistance to this insect pest and also molecular interactions between the insect pest and the cashew tree are being studied to devise novel strategies for management of this pest.

Among the cashew diseases, anthracnose and black mold are the major fungal diseases in the world causing huge losses in yield and quality of nut and cashew apple. In India, no economically damaging diseases are reported. But in Brazil, the diseases are reported to cause severe losses and several disease resistant varieties have been developed over a period of time. However, anticipatory breeding may be initiated in India to breed for varieties with increased resistance to diseases keeping in mind the changing climatic scenario in the world as well as world trade and exchange of germplasm.

A high shelling percentage, i.e., higher kernel recovery from the raw cashew nuts is another important breeding objective as kernels are the main economical part of cashew. A new variety should have a shelling percentage of > 28% to enhance recovery and processing efficiency, which is a costly affair in the cashew industry. With increasing health awareness and per capita income, consumers prefer more nutritious and less allergens or anti-nutrients in the food consumed. Therefore, breeders should start tweaking breeding programs to include nutritional quality of both nuts and apples because the need has arisen to develop new varieties with nutrition-rich kernels especially protein and minerals, and less in sugars and free-from-nut allergens. The Cashew apple is another edible and major product of cashew. Currently, most of the cashew apple goes to waste due to its high perishability and astringency-causing antinutrients. Therefore, these cashew apple parameters along with high TSS, etc. must be considered in cashew breeding as an ancillary trait considering the huge economic demand in the food and beverage industry for cashew apples. CNSL is an industrially valued product and is widely used in paints and varnishes, resins, industrial and decorative laminates, brake linings, and rubber resins; recently its components anacardic acid, cardol, and cardanol have been shown to have great pharmacological value especially in treating cancers. However, the corrosive nature of CNSL hinders its use of young nuts in culinary. Therefore, genotypes with high CNSL and its varied components should be bred for industrial purposes while free types should be bred for free types. For instance, a Fruit research station at Vengurle, Maharashtra, India has identified a high yielding CNSL free cashew genotype, viz., RFRS 195 with potential for commercial cultivation, which serves culinary purposes as well as small-scale domestic processing. ICAR-DCR, Puttur, the coordinating center for the AICRP, and its AICRP centers have been working on most of these breeding aspects in cashew for decades and many varieties have been released for cultivation, and new varieties with improved quality and special traits are expected in the future over the period of time.

Cashew breeding is possible through either direct selection from the germplasm or population improvement using germplasm, followed by the selection and clonal evaluation of progenies with desired traits (Paiva et al., 2009). Paiva et al. (2009) suggested considering diverse genetic material to avoid the genetic vulnerability of varieties and to enhance genetic gains in breeding programs. In trees, utilizing diverse genetic plant material is possible through the polycross method. This method was used in cashew breeding in 1978 to develop dwarf, precocious, and high-yielding varieties (Barros et al., 1984). Hybridization is the widely used breeding method in cashew resulting in significant genetic variations in important yield and agronomic traits such as tree height, number of nuts per panicle, nut yield, canopy diameter, and nut and kernel weight (Sethi et al., 2016; Hawerroth et al., 2019). Recent advances in molecular biology and genomics of cashew will provide more tools to cashew breeders for expediting the breeding process through faster evaluation of progeny (Paiva et al., 2009; Savadi et al., 2020).

In India, several selection varieties and hybrids of cashews have been developed and released by AICRP on cashews using the germplasm collections at NAGS, ICAR-Directorate of Cashew Research (DCR), Puttur (Saroj and Mohana, 2016). Further, about 155 germplasm accessions with desired traits have been utilized in different hybridization programs at the ICAR-DCR, Puttur, and other cashew research centers under the AICRP system for breeding and other studies (Saroj and Mohana, 2016). In Brazil, germplasm banks have been utilized to develop early dwarf cashew clones since the 1980s (Castro et al., 2010). The dwarf cashew varieties were developed by hybridization of the dwarf germplasm with regular cashew genotypes to increase their nut size. Further, interspecific hybridizations of *A. occidentale, A. othonianum, A. occidentale,* and *A. microcarpum* were made to successfully transfer anthracnose resistance and apple quality traits in cultivated cashew (Castro et al., 2010).

Germplasm users find it difficult to handle or work with or to make decisions on the choice of diverse material from large germplasm collections. Therefore, the concept of core collections of germplasm was proposed in 1984 to reduce the number of accessions to be handled or worked with and at the same time ensure the representation of the maximum diversity within a set of accessions.

In cashew, a core collection of 49 accessions has been formed out of the germplasm collections of 478 accessions available at NAGS, ICAR-DCR using the advanced M strategy with a heuristic approach with 68 morphological characters (Mohana and Nayak, 2018a). The core set in cashews is of convenient size and allows comprehensive evaluation and enhanced use of germplasm in breeding programme. Further, the utilization of such diverse accessions will enrich breeding material through the infusion of new genetic diversity and allow the generation of broad-based cashew cultivars needed for sustainable cashew production in the face of changing climate.

Species and Cultivars were Released

In India, 63 cashew varieties have been released for commercial cultivation by the ACIRP centers working on cashew research through selections and hybridizations of evaluated germplasm. Among the 63 varieties released, 35 are selections and 28 are hybrids (Eradasappa and Mohana, 2023). In Brazil, the development of early dwarf cashew cultivars has enhanced yield and production because they are more productive, easy to grow, and more disease-resistant compared to the regular giant varieties (de Figueirêdo et al., 2016). Cashew dwarf clones, viz., CPC 06, CPC 09, CPC 76, and CPC 1001, were derived through selections from natural populations existing in the northeast coastal region of Brazil. Further, dwarf cashew clones, viz., Embrapa 50, Embrapa 51, BRS 189, BRS 226, 265, 274, and 275, were developed through different breeding methods, such as the polycross method and intra-specific hybridization (Paiva et al., 2008). Like India and Brazil, the major cashew producers, viz., Benin, Côte d'Ivoire, Mozambique, Nigeria, Tanzania, and Vietnam have developed cashew varieties adaptable to their country's climatic conditions. A list of cashew varieties and their tree-type characteristic is mentioned in Table 2.

Conservation of Useful Germplasm/Genetic Resources

The Convention on Biological Diversity (CDB) is an international legal establishment, created by the United Nations in 1998, for the conservation and sustainable use of biological diversity on earth, and for sharing of the benefits from the outcomes of the utilization of genetic resources fairly and equitably. CDB is signed by 196 nations, and it provides access and technology transfer for genetic resources' conservation, protection of traditional knowledge associated with genetic resources, and benefit sharing with the locals involved in the conservation of genetic resources over the centuries (Alves et al.,2018).

Germplasm conservation refers to the protection of genetic diversity present in crop plants' germplasm from genetic erosion. Two types of methods are used for germplasm conservation: i) *in situ* conservation, and ii) *ex situ* conservation. Conservation of germplasm accessions in natural habitats is called *in situ* conservation. Conservation of germplasm in gene banks outside their natural habitat is called *ex situ* conservation. This is the most practical method of conserving and utilizing germplasm. In cashews, comprehensive germplasm bank creation and conservation have been reported in Brazil and India. A schematic representation of the cashew germplasm conservation, utilization, evaluation, data documentation, and registration of unique accessions is given in Figure 3.

Germplasm Bank in Brazil

The cashew germplasm bank of Brazil has a collection of 621 cashew and related *Anacardium* species accessions. Cashew germplasm exhibited significant variability with respect to morphological characters, especially shoot and fruit attributes. Besides, Brazil, being the center of origin of cashews, has huge genetic variability in the wild (Castro et al., 2010).

The genetic diversity available in the cashew germplasm collection has permitted the development of novel cultivars, especially the early dwarf cashew clones, which are highly suitable for commercial

Table 2. List of cashew varieties and hybrids developed and cultivated in different countries (adopted from Savadi et al. 2020 and modified).

Country	Varieties (Features)	Tree Type	References
Brazil	CCP 06 and CCP1001, CP 09, CCP 76, Embrapa 50, Embrapa 51, BRS189, Bahia 12 and BRS 226	Early dwarf clones	Castro et al. (2010)
	BRS 274	Giant type	Kapinga et al. (2019)
Benin	E11-4, C18-1, O24-4, E2-4, E26-1, E15-1, E9-1, E34-2, E21-2, O17-1, C21-2, E32-2, C39-2, E6-1, C40-1, C22-2, C1-1, E39-4, E28-4, O12-4, E25-2, C28-2, C8-4, E31-2, C29-1, E13-4, C14-2, E5-1, E37-1, C24-1, E35-2, O32-1, C11-2, C31-4, C20-1 C25-1, C5-1, C9-3, C4-1, O16-4, C23-2, C16-1, C12-4, O27-1, O29-1, C10-1, C6-1 O20-1, O11-1, O14-3, E19-4, E18-4, O19-1, C43-3, E7-2, C26-2, C17-4, C33-3, O21-2, C3-2	Dwarf types Giant type	Sika et al. (2015a) Sika et al. (2015b)
Côte d'Ivoire	National Center for Agronomic Research (CNRA) selections: LAX3264, LAX4297, LAZ330 Jumbo variety from Brazil	Giant type	Kouakou et al. (2020)
India	NRCC Sel2, Bhaskara, Vengurla-1, Vengurla-3,4,7,8, Madakkathara -2, Ullal-1,2,3, VRI-3	Giant type	Bhat et al. (2010)
Mozambique	H1	Giant type	Uaciquete (2006)
	1.12PA, 12.8PA and 1.18PA, 11.9PA and 2.3BG	Dwarf type	Uaciquete et al. (2013)
	NA7, MB77, 1.5R and MCH-2, IM1 and MU3	Giant type	
Nigeria	G-series by CRIN	Giant type	Aliyu (2012)
Tanzania	AZA 2, AZA 17, AC 43, AC 34	Giant type	Kapinga et al. (2019)
Vietnam	TL2/11, TL6/3, and TL11/2 PN1, CH1, LG1, MH2/7, MH3/5, MH4/5, MH5/4, MH6/2, DDH66-14 and DDH67-15 B01, KP11, BD 01	Giant type	Van Bien et al. (2005)

cultivation. Natural and artificial hybridization among the cashew accessions from the germplasm allowed a significant increase in the nut weight and kernel size. Further, interspecific hybrids derived from crossing cashew with wild relatives, viz., *A. othonianum* and *A. microcarpum,* have helped in transferring anthracnose resistance alleles and cashew apple quality traits into cashew (Castro et al., 2010). The evaluation and utilization of germplasm resources in the Brazilian genebank are being done on a regular basis.

Indian Cashew Genebank

The survey and collection of cashew accessions carried out in cashew-growing Indian states and exotic collections have resulted in 558 accessions of cashew and 3 related *Anacardium* species accessions in the National Cashew Field Gene Bank (NCFGB) at ICAR-DCR, Puttur. Likewise, genebanks established at AICRP Centres referred to as Regional Cashew Gene Banks (RCGBs), hold a total of 1,417 accessions. The conserved cashew accession in the Indian germplasm banks has been characterized using the "Cashew Descriptors" (IBPGR, 1986) over the period of years since the

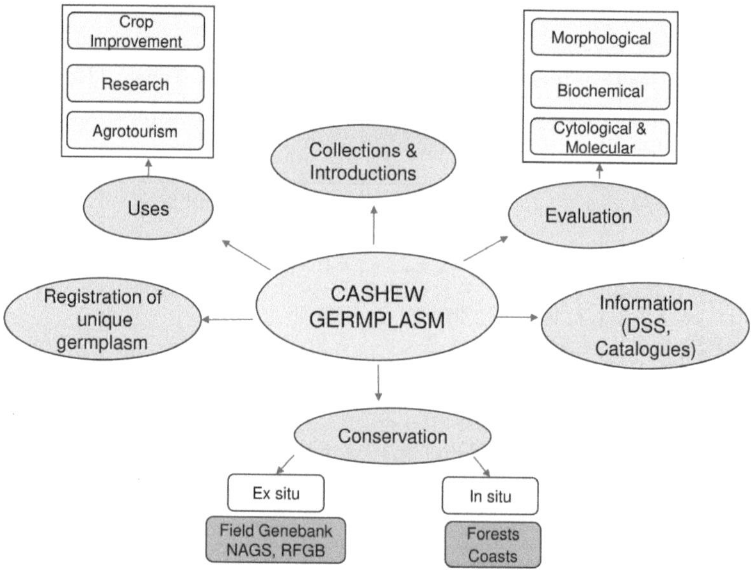

Figure 3. A schematic representation of the germplasm conservation, uses, evaluation, information documentation and registration of unique accessions.

inception of germplasm collection. The germplasm accessions with unique and potential (verified or verifiable) attributes of scientific or commercial value are being registered at NBPGR, New Delhi. Further, for better management and utilization of the germplasm resources, a Decision Support System (DSS) has been developed for cashew accessions collected in the 68 characters (NCFGB) at ICAR-DCR, Puttur. Using DSS, it is possible to select multiple character combinations to select germplasm accessions (Mohana and Nayak, 2018b).

In India, the early efforts for cashew germplasm collections were started during the 1950s and the research stations in Kerala (Kottarakkara), Karnataka (Ullal), Tamil Nadu (Vridhachalam), Andhra Pradesh (Bapatla), and Maharashtra (Vengurle) were involved in collection of locally available elite trees for conservation and evaluation. National Research Centre for Cashew (NRCC) at Puttur (Karnataka) was established in 1986 to carry out strategic research in cashew and the germplasm collection through seeds was discontinued and only the vegetatively propagated clonal material was started and conserved in the National Cashew Gene Bank (NCGB). Further, a coordinated and systematic approach was started through joint survey teams comprising scientists of NRCC (currently DCR, Puttur) and the AICRP on Cashew centers under State Agricultural Universities (SAUs). During the last six to seven decades of cashew research in India, a large number of novel undescribed cashew germplasm of cashew were collected. The cashew accessions were collected from forest plantations, farmers' fields, and wastelands, etc. from Tamil Nadu, Kerala, Karnataka, Andhra Pradesh, and a few other countries. National genebank exists at ICAR-DCR, Puttur, and the Regional Cashew Gene Banks (RCGB) exist at the AICRP centres. On a regular basis the collected germplasm has been characterized as per the IBPGR descriptors and cataloging was done for most of the accessions collected. The catalogs contain characterization data for 68 morphometric characters. The collections at both National genebanks and regional genes show wide morphological differences in growth and reproductive characteristics. For instance, 85% of germplasm accessions at RCGB, Vridachalam, Tamil Nadu have upright and compact tree habit and 76% showed extensive type branching habit. With respect to flowering, the majority of accessions were mid-season flowering and 60% of accessions showed compact inflorescence. The nut weight varied among the accessions and two accessions were boldnut type, while 59 were intermediate type and 64 were small nut size accessions (Jeeva et al., 2011).

Conclusion

Cashew is a nut crop of global interest, and there has been increasing demand in international and domestic markets in recent times. In the world of cashew production, developing countries like Vietnam, India, Nigeria, and Ghana have substantial reliance on the cashew industry. In cashews, there is a need to enhance productivity while maintaining ecological balances. Several breeding technologies have evolved in cashews. The collection and conservation of diverse germplasm play an important role in cashew breeding and sustained cashew production. The documentation of information on the characterization data of cashew germplasm accessions over the years would be useful for designing appropriate crop improvement programmes in cashew to cater to the needs of future research programmes. Further, the formation of core germplasm would facilitate maximum utilization of diverse germplasm in breeding. Therefore, more efforts are needed for systematic conservation and comprehensive characterization using high-throughput technologies, which would facilitate maximum utilization of germplasm in cashew breeding.

References

Adiga, J.D., B.M. Muralidhara, P. Preethi, and S. Savadi. 2019. Phenological growth stages of the cashew tree (*Anacardium occidentale* L.) according to the extended BBCH scale. *Annals of Applied Biology*, *175*(2): 246–252.

Adu-Gyamfi, P.K., A. Akpertey, M. Barnnor, A. Ofori, and F. Padi. 2020. Genotypic characterization of cashew (*Anacardium occidentale* L.) clones using agro-morphological traits. *Plant-Environment Interactions*, *1*(3): 196–206.

Akinhanmi, T.F., V.N. Atasie, and P.O. Akintokun. 2008. Chemical composition and physicochemical properties of cashew nut (*Anacardium occidentale*) oil and cashew nut shell liquid. *Journal of Agricultural Food and Environmental Sciences*, *2*(1): 1–10.

Aliyu, O.M. 2012. Genetic diversity of Nigerian cashew germplasm. *Genetic Diversity in Plants*, *498*: 163–184.

Aliyu, O.M. 2014. Analysis of absolute nuclear DNA content reveals a small genome and intra-specific variation in Cashew (*Anacardium occidentale* L.) *Anacardiaceae Silvae Genetica*, *63*(1–6): 285–292.

Aliyu, O.M., and J.A. Awopetu. 2007. Chromosome studies in cashew (*Anacardium occidentale* L.). *African Journal of Biotechnology*, *6*(2): 131–136.

Alves, R.J.V., M. Weksler, J.A. Oliveira, P.A. Buckup, H.R. Santana, A.L. Peracchi, and U. Caramaschi. 2018. Brazilian legislation on genetic heritage harms Biodiversity Convention goals and threatens basic biology research and education. *Anais da Academia Brasileira de Ciências*, *90*: 1279–1284.

Archak, S., A.B. Gaikwad, D. Gautam, F.B. Rao, K.R.M. Swamy, and J.L. Karihaloo. 2003a. DNA fingerprinting of Indian cashew (*Anacardium occidentale* L.) varieties using RAPD and ISSR techniques. *Euphytica*, *130*(3): 397–404.

Archak, S., A.B. Gaikwad., D. Gautam., E.V.V. Bhaskara Rao., K.R.M. Swamy, and J.L. Karihaloo. 2003b. Comparative assessment of DNA fingerprinting techniques (RAPD, ISSR and AFLP) for genetic analysis of cashew (*Anacardium occidentale* L.) accessions of India. *Genome*, *46*: 362–369.

Archak, S., A.B. Gaikwad., K.R.M. Swamy, and J.L. Karihaloo. 2009. Genetic analysis and historical perspective of cashew (*Anacardium occidentale* L.) introduction into India. *Genome*, *52*(3): 222–230.

Aremu, M.O., A. Olonisakin, D.A. Bako, and P.C. Madu. 2006. Compositional studies and physicochemical characteristics of cashew nut (*Anarcadium occidentale*) flour. *Pakistan Journal of Nutrition*, *5*(4): 328–333.

Asati, R., M.K. Tripathi, R.K. Yadav, S. Tiwari, S. Chauhan, N. Tripathi, and M. Yasin. 2023. Morphological description of chickpea (*Cicer arietanum* L.) genotypes using DUS characterization. *International Journal of Environment and Climate Change*, *13*(9): 1321–1341.

Barros, L.M., F.E. Araújo, J.I.L. Almeida, and L.M.S. Teixeira. 1984. A cultura do cajueiro anão, 67pp. EPACE Fortaleza (EPACE Documentos 3).

Berry, A.D., and S.A. Sargent. 2011. Cashew apple and nut (*Anacardium occidentale* L.). *In*: *Postharvest Biology and Technology of Tropical and Subtropical Fruits*, 414–423. Woodhead Publishing, Cambridge, United Kingdom.

Bhat, M.G., K.V. Nagaraja, and T.R. Rupa. 2010. Cashew research in India. *Journal of Horticultural Sciences*, *5*(1): 1–16.

Borges, A.N.C., A.C.A. Lopes, F.B. Britto, L.F.L. Vasconcelos, and P.D.C Lima. 2018. Genetic diversity in a cajuí (*Anacardium* spp.). germplasm bank as determined by ISSR markers. *Genet. Mol. Res.*, *17*: 1–14.

Bradtke, B. 2007. Cashew growing: How to grow cashew trees nuts and apples. *Tropical Permaculture*. http://wwwtropicalpermaculturecom/growing-cashews.html.

Castro, A.C.R., P.N. Bordallo, J.J.V. Cavacanti, and L.M. Barros. 2010. Brazilian cashew germplasm bank. *In*: *XXVIII International Horticultural Congress on Science and Horticulture for People (IHC2010): III International Symposium on 918*: 857–861.

Chabi Sika, K., H. Adoukonou-Sagbadja, L.E. Ahoton, I. Adebo, F.A. Adigoun, A. and Saidou, L. Baba-Moussa. 2015. Morphological characterization and agronomic performances of cashew (*Anacardium occidentale* L.) accessions from Benin. *Journal of Agricultural and Crop Research, 3*(2): 27–40.

Chavan, S.P., and U.A. Raut. 2013. Genetic diversity based on morphological and molecular markers in cashew (*Anacardium occidentale* L.) genotypes. *Vegetos, 26*(2): 255–263.

Chipojola, F.M., W.F. Mwase., M.B. Kwapata., J.M. Bokosi., J.P. Njoloma., and M.F. Maliro. 2009. Morphological characterization of cashew (*Anacardium occidentale* L.) in four populations in Malawi. *African Journal of Biotechnology, 8*(20): 5173–5181.

Costa, T.D.S.A., V.A. Padilha, M.V. Lima, and K.A. Jales. 2003. Avaliacąo Pedunculos de Quatro Genótipos de Caju. *In*: *Proc. Inter Amer. Soc. Trop. Hort. l47*: 137–140.

Cota, L.G., P.A. Moreira, E.V. Menezes, A.S. Gomes, A.R.O. Ericsson, D.A. Oliveira, and A.F. Melo Jr. 2012. Transferability and characterization of simple sequence repeat markers from *Anacardium occidentale* to *A. humile* (*Anacardiaceae*). *Genetics and Molecular Research, 11*(4): 4609–4616.

Croxford, A.E., M. Robson, and M.J. Wilkinson. 2006. Characterization and PCR multiplexing of polymorphic microsatellite loci in cashew (*Anacardium occidentale* L.) and their cross-species utilization. *Molecular Ecology Notes, 6*(1): 249–251.

Dasmohapatra, R., S. Rath., B. Pradhan., and G.R. Rout. 2014. Molecular and agromorphological assessment of cashew (*Anacardium occidentale* L.) genotypes of India. *Journal of Applied Horticulture, 16*(3): 215–221.

De Castro, P. 1994. Summary of the study. *In*: Delogu, A.M. and Haeuster, G. [Eds.], *The World Cashew Economy*, 11–12. Nomisma L'Inchiostroblu, Bologna, Italy.

de Figueirêdo, M.C.B., J. Potting, L.A.L. Serrano, M.A. Bezerra, V. da Silva Barros, R.S. Gondim, and T. Nemecek. 2016. Environmental assessment of tropical perennial crops: The case of the Brazilian cashew. *Journal of Cleaner Production, 112*: 131–140.

Deckers, J., E. Cundall, S.H. Shomari, A. Ngatunga and G. Bassi. 2001. Nut crops. Raemaekers RH. *In*: *Crop Production in Tropical Africa*, 691–715.

Dhanaraj, A.L., E.B. Rao, K.R.M. Swamy, M.G. Bhat, D.T. Prasad, and S.N. Sondur. 2002. Using RAPDs to assess the diversity in Indian cashew (*Anacardium occidentale* L.) germplasm. *The Journal of Horticultural Science and Biotechnology, 77*(1): 41–47.

dos Santos, J.O., S.J. Mayo, C.B. Bittencourt, and I.M. de Andrade. 2019. Genetic diversity in wild populations of the resting ecotype of the cashew (*Anacardium occidentale*) in coastal Piauí Brazil. *Plant Systematics and Evolution, 305*: 913–924.

Dwivedi, S.L., S. Ceccarelli, M.W. Blair, H.D. Upadhyaya, A.K. Are, and R. Ortiz. 2016. Landrace germplasm for improving yield and abiotic stress adaptation. *Trends in Plant Science, 21*(1): 31–42.

Eradasappa, E., and G.S. Mohana. 2023. Improved cashew varieties for doubling farmers' income. *In*: *Training Manual on Improved Cashew Production Technologies*, 8–18. ICAR-Directorate of Cashew Research, Puttur, Karnataka, India.

FAOSTAT. 2019. *Food and Agricultural Organization Statistics*. FAO Publication Catalog March 2019.

Hawerroth, M.C., P.D.N. Bordallo., L.C.P. Oliveira., E.H. Vale., F.D.C. Vidal Neto., and D.S. Melo. 2019. Genetic variability among cashew hybrids and prediction of superior combinations based on agronomic performance. *Pesquisa Agropecuária Brasileira, 54*: e00725.

IBPGR. 1986. International Board for Plant Genetic Resources. Descriptors for cashew (*Anacardium occidentale* L.), 33pp. Rome, Italy.

Ikabanga, D.U., T. Stevart, K.G. Koffi, F.K. Monthe, E.C.N. Doubindou, G. Dauby, and O.J. Hardy. 2017. Combining morphology and population genetic analysis uncover species delimitation in the widespread African tree genus *Santiria* (Burseraceae). *Phytotaxa., 321*(2): 166–180.

Jeeva, S., M.S. Aneesa Rani., V. Ambethgar., and R. Vaidyanathan. 2011. Cashew germplasm of Tamil Nadu: Characterization and cataloguing. *In*: *I International Symposium on Cashew Nut*, 151–154. Cochin, India.

Jena, R.C., K.C. Samal, A. Pal, B.K. Das, and P.K. Chand. 2016. Genetic diversity among some promising Indian local selections and hybrids of cashew nut based on morphometric and molecular markers. *International Journal of Fruit Science, 16*(1): 69–93.

Kahlmann, K., and M. Kohn. 2018. *USDA/FAS Food for Progress LIFFT-CashewSeGaBi Cashew Value Chain Study*, 1–164. https://www.climatefinancelab.org/wp-content/uploads/2018/12/SeGaBi-study_final_18.03.02_pub.pdf.

Kouakou, C.K., A. N'da Adopo., A.J.B. Djaha., D.P. N'da., H.A. N'da., and A.É. N'Guessan. 2020. Genetic characterization of promising high-yielding cashew (*Anacardium occidentale* L.) cultivars from Côte d'Ivoire. *BASE, 24*(1): 46–58.

Kapinga, F.A. L.J.F. Kasuga., and E.M. Kafiriti. 2019. Growth and production of cashewnut. *In*: *Soils Plant Growth and Production. Encyclopaedia of Life Support Systems* (*EOLSS*). https://www.eolss.net/Sample-Chapters/C10/E1-05A-45.pdf.

Klee, H.J. 2010. Improving the flavor of fresh fruits: Genomics, biochemistry, and biotechnology. *New Phytologist, 187*(1): 44–56.

Kshirsagar, H.H., P. Fajer, G.M. Sharma, K.H. Roux, and S.K. Sathe. 2011. Biochemical and spectroscopic characterization of almond and cashew nut seed 11S legumins, amandin, and anacardein. *Journal of Agricultural and Food Chemistry, 59*(1): 386–393.

Lubi, M.C., and E.T. Thachil. 2000. Cashew nut shell liquid (CNSL), a versatile monomer for polymer synthesis. *Designed Monomers and Polymers*, *3*(2): 123–153.

Malhotra, S.K., V.N. Hubballi, and M.G. Nayak. 2017. Cashew: Production, Processing, and Utilization of By-products, 1–142. Directorate of Cashew Nut and Cocoa Development Cochin Kerala India.

Mangesh, S.N., A.G. Patil, K.S. Tippanna, and S.V. Patil. 2022. Osmotic dehydration influence on marking nut (*Semecarpus anacardium*) *International Journal of Agricultural and Applied Sciences*, *3*(2): 55–59.

Mariante, A., M. Albuquerque, A. Egito, C. McManus, M.A. Lopes, and S.R. Paiva. 2009. Present status of the conservation of livestock genetic resources in Brazil: Status of the conservation of livestock genetic resources in Brazil. *Livestock Science*, *120*: 204–212.

Matos, C.H.A., J.A.R. Nunes, Â.C.D.A. Lopes, and R.L.F. Gomes. 2019. Selection of common cashew tree genotypes in commercial growing areas in municipalities of Piauí Brazil. *Crop Breeding and Applied Biotechnology*, *19*: 245–252.

Mneney, E., S. Mantell, and M. Bennett. 2001. Use of random amplified polymorphic DNA (RAPD). markers to reveal genetic diversity within and between populations of cashew (*Anacardium occidentale* L.). *The Journal of Horticultural Science and Biotechnology*, *76*(4): 375–383.

Mohana, G.S., and M.G. Nayak. 2018a. Development of the core collection through advanced maximization strategy with heuristic approach in cashew (*Anacardium occidentale* L.). *Plant Genetic Resources*, *16*: 367–377.

Mohana, G.S., and M.G. Nayak. 2018b. Unique decision support system for cashew germplasm management. *Indian Society for Plantation Crops*, *46*(2): 128.

Ohler, J.G. 1979. *Cashew*, 260pp. Koninklijk Instituut voor de Tropen Amsterdam, Netherlands.

Paiva, J.R.D., L.D.M. d Barros, and J.J.V. Cavalcanti. 2009. Cashew (*Anacardium occidentale* L.) breeding: A global perspective. *In*: Jain, S.M.S. and Priyadarshan, P.M. [Eds.], *Breeding Plantation Tree Crops: Tropical Species*, 287–321. Springer, New York.

Palei, S., R. Dasmohapatra, S. Samal, and G.R. Rout. 2019. Cashew nut (*Anacardium occidentale* L.). breeding strategies. *Advances in Plant Breeding Strategies: Nut and Beverage Crops*, *4*: 77–104.

Pereira, L.D., D.F. Silva., E.F. Reis., J.F. Pinto., H.F. Assunção., C.G. Machado., and C.H. Costa. 2019. Characterization of bushy cashew (*Anacardium humile* A. St.-Hil.) in the State of Goiás, Brazil. *Journal of Agricultural Science*, *11*(5): 183.

Queiroz, M.A.D., L.D.M. Barros, L.P.D. Carvalho, J.D.A. Candeia, and E. Ferraz. 2012. Plant breeding in the semiarid region of Brazil: examples of success. *Crop Breeding and Applied Biotechnology*, *12*: 57–66.

Ranathunge, R.A.C.A., D.P.S.T.G. Attanayaka, D.A D.S. Samarajeewa, and S. Jayasekera. 2010. Genetic diversity of cashew (*Anacardium occidentale* L.) germplasm collected from five districts of Sri Lanka as revealed by Random Amplified Polymorphic DNA (RAPD). *In*: *Proceedings of 10th Agricultural Research Symposium*, 35–39.

Razzaq, A., B. Sadia, A. Raza, M. Khalid Hameed, and F. Saleem. 2019. Metabolomics: A way forward for crop improvement. *Metabolites*, *9*(12): 303.

Rico, R., M. Bulló, and J. Salas-Salvadó. 2016. Nutritional composition of raw fresh cashew (*Anacardium occidentale* L.) kernels from different origin. *Food Science and Nutrition*, *4*(2): 329–338.

Salifou, M.I., and P.A.L. Masawe. 2018. Guide on Developing Cashew Varieties and Improved Planting Materials. *GIZ/Com. Cashew Accra*, *43*.

Saroj, P.L., and G.S. Mohana. 2016. Cashew improvement in India: Retrospect and prospects. *International Journal of Innovative Horticulture*, *5*(1): 14–22.

Savadi, S., B.M. Muralidhara, and P. Preethi. 2020. Advances in genomics of cashew tree: molecular tools and strategies for accelerated breeding. *Tree Genetics and Genomes*, *16*: 1–15.

Savadi, S., K. Sowmya., V.S. Megha, B.M. Muralidhara., and G.S. Mohana. 2021. Genetic diversity and identification of interspecific hybrids of *Anacardium* species using microsatellites. *Brazilian Journal of Botany*, *44*: 139–148.

Savadi, S., B.M. Muralidhara, V. Venkataravanappa, and J.D. Adiga. 2023b. Genome-wide survey and characterization of microsatellites in cashew and design of a web-based microsatellite database: CMDB. *Frontiers in Plant Science*, *14*: 1242025.

Savadi, S., B.M. Muralidhara, V. Venkataravanappa, J.D. Adiga, K. Manjunatha, and B. Patil. 2022. *De novo* transcriptome assembly and its utility in development and characterization of the first set of genic SSR markers in cashew. *Industrial Crops and Products*, *189*: 115734.

Savadi, S., J.D. Adiga, B.M. Muralidhara, P. Prasad, K. Manjunatha, K. Ashwitha, and K. Manoj. 2023a. Discovery of genome-wide genetic variations and development of first set of InDel markers for genetics research in cashew. *Scientia Horticulturae*, *320*: 112233.

Schut, J.W., X. Qi, and P. Stam. 1997. Association between relationship measures based on AFLP markers pedigree data and morphological traits in barley. *Theoretical and Applied Genetics*, *95*: 1161–1168.

Sethi, K., S.K. Tripathy, and P.C. Lenka. 2016. Assessment of genetic diversity and identification of elite cashew hybrids. *Adv. Life Sci.*, *16*: 6200–6205.

Sika, K.C., H. Adoukonou-Sagbadja., L. Ahoton., A. Saidou., and A. Ahanchede. 2015a. Genetic characterization of cashew (*Anacardium occidentale* L.) cultivars from Benin. *J. Hortic.*, 2: 153.

Sika, K. C., H. Adoukonou-Sagbadja., L. Ahoton., I. Adebo., F.A. Adigoun., A. Saidou., and L. Baba-Moussa. 2015. Morphological characterization and agronomic performances of cashew (*Anacardium occidentale* L.) accessions from Benin. *Journal of Agricultural and Crop Research*, *3*(2): 27–40.

Silva, W.C., S.I.C. de Carvalho, and J.B. Duarte. 2013. Identification of minimum descriptors for characterization of Capsicum spp. germplasm. *Horticultura Brasileira, 31*: 190–202.
Singh, K., K. Gupta, V. Tyagi, and S. Rajkumar. 2020. Plant genetic resources in India: Management and utilization. *Vavilov Journal of Genetics and Breeding, 24*(3): 306.
Soares, T.N., L.D.L. Sant'ana, L.K.D. Oliveira, M.P.D.C. Telles, and R.G. Collevatti. 2013. Transferability and characterization of microsatellite loci in *Anacardium humile* AStHil. (*Anacardiaceae*). *Genet. Mol. Res., 12*(3): 3146–3149.
Stebbins, G.L. 1971. *Chromosomal Evolution in Higher Plants*, 1–216. Edward Arnold Ltd., London.
Straits Research Report. 2022. *Global Report on Cashew Market.* https://straitsresearch.com/reports.
Thimmappaiah, D. Shobha, and S.R. Anil. 2009. Assessment of cashew species for molecular diversity. *Journal of Plantation Crops, 37*(2): 146–151.
Thimmappaiah, D. Shobha., G.S. Mohana., J.D. Adiga, and P.G. Bhat. 2016. Fingerprinting of released varieties of cashew based on DNA markers. *Vegetos, 29*(4): 89–95.
Uaciquete, A. 2006. *Epidemiology and Control of Powdery Mildew (Oidium anacardii Noack) on Cashew (Anacardium occidentale* L.) *in Mozambique*. Doctoral dissertation, University of Pretoria.
Uaciquete, A., L. Korsten, and J.E. Van der Waals. 2013. A search for anthracnose resistant cashew cultivars in Mozambique. *Crop Protection, 50*: 6–11.
UNCTAD 2021. *Commodities at a Glance.*—Special Issue on Cashew Nuts, 1–45. United Nations publication issued by the United Nations Conference on Trade and Development. UNCTAD/DITC/COM/2020/1.
Valeriano, C. 1972. The cashew tree. [O cajueiro.]. *Bol. Inst. Biol. Bahia., 11*: 19–58.
Van Beuningen, L.T., and R.H. Busch. 1997. Genetic diversity among North American spring wheat cultivars: III Cluster analysis based on quantitative morphological traits. *Crop Science, 37*(3): 981–988.
Van Bien, P., N.T. Binh., N.T. Hoc., P. Van Dau., T.N.T. Nam, and N.B. Phuong. 2005. Results of selection and development of cashews and pepper varieties. *AGRIS, FAO*, 130–145.
Vanitha, K., and T.N. Raviprasad. 2019. Pollinators of cashew, their forage behavior, and conservation measures. *Technical Bulletin 20*. ICAR-Directorate of Cashew Research, Puttur, Karnataka, India.
Volkova Ne, D.S. 2011. Molecular markers for the examination of distinctness uniformity and stability in UPOV effects of *hg* gene alleles on economic characters of winter bread wheat in the steppes of Ukraine, 41–46.
Wang, C., S. Hu, C. Gardner, and T. Lübberstedt. 2017. Emerging avenues for utilization of exotic germplasm. *Trends in Plant Science, 22*(7): 624–637.
Wickens, G.E. and N. Chomsky. 1995. Cashew or Monkey nut. Edible nuts, *5*: 13–17.
Wunnachit, W., and M. Sedgley. 1992. Floral structure and phenology of cashew in relation to yield. *Journal of Horticultural Science, 67*(6): 769–777.

20

Genetic Diversity of Macadamia Nut (*Macadamia integrifolia* Maiden & Betche) and Sustainable Utilization

*Anusha, Sakthivel, T.** and *Sankaran, M.*

Introduction

The Macadamia nut is native to Australia specifically to northeastern New South Wales and central and southeastern Queensland, it is renowned as the world's finest nut, known by various names such as Bush nut, Bauple nut, Queensland nut, Australian nut, popple nut, kindal, boombera, burrawang (Hardner et al., 2004). The German-Australian botanist Ferdinand von Mueller gave the name Macadamia in 1857 in honor of John Macadam who was a chemist, medical teacher, politician, and Secretary of the Philosophical Institute of Victoria (Hardner et al. 2009). It is a fast-growing semi-hardwood evergreen tree that can reach a height of up to 20 meters and is commercially cultivated globally for its premium kernel quality which is the basis for the expansion of world industry. The genus includes 10 species but only two, *Macadamia integrifolia* and *Macadamia tetraphylla*, are commercially cultivated for their edible nuts which belong to the family Proteaceae with somatic chromosome number of 2n = 28 (McHargue, 1996). The macadamia industry has undergone rapid global expansion in the last 50 years where it is primarily cultivated in tropical regions such as Australia, Indonesia, Brazil, Kenya, South Africa, and New Zealand. Australia and Hawaii hold the position of being the largest producers whereas the nuts in Hawaii are renowned for their exquisite taste, characterized by a delicate flavor and a crunchy texture, making them a delightful choice for consumption. It is considered the world's most delectable nut, featuring a crisp, buttery flavor that is sweet. These nuts can be enjoyed in various forms, including raw from the shell, dry roasted, or cooked in oil. The kernels are used in roasted snacks, chocolate-coated confectionery, bakery products, ice cream, and most notably, as oil. This positions them as luxury goods, with demand being elastic with income. Beyond their deliciousness, these nuts serve as excellent sources of protein and minerals

---

Division of Fruit Crops, ICAR – Indian Institute of Horticultural Research, Hesaraghatta Lake Post, Bangalore – 560 089, India.

*Corresponding author: sakthivel.t@icar.gov.in

(Verma et al., 2017). Macadamia nuts stand out as one of the few natural food sources containing the highest levels of palmitoleic acid. This beneficial fatty acid, found in higher concentrations in the human liver, enhances insulin sensitivity by suppressing inflammation and inhibiting the destruction of insulin-secreting pancreatic beta cells (Khairina et al., 2016). Understanding genetic resources and the relationships between domesticated and wild germplasm is crucial for directing the incorporation of novel germplasm into selective breeding populations and prioritizing the conservation of potentially valuable wild germplasm for future use (Brozynska et al., 2016; Chen et al., 2017). Unfortunately, this knowledge is inadequately documented in the context of Macadamia. In the forthcoming chapter, an assessment of literature from various sources is conducted to outline the distribution, structure, and status of genetic diversity. Additionally, the chapter explores the origin, key selection criteria, conservation, and utilization of genetic resources in Macadamia nut.

Taxonomic Background

It is classified within the Plantae kingdom, Magnoliophyta phylum, Magnoliopsida class, Proteales order, Proteaceae family, and *Macadamia* genus. Belonging to the Proteaceae family, it is a significant component of early angiosperm-dominated rainforests that once covered most of Australia. Macadamia likely had widespread distribution in these early forests, supported by fossil pollen evidence found in sediments in southeast Australia, central coastal Queensland, and New Zealand. The Proteaceae family is renowned for other genera like Banksia, Grevillea, and Hakea. Despite its evolutionary roots in rainforests, Proteaceae is now less represented in these ecosystems (Johnson and Briggs, 1963). Macadamia trees were introduced to Hawaii in the 1880s for cultivation as windbreaks and in 1930, the University of Hawaii recognized their commercial potential as dessert nuts.

Collection and Assessment of Genetic Resources

Genetic diversity is crucial for agriculture and the significance of a diverse gene pool in plant breeding can be overstated. The ability to recognize genetic variations is essential for the effective management and utilization of genetic resources. Additionally, the genetic diversity present within and among populations plays a pivotal role in conserving plant genetic resources for both current and future applications (Quedraogo, 2001). Understanding the extent and structure of genetic diversity enables the identification of populations meriting conservation due to their distinctiveness or diversity (Thormann et al., 1994; Chamberlain, 2001). As stated by Costello et al. (2008), the genus comprises nine species. The species native to Australia include *Macadamia integrifolia*, *M. tetraphylla*, *M. ternifolia*, *M. jansenii*, *M. whelani*, *M. claudiensis*, and *M. grandis*, while *M. neurophylla* is indigenous to New Caledonia and *M. hildebrandii* is native to Sulawesi in Indonesia. However, only the smooth-shelled (*Macadamia integrifolia*) and the rough-shelled (*Macadamia tetraphylla*) are cultivated for their edible nuts, as noted by Peace et al. (2003), whereas *M. ternifolia* produces small nuts with a bitter-flavored kernel (McConachie, 1980). The natural hybridization between *M. integrifolia* and *M. tetraphylla* occurs readily where the species coexist (Storey, 1959; Barry and Thomas, 1994; Hardner et al., 2000; Peace et al., 2005). Natural hybrids between *M. integrifolia* and *M. ternifolia* are also widely observed in areas of sympatry (Peace et al., 2005). Some of the varieties and hybrids have been developed, contributing to the overall gene pool of macadamia nuts. Despite the macadamia's Australian origin, its initial breeding program was initiated by the Hawaii Agricultural Experiment Station (HAES) in the United States (Aradhya et al., 1998; Hardner and Topp, 2018; Nock et al,. 2019; Topp et al., 2019). As the Hawaiian germplasm originated from seeds of a limited number of trees, the narrow genetic foundation may pose constraints on cultivar development (Aradhya et al., 1998; Steiger et al., 2003). This underscores the global significance of characterizing germplasm, shedding light on the relationships between cultivars and wild individuals.

Characterization and Evaluation of Genetic Resources

The assessment and analysis of germplasm can ascertain the potential of wild germplasm for upcoming breeding programs. Diversity develops over time through various processes like migration, selection, mutation, genetic drift, and the mode and method of reproduction. Macadamia exhibits a preference for out-crossing, with over 75% of the instances being out-crossing (Sedgley et al., 1990). Trees that stem from open-pollinated seeds (seedling trees) display variability in both nut yield and quality. A species with diverse characteristics provides the chance to select the most advantageous alleles, whereas low diversity increases the risk of extinction. Recognizing the genetic connection between parent plants is crucial in preventing inbreeding depression, especially when aiming to enhance complex traits. Consequently, understanding genetic divergence is a fundamental requirement for advancing crop improvement. Characterization involves providing a detailed description of plant germplasm and assessing the expression of highly heritable traits that span from morphological or agronomic features to seed proteins or molecular markers. The history of global macadamia breeding is relatively brief and the current cultivars are only a few generations away from the wild germplasm found naturally in rainforests. Given the predominant reliance on Hawaiian cultivars in present macadamia plantations, the limited diversity in sampling implies a potential avenue for future genetic enhancement by tapping into diverse genetic resources. Wild genetic resources possess the capacity to offer favorable traits such as compact tree size, nuts with thinner shells, and resistance or tolerance to biotic and abiotic factors. Investigating the genetic potential of wild germplasm will aid in effectively harnessing these resources for trait enhancement. Genotypic variability can arise from variations at the genome, chromosome, gene, and cytoplasmic element levels. This variation can be assessed by morphological traits, biochemical analysis, cytological evaluation, or the utilization of molecular markers. These genetic markers are chromosomal tag or alleles that enables the tracking of a specific DNA region, a known-position DNA fragment in the genome, a gene with easily distinguishable phenotypic expression, used for identifying individuals or cells (tissues, organs) carrying it, or serving as a probe for labeling cellular organelles (nucleus, mitochondria, and chloroplasts), chromosomes, or chromosomal loci. According to established terminology, genetic markers are typically categorized into three main classes: morphological or phenotypic markers (observable at the organism's phenotype level), molecular markers (detectable at the nucleic acid level), and biochemical markers (comprising various proteins, including enzymes, and metabolites). Additionally, cytogenetic markers, possessing properties of both biochemical and molecular markers, are also classified as genetic markers (Table 1).

An ideal genetic marker for diversity analysis of Macadamia should exhibit polymorphism, as the gene pool inherently contains variability and polymorphism serves as a reflection or manifestation of this diversity. Furthermore, it should be multi-allelic to enhance the frequency and spectrum of

Table 1. List of genetic markers and their salient features.

Properties	Markers			
	Morphological	**Biochemical**	**Molecular**	**Cytogenetic**
Identification	Easily visible	Staining gel using different dyes and substrates.	Staining gel with ethidium bromide, fluorescent or radioactive tags	By using fluorescent or radioactive tags
Level of identification	Morphology	Proteins and metabolites	Nucleic acids	Nucleic acids and proteins
Inheritance pattern	Dominant, recessive	Dominant, codominant	Dominant, codominant	Dominant
Genome coverage	Low	Low	High	Average
Use of automation	No	No	Yes	No
Cost	Low	Average	High	High

(Data from Chesnokov et al., 2020)

polymorphism. The preference is for a codominant marker, as the heterozygous hybrid simultaneously expresses the characteristics of both parents during natural crossbreeding, facilitating the distinction between homozygotes and heterozygotes. Non-epistatic markers are desirable, allowing for the visual detection of an individual genotype's marker manifestation, regardless of its location in the genome. Additionally, a neutral marker is preferred, ensuring that the replacement of alleles at the marker locus has no discernible phenotypic or selective impact, and it should be free from environmental influences.

Morphological Characterization

Morphological characteristics represent one of the initial indicators employed in the management of germplasm (Smith and Smith, 1992). The descriptors utilized for the characterization of macadamia trees in phenotyping encompass a range of features such as Vegetative part: tree shape, branch distribution, and branch angle of insertion; Bark: Surface, color, and thickness; Leaf: Length, width, shape, color, apex, margin, base, petiole and transversal section of lamina; Flowers: number of flowers per inflorescence, color and size of inflorescence, size of flower, petal, and sepal; Fruit: size, number of fruits per stalk, fruit shape, and color, pericarp texture, endocarp texture, size, and thickness; Seed: shape, size, and color. Descriptors for evaluation encompass data related to tree height and diameter, as well as yield parameters like fruit weight, diameter, and the count of polyembryonic fruits. Additionally, oil content and data pertaining to susceptibility to both biotic and abiotic stresses are taken into consideration (Domingo et al., 2002). Storey and Salleeb (1966) utilized four sets of vegetative taxonomic characters for the classification of macadamia parental species and F1 hybrids. These characters encompassed the count of leaves in the nodal whorls (phyllotaxy), leaf type indicating whether the leaves are petiolated or sessile, the number of marginal serrations (spines), and the color of a new leaf. In a study conducted by Gitonga et al. (2008), leaf and fruit morphological characteristics were documented for 23 Macadamia cultivars in Kenya. The analysis of variance uncovered notable distinctions in leaf length, width, petiole length, and leaf marginal serrations. Similarly, significant variations were observed in fruit cluster length, the number of fruits per cluster, and fruit length, although not in fruit width. *Macadamia integrifolia* and *Macadamia tetraphylla* exhibited the highest dissimilarity, with the hybrids occupying an intermediate position. Among the markers studied, leaf petiole and marginal serrations proved to be easily assessable and effective in distinguishing between the macadamia species. These markers were identified as reliable tools for distinguishing between different macadamia cultivars in Kenya. Morphological markers offer several advantages, including ease of use, visual identification, cost-effectiveness compared to alternative markers, no specific instruments are required, and simultaneous disadvantages such as lower polymorphism, susceptibility to environmental influences, limited number, sensitivity to plant growth stages and cover only restricted portion of the genome. For morphological characterization, it is important to understand the phenotype of different species available in the germplasm. Hence, the following information provides detailed morphology of different species of macadamia which may be helpful for further diversity analysis.

Macadamia integrifolia, also known as the Queensland/Australian nut, bush nut, nut oak, Bauple nut, and smooth shell nut, is characterized by vigorous trees featuring dark green foliage, robust crotches, and an ascending branch structure rather than spreading. Typically, individuals are multi-stemmed with small crowns. The leaves, arranged in whorls of three or opposite, are simply obovate to narrowly oblong, measuring 5.5 cm–14 cm in length and 2.5 cm–6 cm in width (Stanley and Ross, 2002). The blade tips are rounded and terminate in a short sharp point, while the base tapers to petioles ranging from 5–10 mm in length. The flowers display axillary creamy white blossoms arranged in brush-like hanging racemes measuring 10–30 cm long. The fruits are green and rounded, with a diameter of 2.5–3.5 cm, featuring a hard brown inner shell that protects the edible nut.

Macadamia tetraphylla, commonly known as the Spiny leaf macadamia, rough shell Queensland nut, or bush nut, is a perennial tree that reaches a height of 18 meters. The tree features greyish-brown branchlets adorned with pale elongated lenticels (Hauser and Blok, 1992). Its leaves are simple, oblong-lanceolate, arranged in whorls of three to four, measuring 6–20 cm in length and 2–4 cm in width. The blade tips are pointed, margins sharply serrated, and petioles extend to 28 mm in length (Stanley and Ross, 2002). Notably, the new leaves of *M. tetraphylla* are characterized by a bright red color, distinguishing them from the light green leaves of *M. integrifolia*. The tree produces axillary pinkish-purple flowers organized in brush-like hanging racemes, ranging from 15–45 cm in length. The fruits are compressed and rounded, displaying a greyish-green color and measuring 2–3.5 cm in width. They have a hard inner shell that protects the nut. Importantly, the seed kernel is edible and does not contain cyanogenic compounds, making it safe for consumption.

Macadamia ternifolia, also known as the Gympie nut, Marcoochy nut, and fruited Queensland nut, are perennial trees reaching a height of 6 meters, characterized by brown branchlets adorned with raised lenticels (Hauser and Blok, 1992). The leaves are simple, featuring narrow oblong to narrow elliptical shapes arranged in whorls of three, with new growth displaying a pinkish-red hue and measuring 10–12 cm in length. The blade tips are pointed and the base tapers to petioles ranging from 3–13 mm in length (Stanley and Ross, 2002). These trees produce axillary pinkish flowers organized in brush-like hanging racemes, typically measuring 4–20 cm in length. The fruits appear compressed, rounded, and greyish, with a length of 1.5–2 cm. They have a hard inner shell that protects the edible nut. It is important to note that the seed kernel contains cyanogenic compounds, rendering it unsafe for consumption due to the presence of poisonous cyanide radicals (Verma et al., 2017).

Macadamia jansenii, commonly known as the Bulburin nut tree, is characterized by small plants, either single or multi-stemmed trees reaching a height of 6–9 meters, exhibiting generally smooth bark punctuated with prominent lenticels (Halford, 1997). The leaves are oblanceolate to oblong-elliptic, typically arranged in whorls of three, measuring 10–18 cm in length, featuring an acute apex, tapered base, and undulating margins (Harden et al., 2006). The distinct net venation on the leaf blades is observable on both surfaces, particularly when held up to the light. The petioles range from 2–14 mm in length. The flowers, colored creamy brownish, possess sepals that are 7–9 mm long. The fruits are globose, with a diameter of 20–25 mm. It's worth noting that the kernel is mildly cyanogenetic and as such, is not suitable for consumption. The Genus *Macadamia*, as presently understood, comprises ten species of tropical and subtropical evergreen trees (Table 2).

Table 2. List of Macadamia species.

Sl. No.	Species	Origin
1.	*M. hildebrandii* Steenis	Celebes
2.	*M. Rousselii* (Veill.) Sleumer	New Caledonia
3.	*M. veillardii* (Brongn. and Gris.) Sleumer	New Caledonia
4.	*M. francii* (Guill.) Sleumer	New Caledonia
5.	*M. whelani* F.M. Bail	Eastern Australia
6.	*M. ternifolia* F. Muell.	Eastern Australia
7.	*M. integrifolia* Maiden and Betche	Eastern Australia
8.	*M. tetraphylla* LAS. Johnson	Eastern Australia
9.	*M. prealta* (F. Muell.) F. M. Bail	Eastern Australia
10.	*M. heyana* (F. M. Bail.) Sleumer	Eastern Australia

(Data from Storey, 1965)

Cytological Characterization

Cytogenetic markers are associated with variations in chromosome numbers, banding patterns, size, shape, order, and position. Analyzing the morphology, quantity, and structure of chromosomes in diverse species allows the identification of specific cytogenetic characteristics, including various forms of aneuploidy, structural variations in chromosomes, and abnormal chromosomes. These features serve as genetic markers for locating particular genes on chromosomes, determining their relative positions, or aiding in genetic mapping through chromosomal manipulations, such as replacing a chromosome or its fragment. The karyotype of chromosomes and specific bands or changes in thickness on chromosomes represent the structural features of chromosomes. Band patterns are typically distinguished by specific colors, widths, sequences, and their positions on chromosomes, thereby indicating variations in the distribution of euchromatin and heterochromatin. Q-groups (revealed through staining with quinacrine hydrochloride), G-discs (stained using the Romanovsky–Giemsa method), and R-discs (resistant to the Romanovsky–Giemsa stain) are examples of these chromosomal features. These landmarks are valuable not only for characterizing a normal chromosome but also for detecting chromosomal mutations. Cytological markers play a crucial role in identifying linkage groups within specific chromosomes and are commonly employed in physical mapping. Indeed, the category of cytogenetic markers encompasses genomic mutations (alterations in the number of chromosome sets), aneuploidy (modifications in the number of individual chromosomes), and segmental mutations (rearrangements of chromosomes), identified through cytological research techniques. This classification also encompasses gene mutations, structural rearrangements, and allelic variants of genes, which can be detected using cytogenetic or cytobiochemical analysis methods, such as GISH (Genomic *In Situ* Hybridization) and FISH (Fluorescence *In Situ* Hybridization) Cytogenetic markers of this nature exhibit dominant inheritance due to their distinctive determination characteristics.

All Macadamia species examined so far have been identified as diploid, with a haploid chromosome count of 14. For instance, studies have confirmed this for *M. integrifolia* (Ramsay, 1963; Storey and Saleeb, 1970), *M. integrifolia*, *M. tetraphylla*, and *M. ternifolia* (Storey, 1965), as well as *M. integrifolia* and *M. tetraphylla* (Storey and Saleeb, 1970). While there is a single mention of polyploidy in a report no specific details were provided for verification. Notably, hybridization between *M. tetraphylla* and *M. integrifolia* does not seem to disrupt normal chromosome pairing or disjunction, as the chromosome count in F1 progeny remains 14 (Storey and Saleeb, 1970). In the most recent examination of cytological data encompassing 188 species across 65 genera of Proteaceae, Stace et al. (1998) propose that the genera within the Grevilleoideae subfamily are predominantly diploid. These genera exhibit chromosome base numbers of $n = x = 10, 11, 12, 13$, and 14, with only two documented instances of triploidy. The authors counter a prior hypothesis suggesting that this chromosome series indicates paleo-polyploidy originating from an ancestral genome with $x = 5$ or 7 (Venkata Rao, 1970; Johnson and Briggs, 1975). Instead, they propose that Proteaceae members derive from an ancestral genome with $x = 12$ or 21, featuring 24 chromosome arms (fundamental number [FN] = 24). The presumed ancestral grevilleoid genome with $x = 14$ likely has Gondwanan or Several members of the Macadamieae tribe appear to have retained the original genome, with a chromosome count of $n = x = 14$ observed in Macadamia, Brabeium, and Floydia. Hicksbeachia and Gevuina, on the other hand, exhibit a count of $n = x = 13$, resulting from the fusion of a telocentric and a metacentric chromosome (Stace et al., 1998). The authors suggest the possibility of additional chromosomal evolution following this initial state. However, the chromosome numbers for Athertonia, Panopsis, and various other members of the Macadamieae tribe remain unreported. A study of five other genera within the Grevilleoideae subfamily reveals relatively small chromosomes (Stace et al., 1998). Notably, Brabeium stellatifolium, belonging to the same subtribe as Macadamia, has the smallest reported genome size, with a mean chromosome length of 1.0 mm (Stace et al., 1998). The presence of small grevilleoid chromosomes in conjunction with the absence of additional isozyme loci in the Australian genera of Proteaceae argues against a paleo-polyploid origin, as noted by Stace et al. (1998). The reported number of isozyme loci in Macadamia by Vithanage and Winks (1992) and Aradhya et al. (1998) also aligns with a diploid rather than an ancient tetraploid origin. Stace et al. (1998) propose that if paleo-polyploidy did occur in ancestral Proteaceae, molecular genetic investigations in genera like Macadamia may reveal substantial gene silencing, a result of the diploidization process.

The challenges associated with identifying cytogenetic markers at the cytological level restrict their accessibility, relying on the morpho-anatomical structure of the studied organism and specialized, costly equipment and chemical reagents. Consequently, cytogenetic markers have not gained widespread adoption or the potential for extensive use compared to other marker classes. It is worth mentioning that due to their limited quantity and relatively modest resolution, cytogenetic markers have restricted utility in analyzing genetic diversity, genetic mapping, and marker-assisted selection (MAS).

Biochemical Characterization

Biochemical markers, also known as isozymes, represent diverse molecular forms of enzymes encoded by different genes yet serving identical functions. These markers are allelic variations of enzymes, allowing the estimation of gene and genotypic frequencies. Biochemical markers have proven effective in discerning genetic diversity, population structure, gene flow, and population subdivision. They possess codominant characteristics, are user-friendly, and offer cost-effective solutions.

The initial investigation, as reported by Vithanage and Winks (1992), utilized nine isozyme systems, with seven of them incorporating nine loci. These systems were employed to differentiate between 73 Macadamia genotypes from the major species Macadamia integrifolia and *Macadamia tetraphylla*. Additionally, *Macadamia ternifolia* was included in the analysis. While no specific records of band presence or absence were documented, there was sufficient variation in the band distribution to facilitate the identification of each cultivar. Among the isozyme systems, phosphoglucoisomerase (PGI) exhibited the highest degree of polymorphism. Members of the Macadamia genus contain cyanogenic compounds that release hydrogen cyanide (HCN) upon hydrolysis. The concentrations of releasable cyanide were assessed in tissues of mature nuts and seedlings from *M. integrifolia, M. tetraphylla*, and *M. ternifolia* by Dahler et al. (1995). In mature seeds of *M. integrifolia* and *M. tetraphylla*, the concentrations were low (0.15 kmol/g fresh weight), aligning with the edibility of seeds in these commercially significant species. In contrast, the inedible seeds of *M. ternifolia* exhibited significantly higher concentrations (9.6 kmol/g) of releasable cyanide. A genetic variability analysis was conducted on 45 accessions of Macadamia, encompassing four species—M. integrifolia, *M. tetraphylla, M. ternifolia, M. hildebrandii,* and a wild relative, *Hicksbeachia pinnatifolia*. This study employed eight enzyme systems encoded by 16 loci (Gpi-1 and 2, Idh-1 and 2, Lap, Mdh-1, 2, and 3, 6Pgd-2, Pgm-2 and 3, Tpi-1 and 2, Ugpp-1, 2, and 3). Results revealed that 43 accessions exhibited distinct isozyme fingerprints, indicating a substantial level of genetic variation (Aradhya et al., 1998). A novel antimicrobial peptide, exhibiting no substantial amino acid sequence similarity to previously documented peptides, has been extracted from the nut kernels of *Macadamia integrifolia*. This peptide, known as MiAMP1, along with the MiAMP1 gene, holds potential for applications in genetic manipulations aimed at enhancing disease resistance in transgenic plants (Marcus et al., 1997). Moreno-Perez et al. (2011) explored the *potential* impact of acyl-ACP thioesterase activity on the composition of macadamia kernel oil. Acyl–acyl carrier protein (ACP) thioesterases (EC 3.1.2.14) are enzymes located within plastids, playing a role in concluding the synthesis of fatty acids in plants. They facilitate the transportation of acyl moieties to the endoplasmic reticulum, where these components can be utilized in glycerolipid production. Hu et al. (2022) analyzed the fatty acid profiles of 197 macadamia accessions, encompassing commercial cultivars, wild accessions, and elite breeding lines derived from *M. integrifolia, M. ternifolia*, and *M. tetraphylla*. The study involved the determination and quantification of these profiles. Ultimately, differences in fatty acid profiles were compared among accessions and species, revealing significant distinctions, particularly in terms of C16:1 and C18:1 concentration, indicating a broader germplasm diversity in macadamia. However, biochemical markers, being limited in number, exhibit less polymorphism and are influenced by diverse extraction methodologies, plant tissues, and different stages of plant growth. Consequently, their application in the analysis of plant diversity is constrained.

Molecular Characterization

Molecular markers are widely regarded as the most appropriate method for assessing genetic diversity, given their polymorphic nature and independence from environmental influences. Various molecular marker systems have been created for investigating macadamia, with only a limited number employed for the genetic characterization of wild germplasm. Mast et al. (2008) explored the relationships among the four Macadamia species and their wild counterparts, employing three cpDNAs and three nDNA genomic regions in their study. Peace et al. (2004) conducted a study to examine the genetic structure of a substantial number of wild germplasm accessions of Macadamia. They utilized low-throughput RAF (randomly amplified DNA fingerprinting, dominant) and RAMiFi (randomly amplified microsatellite fingerprinting, co-dominant) markers. Further, the analysis involved 105 AFLP markers, assessing 26 macadamia accessions that represented four species: *M. integrifolia, M. tetraphylla, M. ternifolia*, and *M. hildebrandii*, along with a wild relative, *Hicksbeachia pinnatifolia* (rose nut). Each macadamia accession displayed distinctive AFLP (Amplified fragment length polymorphism) fingerprints, indicating a substantial level of genetic variation within the macadamia germplasm collection (Steiger et al., 2003). SSRs (Single Sequence Repeats) were employed in the investigation of genetic diversity within wild populations of *M. integrifolia* (Neal et al., 2007) and *M. tetraphylla* (O'Connor et al., 2015). However, the availability of SSR primers for macadamia is limited, especially those capable of successful amplification across species. Consequently, they may not be suitable for large-scale genetic studies encompassing all four wild species. The rapid progress of next-generation sequencing technology (NGS) has facilitated the development of high-throughput and cost-effective molecular marker systems. Utilizing NGS, Diversity Array Technology (DArT) has created a marker system that allows for affordable whole-genome-level genetic characterization. DArT has proven effective in conducting genetic diversity, population structure, and genetic mapping studies in numerous crop species, as demonstrated by Roorkiwal et al. (2014) and Grzebelus et al. (2014). In a recent study, Alam et al. (2018) utilized 11,526 silico DArTs and 3,956 SNPs to investigate the genetic diversity and population structure of 80 macadamia cultivars. Another study by O'Connor et al. (2019) explored the genetic diversity, population structure, and linkage disequilibrium of 295 seedling progenies derived from 29 selected parents, employing 16,171 silico DArTs and 4113 SNPs. These investigations indicate the potential application of DArTseq markers for genomic studies in the wild germplasm of macadamia. Recent investigations utilizing chloroplast genome sequencing provide additional support for the earlier observed strong latitudinal population structure in *M. integrifolia* (Nock et al., 2019; Ahmad Termizi et al., 2014; Peace et al., 2005). Nock et al. (2019) specifically identified extensive variability and structuring in the chloroplast genomic variation of *M. integrifolia*. The mitochondrial genomes of three economically significant Macadamia species, namely, *Macadamia integrifolia, M. ternifolia*, and *M. tetraphylla*, were assembled using the Illumina sequencing platform (Niu et al., 2022). Examining the structure and genetic diversity of 28 macadamia cultivars from the United States, Australia, and Brazil, Sobierajski et al. (2022) utilized 29 new microsatellite loci. These microsatellite loci exhibited high genetic diversity, with an average of 3.65 alleles per locus.

Utilization of Germplasm

Over the past century, global research on crop germplasm resources has experienced significant development. Changes in international, social, and political environments, coupled with advancements in science and technology, have given rise to several new trends and features in germplasm research. These include more extensive globalized germplasm collection, diverse approaches to germplasm preservation and conservation, precise methods for germplasm identification and evaluation, large-scale gene discovery, objective-driven germplasm enhancement, and proactive sharing of germplasm. Substantial progress and breakthroughs have been achieved in the field of crop germplasm research. The effective use of Plant Genetic Resources (PGR) in crop improvement programs relies on the identification of promising accessions. The gathered or introduced germplasm undergoes characterization and evaluation to assess its potential, involving the recording of data on agronomic traits like yield, quality, and tolerance to both biotic and abiotic stresses. There exists a significant gap between the availability of germplasm and its actual utilization, a situation evident in both international and national programs. The widespread use of a limited number of closely related parents in crop improvement may render cultivars vulnerable to pests and diseases. Over the years, the distribution of germplasm has shown a decline in demand, with the demand focusing on a limited number of accessions. Breeders commonly rely on working collections with reduced diversity, contributing to a narrow genetic base in improved cultivars. To achieve sustainable crop improvement, safeguard natural ecosystems, and enhance food security through increased production, greater utilization of genetic and genomic resources is imperative.

Several factors contribute to the limited utilization of germplasm, including the large size of collections, insufficient data on traits relevant to users, restricted access to accessions, inadequate linkages between genebanks and users, and the limited capacity of breeding programs to incorporate new material. To enhance germplasm utilization, it is crucial to conduct an extensive evaluation of collections for agronomic and nutritional traits, as well as resistance to both biotic and abiotic stresses. Identifying numerous promising sources for incorporation into crop improvement programs, characterizing approximately 98% of cultivated accessions, organizing field days to facilitate the selection of promising germplasm by researchers, developing trait-specific gene pools, and establishing composite collections and reference sets of genetically diverse accessions are essential.

Utilization in Breeding and Improvement Programs

Studying genetic diversity plays a crucial role in assessing the future breeding potential of wild germplasm. To expand the genetic foundation of breeding material, it is necessary to identify diverse parents for crossbreeding with the cultivated crop. Understanding the genetic relationships among these parents is vital to prevent inbreeding depression, especially when enhancing complex traits. The domesticated germplasm's limited selection history, low accuracy in selection, and restricted genetic diversity indicate substantial opportunities for gains by tapping into new genetic diversity. Furthermore, the untapped potential of related species adds to these opportunities. Various traits are important for macadamia breeding, such as tree structure, flowering patterns, fruit set and arrangement, precocity, yield, resistance to pests and diseases, nut drop phenology, stick-tights, nut size and shape, kernel recovery, the visual appearance of kernels, oil content and profile, kernel size, percentage of whole kernels, shelf life, roasting response, and environmental sensitivity (Hardner et al., 2009; Table 3).

Table 3. List of key traits of Macadamia germplasm for breeding advancement.

Traits	**Genetic resources**
Early flowering	HAES 842, HAES 814, Kau, Keauhou, Beaumont, HAES 814, and Fuji
Dwarfness	*M. ternifolia*, Fuji
Late flowering	A4, A16, and NG8
Less fruit drop	*M. tetraphylla*
Resistant to sticking tights	Own choice, Keauhou, Makai
Resistant to tropical nut borer	Ikaika, Makai
Resistant to southern stink bug	Purvis, Ikaika, Makai
Resistant koa seed worm	Pahal
Large nut size	Own Venture, A4, Makai, and Kau
Small nut size	HAES 814, Keaau, and NG18
High kernel recovery	*M. tetraphylla*, A4, HAES 849, HAES 816, A16 Pahala, and Keaau
High kernel oil recovery	A4, Makai
Performs well in a cooler environment	*M. tetraphylla*, Beaumont, Own Choice, and Hinde

Data from (Hardner et al., 2009)

Cultivars

Table 4. List of cultivars with their respective traits.

Sl. No.	Cultivar	Characteristics
1.	Beaumont	Hybrid (*Macadamia integrifolia* x *M. tetraphylla*) extensively cultivated in Australia and New Zealand. It has high oil content, early fruiting, and yields up to 18 kg /tree. The shell is of medium thickness and kernel recovery is 40% and produces Grade A kernels.
2.	Burdick	Classified under *M. tetraphylla*, originated in Encinitas, California. It features large nuts, averaging 40/pound, with a thin shell. The kernel recovery is 34% and used as rootstock.
3.	Cate	Belongs to *M. tetraphylla* and originated in California. It exhibits a medium to large size nut, with an average shell thickness; 40% kernel recovery, and is precocious in bearing.
4.	Dorado	Classified under *M. integrifolia* and originating in Hawaii, features medium-sized. With a kernel recovery rate of 35%, it has 75% of oil content and exhibits cold tolerance.
5.	Elimbah	Hailing from Australia, Elimbah is early-bearing, characterized by thin shells and kernel recovery of 45–50%.
6.	James	Belongs to *M. integrifolia* and originated in California, is known for its early fruiting and medium-sized nuts. It has a kernel recovery of 40–42% with 75% oil recovery and possesses the unique ability of self-harvesting.
7.	Keaau	A member of *M. integrifolia* from Hawaii, it features medium-sized nuts, average of 80 nuts/pound. It sports a smooth shell and demonstrates a kernel recovery of 42–46%.
8.	Keauhou	A member of *M. integrifolia* originated in Hawaii, characterized by medium to large nuts, averaging around 54 nuts/pound. The shell has a slightly pebbled texture and is of medium thickness. It has a kernel recovery of 37–40% and exhibits resistance to anthracnose.
9.	Vista	It features small to medium-sized nuts, with a remarkable kernel recovery of 46%. The nuts have excellent flavor and the variety is distinguished by a very thin shell. Additionally, possesses the unique ability of self-harvesting.
10.	Waimanalo	A variety of *M. integrifolia* originated in Hawaii, and produces large nuts, sometimes with twin halves. The shell is moderately thick, and it has a kernel recovery rate of 38%, with a high 75% oil content. This variety exhibits resistance to frost and disease.
11.	PA39	Yields up to 33 kg with 95% Grade A kernel, and pairs well with Beaumont as a pollinator. It is highly susceptible to the green shield bug.
12.	GT1	An effective pollinator can bear a substantial crop, whether through self-pollination or cross-pollination, producing smaller nuts. Despite their size, the nuts are of high quality, featuring a rich oil content. The kernels are clean and appealing, and the cracking efficiency is high. It is susceptible to wind damage.
13.	GT2	Derived from Gordon Titirangi selection, it has a thin-shell, susceptible to both rat depredation and shield bug damage. Despite these challenges, it can yield up to 8 kg/tree.
14.	GT201	Derived from Gordon Titirangi selection, serves as a pollinator for Beaumont and possesses commendable nut quality.
15.	GT207	A product of Gordon Titirangi selection, GT207 stands out with a remarkable 100% grade 1 nut quality.
16.	Own Choice	Hawaiian variety, yields up to 22 kg/tree, with the highest quality nuts. Notably, the nuts remain on the tree when ripe and are an excellent pollinator for Beaumont.
17.	Maroochy	A pure *M. tetraphylla* variety originating from Australia is grown for its high crop yield, and flavorful nuts and is well-suited for pollinating Beaumont. It can yield up to 75 kg/tree.
18.	Nutty Glen	A hybrid selection from Australia recognized for its large and high-quality nuts. Additionally, it is popularly used as rootstock.
19.	Nelmac I	Characterized by late flowering. It produces an average yield of 8 kg/tree. It produces high-quality kernels with favorable oil content.
20.	Nelmac II	This variety is widely favored as it serves as a pollinator for Beaumont, contributing to yields of up to 22 kg/tree.
21.	Renown	Recognized for its high yield of up to 17 kg/tree and possessing quality kernels. Renown is frequently utilized as a parent in the breeding of new varieties.

Cultivars Released by Different Institutes

Department of Agriculture and Fisheries (Queensland): MIV1-G, MIV1-J, MIV1-P and MIV1-R.

Hawaiian Agricultural Experiment Station: HAES 246 (Keauhou), HAES 333(Ikaika), HAES 344 (Kau), HAES 508 (Kakea), HAES 660 (Keaau), HAES 741 (Mauka), HAES 781, HAES783, HAES 791 (Fuji), HAES 800 (Makai), HAES 814, HAES 816, HAES 842, HAES 849, Pahala, Greber hybrid, Purvis and Probert 1 (USDA ARS, 2023; Figure 1).

Hidden Valley Plantations (Queensland): A4, A16, A29, A38, A203, A268, A376, A403, A447 and A538.

Conservation of Germplasm

The Macadamia Nut is categorized as vulnerable and qualifies for vulnerable status under the Environment Protection and Biodiversity Conservation Act 1999 (EPBC Act). Before the enactment of the EPBC Act, it held vulnerable status under Schedule 1 of the Endangered Species Protection Act 1992. Additionally, the Macadamia Nut is designated as vulnerable under the Nature Conservation Act 1992 in Queensland. The main threats to this species include urban and agricultural clearing, weed invasion, and inappropriate fire regimes, as highlighted by the Queensland CRA/RFA Steering Committee in 1997. Land clearing poses a significant risk by causing habitat fragmentation and reducing pollination success. The dominance of invasive species such as Lantana (*Lantana camara*) and Camphor Laurel (*Cinnamomum camphora*) alters the habitat and affects fire patterns. The Macadamia Nut is particularly susceptible to fire damage. Another potential threat is the genetic homogenization of wild stock due to intermingling with widely cultivated orchard strains. Often, cultivated varieties of threatened species are represented by only a few genotypes selected for specific traits. The introduction of these genes into wild populations could lead to genetic homogenization and disrupt local genetic adaptations. The conservation of this iconic species holds significance for biodiversity maintenance and the preservation of wild germplasm.

Macadamia conservation can be achieved through various strategies, including monitoring established populations to identify key threats, pinpointing high-priority conservation areas, minimizing adverse impacts from land use in known locations, and creating and implementing a management plan to control the spread of Lantana and Camphor Laurel in the region. Additionally, it is crucial to develop and implement an effective fire management strategy specifically tailored for Macadamia Nut. Other approaches, such as *in situ* and *ex situ* conservation, can also play a significant role in the comprehensive conservation efforts for this species. To enhance conservation, supplementary planting in small fragments or enhancing connectivity between remnants through habitat restoration is anticipated to counteract some of the observed fragmentation effects. This approach stands out as one of the most effective strategies for preserving the species *in situ*, especially in landscapes that are highly fragmented, as noted by Neal in 2007. Regarding *ex situ* conservation, various methods such as the establishment of a field genebank, seed bank, in vitro conservation, and cryopreservation are employed. A notable example of *ex situ* conservation is the National Macadamia Germplasm Collection, created by utilizing cuttings from over 370 trees across more than 70 sites. These sites include native populations, old planted populations, and stands of unknown origin, forming an extensive core collection of the major species in the southern Macadamia clade (Hardner et al., 2004; Peace et al., 2005). Orchard trials have been conducted using this collection to preserve a diverse range of genetic variations and assess the material for potential inclusion in future breeding programs.

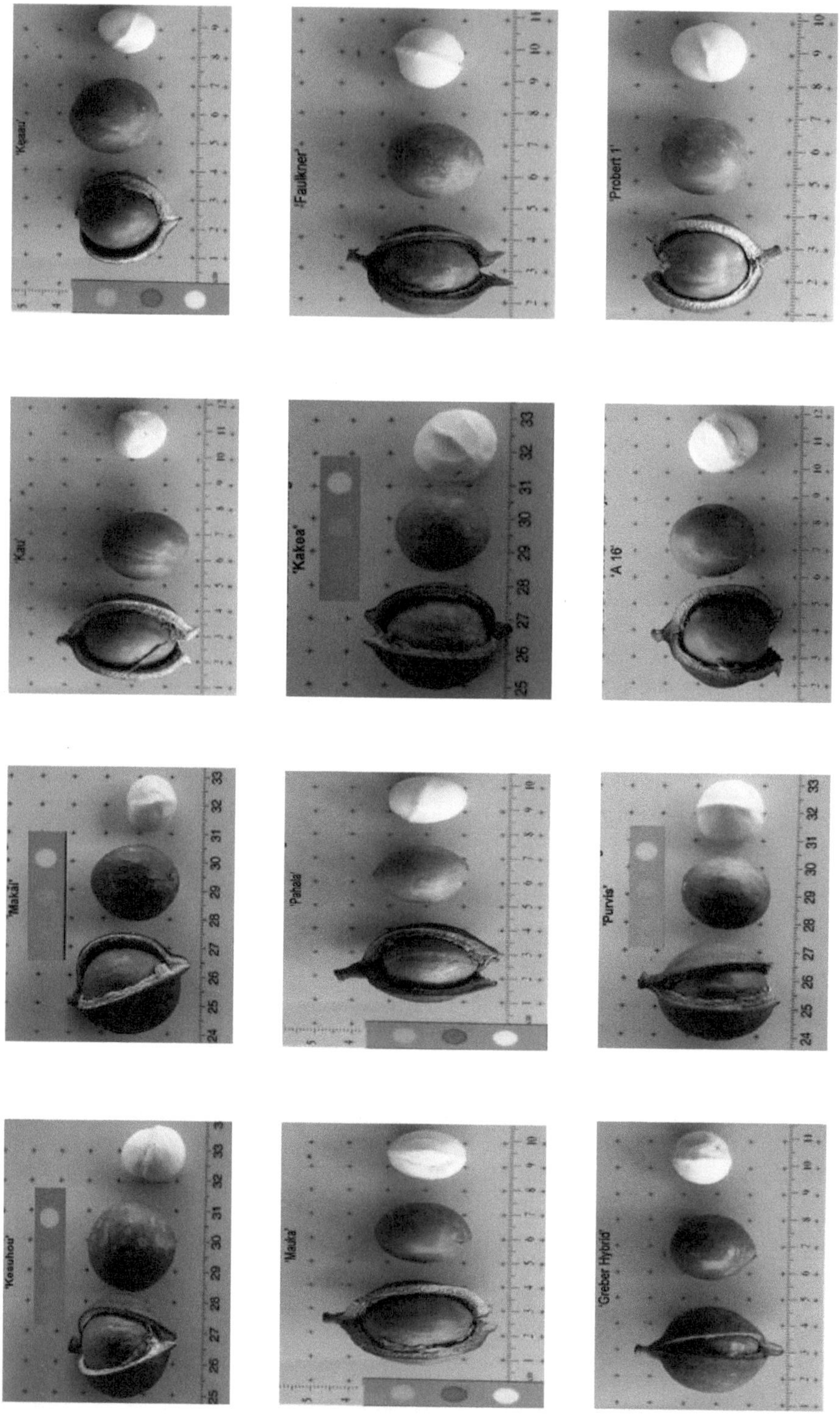

Figure 1. Important released varieties of Macadamia.

Conclusion

Macadamia nuts are in high demand globally due to their distinctive flavors, versatile applications, and extended shelf life. It is essential to comprehend genetic resources and the connections between cultivated and wild germplasm to guide the integration of new germplasm into selective breeding programs. Prioritizing the conservation of potentially valuable wild germplasm for future use is also crucial because the vulnerability classification of the Macadamia nut highlights its eligibility for vulnerable status. Hence, this iconic germplasm should be conserved and utilized for further breeding evolution.

References

Ahmad T.A., C.M. Hardner, J. Batley, C.J. Nock, S. Hayashi, J.D. Montenegro, and D. Edwards. 2014. SNP analysis of *Macadamia integrifolia* chloroplast genomes to determine the genetic structure of wild populations. *Acta Hortic., 1109*: 175–180.

Alam, M., J. Neal, K. O'Connor, A. Kilian, and B. Topp. 2018. Ultra-high-throughput DArTseq-based silicoDArT and SNP markers for genomic studies in macadamia. *PloS One*, *13*(8): 0203465.

Aradhya, M.K., L.K. Yee, F.T. Zee, and R.M. Manshardt. 1998. Genetic variability in macadamia. *Genet. Resour. Crop Evol.*, *45*: 19–32.

Barry, S.J., and G.T. Thomas. 1994. *Threatened Vascular Rainforest Plants of South-East Queensland: A Conservation Review*. Queensland Department of Environment and Heritage/ANCA, Canberra.

Brozynska, M., A. Furtado, and R.J. Henry. 2016. Genomics of crop wild relatives: Expanding the gene pool for crop improvement. *Plant Biotechnol. J.*, *14*(4): 1070–1085.

Chamberlain J.R. 2001. Genetic markers and their application to forest genetic resources conservation and use. Recent research and development in Forest Genetic Resources. *In*: *Proceedings of the Training Workshop on the Conservation and sustainable Use of Forest Genetic Resources in Eastern and Southern Africa 6–11 December 1999*. 59–66. Nairobi, Kenya.

Chen, Y. H., L. R. Shapiro, B. Benrey, and A. Cibrian-Jaramillo. 2017. Back to the origin: *in situ* studies are needed to understand selection during crop diversification. *Front. Ecology Evol.*, *5*: 125.

Chesnokov, Y.V., V.M. Kosolapov, and I.V. Savchenko. 2020. Morphological genetic markers in plants. *Russ. J. Genet.*, *56*: 1406–1415.

Costello, G., M. Gregory, and P. Donaitu. 2008. *Southern Macadamia Species Recover Plan 2008–2012*. Report to Department of the Environment, Water, Heritage, and the Arts, Canberra by Horticulture Australia Limited, Sydney.

Dahler, J.M., C. Mcconchie, and C.G.N. Turnbull. 1995. Quantification of cyanogenic glycosides in seedlings of three Macadamia (Proteaceae) species. *Aust. J. Bot.*, *43*(6): 619–628.

Domingo, R.M., M.A. Gutierrez-E, P. Ramirez-V, and J. Rodriguez-A. 2002. Descriptor for macadamia (*Macadamia* Spp.). *In*: *XXVI International Horticultural Congress: IV International Symposium on Taxonomy of Cultivated Plants*, Article 634: 203–208.

Gitonga L., E. Kahangi, A. Muigai, K. Ngamau, S. Gichuki, E. Mutuma,W. Cheluget, and B. G. Watiki. 2008. Farmers knowledge on macadamia genetic diversity in Kenya as a means for *in situ* conservation. *Int. J. Environ. Sci.*, *3*(1): 55–60.

Grzebelus, D., M. Iorizzo, D. Senalik, S. Ellison, P. Cavagnaro, A. Macko-Podgorni, K. Heller-Uszynska, A. Kilian, T. Nothnagel, and C. Allender. 2014. Diversity, genetic mapping, and signatures of domestication in the carrot (Daucus carota L.) genome, as revealed by Diversity Arrays Technology (DArT) markers. *Mol. Breed.*, *33*: 625–637.

Halford, D. 1997. *Macadamia jansenii*. Queensland Herbarium, Environmental Protection Agency, Brisbane. *Exp. Sta. Prog.*, *51*: 6.

Harden, G., B. McDonald, and J. Williams. 2006. *Rainforest Trees and Shrubs: A Field Guide to their Identification*. Gwen Harden Publishing, Nambucca Heads.

Hardner, C., and B. Topp. 2018 Macadamia integrifolia and *M. tetraphylla*. *In*: *Trees for Life in Oceania: Conservation and Utilisation of Genetic Diversity*, 147–151. Australian Centre for International Agricultural Research: Canberra, Australia.

Hardner, C., C. Peace, J. Henshall, and J. Manners. 2004. Opportunities and constraints for marker-assisted selection in macadamia breeding. *In*: *International Symposium on Harnessing the Potential of Horticulture in the Asian-Pacific Region*. Article 694: 85–90.

Hardner, C.M., C.A. McConchie, A. Vivian-Smith, and S.J. Boyton. 2000. Hybrids in macadamia improvement. *In*: *Hybrid Breeding and Genetics of Forest Trees Proceedings of QFRI/CRC-SPF Symposium*, Noosa, 9–14.

Hardner, C., and B. Topp. 2018. Macadamia integrifolia and *M. tetraphylla. In*: *Trees for Life in Oceania: Conservation and Utilisation of Genetic Diversity*, 147–151. Australian Centre for International Agricultural Research: Canberra, Australia.

Hardner, C.M., C. Peace, A.J. Lowe, J. Neal, P. Pisanu, M. Powell, A. Schmidt, C. Spain, and K. Williams. 2009. 1 Genetic Resources and Domestication of Macadamia. *Horticultural Reviews, 35*(4): 1–125.

Hauser, J., and J. Blok. 1992. *Fragments of Green*. Australian Rainforest Conservation Society, Brisbane.

Hu, W., M. Fitzgerald, B. Topp, B, M. Alam, and T.J. O'Hare. 2022. Fatty acid diversity and interrelationships in macadamia nuts. *LWT., 154*: 112839.

Johnson, L.A.S., and B.G. Briggs. 1963. Evolution of Proteaceae. *Australian J. Bot.*, 11: 21–61.

Johnson, L.A.S., and B.G. Briggs. 1975. On the Proteaceae: The evolution and classification of a southern family. *Bot. J. Lin., 70*(2): 83–132.

Khairina, R., Y. Fitrial, H. Satrio, and N. Rahmi. 2016. Physical, chemical, and microbiological properties of 'Ronto', a traditional fermented shrimp from South Borneo, Indonesia. *Aquatic Procedia,, 7*: 214–220.

Marcus, J.P., K.C. Goulter, J .L. Green, S. J. Harrison, and J.M. Manners. 1997. Purification, characterisation and cDNA cloning of an antimicrobial peptide from *Macadamia integrifolia. European J. Mol. Biol. Biochem., 244*(3): 743–749.

Mast, A.R., C.L. Willis, E.H. Jones, K.M. Downs, and P.H. Weston. 2008. A smaller Macadamia from a more vagile tribe: Inference of phylogenetic relationships, divergence times, and diaspore evolution in Macadamia and relatives (tribe Macadamieae; Proteaceae). *Am. J. Bot., 95*(7): 843–870.

McConachie, I. 1980. *The Macadamia Story*. Macadamia Consultants Pty Limited, Brisbane, Queensland, Australia.

McHargue L.T. 1996. Macadamia production in southern California. *In*: J. Janick [Ed.], *Progress in New Crops*, 458–462. ASHS Press, Arlington, VA.

Moreno-Perez, A.J., A. Sanchez-Garcia, J.J. Salas, R. Garces, and E. Martinez-Force. 2011. Acyl-ACP thioesterases from macadamia (*Macadamia tetraphylla*) nuts: Cloning, characterization and their impact on oil composition. *Plant Physiol. Biochem., 49*(1): 82–87.

Neal, J.M. 2007. *The Impact of Habitat Fragmentation on Wild Macadamia Integrifolia Maiden and Betche (Proteaceae) Population Viability*. University of New England: Armidale, Australia.

Neal, J.M., C.M. Hardner, and C. L. Gross. 2010. Population demography and fecundity do not decline with habitat fragmentation in the rainforest tree *Macadamia integrifolia* (Proteaceae). *Biol. Conserv., 143*(11): 2591–2600.

Niu, Y., Y. Lu, W. Song, X. He, Z. Liu, C. Zheng, S. Wang, C. Shi, and J. Liu. 2022. Assembly and comparative analysis of the complete mitochondrial genome of three Macadamia species (*M. integrifolia, M. ternifolia* and *M. tetraphylla*). *Plos One, 17*(5): 0263545.

Nock, C.J., C.M. Hardner, J.D. Montenegro, A.A. Ahmad Termizi, S. Hayashi, J. Playford, D. Edwards, and J. Batley. 2019. Wild origins of macadamia domestication identified through intraspecific chloroplast genome sequencing. *Front. Plant Sci., 10*: 1–15.

O'Connor, K., A. Kilian, B. Hayes, C. Hardner, C. Nock, A. Baten, M. Alam, and B. Topp. 2019. Population structure, genetic diversity, and linkage disequilibrium in a macadamia breeding population using SNP and silicoDArT markers. *Tree Genet. Genomes, 15*Article No. 24.

O'Connor, K., M. Powell, C. Nock, and A. Shapcott. 2015. Crop to wild gene flow and genetic diversity in a vulnerable Macadamia (Proteaceae) species in New South Wales, Australia. *Biol. Conserv., 191*: 504–511.

Peace C.P., P. Allan, and V. Vithanage. 2005. Genetic relationships amongst macadamia varieties grown in South Africa as assessed by RAF markers. *S. Afr. J. Plant Soil, 22*(2): 71–75.

Peace, C.P., V. Vithanage, J. Neal, C.G.N. Turnbull, and B.J. Carroll. 2004. A comparison of molecular markers for genetic analysis of macadamia. *J. Hortic. Sci. Biotechnol,. 79*(6): 965–970.

Peace, C.P., V. Vithanage, C.G.N. Turnbull, and B.J. Carroll. 2003. A genetic map of macadamia based on randomly amplified DNA fingerprinting (RAF) markers. *Euphytica, 134*(1): 1726.

Quedraogo A.S. 2001. Conservation, management and use of forest genetic resources. Recent Research and Development in Forest Genetic Resources. *Proceedings of the Training Workshop on the Conservation and Sustainable Use of Frest Genetic Resources in Eastern and Southern Africa*, 1–14.

Ramsay, H.P. 1963. Chromosome number in the Proteaceae. *Australian J. Bot., 11*: 1–20.

Roorkiwal, M., E.J. von Wettberg, H.D. Upadhyaya, E. Warschefsky, A. Rathore, and R.K. Varshney. 2014. Exploring germplasm diversity to understand the domestication process in Cicer spp. using SNP and DArT markers. PLoS, *9*: 102016.

Sedgley, M., F.D.H. Bell, D. Bell, C.W. Winks, S.J. Pattison, and T.W. Hancock. 1990. Self-and cross-compatibility of macadamia cultivars. *Int. J. Hortic. Sci., 65*(2): 205–213.

Smith, J.S.C., and O.S. Smith. 1992. Fingerprinting crop varieties. *Adv. Agron., 47*: 85–140.

Sobierajski, G.D.R., G.C. Blain, A.C.B. Sousa, L.J. Cancado, G. Pereira, A.D. Souza, and A.A.F. Garcia. 2022. Analysis of the genetic structure and diversity of a Brazilian macadamia nut (*Macadamia integrifolia*) germplasm. *Crop Breed Appl Biotechnol., 22*.

Stace, H.M., A.W. Douglas, and J.F. Sampson. 1998. Did 'paleo-polyploidy' really occur in Proteaceae. *Syst. Bot., 11*: 613–629.

Stanley, T.D., and E.M. Ross. 2002. *Flora of South Eastern Queensland* (Vol. 2). Queensland Department of Primary Industries, Brisbane.

Steiger, D.L., P.H. Moore, F. Zee, Z. Liu, and R. Ming. 2003. Genetic relationships of macadamia cultivars and species revealed by AFLP markers. *Euphytica, 132*: 269–277.

Storey, W.B. 1959. History of the systematic botany of the Australian species of macadamia. *California Macadamia Soc. Yearbook, 5*: 68–78.

Storey, W.B. 1965. The ternifolia group of Macadamia species, 507–514.

Storey, W.B., and W.F. Salleeb. 1966. Genetics of four vegetative characters in an interspecific Macadamia Hybrid. *California Macadamia Society Yearbook, 12*: 77–88.

Storey, W.B., and W.F. Saleeb. 1970. Interspecific hybridization in macadamia. *California Macadamia Soc. Yearbook, 16*: 75–85.

Thormann C.E., M.E. Ferreira, L.E.A. Camargo, J.G. Tivanga, and T.C. Osborn. 1994. Comparison of RFLP and RAPD markers to estimating genetic relationships within and among cruciferous species. *Theor. Appl. Genet., 88*: 973–980.

Storey, W.B. 1959. History of the systematic botany of the Australian species of macadamia. *California Macadamia Soc. Yearbook, 5*: 68–78.

Topp, B.L., C.J. Nock, C.M. Hardner, M. Alam, and K.M. O'Connor. 2019. Macadamia (*Macadamia* spp.) Breeding. *In*: *Advances in Plant Breeding Strategies: Nut and Beverage Crops*, 221–251. Springer International Publishing: Cham, Switzerland.

USDA ARS 2023. https://www.ars.usda.gov/pacific-west-area/hilo-hi/daniel-k-inouye-us-pacific-basin-agricultural-research-center/tropical-plant-genetic-resources-and-disease-research/docs/macadamia-collection/

Venkata Rao, C. 1970. Studies in the Proteaceae. *Proc. Indian Natl. Sci. Acad., 36*: 345–363.

Verma, M. K., A. Yadav, N. Deepak, K. Usha, and S. Kumar. 2017. Macadamia Nut (*Macadamia integrifolia*). *In*: S.N. Ghosh [Ed.], *Minor Fruits: Nutraceutical Importance and Cultivation*, 583–605.

Vithanage, V., and C.W. Winks. 1992. Isozymes as genetic markers for macadamia. *Scientia. Hort., 49*: 103–115.

Index